Classes of Organic Compounds

Class	Formula	Functional Group	IUPAC Prefix or Suffix
Acyl halide	$\overset{\displaystyle O}{\overset{\|}{RCX}}$	$\overset{\displaystyle O}{\overset{\|}{-CX}}$	-oyl halide
Alcohol	ROH	—OH	-ol
Aldehyde	$\overset{\displaystyle O}{\overset{\|}{RCH}}$	$\overset{\displaystyle O}{\overset{\|}{-CH}}$	-al
Alkane	RH		-ane
Alkene	$R_2C{=}CR_2$	$\text{C}{=}\text{C}$	-ene
Alkyne	$RC{\equiv}CR$	$-C{\equiv}C-$	-yne
Amide	$\overset{\displaystyle O}{\overset{\|}{RCNR_2}}$	$\overset{\displaystyle O}{\overset{\|}{-CN-}}$	-amide
Amine	R_3N	$-N-$	amino-
~~Arene~~ *Aromatic*	ArH		
Carboxylic acid	$\overset{\displaystyle O}{\overset{\|}{RCOH}}$	$\overset{\displaystyle O}{\overset{\|}{-COH}}$	-oic acid
Ester	$\overset{\displaystyle O}{\overset{\|}{RCOR}}$	$\overset{\displaystyle O}{\overset{\|}{-CO-}}$	
Ether	R_2O	—O—	alkoxy-
Halide	RX	—X	halo
Ketone	$\overset{\displaystyle O}{\overset{\|}{RCR}}$	$\overset{\displaystyle O}{\overset{\|}{-C-}}$	-one
Nitrile	$RC{\equiv}N$	$-C{\equiv}N$	-nitrile
Nitro compound	RNO_2	$-NO_2$	nitro-
Phenol	ArOH	—OH	
Sulfide	R_2S	—S—	alkylthio-
Thiol	RSH	—SH	-thiol

Introduction to Organic
and Biological
Chemistry

Stuart J. Baum

STATE UNIVERSITY OF NEW YORK, COLLEGE AT PLATTSBURGH

Introduction to Organic and Biological *Chemistry*

FOURTH EDITION

MACMILLAN PUBLISHING COMPANY
NEW YORK
COLLIER MACMILLAN PUBLISHERS
LONDON

DEDICATED TO MY FAMILY
Sharon, Derek, and Kym

PRINTED IN THE UNITED STATES OF AMERICA

Macmillan Publishing Company
866 Third Avenue, New York, New York 10022

Collier Macmillan Canada, Inc.

LIBRARY OF CONGRESS CATALOGING-IN-PUBLICATION DATA

Baum, Stuart J.
Introduction to organic and biological chemistry.

Includes index.
1. Biological chemistry. 2. Chemistry, Organic.
I. Title.
QP514.2.B37 1987 574.19′2 86-5350
ISBN 0-02-317380-7

Printing: 1 2 3 4 5 6 7 8 Year: 7 8 9 0 1 2 3 4 5 6

ISBN 0-02-317380-7

Preface

The fourth edition of this text maintains the purpose of the previous editions, namely, to present an introduction to organic chemistry and biochemistry. Again, students are assumed to have had at least one prior course in general chemistry. The text is designed to assist the student in acquiring a sound background in the subject without presenting an overwhelming amount of factual material. It is intended for those who are mainly interested in the application of the principles of organic chemistry and biochemistry to related areas of science (such as genetics, microbiology, pharmacology, physiology, and nutrition.)

The book is divided into two main sections. The first part is devoted to establishing the fundamental principles of organic chemistry. Because of the time limit imposed by a one-semester course, much of the material normally covered in an introductory organic chemistry text has been condensed, and many interesting peripheral aspects of the field have been omitted. The first nine chapters deal with the structure and properties of the different classes of organic compounds, with emphasis on the characteristic reactions of the various functional groups. A few selected reaction mechanisms have been presented in order to familiarize the student with such terms as free radical, carbocation, electrophile, and nucleophile. Chapter 10 deals with stereoisomerism, and its importance to an understanding of enzyme specificity is stressed. Systematic names for the organic compounds are used repeatedly. I have found that students enjoy learning organic nomenclature. They derive much the same satisfaction as is gained from learning a foreign language, and are well disposed to practice this language.

Biochemistry is the study of the particular organic molecules that occur in living organisms. Thus, Chapters 11–19 present the fundamental concepts of biochemistry in a form that is understandable to students having only a limited background in organic chemistry. The major emphasis is placed on the dynamic nature of biochemistry and the interrelationships of the various metabolic pathways. Nothing of importance is consciously omitted. Chapters 11–13 discuss the chemistry of the three major classes of foodstuffs—carbohydrates, lipids, and proteins. Enzymes (Chapter 14) are presented in a separate chapter as a special class of proteins. Because of the limited mathematics background of most of the students using this text, statements have to be qualitative rather than mathematical. Thus, for example, no extensive treatment of enzyme kinetics is attempted. In Chapter 15, the molecular basis of life is sketched by a clear and concise discussion of nucleic acid structure and replication and of the role of nucleic acids in protein synthesis. Chapters 16–18 present the basic metabolic reaction sequences that occur within the cells from the point of view of their interrelationships and integration into a fundamental whole. Metabolism is concerned with the production and utilization of energy, and it is here that the student is able to tie together the chemical principles learned in general chemistry with organic chemistry and biochemistry. The human body is viewed as an intricate machine that utilizes the energy of foods to run chemical reactions to meet its own needs. Chapter 19 discusses the blood, and it is especially well suited to those students who are majoring in the health sciences.

Although relevance is a much overused term, I believe that we teachers ought to stress the connection between classroom learning and the world outside. Hence, numerous biochemical and biomedical examples have been integrated through the text. Still other topics appear within the chapters as separate sections, set in two columns and in smaller type (listed as optional in the Contents). I view these as enrichment sections that add flesh to the organic/biological skeleton. Among these "relevant" topics are octane rating, pesticides, effects of alcohol, DMSO, neurotransmitters, hormones, carcinogens, designer drugs, artificial sweeteners, anabolic steroids, estrogen therapy, prostaglandins, genetic diseases, viruses, recombinant DNA, the immune response, and cholesterol and cardiovascular disease.

Users of the previous edition will notice the omission of the chapters "Organic Compounds of Sulfur and Phosphorus" and "Nutrition and Health." Thiols are now included in Chapter 6, and phosphate esters appear in Chapter 8. The vitamin sections have been expanded and now appear in Chapters 12 and 14. The Exercises at the end of each chapter have been revised and substantially increased. A study guide, which accompanies this edition of the text, contains the answers to all these Exercises. In addition, the study guide contains numerous multiple choice, fill-in-the-blanks, and true/false questions, answers to which are also provided. It is my contention that students learn chemistry only through repeated practice. The more questions that they are able to answer, the more confident they will be about their understanding of the subject.

As in previous editions, feedback from my students at Plattsburgh and comments from users and reviewers have greatly added to the worthiness of this text. I wish to thank James Beier, Ferrum College; David Erwin, Loyola University; H. M. Ghose, Cuyahoga Community College; Donald Langr, North Iowa Area Community College; Elva Mae Nicholson, Eastern Michigan University; Alan Price, University of Michigan; Donald Renn, Edinboro University of Pennsylvania; and Karen Wiechelman and Linda Woodward, University of Southwestern Louisiana. I am especially grateful to Chemistry Editor Peter Gordon for selecting the reviewers and capably overseeing the project, to Noella Watts for typing the revisions, to one of my students, Terrence O'Hanlon, for his help with the study guide, and to my daughter Kym for her tireless efforts of proofreading the galleys and checking the answers to the exercises. By her informed and intelligent copyediting, Roberta Gellis has contributed significantly to the readability and currency of this edition and to the effectiveness of the two-color presentation.

Finally, I wish to acknowledge the person who has been the most instrumental in the evolution of this text. Production Supervisor Elisabeth Belfer has worked with me for the past dozen years. Just about every page of the manuscript has contained a suggestion for rewording or the question "Will the student understand this?" Thank you, Liz, for your conscientiousness and dedication.

S. J. B.

Contents

17 *Lipid Metabolism* 459

18 *Protein Metabolism* 473

19 *The Blood* 489

Index 509

1

Introduction to Organic Chemistry

In 1685, Nicolas Lémery, a French chemist, published a book entitled *Cours de Chyme* in which he classified substances as animal, vegetable, or mineral on the basis of their origin. This was probably the first attempt made to distinguish between substances derived from plant or animal sources and those obtained from mineral constituents. The term **organic** was later applied to compounds derived from living matter, and substances having nonliving sources were accordingly referred to as **inorganic.** Furthermore, it was universally believed that living organisms contained some mysterious vital force necessary to the formation of organic substances. Throughout the ensuing century, scientists were thwarted in all their attempts to synthesize organic substances from inorganic materials. Their failures served to entrench more firmly the vital force theory, which eventually achieved a status akin to religious dogma.

In 1814, a Swedish chemist, J. J. Berzelius, dealt the vital force theory a serious blow when he proved that the basic laws of chemical change (the law of definite

1

composition and the law of multiple proportions) applied to organic as well as to inorganic compounds. Fourteen years later, the erroneous theory suffered a crippling blow by a stroke of chemical serendipity.

The birth of modern organic chemistry is generally placed in the year 1828. It was in that year that Friedrich Wöhler, while a medical student at the University of Heidelberg, attempted to prepare ammonium cyanate by heating a mixture of two inorganic substances, lead cyanate and ammonium hydroxide. To his surprise, instead of ammonium cyanate, he obtained crystals of the well-known organic compound urea.

$$\underset{\substack{\text{Lead} \\ \text{cyanate}}}{Pb(OCN)_2} + \underset{\substack{\text{Ammonium} \\ \text{hydroxide}}}{NH_4OH} \longrightarrow \underset{\substack{\text{Ammonium} \\ \text{cyanate}}}{[NH_4OCN]} \xrightarrow[\text{upon} \atop \text{heating}]{\text{rearranges}} \underset{\text{Urea}}{H_2N-\overset{\overset{\displaystyle O}{\|}}{C}-NH_2}$$

Urea had been isolated from human urine in 1780. (As we shall see in Section 18.5, urea is synthesized in the liver, transported to the kidneys, and excreted in the urine.) Wöhler correctly concluded that ammonium cyanate is formed, but then rearranges under the influence of heat to yield urea. Urea contains the same number and kinds of atoms as ammonium cyanate, but these atoms are arranged differently.

The next several decades witnessed a renewed effort on the part of chemists to synthesize organic compounds from inorganic starting materials. As a result of the enlightenment of Wöhler's discovery, many other organic compounds were synthesized in chemical laboratories. Moreover, while many of these compounds were identical to compounds found in nature, many others were entirely new, having no known counterpart in nature.

1.1   *The Nature of Organic Compounds*

By 1850, the vital force theory was essentially dead, and the relationship between the two branches of chemistry was clearly recognized. Table 1.1 contrasts the general properties of organic and inorganic compounds. It must be understood, however, that there are exceptions to every entry in this table.

The one constituent common to all organic compounds is the element carbon. Today, the term organic chemistry, although no longer descriptive, implies the study of carbon-containing compounds.[1] There are about 6 million known organic compounds (isolated from nature or synthesized in the laboratory) and several thousand new compounds are synthesized and described each year. Over 95% of all known chemical compounds are compounds of carbon.

What is so special about carbon that differentiates it from all of the other elements in the periodic table? Carbon has the ability to bond successively to other carbon atoms to form chains and rings of varying sizes. As the number of carbon atoms in a

---

[1] This definition is not adhered to strictly; several of the following compounds of carbon properly belong to the domain of inorganic chemistry.

Carbon monoxide (CO)	Carbonates (e.g., Na_2CO_3)	Thiocyanates (e.g., NaSCN)
Carbon dioxide (CO_2)	Bicarbonates (e.g., $NaHCO_3$)	Cyanates (e.g., KOCN)
Carbon disulfide (CS_2)	Cyanides (e.g., KCN)	Carbides (e.g., CaC_2)

TABLE 1.1 Some Contrasting Properties of Organic and Inorganic Compounds

Organic	Inorganic
1. Low melting points	1. High melting points
2. Low boiling points	2. High boiling points
3. Low solubility in water; high solubility in nonpolar solvents	3. High solubility in water; low solubility in nonpolar solvents
4. Flammable	4. Nonflammable
5. Solutions are nonconductors of electricity	5. Solutions are conductors of electricity
6. Chemical reactions are usually slow	6. Chemical reactions are rapid
7. Exhibit isomerism	7. Isomers are limited to a few exceptions (e.g., the transition elements)
8. Exhibit covalent bonding	8. Exhibit ionic bonding
9. Compounds contain many atoms	9. Compounds contain relatively few atoms

chain increases, the number of ways that these atoms may arrange themselves increases, yielding compounds with the same chemical composition but with different structures. Finally, carbon can form strong bonds with a number of different elements. The elements most frequently encountered in organic compounds are hydrogen, oxygen, nitrogen, sulfur, phosphorus, and the halogens.

1.2 Chemical Bonds

Chemical bonds are *the forces or interactions that cause atoms to be held together as molecules or cause atoms, ions, and molecules to be held together as more complex aggregates.* If no such forces existed, no atoms or molecules would condense to liquids or solids; all atoms would exist as monatomic gases, even at temperatures approaching absolute zero. Chemical bonds range from strong ionic and covalent bonds to weaker metallic bonds to very much weaker dipole–dipole interactions, hydrogen bonds, and dispersion forces. All of these types of chemical bonds can be understood at least partially in terms of electrostatic interactions (either attractive or repulsive) between charged bodies such as positively and negatively charged ions, electrons and proton-containing nuclei, polar molecules, or instantaneous dipoles formed by atoms or nonpolar molecules. No attempt is made here to review all the characteristics of the different types of chemical bonds. It is assumed that the student has taken a previous chemistry course and is familiar with such terms as cation, anion, ionic bond, Lewis structure, and dispersion forces (van der Waals forces). We briefly discuss the covalent bond because of its special relevance to the structure of carbon compounds.

1.3 Covalent Bonds

When elements with a similar hold on outer shell electrons combine, a complete transfer of electrons to form ions is unreasonable. Both atoms try to maintain a hold on the electrons involved. A **covalent bond** arises *from the attraction between two atoms as a result of the sharing of one or more pairs of electrons by the two atoms.* A single covalent bond results when two atoms are held together by the sharing of a pair of electrons to which each atom normally contributes one electron.

The hydrogen molecule is the simplest case involving a covalent bond.

$$\text{H} \cdot \ + \ \text{H} \cdot \ \longrightarrow \ \text{H}:\text{H or H—H}$$

The shared electron pair forming the covalent bond is shown as two dots (or as a line) *between* the two H atoms.

The covalent bond in the diatomic H_2 molecule has specific properties. It is a *single* bond because the two H atoms share *one* electron pair. Its *bond length* is the center-to-center distance between the two nuclei. Its *bond energy* is the energy required to break a gaseous H_2 molecule apart into two gaseous H atoms (104 kcal/mole, about enough energy to raise the temperature of 1 L of water from 0 to 100°C). It is a *nonpolar* covalent bond because both atoms are identical, both have the same attraction for the electron pair, and both share the electron pair equally. There is no separation of positive and negative charge between the two H atoms.

1.4   *Multiple Covalent Bonds*

Multiple covalent bonds occur when more than one electron pair is shared between two atoms. When *two electron pairs are shared between two atoms,* a **double bond** is formed. It is stronger than a single bond, but not twice as strong because the two electron pairs repel each other and cannot become fully concentrated between the two atoms. Its bond length is shorter than that of a single bond. When *three electron pairs are shared between two atoms,* a **triple bond** is formed. It is stronger than a single bond and a double bond, but not three times as strong as a single bond. Its bond length is shorter than that of a double bond. *Saturated* molecules possess only single bonds; *unsaturated* molecules possess one or more double or triple bonds. Table 1.2

TABLE 1.2 Compounds Illustrating Single and Multiple Carbon–Carbon Bonds

Name	Formula	Bond Energy (kcal/mole)	Bond Distance (Å)
Ethane	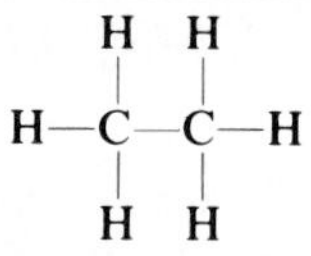 H—C—C—H (with H H above and H H below)	88[a]	1.54
Ethylene	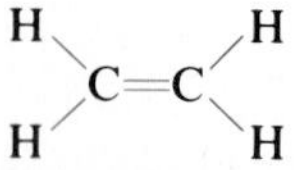 C=C (with H H on left, H H on right)	172	1.33
Acetylene	H—C≡C—H	230	1.20

[a]The carbon–carbon single bond is stronger than any other single bonds between identical atoms, with the exception of H—H.

4

gives examples of some carbon compounds that illustrate the increase in bond energy and decrease in bond distance with the formation of multiple bonds.

Multiple bonds are formed commonly by carbon, nitrogen, and oxygen atoms and occasionally by phosphorus, sulfur, and other atoms. Multiple bonds arise whenever there are too few outer shell electrons to provide each atom with a stable noble gas electron configuration using a completely single-bonded structure. Lewis structures can be used to predict whether or not multiple bonds exist in simple molecules (see Exercise 1.7). Each carbon atom in the molecules in Table 1.2 forms four bonds—by means of four single bonds in ethane, two single bonds and one double bond in ethylene, and one single bond and one triple bond in acetylene.

1.5   The Structure of Carbon Compounds

Lewis structures provide useful representations of the nature of bonding in atoms, but provide little direct information about the distribution of electron density within molecules or the geometry of atoms within molecules.

A number of bonding theories can be used for predicting shapes of simple molecules.[2] We shall use the *valence shell electron pair repulsion* (VSEPR) theory. This model has the advantages of being relatively simple and nonmathematical while still providing accurate qualitative shapes for most of the compounds that we encounter in this text.

The basic premise of VSEPR theory is that *electron pairs, having like charge, repel each other.* Other factors being equal, outer shell or valence shell electron pairs around an atom will try to spread as far apart as possible. Furthermore, interactions between electron clouds of atoms bonded to a central atom are assumed to be negligible. Therefore, orientation of several atoms around a given central atom is dependent only upon interactions between electron pairs around the central atom. A corollary to the basic premise of VSEPR theory is that unshared electron pairs require more space (that is, exhibit greater repulsions) than shared electron pairs.

a. Compounds Containing Single Bonds

The simplest organic molecule, methane, contains one carbon atom and four hydrogen atoms. Its formation may be represented as follows.

$$\cdot \overset{\cdot}{\underset{\cdot}{C}} \cdot \;+\; 4\,H \cdot \;\longrightarrow\; H : \overset{\overset{\displaystyle H}{\cdot\cdot}}{\underset{\underset{\displaystyle H}{\cdot\cdot}}{C}} : H$$

Methane

There are four shared electron pairs around carbon. They can be kept as far apart as possible by placing them at the corners of a tetrahedron with the carbon atom at the center of the tetrahedron. The **bond angle**, which is *the angle formed by two covalent bonds,* is 109.5° (Figure 1.1). This is the approximate configuration around all

---

[2] Many other texts invoke the concept of hybridization of *s* and *p* atomic orbitals to describe the bonding and the shapes of organic molecules.

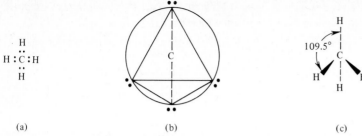

(a)   (b)   (c)

FIGURE 1.1 Various representations of the methane molecule. (a) Lewis structure; (b) structure showing electron pair distributions; (c) structure showing bond angle.

carbon atoms in which the carbon atoms are bonded to four other atoms, as in alkanes (Chapter 2), alkyl halides (Chapter 5), alcohols and ethers (Chapter 6), and amines (Chapter 9).

b. Compounds Containing Multiple Bonds

Double or triple bonds involving two or three shared electron pairs can be considered as just another region of electron density and do not affect the gross shape of a molecule. However, they have larger electron clouds than a single shared electron pair. Therefore, like unshared electron pairs, double and triple shared electron pairs cause greater repulsion and occupy more space than single shared electron pairs.

The Lewis structure for C_2H_4 shows three regions of electron density around each carbon atom (Figure 1.2). They can be kept as far apart as possible by placing them at the corners of an equilateral triangle with a carbon atom in the center of, and thus in the plane of, the equilateral triangle. However, the double shared electron pair occupies more space than the single shared electrons pairs, thus opening the H—C—C bond angles and closing the H—C—H bond angle. These predictions are verified by the experimentally determined bond angles. Similar bond angles around a doubly bonded carbon atom are typical for all alkenes (Chapter 3), as well as for aldehydes and ketones (Chapter 7), acids and esters (Chapter 8), and amides (Chapter 9).

FIGURE 1.2 (a) Lewis structure, (b) structure showing electron pair distribution, and (c) structure showing bond angles for ethylene. All six atoms of ethylene lie in the same plane.

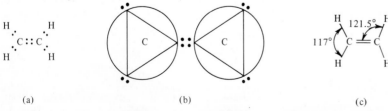

(a)   (b)   (c)

1.6 *Molecular Models and Chemical Formulas*

√

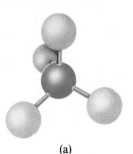

(a)

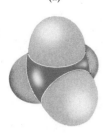

(b)

FIGURE 1.3 Molecular models of methane. (a) Ball-and-stick model; (b) space-filling model.

Molecular models are often employed to illustrate organic molecules. Two widely used methods of representation are

1. The ball-and-stick model in which different colored balls represent the different atoms, and sticks connecting them are used to represent the covalent bonds (Figure 1.3-a).
2. The space-filling model, which shows the shape of the molecule and the relative sizes of the atoms (Figure 1.3-b).

Most organic compounds contain more than one carbon atom and thus are considerably more complex than methane. The simplest compound that illustrates the covalent bonding of one carbon atom to another is ethane. Various representations of the ethane molecule are given in Figure 1.4. However, it is extremely difficult and very inconvenient to draw three-dimensional diagrams each time we wish to discuss a particular compound. Therefore, chemists have chosen to project the three-dimensional ball-and-stick model of a compound onto a two-dimensional surface. The resultant formulas for methane and for ethane are called structural formulas.

$$H-\underset{\underset{\displaystyle H}{|}}{\overset{\overset{\displaystyle H}{|}}{C}}-H \qquad H-\underset{\underset{\displaystyle H}{|}\ \underset{\displaystyle H}{|}}{\overset{\overset{\displaystyle H}{|}\ \overset{\displaystyle H}{|}}{C-C}}-H$$

Methane   Ethane

A **formula** conveys information about the composition and the structure of a molecule. The **empirical formula** (simplest formula) gives a minimum of information since it *expresses only the relative number of atoms in a molecule*. The **molecular formula** gives more information since it *expresses the actual number of atoms in a molecule*. In some cases, the molecular formula and the empirical formula are identical. The **structural formula**, as we have seen, is even more informative. It clearly *shows the arrangement of constituent atoms in a molecule*. A fourth formula, the

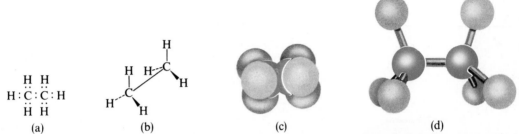

(a)   (b)   (c)   (d)

FIGURE 1.4 Representations of the ethane molecule. (a) Lewis structure; (b) perspective formula; (c) space-filling model; (d) ball-and-stick model.

TABLE 1.3 Chemical Formulas of Some Organic Compounds

Name of Compound	Class of Compound	Empirical Formula	Molecular Formula	Structural Formula	Condensed Structural Formula
Ethane	Alkane	CH_3	C_2H_6	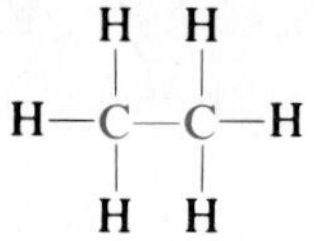 H–C–C–H (with H's above and below each C)	CH_3CH_3
Ethylene	Alkene	CH_2	C_2H_4	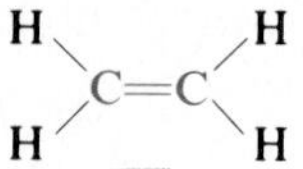 $H_2C{=}CH_2$ structure	$H_2C{=}CH_2$
Acetylene	Alkyne	CH	C_2H_2	 $H–C{\equiv}C–H$	$HC{\equiv}CH$
Ethyl alcohol	Alcohol	C_2H_6O	C_2H_6O	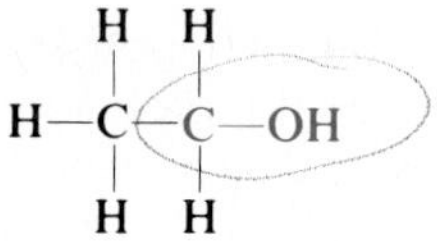 H–C–C–OH (with H's)	CH_3CH_2OH
Dimethyl ether	Ether	C_2H_6O	C_2H_6O	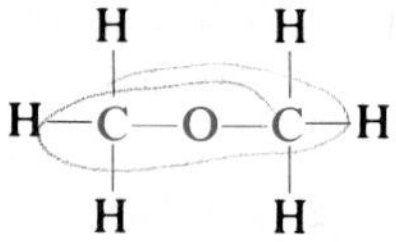 H–C–O–C–H (with H's)	CH_3OCH_3
Acetaldehyde	Aldehyde	C_2H_4O	C_2H_4O	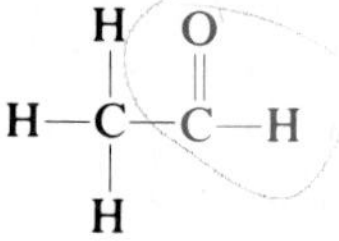 H–C–C(=O)–H (with H's)	CH_3CHO
Acetone	Ketone	C_3H_6O	C_3H_6O	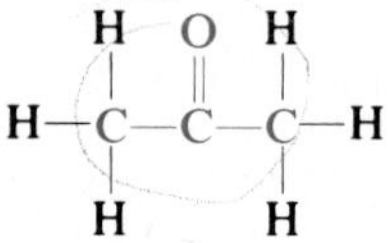 H–C–C(=O)–C–H (with H's)	CH_3COCH_3
Acetic acid	Carboxylic acid	CH_2O	$C_2H_4O_2$	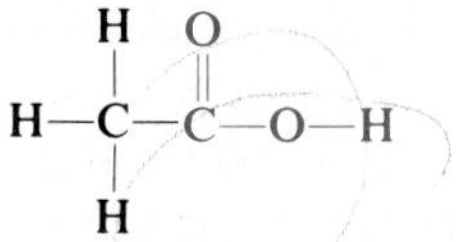 H–C–C(=O)–O–H (with H's)	CH_3COOH
Methyl formate	Ester	CH_2O	$C_2H_4O_2$	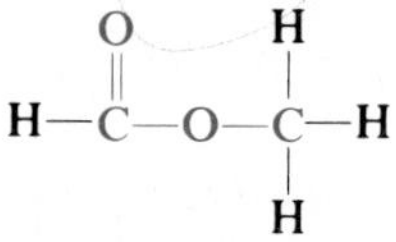 H–C(=O)–O–C–H (with H's)	$HCOOCH_3$
Acetamide	Amide	C_2H_5ON	C_2H_5ON	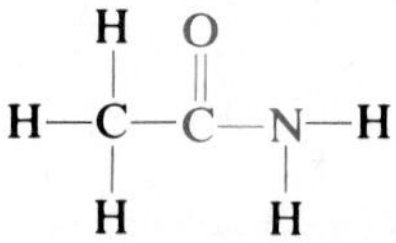 H–C–C(=O)–N–H (with H's)	CH_3CONH_2
Ethylamine	Amine	C_2H_7N	C_2H_7N	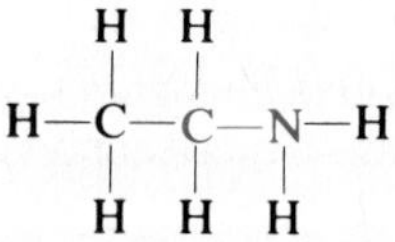 H–C–C–N–H (with H's)	$CH_3CH_2NH_2$

condensed structural formula, is perhaps the most widely employed by organic chemists. As the name implies, it is a shorthand (and less descriptive) method of representing the structural formula. The convention is to omit the single bonds that connect the atoms within the molecule. Exceptions are made when we wish to emphasize a specific bond. Table 1.3 illustrates the differences that exist among all four types of chemical formulas, using as examples specific organic compounds that are studied later in this text. (The functional group of each class of compounds is shown as the shaded portion of the structural formula—see Section 1.8.)

1.7 *Isomers*

For the most part, the empirical and molecular formulas contain enough information for the inorganic chemist, since rarely do two different inorganic compounds have the same formula. They are, however, of limited use in describing organic compounds, as is clearly illustrated in Table 1.3. Notice that the molecular formula for both ethyl alcohol and dimethyl ether is C_2H_6O, and the molecular formula for both acetic acid and methyl formate is $C_2H_4O_2$. These are not exceptional cases. Commonly, several different organic compounds will have identical chemical composition but widely different chemical and physical properties. (This fact was cited in Section 1.1 as being partly responsible for the great abundance of organic compounds.) *Compounds that have the same molecular formula but different structural formulas* are **isomers** (Greek, *isos,* equal; *meros,* part). The phenomenon that describes the existence of such compounds is **isomerism.** In our study of organic and biological compounds, we shall refer to isomerism on many occasions.

1.8 *Homologous Series and Functional Groups*

The total number of possible organic compounds is infinite. However, these compounds may be divided into a comparatively small number of classes on the basis of similarities in structure. Therefore, thorough knowledge of a few members of each class familiarizes one with the chemical and physical properties of almost all the other members of that class. *A family of similar compounds* is a **homologous series** (Greek, *homos,* same). *The individual members of the series* are **homologs.** Characteristics of a homologous series include the following.

1. All compounds in the series contain the same elements and can be represented by a single general formula.
2. The molecular formula of each homolog differs from the one above it and the one below it by a "—CH_2—" increment.
3. There is a gradual variation in physical properties with increasing molecular weight.
4. All compounds in the series have similar chemical properties.

Table 1.3 shows that each of the classes of compounds (alkene, alkyne, aldehyde, ketone, etc.) is distinguished from each of the other classes by some unique structural feature. Such a structural feature is a functional group. A **functional group** is defined as *any particular arrangement of a few atoms that bestows characteristic properties on an organic molecule*. Each member of a homologous series contains the same functional group, and the characteristic differences between the families of organic compounds are a result of differing functional groups. The following chapters treat each of the important homologous series of organic compounds. We shall see that most organic reactions involve the transformation of one functional group into another.

Exercises

1.1 What was the major deterrent to the development of organic chemistry prior to the nineteenth century? What was the significance of Wöhler's experiment?

1.2 The number of different carbon-containing compounds appears to be unlimited, whereas there are relatively many fewer compounds of the other elements. Explain.

1.3 Briefly discuss how organic and inorganic compounds differ.

1.4 Suggest a test that one might use to determine if a newly discovered compound is organic or inorganic.

1.5 What property of the carbon atom accounts for the occurrence of so many carbon compounds?

1.6 Why is it preferable to write structural formulas rather than molecular formulas to designate organic compounds?

1.7 Write Lewis structures for the following molecules.

(a) F_2
(b) CO_2
(c) SO_2
(d) H_2O
(e) NH_3
(f) H_2CO
(g) HCN
(h) HOCl
(i) CH_3Cl
(j) HONO
(k) C_3H_6
(l) C_3H_8
(m) PCl_5
(n) $HClO_4$
(o) $MgCl_2$
(p) LiBr

1.8 Which of the following molecules are polar and which are nonpolar?

(a) HCl
(b) H_2O
(c) CH_3F
(d) $O\!=\!C\!=\!O$
(e) CCl_4
(f) CH_3OH
(g) CH_3NH_2
(h) CH_3CH_3

1.9 What do the melting points of LiH (680°C) and methane (−182°C) suggest as to the type of bonding in the molecules of each substance?

1.10 Describe the physical properties (melting point, boiling point, solubility in water) of (a) a typical ionic compound and (b) a typical covalent compound.

1.11 List five substances that are composed of organic molecules and list five inorganic substances.

1.12 What are the approximate bond angles of the colored carbons in each of the following compounds?

(a) CH_3CH_2OH
(b) $CH_3C\!\equiv\!CH$
(c) $CH_3CH_2CH_2NH_2$
(d) $CH_3CH_2CH_2OCH_3$
(e) $CH_3CH_2CH\!=\!CHCH_3$
(f) $CH_3CH_2CH_2CH_3$

(g) $CH_3CH_2\overset{\displaystyle O}{\overset{\|}{C}}CH_3$
(h) $CH_3\overset{\displaystyle O}{\overset{\|}{C}}\!-\!OCH_3$

(i) $CH_3CH_2CH_2\overset{\displaystyle O}{\overset{\|}{C}}\!-\!H$
(j) $CH_3CH_2\overset{\displaystyle O}{\overset{\|}{C}}\!-\!OH$

(k) $H_2N\!-\!\overset{\displaystyle O}{\overset{\|}{C}}\!-\!NH_2$
(l) $CH_3CH_2\overset{\displaystyle O}{\overset{\|}{C}}\!-\!NH_2$

1.13 Identify the class to which each of the compounds in Exercise 1.12 belongs. (Refer to Table 1.3.)

1.14 Refer to Table 1.3 and draw structural formulas for four organic compounds that do not have isomers.

1.15 Write the molecular formula for the following compounds.

(a) $CH_3CH_2CH_2CH_2CH_3$
(b) $CH_3C\!\equiv\!CCH_2CH_3$
(c) $CH_3CH_2CH\!=\!C(CH_3)_2$
(d) $CH_3CH_2CHBr_2$
(e) $CH_3CH_2CH_2OCH(CH_3)_2$
(f) $CH_3CH_2NH_2$

(g) $CH_3CH_2C(CH_3)_2\overset{\displaystyle O}{\overset{\|}{C}}\!-\!H$
(h) $CH_3\overset{\displaystyle O}{\overset{\|}{C}}CH_2CH_3$

(i) $CH_3CH_2\overset{\displaystyle OH}{\overset{|}{C}H}CH_3$
(j) $CH_3\overset{\displaystyle O}{\overset{\|}{C}}\!-\!NH_2$

(k) $CH_3CH_2CH_2\overset{\displaystyle O}{\overset{\|}{C}}\!-\!OCH_3$
(l) $CH_3CH_2\overset{\displaystyle O}{\overset{\|}{C}}\!-\!OH$

1.16 Write condensed structural formulas for the following compounds.

(a)
```
     H  H  Br
     |  |  |
H —  C— C— C —H
     |  |  |
     H  H  H
```

(b)
```
     H  H  H  H
     |  |  |  |
H —  C— C— C— C —OH
     |  |  |  |
     H  H  H  H
```

(c)
```
     H           H  H
     |           |  |
H —  C— C≡C— C— C —H
     |           |  |
     H           H  H
```

(d)
```
     H  H     H
     |  |     |
H —  C— C— O— C —H
     |  |     |
     H  H     H
```

(e)
```
     H  H  H
     |  |  |
H —  C— C— N —H
     |  |
     H  H
```

(f)
```
     H  H  H  H       H
     |  |  |  |        |
H —  C— C— C— C= C— C —H
     |  |  |       |  |
     H  H  H       H  H
```

1.17 Write structural formulas to represent the different isomers of the following compounds.

(a) C_3H_8O (3)   (b) $C_2H_4Cl_2$ (2)
(c) $C_2H_3Cl_3$ (2)   (d) C_3H_6O (2)
(e) $C_2H_4O_2$ (3)   (f) C_3H_7Cl (2)
(g) C_3H_6BrCl (5)   (h) C_4H_9Br (4)
(i) $C_2H_6O_2$ (5)   (j) C_2H_7N (2)

1.18 Which of the following pairs of compounds are isomers?

(a) $CH_3CH_2CH_3$ and $CH_3CH_2CH_2CH_3$

(b) $CH_3CH=CH_2$ and H_2C-CH_2 / CH_2 (cyclopropane)

(c) $CH_3CH=CH_2$ and $CH_3C≡CH$

(d) H_2C-CH / CH_3 and H_2C-CH_2 / H_2C-CH_2

(e) CH_3CH_2OH and $CH_3C(=O)H$

(f) $CH_3CH_2C(=O)H$ and CH_3CCH_3 (with =O)

(g) $CH_3CH_2C(=O)OH$ and $CH_3C(=O)OCH_3$

(h) $CH_3-C(=O)OCH_3$ and $H-C(=O)OCH_2CH_3$

(i) $CH_3CH_2OCH_2CH_3$ and $CH_3CHCH_2CH_3$ (with OH)

(j) $CH_3CH_2C(=O)-NH_2$ and $CH_3C(=O)-N(H)-CH_3$

(k) $CH_3CH_2-N(H)-CH_3$ and $CH_3CH_2CH_2-NH_2$

(l) $CH_2BrCH_2CH_2Cl$ and $CH_2ClCH_2CH_2Br$

1.19 Can homologs be isomers? Explain.

1.20 Define the following terms and give an example for each.

(a) cation   (b) anion
(c) chemical bond   (d) ionic bond
(e) covalent bond   (f) single bond
(g) double bond   (h) polar bond
(i) nonpolar bond   (j) bond length
(k) bond energy   (l) bond angle
(m) empirical formula   (n) molecular formula
(o) structural formula   (p) isomers
(q) homologous series   (r) functional group

1.21 A hydrocarbon was shown by analysis to contain 80% carbon and 20% hydrogen. At STP (standard temperature, 0°C, and pressure, 760 mmHg) 1 L of the hydrocarbon weighed 1.34 g. What is the molecular formula of the hydrocarbon?

1.22 Quantitative combustion of an organic compound indicated that it contained 82.8% carbon and 17.2% hydrogen.
(a) What is the empirical formula of this compound?
(b) If the molecular weight of the compound is 58, what is the molecular formula of the compound?
(c) Write structural formulas for the two isomers of this compound.

2

The Saturated Hydrocarbons

The saturated hydrocarbons are often referred to as the *parent* compounds of organic chemistry since all other known organic compounds can be considered to be derivatives of them. **Hydrocarbons,** as the name implies, are *compounds composed only of carbon and hydrogen atoms.* The adjective, **saturated,** describes the type of bonding within the hydrocarbon molecule. It signifies that *each carbon atom is covalently bonded to four other atoms by **single** bonds.* The compounds are also referred to as paraffin hydrocarbons or alkanes. The name *paraffin* (Latin, *parum affinis,* little activity) alludes to their unreactive nature. The name **alkane** is the generic name for this class of compounds based upon the IUPAC system of nomenclature.

The saturated hydrocarbons belong to the *aliphatic* series of compounds. (The two other main series, aromatic and heterocyclic, are mentioned later in the text.) Since many of the first hydrocarbons to be studied were derived from fatty acids, the name aliphatic (Greek, *aleiphatos,* fat) was applied to them to indicate their source. Today, aliphatic compounds comprise all the open-chain (acyclic) hydrocarbons and cyclic (also called alicyclic) hydrocarbons as well as their derivatives.

2.1 Nomenclature of Alkanes

The general formula for the alkane homologs is C_nH_{2n+2}. Table 2.1 contains the names and formulas for the first ten saturated hydrocarbons. Also included is a listing of the number of possible isomers that correspond to each molecular formula. The number of possible isomers increases extremely rapidly with each addition of a carbon atom to the chain. For example, addition of four carbon atoms and eight hydrogen atoms to hexane, to form decane, causes a fifteen-fold increase in the number of possible isomers. Even more striking is the calculation that there are 366,319 possible isomers of the alkane whose molecular formula is $C_{20}H_{42}$ and there are 62,491,178,805,831 possible isomers of the alkane whose molecular formula is $C_{40}H_{82}$. These numbers are given just by way of illustration, and in no way are meant to imply that all of the isomers occur in nature or have been synthesized.

a. Common System

There is no difficulty in naming the first three alkanes, since there is only one structure that can be written for each of these compounds. However, it is possible to write two (and only two) structural formulas that correspond to the molecular formula C_4H_{10} (Figure 2.1). The important point here is that not only can two separate structures be drawn that correspond to this formula, but two distinct compounds actually exist. Each is found in nature and each can be synthesized in the laboratory. They are structural isomers of each other and thus have different physical properties.

Compound I is a straight-chain hydrocarbon. Its four carbon atoms are arranged in one continuous chain. Compound II is a branched-chain hydrocarbon. Three of its carbon atoms are arranged in a straight chain, and the fourth is branched off from the middle carbon atom. The straight-chain hydrocarbon is sometimes designated by prefixing the letter *n* (for normal) to the name of the compound. Thus, compound I is *n*-butane, or simply butane. The branched-chain compound is designated by the prefix *iso-*; compound II is isobutane.

There are only three different structural formulas that can be drawn to represent the compound with the molecular formula C_5H_{12} (Figure 2.2).

TABLE 2.1   Straight-Chain Alkanes

Name	Molecular Formula	Condensed Structural Formula	Number of Possible Isomers
Methane	CH_4	CH_4	1
Ethane	C_2H_6	CH_3CH_3	1
Propane	C_3H_8	$CH_3CH_2CH_3$	1
Butane	C_4H_{10}	$CH_3(CH_2)_2CH_3$	2
Pentane	C_5H_{12}	$CH_3(CH_2)_3CH_3$	3
Hexane	C_6H_{14}	$CH_3(CH_2)_4CH_3$	5
Heptane	C_7H_{16}	$CH_3(CH_2)_5CH_3$	9
Octane	C_8H_{18}	$CH_3(CH_2)_6CH_3$	18
Nonane	C_9H_{20}	$CH_3(CH_2)_7CH_3$	35
Decane	$C_{10}H_{22}$	$CH_3(CH_2)_8CH_3$	75

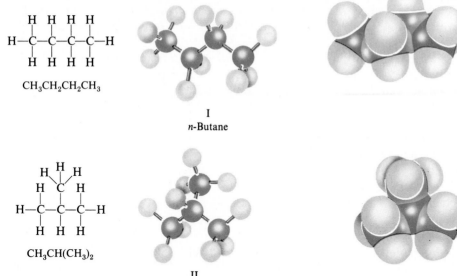

CH₃CH₂CH₂CH₃

I
n-Butane

CH₃CH(CH₃)₂

II
Isobutane

FIGURE 2.1 Structures of the butane isomers.

FIGURE 2.2 Structures of the pentane isomers.

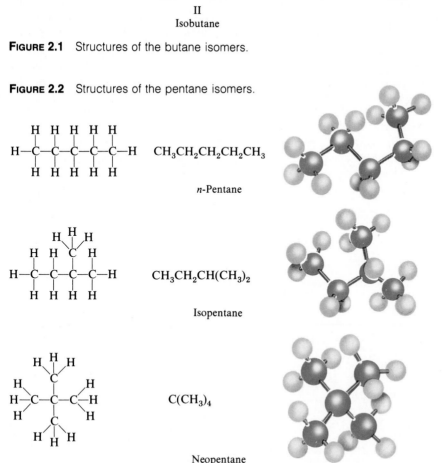

CH₃CH₂CH₂CH₂CH₃

n-Pentane

CH₃CH₂CH(CH₃)₂

Isopentane

C(CH₃)₄

Neopentane

b. Alkyl Groups

The group of atoms that results when a hydrogen atom is removed from an alkane is called an **alkyl group.** Thus, the general formula for an alkyl group is C_nH_{2n+1}. The group is named by replacing the *-ane* suffix of the parent hydrocarbon with *-yl* (Table 2.2). It is important to note that alkyl groups are not independent molecules. Rather, they exist as parts of molecules.

Notice that not all of the hydrogen atoms of the propane molecule are equivalent. Six of them, three attached to each terminal carbon, are equivalent. The two attached to the middle carbon, however, are different. In Section 6.1 this difference is more fully explained. Let it suffice for now to say that if either of the hydrogen atoms on the middle carbon is removed, a different alkyl group results.

For convenience, the organic chemist often represents an alkyl group by the symbol **R.** This symbol is used whenever it is desirable to represent a general class of compounds. Thus, the symbol for the family of saturated hydrocarbons is **R—H.**

c. IUPAC System

For the higher homologs of this series, as well as for other classes of organic compounds, the system of using prefixes is unsatisfactory. It would overtax the memory of anything short of a computer, and language simply cannot provide meaningful prefixes. To tackle this problem, an international meeting of chemists was held at Geneva, Switzerland, in 1892. The meeting resulted in a simple, unequivocal system for naming organic compounds. The system was modified by the International Union of Chemists and is kept up to date by its successor, the International Union of Pure and Applied Chemistry (IUPAC). What has evolved is

TABLE 2.2 Common Alkyl Groups

Parent Hydrocarbon	Alkyl Group	Condensed Formula
Methane 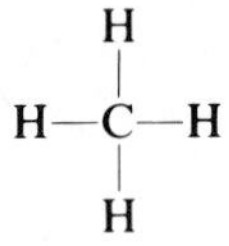	Methyl 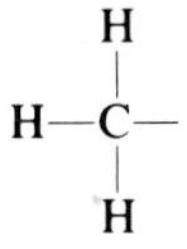	CH_3-
Ethane 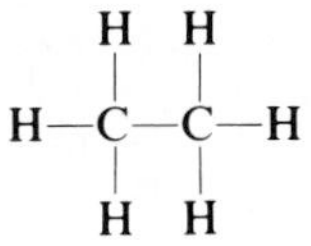	Ethyl 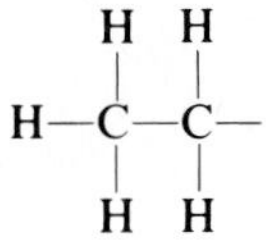	CH_3CH_2- or C_2H_5-
Propane 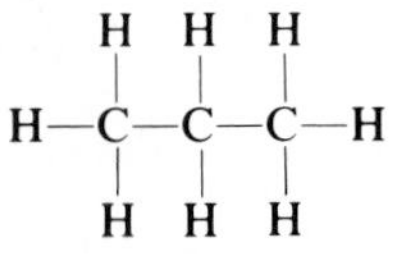	Propyl 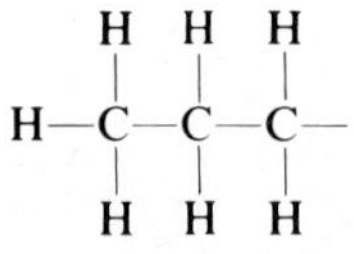	$CH_3CH_2CH_2-$ or C_3H_7-
	Isopropyl 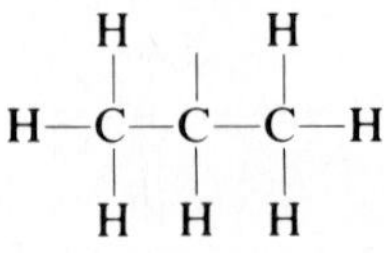	CH_3CHCH_3 or $(CH_3)_2CH-$

a set of rules known as the IUPAC System of Nomenclature. The rules for the alkanes are as follows.

1. Saturated hydrocarbons are named according to the longest continuous chain of carbons in the molecule (rather than the total number of carbon atoms). This is the parent compound.
2. The suffix -ane indicates that the molecule is a saturated hydrocarbon.
3. The name of the parent hydrocarbon is modified by noting what alkyl groups are attached to the chain.
4. The chain is numbered, and the position of each substituent alkyl group is indicated by the number of the carbon atom to which it is attached. The chain is numbered in such a way that the substituents occur on the carbon atoms with the lowest numbers.
5. Names of the substituent groups are placed in alphabetical order before the name of the parent compound.
6. If the same alkyl group appears more than once, the numbers of all the carbons to which it is attached are expressed. If the same group appears more than once on the same carbon, the number of the carbon is repeated as many times as the group appears.
7. Hyphens are used to separate numbers from names of substituents; numbers are separated from each other by commas. The number of identical groups is indicated by the Greek prefixes *di-, tri-, tetra,* etc. These prefixes are **not** considered in determining the alphabetical order of the substituents (e.g., ethyl precedes dimethyl).
8. The last alkyl group to be named is prefixed to the name of the parent alkane forming one word.

The application of these rules may appear to be very difficult, but in reality, learning the systematic nomenclature is probably the easiest part of organic chemistry. All that is required is a good deal of practice. (Students are encouraged to do many nomenclature problems early in the semester. It has been the author's experience that through repeated practice students quickly learn how to apply these rules, and organic nomenclature becomes a game rather than a chore.)

2.2 Application of the IUPAC Rules

Table 2.3 shows the isomers of butane and pentane. Examination of the table raises the following questions: If compounds I and II are both isomers of butane, why is one called butane and the other methylpropane? Similarly, if compounds III, IV, and V are all isomers of pentane, C_5H_{12}, why is one designated as pentane, one as a derivative of butane, and one as a derivative of propane? The answer to both questions is that under the IUPAC system, a compound may be an isomer of another compound without having the same parent name. The only criterion that two compounds must meet in order to be isomers is that they have the same molecular formula. It is to be emphasized that according to the IUPAC system, the name of the parent hydrocarbon corresponds to the number of carbons in the

TABLE 2.3   Isomers of Butane and Pentane

Isomer	Formula	IUPAC Name
I	$CH_3CH_2CH_2CH_3$	Butane
II	$\underset{\displaystyle}{CH_3}$ $CH_3\overset{\displaystyle CH_3}{\underset{\displaystyle \vert}{C}}HCH_3$	2-Methylpropane
III	$CH_3CH_2CH_2CH_2CH_3$	Pentane
IV	$CH_3CH_2\overset{\displaystyle CH_3}{\underset{\displaystyle \vert}{C}}HCH_3$	2-Methylbutane
V	$CH_3\!-\!\overset{\displaystyle CH_3}{\underset{\displaystyle \underset{\textstyle CH_3}{\vert}}{\overset{\vert}{C}}}\!-\!CH_3$	2,2-Dimethylpropane

longest continuous chain, not to the total number of carbon atoms in the molecule. This point is well illustrated by the IUPAC names for the five isomers of hexane, C_6H_{14} (Table 2.4).

There are several aspects of nomenclature that invariably trouble the beginning student in organic chemistry. Repeated practice will insure that the student will avoid the following common pitfalls.

1. *Thinking that a substituent positioned above the chain is different from the substituent positioned below the chain.* There is free rotation about any of the single bonds in the molecule, so above and below have no meaning. The

TABLE 2.4   Isomers of Hexane

Formula	IUPAC Name
$CH_3CH_2CH_2CH_2CH_2CH_3$	Hexane
$CH_3CH_2CH_2\overset{\displaystyle CH_3}{\underset{\displaystyle \vert}{C}}HCH_3$	2-Methylpentane
$CH_3CH_2\overset{\displaystyle CH_3}{\underset{\displaystyle \vert}{C}}HCH_2CH_3$	3-Methylpentane
$CH_3CH_2\overset{\displaystyle CH_3}{\underset{\displaystyle \underset{\textstyle CH_3}{\vert}}{\overset{\vert}{C}}}CH_3$	2,2-Dimethylbutane
$CH_3\overset{\displaystyle CH_3}{\underset{\displaystyle \underset{\textstyle CH_3}{\vert}}{\overset{\vert}{C}}}HCHCH_3$	2,3-Dimethylbutane

Chapter 2 · The Saturated Hydrocarbons

significant thing about the position of a substituent group is the number of the carbon atom to which it is attached.

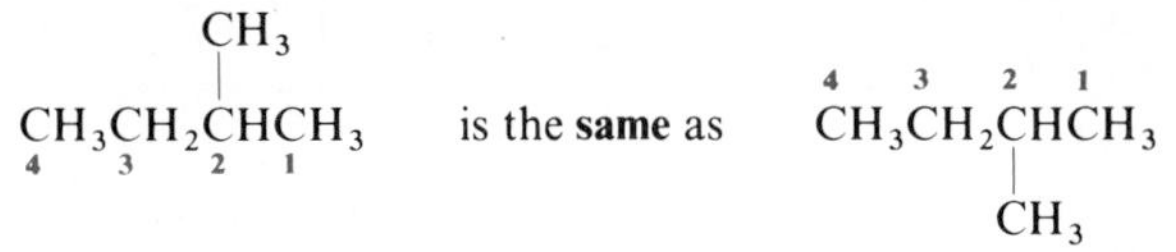

$$\underset{4}{CH_3}\underset{3}{CH_2}\underset{2}{CH}\underset{1}{CH_3} \quad \text{with } CH_3 \text{ branch} \qquad \text{is the same as} \qquad \underset{4}{CH_3}\underset{3}{CH_2}\underset{2}{CH}\underset{1}{CH_3} \quad \text{with } CH_3 \text{ branch}$$

2. *Failing to identify the longest continuous chain.* Nothing but convention dictates that an organic molecule must be represented in the form of a linear chain. Examine curiously branched structures and seek out the longest continuous chain.

is the same as

2,2,4-Trimethylheptane

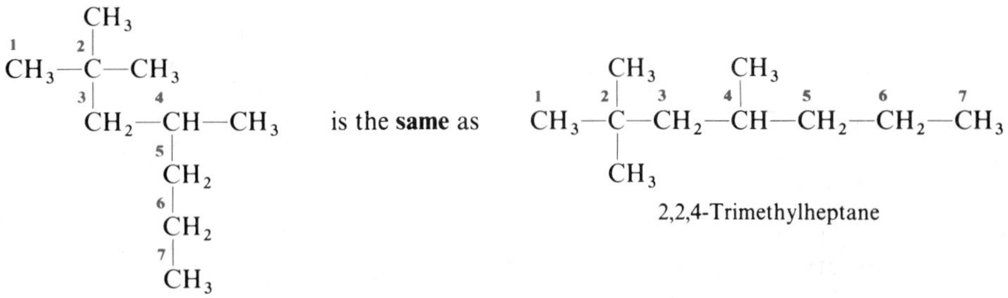

3. Being fooled by an *apparent* alkyl substituent on a terminal carbon.

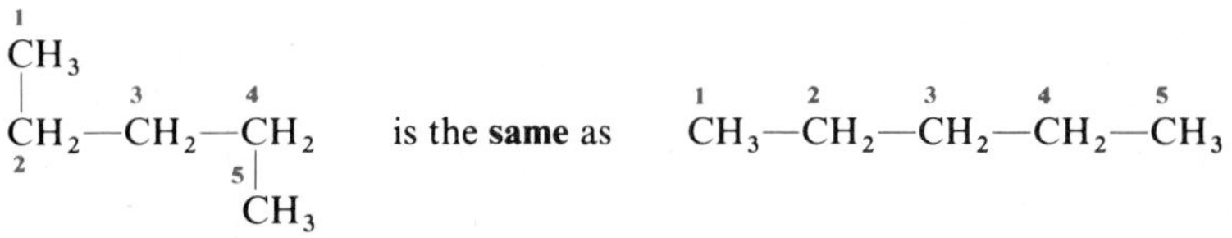

is the same as $\quad \underset{1}{CH_3}-\underset{2}{CH_2}-\underset{3}{CH_2}-\underset{4}{CH_2}-\underset{5}{CH_3}$

4. *Numbering from the wrong end.*

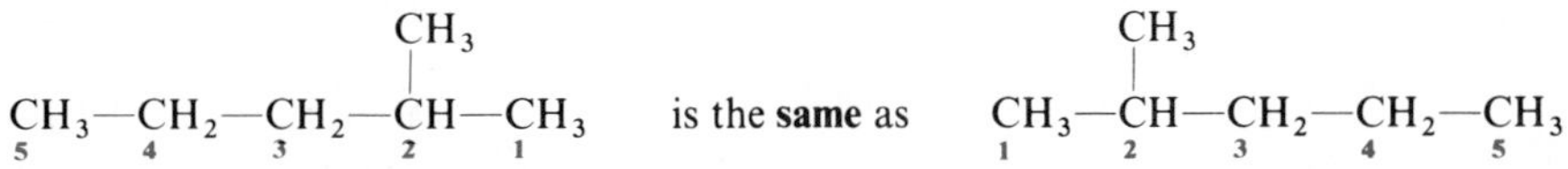

$$\underset{5}{CH_3}-\underset{4}{CH_2}-\underset{3}{CH_2}-\underset{2}{CH}-\underset{1}{CH_3} \quad (CH_3) \qquad \text{is the same as} \qquad \underset{1}{CH_3}-\underset{2}{CH}-\underset{3}{CH_2}-\underset{4}{CH_2}-\underset{5}{CH_3} \quad (CH_3)$$

5. *Failing to write the same number twice and/or failing to use the prefix di-, tri-, etc., when two identical substituents are bonded to the same carbon atom.*

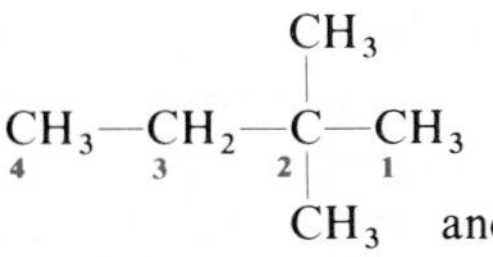

$$\underset{4}{CH_3}-\underset{3}{CH_2}-\underset{2}{C}-\underset{1}{CH_3} \quad (CH_3) \qquad \text{is 2,2-dimethylbutane}$$

and **not** 2-dimethylbutane or 2,2-methylbutane

Table 2.5 contains some further examples of the application of the IUPAC system for naming organic compounds.

TABLE 2.5 Examples of IUPAC Nomenclature

Condensed Structural Formula	Rewritten and Numbered	IUPAC Name
$CH_3CH_2CH(CH_3)CH_2CH(CH_3)_2$	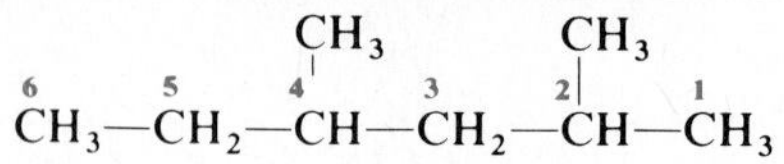 (see structure below)	2,4-Dimethylhexane **not** 3,5-Dimethylhexane
$(C_2H_5)_2CHCH(CH_3)CH_2CH_2CH_3$	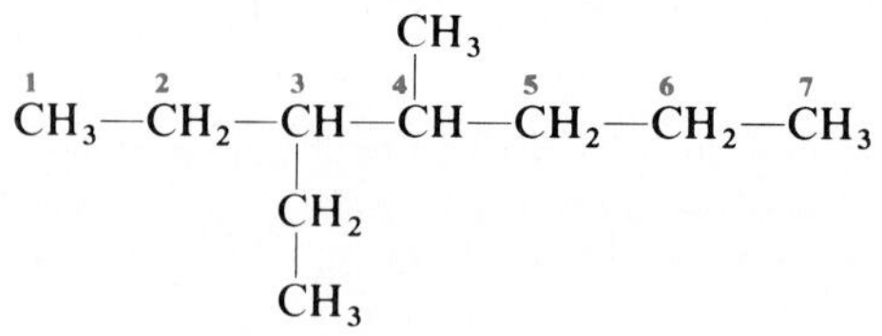 (see structure below)	3-Ethyl-4-methylheptane **not** 4-Methyl-5-ethylheptane
$(CH_3)_2CHCH(C_3H_7)_2$	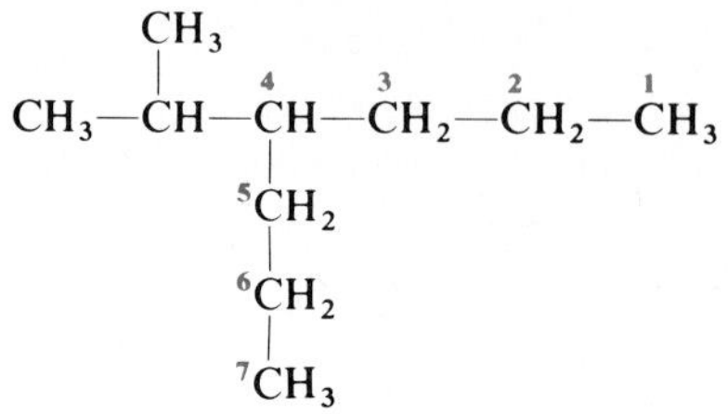 (see structure below)	4-Isopropylheptane **not** 2-Methyl-3-propylhexane
$(CH_3)_2CHCH_2CH_2C(CH_3)_3$	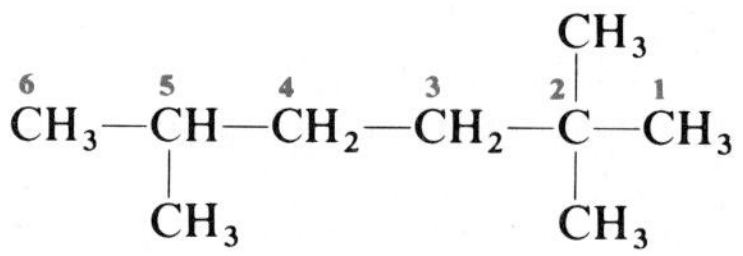 (see structure below)	2,2,5-Trimethylhexane **not** 2,5-Trimethylhexane
$(C_2H_5)_2CHC(CH_3)(C_2H_5)_2$	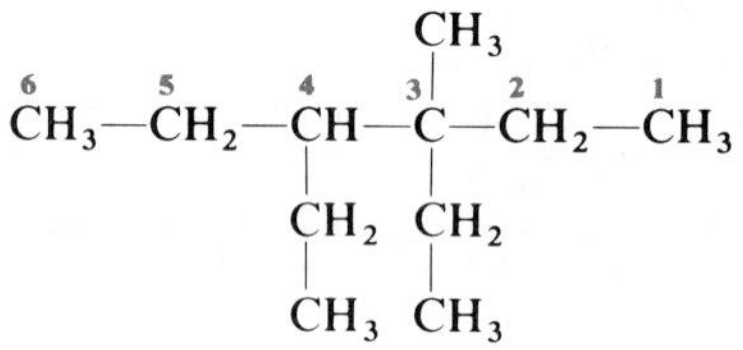 (see structure below)	3,4-Diethyl-3-methylhexane **not** 3,4-Ethyl-3-methylhexane

Structure 1:
$$\overset{6}{CH_3}-\overset{5}{CH_2}-\overset{4}{CH}(CH_3)-\overset{3}{CH_2}-\overset{2}{CH}(CH_3)-\overset{1}{CH_3}$$

Structure 2:
$$\overset{1}{CH_3}-\overset{2}{CH_2}-\overset{3}{CH}(CH_2CH_3)-\overset{4}{CH}(CH_3)-\overset{5}{CH_2}-\overset{6}{CH_2}-\overset{7}{CH_3}$$

Structure 3:
$$CH_3-\overset{4}{CH}(CH_3)-\overset{3}{CH}(CH_2CH_2CH_3)-\overset{2}{CH_2}-\overset{1}{CH_3}$$

Structure 4:
$$\overset{6}{CH_3}-\overset{5}{CH}(CH_3)-\overset{4}{CH_2}-\overset{3}{CH_2}-\overset{2}{C}(CH_3)_2-\overset{1}{CH_3}$$

Structure 5:
$$\overset{6}{CH_3}-\overset{5}{CH_2}-\overset{4}{CH}(CH_2CH_3)-\overset{3}{C}(CH_3)(CH_2CH_3)-\overset{2}{CH_2}-\overset{1}{CH_3}$$

2.3 Physical Properties of Alkanes

The alkanes are nonpolar molecules. They are, therefore, insoluble in water, but they are soluble in organic solvents such as benzene and carbon tetrachloride. (Indeed, many of the hydrocarbons are commonly used as nonpolar solvents.) The hydrocarbons are generally less dense than water, their densities being less than 1.0 g/mL. They are colorless and tasteless, and many of them are odorless. The odor associated with natural gas is not from methane or any of the other alkanes, but rather from an odorant (usually CH_3SH) purposely added to the gas in sufficient quantities to allow for the detection of leaks. Table 2.6 summarizes the properties of the first ten straight-chain alkanes.

All organic chemistry textbooks include many tables of physical properties. The figures in these tables are not meant to be memorized, but are given in order to present a numerical description of the physical characteristics of organic molecules and to serve as standards of purity. The physical states of substances at room

TABLE 2.6 Physical Properties of Some Alkanes

Name	Molecular Formula	MP (°C)	BP (°C)	Density (g/mL)	Normal State
Methane	CH_4	−182	−164	—	Gas
Ethane	C_2H_6	−183	−89	—	Gas
Propane	C_3H_8	−190	−42	—	Gas
n-Butane	C_4H_{10}	−138	−1	—	Gas
n-Pentane	C_5H_{12}	−130	36	0.63	Liquid
n-Hexane	C_6H_{14}	−95	69	0.66	Liquid
n-Heptane	C_7H_{16}	−91	98	0.68	Liquid
n-Octane	C_8H_{18}	−57	125	0.70	Liquid
n-Nonane	C_9H_{20}	−51	151	0.72	Liquid
n-Decane	$C_{10}H_{22}$	−30	174	0.73	Liquid

temperature are obtained directly from a knowledge of their melting points and boiling points. Room temperature is considered to be approximately 67°F or about 20°C. If the melting point of a substance is above 20°C, the substance will exist as a solid at room temperature. (For example, the first member of the alkane series that is a naturally occurring solid is n-octadecane, $C_{18}H_{38}$, mp 28°C.) If the boiling point of a substance is above 20°C, the substance will exist as a liquid at room temperature. (Note that n-pentane, bp 36°C, is the first liquid alkane.) Similarly, knowledge of the density of a substance indicates whether the substance is lighter or heavier than water. (The density of water is 1.00 g/mL.) We observe that oil and grease do not mix with water, but rather float on the surface. They do not mix with water because they are insoluble in water; they float on top of the water because they are less dense than water. Densities are also useful in calculations that call for the conversion of a certain mass of a liquid into the corresponding volume of that liquid, or vice versa. (Recall that density equals the mass of a substance divided by its volume.)

As indicated in Table 2.6, the first four members of the alkane series are gases. A comparison of the densities of these gases with the density of air allows one to predict whether or not they will rise in the air. Natural gas is composed chiefly of methane, which has a density of about 0.65 g/L. The density of air is about 1.29 g/L. Natural gas, then, is much lighter than air, and it will rise in a room in which there may be a natural gas leak. Once the leak is detected and eliminated, the gas can be removed from the room by opening an upper window. On the other hand, the three constituents of bottled gas[1] are much heavier than air. Propane has a density of 1.6 g/L, and the butane isomers have densities of about 2.0 g/L. If bottled gas escapes into a room, it collects near the floor and a much more serious fire hazard develops because it is then more difficult to rid the room of the gas.

As shown in Table 2.6, the boiling points of the straight-chain alkanes increase with increasing molecular weight. This general rule holds true within all the families of organic compounds whenever the straight-chain homologs of the family are

[1]Bottled gas consists chiefly of propane (with some butane and isobutane) that, through pressure, has been condensed to a liquid (liquefied petroleum gas, LPG) so that it may be compactly stored and shipped in metal containers.

TABLE 2.7 Physical Properties of the Isomers of Butane and Pentane

IUPAC Name	Condensed Structural Formula	MP (°C)	BP (°C)
Butane	$CH_3(CH_2)_2CH_3$	-138	-1
2-Methylpropane	$CH_3CH(CH_3)_2$	-159	-11
Pentane	$CH_3(CH_2)_3CH_3$	-130	36
2-Methylbutane	$CH_3CH_2CH(CH_3)_2$	-160	30
2,2-Dimethylpropane	$(CH_3)_4C$	-17	9

considered. Larger molecules are able to wrap around and interact with one another, and thus more energy is required to separate them. In general, the straight-chain isomers have higher boiling points than the branched-chain isomers. The former compounds can be likened to strands of spaghetti. The molecules can be very closely packed together, resulting in relatively strong intermolecular van der Waals forces of attraction. Van der Waals forces depend on the total area of contact available between two molecules. The greater this area of contact, the greater is the attractive force. Branched-chain hydrocarbons are more compact than their straight-chain isomers. Consequently, there is less surface area to interact. The van der Waals forces between molecules are weaker and the molecules can more easily escape from the liquid. Table 2.7 lists the melting points and boiling points of the isomers of butane and pentane.

Notice that there is no obvious trend that enables us to predict melting points. In general, the more symmetrical isomers tend to have higher melting points than the less symmetrical isomers (because symmetrical molecules can pack closer together

FIGURE 2.3 Tripalmitin, a typical fat molecule.

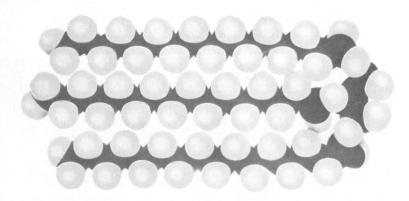

Chapter 2 · The Saturated Hydrocarbons

in the crystal lattice). Contrast 2,2-dimethylpropane with the less symmetrical pentanes in Table 2.7. Contrast, also, the very symmetrical octane isomer $(CH_3)_3CC(CH_3)_3$ with n-octane. The former is a solid and melts at $101°C$; the latter is a liquid whose melting point is $-57°C$.

An extensive review of the physical properties of the alkanes has been given here, not because these compounds are so important in and of themselves, but rather because of their contributions to the structures of the other families of organic and biological compounds. A knowledge of alkane properties is also vital to an understanding of the functions of lipids because large portions of their structures consist of segments of alkane moieties (Figure 2.3). One of the major functions of phospholipids (Section 12.9) and glycolipids (Section 12.10) is to serve as structural components of living tissues. Their biological importance here is dependent upon the presence of both polar and nonpolar groups, which enable them to bridge the gap between water-soluble and water-insoluble phases. This is a requisite in maintaining the selective permeability of cell membranes.

2.4 Chemical Properties of Alkanes

The alkanes are the most unreactive of all classes of organic compounds. At room temperature, alkanes do not react with such strong reagents as concentrated sulfuric acid, concentrated sodium hydroxide, sodium metal, and potassium permanganate. Under favorable reaction conditions, they can be made to react with oxygen (combustion), nitric acid, and chlorine and bromine. The reactions with nitric acid and with the halogens occur by a process known as **substitution.** In a substitution reaction involving an alkane, *a part of the reagent molecule substitutes for one of the hydrogen atoms* of the alkane.

Halogenation of Alkanes (Substitution of Halogen for Hydrogen) The student is probably aware that a mixture of natural gas and air can exist together indefinitely without entering into any reaction. As soon as a match is lit, however, the gas begins to burn. Similarly, mixtures of alkanes and either chlorine or bromine are quite stable when kept in dark containers. When the reaction mixtures are heated or exposed to the sunlight, rapid exothermic reactions occur. Therefore, heat and light must be involved in activating an otherwise unreactive mixture. Through considerable research, this reaction, as well as many of the others we shall study, has been elucidated. When we have arrived at that point in research where we know the pathway by which the reactants are converted to the products, we say that we know the *reaction mechanism* for the reaction under study.

The first step in the reaction mechanism for the chlorination or bromination of an alkane involves the splitting of a halogen molecule into two highly energetic halogen atoms. (Halogen atoms are much more reactive than the corresponding halogen molecules.) This initiation step is followed by attack of the energetic halogen atom upon a molecule of the alkane. The halogen atom pulls a hydrogen atom from the alkane to form a molecule of hydrogen halide and a new highly reactive species, referred to as a free radical. (**Free radicals** are *species having odd numbers of electrons;* they are important intermediates in many organic reactions.) This newly formed free radical then attacks an unreacted halogen molecule to form the halogenated alkane and another energetic halogen atom. The halogen atom then attacks another alkane molecule and the process continually repeats itself. (This reaction is not applicable to iodine or fluorine. Iodine is unreactive, whereas fluorine is too reactive to control.)

This particular type of repetitive sequence of two or more reactions is called a **chain reaction** because once it is initiated, it is self-perpetuating. Once initiated, the reaction will proceed rapidly in the absence of heat or light because it is highly exothermic. The process is terminated only by the chance combination of two radicals. The following is a general scheme for the chlorination of alkanes.

Initiation

$$Cl:Cl \xrightarrow[\text{or heat}]{\text{ultraviolet light}} 2\ Cl\cdot$$

Propagation

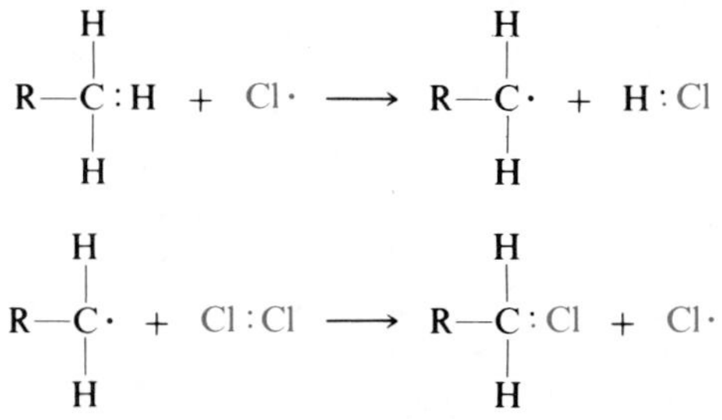

$$R-\underset{\underset{H}{|}}{\overset{\overset{H}{|}}{C}}:H + Cl\cdot \longrightarrow R-\underset{\underset{H}{|}}{\overset{\overset{H}{|}}{C}}\cdot + H:Cl$$

$$R-\underset{\underset{H}{|}}{\overset{\overset{H}{|}}{C}}\cdot + Cl:Cl \longrightarrow R-\underset{\underset{H}{|}}{\overset{\overset{H}{|}}{C}}:Cl + Cl\cdot$$

Termination

$$R-\underset{\underset{H}{|}}{\overset{\overset{H}{|}}{C}}\cdot + Cl\cdot \longrightarrow R-\underset{\underset{H}{|}}{\overset{\overset{H}{|}}{C}}:Cl$$

$$R-\underset{\underset{H}{|}}{\overset{\overset{H}{|}}{C}}\cdot + R-\underset{\underset{H}{|}}{\overset{\overset{H}{|}}{C}}\cdot \longrightarrow R-\underset{\underset{H}{|}}{\overset{\overset{H}{|}}{C}}:\underset{\underset{H}{|}}{\overset{\overset{H}{|}}{C}}-R$$

$$Cl\cdot + Cl\cdot \longrightarrow Cl:Cl$$

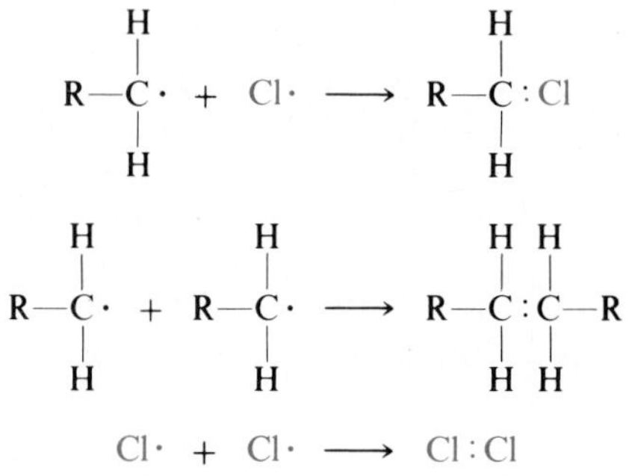

Free radical halogenation reactions are quite difficult to control. The halogenation of an alkane usually results in a mixture of products, which can be separated from one another by fractional distillation. The composition of the mixture depends on the relative amounts of the initial reagents and the reaction conditions.

$$CH_4 + Cl_2 \xrightarrow[\text{light}]{\text{heat or}} CH_3Cl + CH_2Cl_2 + CHCl_3 + CCl_4 + C_2H_6 + \text{Other products}$$

2.5 Free Radical Theory of Aging

Over the years, numerous theories have been advanced to explain the causes of human aging. The free radical theory is just one of these; although there is supportive evidence based upon experimental observations, this theory, like a half dozen others, is controversial and speculative.

The basic concept of the free radical theory of aging is that a steady flux of radicals exists in the body cells and these radicals can react with critical cellular components. Free radicals may be especially harmful when they attack the fatty acids present in the lipids of cell membranes (see Figure 12.6). The resulting altered lipids may seriously impede the diffusion of substances through the membrane—the flow of nutrients and oxygen into the cell and the flow of waste products out of the cell—and thus lead to death of the cell. Some of the damage to the membranes is probably repaired by cellular processes. Aging, according to this theory, could involve either an increased rate of membrane damage with time, or a decrease in the efficiency of repair, or both. Free radicals also cause degradation of DNA (Chapter 15) and destruction of the endothelial cells lining blood vessels. Both of these effects have been implicated in the aging process.

What is the source of these biological free radicals? Since radiation causes a cleavage of chemical bonds, one possibility is the dosage received from the background radiation of cosmic rays and radioactive elements. Although this radiation is partially responsible for cell damage and the low level of natural mutations, it probably has no effect on the aging process per se.[2] It is more likely that free radicals arise from the hydrogen peroxide (HOOH) produced in the oxidative deamination of amino acids (Section 18.3-b) and from the organic hydroperoxides (ROOH) formed by the oxidation of unsaturated lipids (Section 12.6).

It is well known that normal cleavage of a peroxide molecule to form radicals is an extremely slow process at the relatively low temperatures (37°C) of living systems.

$$H—O:O—H \longrightarrow 2\,OH\cdot \quad \text{(extremely slow)}$$

However, cellular fluid contains several transition metal ions (such as iron, copper) that can readily undergo one-electron oxidation-reduction reactions. Ferrous ions, for example, react with hydrogen peroxide or with organic hydroperoxides by a redox process to produce radicals at an exceedingly rapid rate. These radicals then attack the carbon–hydrogen bonds in membrane lipids, producing alkyl radicals, which in turn can attack other lipid molecules.

[2] William A. Pryor, *Chemical and Engineering News*, June 7, 1971, p. 34.

$$HO—OH + Fe^{2+} \longrightarrow HO\cdot + HO^- + Fe^{3+}$$
$$RO—OH + Fe^{2+} \longrightarrow RO\cdot + HO^- + Fe^{3+}$$
$$R—H + HO\cdot (or\ RO\cdot) \longrightarrow R\cdot + HOH\ (or\ ROH)$$

Ozone is another source of radicals. It reacts with unsaturated fatty acids to produce organic radicals. The normal ozone content of air is about 0.02 parts per million (ppm). (Ozone levels in smog normally reach 0.2–0.3 ppm, and a value as high as 0.6 ppm has been recorded.) If 0.02 ppm of ozone were entirely converted to radicals, then about 10^{-6} mole of radicals per human body per day would be formed. (See Section 5.5 for a discussion of stratospheric ozone depletion by chlorine atoms.)

The free radical theory of aging received its impetus from reports that antioxidants[3] had been found to increase the average life span of mice when added to their daily diet. The synthetic antioxidants 2-aminoethanethiol and 3,5-di-*tert*-butyl-4-hydroxytoluene (butylated hydroxytoluene) were found to be the most effective. These compounds preferably react with, and intercept, radicals before the radicals can attack cell membranes.

2-Aminoethanethiol
(2-Mercaptoethylamine)

3,5-Di-*tert*-butyl-4-hydroxytoluene
(Butylated hydroxytoluene, BHT)

The natural antioxidant vitamin E (see Section 12.16-c) has been found to extend the life of fruit flies by about 13%. However, this does not prove that the antioxidant property of these compounds is responsible for their life-extending action. Pryor notes that the following alternative explanations have been postulated.

… these antioxidants induce chemical stress. Certain kinds of stress itself lengthen life. Another possibility is that these antioxidants prolong life because they are enzyme inducers; that is, they increase the formation of [certain] enzymes. The final possibility is that these compounds restrict either appetite or food absorption; it is known that restriction of calories prolongs life.

[3] Antioxidants are substances that inhibit oxidation by atmospheric oxygen (*autoxidation*). Autoxidation occurs via a free radical chain reaction, and this can be retarded by antioxidants at the initiation and propagation steps. Free radical inhibitors are added to rubber, gasoline (Section 2.7), and certain foods (Section 12.6).

In summation, then, experimental observations suggest that radicals may be involved in several pathological conditions and that they play a role in the aging of mammals. There is, however, no evidence that synthetic antioxidants or additional vitamin E (beyond that required for nutritional needs) increases the life span of humans. It is probable that human aging is caused by several different mechanisms operating simultaneously. Clearly, much more extensive research is required in this area.

2.6 Sources and Uses of Alkanes

Petroleum is a complex mixture of hydrocarbons produced by the decomposition of animal and vegetable matter that has been entrapped in the earth's crust for long periods of time. Refining of petroleum involves distillation into fractions of different boiling ranges (Table 2.8). Crude oil and natural gas are the liquid and gaseous components of petroleum. Natural gas contains about 80% methane and 10% ethane, and the remaining 10% is a mixture of the higher alkanes. Methane is a product of the bacterial decay of plant and marine organisms that have become buried beneath the earth's surface. It is also produced by the microbial decomposition of organic matter in sewage treatment plants. Because methane was first isolated in marshes, it was given the name "marsh gas." Methane is used primarily as a cooking fuel (and in bunsen burners). Natural gas is the cleanest of the fossil fuels because it contains the least amount of sulfur compounds. Therefore, there is very little sulfur dioxide produced when natural gas is combusted. When methane burns in an ample supply of oxygen, the following reaction takes place.[4]

TABLE 2.8 Approximate Boiling Ranges of Various Petroleum Fractions

Fraction	Boiling Range (°C)
Natural gas (C_1 to C_4)	<20
Petroleum ether (C_5 to C_6)	30–60
Gasoline (C_6 to C_{12})	85–200
Kerosene (C_{12} to C_{15})	200–300
Fuel oil (C_{15} to C_{18})	300–400
Lubricating oil, paraffin wax, greases, asphalt, mineral oil (C_{16} to C_{40})	>400

$$CH_4 + 2\,O_2 \longrightarrow CO_2 + 2\,H_2O + 210\ \text{kcal/mole}$$

Notice that the reaction is exothermic. The basic source of this released energy is the capacity of these carbon-containing compounds to form stable chemical bonds with oxygen. The larger the alkane molecule, the greater the amount of heat produced. The complete oxidation of 1 g of an alkane, to carbon dioxide and water, releases about 12 kcal of heat. This is a very good ratio of mass of fuel to energy produced. The general equation for the reaction may be written as follows.

$$C_nH_{2n+2} + \frac{3n+1}{2}O_2 \qquad n\,CO_2 + (n+1)\,H_2O + \text{Energy}$$

Although the hydrocarbons have numerous commercial and industrial uses (for example, lubricating oils, greases, and in the manufacture of synthetic chemicals),

[4]A limited supply of oxygen within a gasoline engine results in the incomplete combustion of hydrocarbons, producing elemental carbon (which can form soot deposits on automobile pistons) or the poisonous air pollutant carbon monoxide. (For every gallon of gasoline, more than 1 lb of carbon monoxide is produced.) Sulfur dioxide, unburned hydrocarbons, and reaction by-products such as aldehydes are also released into the atmosphere, significantly contributing to the air pollution.

$$2\,C_8H_{18} + 17\,O_2 \longrightarrow 16\,CO + 18\,H_2O + \text{Energy}$$
$$2\,C_8H_{18} + 9\,O_2 \longrightarrow 16\,C + 18\,H_2O + \text{Energy}$$

the largest use of hydrocarbons ($\sim 85\%$) is as a fuel.[5] The combustion of hydrocarbons provides the energy that propels our motor vehicles and generates our electricity. In general, the combustion of any fuel releases energy. In Chapter 16, we shall see that the equation for the energy-releasing combustion of hydrocarbons is similar to the equation for the burning of food in our body to produce energy.

$$C_6H_{12}O_6 + 6\,O_2 \longrightarrow 6\,CO_2 + 6\,H_2O + 686\ \text{kcal/mole}$$

2.7 Octane Rating

Gasoline is the most important component of crude oil and is itself a mixture of hydrocarbons. In rating the various hydrocarbon fuels, a standard one-cylinder test engine was employed. It was discovered that pure isooctane burned very smoothly (without vibration or "knocking"), and isooctane was assigned an octane number of 100. *n*-Heptane caused very severe knocking and was given an octane number of 0. Mixtures of these two compounds give knocking properties between the two extremes. Knocking occurs when the gasoline–air mixture in the cylinder explodes before it is ignited by the spark plug. This is caused either by too high a compression or the wrong kind of fuel. Constant knocking results in reduced power and in engine wear. In general, branched-chain alkanes have less tendency to knock than straight-chain alkanes. Unfortunately, distillation of petroleum yields mainly straight-chain hydrocarbons.

$$CH_3CH_2CH_2CH_2CH_2CH_2CH_3$$

Heptane

"Isooctane"
2,2,4-Trimethylpentane

The octane number of any fuel is determined by the percentage of isooctane in the heptane–isooctane mix. For example, if a fuel has the same knocking properties as a 6% heptane–94% isooctane mix, it would have an octane number of 94. Technological advances have resulted in engines that have greater power and compression ratios. This has necessitated a higher quality fuel. Some fuels containing more highly branched

hydrocarbons, unsaturated hydrocarbons, or aromatic hydrocarbons may have octane numbers greater than 100 if they burn more smoothly than isooctane. Most regular grades of gasoline have octane ratings of 85–90, and premium grades of gasoline average 97–100.

A method used to minimize engine knocking is to add small quantities of a so-called antiknocking ingredient such as tetraethyllead, $(C_2H_5)_4Pb$. Its function is to control the concentration of free radicals and prevent the premature explosions that are characteristic of knocking. Addition of 3 mL of tetraethyllead per gallon of gasoline can raise the octane number by 15 units. Tetraethyllead adds another air pollutant (lead) to those already produced by the automobile. In addition, the catalytic converters found in most of today's cars are deactivated by lead. Leaded gas may some day be eliminated entirely because the majority of cars now being manufactured use unleaded gasoline. The major oil companies now add aromatic compounds such as toluene and xylene (Chapter 4) or methyl *tert*-butyl ether to their unleaded gas to increase the octane number. In addition, they utilize various "reforming" processes, in which straight-chain alkanes are chemically converted to branched-chain or cyclic compounds. The increased cost of the additives and/or the reforming process has substantially increased the cost of unleaded gas compared to the leaded variety. There is a continuous search for other blending agents that can inexpensively boost octane ratings.

Methyl *tert*-butyl ether

[5] In today's energy-conscious world, the unit of commerce on the oil market is the **barrel**. *One barrel of crude oil equals* 42 *gallons.* Of this, approximately 45% is converted into gasoline, 30% into fuel oil for heating purposes, and 10% into jet fuel. Only about 5% of this precious liquid is used in the petrochemical industry to manufacture detergents, dyes, fertilizers, pesticides, plastics, etc. The remainder is sold as aviation gasoline, lubricating oils, greases, and asphalt.

2.8 Cycloalkanes

The general formula for cycloalkanes is C_nH_{2n}; they contain two fewer hydrogen atoms than the open-chain hydrocarbons. The cycloalkanes can be imagined to arise from the removal of one hydrogen atom from each of the terminal carbons of the corresponding open-chain hydrocarbon. These carbons are then joined together.

The simplest cycloalkane, cyclopropane, contains three carbons. The names of the cycloalkanes are obtained by attaching the prefix *cyclo-* to the name of the parent alkane. If there is only one substituent on the cycloalkane ring, the substituent does not have to be numbered, since all the positions of the ring are equivalent. When there is more than one substituent, numbers are required. The ring carbons are numbered so that the carbons bearing the substituents have the lowest numbers. It is a convention in the writing of cycloalkane formulas that the ring carbon atoms and their attached hydrogens need not be shown. Only the appropriate geometric figure is drawn; the carbon atom at each corner of the figure is understood, and substituent groups are shown attached to the proper carbon atoms of the ring.

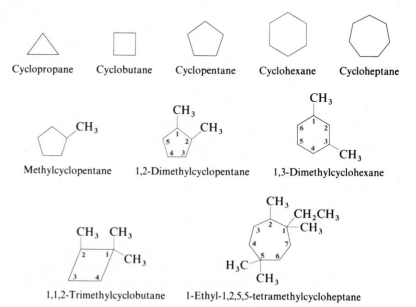

Cyclopropane   Cyclobutane   Cyclopentane   Cyclohexane   Cycloheptane

Methylcyclopentane   1,2-Dimethylcyclopentane   1,3-Dimethylcyclohexane

1,1,2-Trimethylcyclobutane   1-Ethyl-1,2,5,5-tetramethylcycloheptane

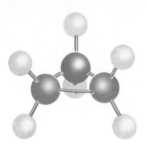
Cyclopropane

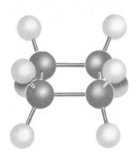
Cyclobutane

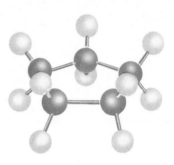
Cyclopentane

FIGURE 2.4 Ball-and-stick models of some cycloalkanes.

In most respects, the chemical and physical properties of the cycloalkanes are similar to those of their noncyclic counterparts. There are, however, some variations that deserve special consideration.

The major distinction between the cyclic and noncyclic alkanes arises from the differences in the geometric configuration of the two types of compounds. As mentioned previously, there is free rotation about all carbon–carbon bonds of open-chain alkanes. This is not the case with the cycloalkanes. Here the carbons are held rigidly in place (Figure 2.4). Free rotation is impossible without disruption of the ring structure. For these compounds, the position of any substituent relative to the ring becomes extremely important. (The substituent will be situated either above or below the ring.) A new form of isomerism is possible, which we discuss in Chapter 10. For now, let it suffice to say that the compounds at the left are not the same, but are isomers.

Cyclopropane is planar, cyclobutane is slightly bent, and cyclic compounds containing more than four carbons in the ring are decidedly nonplanar molecules. Four of the carbon atoms in cyclopentane are coplanar, but the fifth carbon is puckered out of the plane. If cyclohexane were planar, the internal bond angles would have to be 120° and the molecule would be strained. By assuming a bent geometric shape, cyclohexane and the higher cycloalkanes can achieve the unstrained tetrahedral angle (109.5°). Molecular models indicate that a six-membered ring may assume two distinct nonplanar **conformations**.[6] One form is shaped like a chair and is called the *chair* conformation; the other form resembles a boat and is called the *boat* conformation (Figure 2.5). In the boat form, the hydrogen atoms on carbons 1 and 4 are close enough to repel each other, causing a further twisting to occur, to give the so-called twist-boat structure. In the chair form, all of the hydrogen atoms are staggered and are maximally separated. The chair form is the most stable form of cyclohexane, and it is the preferred conformation.

The carbon atoms in the puckered cyclohexane ring approximate a plane. Six of the twelve hydrogen atoms lie in this plane and are referred to as **equatorial**

[6]Structures that can be converted into one another by rotation about one or more single bonds are referred to as *conformational isomers* or *conformers*.

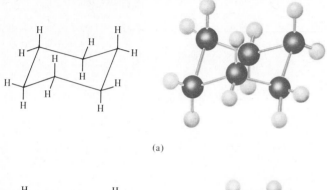

(a)

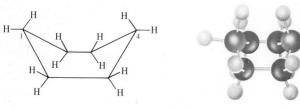

(b)

FIGURE 2.5 Conformations of cyclohexane. (a) "Chair" form; (b) "boat" form.

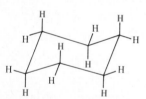

FIGURE 2.6 Axial and equatorial positions in cyclohexane. Colored hydrogens are axial.

hydrogens. The other six hydrogen atoms, which lie roughly at right angles above or below this plane, are called **axial** hydrogens (Figure 2.6). Substituent groups usually occupy equatorial positions so that they are as far away as possible from ring hydrogens and/or other groups.

In summary, it should be emphasized that the alkanes as a class are generally unreactive toward most chemicals. This characteristic is usually imparted to the alkane portion of more complex molecules. In other words, although a compound may contain one or more highly reactive functional groups, the alkane portion of that molecule will generally remain unchanged despite transformations elsewhere in the molecule.

Exercises

2.1 Write structural formulas and give IUPAC names for all the isomeric heptanes, C_7H_{16}.

2.2 Which of the following formulas are equivalent?

(a) $CH_3—\overset{\displaystyle CH_3}{\underset{\displaystyle CH_3}{C}}—CH_3$

(b) $CH_3CH_2C(CH_2CH_3)_2CH_2CH_3$

(c) $(CH_3)_2CHCH_2CH(CH_3)_2$

(d) $CH_3\overset{\displaystyle CH_3}{\underset{}{CHCH_2(C_2H_5)}}$

(e) $(CH_3)_3CCH_2CH(CH_3)_2$

(f) $CH_3\overset{\displaystyle CH_3}{\underset{}{CHCH_2C(CH_3)_3}}$

(g) $CH_3\overset{\displaystyle CH_3}{\underset{}{CHCH_2}}\overset{\displaystyle CH_3}{\underset{}{CHCH_3}}$

(h) $C(CH_3)_4$

(i) $CH_3(CH_2)_2CH(CH_3)_2$

(j) $C(CH_2CH_3)_4$

2.3 Name the following compounds.

(a) $CH_3CH_2CH(CH_3)CH_2CH_3$ *3-methyl 5 pentane*

(b) $(CH_3)_2CHCH(CH_3)_2$

(c) $CH_3CH(CH_2CH_3)_2$

30

(d) $(CH_3)_2CHC(CH_3)_2C(CH_2CH_3)_3$
(e) $(CH_3)_3CCH_2CH_3$
(f) $(CH_3)_2CHCH_2CH(CH_3)_2$
(g) $(CH_3CH_2)_2CHCH_2CH_2CH_3$
(h) $CH_3CH_2CH_2CH_2CH(n\text{-}C_3H_7)CH(CH_3)_2$
(i) $(CH_3)_2CHCH_2C(CH_3)_2CH(CH_2CH_3)C(CH_3)_3$
(j) $CH_3CH(CH_2CH_3)CH(CH_2CH_3)CH_3$

(k) 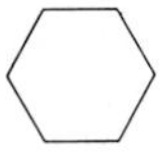

(l) 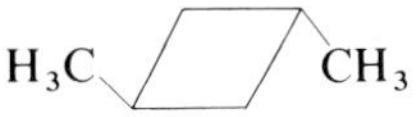

(m) 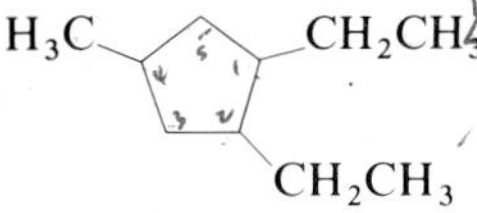

(n) 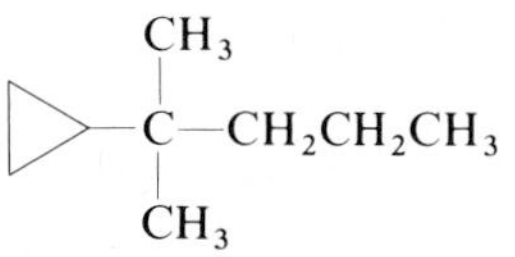

2.4 Write structural formulas for the following compounds.
(a) 2-methylpentane
(b) 4-ethyl-2-methylhexane
(c) 2,2,3,3-tetramethylbutane
(d) 4-ethyl-3-isopropyloctane
(e) 2,2,4-trimethylpentane
(f) 3-ethyl-2,7-dimethyl-5-propyloctane
(g) 2,4-dimethylhexane
(h) 4-ethyl-2,2-dimethylheptane
(i) 2,3,4-trimethylpentane
(j) 4,4-diethyl-2,3-dimethylheptane
(k) 1-ethyl-3-methylcyclohexane
(l) 2-cyclopentyl-2-methylpropane
(m) 1,1-dimethylcyclobutane
(n) 3-cyclobutyl-2,5-dimethylhexane

2.5 Give another name for 3-ethyl-2,3,4-trimethylpentane.

2.6 What is wrong with each of the following names? Give the structure and the correct name for each compound.
(a) 2-dimethylpropane
(b) 2,3,3-trimethylbutane
(c) 2,4-diethylpentane
(d) 3,4-dimethyl-5-propylhexane
(e) 4-isopropyl-5-methylheptane
(f) 3-ethyl-4-methylpentane
(g) 3,3-diethyl-5,5-dimethylhexane
(h) 2,2-diethyl-4,4-dimethylheptane
(i) 2,8-dimethyloctane
(j) 1,4-dimethylcyclobutane
(k) 2,5-diethylcyclohexane
(l) 2-isopropylcyclopentane
(m) 3,3-dimethyl-1-ethylcycloheptane
(n) 3-cyclopropylbutane

2.7 What compounds contain fewer carbons than propane and are homologs of propane?

2.8 Which member of each pair will have the higher boiling point?
(a) pentane or butane
(b) isopentane or pentane
(c) $(CH_3)_2CHCH(CH_3)_2$ or $CH_3(CH_2)_4CH_3$
(d) butane or isohexane
(e) cyclopentane or cyclohexane
(f) $CH_3(CH_2)_5CH_3$ or $CH_3(CH_2)_7CH_3$

2.9 Which member of each pair will have the higher melting point?
(a) pentane or butane
(b) neopentane or pentane
(c) $(CH_3)_2CHCH(CH_3)_2$ or $CH_3(CH_2)_4CH_3$
(d) butane or isohexane
(e) cyclopentane or cyclohexane
(f) $CH_3(CH_2)_5CH_3$ or $CH_3(CH_2)_7CH_3$

2.10 Write structures for the nine possible brominated derivatives of ethane that can result from the free radical bromination of ethane. Indicate which products are isomers.

2.11 **(a)** If *n*-pentane reacts with chlorine in the presence of light to form monochloropentane, how many isomers will be formed? Draw their structural formulas.
(b) How many monochloro isomers would you expect from the chlorination of isopentane and neopentane respectively?

2.12 Comparatively few alkanes are found in living organisms, yet they are vital to our way of life.
(a) What are the chief sources of alkanes.
(b) List several uses of alkanes.

2.13 Using appropriate examples, distinguish between the terms in the following pairs.
(a) alkyl group and alkane
(b) alkyl anion and alkyl free radical
(c) common name and IUPAC name
(d) propyl group and isopropyl group
(e) straight-chain alkane and branched-chain alkane
(f) acyclic alkane and alicyclic alkane
(g) chair form and boat form
(h) axial position and equatorial position
(i) polar group and nonpolar group
(j) antioxidant and antiknocking agent

2.14 Distinguish between the terms in each pair.
(a) natural gas and bottled gas
(b) physical properties and chemical properties
(c) exothermic reaction and endothermic reaction
(d) crude oil and natural gas

(e) chain-initiating step and chain-terminating step

2.15 What is meant by an octane rating of 90? What, in general, is the relationship between the structures of various alkanes and their octane ratings?

2.16 Predict the relative octane ratings of the following compounds. Begin with the one that has the lowest octane number.
(a) 2-methylbutane
(b) hexane
(c) 2-methylhexane
(d) 2,2,4-trimethylpentane

2.17 Complete the following equations for those reactions that will occur.
(a) $CH_3CH_2CH_3 + NaOH \longrightarrow$
(b) $CH_3CH_2CH_2CH_3 + O_2 \longrightarrow$

(c) 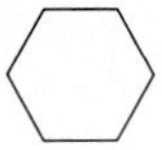 $+ Br_2 \xrightarrow{light}$

(d) $CH_3(CH_2)_3CH_3 + KMnO_4 \longrightarrow$
(e) $CH_3CH_2CH_2CH_2CH_3 + H_2SO_4 \longrightarrow$

(f) 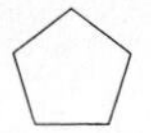 $+ Na \longrightarrow$

2.18 An alkane having a molecular weight of 72 forms only one monosubstituted bromine derivative upon its bromination in the presence of sunlight.
(a) What is the structural formula of the alkane?
(b) Bromination of an isomer of this alkane yields a mixture of four isomeric monobromo derivatives. What is the structural formula of the isomer?

2.19 The density of gasoline is 0.69 g/mL. On the basis of the complete combustion of octane, calculate the amount of carbon dioxide and water formed per gallon (3.78 L) of gasoline used in an automobile.

2.20 Air is approximately 20% oxygen by volume. How many liters of air would be needed to supply just enough oxygen for the complete combustion of 2.24 L of ethane at STP?

2.21 A hydrocarbon was found to contain 85.7% carbon and 14.3% hydrogen, and it had a density of 1.25 g/L (at STP). Write a structural formula for the compound.

3

The Unsaturated Hydrocarbons

Two families of homologous compounds are included under the general classification of unsaturated hydrocarbons. In both classes, each compound contains fewer hydrogen atoms than the corresponding alkane. Because these *compounds are deficient in hydrogen,* they are said to be **unsaturated** and must therefore contain multiple bonds. One family of compounds contains carbon–carbon double bonds; that is, two carbon atoms are joined together by two pairs of shared electrons. These compounds are called **alkenes** or **olefins.**[1] The other family of compounds contains carbon–carbon triple bonds (three pairs of shared electrons) and are known as **alkynes** or **acetylenes.**

[1] The action of chlorine gas on compounds that contain double bonds produces an oily liquid. The name olefin comes from the Latin *oleum,* oil, and *ficare,* to make.

33

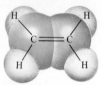

FIGURE 3.1 Structure of ethylene.

In the following sections, we deal mainly with *those compounds that contain only one double bond per molecule,* the **alkenes.** The general formula for such a homologous series is C_nH_{2n}.[2] Since there can be no alkene with only one carbon atom, the first member of the series has the molecular formula C_2H_4 (ethylene), and its structural formula, superimposed on its space-filling model, is shown in Figure 3.1.

3.1 *Nomenclature of Alkenes*

The IUPAC rules for alkane nomenclature have the following modifications when applied to the alkenes.

1. Locate the longest chain of carbon atoms that contains the functional group (in this case, the carbon–carbon double bond). Again, the longest chain determines the parent hydrocarbon.
2. The name of the parent hydrocarbon is modified by replacing the *-ane* suffix with the ending *-ene*. The *-ene* ending signifies that the molecule is an alkene.
3. Number the parent hydrocarbon so that the first carbon of the double bond will have the smallest possible number (for example, $CH_3CH_2CH{=}CH_2$ is 1-butene and not 1,2-butene or 3-butene).

Regardless of all the rules of nomenclature that have been agreed upon by chemists throughout the world, there is still a problem of nomenclature of which the student should be aware. This problem was alluded to in the previous chapter, when it was stated that many of the common names of the more familiar organic compounds still persist in the chemical literature. Chemists often use these common names, especially when referring to the lower homologs of each series of organic compounds. The student must be familiar with these common names as well as with the IUPAC names. Table 3.1 contains the names and formulas of some of the lower members of the alkene series. Table 3.2 contains some examples of the application of the IUPAC rules of nomenclature to the alkenes and cycloalkenes.

Before leaving the subject of alkene nomenclature, one additional aspect must be mentioned. The double bond of alkene molecules, like the ring structure of cycloalkanes, imposes geometric restrictions on the molecules. Recall that according to VSEPR theory (Section 1.5-b), each doubly bonded carbon atom lies in the center

---

[2]Notice that the alkenes and the cycloalkanes have the same general formula. Therefore alkenes and cycloalkanes that contain the same number of carbon atoms are isomers (e.g., $CH_3{-}CH{=}CH_2$ is an isomer of cyclopropane).

TABLE 3.1 Nomenclature of Alkenes

Molecular Formula	Structural Formula	IUPAC Name	Common Name
C_2H_4	$\begin{array}{c} \text{H} \quad \text{H} \\ \mid \quad \mid \\ \text{H—C=C—H} \end{array}$	Ethene[a]	Ethylene
C_3H_6	$\begin{array}{c} \text{H} \quad \text{H} \\ \mid \quad \mid \\ \text{CH}_3\text{—C=C—H} \end{array}$	Propene[a]	Propylene
C_4H_8	$\begin{array}{c} \text{H} \quad \text{H} \\ \mid \quad \mid \\ \text{CH}_3\text{—CH}_2\text{—C=C—H} \end{array}$	1-Butene	
C_4H_8	$\begin{array}{c} \text{H} \quad \text{H} \\ \mid \quad \mid \\ \text{CH}_3\text{—C=C—CH}_3 \end{array}$	2-Butene	
C_4H_8	$\begin{array}{c} \text{CH}_3 \\ \mid \\ \text{CH}_3\text{—C=C—H} \\ \mid \\ \text{H} \end{array}$	2-Methylpropene	

[a] These compounds need not be named 1-ethene and 1-propene since it is understood that the double bond can only be between carbon-1 and carbon-2. For all higher homologs, the position of the double bond must be indicated by number.

of an equilateral triangle (to minimize the repulsive forces among the three regions of electron density). The carbon atoms of a double bond and the two atoms bonded to each carbon all lie in a single plane (Figure 3.2). Free rotation about doubly bonded carbon atoms is *not* possible without rupturing the bond. Therefore, the relative positions of substituent groups located above or below the double bond become very significant. The nomenclature employed in situations such as this, as well as the consequences of restricted rotation, are discussed in Chapter 10.

FIGURE 3.2 (a) Planar configuration of the double bond. (b) Ball-and-spring model of ethylene.

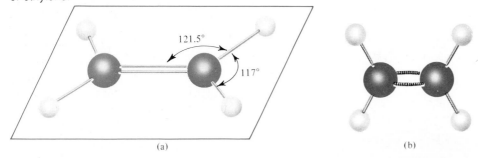

(a)   (b)

TABLE 3.2 Application of the IUPAC Rules to the Naming of Alkenes and Cycloalkenes

Condensed Structural Formula	Rewritten and Numbered	IUPAC Name
$CH_3(CH_2)_2CH{=}CH_2$	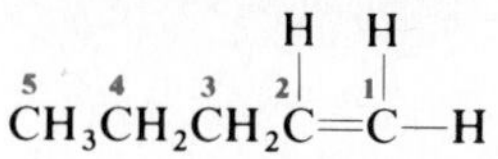	1-Pentene
$CH_3CH{=}CHCH_2CH_3$	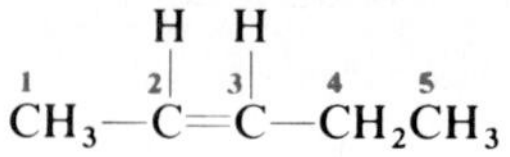	2-Pentene
$CH_3CH_2CH{=}C(CH_3)_2$	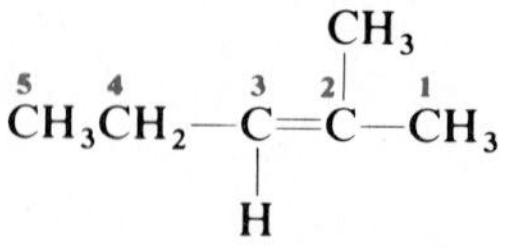	2-Methyl-2-pentene
$(CH_3)_2CHC(CH_3){=}CHCH_3$	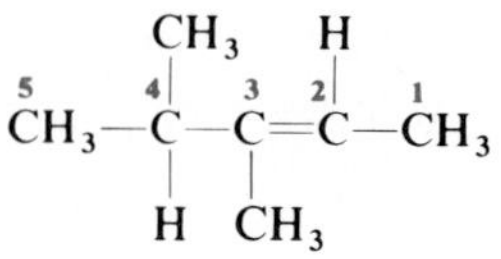	3,4-Dimethyl-2-pentene
$CH_3CH_2CH(CH_3)C(C_3H_7){=}CH_2$	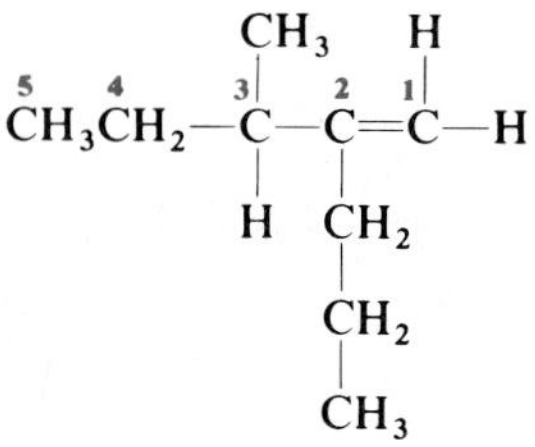	3-Methyl-2-propyl-1-pentene
		Cyclopentene
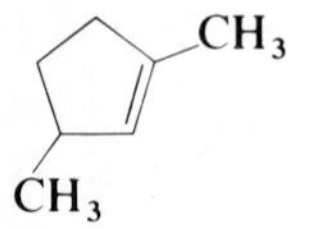	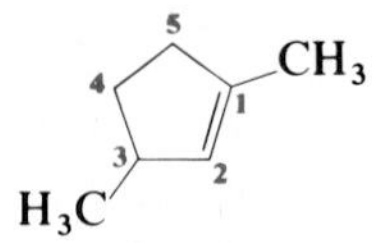	1,3-Dimethylcyclopentene
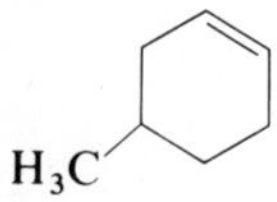	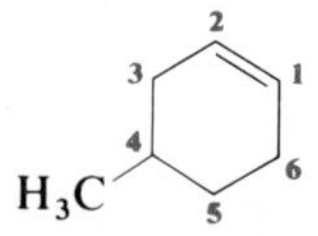	4-Methylcyclohexene
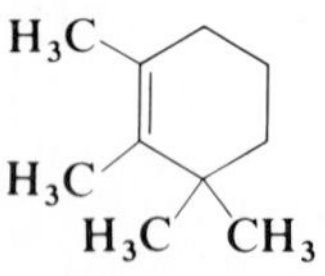	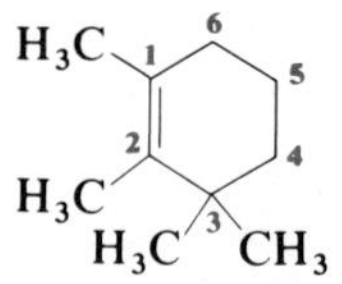	1,2,3,3-Tetramethylcyclohexene
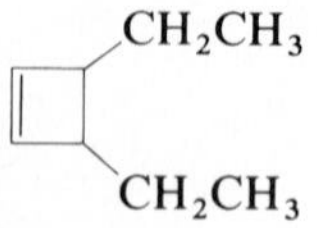	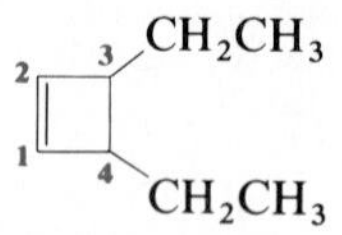	3,4-Diethylcyclobutene

3.2 *Preparation of Alkenes*

Ethylene and, to a lesser extent, the higher homologs of the alkene series can be obtained from the cracking of petroleum. **Cracking** (also called **pyrolysis**) is a *process in which the saturated hydrocarbons are subjected to very high temperatures* in the presence of a silica–alumina catalyst. The result is the elimination of hydrogen and the smaller hydrocarbons, such as methane and ethane. However, the process always yields a mixture of products and is thus of limited use in a chemical laboratory.

$$CH_3CH_2CH_2CH_3 \xrightarrow[\text{catalyst}]{>400°C} \begin{array}{l} CH_2{=}CH_2 + CH_3CH_3 \\ CH_3CH{=}CH_2 + CH_4 \\ CH_3CH{=}CHCH_3 + H_2 \end{array}$$

a. Dehydration of Alcohols

Dehydration reactions are extremely common in both organic chemistry and biochemistry. As the name implies, **dehydration** *involves the elimination of a molecule of water* from a reactant molecule. (See Section 6.4-b for a discussion of the mechanism of the dehydration reaction.)

The reaction can be brought about by heating the alcohol in the presence of concentrated sulfuric acid or phosphoric acid. High temperatures favor the elimination reaction, and the acids serve as effective catalysts. This procedure works best for secondary and tertiary alcohols. (See Section 6.1 for a definition of these terms.) Note that if the Greek letter delta (Δ) appears above or below the arrow in a chemical equation, it indicates that the reaction is endothermic and requires heat to occur.

General Equation

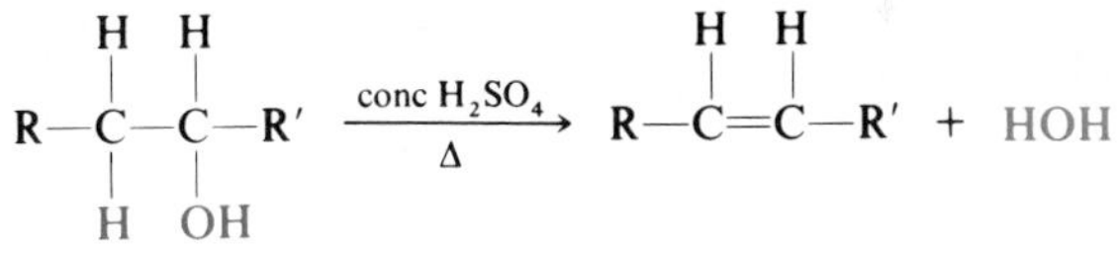

Specific Equations

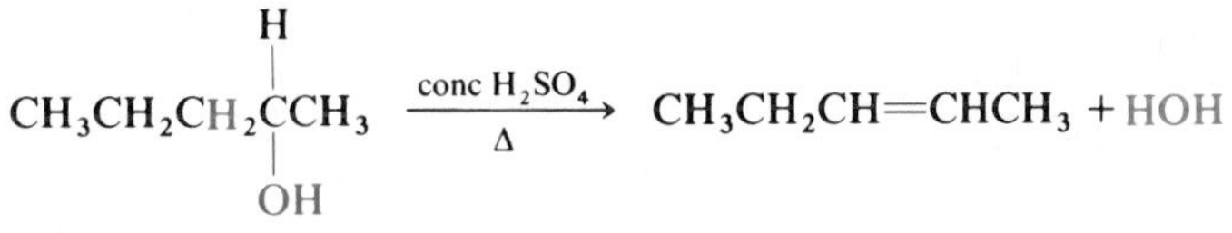

2-Pentanol   2-Pentene

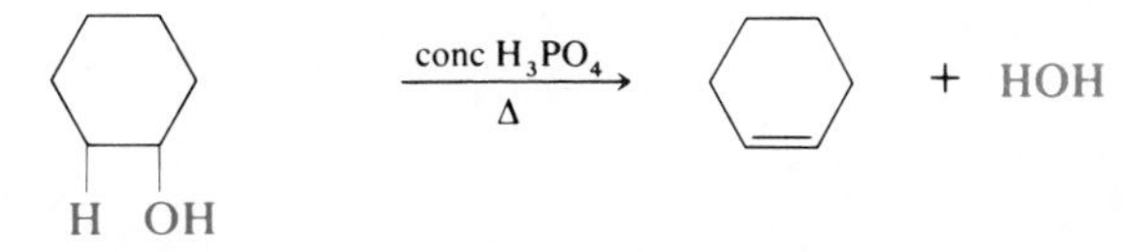

Cyclohexanol   Cyclohexene

b. Dehydrohalogenation of Alkyl Halides

Dehydrohalogenation *involves the elimination of the elements of hydrogen halide (HX) from adjacent carbon atoms.* The action of a strong base upon the alkyl halide is required to bring about dehydrohalogenation. Since the reaction is carried out in solution, it is necessary to find a solvent that will dissolve both the alkyl halide and the strong base. Water, the common solvent in inorganic chemistry, is of no value here since alkyl halides are water insoluble. The best solvent for this reaction is alcohol. (See Section 5.4-a for a discussion of the mechanism of the dehydrohalogenation reaction.) Dehydrohalogenation is not a widely used preparative reaction because the alkenes that would be formed are more readily available by the dehydration of an alcohol.

General Equations

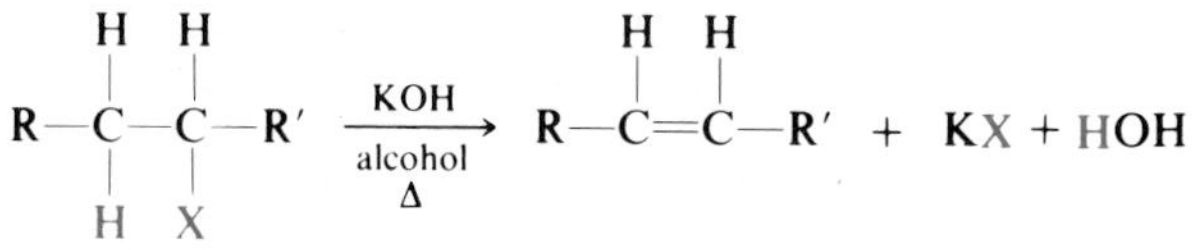

$$
\underset{\substack{| \quad | \\ \text{H} \quad \text{X}}}{\text{R}-\overset{\text{H}}{\underset{}{\text{C}}}-\overset{\text{H}}{\underset{}{\text{C}}}-\text{R}'} \quad \xrightarrow[\substack{\text{alcohol} \\ \Delta}]{\text{KOH}} \quad \text{R}-\overset{\text{H}}{\text{C}}=\overset{\text{H}}{\text{C}}-\text{R}' \;+\; \text{KX} \;+\; \text{HOH}
$$

Specific Equation

$$
\underset{\substack{| \quad | \\ \text{Br} \quad \text{H}}}{\text{CH}_3-\overset{\text{H}}{\underset{}{\text{C}}}-\overset{\text{H}}{\underset{}{\text{C}}}-\text{H}} \quad \xrightarrow[\substack{\text{alcohol} \\ \Delta}]{\text{KOH}} \quad \text{CH}_3-\overset{\text{H}}{\text{C}}=\overset{\text{H}}{\text{C}}-\text{H} \;+\; \text{KBr} \;+\; \text{HOH}
$$

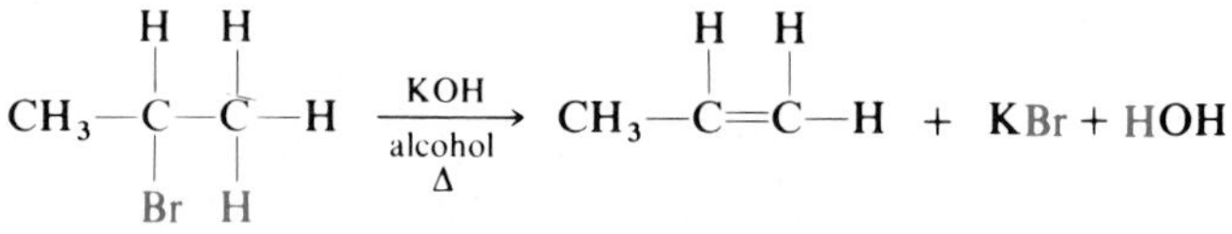

2-Bromopropane   Propene

3.3 Physical Properties of Alkenes

Table 3.3 lists the physical constants for several members of the alkene series. Since an alkene is only two atomic mass units lighter than its corresponding alkane, it is not surprising to find that alkenes have slightly lower boiling points. Notice once again that isomers (e.g., C_4H_8) have similar boiling points but significantly different melting points. The different shapes strongly influence how the molecules pack into

TABLE 3.3 Physical Constants of Some Alkenes

Molecular Formula	Structural Formula	IUPAC Name	MP (°C)	BP (°C)
C_2H_4	$CH_2{=}CH_2$	Ethene	−169	−104
C_3H_6	$CH_3CH{=}CH_2$	Propene	−185	−47
C_4H_8	$CH_3CH_2CH{=}CH_2$	1-Butene	−130	−6
C_4H_8	$(CH_3)_2C{=}CH_2$	2-Methylpropene	−141	−7
C_5H_{10}	$CH_3(CH_2)_2CH{=}CH_2$	1-Pentene	−138	30
C_6H_{12}	$CH_3(CH_2)_3CH{=}CH_2$	1-Hexene	−140	63

38

a crystal lattice. Like all hydrocarbons, the alkenes are insoluble in water but soluble in organic solvents.

3.4 *Chemical Properties of Alkenes*

The major distinction between the alkanes and the alkenes is apparent from a comparison of their chemical properties. Alkanes are inert. The few reactions that they do undergo occur by substitution. The alkenes, on the other hand, are very reactive, and the majority of their reactions are characterized by **addition** to the double bond.

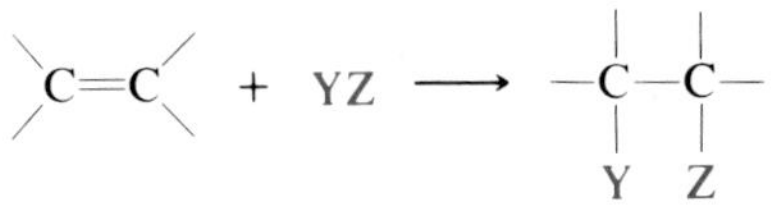

$$\text{>C=C<} + \text{YZ} \longrightarrow -\underset{Y}{\overset{|}{C}}-\underset{Z}{\overset{|}{C}}-$$

a. Addition of Symmetrical Reagents

1. Addition of Halogen (Halogenation) Alkenes react with chlorine or bromine. With the less reactive iodine molecule, the reaction is very slow. Illumination speeds up this latter reaction considerably. At the opposite extreme, the reaction of fluorine with alkenes is too violent to control. For reasons of safety and ease of handling, the reactions between the halogens and the alkenes are usually carried out in some inert solvent such as carbon tetrachloride. In fact, this reaction is the basis for a very useful test for the presence of an unsaturated compound. A few drops of a 1% solution of bromine in carbon tetrachloride are added to the unknown. If the unknown contains a double bond, the red-brown color of the bromine will disappear as a result of the formation of a colorless dibromoalkane.

General Equation

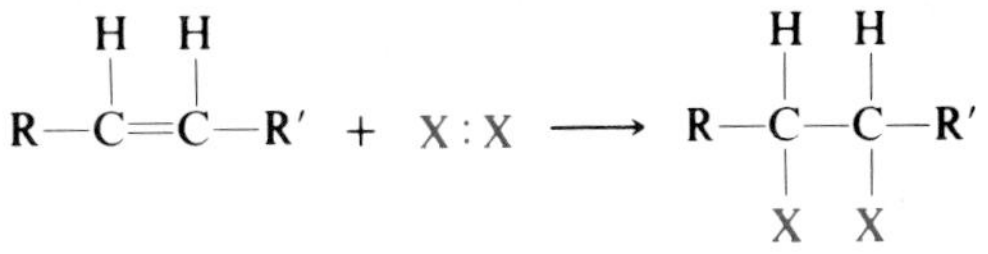

$$R-\underset{}{\overset{H}{\underset{}{C}}}=\underset{}{\overset{H}{\underset{}{C}}}-R' + X:X \longrightarrow R-\underset{X}{\overset{H}{\underset{|}{C}}}-\underset{X}{\overset{H}{\underset{|}{C}}}-R'$$

Specific Equations

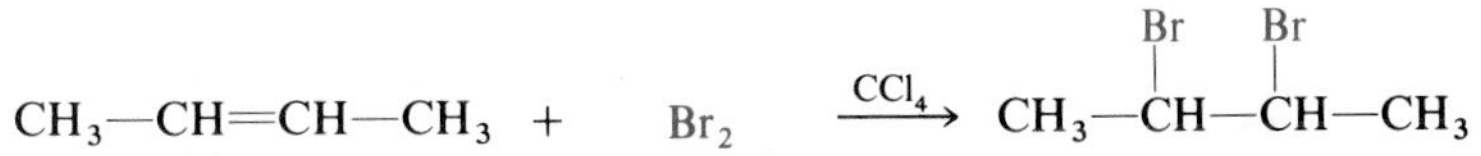

$$CH_3-CH=CH-CH_3 + Br_2 \xrightarrow{CCl_4} CH_3-\underset{Br}{\overset{Br}{\underset{|}{CH}}}-\underset{Br}{\overset{Br}{\underset{|}{CH}}}-CH_3$$

2-Butene   (red-brown   2,3-Dibromobutane
(colorless liquid)   liquid)   (colorless oil)

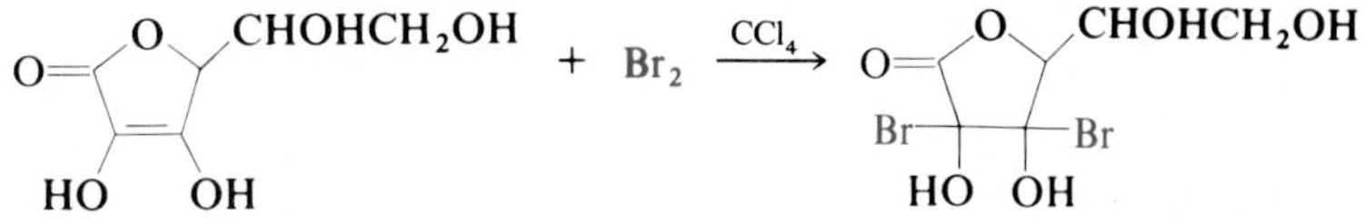

Ascorbic acid
(Vitamin C)
           2,3-Dibromoascorbic acid

2. Addition of Hydrogen (Hydrogenation) When hydrogen adds to an alkene, the corresponding alkane is produced. The process requires a catalyst and is thus termed catalytic hydrogenation.[3] The best catalysts for the reaction are platinum and palladium, but because of the high cost of these metals, finely divided nickel is often used in the laboratory. The metals adsorb hydrogen molecules on their surfaces and promote the dissociation of hydrogen gas into hydrogen atoms. These metals also adsorb the alkene, so the reaction occurs on the surface of the catalyst.

General Equation

$$R-\underset{\underset{H}{|}}{\overset{\overset{H}{|}}{C}}=\underset{}{\overset{\overset{H}{|}}{C}}-R' \ + \ H:H \ \xrightarrow[\text{catalyst}]{\text{Pt, Pd, or Ni}} \ R-\underset{\underset{H}{|}}{\overset{\overset{H}{|}}{C}}-\underset{\underset{H}{|}}{\overset{\overset{H}{|}}{C}}-R'$$

Specific Equation

$$CH_3CH-\underset{\underset{CH_3}{|}}{\overset{\overset{H}{|}}{C}}=\underset{}{\overset{\overset{H}{|}}{C}}-H \ + \ H_2 \ \xrightarrow{\text{Pt}} \ CH_3CH-\underset{\underset{CH_3}{|}}{\overset{\overset{H}{|}}{C}}-\underset{\underset{H}{|}}{\overset{\overset{H}{|}}{C}}-H$$

3-Methyl-1-butene   2-Methylbutane

The catalytic hydrogenation of the alkenes provides a method for the preparation of pure alkanes. The reaction is also of considerable commercial importance. It is used in the manufacture of high-octane fuels and in the preparation of synthetic detergents. Margarine and cooking shortenings are produced by the catalytic hydrogenation of polyunsaturated vegetable oils (see Section 12.5-c).

b. Addition of Unsymmetrical Reagents

The majority of alkene addition reactions are initiated by the attack of a proton or a positive ion on the electrons of the double bond. The *electron-seeking reagents* that are capable of forming a covalent bond with carbon are referred to as **electrophilic reagents** or **electrophiles** (electron-loving). The electrophile reacts with the alkene to form an intermediate carbocation, which subsequently combines with the anionic species to yield the addition product.

$$\overset{}{\underset{}{C}}=\overset{}{\underset{}{C} } \ + \ H^+A^- \ \longrightarrow \ \left[-\underset{\underset{H}{|}}{\overset{|}{C}}-\overset{|}{\underset{+}{C}}- \right] \ + \ A^-$$

A carbocation

$$-\underset{\underset{H}{|}}{\overset{|}{C}}-\overset{|}{\underset{+}{C}}- \ + \ A^- \ \longrightarrow \ -\underset{\underset{H}{|}}{\overset{|}{C}}-\underset{\underset{A}{|}}{\overset{|}{C}}-$$

[3] When hydrogens are added to the unsaturated carbon atoms in an alkene, the oxidation number of the carbons is decreased. Therefore these hydrogenation reactions are also referred to as *reduction* reactions.

Carbocations are highly reactive intermediates in many organic and biochemical reactions. A **carbocation** is defined as *an electron-deficient carbon with a sextet of electrons.* These organic cations may be described as *primary carbocations* ($RCH_2{}^+$), *secondary carbocations* (R_2CH^+), and *tertiary* carbocations (R_3C^+). (See Section 6.1 for an explanation of the terms primary, secondary, and tertiary.) Lewis structures for some carbocations are

$$
\begin{array}{ccc}
\overset{\displaystyle H}{\underset{\displaystyle H}{CH_3:\overset{\cdot\cdot}{C}{}^+}} & \overset{\displaystyle H}{\underset{\displaystyle CH_3}{CH_3:\overset{\cdot\cdot}{C}{}^+}} & \overset{\displaystyle CH_3}{\underset{\displaystyle CH_3}{CH_3:\overset{\cdot\cdot}{C}{}^+}}
\end{array}
$$

Ethyl cation	Isopropyl cation	*tert*-Butyl cation
A primary carbocation	A secondary carbocation	A tertiary carbocation

Reaction between a symmetrical alkene (such as $H_2C{=}CH_2$) and an unsymmetrical reagent (HA) presents no problem since only one product results no matter which carbon atom accepts the H and which accepts the A. For example, consider the reaction of a symmetrical alkene with hydrogen chloride.

$$
\underset{H}{\overset{H}{H-C}}{=}\underset{\ }{\overset{H}{C}}-H \;+\; HCl \;\longrightarrow\; \underset{\underset{H\quad Cl}{|\quad|}}{H-C-C-H} \quad \text{or} \quad \underset{\underset{Cl\quad H}{|\quad|}}{H-C-C-H}
$$

Identical compounds
(Chloroethane)

However, in the case of a reaction between an unsymmetrical alkene and an unsymmetrical reagent, one might predict that two possible products would be formed. For example, the addition of hydrogen bromide to propene might be expected to yield both 1-bromopropane and 2-bromopropane. If the reaction were governed solely by the statistical addition of the fragments H and Br, a mixture of the two isomeric alkyl halides would be expected.

$$
CH_3-\overset{H}{\underset{H}{C}}{=}\overset{H}{C}-H \;+\; HBr \;\longrightarrow\; CH_3-\underset{\underset{H\quad Br}{|\quad|}}{C-C-H} \quad \text{and/or} \quad CH_3-\underset{\underset{Br\quad H}{|\quad|}}{C-C-H}
$$

Propene	1-Bromopropane	2-Bromopropane
	I	II

Actually, when this reaction is carried out in the laboratory, only one product is formed, and that product is compound II. This phenomenon was observed in all reactions between unsymmetrical reagents and alkenes, and in 1871, a Russian chemist, Vladimir Markovnikov, suggested a generalization that enables us to predict what the outcome of such a reaction will be. This empirical generalization is known as **Markovnikov's rule** and can be stated as follows: *In the addition of unsymmetrical reagents to alkenes, the positive portion of the reagent (which is usually* *hydrogen) adds to the carbon atom that already has the most hydrogen atoms* (*or "them that has, gets,"* in the vernacular).

The direction of addition according to Markovnikov's rule can be rationalized by the order of relative stability of the intermediate carbocations. The relative stabilities of the various carbocations are tertiary > secondary > primary.

This observed order of stability of carbocations is explained, in part, by the effects of electrostatic interactions. *These electrostatic factors, in which electrons are either attracted to or repelled by one atom* (*or group of atoms*), are referred to as **inductive effects.** Atoms or groups of atoms that have high electronegativities (halogens, oxygen, nitrogen) are said to be *electron-attracting* groups. As we shall see in Section 8.6, these groups tend to stabilize carboxylate anions by withdrawing and spreading out the negative charge.

Alkyl groups, on the other hand, are *electron-donating* species. They stabilize carbocations by releasing electron density to the positively charged carbon atom. This results in spreading the positive charge over a larger volume (delocalizing the charge). Ions with concentrated charge are less stable than ions in which the charge is spread over a greater volume. The greater the number of alkyl groups bonded to the carbocation carbon, the greater the charge delocalization, and the greater the stability of the carbocation.

In the addition of unsymmetrical reagents to unsymmetrical alkenes, the initial protonation occurs so as to yield the more stable carbocation.

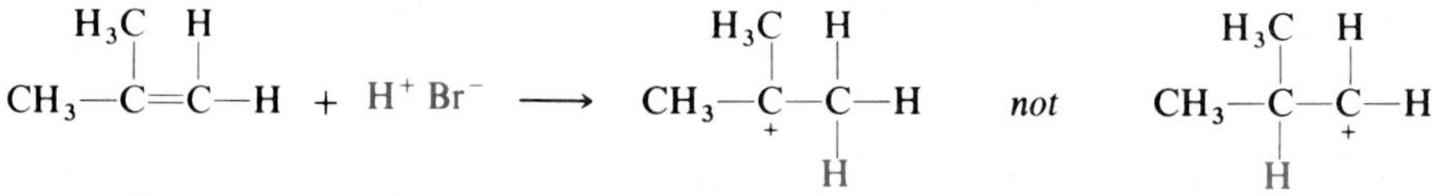

2-Methylpropene   A tertiary carbocation   *not*   A primary carbocation

The intermediate carbocation is stabilized by the electron-donating effects of the adjacent methyl groups and is subsequently attacked by the anionic portion of the reagent.

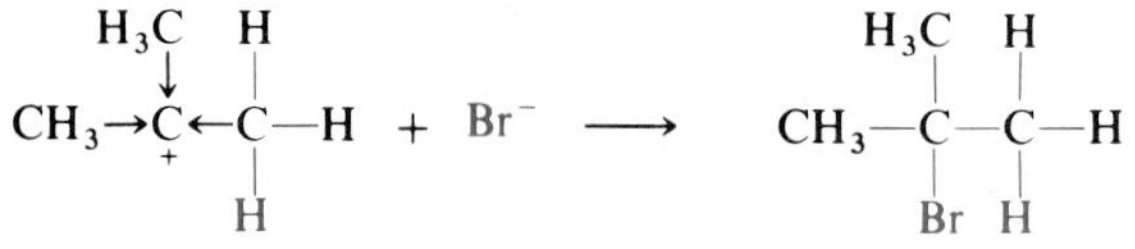

2-Bromo-2-methylpropane
(*tert*-Butyl bromide)

General Equation

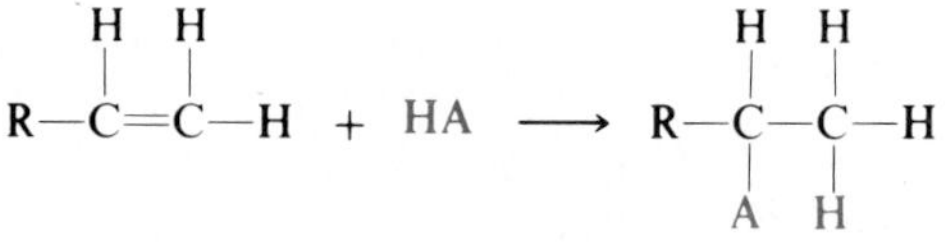

where R is not hydrogen and A = Cl, Br, I, OSO_3H, OH, etc.

Chapter 3 · The Unsaturated Hydrocarbons

Specific Equations

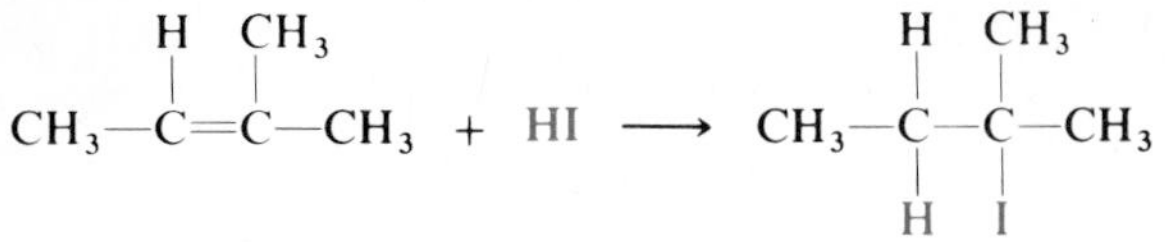

$$\underset{\text{2-Methyl-2-butene}}{CH_3-\overset{\displaystyle H}{\underset{\displaystyle }{C}}=\overset{\displaystyle CH_3}{\underset{\displaystyle }{C}}-CH_3} + HI \longrightarrow \underset{\text{2-Iodo-2-methylbutane}}{CH_3-\overset{\displaystyle H}{\underset{\displaystyle H}{C}}-\overset{\displaystyle CH_3}{\underset{\displaystyle I}{C}}-CH_3}$$

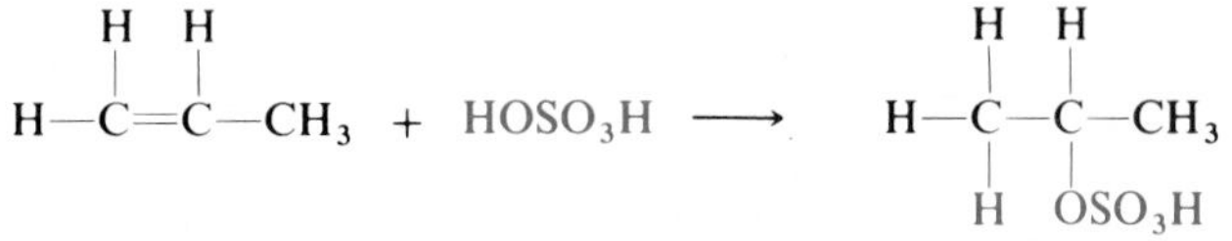

$$\underset{\text{Propene}}{H-\overset{\displaystyle H}{\underset{\displaystyle }{C}}=\overset{\displaystyle H}{\underset{\displaystyle }{C}}-CH_3} + HOSO_3H \longrightarrow \underset{\text{Isopropyl hydrogen sulfate}}{H-\overset{\displaystyle H}{\underset{\displaystyle H}{C}}-\overset{\displaystyle H}{\underset{\displaystyle OSO_3H}{C}}-CH_3}$$

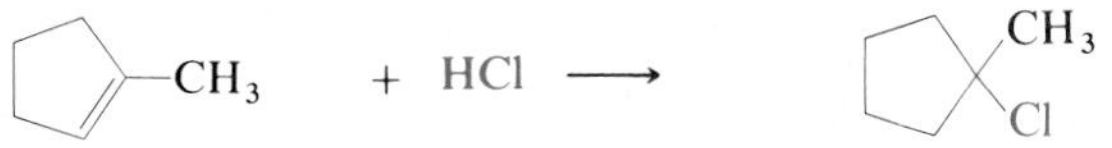

1-Methylcyclopentene ⟶CH$_3$ + HCl ⟶ 1-Chloro-1-methylcyclopentane

c. Oxidation

Alkenes, like alkanes, can be completely oxidized (via combustion) to carbon dioxide and water. They release slightly less energy than the corresponding alkanes.

$$CH_2{=}CH_2 + 3\,O_2 \longrightarrow 2\,CO_2 + 2\,H_2O + 332\,\text{kcal}$$

However, alkenes are much more susceptible to chemical oxidation than alkanes, and they are readily oxidized by a solution of potassium permanganate. The products of this reaction are dependent upon reaction conditions. This is not at all unusual in organic chemistry. Therefore it is very important that the reaction conditions be specified as part of the equation. When the oxidation occurs with cold dilute potassium permanganate solution, the alkenes are oxidized to **glycols** (*compounds containing two hydroxyl groups*).

General Equation

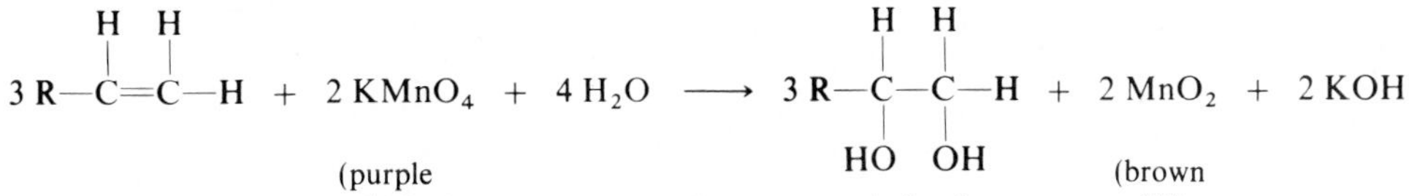

$$3\,R-\overset{\displaystyle H}{\underset{\displaystyle }{C}}=\overset{\displaystyle H}{\underset{\displaystyle }{C}}-H + 2\,KMnO_4 + 4\,H_2O \longrightarrow 3\,R-\overset{\displaystyle H}{\underset{\displaystyle HO}{C}}-\overset{\displaystyle H}{\underset{\displaystyle OH}{C}}-H + 2\,MnO_2 + 2\,KOH$$

(purple solution)   A glycol   (brown solid)

Notice that the equation given is a balanced net equation. It is necessary to write the complete equation whenever the reaction is to be carried out in the laboratory. A knowledge of the stoichiometry of a reaction enables one to calculate the necessary quantities of each reagent and the expected mass of the product to be formed. Often, however, we are only interested in knowing what organic products are obtained. In these cases, it is sufficient just to indicate the inorganic reagents above the arrow.

$$\underset{\text{Ethylene}}{H-\overset{\overset{\displaystyle H}{|}}{C}=\overset{\overset{\displaystyle H}{|}}{C}-H} \xrightarrow{\text{dil KMnO}_4} \underset{\text{Ethylene glycol}}{H-\overset{\overset{\displaystyle H}{|}}{\underset{\underset{\displaystyle HO}{|}}{C}}-\overset{\overset{\displaystyle H}{|}}{\underset{\underset{\displaystyle OH}{|}}{C}}-H}$$

The reaction of alkenes with dilute potassium permanganate is the basis of a second test (the Baeyer test) for unsaturation. If an unknown compound contains no other easily oxidized groups, then a color change from purple to brown (from purple $KMnO_4$ to brown MnO_2) indicates the presence of an unsaturated linkage in the compound.

If the oxidation of alkenes is carried out with concentrated $KMnO_4$ at low pH and high temperature, the initially formed glycol is cleaved. The product is a mixture of ketones and/or carboxylic acids, depending on the extent of substitution of the double bond.

$$R-\overset{\overset{\displaystyle H}{|}}{C}=\overset{\overset{\displaystyle R'}{|}}{C}-R'' \xrightarrow[H^+,\ \Delta]{\text{conc KMnO}_4} \left[R-\overset{\overset{\displaystyle H}{|}}{\underset{\underset{\displaystyle HO}{|}}{C}}-\overset{\overset{\displaystyle R'}{|}}{\underset{\underset{\displaystyle OH}{|}}{C}}-R'' \right] \longrightarrow \underset{\substack{\text{Carboxylic}\\\text{acid}}}{R-C\overset{\displaystyle O}{\underset{\displaystyle OH}{\diagdown}}} + \underset{\text{Ketone}}{O=C\overset{\displaystyle R'}{\underset{\displaystyle R''}{\diagup}}}$$

d. Polymerization

Polymerization (Greek, *poly,* many; *meros,* parts) *is the process whereby a small molecule* (*a monomer*) *reacts with itself in such a way as to form one large molecule* (a macromolecule or *polymer*—MW $\sim 10^5$). From an industrial viewpoint, this is the most important type of reaction of the alkenes. Polymers are classified into two main groups based on a comparison of the structure of their repeating units. The two general divisions are **addition polymers** and **condensation polymers.** In an addition polymer, the repeating unit is identical with that of the monomer. The polymerization reactions of alkenes are examples in this category. A condensation polymerization yields a product in which the repeating unit contains fewer atoms than the monomer or monomers (see pages 161 and 180).

Polymers are readily distinguishable from the small molecules from which they are synthesized. Besides the fundamental differences in molecular weight, structure, and bonding, polymers usually exhibit different physical and chemical properties. The small molecules tend to be volatile liquids of low viscosity, or even gases at room temperature. Polymers, on the other hand, tend to be nonvolatile, highly viscous liquids, glasses, or solids that soften only at high temperatures. Natural materials such as proteins (Chapter 13), natural rubber, cellulose, starch (Chapter 11), and complex silicate minerals are polymers. Artificial materials such as fibers, films, plastics, semisolid resins, and synthetic rubbers are also polymers. Synthetic polymers represent more than half of the compounds produced by the chemical industry (Table 3.4).

TABLE 3.4 Polymers and Some of Their Uses

Monomer	Polymer	Uses

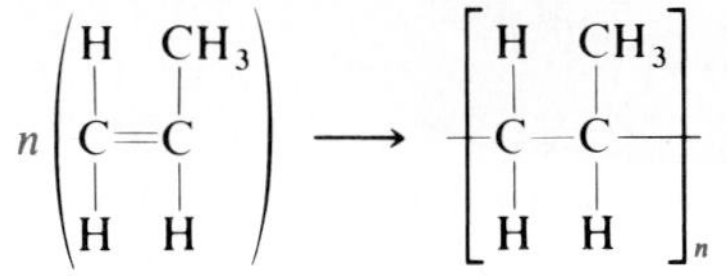

Propylene → Polypropylene — Plastic bottles, biomedical implants, carpet fibers

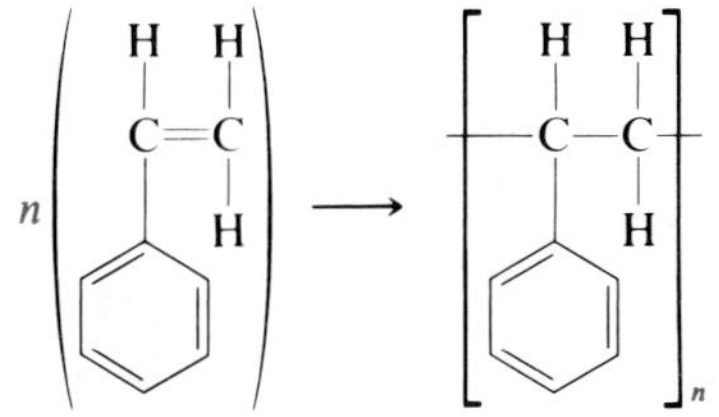

Styrene → Polystyrene — Electrical insulators, foamed plastic fabrication (styrofoam)

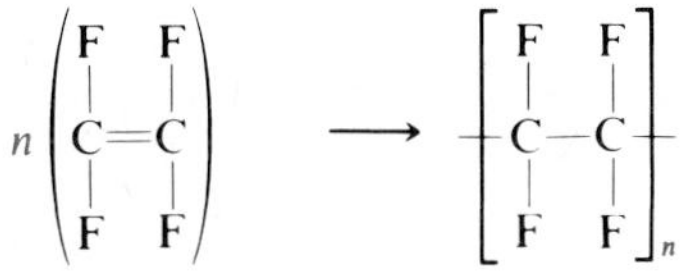

Tetrafluoroethylene → Teflon — Lubricant, bearings, nonstick surface for kitchen utensils

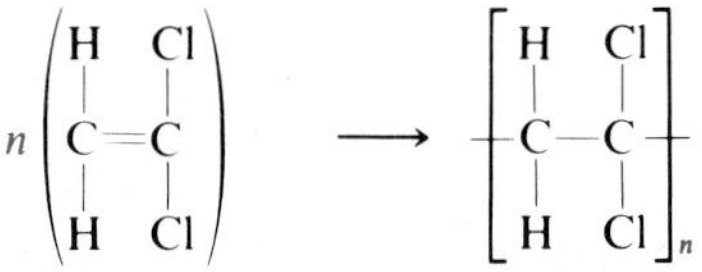

1,1-Dichloroethylene → Saran — Self-adhering food wrappers, seatcovers

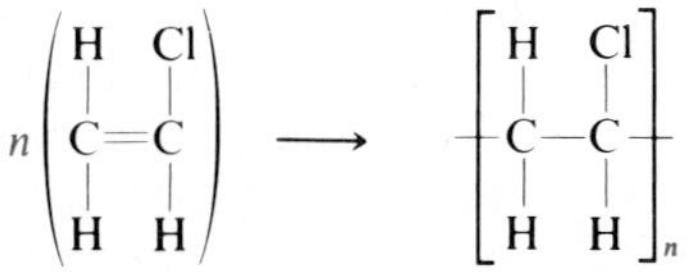

Vinyl chloride → Polyvinyl chloride — Furniture and floor covering, phonograph records, electrical insulators, rainwear, garden hoses

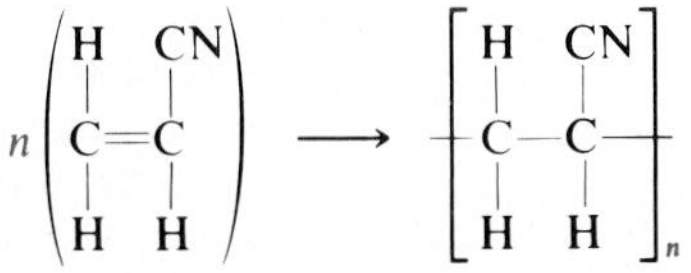

Cyanoethylene → Orlon — Fibers for clothing

Perhaps the most familiar polymer is polyethylene, which is made from the polymerization of thousands of ethylene monomers. Annual production of polyethylene in the United States is well over 1 million tons. It is used for electrical insulation and for making plastic cups, refrigerator trays, squeeze bottles, etc. The repeating unit of the polymer is

$$\left[\begin{array}{cc} H & H \\ | & | \\ -C & -C- \\ | & | \\ H & H \end{array}\right]$$

Therefore the equation for the overall process can be written as

$$n\left(\begin{array}{cc} H & H \\ | & | \\ C = C \\ | & | \\ H & H \end{array}\right) \xrightarrow[\text{heat, pressure}]{\text{initiator,}} \left[\begin{array}{cc} H & H \\ | & | \\ -C & -C- \\ | & | \\ H & H \end{array}\right]_n$$

Ethylene $n \approx 10{,}000$ Polyethylene

Many other alkenes serve as monomer units to provide, among others, the well-known polymers shown in Table 3.4.

3.5 The Isoprene Unit

Complex alkenes containing multiple double bonds within the molecules occur widely in nature. One of our important commodities, natural rubber, is a linear polymer of the diolefin 2-methyl-1,3-butadiene, **isoprene.**

Natural rubber

Isoprene
(2-Methyl-1,3-butadiene)

Isoprene is an example of a **conjugated diene,** *a diene in which the two double bonds are separated by a single bond.* Conjugated double bonds are particularly stable, and we shall see that this structural feature occurs in many biochemical molecules.

Compounds derived from a single isoprene unit are rare in nature. However, compounds in which there are two or more isoprene units are common.

The odoriferous constituents of many plants (such as cedar, clove, pine, eucalyptus, peppermint, and rose) are volatile compounds referred to as **essential oils.** The essential oils are extracted from the plants by distilling with water. (The oils are then separated from the distillate.) These oils are widely used in cosmetics, food flavorings, medicines, and perfumes. They all contain isomeric hydrocarbons of the composition $C_{10}H_{26}$, which are called **terpenes** (because some of the first members were isolated from oil of turpentine).

The compound responsible for the color of carrots, tomatoes, and autumn leaves is a tetraterpene called β-carotene. The carotenes are important intermediates in photosynthesis, in the biosynthesis of vitamin A (the reason that carrots are a good source of vitamin A), and in other cellular processes. In animals, β-carotene is split into two molecules of vitamin A in the small intestine during digestion. β-Carotene is added to margarine, butter, cheese, salad dressings, fruit juices, and carbonated beverages to impart a yellow color. Vitamin E (Section 12.16-c) and vitamin K (Section 12.16-d) also contain isoprene units.

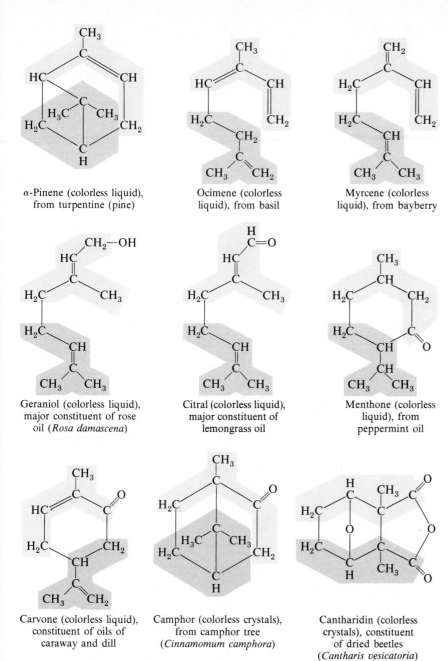

α-Pinene (colorless liquid),
from turpentine (pine)

Ocimene (colorless
liquid), from basil

Myrcene (colorless
liquid), from bayberry

Geraniol (colorless liquid),
major constituent of rose
oil (*Rosa damascena*)

Citral (colorless liquid),
major constituent of
lemongrass oil

Menthone (colorless
liquid), from
peppermint oil

Carvone (colorless liquid),
constituent of oils of
caraway and dill

Camphor (colorless crystals),
from camphor tree
(*Cinnamomum camphora*)

Cantharidin (colorless
crystals), constituent
of dried beetles
(*Cantharis vesicatoria*)

FIGURE 3.3 Some natural C_{10} products of diverse origin that obey the isoprene rule. The plant origin of each substance is given, and the isoprene units are distinguished by the grey and colored areas.

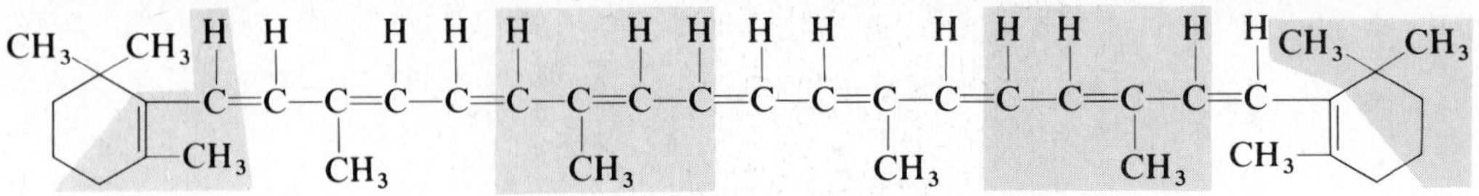

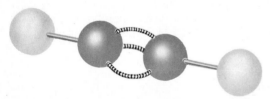

β-Carotene
(8 isoprene units)

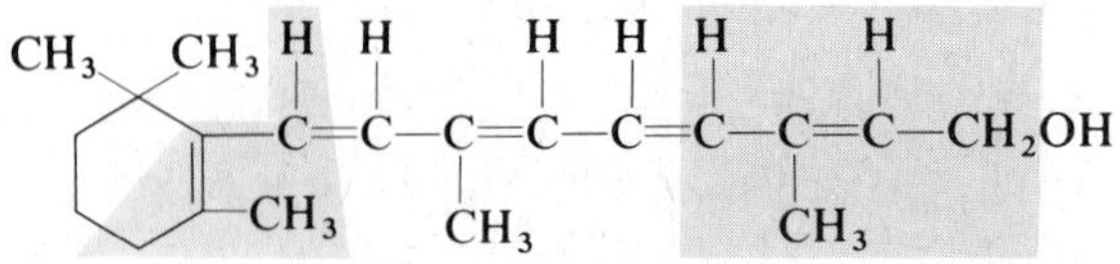

Vitamin A
(4 isoprene units)

The great abundance of naturally occurring plant products whose carbon skeletons contain isoprene units has resulted in the formulation of the **isoprene rule:** *All terpenes should be* *formally divisible into isoprene units.* Some examples of terpenes that obey the rule are shown in Figure 3.3.

Alkynes

The **alkynes** are unsaturated hydrocarbons that *contain a triple bond between adjacent carbon atoms.* The general formula for this class of compounds is C_nH_{2n-2}. The first and most important member in the series is called **acetylene,** and has the molecular formula C_2H_2 (Figure 3.4). Because of the importance of this simple compound, the higher homologs are often referred to as acetylenic hydrocarbons or acetylenes.

FIGURE 3.4 Ball-and-spring model of acetylene showing the linear configuration of the triple bond.

3.6 Nomenclature of Alkynes

The procedure used in naming the alkynes is identical to that for naming the alkenes, but the generic ending for this class of compounds is, of course, different. The suffix *-yne* is used to indicate the presence of the triple bond in the compound, and must be added to the stem name of the parent alkane. Once again, we find that the lower homologs of the series are very often referred to by their common names rather than by their IUPAC names. Table 3.5 lists formulas and names for some of the alkynes.

TABLE 3.5 Formulas and Nomenclature for Some Alkynes

Molecular Formula	Structural Formula	Common Name	IUPAC Name
C_2H_2	$H-C\equiv C-H$	Acetylene	Ethyne
C_3H_4	$CH_3-C\equiv C-H$	Methylacetylene	Propyne
C_4H_6	$CH_3CH_2-C\equiv C-H$	Ethylacetylene	1-Butyne
C_4H_6	$CH_3-C\equiv C-CH_3$	Dimethylacetylene	2-Butyne
C_5H_8	$CH_3CH_2CH_2-C\equiv C-H$	Propylacetylene	1-Pentyne
C_5H_8	$CH_3CH_2-C\equiv C-CH_3$	Methylethylacetylene	2-Pentyne
C_5H_8	$CH_3-\underset{\underset{CH_3}{\vert}}{CH}-C\equiv C-H$	Isopropylacetylene	3-Methyl-1-butyne (**not** 2-Methyl-3-butyne)

3.7 Preparation of Acetylene

Despite the reactivity of the triple bond, there is little difficulty in synthesizing the alkynes. The $C\equiv C$ bond is found in compounds that are constituents of birth control pills (see Section 12.12-d) and in a variety of living organisms, particularly molds. Acetylene can be prepared easily and inexpensively from calcium carbide and water.

$$CaC_2 \ + \ 2\,H_2O \ \longrightarrow \ HC\equiv CH \ + \ Ca(OH)_2$$

Calcium carbide   Acetylene

Acetylene is also prepared by the cracking of methane in an electric arc. This process was perfected and used extensively by Germany during World War II. In the United States about 80% of acetylene usage is as a raw material for the synthesis of various organic compounds (e.g., acetaldehyde, acetic acid, plastics, and synthetic rubber). The remaining 20% is used for metal welding or cutting.

3.8 Physical Properties of Acetylene

Acetylene is a colorless gas, insoluble in water, and it has a rather inoffensive odor when pure. The very disagreeable odor usually associated with acetylene is due to the presence of phosphine (PH_3) impurities. The gas can be condensed to a liquid at $-84°C$ and a pressure of 1 atmosphere (atm). In the liquid state acetylene is very sensive to shock and is highly explosive. Therefore, for commercial purposes, it can be safely transported dissolved in acetone within pressurized cylinders. Acetylene burns with a highly luminous flame; one of its earliest uses was a fuel for miners' lamps (carbide lamps). It was later utilized in bicycle and automobile lamps. The combustion of acetylene in oxyacetylene torches generates enough heat to cut and weld metals.

$$2\,C_2H_2 \ + \ 5\,O_2 \ \xrightarrow{3000°C} \ 4\,CO_2 \ + \ 2\,H_2O \ + \ 620\,kcal$$

3.9   *Chemical Properties of Alkynes*

The chemical reactions of the alkynes are, for the most part, analogous to the reactions of the alkenes. For example, 2 moles of the reagents that add to the alkenes will add to the alkynes. The reactions may be visualized as occurring in two steps.

Symmetrical Reagents

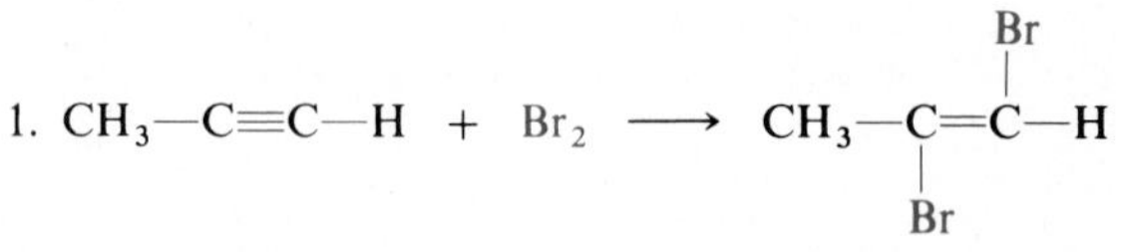

1. $CH_3-C\equiv C-H$ + Br_2 $\longrightarrow$ $CH_3-\overset{\displaystyle Br}{\underset{\displaystyle Br}{C}}=C-H$

   Propyne   1,2-Dibromopropene

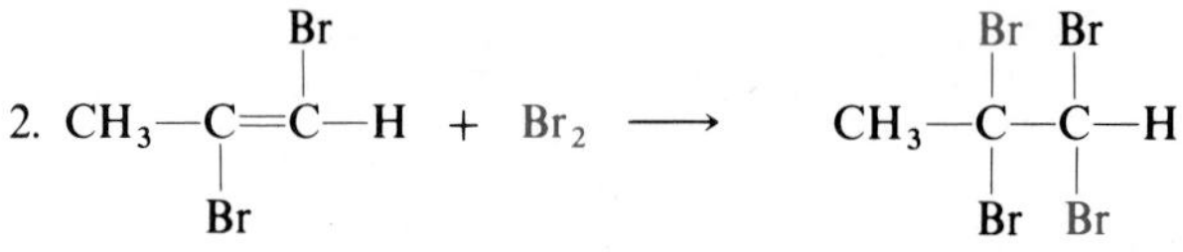

2. $CH_3-\overset{\displaystyle Br}{\underset{\displaystyle Br}{C}}=C-H$ + Br_2 $\longrightarrow$ $CH_3-\overset{\displaystyle Br}{\underset{\displaystyle Br}{C}}-\overset{\displaystyle Br}{\underset{\displaystyle Br}{C}}-H$

                        1,1,2,2-Tetrabromopropane

Unsymmetrical Reagents (Follow Markovnikov's Rule)

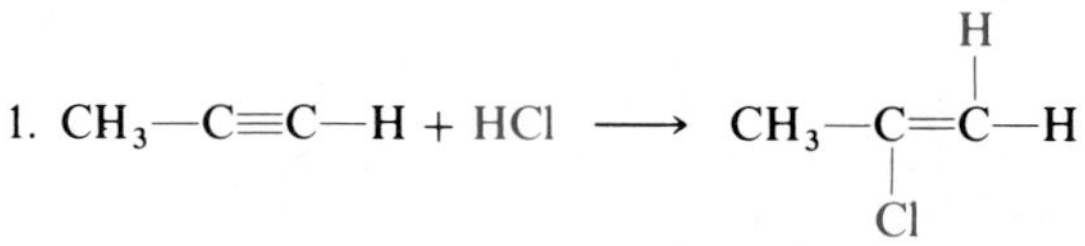

1. $CH_3-C\equiv C-H$ + HCl $\longrightarrow$ $CH_3-\overset{\displaystyle H}{\underset{\displaystyle Cl}{C}}=C-H$

                2-Chloropropene

2. $CH_3-\overset{\displaystyle H}{\underset{\displaystyle Cl}{C}}=C-H$ + HCl $\longrightarrow$ $CH_3-\overset{\displaystyle Cl}{\underset{\displaystyle Cl}{C}}-\overset{\displaystyle H}{\underset{\displaystyle H}{C}}-H$

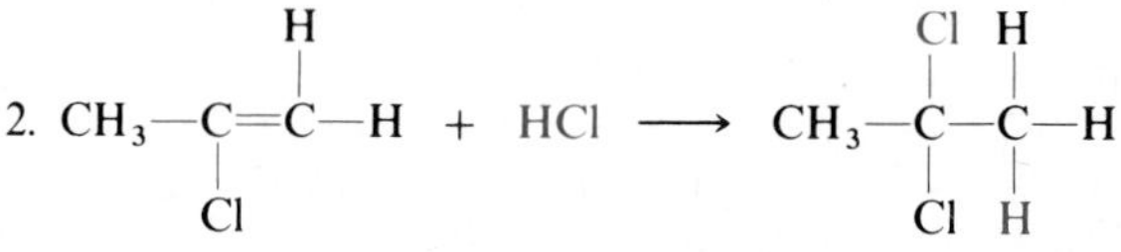

                2,2-Dichloropropane

   In addition, certain reactions are unique to the alkynes. Under the same conditions, alkynes will react with several reagents toward which the alkenes are inert. Acetylene, in the presence of certain catalysts, will react with hydrogen cyanide, whereas ethylene will not. The product of the reaction is cyanoethylene and is an important industrial chemical from which Orlon is produced (Table 3.4).

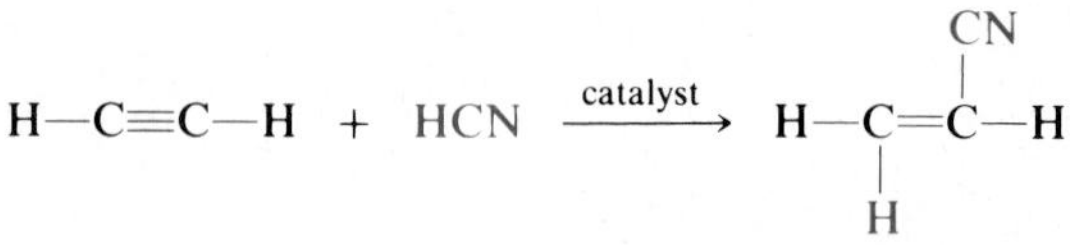

$H-C\equiv C-H$ + HCN $\xrightarrow{\text{catalyst}}$ $H-\overset{\displaystyle CN}{\underset{\displaystyle H}{C}}=C-H$

   Acetylene   Cyanoethylene
                                (Acrylonitrile)

Acetylene reacts, by addition, with water, hydrochloric acid, and acetic acid. The hydration reaction leads to an unstable product that rearranges to form acetaldehyde. The vinyl chloride and vinyl acetate, obtained from the reaction of acetylene and the corresponding acids, are used as monomers in the production of a wide variety of polymers.

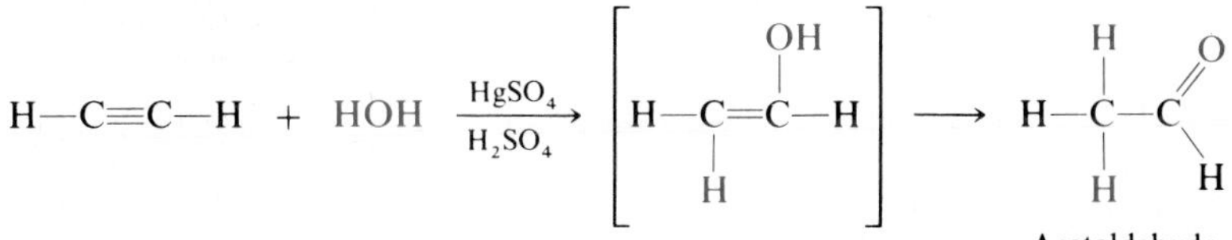

Acetaldehyde

Vinyl chloride

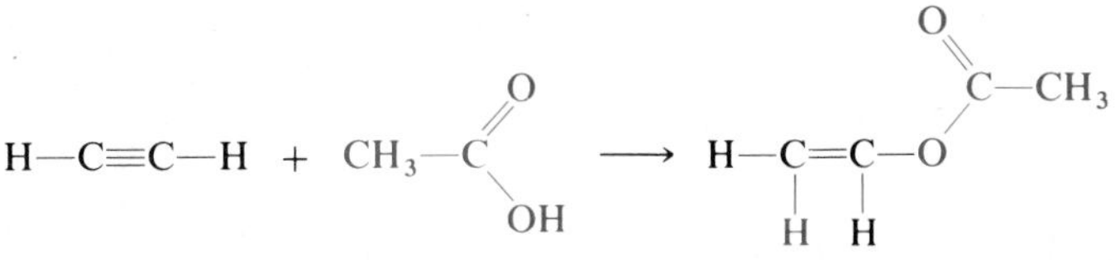

Acetic acid   Vinyl acetate

(Note that H—C=C— is called the *vinyl group*)

Exercises

3.1 Draw and name all the noncyclic structural isomers of hexene, C_6H_{12}.

3.2 Draw and name all the cyclic isomers of C_6H_{12}.

3.3 Draw and name all the noncyclic isomers of C_5H_8.

3.4 Name the following compounds.
(a) $CH_2\!=\!CHCH_2CH_2CH_3$
(b) $CH_3CH_2CH\!=\!C(CH_3)_2$
(c) $(CH_3)_3CCH\!=\!C(CH_3)CH_2CH_3$
(d) $(CH_3)_2C\!=\!CHCHBrCH_3$
(e) $CH_3CHClCH_2CH\!=\!CHCH_2CH_3$

(f) 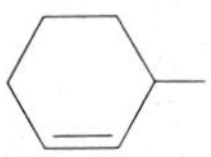 — CH_2CH_3

(g)

(h) $CH_3CH_2CHClCH\!=\!CHCHCl_2$
(i) $CH_3CH\!=\!C\!=\!CH_2$
(j) $(CH_3)_2CHC\!\equiv\!CCH(CH_3)_2$
(k) $CH_3CH_2CH_2C(CH_2CH_3)_2C\!\equiv\!CH$

(l) $(CH_3CH_2CH_2)_2CHC\!\equiv\!CCH_3$

(m) 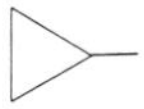 —$CH_2CH_2CH\!=\!CH_2I$

(n)

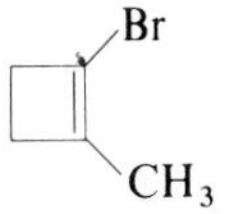

3.5 Write structural formulas for the following compounds.
(a) 2-methyl-2-pentene
(b) 5-methyl-1-hexene
(c) 1,3-butadiene
(d) 2-ethyl-1-butene
(e) 2,4,6,6-tetramethyl-2-heptene
(f) 1-cyclohexyl-2-methylpropene
(g) 3-methylcyclopentene
(h) 4-ethyl-2-methyl-3-hexene
(i) 2,2-dimethyl-4-octene
(j) 5-ethyl-5-methyl-2-heptene
(k) 3,5-dimethyl-1-hexyne
(l) 4-methyl-2-pentyne
(m) 2-ethyl-1,3-cyclohexadiene
(n) 3,4-dipropylcyclobutene

3.6 What is wrong with each of the following names? Give the structure and correct name for each compound.
  (a) 2-methyl-4-heptene
  (b) 2-ethyl-3-hexene
  (c) 2-methyl-2,5-hexadiene
  (d) 2,2-dimethyl-3-pentene
  (e) 2-ethylcyclopentene
  (f) 2-propyne
  (g) 4-iodo-2-butyne
  (h) 2-chloro-4-pentyne
  (i) 2,4-cyclopentadiene
  (j) 4-bromocyclobutene

3.7 Complete the following equations, giving only the organic products (if any).

(a) $CH_3CH_2CHOHCH_2CH_3 \xrightarrow[\Delta]{conc\ H_2SO_4}$

(b) $(CH_3)_3COH \xrightarrow[\Delta]{conc\ H_2SO_4}$

(c) 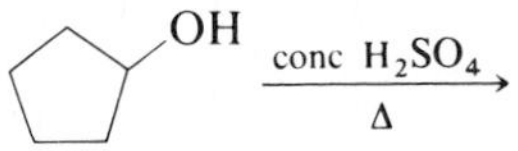 $\xrightarrow[\Delta]{conc\ H_2SO_4}$

(d) $CH_3CH_2Br \xrightarrow[alcohol,\ \Delta]{KOH}$

(e) 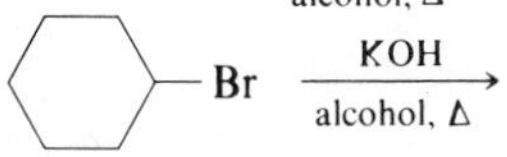 $\xrightarrow[alcohol,\ \Delta]{KOH}$

(f) $(CH_3)_3CCHClCH(CH_3)_2 \xrightarrow[alcohol,\ \Delta]{KOH}$

(g) $(CH_3)_2C\!=\!CH_2 + Br_2 \longrightarrow$

(h) 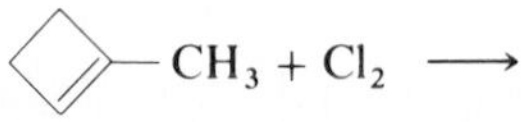 $-CH_3 + Cl_2 \longrightarrow$

(i) $CH_2\!=\!C(CH_3)CH_2CH_3 + H_2 \xrightarrow{Ni}$

(j) $CH_2\!=\!CHCH\!=\!CH_2 + 2\,H_2 \xrightarrow{Ni}$

(k) $(CH_3)_2C\!=\!CHCH_3 + HBr \longrightarrow$

(l) 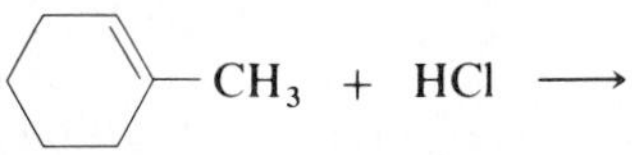 $-CH_3 + HCl \longrightarrow$

(m) $(CH_3)_2C\!=\!CH_2 + HOSO_3H \longrightarrow$

(n) $CH_3CH_2CH\!=\!C(CH_3)CH_2CH_3 + HI \longrightarrow$

(o) $(CH_3)_2C\!=\!C(CH_3)_2 + KMnO_4 \xrightarrow[cold]{dil}$

(p) $(CH_3)_2C\!=\!CHCH_3 + KMnO_4 \xrightarrow[\Delta]{conc}$

(q) $CH_3CH_2C\!\equiv\!CH + Br_2 \longrightarrow$

(r) $CH_3C\!\equiv\!CCH_3 + 2\,H_2 \xrightarrow{Ni}$

(s) $CH_3C\!\equiv\!CH + HCl \longrightarrow$

(t) $CH_3C\!\equiv\!CH + HOH \xrightarrow[H_2SO_4]{Hg^{2+}}$

3.8 Write the structural formula(s) of the product(s), if any,

that would be formed from the reaction of propene with each of the following reagents.
  (a) H_2, nickel catalyst
  (b) Br_2
  (c) HCl
  (d) H_2SO_4
  (e) cold, dil $KMnO_4$
  (f) hot, conc $KMnO_4$

3.9 What reagents would you use to effect the following reactions?
  (a) $CH_2\!=\!CH_2 \longrightarrow CO_2 + H_2O$
  (b) $CH_2\!=\!CHCH\!=\!CH_2 \longrightarrow$
$$CH_2BrCHBrCHBrCH_2Br$$
  (c) $CH_3CH_2CH\!=\!CH_2 \longrightarrow$
$$CH_3CH_2CH(OSO_3H)CH_3$$
  (d)  $\longrightarrow$
  (e) $CH_3CH_2C\!\equiv\!CH \longrightarrow CH_3CH_2CBr\!=\!CH_2$
  (f) $CH_3C\!\equiv\!CCH_3 \longrightarrow CH_3CCl_2CCl_2CH_3$
  (g) $CH_3CH_2CH\!=\!CHCH_3 \longrightarrow$
$$CH_3CH_2CHOHCHOHCH_3$$
  (h) $CH_3CH_2C(CH_3)\!=\!CH_2 \longrightarrow$

$$CH_3CH_2\overset{\displaystyle O}{\overset{\|}{C}}CH_3 + H\!-\!\overset{\displaystyle O}{\overset{\|}{C}}\!-\!OH$$

3.10 By means of equations, distinguish between the terms in the following pairs.
  (a) an addition reaction and a substitution reaction
  (b) a dehydration reaction and a dehydrohalogenation reaction
  (c) an oxidation reaction and a reduction reaction

3.11 Define and illustrate each of the following terms.
  (a) electrophile
  (b) carbocation
  (c) polymerization
  (d) cracking
  (e) electron-attracting group
  (f) electron-donating group
  (g) Markovnikov's rule
  (h) isoprene
  (i) isoprene rule
  (j) terpene
  (k) halogenation
  (l) hydrogenation
  (m) inductive effect
  (n) glycols
  (o) monomer
  (p) polymer
  (q) conjugated diene
  (r) vinyl group

3.12 (a) In the reaction of HBr with an alkene, which portion of the reagent makes the initial attack?

(b) If the alkene is unsymmetrical, where will the initial attachment occur?

(c) Illustrate this reaction by writing an equation for the addition of HBr to an unsymmetric cycloalkene.

3.13 Pentane and 1-pentene are both colorless, low-boiling liquids. Give two simple tests that would distinguish the two compounds. Indicate what you would observe.

3.14 Predict the order of stability of the following carbocations.

$$CH_3CH_2\overset{+}{C}HCH_3 \qquad (CH_3)_3C^+ \qquad CH_3CH_2CH_2CH_2{}^+$$

3.15 Compare the carbon–carbon bond lengths of single bonds, double bonds, and triple bonds.

3.16 Since acetylene will explode if subjected to pressures in excess of 2 atm, how is it possible to store and ship acetylene in steel containers?

3.17 What are the chief uses of acetylene? How is it produced industrially?

3.18 Write equations for two reactions of acetylene that are similar to those of ethene and two reactions that are unique to acetylene.

3.19 Name three polymers and give one use for each.

3.20 Vitamin E and vitamin K are two "fat-soluble" vitamins that play important roles in the maintenance of human health. The formulas of vitamin E and vitamin K, with condensed isoprene units, are shown in Sections 12.16-c and 12.16-d, respectively. Draw the formulas of the vitamins and write the full structures of their isoprene units. Circle each separate unit.

3.21 Limonene is a terpene that gives lemon and orange peels their characteristic odors. Circle the isoprene units.

Limonene

3.22 56 mg of a monoalkene requires 22.4 mL of hydrogen (at STP) to convert it to an alkane.

(a) What is the molecular weight of the alkene, and what is its molecular formula?

(b) Draw all the possible alkenes that correspond to this molecular formula.

3.23 A hydrocarbon whose molecular formula is C_4H_6 reacts with 2 moles of chlorine to yield $C_4H_6Cl_4$. Is this information sufficient to determine the structure of the original hydrocarbon? Explain.

3.24 A hydrocarbon of molecular formula C_5H_8 combines with 2 moles of hydrogen to yield 2-methylbutane. What is the structural formula of the original compound?

3.25 An unknown compound has the molecular formula C_6H_{10}, yet it only adds 1 mole of hydrogen or of bromine per mole of compound. What is a possible structural formula?

3.26 Three isomeric pentenes, X, Y, and Z, can be hydrogenated to 2-methylbutane. Addition of chlorine to Y gives 1,2-dichloro-3-methylbutane, and 1,2-dichloro-2-methyl-butane is obtained from Z. Write the structural formulas for the three isomers.

3.27 How much HBr is necessary to convert 13 g of acetylene into 1,1-dibromoethane?

3.28 Excess bromine was added to 1 mole of ethene and 150 g of product were formed. What percent of the theoretical yield was obtained?

4

The Aromatic Hydrocarbons

Previously, distinctions among classes of hydrocarbons were made according to the type of bonding that occurred between the carbon atoms in the molecule. The alkanes were characterized by the occurrence of single bonds, the alkenes by double bonds, and the alkynes by triple bonds.

It is not so easy to characterize the aromatic hydrocarbons in this manner. The compounds of this family were called aromatic originally because the earliest known derivatives had spicy or sweet-smelling odors. However, not all aromatic compounds are odorous, and not all fragrant organic compounds are aromatic. The name is used today to denote a particular type of chemical structure that sets these compounds apart from aliphatic compounds. *The majority of* **aromatic** *compounds resemble benzene in structure and reactivity.*

4.1   *Structure of the Aromatic Ring*

In 1825 Michael Faraday isolated a hydrocarbon from a sample of lighting fuel (illuminating gas). He determined that the compound had an empirical formula of CH_2 (based upon the then current belief that the atomic weight of carbon was 6), and he named the compound "bicarburet of hydrogen." Later the compound was

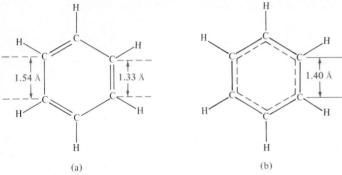

(a)　　　　　　　　　　　(b)

FIGURE 4.1 (a) The hypothetical 1,3,5-cyclohexatriene would be a slightly lopsided hexagon because double bonds are shorter than single bonds. (b) The actual benzene molecule is a regular hexagon with all sides 1.40 Å.

given the name benzene because it could be obtained by distilling benzoic acid with calcium oxide. During the ensuing decades the concept of atomic weight became more clearly defined, and vapor density measurements established the molecular formula of benzene to be C_6H_6.

One of the many possible structural formulas corresponding to this molecular formula is 1,3,5-cyclohexatriene. The compound would be a slightly lopsided hexagon, having three double bonds in a six-membered ring (Figure 4.1-a). The high degree of unsaturation would imply a very high reactivity. Chemists soon discovered, however, that benzene is chemically stable and that it behaves more like an alkane than an alkene or alkyne. It is unreactive with respect to both the bromine and permanganate tests for unsaturation (Section 3.4). Under special conditions, it can be made to react with bromine, but the reaction is a substitution reaction and not an addition.

$$C_6H_6 \ + \ Br_2 \ \xrightarrow[\text{catalyst}]{\text{FeBr}_3} \ C_6H_5Br \ + \ HBr$$

Benzene　　　　　　　　　　　　　　Bromobenzene

This reaction proved significant in elucidating the structure of benzene. It was found that one and only one monobromobenzene is formed in the reaction. There are no isomers of this compound, and therefore the six hydrogens of benzene must be equivalent.

Finally, in 1865, the German chemist Friedrich August Kekule proposed a structure that could account for all of the known chemical properties of the compound.[1] His theory was that the benzene molecule consisted of a cyclic, hexagonal, planar structure of six carbon atoms with alternate single and double bonds. Each carbon atom was bonded to only one hydrogen atom. He accounted for

---

[1] There is a story told that one evening Kekule fell asleep while sitting in front of a fire. He dreamed about chains of atoms having the forms of twisting snakes. Suddenly one of the snakes caught hold of its own tail, forming a whirling ring. He awoke, freshly inspired, and spent the remainder of the night working on his now famous hypothesis. Kekule is said to have written "Let us learn to dream, gentlemen, and then perhaps we shall learn the truth."

the equivalance of all six carbon atoms by suggesting that the double bonds are not static but rather are mobile, and that they oscillate from one position to another.

Structures I and II differ from each other only in the positions of the double and single bonds, and although they satisfy the requirements for equivalent hydrogens, they do not explain why benzene does not behave like an unsaturated hydrocarbon. Furthermore, x-ray diffraction measurements of the bond lengths indicate that all the carbon–carbon bonds in benzene are the same length, 1.40 Å (Figure 4.1-b). This value falls between the length of a carbon–carbon double bond (1.33 Å), and a carbon–carbon single bond (1.54 Å).

To accommodate these discrepancies, chemists have postulated that benzene exhibits resonance. **Resonance**[2] is a word used to describe *the phenomenon in which no single classical Lewis structure adequately accounts for the experimentally observed properties of a molecule* (such as bond energies and bond distances). According to the resonance theory, the actual benzene molecule is a resonance hybrid of structures I and II. That is, neither of the two structures actually exists, but taken together, they represent the true structure of the molecule. We depict this situation with a double-headed arrow (↔) that connects the contributing structures (Figure 4.2-a). This resonance arrow should be clearly distinguished from the pair of half-headed arrows (⇌) that are used to indicate an equilibrium condition.

Some authors combine the two contributing forms into a single structure as depicted in Figure 4.2-b. The inner circle indicates that the valence electrons are shared equally by all six carbon atoms (that is, the electrons are *delocalized,* or spread out, over all the carbon atoms). This method is a valid, short-cut device for writing benzene, but it does not adequately account for all the electrons in the molecule. The representation of benzene that contains alternating single and double bonds (*resonance* or *Kekule forms*) is the best model for keeping track of the electrons within the molecule. The circle within a hexagon better describes the molecule by indicating the equal sharing of the electrons and the identical bond lengths within the molecule. We shall use both structures: hexagon with inscribed circle when we discuss nomenclature, but one of the Kekule forms when we are talking of resonance structures and/or chemical reactions. When we show either representation, it is understood that each corner of the hexagon is occupied by one carbon atom. Attached to each carbon atom is one hydrogen atom, and all the atoms lie in the same plane. Whenever any other atom, or group of atoms, is

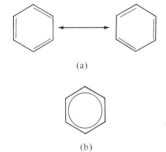

FIGURE 4.2 Two methods of representing the structure of benzene. (a) Kekule structures, (b) hexagon with inscribed circle.

---

[2] Although the term resonance is utilized a great deal throughout organic chemistry, it is not unique to organic compounds. Many inorganic compounds (e.g., O_3, SO_2, NO_3^-) also exhibit resonance.

substituted for a hydrogen atom, it must be shown to be bonded to a particular corner of the hexagon.

One final note: taken literally, the term resonance is a misnomer. It indicates oscillation from one structure to another or oscillation of an electron pair from one bond to another. *There is no oscillation.* A mule, the offspring of a male donkey and a mare, can be considered a hybrid of a donkey and a horse. However, it is not a horse at one instant and a donkey at another; it is always a mule. Likewise, any molecule that exhibits resonance has *one* real structure, which is never any of the rather unsophisticated Lewis structures used to describe it. The difficulty arises because we cannot adequately portray the molecule—not from the molecule itself.

4.2 *Nomenclature of Aromatic Hydrocarbons*

There is more than one system for naming aromatic hydrocarbons. Both common names and systematic names are encountered. Some aromatic compounds are referred to exclusively as derivatives of benzene, whereas others are more frequently denoted by their common names. Note that in the following structures, it is immaterial whether the substituent is written at the top, side, or bottom of the ring: a hexagon is symmetrical, and all positions are equivalent.

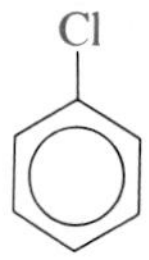

Chlorobenzene	Bromobenzene	Nitrobenzene	Ethylbenzene

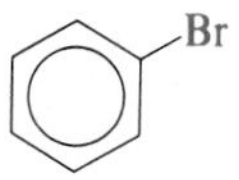

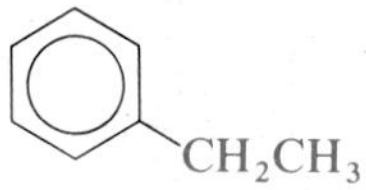

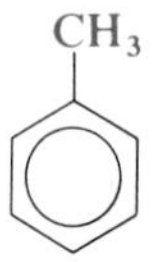

Toluene (Methylbenzene)	Phenol (Hydroxybenzene)	Aniline (Aminobenzene)	Styrene (Vinylbenzene)

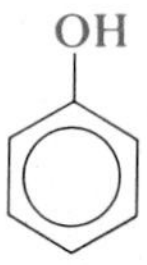

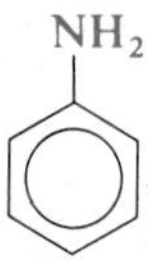

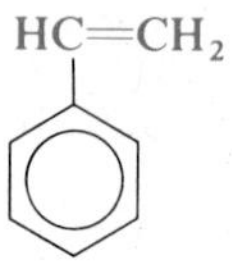

A further complication arises when there is more than one substituent attached to a benzene ring. When this occurs, all the positions on the hexagon are no longer equivalent, and the relative positions of the substituents must be designated. In the case of a disubstituted benzene, one nomenclature system uses the prefixes **ortho** (*o-*), **meta** (*m-*), and **para** (*p-*). Ortho designates 1,2-disubstitution, meta designates 1,3-disubstitution, and para designates 1,4-disubstitution.

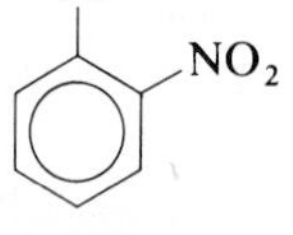

o-Chloronitrobenzene 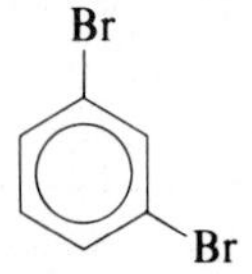   *m*-Dibromobenzene 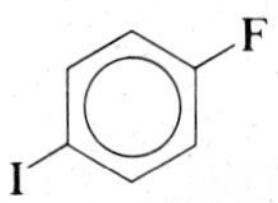   *p*-Fluoroiodobenzene

Alternatively, the ring is numbered and the substituent names are listed in alphabetical order. The first substituent is given the lowest number. When a common name is used, the carbon atom that bears the group responsible for the name is considered to be carbon-1.

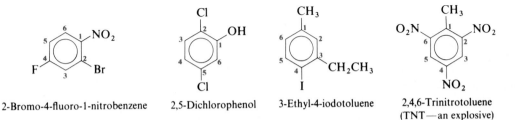

m-Chloroethylbenzene   *o*-Bromotoluene   *p*-Nitrophenol
1-Chloro-3-ethylbenzene   2-Bromotoluene   4-Nitrophenol

If more than two substituents occur, the numbering is determined by the requirement that the numbers give the smallest possible sum (substituents are listed in alphabetical order).

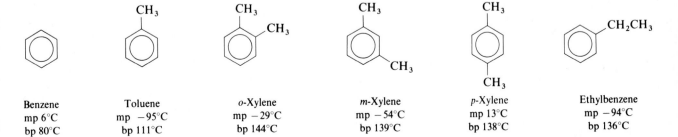

2-Bromo-4-fluoro-1-nitrobenzene   2,5-Dichlorophenol   3-Ethyl-4-iodotoluene   2,4,6-Trinitrotoluene
                                                                                      (TNT—an explosive)

There are three isomers corresponding to the formula $C_6H_4(CH_3)_2$. Like toluene, they are homologs of benzene. The parent name of all three compounds is xylene.

Benzene	Toluene	*o*-Xylene	*m*-Xylene	*p*-Xylene	Ethylbenzene
mp 6°C	mp −95°C	mp −29°C	mp −54°C	mp 13°C	mp −94°C
bp 80°C	bp 111°C	bp 144°C	bp 139°C	bp 138°C	bp 136°C

The three xylenes are also isomeric with ethylbenzene, $C_6H_5(C_2H_5)$, since they all have the molecular formula C_8H_{10}. These four isomers again illustrate that the boiling points of compounds can be very nearly identical, whereas their melting points can differ over a wide range, with the symmetrical isomer (*p*-xylene) melting at the highest temperature.

Occasionally, an aromatic group is a substituent that is bonded to an aliphatic compound or to another aromatic ring. The general designation for an aromatic (**aryl**) group as a substituent is **Ar** just as **R** is used to represent an alkyl group. The most common aryl group is the one derived from benzene (C_6H_5—). It was given the name *phenyl*, derived from *pheno*, an old name for benzene.[3]

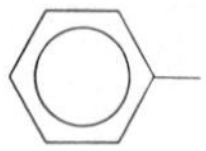

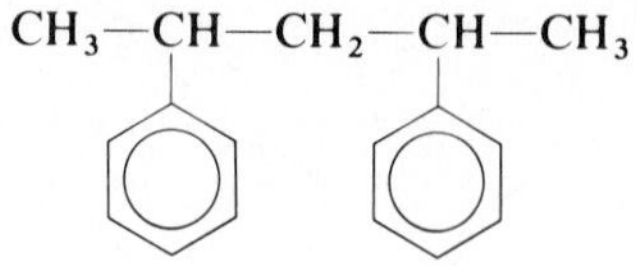

$$CH_3-CH-CH_2-CH-CH_3$$

| Phenyl | Biphenyl | 2,4-Diphenylpentane |

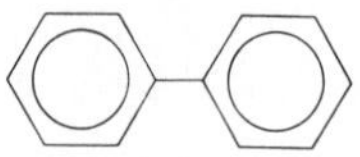

Mention should be made of some common aromatic hydrocarbons that are not substituted benzenes but are condensed benzene rings.

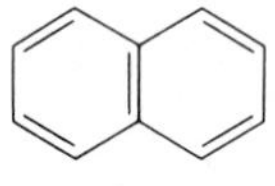

Naphthalene	Anthracene	Phenanthrene
mp 80°C	mp 218°C	mp 101°C
bp 218°C	bp 342°C	bp 340°C

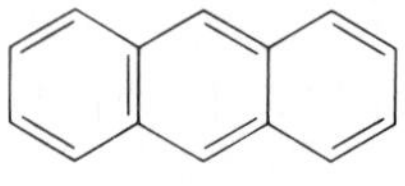

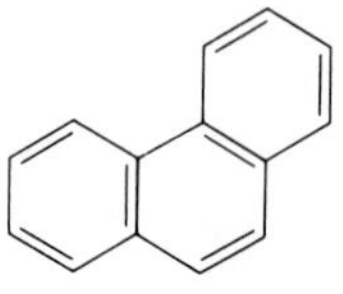

These three substances are colorless, crystalline solids that are obtained from coal tar. Naphthalene has a pungent odor and is commonly used in moth balls and in moth flakes. Anthracene is an important starting material in the manufacture of certain dyes. A large group of naturally occurring substances, the steroids, contain the hydrogenated phenanthrene structure (see Section 12.12).

These polycyclic aromatic compounds do not exist in coal itself, but are formed by the intense heating involved in the distillation of coal tar. For many years, it has been known that workers in coal-tar refineries are susceptible to a type of skin cancer known as tar cancer. Investigation has shown that a number of these polycyclic aromatic hydrocarbons have the ability to cause cancer when applied to the skin. Such compounds are called *carcinogens* (cancer-producers). One of the most active carcinogenic compounds, benzo[a]pyrene (Figure 4.3), occurs in coal tar to the extent of 1.5%, and it has been isolated from cigarette smoke, automobile exhaust gases, and charcoal-broiled steaks. It is estimated that more than 1000 tons of benzo[a]pyrene is emitted into the air over the United States each year. Only a few milligrams of benzo[a]pyrene is required to induce cancer in experimental animals.

The mechanism by which these compounds cause cancer has not yet been elucidated. One hypothesis is that the active carcinogens are not the polycyclic

---

[3]This terminology is confusing because it would seem that the group derived from benzene should be called benzyl. The problem is compounded by the fact that another group is called benzyl. Replacement of one of the methyl hydrogens of toluene gives the *benzyl* group, $C_6H_5CH_2-$.

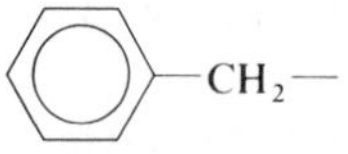

| Benzyl | Benzyl bromide |

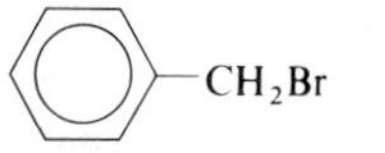

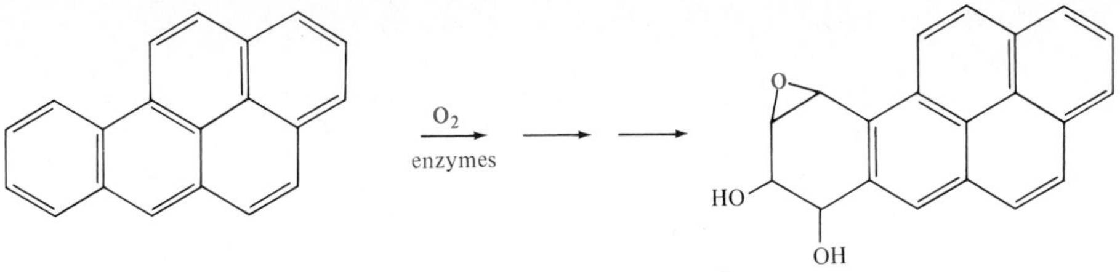

Benzo[a]pyrene · A diolepoxide

FIGURE 4.3 Benzo[a]pyrene can be metabolized in the body to produce the active carcinogen.

hydrocarbons themselves but one or more of their metabolites. (As we shall learn in Chapter 16, metabolites are the products of chemical transformations in living cells.) Figure 4.3 indicates the conversion of benzo[a]pyrene via a multistep oxidation sequence to yield a highly carcinogenic diolepoxide metabolite. To a certain extent, fused polycyclic hydrocarbons are formed whenever organic molecules are heated to high temperatures. It is the current belief that lung cancer is caused by the formation of carcinogenic compounds in the burning of cigarettes.

The aromatic hydrocarbons containing only one benzene ring are generally liquids, whereas the polyring aromatics are generally solids. They are all insoluble in water, but are soluble in organic solvents. Aromatic hydrocarbons are readily combustible. Unlike aliphatic hydrocarbons, which burn with a relatively clean flame, the aromatic compounds burn with a very sooty flame.

4.3 Reactions of Aromatic Hydrocarbons

Although benzene can be made to undergo addition reactions under special conditions, its principle mode of reaction is substitution. Aromatic substitution reactions, unlike the substitution reactions of the alkanes, are rather easily controlled. The general equation for the substitution of a ring hydrogen is as follows.

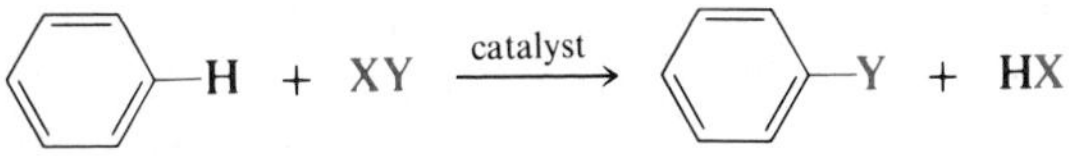

Following are specific illustrative reactions.

Halogenation—the substitution of hydrogen by chlorine or bromine

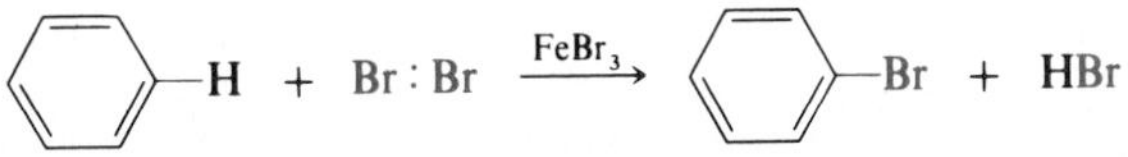

Nitration—the substitution of hydrogen by the nitro group of nitric acid

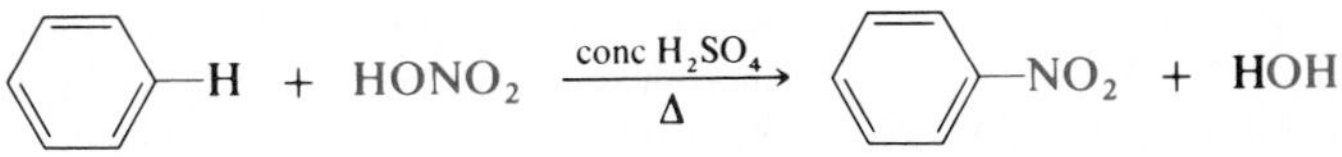

Sulfonation—the substitution of hydrogen by the sulfonic acid group

$$\text{C}_6\text{H}_5\text{—H} + \text{HOSO}_3\text{H} \xrightarrow[\Delta]{\text{SO}_3} \text{C}_6\text{H}_5\text{—SO}_3\text{H} + \text{HOH}$$

Alkylation—the substitution of hydrogen by an alkyl group

$$\text{C}_6\text{H}_5\text{—H} + \text{ClCH}_2\text{CH}_3 \xrightarrow{\text{AlCl}_3} \text{C}_6\text{H}_5\text{—CH}_2\text{CH}_3 + \text{HCl}$$

This last reaction is also known as the Friedel–Crafts reaction after its discoverers.

4.4 *Uses of Benzene and Benzene Derivatives*

Most of the benzene used commercially comes from petroleum. Benzene is employed industrially as a starting material for the production of many other products (e.g., detergents, drugs, dyes, insecticides, plastics). Benzene had been widely used as an organic solvent, but we now recognize benzene to be a poisonous substance with acute and chronic effects. Inhalation of large concentrations of benzene can cause nausea and even death due to respiratory or heart failure. Repeated exposure to benzene leads to a progressive disease in which the ability of the bone marrow to make new blood cells is eventually destroyed. This results in a condition called *aplastic anemia,* in which there is a decrease in both the red and white blood cells. Because of its hazardous nature, many chemical laboratories have replaced benzene with toluene as a general solvent.[4] Toluene is utilized in the production of dyes, drugs, and explosives and as a solvent in lacquers. It is commonly used as a preservative for urine specimens, and it is added to fuels to improve their octane numbers. Trinitrotoluene (TNT), unlike nitroglycerin (see Section 6.5-e), is not sensitive to shock on jarring and must be exploded by a detonator. The xylenes are good solvents for grease and oil and are used for cleaning slides and optical lenses of microscopes. *p*-Xylene is oxidized to terephthalic acid, which is then utilized in the production of Dacron (see page 161). The xylenes are also used to raise the octane number of unleaded gasolines.

Nitrobenzene is used extensively in the manufacture of aniline, the parent compound of many dyes and drugs. Phenol containing a small amount of water is a liquid, and in this form is referred to as carbolic acid. It is a good antiseptic and germicide, but its use is limited because of its toxicity (see Section 6.9).

---

[4]The maximum allowed concentration for an 8-hour exposure to toluene is 100 ppm, whereas the allowable exposure to benzene for the same period of time is 1 ppm. There are no toxic symptoms attributable to toluene until concentrations reach 200 ppm. Care must be taken in using toluene because most commercial toluene contains benzene as an impurity. If the concentration of benzene in toluene is less than 1%, the toluene is safe to use. *However,* toluene has been shown to cause birth defects, so pregnant women should avoid breathing toluene vapors. Prolonged inhalation of toluene should be avoided by everyone. See footnote 1 (page 141) for an explanation of the greater safety of toluene compared to benzene.

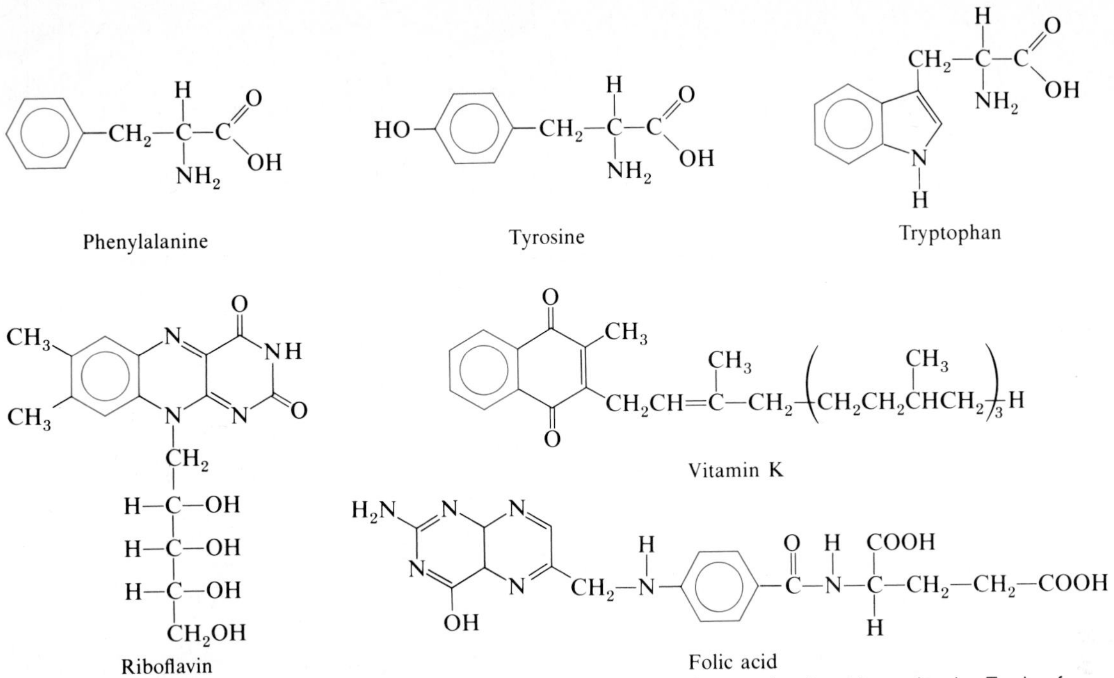

FIGURE 4.4 Some biologically important compounds that cannot be synthesized by animals. Each of these compounds contains a benzene ring that must be supplied from substances in the diet.

Substances containing the benzene ring are commonly found in both the animal and plant kingdoms, although they are more abundant in the latter. Plants have the ability to synthesize the benzene ring from carbon dioxide, water, and inorganic materials. Animals, on the other hand, are incapable of this synthesis but are dependent on benzenoid compounds for their survival. Therefore, the animal must obtain the compounds from the food that it ingests. Included among the aromatic compounds necessary for animal metabolism are the amino acids phenylalanine, tyrosine, and tryptophan and certain vitamins such as vitamin K, riboflavin, and folic acid (Figure 4.4). In addition, a great majority of drugs contain the benzene ring (see Section 9.12-a–h).

Exercises

4.1 What characteristic features of aromatic hydrocarbons distinguish them from other hydrocarbons?

4.2 Briefly discuss the concept of resonance.

4.3 Draw all of the resonance structures for **(a)** ozone and **(b)** the nitrate ion (NO_3^-).

4.4 Name the following compounds.

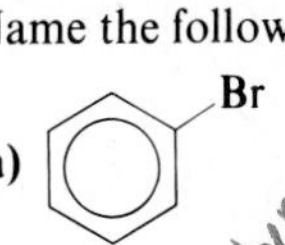

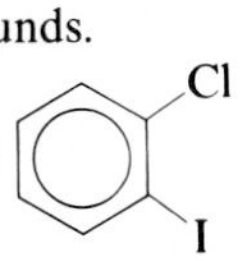

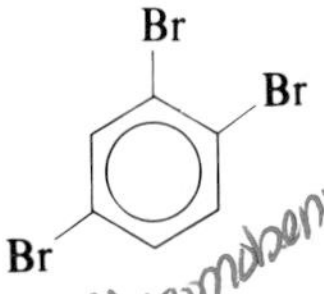

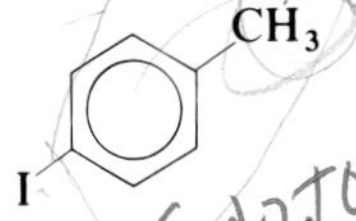

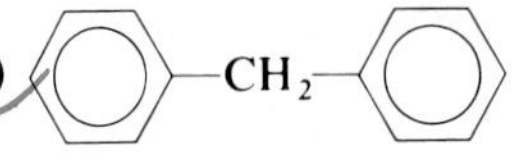

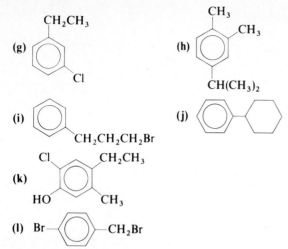

(g) ethylbenzene with Cl (meta position)

(h) 1,2-dimethyl-4-isopropylbenzene

(i) phenyl-CH₂CH₂CH₂Br

(j) phenyl-cyclohexane

(k) chloro, CH₂CH₃, HO, CH₃ substituted benzene

(l) Br—⟨benzene⟩—CH₂Br

4.5 Write structural formulas for the following compounds.
  (a) *p*-diethylbenzene
  (b) 1-ethyl-4-isopropylbenzene
  (c) 1,2,3-trimethylbenzene
  (d) *o*-xylene
  (e) *p*-nitrotoluene
  (f) 4-amino-1,2-dichlorobenzene
  (g) *m*-bromoaniline
  (h) 2,6-dichlorophenol
  (i) 2-iodo-3-nitrostyrene
  (j) 3-ethyl-5-phenyl-1-pentene
  (k) benzyl iodide
  (l) 2-(*p*-nitrophenyl)hexane

4.6 Distinguish between the terms *alkyl group* and *aryl group*. Give an example of each.

4.7 Draw and name the six isomeric dichlorophenols.

4.8 Draw and name all the C_9H_{12} aromatic isomers (there are eight!).

4.9 Draw and name all the isomeric tetramethylbenzenes.

4.10 Draw and name all the possible tribromobenzenes obtained by bromination of (a) *o*-dibromobenzene, (b) *m*-dibromobenzene, (c) *p*-dibromobenzene.

4.11 How many products would one expect to be formed by the mononitration of the following compounds?

(a) 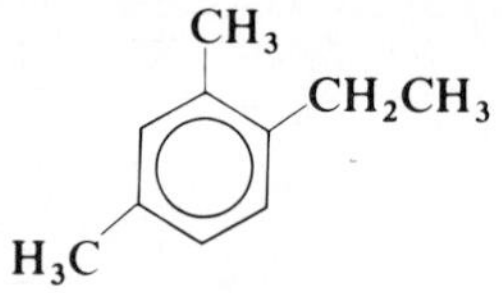 benzene with CH₃, CH₂CH₃, H₃C substituents

(b) 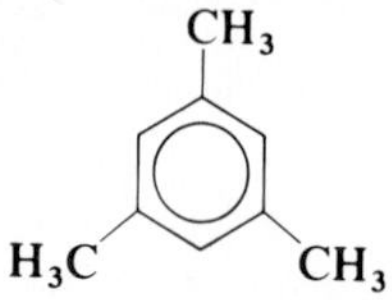 benzene with CH₃, H₃C, CH₃ substituents

(c) 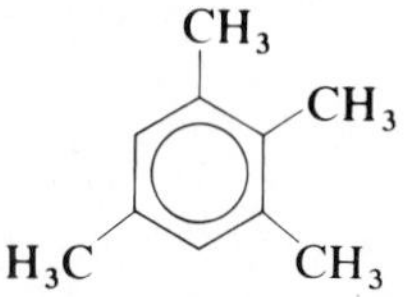 benzene with CH₃, CH₃, H₃C, CH₃ substituents

(d) 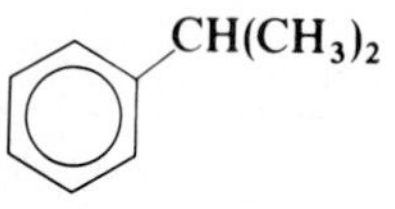 benzene with CH(CH₃)₂

4.12 Write three Kekule formulas for naphthalene, four for anthracene, and five for phenanthrene.

4.13 (a) How many monochloro derivatives of naphthalene are possible? Draw their formulas.
  (b) How many isomers are there for dichloronaphthalene?

4.14 Contrast the bromination of benzene with that of methane.

4.15 What is a quick method to distinguish an aliphatic compound from an aromatic compound?

4.16 Compare the chemical behaviors of benzene and 1,3-cyclohexadiene.

4.17 What chemical tests would you perform to distinguish among cyclohexane, cyclohexene, and benzene?

4.18 (a) What is meant by the term carcinogen?
  (b) Draw the structure of a compound that is carcinogenic.

4.19 List five substances found in your home that contain the benzene ring.

5

Halogen Derivatives of Hydrocarbons

The class of compounds we are about to discuss is of great importance to the industrial chemist but only of limited value to the biochemist. The halogen derivatives of the hydrocarbons result from the replacement of one or more hydrogen atoms of a hydrocarbon with halogen atoms. The letter **X** is often used as a generic symbol for halogen. If *the substituted hydrocarbon is aliphatic,* then the derivative is an **alkyl halide (R—X);** if the *hydrocarbon is aromatic,* then the derivative is an **aryl halide (Ar—X).** Previously, the functional group of the different classes of hydrocarbons was a particular type of bond—single, double, or triple. In the case of the organic halides, the functional group is a single atom, the covalently bonded halogen atom.

5.1 *Nomenclature of Alkyl Halides*

The common names of the alkyl halides are quite prevalent in the chemical literature. The naming system in this case is quite similar to the naming of inorganic halide salts. For example, NaCl is sodium chloride and CH_3Cl is methyl chloride. The alkyl group is always named first and is followed by the name of the halide. The halogenated derivatives of methane are invariably referred to by their common names (CH_2Cl_2, methylene chloride; $CHCl_3$, chloroform; CCl_4, carbon tetra-

TABLE 5.1   Nomenclature of Alkyl Halides

Molecular Formula	Structural Formula	Common Name	IUPAC Name
CH_3I	CH_3I	Methyl iodide	Iodomethane
C_2H_5I	CH_3CH_2I	Ethyl iodide	Iodoethane
C_3H_7Cl	$CH_3CH_2CH_2Cl$	*n*-Propyl chloride	1-Chloropropane
C_3H_7Cl	$CH_3\overset{\displaystyle Cl}{\underset{\displaystyle |}{C}}HCH_3$	Isopropyl chloride	2-Chloropropane
C_4H_9Br	$CH_3CH_2CH_2CH_2Br$	*n*-Butyl bromide	1-Bromobutane
C_4H_9Br	$CH_3CH_2\overset{\displaystyle Br}{\underset{\displaystyle |}{C}}HCH_3$	*sec*-Butyl bromide	2-Bromobutane
C_4H_9Br	$CH_3\overset{\displaystyle CH_3}{\underset{\displaystyle |}{C}}HCH_2Br$	Isobutyl bromide	1-Bromo-2-methylpropane
C_4H_9Br	$CH_3-\overset{\displaystyle CH_3}{\underset{\displaystyle \underset{\displaystyle Br}{|}}{\overset{\displaystyle |}{C}}}-CH_3$	*tert*-Butyl bromide	2-Bromo-2-methylpropane
$C_6H_{11}Cl$	 ⬡—Cl	Cyclohexyl chloride	Chlorocyclohexane
C_7H_7Br	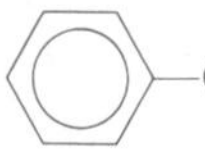 ⬡—CH_2Br	Benzyl bromide	Bromophenylmethane
C_2H_3Cl	$H_2C=CHCl$	Vinyl chloride	Chloroethene
C_3H_5Cl	$H_2C=CH-CH_2Cl$	Allyl chloride	3-Chloropropene
CH_2Br_2	$H-\overset{\displaystyle Br}{\underset{\displaystyle \underset{\displaystyle Br}{|}}{\overset{\displaystyle |}{C}}}-H$	Methylene bromide	Dibromomethane
$C_2H_4Br_2$	$\underset{\displaystyle Br}{\underset{\displaystyle |}{C}H_2}-\underset{\displaystyle Br}{\underset{\displaystyle |}{C}H_2}$	Ethylene bromide	1,2-Dibromoethane

chloride). Under the IUPAC system, the parent compound is the longest continuous chain containing the halogen atom. In designating the halogen substituent, the -*ine* halogen ending is replaced by -*o*. Table 5.1 illustrates the nomenclature of some alkyl halides. (Notice that in the last six entries in the table, the organic groups rather than the halogens are colored. The colored portions correspond to the common names of the organic groups.) Alkyl halides of more than five carbons are most conveniently named according to the IUPAC system.

$$\underset{6\ \ \ 5\ \ \ 4\ \ \ 3\ \ 2\ \ 1}{CH_3CH_2CH_2\overset{\overset{\displaystyle CH_3}{|}}{C}H\overset{\overset{\displaystyle Br}{|}}{C}H\underset{\underset{\displaystyle Cl}{|}}{C}H_2}$$

1-Bromo-2-chloro-3-methylhexane

$$\underset{1\ \ \ 2\ \ 3\ \ \ \ 4\ \ \ \ 5\ \ 6\ \ \ \ 7\ \ \ \ 8}{CH_3\overset{\overset{\displaystyle Cl}{|}}{\underset{\underset{\displaystyle Cl}{|}}{C}}CH_2CH_2\overset{}{C}H\overset{\overset{\displaystyle \bigcirc}{}}{C}H\underset{\underset{\displaystyle CH_3}{|}}{C}H_2CH_3}$$

2,2-Dichloro-5-methyl-6-phenyloctane

$$\underset{7\ \ \ \ 6\ \ \ 5\ \ 4\ \ \ \ 3\ \ \ \ 2\ \ \ \ 1}{CH_3CH_2\overset{\overset{\displaystyle I}{|}}{C}CH_2\overset{\overset{\displaystyle I}{|}}{C}HCH=CH_2}$$
$$\underset{CH_2CH_3}{|}$$

5-Ethyl-3,5-diiodo-1-heptene

1-Bromo-1-ethylcyclohexane

1-Chlorocyclobutene

3-Iodocyclopentene

5.2   *Preparation of Organic Halides*

Organic halides are synthesized in great quantities because they are used extensively in industry as the starting materials for the preparation of a great variety of organic reagents. In addition, many organic halides are useful in their own right, as we see in Section 5.5.

a. From Alkenes

Dihalides are prepared by the addition of halogen molecules to the appropriate alkene, whereas monohalides are prepared by the addition of hydrogen halides to alkenes. Notice that in equation (2) the hydrogen bromide adds to the alkene in accordance with Markovnikov's rule.

$$CH_3-CH=CH_2 + Br_2 \longrightarrow CH_3-\overset{\overset{\displaystyle Br}{|}}{C}H-\overset{\overset{\displaystyle Br}{|}}{C}H_2 \qquad (1)$$

Propene   1,2-Dibromopropane

$$\bigcirc\!\!\!-\overset{\overset{\displaystyle H}{|}}{C}=\overset{\overset{\displaystyle H}{|}}{C}-H + HBr \longrightarrow \bigcirc\!\!\!-\overset{\overset{\displaystyle H}{|}}{\underset{\underset{\displaystyle Br}{|}}{C}}-\overset{\overset{\displaystyle H}{|}}{\underset{\underset{\displaystyle H}{|}}{C}}-H \qquad (2)$$

Styrene   1-Bromo-1-phenylethane
(Phenylethene)

b. From Alcohols

The alcohols are the common laboratory reagents for the synthesis of the alkyl halides. There are several reagents that will effect the replacement of the alcoholic hydroxyl group by a halogen atom (see also Section 6.4-c).

$$3 \ CH_3CH_2OH \ + \ PX_3 \ \longrightarrow \ 3 \ CH_3CH_2X \ + \ H_3PO_3 \qquad (3)$$
$$X = Cl, \ Br, \ I$$

$$CH_3CH_2CH_2OH \ + \ conc \ HX \ \longrightarrow \ CH_3CH_2CH_2X \ + \ HOH \qquad (4)^{[1]}$$
$$X = Br \ or \ I; \ if \ X = Cl, \ ZnCl_2 \ is \ needed \ as \ a \ catalyst$$

$$\bigcirc\!\!\!\!\bigcirc-CH_2CH_2OH \ + \ HBr \ \longrightarrow \ \bigcirc\!\!\!\!\bigcirc-CH_2CH_2Br \ + \ HOH \qquad (5)$$

2-Phenyl-1-ethanol   1-Bromo-2-phenylethane

c. Halogenation of Aromatic Hydrocarbons

Direct halogenation is most frequently applied in the synthesis of aryl halides. Recall that the direct halogenation of alkanes results in the formation of polyhalogenated products (Section 2.4).

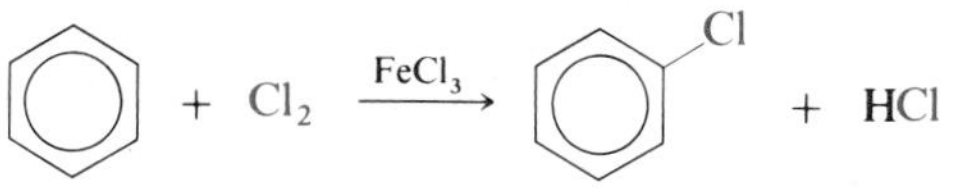

The unpleasant taste and odor of drinking water are sometimes caused by the chlorination process. In particular, chlorination of water containing phenol (a common contaminant of industrial waste water) leads to substitution of the ring hydrogens to produce mono- and dichlorophenols. The offensive taste of 2,6-dichlorophenol can be detected at a concentration of 0.1 part per billion (ppb).

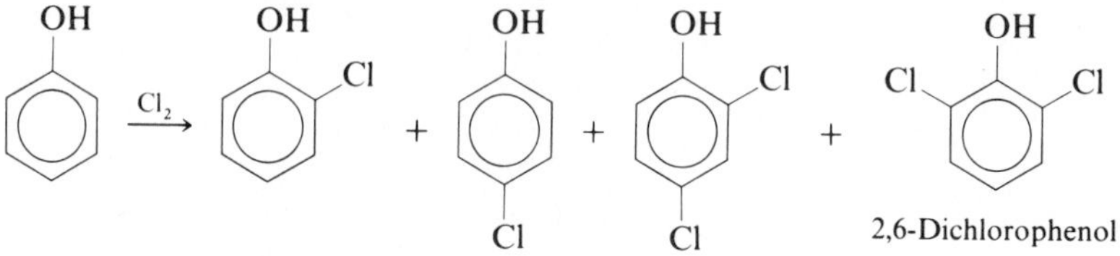

2,6-Dichlorophenol

5.3 Physical Properties of Alkyl Halides

Alkyl halides are insoluble in water. The chlorides are lighter than water, and the bromides and iodides are heavier. Table 5.2 illustrates once again the effect that increasing molecular weight has upon the boiling points of a series of homologous compounds. It is convenient to remember that methyl iodide is the only methyl halide that is not a gas at room temperature. Aside from the rather uncommon ethyl fluoride, ethyl chloride is the only ethyl halide that is a gas at room temperature. Ethyl chloride has an extremely high vapor pressure, and many uses are made of this

[1]Conceptually, the reaction in which a hydrogen halide reacts with an alcohol to form an alkyl halide and water is similar to a neutralization reaction in which an acid and a base react to produce a salt and water. Care must be taken to note that while the reaction appears similar it is *not* the same. Alcohols are not organic bases even though they do possess a hydroxyl group, and alkyl halides are *not* organic salts.

TABLE 5.2 Boiling Points of Alkyl Halides (°C)

Alkyl Group	Chloride	Bromide	Iodide
Methyl	−24	4	43
Ethyl	12	38	72
n-Propyl	47	71	102
Isopropyl	35	60	90

physical property. For example, it is employed in baseball games as a temporary anesthetic whenever a player is hit by a pitched ball. The affected area is sprayed with liquefied ethyl chloride. The halide then begins to evaporate, and as it does so, it abstracts heat from the injured area. The tissue becomes temporarily frozen, and thus insensitive to pain. The same treatment is used in medicine when extremely painful injections must be administered. Caution must be taken in using alkyl halides because they can cause skin damage and almost all of them are toxic. Exposure over prolonged periods may cause liver damage.

Many alkyl halides are light sensitive. Upon exposure to light, they slowly decompose to liberate the free halogen, which then dissolves to yield a colored solution. For this reason liquid halides are generally stored in brown bottles.

5.4   Chemical Properties of Organic Halides

As indicated earlier, the replacement of a hydrogen atom by a halogen atom effects an enormous difference in the reactivity of the molecule. Most of the reactions that the alkyl halides undergo are either substitution or elimination reactions. In some cases, both types of reaction are possible for the same alkyl halide and a given reagent. The mechanism of reaction in such an instance will depend on the structure of the alkyl halide involved, the solvent utilized, and the nature and concentration of the participating reagent.

a. Elimination (Dehydrohalogenation)

The dehydrohalogenation reaction was mentioned in Section 3.2-b as a method for the preparation of alkenes.

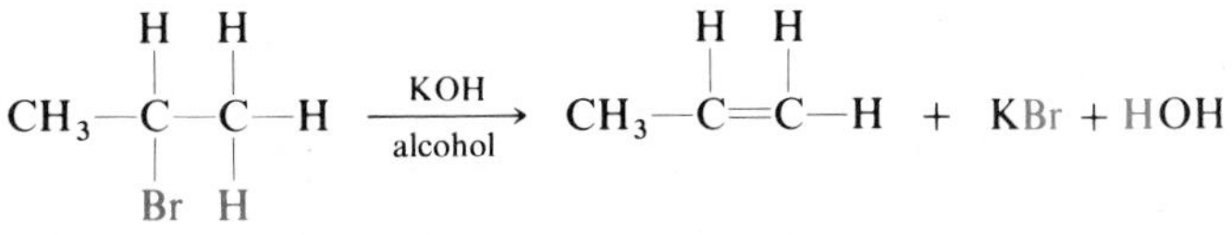

2-Bromopropane   Propene

The reaction is accomplished by the use of a strong base in an alcoholic solution. A concerted mechanism has been proposed; that is, the removal of a proton occurs simultaneously with a shift of the electron pair and ionization of the halide. The overall result is the loss of hydrogen halide with formation of a carbon–carbon double bond. This type of reaction is termed an **elimination reaction**. It is defined as a *reaction that effects the removal of some simple molecule and the formation of a multiple bond.*

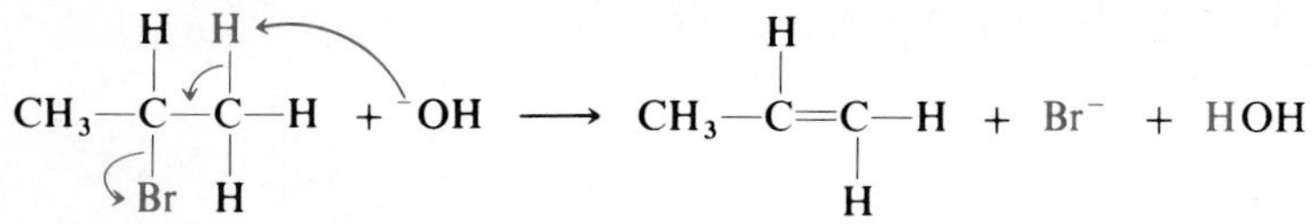

$$\text{CH}_3-\overset{\overset{\displaystyle H}{|}}{\underset{\underset{\displaystyle Br}{|}}{C}}-\overset{\overset{\displaystyle H}{|}}{\underset{\underset{\displaystyle H}{|}}{C}}-H \;+\; {}^{-}OH \;\longrightarrow\; CH_3-C=\overset{\overset{\displaystyle H}{|}}{\underset{\underset{\displaystyle H}{|}}{C}}-H \;+\; Br^{-} \;+\; HOH$$

In the foregoing example, all six of the hydrogen atoms adjacent to the halogen are equivalent; hence only one product is possible. When there are nonequivalent hydrogens, more than one alkene may be formed. The predominant product will be the more highly substituted alkene (see also Section 6.4-b).

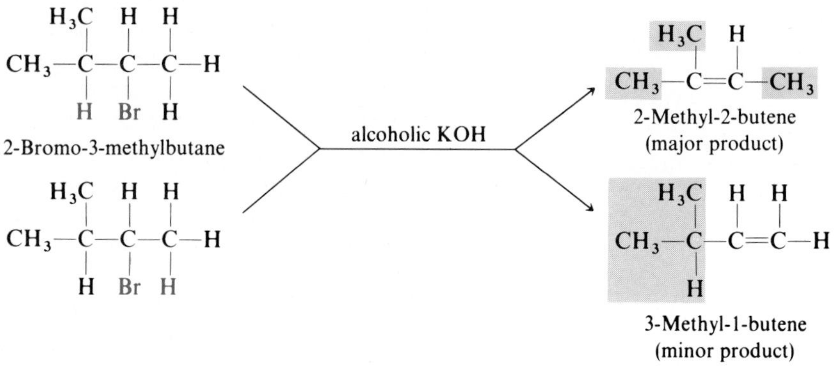

2-Bromo-3-methylbutane

alcoholic KOH

2-Methyl-2-butene
(major product)

3-Methyl-1-butene
(minor product)

The dehydrohalogenation reaction is useful for the preparation of some alkynes. For example, phenylacetylene may be synthesized by the dehydrobromination of styrene dibromide in the presence of a strong base such as potassium hydroxide.

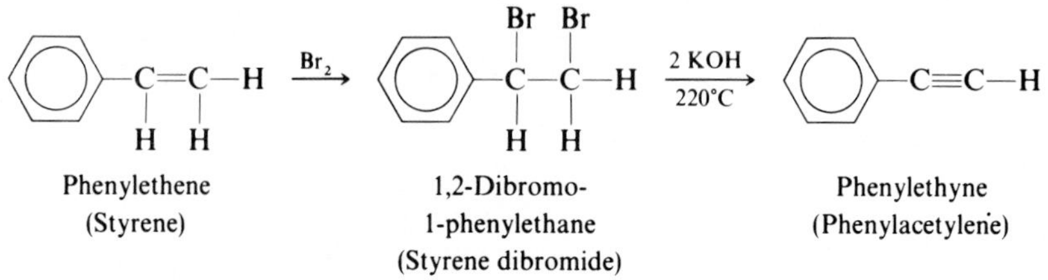

Phenylethene
(Styrene)

1,2-Dibromo-
1-phenylethane
(Styrene dibromide)

Phenylethyne
(Phenylacetylene)

Several insect species have been found to be resistant to DDT (see Section 5.6). These insects have developed an enzyme, *DDT dehydroclorinase,* that effects the dehydrochlorination of DDT and converts DDT into a noninsecticidal ethylene derivative (DDE).

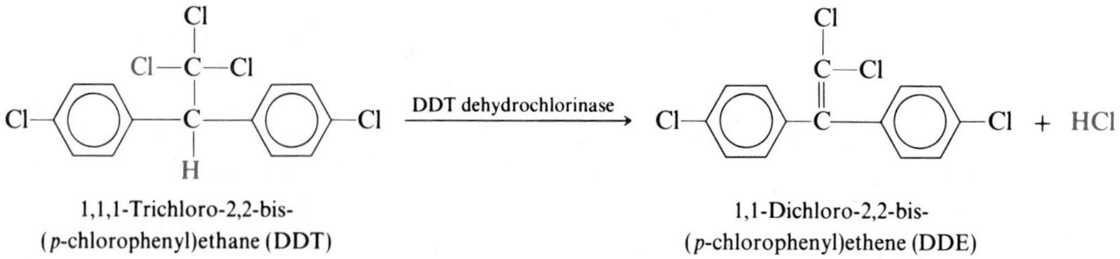

1,1,1-Trichloro-2,2-bis-
(*p*-chlorophenyl)ethane (DDT)

DDT dehydrochlorinase

1,1-Dichloro-2,2-bis-
(*p*-chlorophenyl)ethene (DDE)

b. Substitution

The halogen atom of alkyl halides may be displaced by a number of other negative groups to yield a wide variety of organic derivatives. These attacking groups are referred to as **nucleophiles** (nucleus-loving) or **nucleophilic reagents.** Nucleophilic reagents *seek a center of positive charge, and* therefore *can donate a pair of electrons* (that is, are Lewis bases) *in a chemical reaction*. Common nucleophiles are negative ions or neutral molecules containing unshared electron pairs (for example, $\ddot{N}H_3$, $\ddot{P}(CH_3)_3$, $H_2\ddot{O}{:}$). In contrast, recall that positive ions are designated as electrophiles (Section 3.4-b).

General Equation

$$
\begin{array}{ccc}
\overset{\displaystyle H}{\underset{\displaystyle H}{R-\overset{|}{\underset{|}{C}}-X}} + A^- & \longrightarrow & \overset{\displaystyle H}{\underset{\displaystyle H}{R-\overset{|}{\underset{|}{C}}-A}} + X^-
\end{array}
$$

where A = a nucleophile such as OH^-, OR^-, CN^-, NH_2^-, SH^-

Specific Equations

$$CH_3I \;+\; KCN \xrightarrow{\text{alcohol}} CH_3CN \;+\; KI$$

Methyl iodide   Methyl cyanide (Acetonitrile)

$$CH_3CH_2CH_2Br \;+\; NaNH_2 \xrightarrow{NH_3} CH_3CH_2CH_2NH_2 \;+\; NaBr$$

n-Propyl bromide   *n*-Propylamine

$$\text{(ring)}-CH_2Cl \;+\; NaOH \xrightarrow{HOH} \text{(ring)}-CH_2OH \;+\; NaCl$$

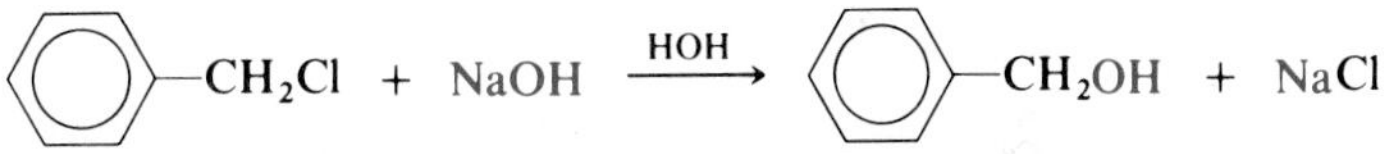

Benzyl chloride   Benzyl alcohol

Aromatic compounds containing halogen atoms in their side chain behave essentially like alkyl halides. On the other hand, the aryl halides, in which the halogen is directly bonded to a carbon of the aromatic ring, are very unreactive. They will not undergo the normal substitution reactions with nucleophilic reagents. However, if strongly electron-attracting groups (for example, the nitro group) are positioned ortho or para to the halogen atom, the halogen may be displaced. The following reaction was used by Frederick Sanger to determine the sequence of amino acids in proteins (see Chapter 13).

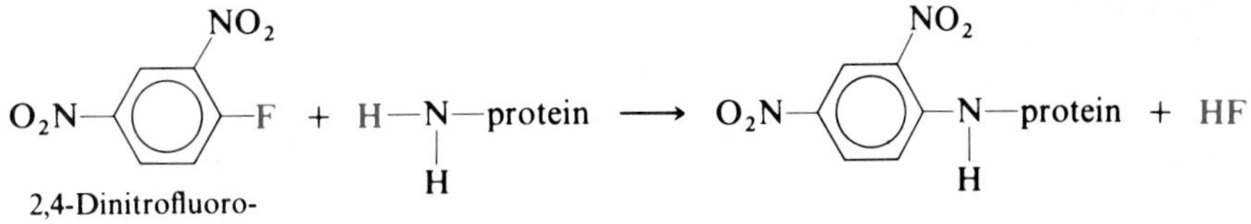

$$O_2N-\text{(ring,}NO_2\text{)}-F \;+\; H-\underset{H}{\overset{}{N}}-\text{protein} \longrightarrow O_2N-\text{(ring,}NO_2\text{)}-\underset{H}{\overset{}{N}}-\text{protein} \;+\; HF$$

2,4-Dinitrofluoro-
benzene

TABLE 5.3 Commercially Important Halogenated Hydrocarbons

Name	Formula	Physical State	Remarks
Methyl chloride	CH_3Cl	Gas	Used in the manufacture of silicones, synthetic rubber, and as a general methylating agent.
Methylene chloride	CH_2Cl_2	Liquid	Effective extraction solvent in the food and pharmaceutical industries. The EPA is reviewing the chemical because it has been shown to cause cancer in laboratory animals. The FDA has proposed that it be banned as a solvent in aerosol hair sprays.
Iodoform	CHI_3	Yellow solid	Formerly used as an antiseptic for wounds; used in treatment of some skin diseases and skin ulcers.
Halothane	$CF_3CHClBr$	Gas	Inhalative anesthetic. It is effective and relatively nontoxic.
Freons	CF_2Cl_2 $CFCl_3$	Gas	Nontoxic, nonodorous, noncombustible; used as a refrigerant, and formerly in aerosol spray propellants. Possible cause of some ozone depletion in the upper atmosphere.
Ethyl chloride	C_2H_5Cl	Gas	External local anesthetic. Used in the manufacture of tetraethyllead.
Teflon	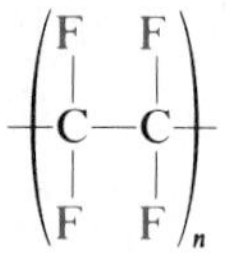	Waxy plastic	A polymer that is resistant to oxidation and corrosion. Used as an electrical insulator and as a liner in frying pans and other utensils to provide a nonsticking surface.
Trichloroethylene	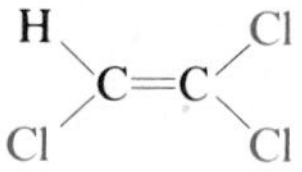	Liquid	They have replaced CCl_4 as drycleaning solvents. However, they are carcinogenic if inhaled for prolonged periods; adequate ventilation is strongly advised.
Tetrachloroethylene	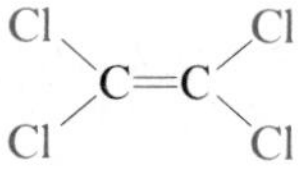	Liquid	
p-Dichlorobenzene		Solid	Larvicide. It is extensively used in place of naphthalene to kill moths and caterpillars.
2-Perfluoropropylfuran	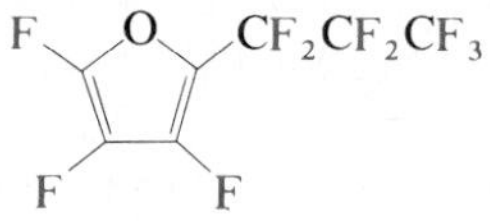	Liquid	A solution containing a mixture of isomers of perfluoropropylfuran is marketed as a blood substitute. The product has been licensed in Japan, but is still undergoing clinical tests in the U.S. (see Figure 5.1).
3-Perfluoropropylfuran	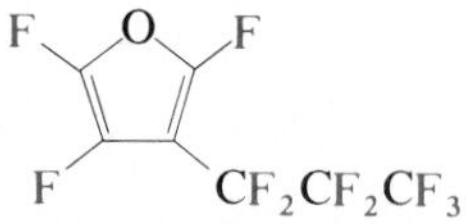	Liquid	

5.5 Uses of Halogenated Hydrocarbons

In addition to their value as intermediates in synthetic processes, organic halides have numerous commercial uses, for example, as solvents, refrigerants, polymers, antiseptics, and anesthetics. Table 5.3 lists some organic halides, their formulas, physical states, and chief uses.

Table 5.4, on the other hand, contains a listing of compounds that have been embroiled in controversy throughout the twentieth century. It includes toxic gases, metabolic poisons, plasticizers, aerosols, and herbicides. The one common feature of all of these compounds is that they contain carbon–halogen bonds. Organic halides, with but a few exceptions (for example, the thyroid hormone thyroxine—see Figure 13.1) are incompatible with living systems.

Special mention should be made of the fluorocarbons that contain two or more fluorine atoms on the same carbon. Not only are these compounds extremely inert, but they are nontoxic, noncorrosive, and nonflammable. Recall that tetrafluoroethylene is polymerized to Teflon (Table 3.4), which is resistant to attack by acids, oxidizing or reducing agents, and organic solvents. In Section 6.14 we discuss some of the fluorocarbons that are used as general anesthetics.

Some chlorofluorocarbons are prepared from carbon tetrachloride by a halogen exchange process.

$$CCl_4 + 2\ HF \xrightarrow{\text{SbF}_5} CF_2Cl_2 + 2\ HCl$$

This equation represents the preparation of Freon 12 (Freon 11 is $CFCl_3$). These gaseous chlorofluoromethanes are readily liquefied under pressure. Hence, they became widely used as refrigerants and aerosol propellants to dispense deodorants, hair sprays, shaving creams, and so on. However, since these compounds are so inert, they do not react with substances with which they come in contact. As a result, they diffuse up through the atmosphere unchanged and eventually reach the stratosphere (altitude of 10–12 miles).

Ozone is an essential component of the stratosphere, where it functions to absorb the high energy ultraviolet radiation from the sun. If this radiation were not filtered out, it would quickly burn exposed skin, causing skin cancer and eye damage. The Freons that reach the stratosphere absorb ultraviolet solar radiation and break down, liberating chlorine atoms ($Cl\cdot$). The chlorine atoms are believed to destroy the ozone layer by a sequence of free-radical chain reactions that yield oxygen.

$$CF_2Cl_2 \xrightarrow[\text{light}]{\text{ultraviolet}} Cl\cdot + \text{Other products}$$

$$Cl\cdot + O_3 \longrightarrow ClO + O_2$$

This ozone-depleting mechanism is based upon theoretical models and laboratory studies and not upon correlations of Freon and ozone concentrations actually measured in the stratosphere. Nevertheless, the public became aroused; the potential for increased incidence of skin cancer was too high a price to pay for consumer comforts. Manufacturers changed

FIGURE 5.1 The fluorocarbon liquid (FC75), a mixture of isomers of perfluoropropylfuran, is being bubbled with oxygen. The mouse has been breathing the liquid for about half an hour. All the normally air-filled spaces in the lung alveoli are now filled with liquid containing dissolved oxygen. When removed, the mouse is held head down, the fluorocarbon liquid drains out, and the mouse is unharmed. Other perfluorinated liquids have also been used. [Courtesy of Leland C. Clark, Jr., Children's Hospital Research Foundation, Cincinnati, OH.]

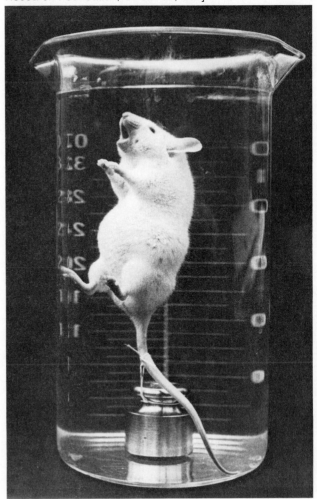

TABLE 5.4 Toxic Organic Halides

Name	Formula	Remarks
Carbon tetrachloride	CCl_4	Highly toxic — an accumulative poison that causes severe liver damage. CCl_4 is restricted to industrial uses (e.g., in manufacture of fluorocarbon refrigerants).
Chloroform	$CHCl_3$	Formerly an inhalative anesthetic in surgery, but no longer used for this purpose because of its toxicity. It has a depressant action on the heart and on the respiratory center. Chloroform has been shown to cause cancer in laboratory mice and rats.
Phosgene	$\begin{matrix} & O \\ & \| \\ Cl - & C - Cl \end{matrix}$	Lethal gas developed during World War I. It attacks hydroxyl groups and amino groups of the proteins of the lung membranes, producing HCl and heat, which are responsible for its toxic effect.
Mustard gas	$Cl - CH_2CH_2 - S - CH_2CH_2 - Cl$	Skin inflammatory agent — causes severe, slow-healing blistering on any part of the body that it contacts.
Diisopropylfluoro-phosphate (DFP)	$\begin{matrix} & & & H \\ & & & \| \\ O & & O - C - CH_3 \\ \| & & \quad\quad\quad \| \\ F - P & & \quad\quad\quad CH_3 \\ & \| & \\ & O - C - CH_3 \\ & \quad\quad\quad \| \\ & H \quad\quad CH_3 \end{matrix}$	Nerve gas — combines with a specific hydroxyl group (of serine) situated at the active site of the enzyme acetylcholine esterase, the enzyme that functions in the transmittal of the nerve impulse. Deactivation of the enzyme by DFP disrupts coordination in the autonomic nervous system and in the muscles, causing tremors, convulsions, and death. DFP activity resembles that of the organophosphate pesticides (see Section 8.15).
Phenacyl chloride	$\begin{matrix} O \\ \| \\ C - CH_2 - Cl \end{matrix}$ (attached to benzene ring)	Lachrymator (tear gas) — used routinely in civilian police work and occasionally by the military.
Iodoacetate	$I - CH_2 - COO^-$	Metabolic poison — combines with sulfhydryl groups ($-SH$) of enzymes, deactivating the enzymes (see Section 14.8-b).
Fluoroacetate	$F - CH_2 - COO^-$	Metabolic poison — reacts with citric acid, an intermediate of the Krebs cycle (see Section 16.8), to form 2-fluorocitric acid, which is not further metabolized, thus blocking the Krebs cycle.
Ethylene dibromide (EDB)	$Br - CH_2CH_2 - Br$	Crop fumigant — expected to vaporize before marketing of crops, but residues have been found on citrus fruits and in processed flour. In 1984, its use as a fumigant was restricted by the EPA.
Vinyl chloride	$\begin{matrix} H & & H \\ \quad \diagdown & & \diagup \\ & C = C \\ \quad \diagup & & \diagdown \\ H & & Cl \end{matrix}$	A potent animal carcinogen. Formerly used as aerosol propellant in household spray paints and finishes, protective and decorative coatings, paint removers, adhesives, and solvents. Vinyl chloride has been linked to a form of liver cancer occurring among workers exposed to it in chemical plants. In 1974 its use in aerosol drugs, cosmetics, pesticides, and household sprays was banned. Still used in the production of polyvinyl chloride plastics.

Polyvinyl chloride
(PVC)

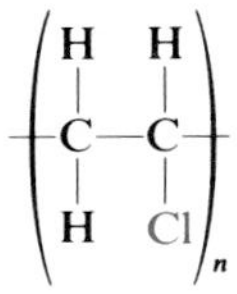

$$\left(\begin{array}{c} H \quad H \\ -C-C- \\ H \quad Cl \end{array}\right)_n$$

Plastic polymer made from vinyl chloride monomer units and used in phonograph records, plastic bottles, garden hoses, etc. Disposal of PVC products presents a problem because they undergo loss of HCl gas when heated in an incinerator. HCl is a corrosive irritant that significantly increases the air pollution.

2,4-Dichlorophenoxy-
acetic acid
(2,4-D)

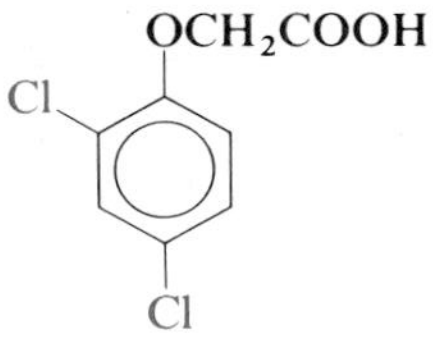

OCH_2COOH

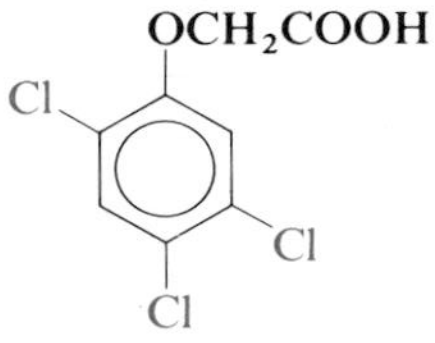

Both 2,4-D and 2,4,5-T were widely used as herbicides. They exert their effect by mimicking natural plant growth hormones. The cell walls take up water at a faster rate than normal, producing elongation in the stem, little or no root growth, and leaves that are deficient in chlorophyll. The cessation of normal physiological functions results in the death of the plant. The chief advantage of these compounds is a preferential attack on broadleaf weeds without harm to grasses or trees. Both substances were used by the U.S. military in Vietnam to defoliate the jungle growth. Controversy arose over reports of increased birth defects in Vietnamese children in the sprayed areas and of cancer deaths in U.S. military men and birth defects in their children. These effects may be due in part to an impurity (a dioxin—TCDD) that is formed in the production of 2,4,5-T. TCDD is lethal at levels as low as $0.6 \ \mu g/kg$ of body weight in experimental animals. Vietnam veterans' suit against the herbicide manufacturers was settled out of court.

2,4,5-Trichlorophenoxy-
acetic acid
(2,4,5-T; Agent Orange)

OCH_2COOH

2,3,7,8-Tetrachloro-
dibenzo-p-dioxin
(TCDD)

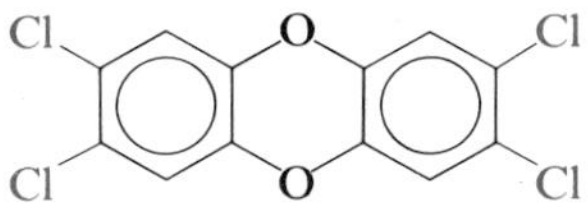

Polychlorinated biphenyls
(PCBs)

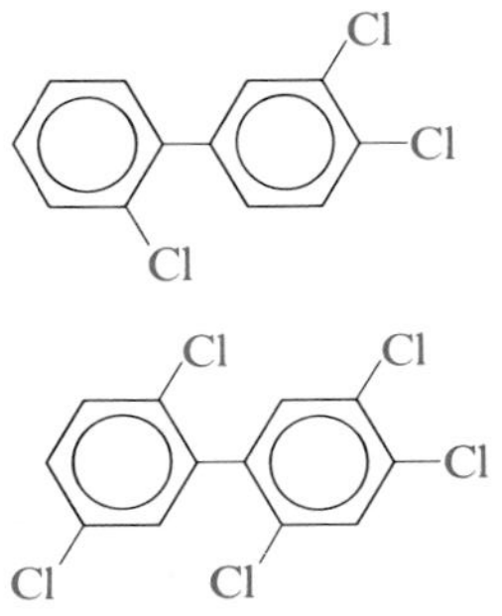

There are 210 possible PCBs depending on the number of Cl atoms bonded to the biphenyl structure. Commercial PCBs are various mixtures that contain an average of 3–5 Cl atoms per molecule. PCBs have been used in lubricants, hydraulic fluids, waxes, paints, inks, and plastics. Their high heat capacities make them useful as heat transfer fluids in electric transformers. Like DDT, PCBs pervade the environment, decompose slowly, and accumulate in the fatty tissue of all organisms in the food chain. In laboratory animals they have caused reproductive problems, gastric disorders, skin lesions, and cancerous tumors. Careful disposal of PCBs is required to avoid leakage into rivers and streams. PCBs are no longer manufactured commercially in the U.S., and the use (and disposal) of remaining supply is regulated by the EPA. EPA regulations ban high voltage PCB transformers by October 1990; this will require costly removal and replacement of equipment.

from chlorofluorocarbon aerosols to spray pump dispensers or to other types of propellants (such as propane and isobutane). The Freons are still used as refrigerants in air conditioning units, freezers, and refrigerators.

A mixture of isomers of perfluoropropylfuran (see Table 5.3) has been utilized as a synthetic blood substitute. These compounds, in which all hydrogen atoms have been substituted by fluorine atoms, can dissolve and transport large quantities of oxygen. Mice were submerged in a beaker containing the liquid fluorocarbons for extended periods. They did not drown (suffocate) because they were able to breathe the dissolved oxygen directly (Figure 5.1). Humans have been treated with this blood supplement in Japan, and clinical trials are underway in the United States. If no adverse effects are discovered, these compounds will be a welcome addition to the dwindling blood supplies.

5.6   *The Polychlorinated Pesticides*

In 1962, Rachel Carson published the highly provocative book *Silent Spring* in which she provided dramatic examples of how the environment had been damaged by the indiscriminate use of chemicals.

For the first time in the history of the world, every human being is now subjected to contact with dangerous chemicals, from the moment of conception until death. In the less than two decades of their use, the synthetic pesticides have been so thoroughly distributed throughout the animate and inanimate world that they occur virtually everywhere. They have been recovered from most of the river systems and even from streams of groundwater flowing unseen through the earth. Residues of these chemicals linger in soil to which they may have been applied a dozen years before. They have entered and lodged in the bodies of fish, birds, reptiles, and domestic and wild animals so universally that scientists carrying on animal experiments find it almost impossible to locate subjects free from such contamination. They have been found in fish in remote mountain lakes, in earthworms burrowing in soil, in the eggs of birds—and in man himself. For these chemicals are now stored in the bodies of the vast majority of human beings, regardless of age. They occur in the mother's milk, and probably in the tissues of the unborn child.

Miss Carson focused her attention on two major groups of synthetic pesticides, the chlorinated hydrocarbons and the organic phosphates (see page 170). We shall discuss the former group here with a view toward understanding their structure and the nature of their deadly action. Figure 5.2 contains the names and formulas of the common chlorinated hydrocarbon pesticides.

FIGURE 5.2   The polychlorinated pesticides.

p,p-DDT   o,p-DDT   Chlordane   Heptachlor

Dieldrin   Aldrin   Endrin   Gammexane

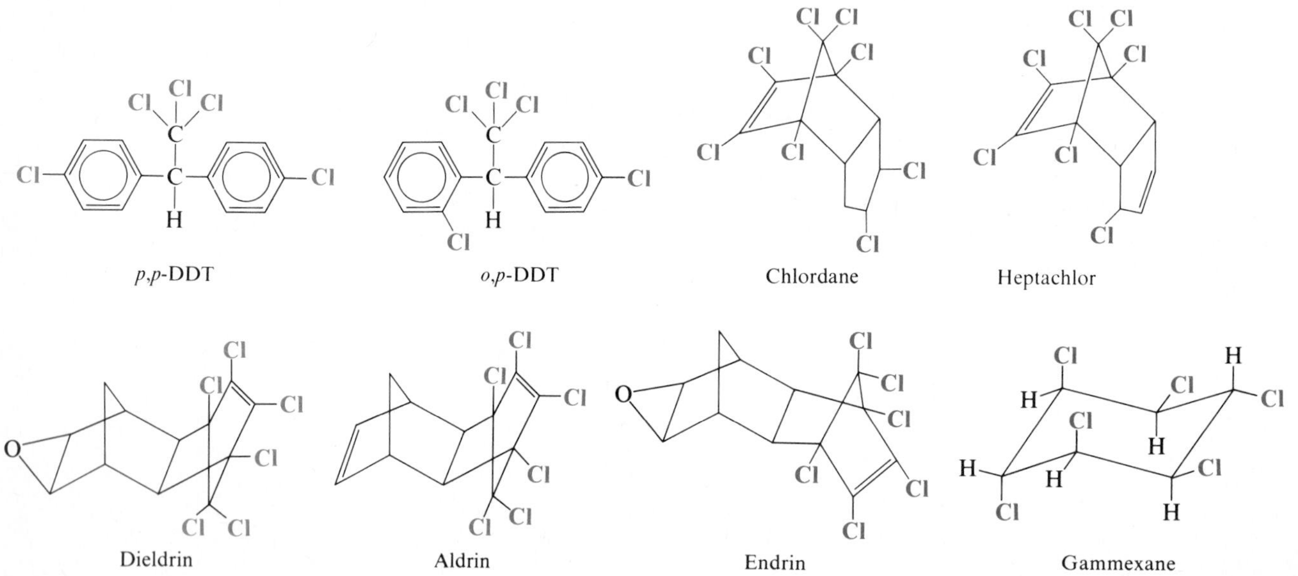

a. DDT (Dichlorodiphenyltrichloroethane[2])

DDT is formed by the reaction of chlorobenzene with chloral hydrate in the presence of concentrated sulfuric acid as a catalyst.

$$2 \ Cl-\bigcirc-H \ + \ HO-\overset{\overset{\displaystyle Cl}{|}}{\underset{\underset{\displaystyle H}{|}}{C}}-OH \ \xrightarrow{\text{conc } H_2SO_4}$$

Chlorobenzene   Chloral hydrate

$$Cl-\bigcirc-\overset{\overset{\displaystyle Cl-\overset{\overset{\displaystyle Cl}{|}}{\underset{\underset{\displaystyle }{}}{C}}-Cl}{}}{\underset{\underset{\displaystyle H}{|}}{C}}-\bigcirc-Cl \ + \ 2 \ HOH$$

DDT

No other pesticide has been as assiduously studied as DDT. It was first synthesized in 1874 by Othmar Zeidler, a German chemist working on his doctoral thesis. Its properties as an insecticide went unnoticed until 1939, when Paul Mueller of Switzerland observed its extreme toxicity to moths and house-flies. (Mueller was awarded the Nobel prize in 1948 for discovering the insecticidal properties of DDT.) Almost immediately it was hailed as a life-saving protector against insects and insect-borne diseases. In 1943, United States military authorities aborted a typhus epidemic in Naples, Italy, by a massive dusting with DDT. The chemical has virtually eliminated the yellow fever mosquito in many South American countries. It has killed the black fly that transmitted a thread-worm responsible for blindness in millions of Africans. According to the World Health Organization, malaria is the chief cause of death in the world (aside from natural causes). Use of DDT on a worldwide basis resulted in a major control over the malarial (Anopheles) mosquito, with a saving of millions of lives.

When Rachel Carson's book appeared, DDT was proclaimed to have permeated all living tissue. It was found in eggs and in vegetables, especially the leafy varieties such as lettuce and spinach. The penguins in Antarctica, pheasants in California, salmon in Lake Michigan, and peregrine falcons in England had all been polluted with DDT and its analogs. There had been

[2]Notice that this name is not a correct application of IUPAC nomenclature (see page 70). Commercial DDT is usually composed of 20–25% of the o,p-isomer and 75–80% of the p,p-isomer.

massive fish kills, the robin population had been severely depleted, and the bald eagle was in danger of extinction.

As yet, there has never been a proved instance of a human death having resulted from the proper use of DDT. Many people, therefore, regard the chemical as harmless to humans. During World War II, thousands of soldiers, refugees, and prisoners were dusted with DDT powder to kill the typhus-carrying louse. These individuals suffered no immediate ill effects of the chemical. The reason for this apparent harmlessness is that DDT, unlike other chlorinated hydrocarbons, is not readily absorbed through the skin when it is in *powder form*. However, it does find its way into the human system either through the nose or the mouth. When DDT is inhaled, it can be absorbed through the lungs; when it is ingested as a contaminant of food, it is slowly absorbed through the digestive tract. Once it has entered the body, it is stored in fatty tissue (because DDT is fat soluble) such as the adrenals, testes, or thyroid. Relatively large amounts are deposited in the brain, liver, and kidneys.

The dangerous characteristic of DDT are

1. The tremendous capability of DDT to spread throughout the world. Wind and water, migrating birds and fish carry it thousands of miles from the point of application.
2. DDT is one of the most persistent pesticides. It is insoluble in water and settles to the bottom of streams and lakes. When sprayed into the environment, it resists decomposition into its simpler constituents. The half-life of DDT in the soil is estimated at five to ten years.
3. DDT concentrates in living tissue. Each successive link in the food chain accumulates a much larger amount of DDT. Small fish absorb ten times more DDT than the plankton they feed upon and birds that eat these fish contain thirty times more DDT than the fish.
4. Numerous insects have developed a resistance to DDT. It is estimated that over 200 insect species have become immune to the chemical. Some of these insects are found to contain an enzyme (*DDT dehydrochlorinase*) that converts DDT to a noninsecticidal ethylene derivative (Section 5.4-a). Other insects are able to form a nonlethal analog by replacing the tertiary hydrogen with a hydroxyl group. (In Section 8.15 we discuss the organophosphate pesticides that have been developed to combat insects that are resistant to DDT.)

DDT is highly toxic to insects because it is readily absorbed through the insect cuticle. It affects peripheral sensory organs, producing violent impulses that cause hyperactivity and convulsions. Paralysis and eventual death probably occur from metabolic exhaustion.

In higher organisms, DDT and the other pesticides exert their effect by blocking the action of enzymes. Since the concentration of enzymes is relatively small, minute amounts of DDT can bring about vast bodily changes. In animal experiments, 3 ppm has been found to inhibit an essential enzyme in heart muscle; only 5 ppm has brought about necrosis or disintegration of liver cells. DDT is highly toxic to fish and only moderately toxic to birds. However, massive bird deaths have resulted from biomagnification of DDT through the food chain. In addition, DDT apparently interferes with the bicarbonate metabolism of birds. As a result, they lay eggs that are too thin-shelled (about 20% thinner than normal) to survive the incubation period.

The question frequently asked is: How reliable is extrapolation of animal data to humans? We do not know the answer to this question. No animal species exists that handles pesticides in every respect as humans do. Little is known about the chronic effects of continued and possibly increasing doses of pesticides over a long period of time. Until the cause and effect relationship between pesticide exposure and human disease is understood, it is necessary that we prevent their excessive use. At present, DDT and other chlorinated hydrocarbons are banned in the United States except for essential purposes, such as the control of disease of epidemic proportions or of a major infestation of crop insects. The reduction in DDT use in the United States has led to marked declines in DDT levels in fish, birds, food, and human fat tissue.

b. Other Chlorinated Hydrocarbons

Chlordane is a more potent pesticide than DDT. It has been used for control of ants, cockroaches, termites, and agricultural pests. Like DDT, chlordane is not readily decomposed and persists for long periods of time in the soil, on foodstuffs, or on surfaces to which it may be applied. Unlike DDT, chlordane can enter the body through all available openings. It is absorbed through the skin, breathed in as a spray, and absorbed from the digestive tract. In common with all the other chlorinated hydrocarbons, chlordane deposits build up in the body in a cumulative fashion. A diet containing only 2.5 ppm of chlordane eventually leads to storage of 75 ppm in the fat of experimental animals.

Heptachlor is a derivative of chlordane that was marketed as a separate agricultural formulation, often mixed with fertilizer to control pests in the soil. Heptachlor was also used against household pests, such as flies and mosquitoes. It has a particularly high capacity for storage in fatty tissue. A diet containing as little as 0.1 ppm will result in the deposit of measurable amounts of heptachlor in the body. In the soil and in the tissues of both plants and animals, it may be converted into another compound known as heptachlor epoxide. Tests on birds indicated that the epoxide was more toxic than the original compound. In 1978, most of the uses of chlordane and heptachlor were cancelled in the United States.

Endrin is the most toxic of all the chlorinated hydrocarbons. It is 15 times as poisonous as DDT to mammals, 30 times as poisonous to fish, and about 300 times as poisonous to some birds. Endrin, like dieldrin and aldrin, is related structurally to chlorinated naphthalenes, a class of compounds that has been shown to cause hepatitis. Endrin had been used to combat soil and foliage insects.

Dieldrin[3] is a stereoisomer of endrin. It is about five times as toxic as DDT when swallowed, but 40 times as toxic when absorbed through the skin in solution. Dieldrin strikes quickly at the nervous system, sending the victim into convulsions. It had been substituted for DDT in malaria-control work in regions where the malaria mosquitoes had become resistant to DDT.

Aldrin[3] produces degenerative changes in the liver and kidneys. A quantity the size of an aspirin tablet can kill over 400 quail. Pheasants and rats that have been fed quantities too small to kill them produced fewer offspring, and the majority of the offspring died within a week. Aldrin has been used effectively in the Middle East for the control of locust plagues. Aldrin and dieldrin have been used in the control of forestry pests, where their long-lasting residual effect was considered advantageous. In 1974, the use of dieldrin and aldrin as pesticides was banned by the Environmental Protection Agency because of studies linking their use with increased liver cancers in humans.

Gammexane (Lindane) is prepared by the addition of 3 moles of chlorine to benzene. Nine different stereoisomeric 1,2,3,4,5,6-hexachlorocyclohexanes are possible, and these are distinguished by the Greek letters α, β, γ, and so on. It is remarkable that only the γ-isomer has insecticidal properties. Its stereochemical structure is given in Figure 5.2. Gammexane, like DDT, is a broad-spectrum insecticide, a poison active against many species of insects. Its primary use in the United States is in seed treatment. In 1980, the EPA proposed a ban on most other uses of Lindane. All home uses of the chemical were stopped, including its use in flea collars, on ornamental plants, many agricultural crops, and on tree and wood products. According to the EPA, Lindane had been shown to cause cancer, birth defects, and nerve damage in animal studies and had been linked to aplastic anemia in humans.

[3]Dieldrin and aldrin are named for two German chemists, Otto Diels and Kurt Alder. Both men received the Nobel Prize in 1950 for discovering a very important organic reaction that bears their names. This reaction is utilized in the synthesis of these insecticides.

Aldrin

Exercises

5.1 Draw and name all the isomers of $C_5H_{11}Br$.

5.2 Draw and name all the dichloro derivatives of butane and isobutane.

5.3 Give common names for (a) CH_2Br_2, (b) CHI_3, and (c) CF_4.

5.4 Name the following compounds.
   (a) $CH_3CH_2CHICH_2CH_3$
   (b) $(CH_3)_2CCICH_2CH_2Cl$
   (c) $(CH_3)_3CCBr_2CH_2C(CH_3)_3$
   (d) $CH_2IC(CH_3)_2CH_2CH_2F$

   (e) ⬡—CH_2I

   (f) Br—⬡—CH_3
       |
       I

   (g) $CH_3CHICH_2C(CH_3)\!=\!CH_2$
   (h) $(CH_3CH_2)_2CHF$
   (i) $CH_3CH\!=\!CICH_3$
   (j) CF_3CF_3

   (k) Cl⟍⬠⟋CH_2CH_3

   (l) ⬡—C(CH_3)(Br)—CH_2CH_3

5.5 Write structural formulas for the following compounds.
   (a) cyclopropyl bromide
   (b) 3-bromo-2,3-dimethylpentane
   (c) 4-iodo-3-methyl-2-heptene
   (d) 1-bromo-3-isopropylcyclopentene
   (e) 1-chloro-3-fluorobenzene
   (f) chlorodiphenylmethane
   (g) 1,1,3-triiodopropene
   (h) DDT
   (i) vinyl bromide
   (j) p-bromotoluene
   (k) 1,3,5-trichlorobenzene
   (l) 1-bromo-3-cyclobutylbutane

5.6 1,2-Dibromo-3-chloropropane (DBCP) was formerly used as a preemergence soil fumigant to kill nematodes. It was shown to be a health hazard when workers in chemical plants became sterile due to exposure to DBCP. The fear that residues were retained in vegetables caused the halt of its production in the United States. Draw the structure of DBCP.

5.7 Why are liquid alkyl halides usually stored in brown bottles?

5.8 Notice in Table 5.2 that the n-propyl halides boil about 10°C higher than the corresponding isopropyl halides. Offer an explanation.

5.9 Which compound in each of the following pairs would have the higher boiling point?
   (a) $CHCl_3$ or CH_3Cl

   (b) ⬡—Br   or   ⬡—F

   (c) $CH_3CH_2CH_2CH_2Br$ or CH_3CH_2I
   (d) $CH_3CH_2CH_2CH_2Cl$ or $(CH_3)_3CBr$

5.10 Complete the following equations, giving only the organic products (if any).
   (a) $CH_3CH_2CH_2Br \xrightarrow[\text{alcohol}]{\text{KOH}}$
   (b) $(CH_3)_2CHCH\!=\!CHCH_3 + Br_2 \longrightarrow$
   (c) $CH_3CH_2CHOHCH_3 + PBr_3 \longrightarrow$
   (d) $CH_3CH_2CH_2CHClCH_3 + NaCN \longrightarrow$
   (e) $CH_3CH_2I \xrightarrow[\text{HOH}]{\text{NaOH}}$

   (f) ⬡ + $Br_2 \xrightarrow{\text{FeBr}_3}$

   (g) $(CH_3)_2C\!=\!CHCH_3 + HCl \longrightarrow$
   (h) $CH_2\!=\!CHCH_2CH_3 + HBr \longrightarrow$
   (i) $CH_3CHOHCH_3 + PI_3 \longrightarrow$
   (j) $(CH_3)_3COH + HBr \longrightarrow$
   (k) $CH_4 + Cl_2 \xrightarrow{\text{sunlight}}$

   (l) ⬡—CH_3 + $Br_2 \xrightarrow{\text{sunlight}}$

5.11 Write equations for two reactions that 2-chlorobutane can undergo with KOH. (One occurs with conc KOH in alcohol at high temperatures; the other with dilute, aqueous KOH at room temperature.)

5.12 Define the following terms and illustrate with a suitable example.
   (a) alkyl halide
   (b) aryl halide
   (c) Freons
   (d) PCBs
   (e) Agent Orange
   (f) elimination reaction
   (g) nucleophile
   (h) herbicide
   (i) biomagnification
   (j) blood substitute

5.13 Show reaction sequences for the following conversions. Indicate reagents and reaction conditions. In some cases, more than one equation is required.

(a) $(CH_3)_2CHCH_2CH_2Br \longrightarrow (CH_3)_2CHCHBrCH_3$

(b) $CH_3CH_2CH_2Br \longrightarrow CH_3CH_2CH_3$

(c) $CH_2{=}CH_2 \longrightarrow CH_3CH_2CN$

(d) 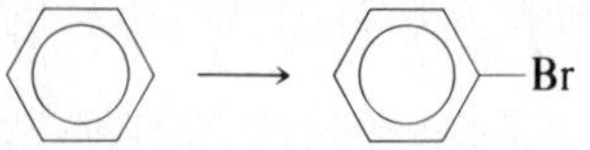 $\longrightarrow$ —Br

5.14 Distinguish between the terms *antiseptic* and *anesthetic*. Give an example of each. What is the disadvantage of using chloroform as an anesthetic?

5.15 Match each of the following with the appropriate number.

(a) chlordane

(b) *p*-dichlorobenzene

(c) diisopropylfluorophosphate

(d) ethyl chloride

1. heat transfer fluid
2. herbicide
3. plastic polymer
4. crop fumigant

(e) ethylene dibromide

(f) Freon

(g) halothane

(h) mustard gas

(i) phenacyl chloride

(j) polychlorinated biphenyls

(k) polyvinyl chloride

(l) 2,4,5-trichlorophenoxy-acetic acid

5. lachrymator
6. nerve gas
7. skin inflammatory agent
8. larvicide
9. topical local anesthetic
10. refrigerant
11. inhalative anesthetic
12. pesticide

5.16 Examine the labels of consumer products found in a supermarket (soaps, toothpastes, cleaning fluids, etc.) and list five that contain organic halogen compounds. Write the names of these compounds and their respective structural formulas.

Alcohols, Phenols, Ethers, and Thiols

Conceptually, alcohols, phenols, and ethers can be thought of as derivatives of water. When *one of the hydrogen atoms of a water molecule is replaced by an alkyl group,* an alcohol results. Substitution of an aryl group for one hydrogen atom of water yields a phenol. The general formulas for alcohols and phenols are **R—OH** and **Ar—OH,** respectively. When both hydrogen atoms of water are replaced by alkyl or aryl groups, the compound is an ether. The general formula for all ethers, then, is **R—O—R′** (where R and R′ can be alkyl and/or aryl groups). Thiols are derivatives of hydrogen sulfide; hence their formula is **R—SH.**

Alcohols

6.1 Classification and Nomenclature of Alcohols

The properties of alcohols depend on the structural arrangement of the carbon atoms in the molecule. Alcohols may be divided into three classes that serve to distinguish these different structural arrangements from one another. The classes are known as primary, secondary, and tertiary, and alcohols are placed into these classes according to the type of carbon atom to which the hydroxyl group is attached.

1. A **primary (1°) carbon atom** is a carbon atom that *is attached directly to only one other carbon.*

Example

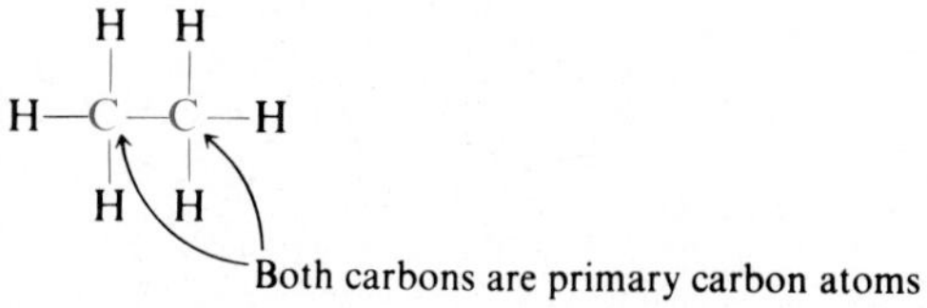

Both carbons are primary carbon atoms

2. A **secondary (2°) carbon atom** *is one that is attached directly to two other carbon atoms.*

Example

A secondary carbon atom

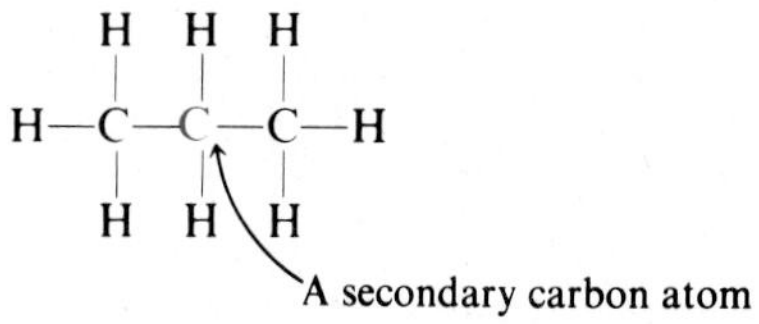

3. A **tertiary (3°) carbon atom** *is one that is attached directly to three other carbon atoms.*

Example

A tertiary carbon atom

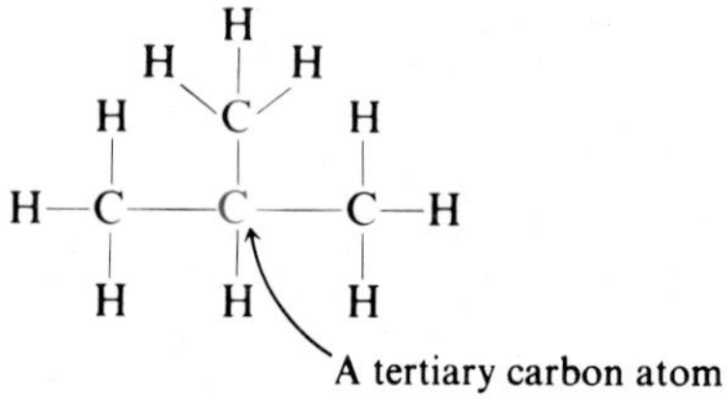

Therefore, a *primary alcohol* is one in which the hydroxyl group is attached to a primary carbon atom, a *secondary alcohol* is one whose hydroxyl group is located on a secondary carbon atom, and a *tertiary alcohol* has its hydroxyl group bonded to a tertiary carbon. Table 6.1 lists the nomenclature and classification of some of the simpler alcohols.

As seen in the table, the common names of the lower members of the alcohol series are formed in a manner analogous to that used for the naming of the alkyl halides. The name of the alkyl group is followed by the word alcohol. The IUPAC system is adapted for use in naming the higher homologs. In the IUPAC system, the designations primary, secondary, and tertiary are unnecessary and have no significance. Alcohols are named under the IUPAC system as follows.

1. The longest continuous chain of carbons containing the —OH group is taken as the parent compound.

TABLE 6.1 Classification and Nomenclature of Some Alcohols

Structural Formula	Type of Alcohol	Common Name	IUPAC Name
CH_3OH	Primary	Methyl alcohol	Methanol
CH_3CH_2OH	Primary	Ethyl alcohol	Ethanol
$CH_3CH_2CH_2OH$	Primary	n-Propyl alcohol	1-Propanol
$CH_3\overset{\underset{\|}{OH}}{C}HCH_3$	Secondary	Isopropyl alcohol	2-Propanol
$CH_3CH_2CH_2CH_2OH$	Primary	n-Butyl alcohol	1-Butanol
$CH_3CH_2\overset{\underset{\|}{OH}}{C}HCH_3$	Secondary	sec-Butyl alcohol	2-Butanol
$CH_3\overset{\underset{\|}{CH_3}}{C}HCH_2OH$	Primary	Isobutyl alcohol	2-Methyl-1-propanol
$CH_3-\overset{\overset{OH}{\|}}{\underset{\underset{CH_3}{\|}}{C}}-CH_3$	Tertiary	tert-Butyl alcohol	2-Methyl-2-propanol
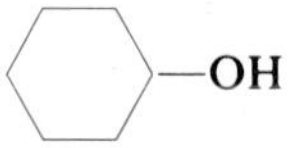 ⬡—OH	Secondary	Cyclohexyl alcohol	Cyclohexanol
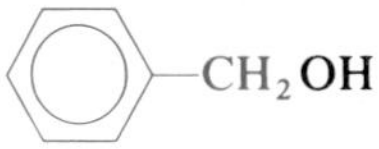 ◯—CH_2OH	Primary	Benzyl alcohol	Phenylmethanol

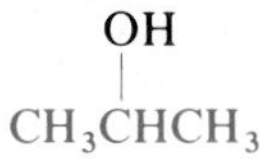

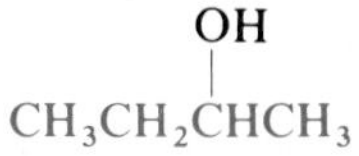

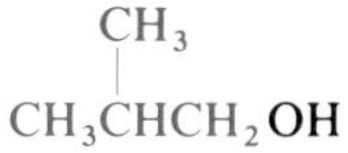

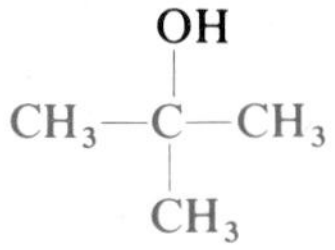

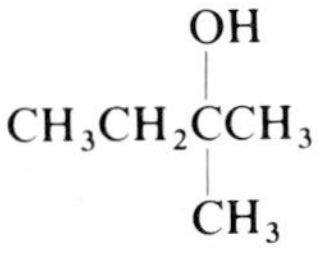

$CH_3CH_2\overset{\overset{OH}{\|}}{\underset{\underset{CH_3}{\|}}{C}}CH_3$

2-Methyl-2-butanol

$CH_3\overset{\overset{CH_3}{\|}}{\underset{6\quad5}{C}}H\underset{4}{C}H_2\overset{\overset{OH}{\|}}{\underset{\underset{CH_3}{\|}}{\underset{3\,2}{C}}}CH_2CH_3$

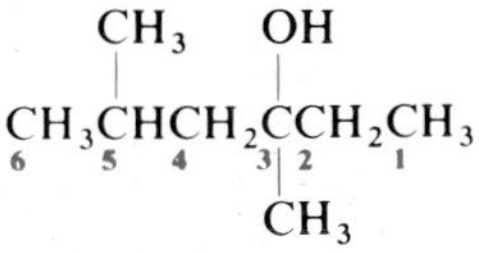

3,5-Dimethyl-3-hexanol

2. The chain is numbered from the end closer to the hydroxyl group. The number that then appropriately indicates the position of the hydroxyl group is prefixed to the name of the parent hydrocarbon.

3. The -e ending of the parent alkane is replaced by the suffix -ol, the generic ending for the alcohols.

4. If more than one hydroxyl group appears in the same molecule (polyhydroxy alcohols), the suffixes -diol, -triol, etc., are used. In these cases, the -e ending of the parent alkane is retained.

5. If a molecule contains a double bond and a hydroxyl group, two suffixes are used, and the -ol suffix has priority. (Notice that the final -e of the alkene suffix -ene is dropped when another suffix follows.)

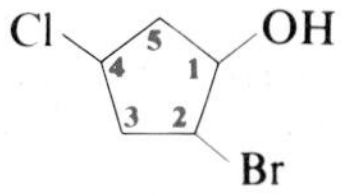

2-Bromo-4-chlorocyclopentanol

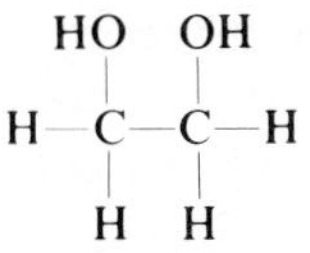

1,2-Ethanediol
(Ethylene glycol)

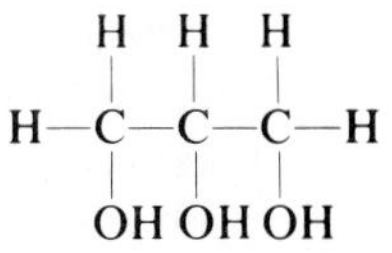

1,2,3-Propanetriol
(Glycerol)

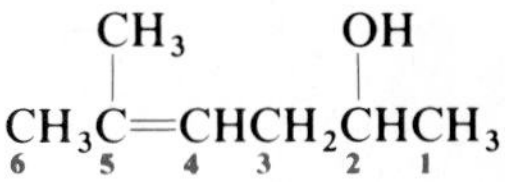

$\underset{6}{C}H_3\underset{5}{C}=\underset{4}{C}H\underset{3}{C}H_2\overset{\overset{OH}{\|}}{\underset{2}{C}}H\underset{1}{C}H_3$

5-Methyl-4-hexen-2-ol
(**not** 2-Methyl-2-hexen-5-ol)

6.2 Preparation of Alcohols

In studying the reactions of the alkenes and the alkyl halides, we have already encountered reactions that are useful for the preparation of alcohols. The following examples, then, should serve as a review for the student. In organic chemistry, a typical reaction of one class of compounds very often is the preparative reaction of another class. In fact, a thorough knowledge of organic chemistry requires a knowledge of the interconversion of the various classes of organic compounds. In addition to the following methods of preparation, special reactions for the preparation of methyl and ethyl alcohols are discussed in Section 6.5.

a. Hydration of Alkenes

In Section 3.4-b the addition of sulfuric acid across the double bond of an alkene was illustrated. If the alkylsulfuric acid that results from this reaction is treated with water, an alcohol is formed. The $-OSO_3H$ group is replaced by the $-OH$ from the water.

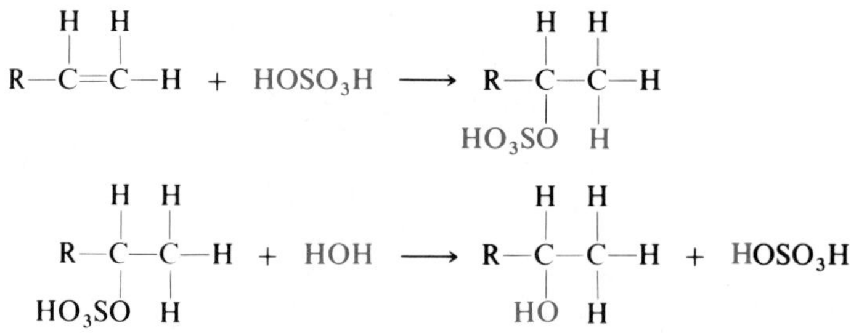

$$R-\underset{\underset{H}{|}}{C}=\underset{\underset{H}{|}}{C}-H \ + \ HOSO_3H \ \longrightarrow \ R-\underset{\underset{HO_3SO}{|}}{C}-\underset{\underset{H}{|}}{C}-H$$

$$R-\underset{\underset{HO_3SO}{|}}{C}-\underset{\underset{H}{|}}{C}-H \ + \ HOH \ \longrightarrow \ R-\underset{\underset{HO}{|}}{C}-\underset{\underset{H}{|}}{C}-H \ + \ HOSO_3H$$

Since the sulfuric acid is regenerated, it acts as a catalyst, and the entire reaction can be represented by the following single equation.

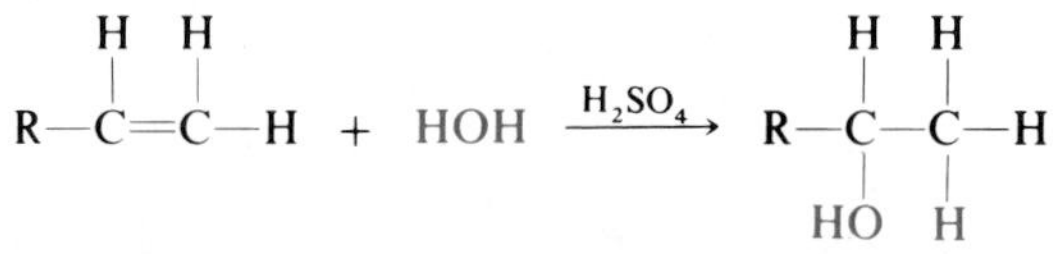

$$R-\underset{\underset{H}{|}}{C}=\underset{\underset{H}{|}}{C}-H \ + \ HOH \ \xrightarrow{H_2SO_4} \ R-\underset{\underset{HO}{|}}{C}-\underset{\underset{H}{|}}{C}-H$$

When the reaction is represented in this manner, it appears as though a direct addition of the H and OH from a molecule of water has taken place across the double bond of the alkene. This particular representation of the reaction is deceiving because the composition of the products does not give a correct indication of the mode (mechanism) of reaction.

Notice that if the reacting alkene in the above reaction were ethylene (if R = H), the resulting alcohol would be ethyl alcohol. Ethyl alcohol is the only primary alcohol that can be prepared by this method. Why? (Hint: Recall Markovnikov's rule.) Because the alkenes are so readily available from petroleum, the hydration reaction accounts for the major production of industrial ethyl alcohol and isopropyl alcohol (rubbing alcohol).

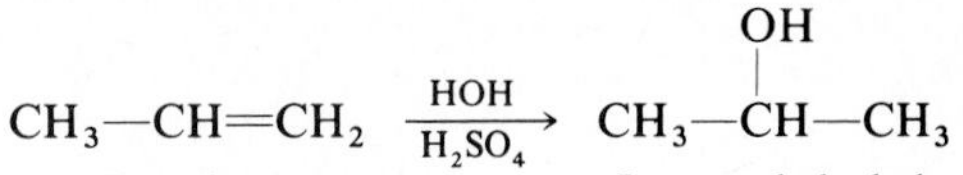

$$\text{CH}_3\text{---CH==CH}_2 \xrightarrow[\text{H}_2\text{SO}_4]{\text{HOH}} \overset{\displaystyle \text{OH}}{\underset{}{\text{CH}_3\text{---CH---CH}_3}}$$

Propylene   Isopropyl alcohol

Although the direct hydration of alkenes is an important industrial process for the preparation of alcohols, it is seldom used as a laboratory procedure because other synthetic methods are more convenient. The hydration reaction, however, is very common in biochemistry; many of the hydroxy compounds that occur in living systems are formed in this manner. The following reaction, for example, occurs as one of the steps in the Krebs cycle (see Figure 16.7).

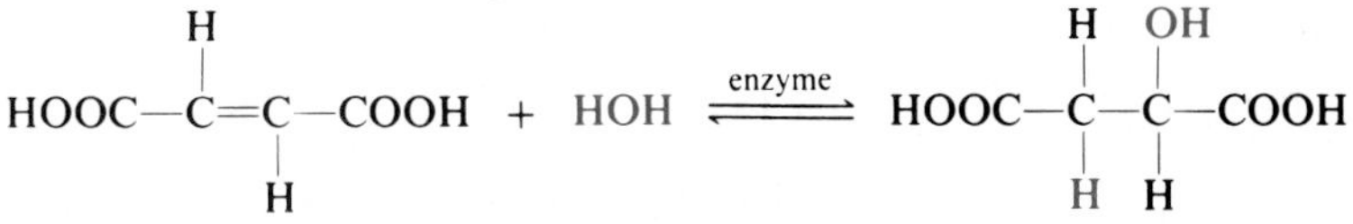

$$\underset{\text{Fumaric acid}}{\overset{\displaystyle \text{H}}{\underset{\displaystyle \text{H}}{\text{HOOC---C==C---COOH}}}} + \text{HOH} \underset{}{\overset{enzyme}{\rightleftharpoons}} \underset{\text{Malic acid}}{\overset{\displaystyle \text{H} \quad \text{OH}}{\underset{\displaystyle \text{H} \quad \text{H}}{\text{HOOC---C---C---COOH}}}}$$

b. Hydrolysis of Alkyl Halides

The products of the hydrolysis of alkyl halides are dependent on many factors, among them, reaction conditions. In order to prepare an alcohol by the hydrolysis of an alkyl halide, experimental conditions must be sought that will maximize the hydrolysis (substitution) reaction and minimize the elimination reaction. One method that accomplishes this involves the use of primary alkyl halides and dilute aqueous solutions of sodium hydroxide. Secondary and tertiary alkyl halides tend to give elimination products (Section 5.4-a).

General Equation

$$\text{R---X} + \text{NaOH} \xrightarrow{\text{H}_2\text{O}} \text{R---OH} + \text{NaX}$$

Specific Equation

$$\underset{\text{1-Chlorobutane}}{\text{CH}_3\text{CH}_2\text{CH}_2\text{CH}_2\text{Cl}} + \text{NaOH} \xrightarrow{\text{H}_2\text{O}} \underset{\text{1-Butanol}}{\text{CH}_3\text{CH}_2\text{CH}_2\text{CH}_2\text{OH}} + \text{NaCl}$$

6.3 Physical Properties of Alcohols

It was earlier stated that the alcohols can be considered to be derivatives of water. This relation becomes particularly apparent, especially for the lower homologs, in a discussion of the physical and chemical properties of the alcohols. Remember that methane, ethane, and propane are gases and are insoluble in water. In contrast, methanol, ethanol, and propanol are liquids and are completely miscible with water. Therefore, replacement of a single hydrogen atom with a hydroxyl group brings about marked changes in solubility and physical state of the molecules. These

TABLE 6.2 Comparison of Boiling Points and
Molecular Weights

Formula	Name	MW	BP (°C)
CH_4	Methane	16	−164
HOH	Water	18	100
C_2H_6	Ethane	30	−89
CH_3OH	Methanol	32	65
C_3H_8	Propane	44	−42
CH_3CH_2OH	Ethanol	46	78
C_4H_{10}	Butane	58	−1
$CH_3CH_2CH_2OH$	1-Propanol	60	97

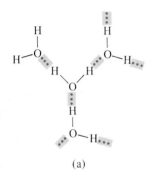

(a)

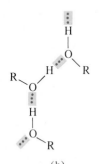

(b)

FIGURE 6.1 Intermolecular
hydrogen bonding in water (a)
and in alcohol (b).

differences result from the hydrogen-bonding capabilities of the alcohols. *Hydro-gen bonding*[1] is responsible for the intermolecular attractions between alcohol molecules, and hence even the lightest homolog of the series exists as a liquid at room temperature. Alcohols can form hydrogen bonds with water molecules as well, and so the lower homologs of the series are water soluble. Table 6.2 lists the molecular weights and the boiling points of some common compounds. The table shows that substances having similar molecular weights do not always have similar boiling points.

In the case of the alcohols, the relatively high boiling points are a direct result of strong intermolecular attractions. Recall that a boiling point is a rough measure of the amount of energy necessary to separate a liquid molecule from its nearest neighbors. If the nearest neighbors of a molecule are attached to that molecule by means of hydrogen bonding, a considerable amount of energy must be supplied to break the hydrogen bonds. Only then can an individual molecule escape from the liquid into the gaseous state.

Figure 6.1 illustrates hydrogen bonding in water and in the alcohols. Reference to this schematic representation reveals the reasons why water boils at a higher temperature than methyl alcohol, even though water is a lighter molecule. The oxygen atom and both hydrogen atoms of the water molecule participate in hydrogen bonding to three, or even four, adjacent water molecules. Alcohols result from the replacement of one hydrogen atom of water with an alkyl group. The alkyl group does not participate in hydrogen bonding, and so the alcohol is associated to only two other alcohol molecules. More energy is required to disrupt three or four intermolecular bonds than two, and thus greater energy is needed to vaporize the water. (The energy required to break a hydrogen bond is about 5 kcal/mole. Although this is distinctly less than the energy required to break any of the intramolecular bonds in water or alcohol, it is still an appreciable amount of energy and shows its significance in the boiling point differences.)

[1]Hydrogen bonding occurs in molecules whenever hydrogen is bonded to a strongly electronegative element such as fluorine, oxygen, or nitrogen. The concept of hydrogen bonding is of extreme importance in organic chemistry as well as in biochemistry. It is vital to understanding the structure and functions of proteins and nucleic acids.

Chapter 6 · Alcohols, Phenols, Ethers, and Thiols

Polarity and hydrogen bonding are significant factors in the water solubility of alcohols. A common expression among chemists is "like dissolves like" (the word *like* refers to relative polarities). This rule of thumb implies the useful generalization that polar solvents will dissolve polar solutes, and nonpolar solvents will dissolve nonpolar solutes. Care must be taken, however, not to apply this generalization haphazardly to all cases. All alcohol molecules are polar, yet not all alcohols are water soluble. On the other hand, all alcohols are soluble in most of the common nonpolar solvents (carbon tetrachloride, ether, benzene). Only the lower homologs of the series have an appreciable solubility in water. As the length of the carbon chain increases, water solubility decreases.

The differences in water solubility can be explained in the following manner. The hydroxyl group confers polarity and water solubility upon the alcohol molecule. The alkyl group confers nonpolarity and water insolubility. Whenever the hydroxyl group comprises a substantial portion of a molecule, the molecule will be water soluble. (The hydroxyl group can be thought of as dragging the remainder of the molecule into the water structure.)

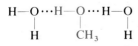

CH_3OH has more than 50% of its mass ($\frac{17}{32}$) located in the hydroxyl portion of the molecule. CH_3CH_2OH has more than 37% of its mass ($\frac{17}{46}$) located in the hydroxyl group. Both have infinite water solubility. As the carbon chain of monohydroxy alcohols increases, the contribution of the hydroxyl group to the total mass of the molecule decreases and soon becomes overwhelmed by the mass of the carbon chain. Consider *n*-decyl alcohol, in which the hydroxyl group is only $\frac{17}{158}$ of the mass of the entire molecule. In this case, the hydroxyl group is almost totally ineffective in dragging the large alkane portion of the molecule into the water structure (Figure 6.2). Therefore, *n*-decyl alcohol behaves more like an alkane than an alcohol in many of its physical properties.

$$\underbrace{CH_3CH_2CH_2CH_2CH_2CH_2CH_2CH_2CH_2CH_2}_{\text{Alkane portion}}OH$$

Consider Table 6.3, which contains the solubilities of the butyl alcohols in water. The large discrepancies in the water solubilities of these isomeric alcohols cannot be attributed to differences in molecular weight or mass percent of the hydroxyl group. The differing solubilities are a result of the different geometric shapes of the alcohols. The very compact *tert*-butyl alcohol molecules experience weaker intermolecular van der Waals attractions and are more easily surrounded by water

TABLE 6.3 Solubilities of the Butyl Alcohols in Water

Alcohol	Formula	Solubility (g/100 g of water)
n-Butyl	$CH_3CH_2CH_2CH_2OH$	8
Isobutyl	$(CH_3)_2CHCH_2OH$	11
sec-Butyl	$CH_3CH_2CH(OH)CH_3$	12.5
tert-Butyl	$(CH_3)_3COH$	Completely soluble

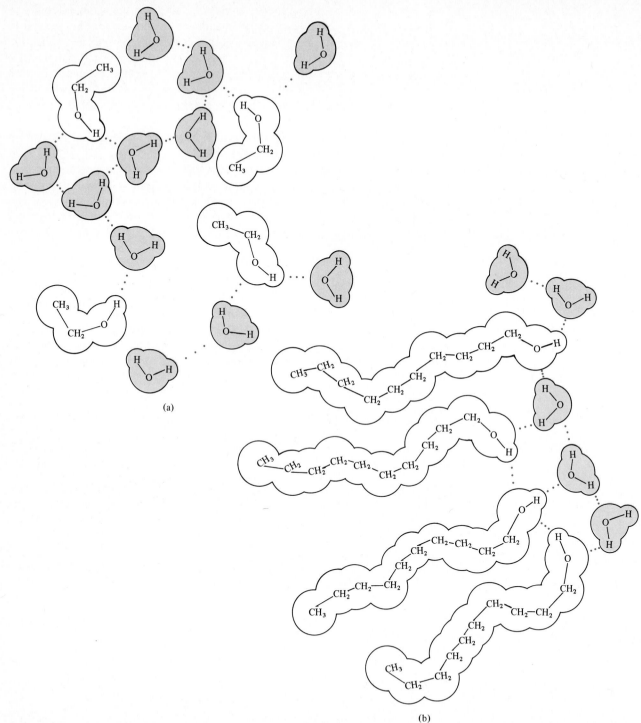

FIGURE 6.2 (a) Hydrogen bonding between ethyl alcohol and water molecules accounts for the solubility of ethyl alcohol in water. (b) Water molecules are unable to surround molecules of *n*-decyl alcohol so it is insoluble in water.

molecules. Hence, *tert*-butyl alcohol has the lowest boiling point (83°C) of any of its isomers (all of which boil above 100°C) and the highest solubility in water.

In summary, solubility considerations involve the balance of polar and nonpolar groups within a molecule, as well as molecular shape. The more polar a molecule and the more compact its shape, the greater will be its water solubility.

6.4 Chemical Properties of Alcohols

a. Acidity and Basicity

The acid–base properties of the aliphatic alcohols are similar to those of water. That is, upon reaction with a strong acid the alcohols will accept protons (that is, they act as bases) to form positively charged analogs of the hydronium ion. *Protonated alcohols* are called **oxonium ions.**

$$H-\overset{\overset{\displaystyle \cdot\cdot}{|}}{\underset{\displaystyle H}{O}}: \ + \ H_2SO_4 \ \rightleftharpoons \ H-\overset{\overset{\displaystyle \cdot\cdot +}{|}}{\underset{\displaystyle H}{O}}-H \ + \ HSO_4^-$$

Hydronium ion

$$CH_3-\overset{\overset{\displaystyle \cdot\cdot}{|}}{\underset{\displaystyle H}{O}}: \ + \ H_2SO_4 \ \rightleftharpoons \ CH_3-\overset{\overset{\displaystyle \cdot\cdot +}{|}}{\underset{\displaystyle H}{O}}-H \ + \ HSO_4^-$$

An oxonium ion

Methanol and ethanol are about as acidic as water; the larger alcohols are weaker acids than water. Alcohols and water do not react with strong bases, but they both react with alkali metals to liberate hydrogen gas. The resulting anion is termed an *alkoxide ion* (RO^-); alkoxides are strong bases, even stronger than hydroxides. Sodium and potassium alkoxides are important reagents in organic chemistry, whenever strong bases are required in nonaqueous conditions.

$$CH_3OH \ + \ HOH \ \rightleftharpoons \ CH_3O^- \ + \ H_3O^+ \qquad K \approx 10^{-16}$$

Methoxide
ion

$$CH_3OH + NaOH \ \longrightarrow \ \text{No reaction}$$

$$HOH \ + \ Na \ \longrightarrow \ Na^+ \ {}^-OH \ + \ \tfrac{1}{2}H_2$$

$$CH_3CH_2OH \ + \ Na \ \longrightarrow \ Na^+ \ {}^-OCH_2CH_3 \ + \ \tfrac{1}{2}H_2$$

Sodium ethoxide

The reactions of active metals with alcohols are not as vigorous as those with water and thus are more easily controlled. (Isopropyl alcohol is often employed in the laboratory to decompose excess pieces of sodium because this reaction is relatively slow and moderate.) The evolution of hydrogen gas is a good qualitative test to indicate the presence of a hydroxyl group in an organic molecule.

b. Dehydration of Alcohols (Preparation of Alkenes or Ethers)

The dehydration of alcohols was cited in Section 3.2-a as a method for the preparation of alkenes. We shall mention it again in our later discussions of carbohydrate and fatty acid metabolism. The reaction involves intramolecular dehydration, represented by the following general equation.

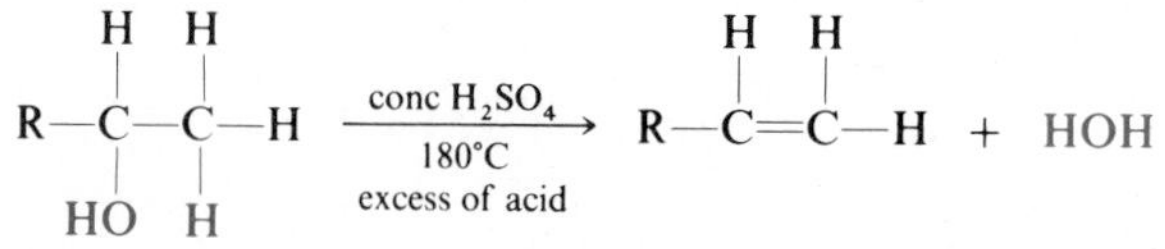

$$\underset{\substack{| \quad | \\ HO \quad H}}{\overset{\substack{H \quad H \\ | \quad |}}{R-C-C-H}} \xrightarrow[\substack{180°C \\ \text{excess of acid}}]{\text{conc } H_2SO_4} \underset{\substack{\\ }}{\overset{\substack{H \quad H \\ | \quad |}}{R-C=C-H}} + HOH$$

The first step in the reaction is the protonation of the alcohol to form an oxonium ion intermediate.

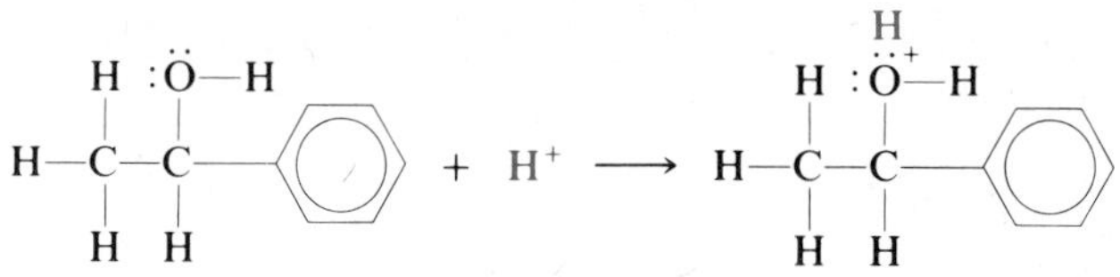

1-Phenylethanol

Next a molecule of water is eliminated, and the positive charge then resides on the carbon atom (a carbocation is formed). Experimental evidence indicates that the controlling factor in the dehydration of alcohols is the ease of formation of the carbocation intermediate. Furthermore, the ease of formation of the carbocation is directly related to the stability of the carbocation (recall page 42). We thus find that tertiary alcohols are dehydrated more readily than secondary alcohols, which are in turn more easily dehydrated than primary alcohols.

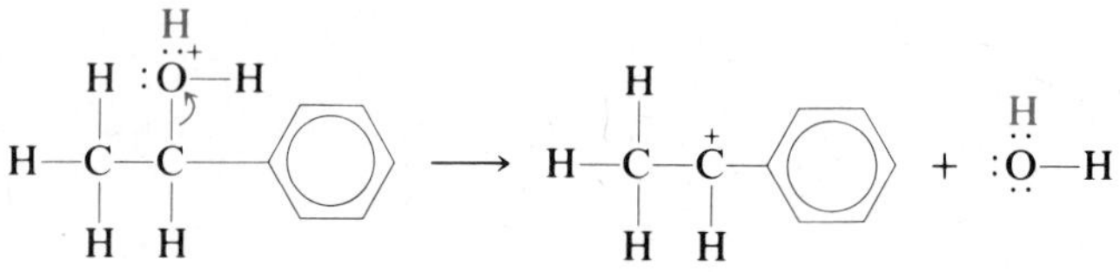

In the last step the acid catalyst is regenerated by loss of a hydrogen ion from the adjacent carbon atom.

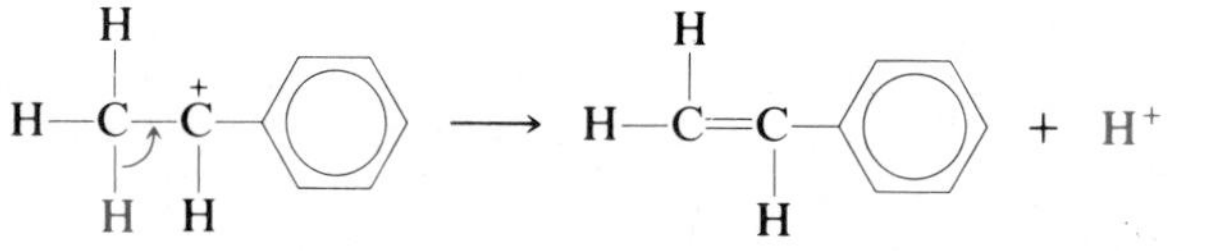

Styrene

Dehydration of 1-phenylethanol can yield only one possible alkene. However, there are many alcohols that present a choice of possible products (such as $CH_3CH_2CHOHCH_3$), and we need to be able to predict the major product. This prediction is based upon a consideration of the relative stabilities of the prepared

90

alkenes. Because alkyl groups are electron-donating, they tend to stabilize multiple carbon–carbon bonds, and thus *the more highly substituted alkene is usually the major product obtained from the acid-catalyzed dehydration of alcohols.* Predominant formation of the more highly substituted isomer is referred to as **Saytzeff's rule** after the Russian chemist Alexander Saytzeff.

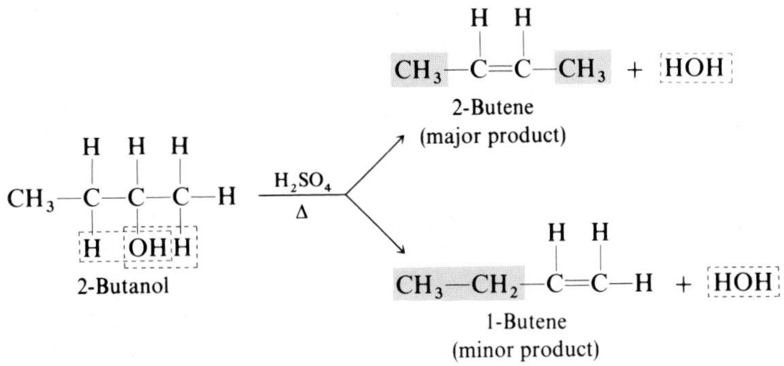

$$CH_3-C=C-CH_3 \ + \ HOH$$

2-Butene
(major product)

$$CH_3-C-C-C-H \xrightarrow[\Delta]{H_2SO_4}$$

2-Butanol

$$CH_3-CH_2-C=C-H \ + \ HOH$$

1-Butene
(minor product)

With primary alcohols, an alteration of reaction conditions, particularly reaction temperature, can bring about an intermolecular dehydration, which yields another class of organic compounds, the ethers (see Section 6.11).

General Equation

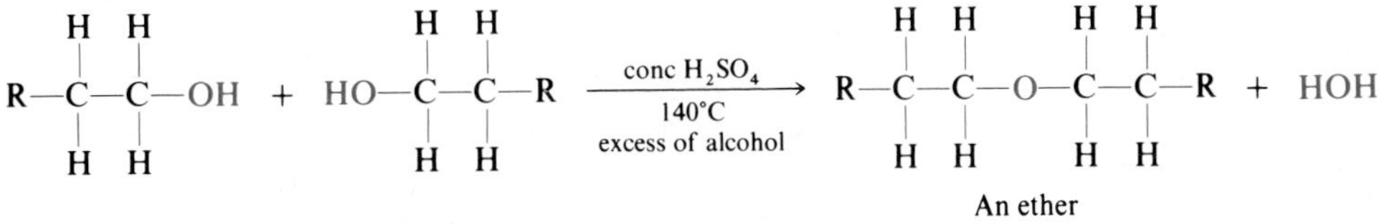

$$R-C-C-OH \ + \ HO-C-C-R \xrightarrow[\substack{140°C \\ \text{excess of alcohol}}]{\text{conc } H_2SO_4} R-C-C-O-C-C-R \ + \ HOH$$

An ether

The first step of the mechanism again involves oxonium ion formation. However, at this lower temperature and in the presence of excess alcohol, displacement of a water molecule by another alcohol molecule takes precedence over carbocation formation. The catalyst is then regenerated in the third step.

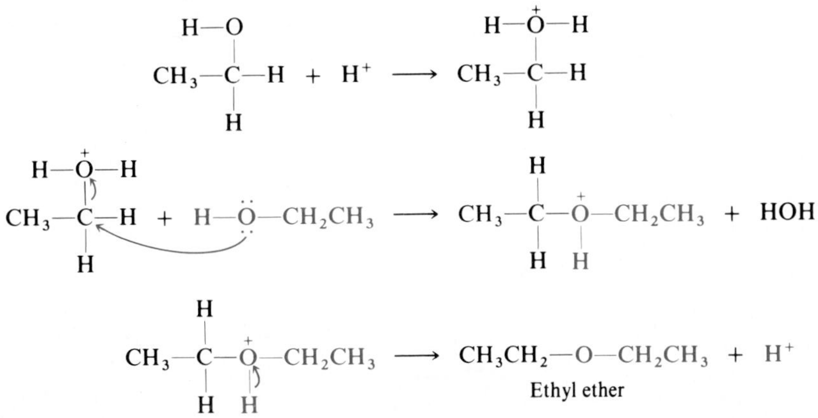

$$CH_3-\overset{\overset{H-O}{|}}{C}-H \ + \ H^+ \longrightarrow CH_3-\overset{\overset{H-\overset{+}{O}-H}{|}}{C}-H$$

$$CH_3-\overset{\overset{H-\overset{+}{O}-H}{|}}{C}-H \ + \ H-\overset{..}{\underset{..}{O}}-CH_2CH_3 \longrightarrow CH_3-\overset{\overset{H}{|}}{C}-\overset{+}{O}-CH_2CH_3 \ + \ HOH$$

$$CH_3-\overset{\overset{H}{|}}{\underset{\overset{|}{H}}{C}}-\overset{+}{O}-CH_2CH_3 \longrightarrow CH_3CH_2-O-CH_2CH_3 \ + \ H^+$$

Ethyl ether

Dehydration (and its reverse, hydration) reactions occur continuously in cellular metabolism. In these biochemical dehydrations, enzymes serve as catalysts instead of acids, and the reaction temperature is 37°C instead of the elevated temperatures required in the laboratory.

$$\underset{\text{2-Phosphoglyceric acid}}{\underset{\displaystyle H-\overset{\displaystyle HO}{\underset{\displaystyle H}{C}}-\overset{\displaystyle H}{\underset{\displaystyle OPO_3{}^{2-}}{C}}-\overset{O}{\underset{OH}{C}}}} \quad \underset{enolase}{\rightleftharpoons} \quad \underset{\text{Phosphoenolpyruvic acid}}{\underset{\displaystyle \overset{H}{\underset{H}{}}C=\overset{\overset{O}{\|}}{\underset{OPO_3{}^{2-}}{C}}-C-OH}} \quad + \quad HOH$$

c. Replacement of the Hydroxyl Group

The hydroxyl group of alcohols can be replaced by various acid anions (Section 5.2-b). Hydrobromic acid and hydriodic acid react readily with all alcohols. The less reactive hydrochloric acid requires the presence of zinc chloride for reaction with primary and secondary alcohols. A concentrated hydrochloric acid solution that has been saturated with zinc chloride is known as the *Lucas reagent*. The rates for the reactions of alcohols with the Lucas reagent decrease in the order tertiary > secondary > primary.

$$\underset{\textit{tert-Butyl alcohol}}{CH_3-\overset{\displaystyle CH_3}{\underset{\displaystyle CH_3}{C}}-OH} \;+\; HCl \;\longrightarrow\; \underset{\textit{tert-Butyl chloride}}{CH_3-\overset{\displaystyle CH_3}{\underset{\displaystyle CH_3}{C}}-Cl} \;+\; HOH$$

The phosphorus halides and thionyl chloride are also utilized to replace the hydroxyl group by halogen. Thionyl chloride is especially effective because the by-products of the reaction, SO_2 and HCl, are volatile and escape from the reaction mixture.

$$3\;\underset{\text{Cyclohexanol}}{\bigcirc\!-OH} \;+\; \underset{\displaystyle Br}{Br-\overset{\displaystyle}{\underset{\displaystyle |}{P}}-Br} \;\longrightarrow\; 3\;\underset{\text{Bromocyclohexane}}{\bigcirc\!-Br} \;+\; \underset{\substack{\text{Phosphorous acid}}}{HO-\overset{\overset{O}{\|}}{\underset{\underset{H}{|}}{P}}-OH}$$

$$\underset{\substack{\text{Allyl alcohol}\\\text{(2-Propen-1-ol)}}}{CH_2{=}CH{-}CH_2OH} \;+\; \underset{\substack{\text{Thionyl}\\\text{chloride}}}{SOCl_2} \;\longrightarrow\; \underset{\substack{\text{Allyl chloride}\\\text{(3-Chloropropene)}}}{CH_2{=}CH{-}CH_2Cl} \;+\; SO_2(g) \;+\; HCl(g)$$

d. Oxidation–Reduction Reactions

The oxidation reactions of alcohols to form aldehydes and ketones are discussed in the next chapter and also in the chapters on carbohydrate metabolism and lipid metabolism. Primary alcohols are oxidized to aldehydes, whereas secondary alcohols are oxidized to ketones. Tertiary alcohols are stable to oxidation under the usual conditions. Many procedures are available for accomplishing these transformations, but the most common oxidizing system is some form of Cr(VI) in an organic solvent. The symbol [O] implies that an oxidizing agent is used without indicating the specific reagent.

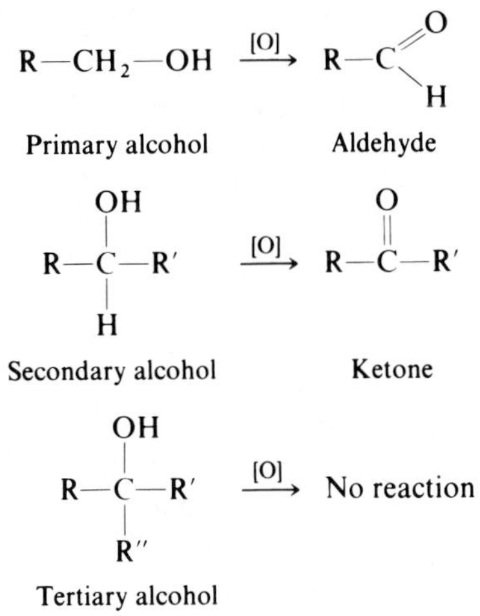

$$R-CH_2-OH \xrightarrow{[O]} R-C\!\!\begin{array}{c} \nearrow O \\ \searrow H \end{array}$$

Primary alcohol   Aldehyde

$$\underset{\underset{H}{|}}{\overset{\overset{OH}{|}}{R-C-R'}} \xrightarrow{[O]} \overset{\overset{O}{\|}}{R-C-R'}$$

Secondary alcohol   Ketone

$$\underset{\underset{R''}{|}}{\overset{\overset{OH}{|}}{R-C-R'}} \xrightarrow{[O]} \text{No reaction}$$

Tertiary alcohol

e. Inorganic Ester Formation

The reaction of primary and secondary alcohols with oxyacids rather than with halogen acids leads to the formation of compounds in which the —OH group of the acid is converted to —OR. Such compounds are called **esters,** and they are very important and useful substances. In Chapter 8, we discuss the organic esters prepared from alcohols and organic acids, and in Chapter 12, we discuss the esters formed from fatty acids and glycerol. The inorganic esters may be formed from alcohols and inorganic acids such as sulfuric, nitrous, nitric, and phosphoric.

$$R-OH \ + \ H-OY \ \longrightarrow \ R-OY \ + \ HOH$$

Alcohol   Oxyacid   Inorganic ester

The reaction of ethanol with sulfuric acid merits special mention because it illustrates the importance of specifying reaction conditions. At a temperature of 180°C and in the presence of excess sulfuric acid, dehydration occurs to form ethylene (Section 6.4-b). At 140°C and with excess of alcohol, the product is ethyl ether (see Section 6.11). At room temperature the alcohol and acid react to yield the ester, ethyl hydrogen sulfate.

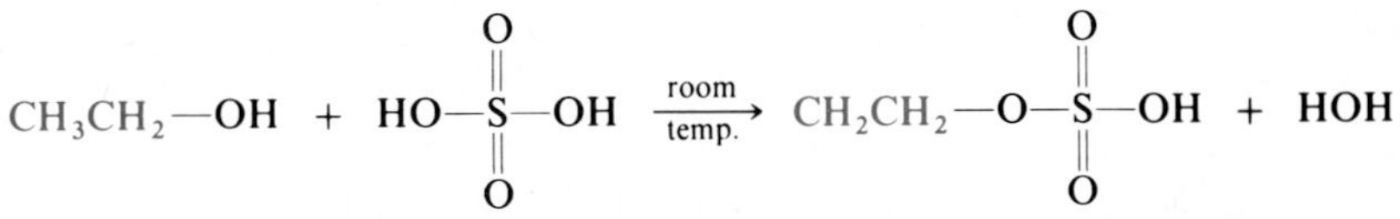

$$CH_3CH_2-OH \ + \ HO-\overset{\overset{O}{\|}}{\underset{\underset{O}{\|}}{S}}-OH \ \xrightarrow[\text{temp.}]{\text{room}} \ CH_2CH_2-O-\overset{\overset{O}{\|}}{\underset{\underset{O}{\|}}{S}}-OH \ + \ HOH$$

Ethyl hydrogen sulfate

Esters of nitrous acid form readily by the direct reaction of an alcohol with the acid. The valuable drug isopentyl nitrite is used clinically to relieve the pain associated with angina pectoris. The severe and disabling chest pain results when blood flow to the heart muscle is restricted, thereby limiting the amount of oxygen delivered to the heart. Isopentyl nitrite and polynitrates such as nitroglycerin (Section 6.5-e) and pentaerythritol tetranitrate function by dilating the smaller blood vessels (vasodilators), thus increasing blood flow to the heart.

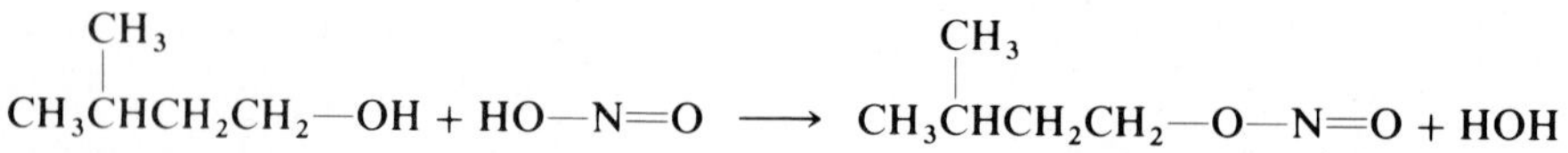

$$CH_3CHCH_2CH_2—OH + HO—N{=}O \longrightarrow CH_3CHCH_2CH_2—O—N{=}O + HOH$$

Isopentyl alcohol (Isoamyl alcohol)	Nitrous acid	Isopentyl nitrite (Isoamyl nitrite)

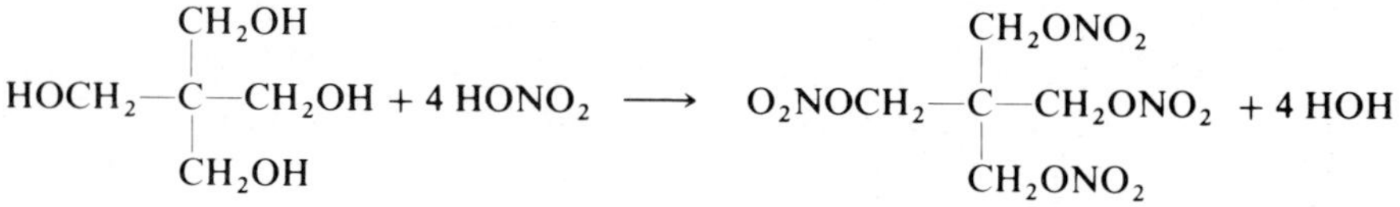

$$HOCH_2—C—CH_2OH + 4\ HONO_2 \longrightarrow O_2NOCH_2—C—CH_2ONO_2\ + 4\ HOH$$

Pentaerythritol	Nitric acid	Pentaerythritol tetranitrate (PETN)

It is possible to replace all three hydroxyl groups of phosphoric acid by alkoxy groups. Phosphate esters play vital roles in all biological systems. We discuss these very important compounds in more detail in Chapters 8 and 16.

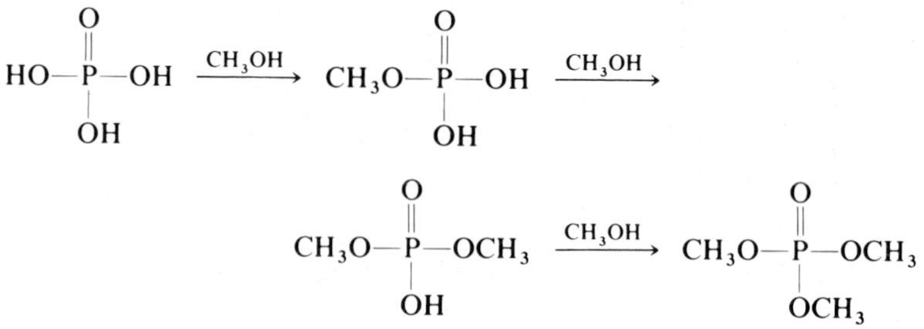

$$HO—\overset{O}{\overset{\|}{P}}—OH \xrightarrow{CH_3OH} CH_3O—\overset{O}{\overset{\|}{P}}—OH \xrightarrow{CH_3OH}$$

$$CH_3O—\overset{O}{\overset{\|}{P}}—OCH_3 \xrightarrow{CH_3OH} CH_3O—\overset{O}{\overset{\|}{P}}—OCH_3$$

6.5 Important Alcohols

a. CH₃OH (Methanol, Methyl Alcohol, Wood Alcohol)

Prior to 1923, methanol was prepared by the destructive distillation of wood—hence its common name, wood alcohol. When wood is heated to a temperature of 450°C in the absence of air, it decomposes to charcoal and a volatile fraction. Two to three percent of this fraction is methanol, and it can be separated from the other components (acetic acid and acetone) by fractional distillation. On the average, 1 ton of wood produces about 35 lb of the alcohol. Presently, methanol is prepared more economically by combining hydrogen and carbon monoxide under conditions of high temperature and pressure in the presence of a zinc oxide–chromium oxide catalyst.

$$2\ H_2 + CO \xrightarrow[ZnO–Cr_2O_3]{200\ atm,\ 350°C} CH_3OH$$

Methyl alcohol, unlike ethyl alcohol, is poisonous. Ingestion of as little as 15 mL of methanol can cause blindness; 30 mL (1 fluid ounce) can cause death. The poisoning effect is brought about by the body's oxidation of methanol to formic acid (HCOOH), and probably to some extent, to formaldehyde (HCHO). Formic acid causes severe acidosis (see Section 17.4) and other characteristic symptoms of methyl alcohol poisoning. The basis of the selective injury to retinal cells in methanol intoxication is not definitely known. Some believe that formaldehyde has a specific toxicity for the cells of the retina, whereas others believe that formic acid is the offending substance. Since the dominant feature of methanol poisoning is acidosis, the antidote is a solution of the weak base sodium bicarbonate (baking soda, $NaHCO_3$). Alternatively, large doses of ethanol can be administered. Ethanol blocks the oxidation of methanol by competing for the oxidation enzyme. Methyl alcohol should never be applied to the body, nor should its vapors be inhaled because methanol is readily absorbed through the skin and respiratory tract.

The largest use of methanol is as the starting material for the commercial synthesis of formaldehyde. It is also used in car windshield washer fluids and it is employed commercially as a solvent for paint, gum, and shellac. Methanol, along with ethanol (see following), is mixed with gasoline and sold as **gasohol** for use as a motor fuel.

b. C_2H_5OH (*Ethanol, Ethyl Alcohol, Grain Alcohol, Alcohol*)

Ethyl alcohol is probably the best known and most important member of the alcohol series. In fact, when the layman uses the term alcohol, he is usually referring to ethyl alcohol.

The production of alcoholic spirits is one of the oldest known chemical reactions. Even in biblical times ethanol was prepared by the fermentation of sugars or starch from various sources (potatoes, corn, wheat, rice, etc.). Biochemical investigations have shown that the fermentation process is catalyzed by enzymes found in yeast, and that it proceeds by an elaborate multistep mechanism. These steps are considered in detail in Chapter 16. The equation for the overall process can be written as follows.

$$(C_6H_{10}O_5)_x \xrightarrow{enzymes} C_6H_{12}O_6 \xrightarrow{enzymes} 2\,C_2H_5OH + 2\,CO_2$$
$$\text{Starch} \qquad\qquad \text{Glucose} \qquad\qquad \text{Ethanol}$$

The greatest use of ethanol is as a beverage. Wines contain about 12% ethanol by volume; champagnes contain 14–20%; beers and ciders contain about 4%; and whiskey, gin, and brandy contain 40–50%. The alcoholic content of a beverage is indicated by a measure known as *proof spirit*.[2] The proof value is twice the alcoholic content by volume, and thus whiskey that is 50% alcohol is said to be 100 proof.

[2]This term has its origins in an early procedure used to test the alcoholic strength of a beverage. The whiskey to be tested was poured over a portion of gunpowder and the mixture was ignited. Since an ethanol–water solution will ignite when the alcohol concentration is about 50%, this solution scored 100 in the test, "proof" of the spirit content in the whiskey. If the powder did not burn, it was due to the presence of too much water in the whiskey.

Ethyl alcohol is potentially toxic. Rapid ingestion of 1 pint of pure alcohol would kill most people. Alcohol freely crosses into the brain where it depresses the respiratory control center, resulting in failure of the respiratory muscles in the lungs and hence suffocation. Alcohol is believed to act on the nerve cell membranes, causing a diminution in speech, thought, cognition, and judgment. Excessive ingestion over a long period of time leads to deterioration of the liver (cirrhosis), loss of memory, and may lead to strong physiological addiction. Addiction to alcohol (alcoholism) is the most serious drug problem in the United States. It has been estimated that there are about 40 times as many alcoholics (about 10 million) as there are heroin addicts in the United States. If the alcohol is diluted (as in alcoholic beverages) and is consumed in small quantities, it is relatively safe.[3] The body possesses enzymes that have the capacity to metabolize ethyl alcohol to carbon dioxide and water (see page 438).

Alcohol not intended for beverage purposes is commercially prepared by the direct hydration of ethylene, which is a by-product of the petroleum industry. The alcohol so produced is 95% alcohol and 5% water. The water that remains in this mixture cannot be removed by ordinary distillation procedures, since 95% ethanol is *a constant boiling mixture* (**azeotrope**).[4] This 95% alcohol is used in chemical laboratories as a solvent. If 100% alcohol is needed, special procedures must be employed to prepare it. One method is to dry the alcohol over calcium oxide for several hours. The calcium oxide has a great affinity for water but will not combine with ethanol. The remaining alcohol is then distilled. Alcohol so prepared is known as *absolute* (100%) alcohol.

Ethanol is used as a solvent for perfumes, medicinal formulations (tinctures), lacquers, varnishes, and shellacs. Alcohol denatures enzymes in bacteria (see Section 13.9-b) and for this reason it is widely used as an antiseptic in mouthwashes and aerosol disinfectants. Some brands of rubbing alcohol consist of a 70% solution of ethanol in water. Ethanol is also employed in the synthesis of other organic compounds. When ethanol is used for such industrial purposes it is not subject to a federal tax. (The tax is greater than $20/gallon in most states.) To insure the legitimate use of tax-free alcohol, the government treats it with certain additives that make it unfit to drink. Such alcohol is known as *denatured* alcohol. Common denaturants are methanol and 2-propanol. These compounds are toxic but do not interfere with the solvent properties of the alcohol.

The use of ethanol (and methanol) as a blend component of gasoline began in the United States in 1979. Favorable results have been achieved with mixtures

---

[3]Contrary to popular belief, alcohol is a depressant of the central nervous system, not a stimulant. The illusionary stimulation comes from its effects of depressing brain areas responsible for judgment. The resulting lack of inhibitions and restraints may cause one to feel "stimulated." People under the influence of alcohol suffer from diminished control of their judgment and actions and hence may endanger themselves and/or others (especially if they are driving). In the United States a blood–alcohol concentration of 0.1% (100 mg of alcohol in 100 mL of blood) is legal evidence of intoxication. A concentration of 0.5–1% leads to coma and death.

[4]An azeotropic mixture is a constant-boiling mixture of two components. The two components are present in a fixed composition. A mixture of 95% ethanol and 5% water is an azeotropic mixture that boils at 78°C.

Chapter 6 · Alcohols, Phenols, Ethers, and Thiols

containing 80–90% unleaded gasoline and 10–20% alcohol. Since ethanol can be obtained easily from grain (and methanol from coal, wood chips, or municipal refuse), this represents one method of augmenting our dwindling fuel supplies. Brazil, which has an abundance of sugarcane, is producing cars that are designed to burn only alcohol as a motor fuel. In addition to its use as a fuel extender, alcohol also increases the octane rating of the gasoline with which it is blended.

c. $CH_3CHOHCH_3$ (Isopropyl Alcohol, 2-Propanol, Rubbing Alcohol)

Isopropyl alcohol is a colorless, bitter-tasting liquid. It is manufactured from propene by hydration with sulfuric acid. A 70% solution of isopropyl alcohol is commonly referred to as rubbing alcohol. It has a high vapor pressure, and its rapid evaporation from the skin produces a cooling effect that helps to lower fever. Rubbing alcohol, like ethanol, denatures bacterial enzymes and is used as an antiseptic to cleanse the skin before a blood sample is taken or an injection is given. Isopropyl alcohol is toxic when ingested, but, unlike methanol, it is not absorbed through the skin. Much of the production of isopropyl alcohol is for the manufacture of acetone (see Section 7.2) and to introduce the isopropyl group into organic molecules.

d. CH_2OHCH_2OH (Ethylene Glycol, 1,2-Ethanediol, Glycol)

Ethylene glycol is a colorless, viscous, sweet liquid that is water soluble and denser than water. When taken internally, it is toxic because glycol is oxidized in the body to oxalic acid (see Section 8.8-d). Approximately 50 deaths each year occur from the accidental ingestion of antifreeze. The most important of the dihydroxy alcohols, it is used as a permanent radiator antifreeze because of its high boiling point (198°C) and its effectiveness in lowering the freezing point of water. A solution of 60% ethylene glycol in water will not freeze until the temperature falls to $-49°C$ ($-56°F$). The color of most of the commercial antifreezes is due to the additives. Ethylene glycol is also used in the manufacture of polyester (Dacron) fiber and film (Mylar) used in tapes for recorders and computers.

e. $CH_2OHCHOHCH_2OH$ (Glycerol, 1,2,3-Propanetriol, Glycerin)

This important trihydroxy alcohol is also viscous, sweet, heavier than water, and water soluble. It is, however, nontoxic, and is an ingredient of all fats and oils. Glycerol is relatively inexpensive because it is obtained as a by-product of the alkaline hydrolysis of fats during the soap-making process (see Section 12.8). Glycerol has widespread industrial use; the following list is not all-inclusive.

1. Preparation of hand lotions and cosmetics.
2. In inks, tobacco products, and plastic clays to prevent dehydration (glycerol is hygroscopic).
3. Constituent of glycerol suppositories.
4. Sweetening agent and solvent for medicines.
5. Lubricant, especially in chemical laboratories.

6. Production of plastics, surface coatings, and synthetic fibers.

7. Production of nitroglycerin.

The last glycerin derivative deserves special attention. Nitroglycerin was first prepared in 1846 by the Italian chemist Sobrero, who was lucky that he lived to tell of his discovery. Sobrero mixed nitric acid and glycerin, and the ensuing explosion nearly killed him. It was not until 15 years later that the famed Swedish chemist and inventor Alfred Nobel discovered a method to prepare and transport the compound safely. He found that a type of diatomaceous earth, *Kieselguhr,* was capable of absorbing the nitroglycerin, thus rendering it insensitive to shock. The mixture was referred to as dynamite. Unless exploded by means of a percussion cap or by a detonator containing mercuric fulminate ($Hg[ONC]_2$), it was quite stable. The production of dynamite was a major breakthrough in the building industry. With dynamite, the construction of canals, dams, highways, mines, and railroads became much easier. Its use as a weapon in warfare greatly disturbed the Nobel family, however, and after Alfred Nobel's death (1896) a trust fund was set up for an annual award for an outstanding contribution toward peace. (A trust fund was also set up to offer annual awards for contributions in the fields of chemistry, physics, economics, literature, and medicine or physiology.) It is surprising that a compound that is so sensitive to shock is also used as a drug to relieve angina pectoris. Nitroglycerin is administered either in tablet form (mixed with nonactive ingredients) or as an alcoholic solution (spirit of glyceryl trinitrate). Nitroglycerin functions in a manner similar to isopentyl nitrite (i.e., as a vasodilator). It relaxes cardiac muscle and smooth muscle in the smaller blood vessels, thus increasing the supply of blood (and hence oxygen) to the heart.

The reaction for the preparation of nitroglycerin is analogous to the nitration of a monohydroxy alcohol, but three molecules of nitric acid are required for every molecule of glycerin. The glycerin must be very pure to ensure stability of the product.

$$
\begin{array}{ccc}
\begin{array}{c}
\text{H} \\
| \\
\text{H}-\text{C}-\text{OH} \\
| \\
\text{H}-\text{C}-\text{OH} \\
| \\
\text{H}-\text{C}-\text{OH} \\
| \\
\text{H}
\end{array}
+ \; 3\,\text{HONO}_2
\xrightarrow[10-20^\circ\text{C}]{\text{H}_2\text{SO}_4}
\begin{array}{c}
\text{H} \\
| \\
\text{H}-\text{C}-\text{ONO}_2 \\
| \\
\text{H}-\text{C}-\text{ONO}_2 \\
| \\
\text{H}-\text{C}-\text{ONO}_2 \\
| \\
\text{H}
\end{array}
+ \; 3\,\text{H}_2\text{O}
\end{array}
$$

Glycerol (Glycerin)   Glycerol trinitrate (Nitroglycerin)

Nitroglycerin is a pale yellow, oily liquid that detonates upon slight impact. The explosive power arises from the extremely rapid conversion of a small volume of liquid into a large volume of hot, expanding gases.

$$4\,\text{C}_3\text{H}_5(\text{ONO}_2)_3 \longrightarrow 6\,\text{N}_2(g) + 12\,\text{CO}_2(g) + 10\,\text{H}_2\text{O}(g) + \text{O}_2(g)$$

The reaction produces temperatures of over 3000°C and pressure above 2000 atm.

The explosion wave caused by such temperatures and pressures is enormous, accounting for the damaging effect of the detonation.

Phenols

Phenols are compounds that *have a hydroxyl group directly bonded to an aromatic ring*. Like alcohols, phenols have a C—OH grouping, but because the carbon is part of an aromatic system, the properties of phenols are different than the properties of alcohols. Phenols are important synthetic intermediates in the chemical industry. Additionally, the phenolic group occurs in many diverse naturally occurring substances such as alkaloids, amino acids, antibiotics, coenzymes, marijuana, etc.

6.6 Nomenclature of Phenols

Phenol (hydroxybenzene) is the name given to the parent compound. The IUPAC system has accepted this name, and compounds containing an aromatic hydroxy group are referred to as phenol derivatives.

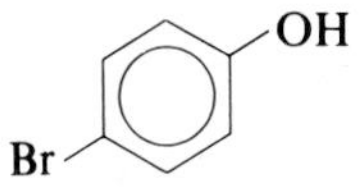

p-Bromophenol 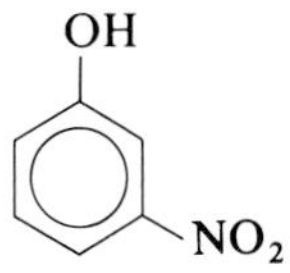   m-Nitrophenol 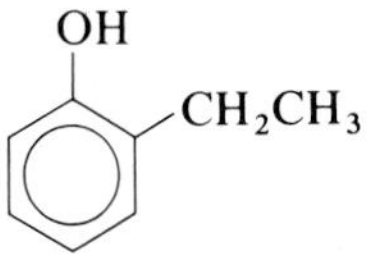   o-Ethylphenol

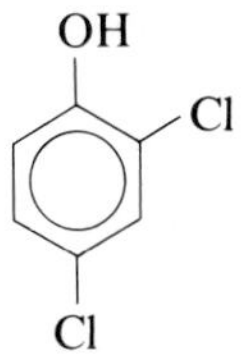

2,4-Dichlorophenol 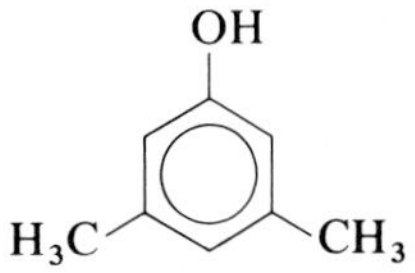   3,5-Dimethylphenol 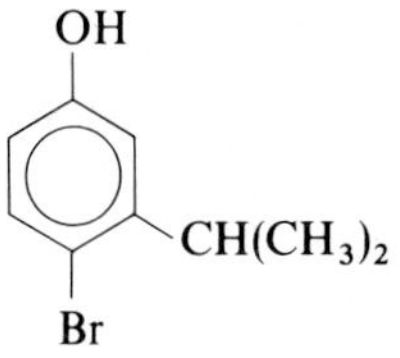   4-Bromo-3-isopropylphenol

Because many of the phenols were named before there was a systematic nomenclature, common names are still prevalent. The methyl derivatives of phenols are called *cresols*. They are important ingredients in the wood preservative creosote. Lysol is a combination of an aqueous soap solution and the cresol isomers. It is used in hospitals as a disinfectant.

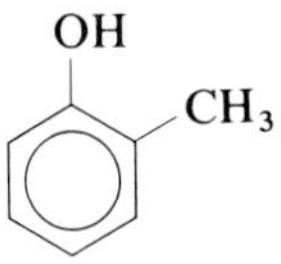

o-Cresol 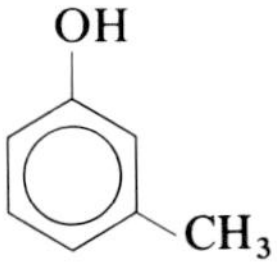   m-Cresol 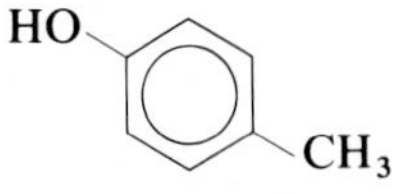   p-Cresol

Unlike the three cresol isomers, the three dihydroxybenzenes have been given individual names. They all have commercial significance, and two of them are important components of biochemical molecules. Hydroquinone occurs in a co-enzyme (page 451) and catechol forms part of the structure of certain neurotransmitters termed catecholamines (see Section 9.12-d).

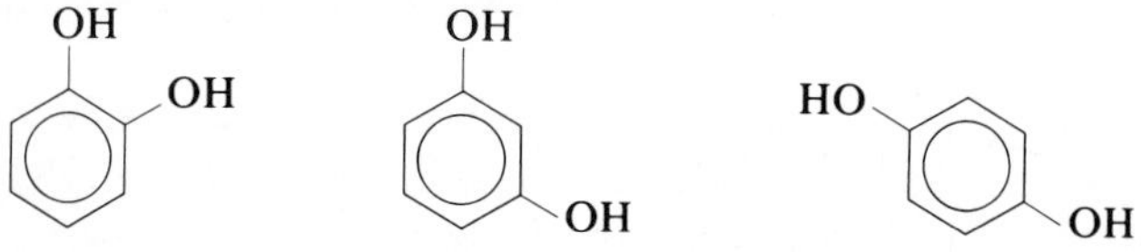

Catechol · Resorcinol · Hydroquinone

6.7 *Properties of Phenols*

Phenol is a white crystalline compound, with a distinctive ("hospital smell") odor. Its freezing point is 41°C and its boiling point is 182°C. Phenol is slightly soluble in cold water (7 g/100 mL) and is miscible in hot water. Phenol is a powerful germicide, and dilute solutions are sometimes used to disinfect hospital equipment, floors, and walls.

One of the major distinguishing characteristics between phenols and alcohols is that *phenols are acidic* ($K_a = 10^{-10}$), whereas *alcohols are neutral*. Phenols can be neutralized by strong bases, but they are too weakly acidic to react with weak bases such as aqueous sodium bicarbonate.

$$CH_3CH_2OH + NaOH \longrightarrow \text{No reaction (recall Section 6.4-a)}$$

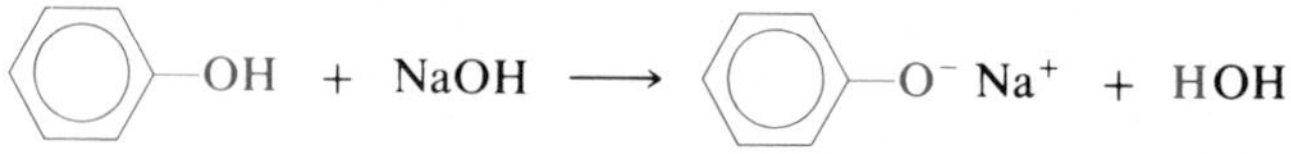

Sodium phenoxide

Phenols are more acidic than alcohols because the negative charge of the phenoxide ion is delocalized by resonance throughout the benzene ring. No such delocalization of charge can take place for the alkoxide ion of alcohols. (See Section 8.6 for further discussion about the effects of molecular structure on the acidity of molecules.)

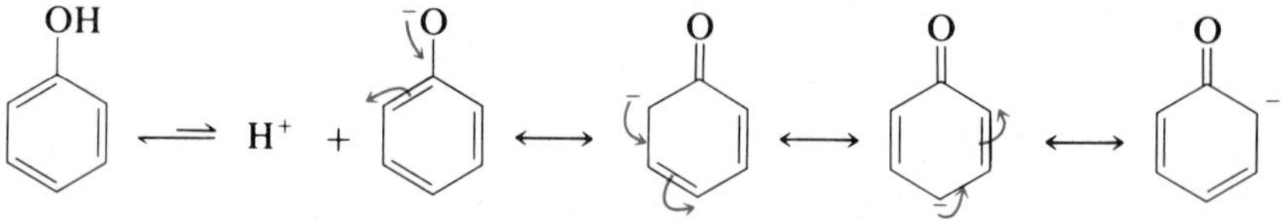

Electron-attracting groups attached to the benzene ring enhance the acidity of phenols. 2,4,6-Trinitrophenol (picric acid), in which all three nitro groups help in delocalizing the negative charge, is almost as strong as hydrochloric acid. It readily reacts with weak bases to form salts.

Picric acid   + NH₄OH ⟶   Ammonium picrate   + HOH

Picric acid is a yellow, crystalline compound that was once used as a synthetic dye. It is used as an antiseptic in burn ointments (see Section 13.9-e). Solid picric acid is an explosive when heated (above 300°C) rapidly. Care should be taken in storing this chemical in the laboratory. The salt ammonium picrate is also an explosive, but it is used only when a high explosive particularly insensitive to shock is required.

6.8 Oxidation of Phenols

Phenols, unlike benzene, are readily oxidized. Oxidation in neutral or basic solutions results in a mixture of products arising from coupling reactions. If the oxidation occurs under acidic conditions, colored substances called *quinones* are formed.

p-Benzoquinone

Although the reaction shown does occur, side products are numerous. A better preparative method for quinones is the oxidation of hydroquinones, which are more easily oxidized.

Hydroquinone

This reaction is the basis for the use of hydroquinone as a photographic film developer. The exposed silver ions in the film are reduced by hydroquinone to black silver metal (and hydroquinone is oxidized to benzoquinone).

The ease of oxidation of phenolic compounds is of considerable importance in biological systems. The bombardier beetle uses an enzymatically catalyzed oxidation of hydroquinone by hydrogen peroxide (H_2O_2) as a defense mechanism. The hydroquinone and hydrogen peroxide are stored in one part of the gland and the oxidative enzyme is located in a sort of vestibular compartment. When attacked, the beetle injects the hydroquinone–hydrogen peroxide mixture into the vestibule, where the ensuing exothermic reaction sprays hot, highly irritating quinone out of the abdomen.

Recall that we mentioned the antioxidant properties of BHT (page 25) and vitamin E (see Section 12.16-c). Both of these compounds are phenols that react preferentially with oxygen and thus protect cell membranes from being oxidized. Vitamin K (Section 12.16-d) also contains a phenolic group. Although its mechanism is uncertain, vitamin K is necessary for the proper clotting of blood.

In Section 16.9 we shall discuss an extremely important system that operates within the mitochondria of most living organisms. This is the respiratory chain, which involves a sequence of oxidation–reduction reactions that result in energy being made available to the cell. One of the compounds in the respiratory chain is a quinone called coenzyme Q or ubiquinone (because it is so ubiquitous in cells). An important aspect of the hydroquinone–quinone reaction is its reversibility. We shall see that in one of the steps of the respiratory chain, the quinone form of coenzyme Q is reduced by accepting two electrons and two hydrogens. Then in a subsequent step, the hydroquinone passes the electrons and hydrogens on to the next carrier in the chain and is oxidized back to the quinone (see page 452)

(yellow)   (colorless)

6.9 Uses of Phenols

We mentioned that phenol and the cresols are widely used in commercial disinfectants. Phenol was first used in 1867 by Joseph Lister to disinfect patients during surgery and to clean surgical instruments. Phenol is still used for the latter purpose, but it is no longer used as an antiseptic because of its extreme toxicity. Because of its acidity, phenol is highly caustic to the skin. (Phenol spills should be washed away with alcohol.) Like methanol, phenol can be absorbed through the skin, and internally its effects may be lethal.

Although phenol has been replaced as an antiseptic by a number of its derivatives, it is still used as a standard of comparison for other germicides. The *phenol coefficient* is an arbitrary measure of the effectiveness of a given germicide. It is obtained by dividing the concentration of phenol needed to destroy a certain organism (*Staphylococcus aureus* or *Salmonella typhosa*) in a given time by the concentration of some other germicide needed to destroy the same organism in the same time. For example, a 1% germicide solution that destroys an organism in the same time as that required by a 5% phenol solution, is assigned a phenol coefficient of 5. Examine the labels on disinfectants, deodorants, soaps, throat lozenges, and muscle rubs in your home. You will find that most of them contain a phenolic derivative as an active ingredient.

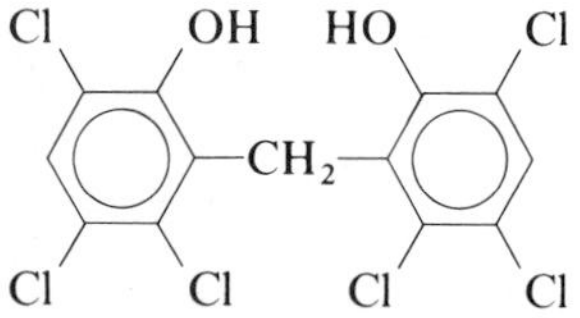

Hexachlorophene is a polysubstituted phenol derivative that is a potent antiseptic (phenol coefficient of 125). It had been used in dilute form in the manufacture of toothpastes, deodorants, and germicidal soaps (such as PhisoHex). PhisoHex was once routinely used in hospitals to clean newborn babies. However, in 1972, studies showed that hexachlorophene is absorbed through the skin of rats, that it caused brain damage in monkeys, and that it may cause certain genetic defects. Consequently, hexachlorophene was withdrawn from the commercial market. The FDA recommended that its use be restricted to preventing the spread of staphylococcus infections, which cannot be treated in any other way.

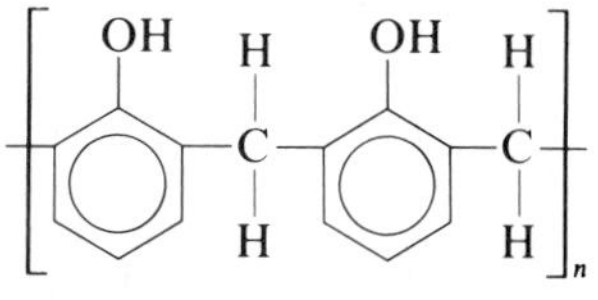

The most important commercial reaction of phenols is the condensation with formaldehyde to yield phenolic polymers (Bakelite). Bakelite was used initially as an electrical insulator and later used to form plastic parts for the automotive and radio industries. Phenol is also used in the production of dyes (e.g., phenolphthalein), drugs, and antiseptic throat lozenges such as Sucrets.

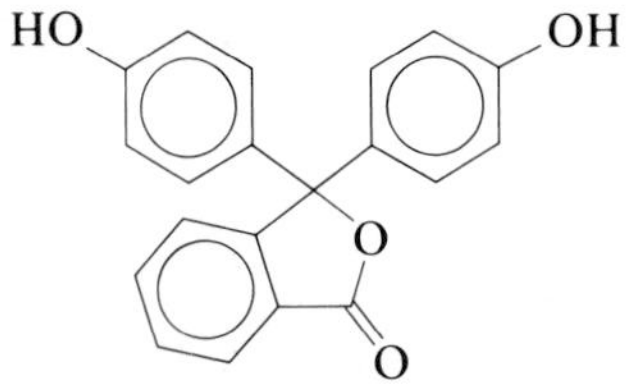

Hexachlorophene

Bakelite

Phenolphthalein

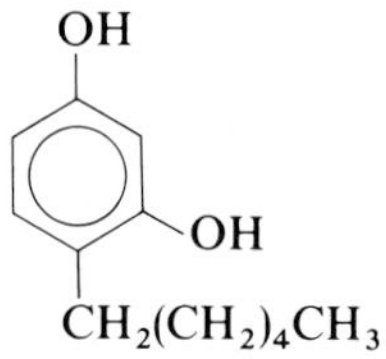

4-Hexylresorcinol
(Sucrets)

Ethers

Ethers may be considered to be *derivatives of water in which both hydrogen atoms have been replaced by alkyl or aryl groups.* They may also be considered as derivatives of an alcohol in which the hydroxyl hydrogen has been replaced by an organic group.

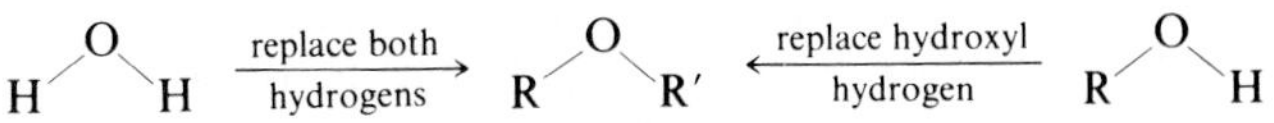

The general formula for the ethers is **R—O—R′**. When both **R—** groups are the same, the compound is a *symmetrical* ether. When **R** and **R′** are different, the ether is said to be an *unsymmetrical* ether.

$$R-O-R \qquad H_3C-O-CH_3 \qquad\qquad R-O-R' \qquad H_3C-O-CH_2CH_3$$

Symmetrical ether   Unsymmetrical ether

TABLE 6.4 Nomenclature of Ethers

Molecular Formula	Structural Formula	Common Name	IUPAC Name
C_2H_6O	CH_3-O-CH_3	(Di)methyl ether	Methoxymethane
C_3H_8O	$CH_3-O-CH_2CH_3$	Ethyl methyl ether	Methoxyethane
$C_4H_{10}O$	$CH_3CH_2-O-CH_2CH_3$	(Di)ethyl ether	Ethoxyethane
$C_4H_{10}O$	$CH_3-O-\overset{1}{C}H_2\overset{2}{C}H_2\overset{3}{C}H_3$	Methyl propyl ether	1-Methoxypropane
$C_4H_{10}O$	$CH_3-O-\overset{2}{C}H\overset{3}{C}H_3$ $\underset{\overset{1}{C}H_3}{}$	Isopropyl methyl ether	2-Methoxypropane
C_7H_8O	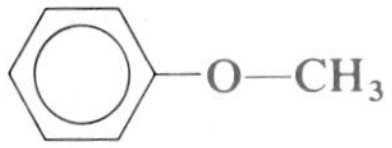 ⬡—O—CH$_3$	Methyl phenyl ether (Anisole)	Methoxybenzene
$C_8H_{10}O$	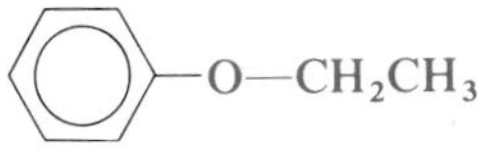 ⬡—O—CH$_2$CH$_3$	Ethyl phenyl ether (Phenetole)	Ethoxybenzene
$C_8H_{10}O$	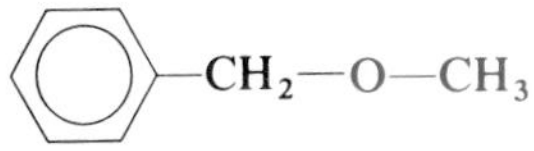 ⬡—CH$_2$—O—CH$_3$	Benzyl methyl ether	Methoxyphenylmethane
$C_{12}H_{10}O$	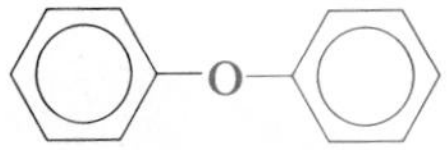 ⬡—O—⬡	(Di)phenyl ether	Phenoxybenzene

6.10 Nomenclature of Ethers

Ethers are generally known by their common names rather than by the IUPAC names. Usually the organic groups are named first in alphabetical order and are followed by the word ether. In naming the symmetrical ethers the prefix di- is unnecessary, although it is often employed to avoid confusion.

Under the IUPAC system one organic group is considered to have been derived from a parent hydrocarbon, and the other group *plus the oxygen* is referred to as an *alkoxy* substituent, —OR. Names of alkoxy substituents are obtained by dropping the *-yl* from the alkyl name and adding *-oxy*. By convention the larger group is considered to have been derived from the parent hydrocarbon and the smaller group is part of the alkoxy substituent. Thus a methyl group attached to the oxygen (OCH_3) of an ether would be called a *methoxy* group, and an ethyl group bonded to an oxygen (OCH_2CH_3), an *ethoxy* group, and so on. Table 6.4 illustrates the rules of nomenclature for several simple and mixed ethers.

6.11 Preparation of Ethyl Ether

The aliphatic ethers, like the alkyl halides, are strictly synthetic compounds. By far the most common and most important ether to be synthesized is ethyl ether. Its preparation has already been mentioned in Section 6.4-b. A general scheme, including significant reaction conditions, for the interactions of ethene, ethanol, and sulfuric acid is

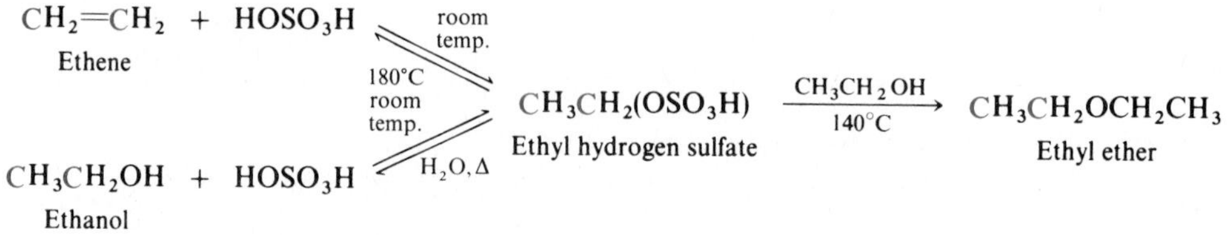

The commercial manufacture of ethyl ether involves continuous addition of ethanol to a mixture of sulfuric acid and ethanol that is maintained at a temperature of 140°C. At this temperature, the ethyl ether distills out of the reaction mixture as it is formed. This is one reaction in which the maintenance of reaction temperature is critical. At temperatures below 130°C, the reaction between acid and alcohol is so slow that unreacted alcohol will distill out of the mixture. At temperatures above 150°C, the ethyl hydrogen sulfate intermediate will decompose to yield ethene and sulfuric acid.

6.12 Physical Properties of Ethers

The carbon–oxygen bond of the ethers is in many respects similar to the carbon–carbon single bond of the alkanes. The ethers, therefore, are colorless, highly volatile, very flammable, and less dense than water. However, they exhibit a

TABLE 6.5 Comparison of Boiling Points of Alkanes, Alcohols, and Ethers

Formula	Name	MW	BP (°C)
$CH_3CH_2CH_3$	Propane	44	−42
CH_3OCH_3	Methyl ether	46	−25
CH_3CH_2OH	Ethyl alcohol	46	78
$CH_3CH_2CH_2CH_2CH_3$	Pentane	72	36
$CH_3CH_2OCH_2CH_3$	Ethyl ether	74	35
$CH_3CH_2CH_2CH_2OH$	n-Butyl alcohol	74	117

solubility in water that is similar to that of the primary alcohols of comparable molecular weight. (For example, methyl ether and ethanol are completely soluble in water, whereas ethyl ether and 1-butanol are soluble to the extent of 8 g/100 mL of water.) The bent shape of the ether molecule allows the hydrogens of the water molecules to form hydrogen bonds with the unshared electrons on the ether oxygen.

Since ethers lack a hydrogen–oxygen bond, intermolecular hydrogen bonding between ethers is impossible, and the ethers, like the alkanes, are unassociated. This is evidenced by their boiling points, which are similar to those of the alkanes of comparable molecular weight and much lower than their isomeric alcohols. Table 6.5 lists comparative boiling points for some ethers, alkanes, and alcohols.

6.13 *Chemical Properties of Ethers*

The chemical behaviors of the ethers and the alkanes are strikingly similar. Ethers are characterized by their inertness toward most of the common reagents under ordinary conditions. They are resistant to attack by metallic sodium, by oxidizing and reducing agents, and by strong bases. However, they are soluble in concentrated sulfuric acid, a characteristic that distinguishes ethers from alkanes. Solubility in sulfuric acid is a result of the basic character of the ether molecule. Ethers are protonated by acids to form oxonium ions.

$$C_2H_5-\ddot{O}-C_2H_5 \;+\; \text{conc } H_2SO_4 \longrightarrow C_2H_5-\overset{..+}{\underset{\underset{H}{}}{O}}-C_2H_5 \; HSO_4^-$$

Diethyloxonium hydrogen sulfate

The ether molecule can be cleaved by heating with strong acids. The acids commonly used for this reaction are hydrobromic (HBr) and hydriodic (HI). If an excess of the acid is employed, the alcohol formed initially is converted to an alkyl halide.

$$CH_3CH_2-O-CH_2CH_3 \;+\; HI \xrightarrow{\Delta} CH_3CH_2OH \;+\; CH_3CH_2I$$

$$CH_3CH_2-O-CH_2CH_3 \xrightarrow[\Delta]{\text{excess HI}} 2\,CH_3CH_2I \;+\; HOH$$

6.14 Uses and Hazards of Ethyl Ether

a. Ether as an Anesthetic

Ethyl ether was first used during surgery in 1842, and for many years it was the most generally used of all the volatile anesthetics. An **anesthetic** is *any substance that causes unconsciousness or insensitivity to pain.* Ether produces unconsciousness by depressing the activity of the central nervous system. It is easy to administer, has very little effect upon the rate of respiration or the blood pressure, and is a good muscle relaxant. The major disadvantages of ether are its flammability, its irritating effects on the respiratory passages, and the occurrence of postanesthetic nausea and vomiting. Many hospitals are now using nitrous oxide (N_2O) and/or halogenated compounds (2–3% in oxygen) such as halothane (Fluothane, $CF_3CHClBr$) or enflurane (Ethrane, CHF_2OCF_2CHFCl) as general anesthetics. They are more potent than ethyl ether, nonflammable, and less irritating to the respiratory passages. Caution must also be exercised in their use, since halothane has been shown to cause a severe liver toxicity in a small percentage of people given this anesthetic. The major disadvantage of halothane is its potential for causing respiratory and cardiovascular depression. The concentration of halothane required to maintain surgical anesthesia is only slightly lower than that needed to cause respiratory arrest.

Modern surgical practice has moved away from a single anesthetic. Generally, the patient is given a strong sedative such as Pentothal by injection to produce unconsciousness. The gaseous anesthetic is then administered through a mask (in a stream of nitrous oxide and oxygen gases) to provide insensitivity to pain and to retain unconsciousness.

Several theories have been presented to explain the biochemical activity of anesthetics. The one property they all share in common is their high solubility in fats and low solubility in water. One theory suggests that anesthetics interfere with electrical activity in nerves by dissolving in the fatty components of the brain cells. In some manner anesthetics inhibit a vital metabolic process, and/or they interrupt oxygen transfer reactions.

There are four stages associated with anesthesia. In the first stage, there is a high degree of **analgesia** (incapacity to feel pain) without loss of consciousness. The second stage is characterized by loss of consciousness, by irregular deep respiration, and often by active and purposeless muscular activity. In the third stage, the necessary muscular relaxation is achieved and surgery can be performed. The fourth stage is characterized by paralysis of the respiratory muscles. Death will follow unless corrective action is initiated. A principal function of the anesthesiologist is to determine the level of anesthesia by monitoring the alterations in respiration and the degree of muscular relaxation in the abdomen.

b. Other Uses

Ether is extremely hygroscopic. A freshly opened can of ether will immediately pick up about 1–2% water from the moisture in the air. Special techniques (dry box, nitrogen atmosphere) are employed in handling the compound when reaction conditions call for anhydrous ether.

Ether serves as an excellent extraction medium[5] because (1) it is a good solvent for many organic substances, (2) it is not miscible with water, (3) it is relatively unreactive, (4) it does not dissolve most inorganic substances, and (5) it has a low boiling point and can be removed easily from the extract by low-temperature distillation. Ether is also employed in separating constituents of foodstuffs. Lipids are soluble in ether and thus can be separated from carbohydrates and proteins, which are ether insoluble.

Ether is also used to start cold gasoline or diesel engines. A cloth soaked with ether is placed near the air intake of the engine. The ether evaporates and its vapors are drawn into the cylinders. Since ether ignites much more readily than gasoline or diesel fuel, the engine usually turns over.

Two hazards must be avoided when working with ether. First, since ether vapor is heavier than air, it remains at the bottom of a room and has the nasty property of "rolling" long distances across bench tops—hence, the necessity of avoiding open flames in a laboratory in which ether is being used. Ether fires *cannot* be extinguished with water, since ether floats on top of water. A carbon dioxide extinguisher is most often used.

Second, by reaction with oxygen in the air, ethers form organic peroxides (e.g., $CH_3CH_2OCH(OOH)CH_3$) when left standing in open containers. These peroxides remain with the residue when ether is distilled or evaporated to dryness. Solid peroxides are dangerous because they are unstable and explode readily. For this reason, ether from a previously opened bottle should be treated with a reducing agent, such as ferrous sulfate, before it is used. Commercial absolute ether contains 0.05 ppm of sodium diethyldithiocarbamate, $(C_2H_5)_2NCS_2^- Na^+$, as an antioxidant.

---

[5] The extracting power of ether has made it the solvent of choice among cocaine users. "Free-basing" first involves separating the cocaine from other substances by extracting it into ether. Enormous quantities of ether are used in extracting cocaine from the coca plant. Drug enforcement officials in the United States have been able to apprehend some cocaine manufacturers by keeping track of the shipments of large quantities of ether.

Sulfur is located under oxygen in the same group of the periodic table; therefore, we should expect to find a similarity among sulfur and oxygen compounds. A principal difference between the two series of compounds arises from the fact that sulfur exhibits a greater number of oxidation states than does oxygen. (For example, in SO_2 and H_2SO_4 the oxidation number of S is $+4$ and $+6$, respectively.) The sulfur analogs of alcohols and ethers, and their oxidation products, are of considerable importance to the organic chemist and biochemist. Sulfur atoms are encountered in such diverse compounds as amino acids, coenzymes, detergents, dyes, mustard gas, penicillin, saccharin, and sulfa drugs. Examples of some organosulfur compounds are represented in Figure 6.3.

The prefix *thio* in the name of a compound indicates that a sulfur atom has been substituted for an oxygen atom. **Thiols,** also known as thioalcohols (**RSH**), are the *monoalkyl derivatives of hydrogen sulfides*, just as alcohols are the derivatives of water. In biochemistry, the —**SH** group is termed a *sulfhydryl group*. We shall see that many proteins contain sulfhydryl groups, and the poisoning effects of arsenic and mercury salts are due to their ability to inactivate these proteins by precipitation.[6]

FIGURE 6.3 Examples of sulfur-containing compounds.

$CH_3CH_2CH_2$—S—H
Propanethiol
(found in onions)

CH_2=$CHCH_2$—N=C=S
Allyl isothiocyanate
(active principle in mustard)

$ClCH_2CH_2$—S—CH_2CH_2Cl
Bis(β-chloroethyl) sulfide
(Mustard gas—used as a
warfare agent in World War I)

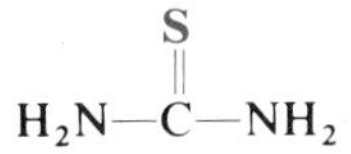

H_2N—C—NH_2 (with S double bonded to C)
Thiourea
(organic reactant)

CH_2=$CHCH_2$—S—CH_2CH=CH_2
Allyl sulfide
(chief constituent of oil of garlic)

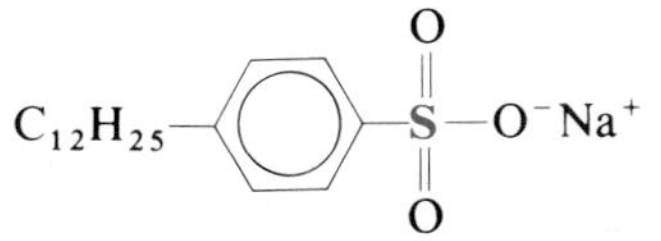

$C_{12}H_{25}$—⟨benzene ring⟩—S—O^-Na^+ (with two O double bonded to S)
Sodium dodecylbenzenesulfonate
(a detergent)

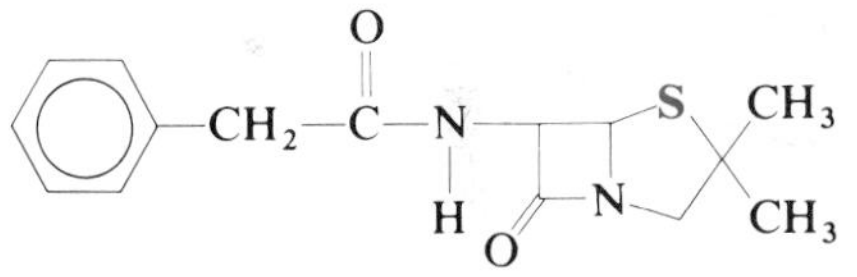

Penicillin G structure with CH_2—C(=O)—N(H)... S, CH_3, CH_3, N, O
Penicillin G
(an antibiotic)

[6] Lewisite (an organoarsenic compound) is a poisonous war gas, named after its developer, W. Lee Lewis. Lewisite exerts its lethal effects by combining with the sulfhydryl groups of proteins. The 1,2-dithiol derived from glycerin is called BAL (British antilewisite). BAL was developed in England during World War II. It is used as an antidote in heavy metal poisoning because the sulfhydryl groups present (in BAL) form strong complexes with arsenic, mercury, silver, and lead, thus preventing these metals from bonding to proteins.

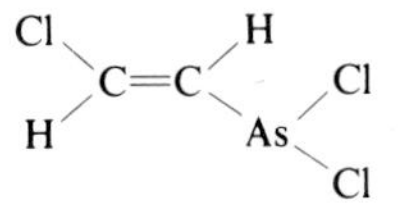

Structure of Lewisite: Cl, H on C=C, then Cl, As, Cl
Lewisite

Structure of BAL: H—C—C—C—H with SH SH OH
BAL

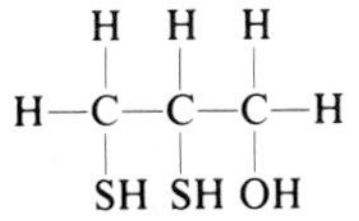

6.15 Nomenclature of Thiols and Thioethers

In the IUPAC system the suffix -thiol is added to the name of the parent hydrocarbon to indicate the —SH substituent. In the common system, the —SH group is designated by adding the word *mercaptan*. This older terminology, which is not often used, derives from the ability of these organosulfur compounds to form insoluble mercury salts (Latin, *mercurium captans,* mercury seizing).

Thioethers are commonly named as sulfides if the organic groups are simple, or as alkylthioalkanes (IUPAC) for complex structures. The *alkylthio substituent* (**RS—**) is analogous to alkoxy (RO—) in ether nomenclature.

$$CH_3CH_2SH$$
Ethanethiol
(Ethyl mercaptan)

$$\overset{\overset{\text{SH}}{|}}{CH_3CHCH_3}$$
2-Propanethiol
(Isopropyl mercaptan)

$$CH_3SCH_3$$
Methylthiomethane
(Dimethyl sulfide)

$$CH_3CH_2SCH_2CH_2CH_2CH_3$$
1-Ethylthiobutane
(Ethyl butyl sulfide)

6.16 Properties of Thiols

The thiols are characterized by their obnoxious odors, and this discourages their use as laboratory reagents. Propanethiol is released from freshly chopped onions, and a mixture of three sulfur-containing compounds is responsible for the scent of skunks (Figure 6.4). The odor of ethanethiol can be detected at a dilution of 1 part per 50 billion parts of air. Small amounts of thiols are added to natural gas so that the odor will make it obvious if a gas leak has occurred. The thiols boil at lower temperatures than the corresponding alcohols (e.g., ethanethiol boils at 37°C compared to 78°C for ethanol). This is indicative of the absence of hydrogen bonding between individual thiol molecules. However, weak hydrogen bonding does occur between the thiol hydrogen and the oxygen of water molecules; hence, the low molecular weight thiols are slightly soluble in water.

One of the chief distinguishing features of the thiols is that they are more acidic than their corresponding alcohols, just as H_2S is a stronger acid than H_2O. The difference in acidities is attributed to a difference in bond strengths; that is, less energy is required to disrupt an S—H bond than an O—H bond.[7] (K_a of ethanethiol is 10^{-10} compared to a K_a of 10^{-16} for ethanol.) As a result, the sodium and potassium salts of thiols can be prepared from aqueous solutions of sodium or potassium hydroxide, whereas the corresponding salts of the alcohols require the use of sodium or potassium metal.

$$CH_3CH_2SH + NaOH \longrightarrow CH_3CH_2S^- Na^+ + HOH$$
Sodium ethyl
mercaptide

This reaction is of practical use in the petroleum industry. Small amounts of organosulfur compounds are often obtained in crude petroleum oils, and these are converted to thiols during the refining process. The thiols, of course, are objectionable because of their odor and the polluting nature of their combustion products (SO_2 and SO_3). They are removed from

[7] For this reason thiols readily react with free radicals. They are termed *radical scavengers* and are utilized to interrupt radical chain processes by combining with radicals. Recall (Section 2.5) that 2-aminoethanethiol was shown to inhibit free radical reactions in mice.

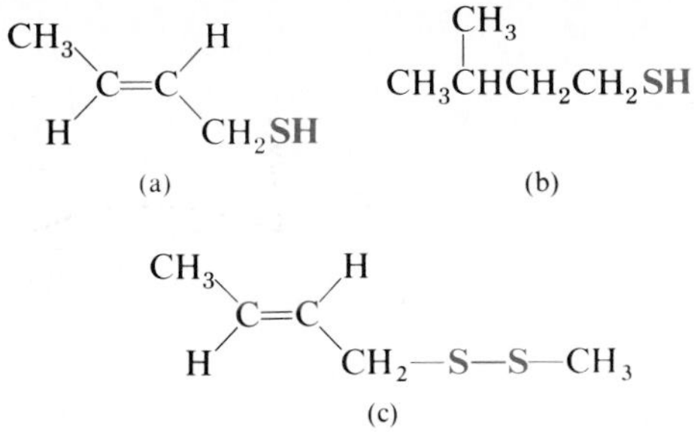

CH₃ structures (a), (b), (c) — skunk scent compounds

FIGURE 6.4 Skunk scent consists of three sulfur-containing compounds: *trans*-2-butene-1-thiol (a), 3-methyl-1-butanethiol (b), and methyl-1-(*trans*-2-butenyl) disulfide (c)—in a ratio of 4:3:3, respectively.

the petroleum products at the refineries by treatment with aqueous alkali. This process is referred to as *scrubbing*. The thiols are converted to nonvolatile, petroleum-insoluble salts that dissolve into the alkaline medium.

Another way that thiols and alcohols differ is in their behavior toward oxidizing agents. Recall that when alcohols are oxidized, it is the carbon bonded to the hydroxyl group, rather than the oxygen, that is oxidized (Section 6.4-d). When thiols are oxidized, the sulfur atom itself undergoes oxidation; the carbon chain remains unchanged. The product that is formed is dependent on the type of oxidizing agent employed. Mild oxidizing agents such as air, ferric chloride, hydrogen peroxide, or iodine yield a disulfide. **Disulfides** are *compounds containing S—S bonds;* they can be considered to be the sulfur analogs of peroxides. The disulfide bond is weak and is easily reduced to regenerate the thiol.

$$2 \ CH_3-S-H + I_2 \longrightarrow CH_3-S-S-CH_3 + 2 \ HI$$

Dimethyl disulfide

$$CH_3-S-S-CH_3 \xrightarrow{\text{LiAlH}_4} 2 \ CH_3-S-H$$

The interconversion of thiols and disulfides by oxidation–reduction reactions is utilized to a considerable extent in cellular metabolism. Two of the amino acids, cysteine and cystine, undergo such a transformation. Cystine is the oxidized form of cysteine, and its disulfide bond plays an extremely important role in the stabilization of the three-dimensional structure of proteins (see Section 13.5-c). The reduction reaction in which the disulfide bond of cystine is cleaved to yield two molecules of cysteine has been utilized in the laboratory to separate polypeptides that are covalently cross-linked by one or more disulfide bonds. The sulfhydryl compound, β-mercaptoethanol (HOCH₂CH₂SH), is the reagent of choice for this procedure.

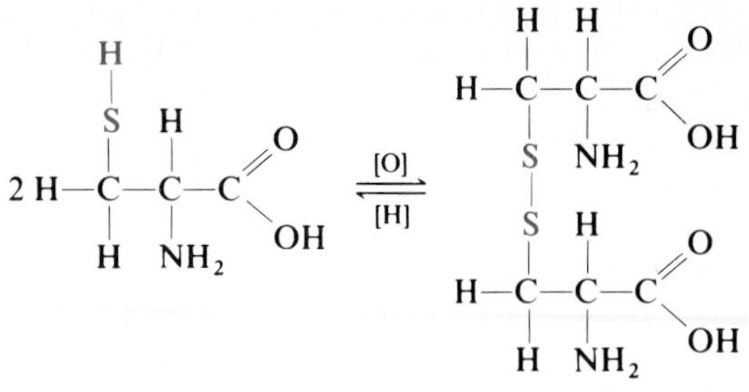

Cysteine ⇌ Cystine

Another example is the oxidation and reduction of the coenzyme lipoic acid. Lipoic acid functions to catalyze certain oxidation reactions (pages 443 and 446), and in the process it is reduced to the corresponding dithiol. Oxidation at a later stage by other components of the cell regenerates the lipoic acid.

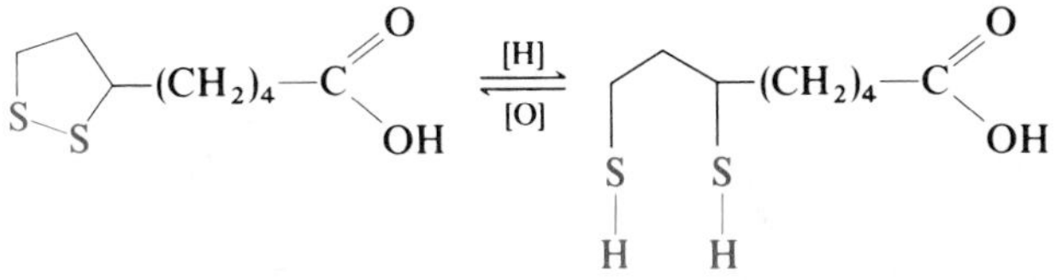

Lipoic acid ⇌ Reduced form of lipoic acid

Sulfides are the sulfur analogs of ethers (i.e., thioethers). Thioethers, unlike ethers, oxidize readily. The oxidation of thioethers with hydrogen peroxide can yield either a *sulfoxide* or a *sulfone*.

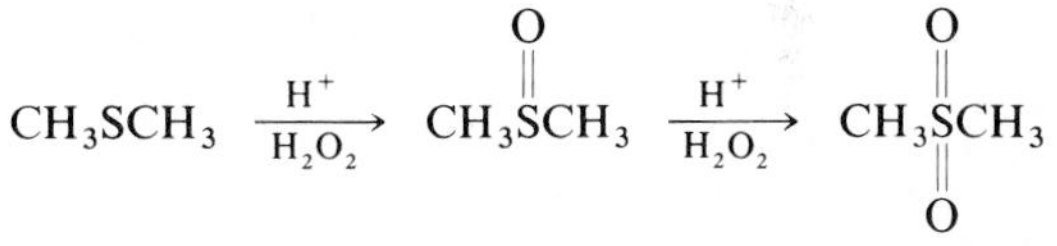

Dimethyl sulfide → Dimethyl sulfoxide → Dimethyl sulfone

Dimethyl sulfoxide (DMSO) is obtained in large quantities as a by-product in the sulfite pulping process for paper manufacture. It is a highly polar solvent that is miscible with both polar and nonpolar compounds (including water, alcohol, acetone, benzene, chloroform, ether). For this reason it has become a useful solvent for a multitude of organic reactions, and it has been utilized in place of water as a solvent in the study of certain enzymes. DMSO has been proclaimed to have potent analgesic properties in the relief of a wide range of painful symptoms (strains and sprains, arthritis, severe head injuries, and certain skin infections). Although this use is still highly controversial, the ability of DMSO to penetrate the skin readily does present a danger to those working with it. That is, since DMSO dissolves almost all solutes, it can serve to transport these solutes into the body if a DMSO solution is accidentally spilled on the skin.

Thiols react with carboxylic acids to produce sulfur analogs of esters, called *thioesters*.

General Equation

$$\underset{\text{Acid}}{R-\overset{\overset{\displaystyle O}{\|}}{C}-OH} + \underset{\text{Thiol}}{R'SH} \underset{}{\overset{H^+}{\rightleftharpoons}} \underset{\text{Thioester}}{R-\overset{\overset{\displaystyle O}{\|}}{C}-S-R'} + HOH$$

Specific Equation

$$\underset{\text{Acetic acid}}{CH_3\overset{\overset{\displaystyle O}{\|}}{C}OH} + \underset{\text{Ethanethiol}}{CH_3CH_2SH} \overset{H^+}{\rightleftharpoons} \underset{\text{Ethyl thioacetate}}{CH_3\overset{\overset{\displaystyle O}{\|}}{C}SCH_2CH_3} + HOH$$

The fact that water is eliminated in this reaction, and not hydrogen sulfide, is consistent with the observation that in esterification reactions the hydroxyl group derives from the acid (see Section 8.10).

Coenzyme A (abbreviated CoA or CoA-SH—see Table 14.2) is a complex sulfur-containing coenzyme that plays a central role in the metabolism of carbohydrates, fats, and proteins as well as in reactions associated with the transmission of nerve impulses and vision. The acetyl derivative of coenzyme A, called *acetyl-CoA*, occurs in all living systems. The thioester linkage between the acetyl group and coenzyme A is a high-energy bond (that is, it has a high energy of hydrolysis, about -7500 cal/mole). Because of its high degree of reactivity, acetyl-CoA is a carrier of acetyl groups in the cell just as ATP is a carrier of phosphate groups (see Section 8.14). Although the biochemical formation of acetyl-CoA involves a series of reactions, we can conceptually visualize the process as a simple thioesterification (see Sections 16.8 and 17.2).

$$\underset{\text{Acetic acid}}{CH_3-\overset{\overset{\displaystyle O}{\|}}{C}-OH} + \underset{\text{Coenzyme A}}{H-S-CoA} \longrightarrow \longrightarrow \longrightarrow$$

$$\underset{\text{Acetyl-CoA}}{CH_3-\overset{\overset{\displaystyle O}{\|}}{C}\sim S-CoA} + HOH$$

Exercises

6.1 Draw and name all the alcohols corresponding to the molecular formula $C_5H_{12}O$. Label each isomer as primary, secondary, or tertiary.

6.2 Draw and name all the isomeric ethers having the molecular formula $C_5H_{12}O$.

6.3 Draw and name all the aliphatic isomers of molecular formula $C_4H_{10}O$.

6.4 Draw and give IUPAC names for all the aliphatic isomers of molecular formula $C_4H_{10}S$.

6.5 Name the following alcohols and classify each as 1°, 2°, or 3°.
   (a) $CH_3CH_2CH_2CHOHCH(CH_3)_2$
   (b) $CH_3CHOHCH_2CHCl_2$
   (c) $(CH_3)_2COHCBr_2CH_3$
   (d) $HOCH_2CH_2CH=CHCH_2I$
   (e) [cyclohexane structure with CH_3 and OH]
   (f) [cyclopentane structure with OH and OH]

   (g) $CH_2BrCHOHCBr_2CH_3$
   (h) $HOCH_2CH_2CH_2OH$
   (i) $CH_3CH_2CH_2COH(CH_2CH_3)_2$
   (j) $CH_3CHOHCH_2CHClCH_2CH_2CH(CH_3)_2$
   (k) $I-\langle\bigcirc\rangle-CH_2OH$
   (l) $HO-\langle\bigcirc\rangle-CH_2CH_3$

6.6 Name the following compounds.
   (a) [benzene ring with OH, O_2N, NO_2 substituents]
   (b) [benzene ring with OH and Br substituents]
   (c) $CH_3OCH_2CH_3$
   (d) $CH_3CH_2CH_2OCH(CH_3)_2$

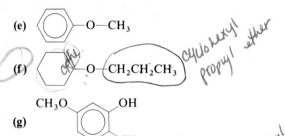

(e) ⬡—O—CH₃

(f) ⬡—O—(CH₂CH₂CH₃)   *cyclohexyl propyl ether*

(g) CH₃O—⬡(OH)(CH₃)

(h) ⬡ with CH₂CH₃ and OH

(i) $CH_3CH(OCH_3)CH(CH_3)CH(OCH_3)CH_3$

(j) $CH_3CH(OCH_2CH_3)CH=CHCH_2CH_3$

(k) $CH_3CHSHCH_3$ *2 propanethiol*

(l) $CH_3CHSHCH_2CHBr_2$

(m) $CH_3SC(CH_3)_3$

(n) $(CH_3CH_2CH_2)_2S$

6.7 **(a)** Menthol is one of the ingredients in mentholated cough drops and nasal sprays. It produces a cooling, refreshing sensation when rubbed on the skin and so is used in shaving lotions and cosmetics. What is its IUPAC name?

(CH₃)₂CH—[cyclohexane ring with OH and CH₃]   Menthol

(CH₃)₂CH—[benzene ring with OH and CH₃]   Thymol

(b) It is interesting to note that the aromatic equivalent of menthol is thymol, the flavoring constituent of thyme. Give two IUPAC names for thymol.

6.8 Write structural formulas for the following alcohols.
(a) 3-methyl-3-heptanol
(b) 4,5-dimethyl-3-hexanol
(c) 2-methyl-2-pentanol
(d) 3-buten-1-ol
(e) 1,2-cyclobutanediol
(f) 2-methyl-1-phenyl-1-butanol
(g) 2-bromo-2-iodo-1-butanol
(h) 3-chloro-5-methoxy-1-hexanol
(i) 1,3-pentanediol
(j) 3-propyl-2-hexanol
(k) 2-cyclopenten-1-ol
(l) *m*-hydroxybenzyl alcohol

6.9 Write structural formulas for the following compounds.
(a) *p*-chlorophenol
(b) *m*-cresol

(c) catechol
(d) *p*-ethoxytoluene
(e) *tert*-butyl *n*-butyl ether
(f) *m*-toluenethiol
(g) ethylthiobenzene
(h) 2,4-dibromophenol
(i) 3,5-dimethylphenol
(j) hydroquinone
(k) 1-methoxy-1-phenyl-2-propanol
(l) isopropyl phenyl ether
(m) *p*-methylbenzenethiol
(n) 1-methylthiohexane

6.10 Coniferyl alcohol, 3-(4-hydroxy-3-methoxyphenyl)-2-propen-1-ol, and sinapyl alcohol, 3-(4-hydroxy-3,5-dimethoxyphenyl)-2-propen-1-ol, are two components of lignin, the woody tissue of plants and a part of the dietary fiber. Draw the structural formulas of these two compounds.

6.11 In addition to the compounds mentioned in Section 6.14, other halogenated hydrocarbons have been used as general anesthetics. For example, methoxyflurane (Penthrane), 2,2-dichloro-1,1-difluoroethyl methyl ether, and isoflurane (Forane), 1-chloro-2,2,2-trifluoroethyl difluoromethyl ether. Draw their structural formulas.

6.12 There is something wrong with each of the following names. Give the structure and correct name for each compound.
(a) 2,3-dimethyl-4-pentanol
(b) 3-methyl-2-propyl-1-butanol
(c) 2,4-butanediol
(d) 4-ethoxyhexane
(e) 3,5-dimethoxycyclohexane
(f) 2-penten-4-ol
(g) *o*-bromo-*m*-nitrophenol
(h) 4,5-dichlorocyclopentanol
(i) 1-propylthioethane
(j) 5-iodothiophenol

6.13 Benzyl alcohol and *p*-cresol are isomers. Write the formulas for each. Compare their solubilities in **(a)** water and **(b)** an aqueous solution of NaOH.

6.14 Arrange the compounds of each group in order of decreasing solubility in water (at room temperature).
(a) $n\text{-}C_4H_9OH$ CH_3OH $n\text{-}C_4H_9Br$

(b) $n\text{-}C_4H_9OH$ $CH_2OHCH_2CH_2CH_2OH$
     $CH_2OHCHOHCH_2OH$

(c) $(CH_3)_3CCH_2OH$ $CH_3(CH_2)_4OH$
     $(CH_3CH_2)_2CHOH$

(d) $(CH_3)_2O$ $CH_3(CH_2)_5OH$ ⬡—OH

(e) $(CH_3CH_2)_2O$ $(CH_3)_3COH$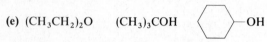

(f) $(CH_3)_2S$ CH_3CH_2SH CH_3SH

6.15 Arrange the compounds of each group in order of decreasing boiling point.

(a) C_4H_{10} CH_3OH C_3H_7OH

(b) $n\text{-}C_4H_9OH$ $(CH_3)_3COH$ [benzene ring]—OH

(c) $n\text{-}C_3H_7OH$ $CH_2OHCH_2CH_2OH$
    $CH_2OHCHOHCH_2OH$

(d) $n\text{-}C_3H_7OH$ $CH_3OCH_2CH_3$ $(CH_3CH_2)_2O$

(e) [benzene ring]—OH   HO—[benzene ring]—OH

    [benzene ring]—O—CH_3

(f) CH_3CH_2SH CH_3CH_2OH CH_3—S—CH_3

6.16 Arrange the compounds of each group in order of decreasing acidity.

(a) CH_3CH_2OH $n\text{-}C_4H_9OH$ CH_3OCH_3

(b) HOH   [benzene ring]—OH   [cyclohexane ring]—OH

(c) [benzene ring]—OCH_3 H_3C—[benzene ring]—OH

    [benzene ring]—CH_2OH

(d) CH_3OH CH_3SH CH_3SCH_3

6.17 Discuss the factors responsible for the differences in water solubility and in boiling points between methyl chloride and methyl alcohol.

6.18 1-Butanol and diethyl ether are isomers. Predict which of these isomers would have (a) the higher vapor pressure and (b) the higher boiling point.

6.19 Assume that you have a mixture of 1-octanol and phenol. How would you separate this mixture?

6.20 Ethyl alcohol, like rubbing alcohol, is often used for sponge baths. What property of alcohols makes them useful for this purpose?

6.21 Write equations for the following preparations.
(a) 1-butanol from an alkyl halide
(b) 2-butanol from an alkene
(c) potassium phenoxide from phenol

(d) 2-propanol from 1-propanol
(e) *n*-pentane from 2-pentanol

6.22 In the preparation of ethyl ether from alcohols, why is it so critical to control the reaction temperature between 130° and 150°C?

6.23 Write equations summarizing the action of sulfuric acid on ethanol. Be sure to include all pertinent reaction conditions.

6.24 Complete the following equations, giving only the organic products (if any).

(a) [cyclohexane ring]—Br + NaOH $\xrightarrow{H_2O}$

(b) $CH_3CHOHCH_3$ $\xrightarrow{KMnO_4}$

(c) $CH_3CH_2CH_2CH_2OH$ + NaOH $\longrightarrow$

(d) $CH_3CH_2CH_2OH$ + Na $\longrightarrow$

(e) $CH_3CH_2CH_2CH_2OH$ + PI_3 $\longrightarrow$

(f) [cyclopentane ring]—OH + H_2SO_4 $\longrightarrow$

(g) [cyclobutane ring]—CH=CH_2 $\xrightarrow[H_2SO_4]{HOH}$

(h) [benzene ring]—CH_2OH + $SOCl_2$ $\longrightarrow$

(i) $CH_3CHOHCH(CH_3)_2$ + HBr $\longrightarrow$

(j) [benzene ring]—OH + $NaHCO_3$ $\longrightarrow$

(k) Cl—[benzene ring]—OH + NaOH $\longrightarrow$

(l) HO—[benzene ring]—OH $\xrightarrow{K_2Cr_2O_7}$

(m) [benzene ring with OH]—OH $\xrightarrow{Na_2Cr_2O_7}$

(n) $CH_3CH_2OCH_2CH_3$ + Na $\longrightarrow$

(o) $CH_3OCH_2CH_2CH_3$ + NaOH $\longrightarrow$

(p) CH_3CH_2—O—CH_2CH_3 + HI $\xrightarrow{\Delta}$

(q) $CH_3CH_2CH_2SH$ + NaOH $\longrightarrow$

(r) CH_3SH $\xrightarrow{KMnO_4}$

6.25 Arrange the following alcohols in order of decreasing reactivity toward the Lucas reagent.

(a) $CH_3CH_2CH_2CHOHCH_3$
$CH_3CH_2CH_2CH_2CH_2OH$
$CH_3CH_2C(CH_3)_2OH$

(b)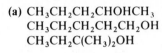

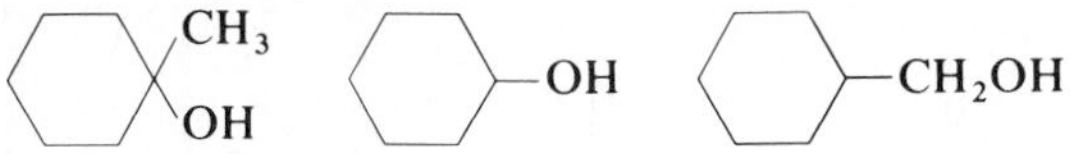

6.26 *o*-Benzoquinone can be prepared by the oxidation of catechol with Ag_2O in ether. Write an equation for this reaction.

6.27 Give the major and minor products that are obtained on dehydration of the following alcohols.
(a) $CH_3CHOHCH_2CH_2CH_3$
(b) $CH_3CH_2CHOHCH_2CH_3$ $CH_3CH_2C=CH_2Ch_3 + H_2O$
(c) $(CH_3)_2COHCH_2CH_3$

(d) 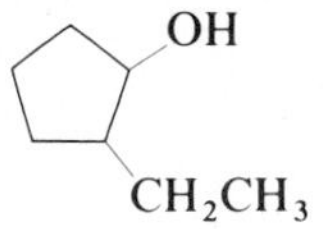
OH
CH₂CH₃

(e) 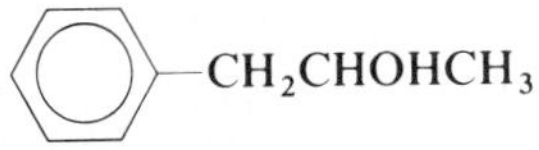 ⬡—$CH_2CHOHCH_3$

(f) $(CH_3)_2CHCHOHCH_3$
(g) $CH_3CBr_2C(CH_3)_2OH$
(h) $(CH_3)_3CCHOHCH_3$

(i) 
CH₃
OH
$◇-CH_3 + H_2O$

(j) 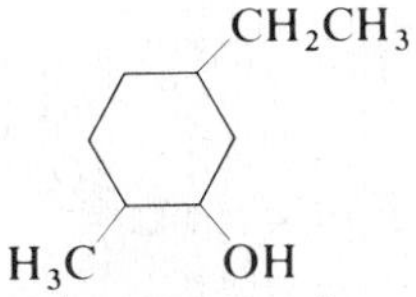
CH₂CH₃
H₃C   OH

6.28 Give the name and one use for each of the following compounds.
(a) CH_3OH
(b) CH_3CH_2OH
(c) $CH_3CHOHCH_3$
(d) CH_2OHCH_2OH
(e) $CH_2OHCHOHCH_2OH$

(f) 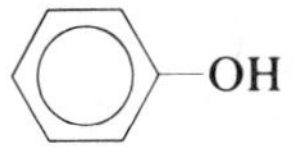 ⬡—OH

(g) 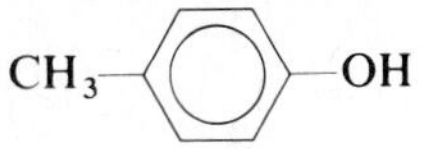 CH_3—⬡—OH

(h) 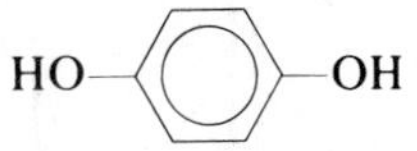 HO—⬡—OH

(i) $CH_3CH_2OCH_2CH_3$
(j) CH_3CH_2SH

6.29 What is meant by the following terms?
(a) hydrogen bonding
(b) absolute alcohol
(c) denatured alcohol
(d) 86 proof
(e) phenol coefficient
(f) tincture
(g) anesthetic
(h) analgesic
(i) alkoxide ion
(j) oxonium ion
(k) azeotrope
(l) gasohol
(m) vasodilator
(n) germicide
(o) mercaptan
(p) DMSO

6.30 Dehydration of a butanol isomer with concentrated sulfuric acid at $180°C$ gave 2-butene as the only product. What was the structural formula of the alcohol?

6.31 1.6 g of an alcohol reacted with excess potassium to liberate 560 mL of hydrogen gas at STP. What is the molecular weight of the alcohol?

6.32 4.6 g of ethanol was treated with excess sulfuric acid. Assuming no side reactions, how many liters (at STP) of ethene could be produced?

6.33 Compound X has the molecular formula C_4H_8. It adds hydrogen bromide to form compound Y which then reacts with AgOH to form a tertiary alcohol. What are the formulas of X and Y?

6.34 An ether containing 68.2% carbon, 13.6% hydrogen, and 18.2% oxygen reacts with hydrogen iodide to yield methanol and a compound containing 69.0% iodine. Treatment of this latter compound with a dilute potassium hydroxide solution yields a secondary alcohol. What is the structural formula of the original ether?

7

Aldehydes and Ketones

The study of aldehydes and ketones brings us for the first time to a consideration of an extremely important chemical grouping, the *carbon–oxygen double bond,*

$$\overset{\text{O}}{\underset{\|}{—C—}}.$$ This unit is referred to as the **carbonyl group.** It is common to both the aldehydes and ketones, as well as to several other classes of organic compounds. The carbonyl carbon of the aldehyde is a terminal carbon and is always bonded to

a hydrogen, $\overset{\text{O}}{\underset{\|}{—C—}}\text{H.}$ The carbonyl carbon of the ketone, on the other hand, is never a terminal carbon since it must be bonded to two other carbon atoms. The

general formula for an aldehyde is $\mathbf{R}—\overset{\text{O}}{\underset{\|}{\mathbf{C}}}—\mathbf{H}$, and the general formula for

a ketone is $\mathbf{R}—\overset{\text{O}}{\underset{\|}{\mathbf{C}}}—\mathbf{R'}$. (Note that the general formula for an aldehyde may be abbreviated to RCHO but not to RCOH since the latter implies an alcohol.)

115

Aldehydes and ketones are thus two different classes of organic compounds. The rationale for including both in the same chapter is based upon their similarity of properties. Most reagents that attack carbonyl groups will react with both aldehydes and ketones. Any observable differences in their properties, with the exception of ease of oxidation, are differences of degree rather than kind.

7.1 Nomenclature of Aldehydes and Ketones

The generic name aldehyde has a significant meaning. The compound can be obtained by the removal of hydrogen from an alcohol, and the name is derived from the two words, *alcohol dehydr*ogenation. Both the common and IUPAC names of the aldehydes are frequently used; however, usage of common names predominates for the lower homologs of the family. The common names are taken from the names of the acids into which the aldehydes are convertible by oxidation.

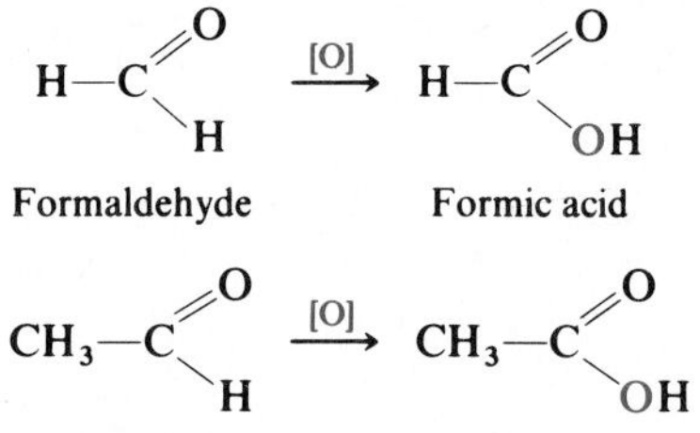

$$H-C\overset{O}{\underset{H}{\big<}} \quad \xrightarrow{[O]} \quad H-C\overset{O}{\underset{OH}{\big<}}$$

Formaldehyde   Formic acid

$$CH_3-C\overset{O}{\underset{H}{\big<}} \quad \xrightarrow{[O]} \quad CH_3-C\overset{O}{\underset{OH}{\big<}}$$

Acetaldehyde   Acetic acid

Naming of the aldehydes according to the IUPAC system follows the previously established rules. The longest chain of carbons is taken to be the one containing the carbonyl group. The final -*e* of the alkane name is replaced by the suffix -*al,* which designates the functional group of the aldehydes. The aldehyde functional group takes precedence over all the groups discussed in previous chapters. Thus, the carbonyl carbon is always considered to be carbon number 1, and it is unnecessary to designate this group by number. Examples of aldehyde nomenclature are provided in Table 7.1.

The common names of the ketones, like those of the ethers, are obtained by naming each of the attached organic groups and adding the word ketone. The lowest homolog of the ketone series is universally called acetone rather than dimethyl ketone. (It was first prepared from acetic acid.) According to the IUPAC system, the longest continuous chain containing the carbonyl group designates the parent hydrocarbon. The -*e* ending of the alkane name is replaced by the suffic -*one,* signifying that the compound is a ketone. Because the carbonyl carbon of a ketone can never be a terminal carbon, the position of the functional group must be indicated by the appropriate number. The chain is numbered so that the carbonyl carbon is assigned the lowest possible number. In cyclic ketones, it is understood that the carbonyl carbon is number one. Table 7.2 illustrates the nomenclature for some of the ketones.

TABLE 7.1 Nomenclature of Aldehydes

Molecular Formula	Condensed Structural Formula	Common Name	IUPAC Name
CH_2O	$H-C{\overset{O}{\underset{H}{}}}$	Formaldehyde	Methanal
C_2H_4O	$CH_3C{\overset{O}{\underset{H}{}}}$	Acetaldehyde	Ethanal
C_3H_6O	$CH_3CH_2C{\overset{O}{\underset{H}{}}}$	Propionaldehyde	Propanal
C_4H_8O	$CH_3CH_2CH_2C{\overset{O}{\underset{H}{}}}$	Butyraldehyde	Butanal
C_4H_8O	$CH_3CHC{\overset{O}{\underset{H}{}}}$ $\ \ \ \ CH_3$	Isobutyraldehyde	2-Methylpropanal
C_4H_6O	$CH_3CH{=}CHC{\overset{O}{\underset{H}{}}}$	Crotonaldehyde	2-Butenal
$C_5H_{10}O$	$CH_3CH_2CH_2CH_2C{\overset{O}{\underset{H}{}}}$	Valeraldehyde	Pentanal
$C_5H_{10}O$	$CH_3CHCH_2C{\overset{O}{\underset{H}{}}}$ $\ \ \ \ CH_3$	Isovaleraldehyde	3-Methylbutanal **not** 2-Methylbutanal
C_7H_6O	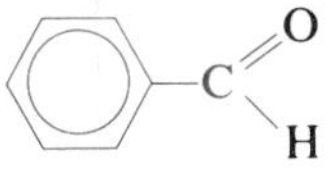 benzene ring $-C{\overset{O}{\underset{H}{}}}$	Benzaldehyde	Benzaldehyde (Phenylmethanal)
C_8H_8O	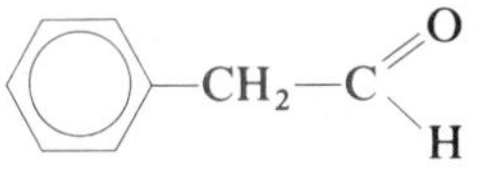 benzene ring $-CH_2-C{\overset{O}{\underset{H}{}}}$	Phenylacetaldehyde	Phenylethanal
$C_8H_8O_3$	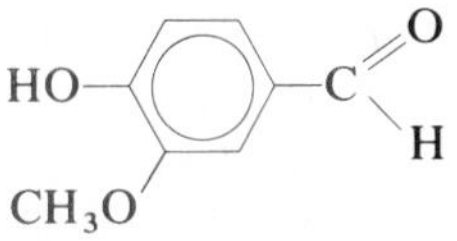 $HO-$ ring $-C{\overset{O}{\underset{H}{}}}$, CH_3O-	Vanillin (odor of vanilla)	4-Hydroxy-3-methoxybenzaldehyde
C_9H_8O	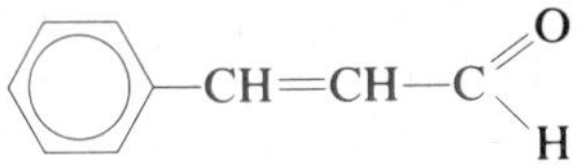 benzene ring $-CH{=}CH-C{\overset{O}{\underset{H}{}}}$	Cinnamaldehyde (odor of cinnamon)	3-Phenyl-2-propenal

TABLE 7.2 Nomenclature of Ketones

Molecular Formula	Condensed Structural Formula	Common Name	IUPAC Name
C_3H_6O	$CH_3\overset{\displaystyle O}{\overset{\|}{C}}CH_3$	Acetone (dimethyl ketone)	Propanone
C_4H_8O	$CH_3\overset{\displaystyle O}{\overset{\|}{C}}CH_2CH_3$	Ethyl methyl ketone	Butanone
C_4H_6O	$CH_2{=}CH\overset{\displaystyle O}{\overset{\|}{C}}CH_3$	Methyl vinyl ketone	3-Buten-2-one **not** 1-Buten-3-one
$C_5H_{10}O$	$CH_3CH_2\overset{\displaystyle O}{\overset{\|}{C}}CH_2CH_3$	Diethyl ketone	3-Pentanone
$C_5H_{10}O$	$CH_3CH_2CH_2\overset{\displaystyle O}{\overset{\|}{C}}CH_3$	Methyl propyl ketone	2-Pentanone
$C_5H_{10}O$	$CH_3\underset{\displaystyle CH_3}{\overset{\displaystyle}{C}}H\overset{\displaystyle O}{\overset{\|}{C}}CH_3$	Isopropyl methyl ketone	3-Methyl-2-butanone **not** 2-Methyl-3-butanone
$C_6H_{10}O$	cyclohexanone structure =O	Cyclohexanone	Cyclohexanone
C_8H_8O	phenyl–$\overset{\displaystyle O}{\overset{\|}{C}}$–$CH_3$	Acetophenone (Methyl phenyl ketone)	Phenylethanone
$C_{13}H_{10}O$	phenyl–$\overset{\displaystyle O}{\overset{\|}{C}}$–phenyl	Benzophenone (Diphenyl ketone)	Diphenylmethanone

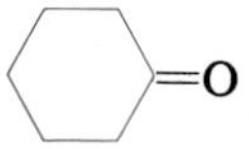

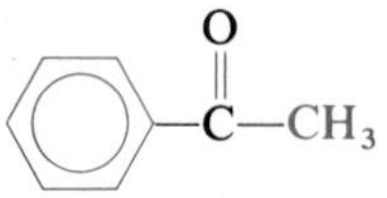

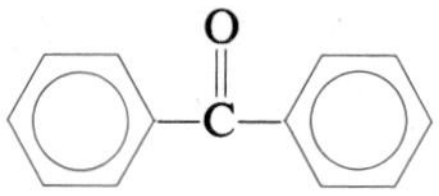

7.2 Preparation of Aldehydes and Ketones

In Section 6.4-d the oxidation of primary and secondary alcohols to form aldehydes and ketones, respectively, was mentioned. However, in aqueous solutions the product aldehyde forms a hydrate that is further oxidized to a carboxylic acid. Therefore, organic solvents are used in the preparation of aldehydes from alcohols. The reagent of choice is chromic oxide in combination with pyridine, methylene chloride, and HCl. In organic solvents chromium(VI) ions are mild oxidizing agents that can oxidize primary alcohols to aldehydes, without further oxidizing the aldehydes to acids. We shall see in Chapter 16 that the enzyme-catalyzed oxidation of alcohols to aldehydes and ketones is of great significance in biological systems.

The oxidation of a primary alcohol results in an aldehyde that contains two fewer hydrogen atoms than the original alcohol. Hence this reaction can also be considered as a dehydrogenation reaction. The student should be aware that oxidation and dehydrogenation both involve the loss of electrons. This loss of electrons may be manifested by either the addition of oxygen or the removal of hydrogen. Similarly, reduction, which may appear to involve only a gain of hydrogen or a loss of oxygen, implies a gain of electrons. The great majority of oxidation–reduction reactions occur through a transfer of electrons, although the balanced net equation shows only a change of atoms within molecules. Since these reactions occur widely in organic chemistry, and because they are of utmost importance in biochemical metabolism, we show the pertinent half-reactions for some of the oxidation–reduction reactions that we study to demonstrate the electron transfers that are involved. The general equation for the oxidation (dehydrogenation) of primary alcohols to produce aldehydes is

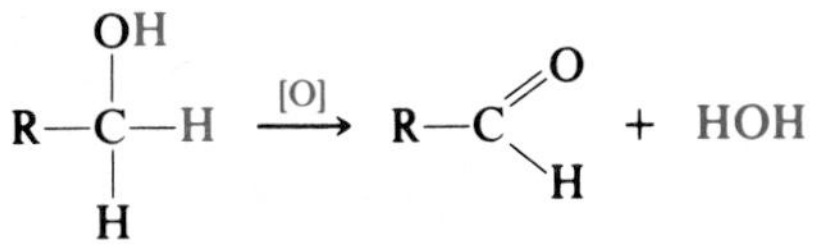

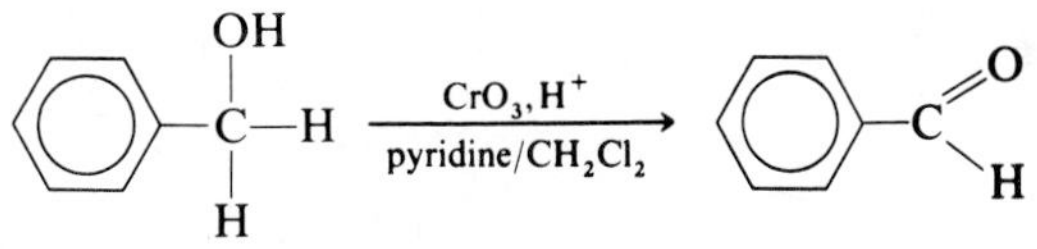

$$
\underset{H}{\overset{OH}{R-\overset{|}{\underset{|}{C}}-H}} \xrightarrow{[O]} R-C\underset{H}{\overset{O}{\diagup}} + HOH
$$

Benzyl alcohol   Benzaldehyde

$$
3\,CH_3CH_2CH_2CH_2OH + CrO_3 + 3\,HCl \xrightarrow[CH_2Cl_2]{pyridine} 3\,CH_3CH_2CH_2C\underset{H}{\overset{O}{\diagup}} + CrCl_3 + 3\,H_2O
$$

1-Butanol   Butanal

Oxidation

$$
CH_3CH_2CH_2CH_2OH \longrightarrow CH_3CH_2CH_2C\underset{H}{\overset{O}{\diagup}} + 2\,H^+ + 2\,e^-
$$

Reduction

$$
6\,H^+ + CrO_3 + 3\,e^- \longrightarrow Cr^{3+} + 3\,H_2O
$$

Like the aldehydes, ketones are obtained by the oxidation of an alcohol. However, the alcohol must be a secondary alcohol, and no special reagents are needed since the ketone is not susceptible to further oxidation. Although many oxidants are used, the most commonly employed are chromium(VI) compounds and sulfuric acid.

General Equation

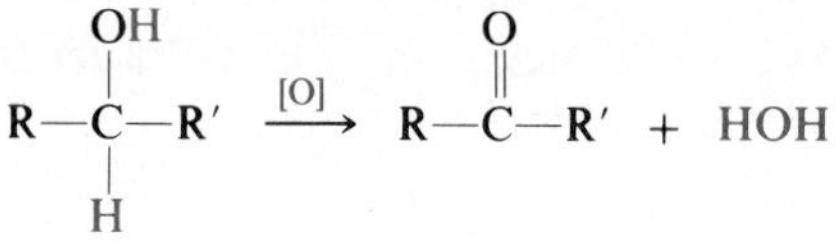

$$\underset{\underset{H}{\overset{OH}{|}}}{R-C-R'} \xrightarrow{[O]} \underset{\overset{O}{\parallel}}{R-C-R'} + HOH$$

Specific Equations

$$3\,\underset{\underset{|}{\overset{OH}{|}}}{CH_3CHCH_3} + K_2Cr_2O_7 + 4\,H_2SO_4 \longrightarrow 3\,\underset{\overset{O}{\parallel}}{CH_3CCH_3} + K_2SO_4 + Cr_2(SO_4)_3 + 7\,H_2O$$

Isopropyl alcohol   (orange)   Acetone   (green)

$$\bigcirc\!\!-OH + 2\,CrO_3 + 3\,H_2SO_4 \longrightarrow \bigcirc\!\!=O + Cr_2(SO_4)_3 + 6\,H_2O$$

Cyclohexanol   Cyclohexanone

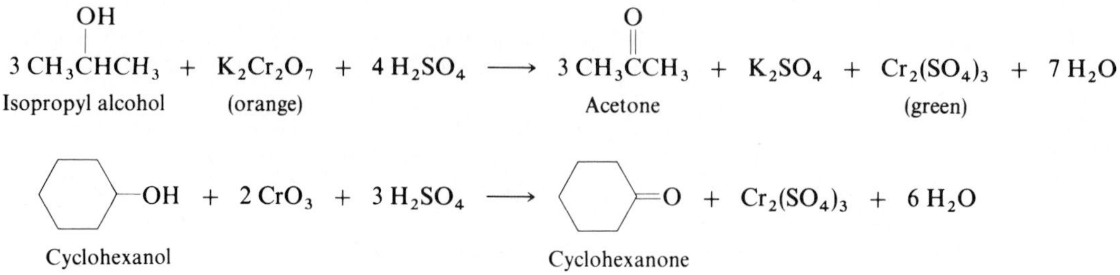

As we shall see in Chapters 16 and 17, the electrons that are released during the oxidation of alcohols to carbonyl compounds are converted into a form of energy that can be utilized by the cell. These biochemical oxidation reactions are carried out at body temperature ($\sim$37°C) and are catalyzed by enzymes. Notice that in the following reaction the enzyme selectively catalyzes the oxidation of the secondary alcohol group to a ketone, but it does not oxidize the primary alcohol group to an aldehyde. We discuss enzyme specificity in Section 14.6.

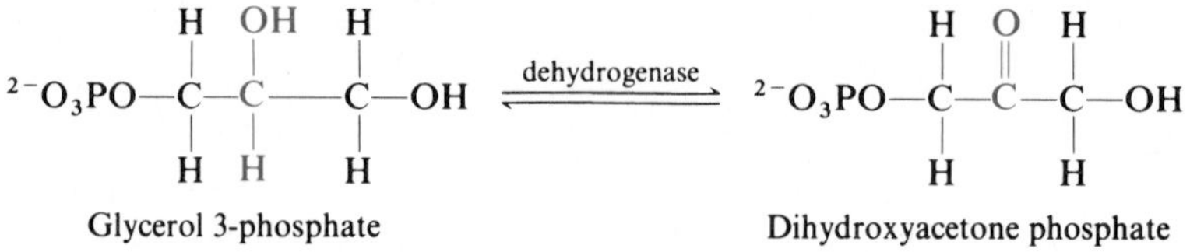

$$^{2-}O_3PO-\underset{\underset{H}{\overset{H}{|}}}{C}-\underset{\underset{H}{\overset{OH}{|}}}{C}-\underset{\underset{H}{\overset{H}{|}}}{C}-OH \underset{\longleftarrow}{\overset{dehydrogenase}{\longrightarrow}} {}^{2-}O_3PO-\underset{\underset{H}{\overset{H}{|}}}{C}-\underset{\overset{O}{\parallel}}{C}-\underset{\underset{H}{\overset{H}{|}}}{C}-OH$$

Glycerol 3-phosphate   Dihydroxyacetone phosphate

7.3   *Physical Properties of Aldehydes and Ketones*

Since aldehydes lack hydroxyl groups, they are incapable of intermolecular hydrogen bonding; thus they have boiling points considerably lower than their corresponding alcohols. On the other hand, their boiling points are somewhat higher than the corresponding alkanes and ethers, indicating the presence of weak intermolecular attractions (dipole–dipole forces). With the exception of the gaseous formaldehyde, the majority of the aldehydes are liquids (Table 7.3). (The physical state of acetaldehyde, bp 21°C, depends on the temperature of the laboratory; in warm rooms acetaldehyde exists as a gas.) Although the lower members of the series have pungent odors, many other aldehydes have pleasant odors and are used in making perfumes and artificial flavorings. The hydrogens of water molecules can form hydrogen bonds with the carbonyl oxygen; thus the solubility of aldehydes is about the same as that of alcohols and ethers. Formaldehyde and acetaldehyde are

120

TABLE 7.3 Physical Constants of Some Aldehydes

Formula	Common Name	MP (°C)	BP (°C)	Solubility (g/100 g of water)
HCHO	Formaldehyde	−92	−21	Miscible
CH_3CHO	Acetaldehyde	−121	21	Miscible
CH_3CH_2CHO	Propionaldehyde	−81	49	20
$CH_3(CH_2)_2CHO$	Butyraldehyde	−99	76	3.6
C_6H_5CHO	Benzaldehyde	−26	178	0.3

soluble in water; as the carbon chain increases, water solubility decreases. All aldehydes are soluble in organic solvents and, in general, are less dense than water.

The physical properties of the ketones are almost identical to those of the corresponding aldehydes. Acetone has a pleasant odor, and it is the only ketone that is completely soluble in water. The higher homologs are colorless liquids, are slightly soluble in water, and, unlike the aldehydes, have rather bland odors. Table 7.4 lists some physical constants for several of the ketones.

TABLE 7.4 Physical Constants of Some Ketones

Formula	Name	MP (°C)	BP (°C)	Solubility (g/100 g of water)
$CH_3\overset{\overset{\displaystyle O}{\|\|}}{C}CH_3$	Acetone	−95	56	Miscible
$CH_3\overset{\overset{\displaystyle O}{\|\|}}{C}CH_2CH_3$	Butanone	−86	80	25
$CH_3CH_2\overset{\overset{\displaystyle O}{\|\|}}{C}CH_2CH_3$	3-Pentanone	−40[a]	102	5
$CH_3\overset{\overset{\displaystyle O}{\|\|}}{C}CH_2CH_2CH_3$	2-Pentanone	−78	102	6
$(CH_3)_2CH\overset{\overset{\displaystyle O}{\|\|}}{C}CH_3$	3-Methyl-2-butanone	−92	94	—

[a]Again note that the symmetrical isomer has the highest melting point.

7.4 The Carbonyl Group

The chemical reactions of aldehydes and ketones are a function of the carbonyl group. Since the carbon–oxygen double bond is a site of unsaturation, it resembles the carbon–carbon double bond in many of its properties. The planar configuration around a double-bonded carbon occurs also for the carbonyl group (Figure 7.1a). The electrons of the carbon–oxygen double bond, like the electrons of the carbon–carbon double bond, are affected to a greater extent by approaching reagents than are the electrons in single bonds.

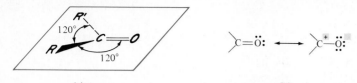

(a)   (b)

FIGURE 7.1 (a) The carbonyl group and the two atoms directly bonded to the carbonyl carbon lie in a plane. (b) Resonance forms of the carbonyl group.

There is, however, one important difference between a C=C bond and a C=O bond. Recall that the electrons of a carbon–carbon double bond are shared equally between the two carbon atoms since they are identical and have identical electronegativities. The electronic distribution is symmetrical and the bond is strictly nonpolar covalent. Oxygen, however, has an appreciably higher electronegativity than carbon, and so the oxygen atom has a tendency to draw the electron pair more closely to itself and away from the carbon atom. Consequently, the electronic distribution of the bond is not symmetrical; the oxygen end is slightly negative and the carbon end is slightly positive. The bond is said to be polarized and can be represented as the sum of two resonance contributors, a neutral form and a doubly charged form (Figure 7.1b). The doubly charged form is a valid way of representing the carbonyl group, and it accounts for the relatively high boiling points of the aldehydes and ketones. Physical measurements indicate that carbonyl compounds have relatively high dipole moments. Thus, if aldehydes or ketones are placed between the poles of an electric field, the molecules will orient themselves so that the oxygen ends of the molecules face the positive pole. Most important is an examination of the products formed when reagents add across the carbon–oxygen double bond. The products always result from the positive fragment of the reagent adding to the oxygen, and the negative fragment adding to the carbon.

7.5 Chemical Properties of Aldehydes and Ketones

a. Addition Reactions

Aldehydes and ketones will react with a variety of compounds by addition. The course of the addition reaction is identical; that is, there is an initial attack by a nucleophilic species on the carbonyl carbon, followed by the addition of a positive species to the oxygen. The initial adducts however, are often unstable and either exist in equilibrium with the reactants or react further to form stable products.

$$R-C\overset{O}{\underset{R'}{\big\langle}} \ + \ Y^+Z \ \longrightarrow \ R-\underset{Z}{\overset{O}{\underset{|}{\overset{|}{C}}}}-R' \ + \ Y^+$$

$$R-\underset{Z}{\overset{O}{\underset{|}{\overset{|}{C}}}}-R' \ + \ Y^+ \ \longrightarrow \ R-\underset{Z}{\overset{OY}{\underset{|}{\overset{|}{C}}}}-R'$$

Chapter 7 · Aldehydes and Ketones

1. **Addition of Water** The reaction of carbonyl compounds with water results initially in the formation of carbonyl hydrates. But with few exceptions,[1] the products so formed are unstable and readily dehydrate to the original aldehyde or ketone.

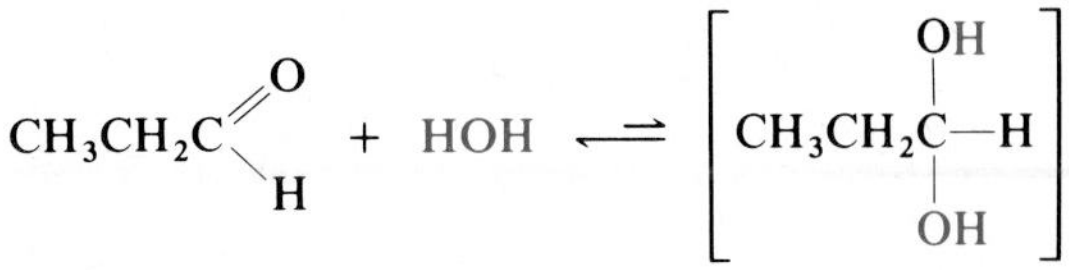

$$CH_3CH_2C \overset{O}{\underset{H}{\diagup}} + HOH \rightleftharpoons \left[CH_3CH_2\overset{OH}{\underset{OH}{C}}-H \right]$$

Propanal hydrate
(unstable)

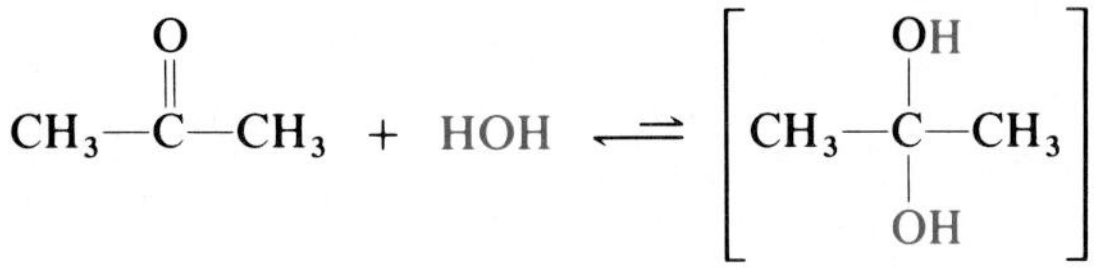

$$CH_3-\overset{O}{\overset{\|}{C}}-CH_3 + HOH \rightleftharpoons \left[CH_3-\overset{OH}{\underset{OH}{C}}-CH_3 \right]$$

Acetone hydrate
(very unstable)

2. **Addition of Alcohol** The addition of 1 mole of an alcohol to an aldehyde or ketone yields a *hemiacetal* or *hemiketal,* respectively. In the presence of an anhydrous acid catalyst, equilibrium is rapidly established and the equilibrium favors the carbonyl compounds (the reactants). As with the hydrates, simple hemiacetals and hemiketals are generally not sufficiently stable to be isolated.

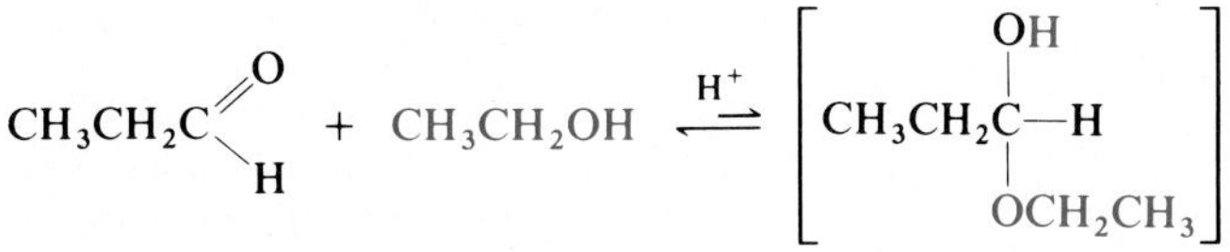

$$CH_3CH_2C \overset{O}{\underset{H}{\diagup}} + CH_3CH_2OH \overset{H^+}{\rightleftharpoons} \left[CH_3CH_2\overset{OH}{\underset{OCH_2CH_3}{C}}-H \right]$$

Unstable hemiacetal

---

[1] Formaldehyde is the only unsubstituted aldehyde that is appreciably hydrated. In general, hydrates are stable if the carbonyl group is adjacent to electron-withdrawing groups.

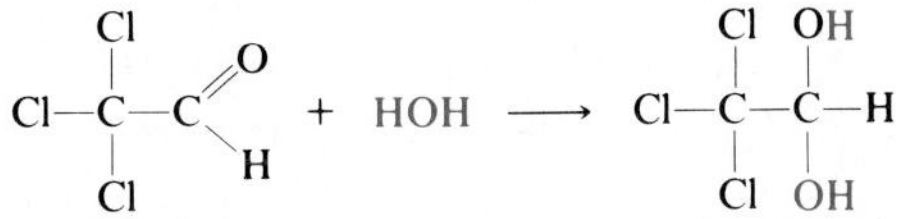

$$Cl-\overset{Cl}{\underset{Cl}{C}}-C\overset{O}{\underset{H}{\diagup}} + HOH \longrightarrow Cl-\overset{Cl}{\underset{Cl}{C}}-\overset{OH}{\underset{OH}{C}}-H$$

Chloral   Chloral hydrate

Chloral hydrate is stabilized by the presence of three electronegative chlorines that further increase the polarity of the carbon–oxygen bond. It is a solid, very soluble in water, and one of the very few organic compounds that possesses two hydroxyl groups on the same carbon atom. It is used in sedatives and is given to patients in hospitals as a nighttime sleeping pill (Noctec). Veterinarians use chloral hydrate as an anesthetic and an analgesic. Chloral hydrate dissolved in alcohol constitutes the so-called knockout drops or Mickey Finns. Recall (Section 5.6) that chloral hydrate is one of the reagents for the synthesis of DDT.

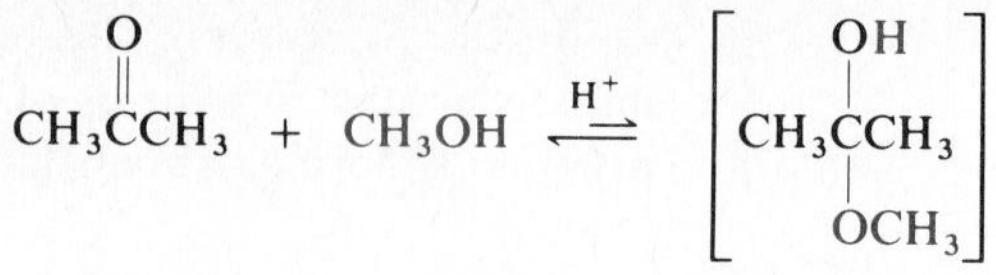

$$\text{CH}_3\overset{\displaystyle\text{O}}{\overset{\|}{\text{C}}}\text{CH}_3 \;+\; \text{CH}_3\text{OH} \;\underset{}{\overset{\text{H}^+}{\rightleftharpoons}}\; \left[\text{CH}_3\overset{\displaystyle\text{OH}}{\underset{\displaystyle\text{OCH}_3}{\overset{|}{\underset{|}{\text{C}}}}}\text{CH}_3 \right]$$

Unstable hemiketal

If, however, the alcohol and carbonyl groups occur within the same molecule, then the equilibrium favors the formation of cyclic hemiacetals and hemiketals. These cyclic compounds result from *intramolecular* interaction between the —**OH** and **C**=**O** groups. The cyclization reactions are of particular significance in our discussion of the structures of monosaccharides (see Section 11.5).

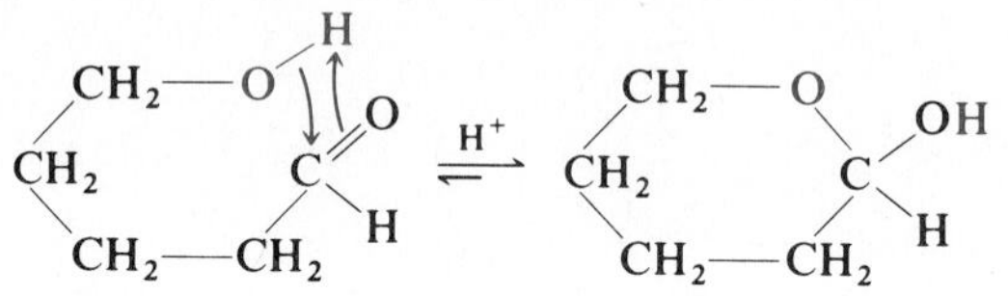

5-Hydroxypentanal   Stable cyclic hemiacetal

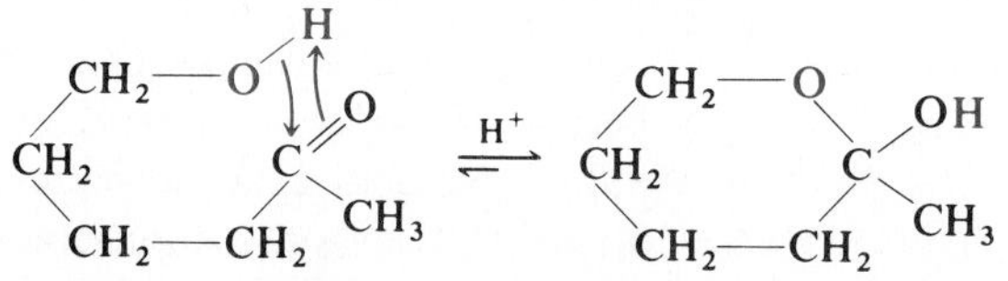

6-Hydroxy-2-hexanone   Stable cyclic hemiketal

In the presence of excess alcohol and an acid catalyst, the hemiacetal and hemiketal intermediates react with a second mole of the alcohol to eliminate water and form stable products called *acetals* and *ketals,* respectively. The formation of acetals is more favorable than that of ketals. We shall see in Section 11.8 that the naturally occurring disaccharides (maltose, lactose, and sucrose) are acetals. The equilibrium can be driven to the right by using alcohol as the solvent and removing the water (via azeotropic distillation).

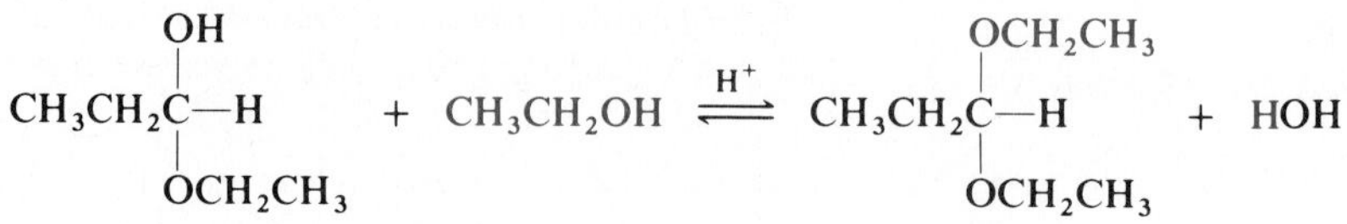

$$\text{CH}_3\text{CH}_2\overset{\displaystyle\text{OH}}{\underset{\displaystyle\text{OCH}_2\text{CH}_3}{\overset{|}{\underset{|}{\text{C}}}}}\text{—H} \;+\; \text{CH}_3\text{CH}_2\text{OH} \;\overset{\text{H}^+}{\rightleftharpoons}\; \text{CH}_3\text{CH}_2\overset{\displaystyle\text{OCH}_2\text{CH}_3}{\underset{\displaystyle\text{OCH}_2\text{CH}_3}{\overset{|}{\underset{|}{\text{C}}}}}\text{—H} \;+\; \text{HOH}$$

A hemiacetal   An acetal

$$\text{CH}_3\overset{\displaystyle\text{OH}}{\underset{\displaystyle\text{OCH}_3}{\overset{|}{\underset{|}{\text{C}}}}}\text{CH}_3 \;+\; \text{CH}_3\text{OH} \;\overset{\text{H}^+}{\rightleftharpoons}\; \text{CH}_3\overset{\displaystyle\text{OCH}_3}{\underset{\displaystyle\text{OCH}_3}{\overset{|}{\underset{|}{\text{C}}}}}\text{CH}_3 \;+\; \text{HOH}$$

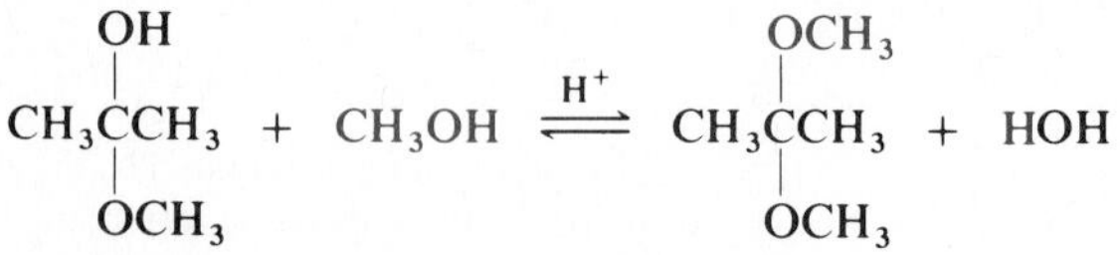

A hemiketal   A ketal

Acetals and ketals are stable in base, but can be hydrolyzed back to carbonyl compounds in acidic solutions.

$$CH_3-\underset{\underset{H}{|}}{\overset{\overset{OCH_3}{|}}{C}}-OCH_3 \xrightarrow[H_2O]{H^+} CH_3-\overset{O}{\overset{\|}{C}}\diagdown H + 2\ CH_3OH$$

3. **Addition of Hydrogen Cyanide** Since HCN is a poisonous gas, it is usually prepared in the reaction mixture from potassium cyanide and sulfuric acid. The products of hydrogen cyanide addition to aldehydes and ketones are called *cyanohydrins* (or hydroxynitriles). They are important intermediates in the preparation of hydroxy acids and amino acids because the —CN group is easily hydrolyzed to a carboxylic acid (see Section 8.2-b). For example, acetaldehyde can be readily converted into the important biological compounds lactic acid and alanine.

$$CH_3-\overset{O}{\overset{\|}{C}}\diagdown H \xrightarrow{HCN} CH_3-\underset{\underset{H}{|}}{\overset{\overset{OH}{|}}{C}}-CN$$

Acetaldehyde cyanohydrin

$$\xrightarrow[HCl,\Delta]{2\ HOH} CH_3-\underset{\underset{H}{|}}{\overset{\overset{OH}{|}}{C}}-\overset{O}{\overset{\|}{C}}-OH + NH_4Cl$$

Lactic acid

$$\xrightarrow{NH_3} CH_3-\underset{\underset{H}{|}}{\overset{\overset{NH_2}{|}}{C}}-CN \xrightarrow[H^+]{HOH} CH_3-\underset{\underset{H}{|}}{\overset{\overset{NH_2}{|}}{C}}-\overset{O}{\overset{\|}{C}}-OH$$

1-Amino-1-cyanoethane   Alanine

$$CH_3-\overset{O}{\overset{\|}{C}}-CH_3 \xrightarrow{HCN} CH_3-\underset{\underset{CH_3}{|}}{\overset{\overset{OH}{|}}{C}}-CN \xrightarrow[HOH]{H^+} CH_3-\underset{\underset{CH_3}{|}}{\overset{\overset{OH}{|}}{C}}-\overset{O}{\overset{\|}{C}}-OH$$

Acetone cyanohydrin   2-Hydroxy-2-methylpropanoic acid

The preceding reaction sequences are examples of *multistep synthesis*. This is the rule, rather than the exception, in both organic chemistry and biochemistry. When organic chemists synthesize a certain compound in the laboratory, or when a necessary component is produced within a living cell, the synthesis usually requires several steps. In these processes the product of the first step becomes the reactant in the second step and so on until the final product is obtained. We shall see numerous examples of this in our studies of metabolic reactions (Chapters 16 and 17).

In an important commercial reaction, acetone cyanohydrin is converted to methyl methacrylate by refluxing with methanol and a strong acid. The methyl methacrylate is then polymerized to form a transparent plastic known as Lucite and Plexiglas.

$$CH_3-\underset{\underset{CH_3}{|}}{\overset{\overset{OH}{|}}{C}}-C\equiv N + CH_3OH \xrightarrow[\Delta]{HCl} CH_2=\underset{\underset{CH_3}{|}}{C}-\overset{O}{\overset{\|}{C}}-OCH_3 \longrightarrow \text{\small$\sim$}CH_2-\underset{\underset{CH_3}{|}}{\overset{\overset{COOCH_3}{|}}{C}}-CH_2-\underset{\underset{CH_3}{|}}{\overset{\overset{COOCH_3}{|}}{C}}\text{\small$\sim$}$$

Methyl methacrylate   Lucite, Plexiglas

4. **Self-Addition (Aldol Condensation)** The hydrogen atoms that are attached to the carbon atom adjacent to the carbonyl carbon have a particular significance because they are more easily released as protons than any of the other hydrogens in the molecule. The explanation for this is given in Section 7.5-b. Such hydrogens are referred to as α-*hydrogens*[2] because they are bonded to the α-carbon.

Any aldehyde molecule that contains α-hydrogens may add to itself or to another aldehyde. The compound resulting from such a process contains both a carbonyl group and a hydroxyl group within the same molecule. The reaction, which is catalyzed by a dilute base, is known as an *aldol condensation.* **Aldol** is a composite word derived from the words *ald*ehyde and alcoh*ol.* A **condensation reaction** is *one in which two molecules combine to form a larger molecule,* usually accompanied by the elimination of a small molecule such as water. The aldol condensation reaction proceeds in the same manner as all other carbonyl addition reactions. An α-hydrogen adds to the carbonyl oxygen; the remainder of the molecule adds to the carbonyl carbon. At a temperature of 5°C and in the presence of a dilute sodium hydroxide solution, two molecules of acetaldehyde will combine.

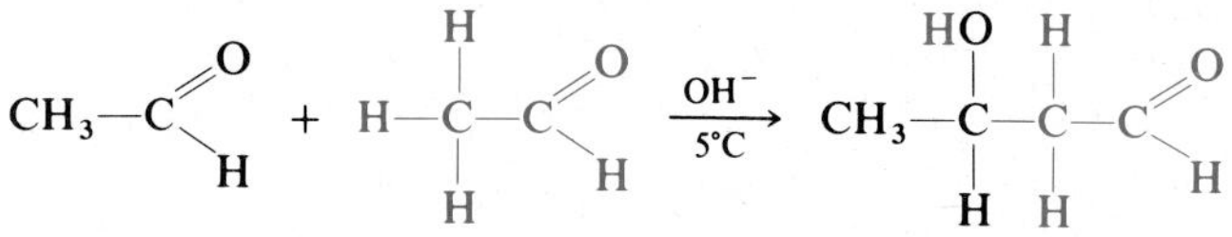

Acetaldehyde   Acetaldehyde   3-Hydroxybutanal

The addition product is very susceptible to dehydration. Heating the solution results in the formation of an α,β-unsaturated aldehyde. The product has a carbon–carbon double bond and a carbon–oxygen double bond in conjugation, resulting in a stable molecule (recall Section 3.5).

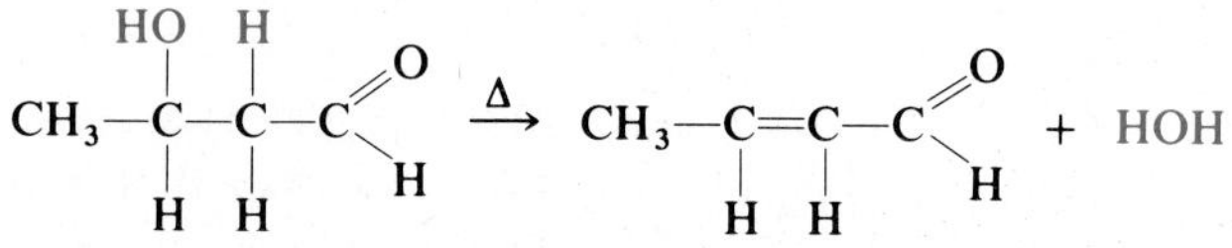

3-Hydroxybutanal   2-Butenal
                                 (Crotonaldehyde)

---

[2]The Greek letters α, β, etc., are sometimes used to designate the positions of carbon atoms with respect to the carbonyl carbon (see Section 8.1). For example,

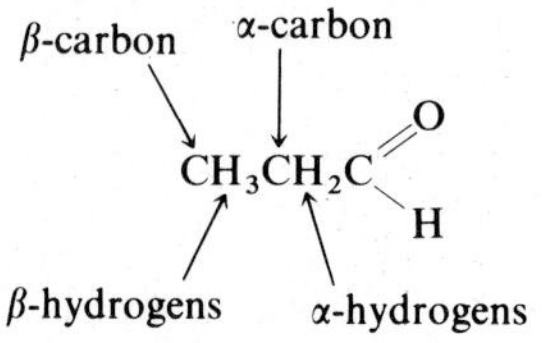

The hydrogens bonded to these carbons are then called α-hydrogens, β-hydrogens, etc.

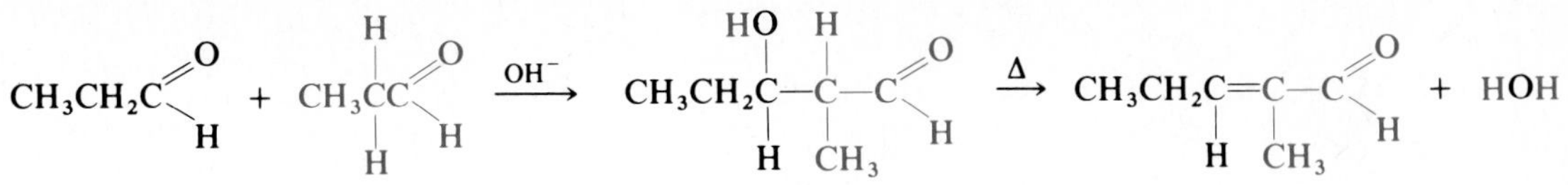

$$CH_3CH_2C\overset{O}{\underset{H}{\diagup}} + CH_3C\overset{O}{\underset{H}{\diagup}} \xrightarrow{OH^-} CH_3CH_2CH_2-\overset{HO\ H}{\underset{H\ \ CH_3}{C}}-C\overset{O}{\diagup}_H \xrightarrow{\Delta} CH_3CH_2C=\overset{}{\underset{H\ \ CH_3}{C}}-C\overset{O}{\diagup}_H + HOH$$

Propanal   Propanal   3-Hydroxy-2-methylpentanal   2-Methyl-2-pentenal

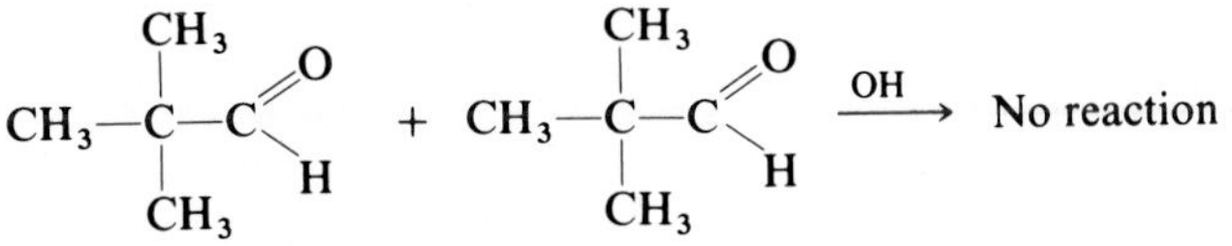

$$CH_3-\overset{CH_3}{\underset{CH_3}{C}}-C\overset{O}{\diagup}_H + CH_3-\overset{CH_3}{\underset{CH_3}{C}}-C\overset{O}{\diagup}_H \xrightarrow{OH} \text{No reaction}$$

2,2-Dimethylpropanal

Aldehydes that lack α-hydrogens do not undergo self-addition reactions. They can, however, combine with other carbonyl compounds by acting as the acceptor molecule. One example is the preparation of cinnamaldehyde from the reaction of benzaldehyde and acetaldehyde. Notice that in cinnamaldehyde the conjugated double bonds extend throughout the entire molecule. The product is so stable that the adduct dehydrates immediately.

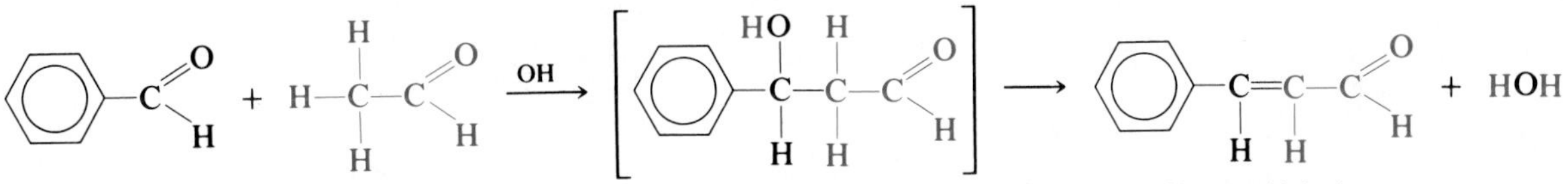

$$C_6H_5-C\overset{O}{\diagup}_H + H-\overset{H}{\underset{H}{C}}-C\overset{O}{\diagup}_H \xrightarrow{OH} \left[C_6H_5-\overset{HO\ H}{\underset{H\ \ H}{C}}-C-C\overset{O}{\diagup}_H \right] \longrightarrow C_6H_5-C=\underset{H\ \ H}{C}-C\overset{O}{\diagup}_H + HOH$$

(No α-hydrogens)   3-Hydroxy-3-phenylpropanal   Cinnamaldehyde

Ketones contain two α-carbon atoms. If either of these carbons bears a hydrogen atom, the ketone can undergo a self-condensation, although much more slowly than aldehydes. Upon addition of acid and heating, the condensation product loses a molecule of water to yield an α,β-unsaturated ketone.

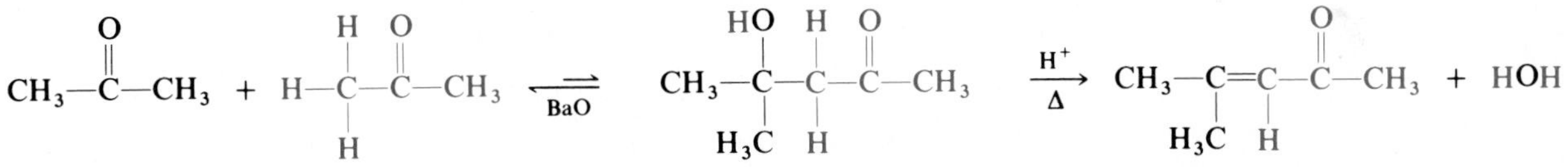

$$CH_3-\overset{O}{\overset{\|}{C}}-CH_3 + H-\overset{H}{\underset{H}{C}}-\overset{O}{\overset{\|}{C}}-CH_3 \underset{BaO}{\rightleftharpoons} CH_3-\overset{HO}{\underset{H_3C\ H}{C}}-\overset{H}{C}-\overset{O}{\overset{\|}{C}}-CH_3 \xrightarrow[\Delta]{H^+} CH_3-\underset{H_3C\ H}{C}=C-\overset{O}{\overset{\|}{C}}-CH_3 + HOH$$

Acetone   Acetone   4-Hydroxy-4-methyl-2-pentanone (Diacetone alcohol)   4-Methyl-3-penten-2-one (Mesityl oxide)

Ketones can also condense with aldehydes to effect the synthesis of a wide variety of organic compounds. A vital reaction in carbohydrate metabolism is the reversible aldol condensation between an aldehyde and a ketone.

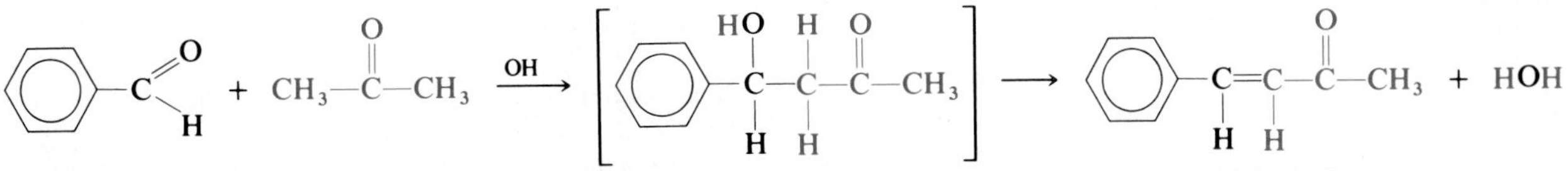

$$C_6H_5-C\overset{O}{\diagup}_H + CH_3-\overset{O}{\overset{\|}{C}}-CH_3 \xrightarrow{OH} \left[C_6H_5-\overset{HO\ H}{\underset{H\ \ H}{C}}-\overset{O}{\overset{\|}{C}}-CH_3 \right] \longrightarrow C_6H_5-C=\underset{H\ \ H}{C}-\overset{O}{\overset{\|}{C}}-CH_3 + HOH$$

Benzaldehyde   Acetone   4-Hydroxy-4-phenyl-2-butanone   4-Phenyl-3-buten-2-one

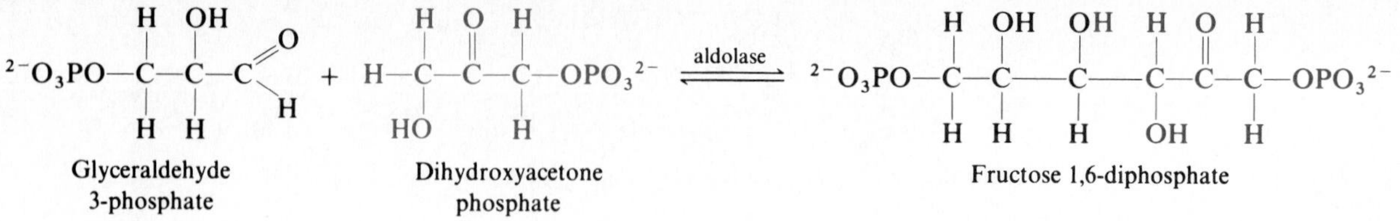

$$^{2-}O_3PO-\underset{\underset{H}{|}}{\overset{\overset{H}{|}}{C}}-\underset{\underset{H}{|}}{\overset{\overset{OH}{|}}{C}}-\overset{O}{\underset{H}{C}} \quad + \quad H-\underset{\underset{HO}{|}}{\overset{\overset{H}{|}}{C}}-\overset{O}{\overset{||}{C}}-\underset{\underset{H}{|}}{\overset{\overset{H}{|}}{C}}-OPO_3{}^{2-} \;\underset{}{\overset{aldolase}{\rightleftharpoons}}\; {}^{2-}O_3PO-\underset{\underset{H}{|}}{\overset{\overset{H}{|}}{C}}-\underset{\underset{H}{|}}{\overset{\overset{OH}{|}}{C}}-\underset{\underset{H}{|}}{\overset{\overset{OH}{|}}{C}}-\underset{\underset{OH}{|}}{\overset{\overset{H}{|}}{C}}-\overset{O}{\overset{||}{C}}-\underset{\underset{H}{|}}{\overset{\overset{H}{|}}{C}}-OPO_3{}^{2-}$$

Glyceraldehyde 3-phosphate   Dihydroxyacetone phosphate   Fructose 1,6-diphosphate

b. Enolization

In the previous section, we mentioned that the α-hydrogens of a carbonyl compound are labile (i.e., they can dissociate). Experimental evidence indicates that aldehydes and ketones exist in solution as an equilibrium mixture of two isomeric forms. These equilibria involve intermediate anions that are stabilized by resonance. For aqueous solutions of acetaldehyde and acetone, we can write the following equations.

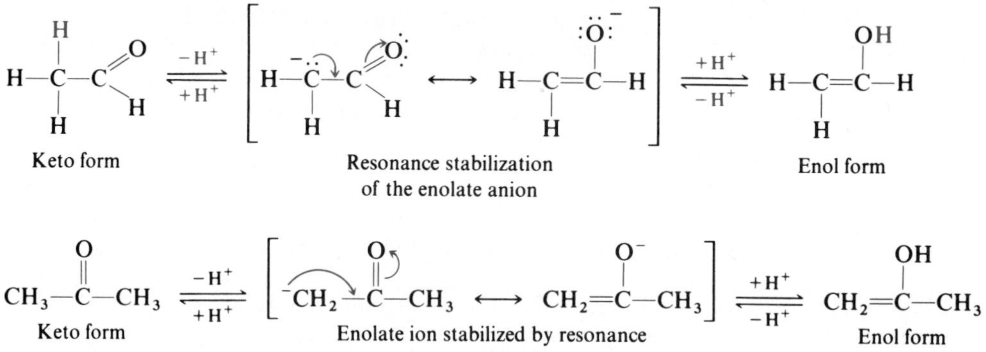

Keto form   Resonance stabilization of the enolate anion   Enol form

Keto form   Enolate ion stabilized by resonance   Enol form

The structure possessing the carbonyl group is known as the **keto** form. *The structure containing the hydroxyl group bonded to an unsaturated carbon* is called the **enol** form. The term enol comes from a combination of the IUPAC endings for an unsaturated alcohol (*ene* + *ol*). **Enolization**[3] *is the process that interconverts the keto*

[3]There is another term, tautomerism, that is often used synonymously with enolization. **Tautomerism** refers to *the rapid interconversion of isomers that differ only in the position of a hydrogen and the corresponding location of a double bond*. The isomers are known as **tautomers.** For example, the two tautomers of the important biological compound adenine can be written as

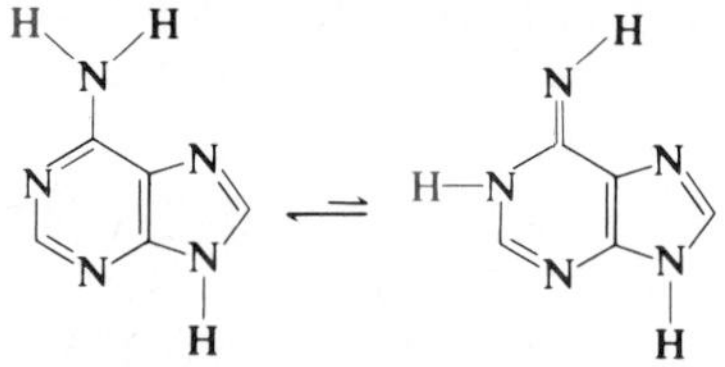

Strictly speaking, enolization is just a specific type of tautomerism that occurs in carbonyl compounds. The keto and enol forms, therefore, are tautomers. Also it is important not to confuse tautomerism with resonance. Tautomers are actual species in equilibrium, and we use a *pair* of half-headed arrows (⇌) to indicate the equilibrium condition. The *unreal* resonance structures that we draw are related by a *single* two-headed arrow (↔).

form and the enol form. The intermediate anion that is formed in the enolization reaction is called the *enolate* ion.

For the simple aliphatic aldehydes and ketones, the keto form is more stable, and there is very little of the enol form (less than 1%) present at equilibrium. However, aqueous solutions of 1,3-dicarbonyl compounds contain substantial amounts of the enol form.

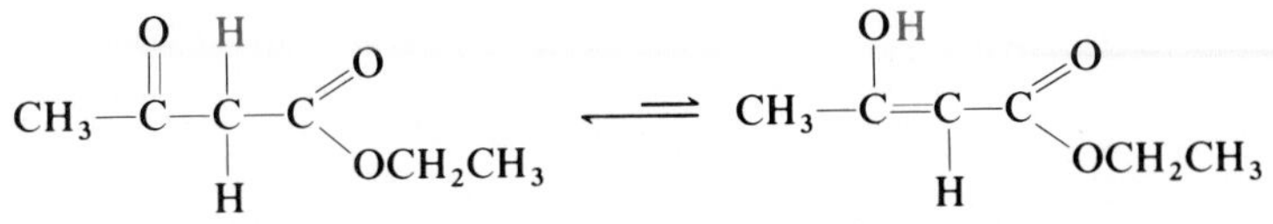

Ethyl acetoacetate
(93% keto form)

Ethyl acetoacetate
(7% enol form)

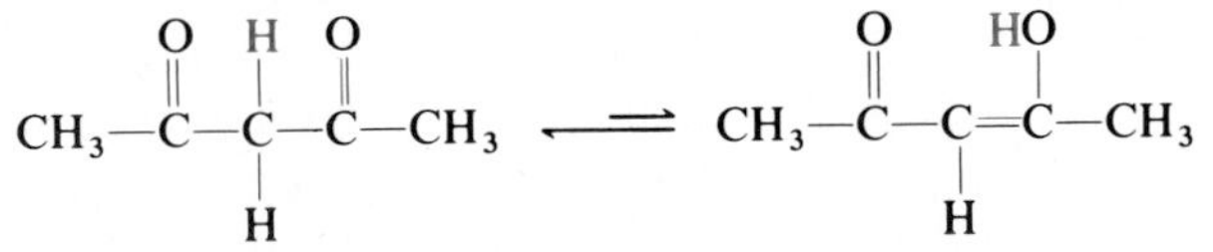

2,4-Pentanedione
(84% keto form)

2,4-Pentanedione
(16% enol form)

Two explanations are given for the greater extent of enolization of these compounds. One is the ability of the enol to form an intramolecular hydrogen bond. Such intramolecular hydrogen bonds are particularly stable when six-membered rings can be formed.

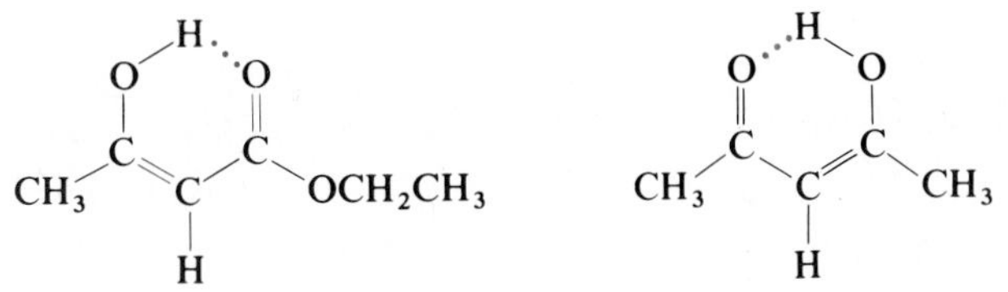

Enol form of
ethyl acetoacetate

Enol form of
2,4-pentanedione

The second explanation is that the intermediate enolate anions are resonance stabilized by the two adjacent carbonyl groups, and therefore the negative charge is more delocalized compared to the enolate ion of monocarbonyl compounds.

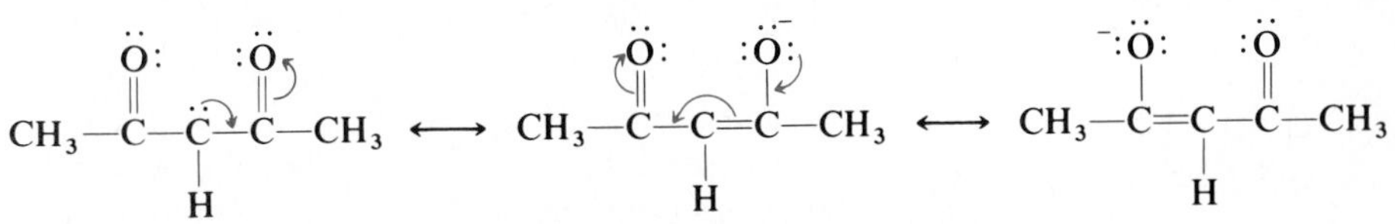

Resonance-stabilized enolate ion of 2,4-pentanedione

We shall have occasion to mention enolization several times in our studies of biochemical compounds. One of the steps in carbohydrate metabolism results in the formation of enolpyruvic acid, which undergoes enolization to the more stable keto form (page 435).

$$H-\overset{\displaystyle H}{\underset{\displaystyle H}{C}}=\overset{\displaystyle HO}{C}-C\overset{\displaystyle O}{\underset{\displaystyle OH}{}} \rightleftharpoons H-\overset{\displaystyle H}{\underset{\displaystyle H}{C}}-\overset{\displaystyle O}{C}-C\overset{\displaystyle O}{\underset{\displaystyle OH}{}}$$

Enolpyruvic acid   Pyruvic acid

Glucose and fructose are interconvertible via a common enol intermediate (page 265). So, too, are glucose 6-phosphate and fructose 6-phosphate (page 432), and glyceraldehyde 3-phosphate and dihydroxyacetone phosphate (page 433).

c. Oxidation

Aldehydes may be characterized by the ease with which they are oxidized to carboxylic acids. Almost any oxidizing agent, including air, will bring about this oxidation. (Recall the experimental precautions employed in the preparation of aldehydes by the oxidation of a primary alcohol.) The general equation for the oxidation of an aldehyde is written as follows.

$$R-C\overset{\displaystyle O}{\underset{\displaystyle H}{}} \xrightarrow{\text{[O]}} R-C\overset{\displaystyle O}{\underset{\displaystyle OH}{}}$$

Aldehyde   Acid

Ketones, on the other hand, are quite resistant to oxidation, and this difference in properties affords a method of distinguishing between the two classes of compounds. Three mild oxidizing solutions are used extensively as test reagents for the presence of the aldehyde group. They are incapable of oxidizing most other organic compounds. Named after the men who discovered their use, they are Benedict's solution, Fehling's solution, and Tollens' reagent. These reagents are commonly employed in testing for the presence of sugar in the blood or in the urine (see Section 11.10-b).

1. **Benedict's Solution and Fehling's Solution** Both of these reagents are alkaline solutions of copper(II) (cupric) ion. The source of the ion is copper(II) sulfate. Because the cupric ion forms an insoluble hydroxide in basic solution, another reagent must be added to the solution to keep the copper ion from precipitating out as the hydroxide. In Benedict's solution, sodium citrate is employed for this purpose; copper remains in solution as the cupric citrate ion. The additional reagent in Fehling's solution is sodium potassium tartrate (Rochelle salt) with which copper forms the water soluble cupric tartrate ion. The blue color of these solutions is due to the presence of the cupric ion complexes. A positive test for the aldehyde group is evidenced by a color change to brick red, indicating the presence of the copper(I) (cuprous) oxide.[4]

---

[4]The cupric ion (+2 oxidation state) is the oxidizing agent and therefore must be the substance that is reduced, in this case to cuprous oxide (+1 oxidation state).

The following equation is written for the reaction of an aldehyde with either Benedict's or Fehling's solution.

$$CH_3C\overset{O}{\underset{H}{\diagdown}} + 2\,Cu^{2+} + 5\,OH^- \longrightarrow CH_3C\overset{O}{\underset{O^-}{\diagdown}} + Cu_2O(s) + 3\,H_2O$$

(blue)   (red)

Oxidation

$$3\,OH^- + CH_3C\overset{O}{\underset{H}{\diagdown}} \longrightarrow CH_3C\overset{O}{\underset{O^-}{\diagdown}} + 2\,H_2O + 2\,e^-$$

Reduction

$$2\,e^- + 2\,OH^- + 2\,Cu^{2+} \longrightarrow Cu_2O + H_2O$$

2. **Tollens' Reagent** Tollens' reagent is an ammoniacal solution of silver nitrate. (In this case, the silver is maintained in solution as the silver ammonia complex ion, $Ag(NH_3)_2^+$.) A positive test is indicated by the appearance of a silver mirror that forms on the inside surface of the reaction vessel due to the reduction of the silver ion to free silver. The deposition of silver on a clean glass surface by the oxidation of formaldehyde is used in the manufacture of mirrors.

$$H-C\overset{O}{\underset{H}{\diagdown}} + 2\,Ag(NH_3)_2^+ + 3\,OH^- \longrightarrow H-C\overset{O}{\underset{O^-}{\diagdown}} + 2\,Ag(s) + 4\,NH_3 + 2\,H_2O$$

Silver
mirror

Oxidation

$$3\,OH^- + H-C\overset{O}{\underset{H}{\diagdown}} \longrightarrow H-C\overset{O}{\underset{O^-}{\diagdown}} + 2\,H_2O + 2\,e^-$$

Reduction

$$2\,Ag(NH_3)_2^+ + 2\,e^- \longrightarrow 2\,Ag + 4\,NH_3$$

d. Reduction

Aldehydes and ketones are readily reduced to the corresponding primary and secondary alcohols, respectively. A wide variety of reducing agents may be used.

$$R-C\overset{O}{\underset{H}{\diagdown}} \xrightarrow{2[H]} R-\overset{H}{\underset{H}{\overset{|}{\underset{|}{C}}}}-OH \qquad \textbf{and} \qquad R-\overset{O}{\overset{||}{C}}-R' \xrightarrow{2[H]} R-\overset{OH}{\underset{H}{\overset{|}{\underset{|}{C}}}}-R'$$

Carbonyl compounds may be reduced to alcohols by hydrogen gas in the presence of a metal catalyst (catalytic hydrogenation). However, this method suffers from the disadvantages that many of the catalysts (Pt, Pd, Ru) are expensive and that other functional groups are also reduced.

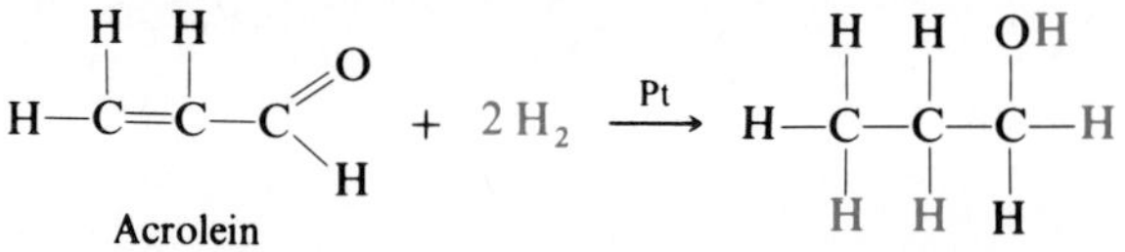

$$
\underset{\substack{\text{Acrolein}\\\text{(2-Propenal)}}}{\overset{\substack{H\;\;H\\|\quad|}}{H-C=C-C}\overset{O}{\diagdown}_H} \;+\; 2\,H_2 \xrightarrow{\;Pt\;} \underset{\substack{\\ \text{1-Propanol}}}{H-\overset{H}{\underset{H}{C}}-\overset{H}{\underset{H}{C}}-\overset{OH}{\underset{H}{C}}-H}
$$

$$
\xrightarrow[\text{Ni}]{2\,H_2}
$$

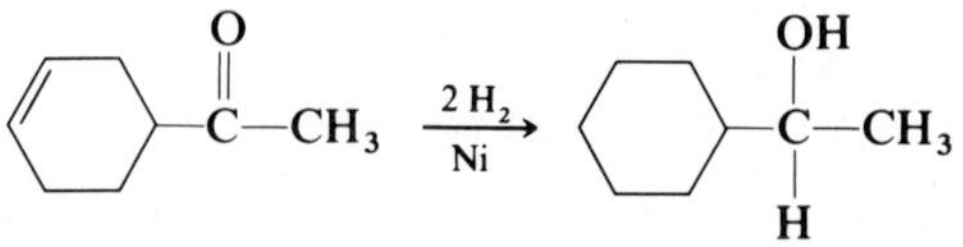

For laboratory applications, reduction with complex metal hydrides has the great advantage of selectivity. Lithium aluminum hydride ($LiAlH_4$) and sodium borohydride ($NaBH_4$) will reduce carbon–oxygen double bonds but not carbon–carbon double bonds.

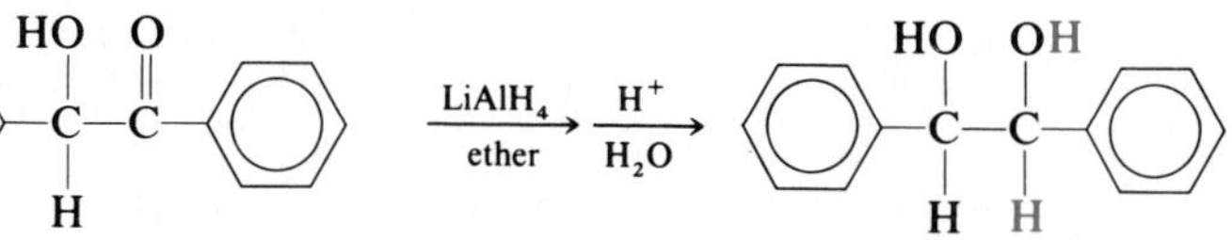

Benzoin
(2-Hydroxy-1,2-diphenylethanone)

$\xrightarrow[\text{ether}]{LiAlH_4}\;\xrightarrow[H_2O]{H^+}$

Hydrobenzoin
(1,2-Diphenylethane-1,2-diol)

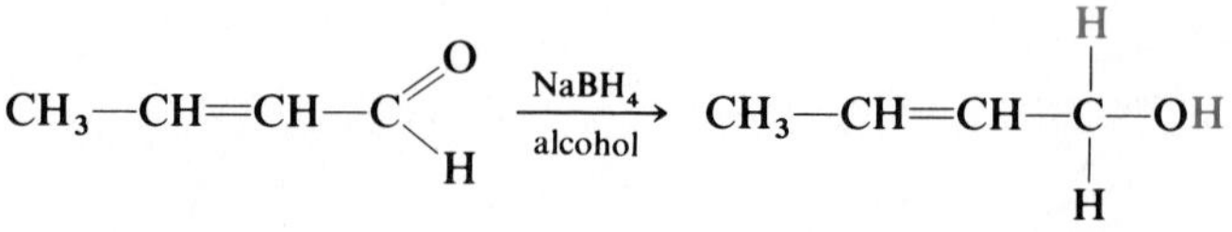

$$
CH_3-CH=CH-C\overset{O}{\diagdown}_H \xrightarrow[\text{alcohol}]{NaBH_4} CH_3-CH=CH-\overset{H}{\underset{H}{C}}-OH
$$

Crotonaldehyde
(2-Butenal)

Crotyl alcohol
(2-Buten-1-ol)

Two extremely important biochemical carbonyl reduction reactions are discussed in Chapter 16. They are the reduction of acetaldehyde to ethyl alcohol and the reduction of pyruvic acid to lactic acid. In each case, the enzyme that catalyzes the reaction contains the coenzyme NADH, which is the reducing agent. (See Table 14.2 for the structure of the coenzyme.)

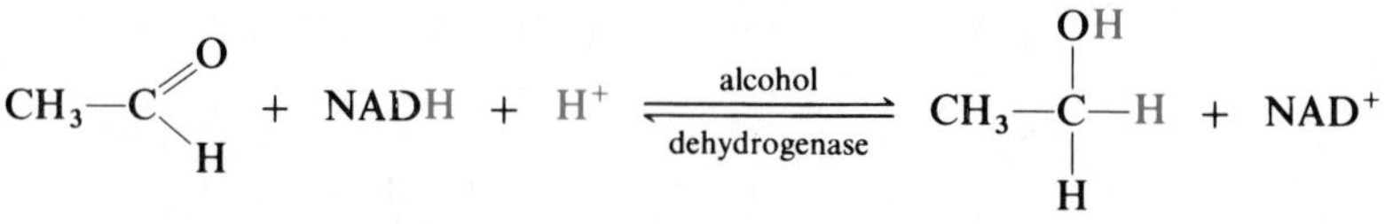

$$
CH_3-C\overset{O}{\diagdown}_H \;+\; NADH \;+\; H^+ \underset{\text{dehydrogenase}}{\overset{\text{alcohol}}{\rightleftharpoons}} CH_3-\overset{OH}{\underset{H}{C}}-H \;+\; NAD^+
$$

Acetaldehyde

Ethyl alcohol

$$CH_3-\overset{\displaystyle O}{\overset{\|}{C}}-\overset{\displaystyle O}{\overset{\diagup}{C}}_{OH} + NADH + H^+ \xrightleftharpoons[\text{dehydrogenase}]{\text{lactic acid}} CH_3-\overset{\displaystyle OH}{\underset{\displaystyle H}{\overset{\|}{C}}}-\overset{\displaystyle O}{\overset{\diagup}{C}}_{OH} + NAD^+$$

Pyruvic acid   Lactic acid

7.6 Important Aldehydes and Ketones

a. Formaldehyde (Methanal)

Formaldehyde is the simplest and industrially the most important member of the aldehyde family. It is manufactured from methanol and oxygen in the air by passing methanol vapors over a copper or silver catalyst at temperatures above 300°C. Formaldehyde is a colorless gas with an extremely irritating odor. Because of its reactivity, formaldehyde cannot be handled easily in the gaseous state and is therefore dissolved in water and sold as a 37–40% aqueous solution (such a solution is called *formalin*).

The largest use of formaldehyde is as a reagent for the preparation of many other organic compounds and for the manufacture of polymers such as Bakelite, Formica, and Melmac. Formaldehyde can denature proteins (see Section 13.9-b), rendering them insoluble in water and resistant to decay bacteria. For this reason it is used in embalming solutions and in the preservation of biological specimens. Formalin is also used as a general antiseptic in hospitals to sterilize gloves and surgical instruments. However, its use as an antiseptic, preservative, and embalming fluid has declined because formaldehyde is suspected of being carcinogenic.

b. Acetaldehyde (Ethanal)

Acetaldehyde is an extremely volatile, colorless liquid. It is prepared by the catalytic (Ag) oxidation of ethyl alcohol or by the catalytic ($PdCl_2$) oxidation of ethylene. It serves as the starting material for the preparation of many other organic compounds such as acetic acid, ethyl acetate, and chloral. Acetaldehyde is formed as a metabolite in the fermentation of sugars and in the detoxification of alcohol in the liver (see page 438).

c. Acetone (Propanone)

Acetone is the simplest and most important of the ketones. It is produced in large quantities by the catalytic (Ag) oxidation of isopropyl alcohol. Because it is miscible with water as well as with most organic solvents, acetone finds its chief use as an industrial solvent (e.g., for paints and lacquers). It is the chief ingredient in some brands of fingernail polish remover. Acetone is also an important intermediate for the preparation of chloroform, iodoform, dyes, methacrylates, and many other complex organic compounds.

Acetone is formed in the human body as a by-product of lipid metabolism. Normally it does not accumulate to an appreciable extent because it is subsequently oxidized to carbon dioxide and water. The normal concentration of acetone in the human body is less than 1 mg/100 mL of blood. In the case of certain abnormalities, such as diabetes mellitus, the acetone concentration rises above this level. The acetone is then excreted in the urine where it can be easily detected. In severe cases, its odor can be noted on the breath (see Section 17.4).

Exercises

7.1 Draw and name all the isomeric carbonyl compounds having the molecular formula $C_5H_{10}O$.

7.2 Name the following compounds.

(a) $Cl_3CCH_2CH_2C{\overset{O}{\underset{H}{\big\backslash}}}$

(b) $GH_3\overset{O}{\overset{\|}{C}}CH_2CH_2CH_3$ *2-pentenone*

(c) $CH_3CH{=}CHC{\overset{O}{\underset{H}{\big\backslash}}}$

(d) $(CH_3)_3C\overset{O}{\overset{\|}{C}}CHBrCH_2CH_3$

(e) $(CH_3)_2CH\overset{O}{\overset{\|}{C}}CH(CH_3)_2$

(f) $(CH_3CH_2)_2CHC{\overset{O}{\underset{H}{\big\backslash}}}$

(g) 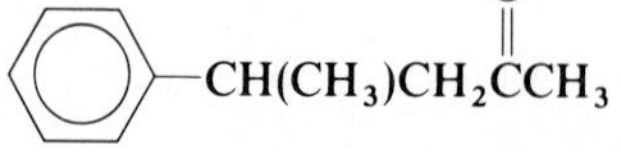 $\text{(phenyl)}{-}CH(CH_3)CH_2\overset{O}{\overset{\|}{C}}CH_3$

(h) $\text{(phenyl)}{-}CH_2CH_2C(CH_3)_2C{\overset{O}{\underset{H}{\big\backslash}}}$

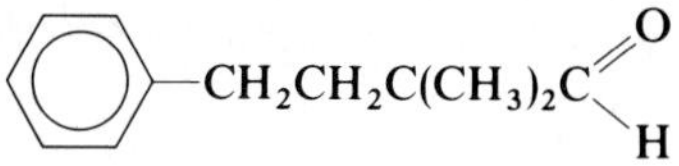

(i) $CH_3CHClCCl_2CH_2CH_2C{\overset{O}{\underset{H}{\big\backslash}}}$

(j) $CH_3\overset{O}{\overset{\|}{C}}CH_2CH_2CH_2I$

(k) $CH_3\overset{O}{\overset{\|}{C}}CH_2\overset{O}{\overset{\|}{C}}CH_2CH_3$

(l) $CH_2OHCH_2C{\overset{O}{\underset{H}{\big\backslash}}}$

(m) $CH_3CHBrCH{=}CHCH_2C{\overset{O}{\underset{H}{\big\backslash}}}$

(n) $CH_3CH{=}CH\overset{O}{\overset{\|}{C}}CH_3$

(o) 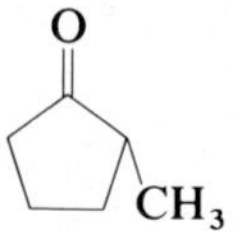 cyclopentanone with CH_3

(p) 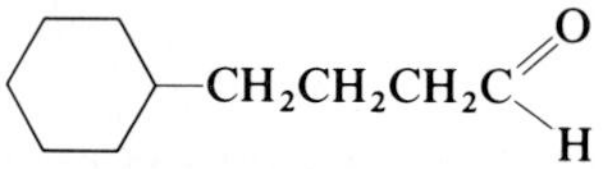 cyclohexyl${-}CH_2CH_2CH_2C{\overset{O}{\underset{H}{\big\backslash}}}$

7.3 Write structural formulas for the following compounds.
  (a) butyraldehyde
  (b) 3-methylbutanal
  (c) 3-hexenal
  (d) 2-hexanone
  (e) 2-chloropropanal
  (f) 3-bromo-2-heptanone
  (g) 3,3-dimethylcyclohexanone
  (h) 1-phenyl-2-butanone
  (i) 4-hexen-2-one
  (j) 2-iodo-2-methyl-4-octanone
  (k) 4-methyl-3-penten-2-one
  (l) 5-ethylheptanal
  (m) 2-hydroxy-4-methyl-3-pentanone
  (n) 2,5-dimethyl-2-hexenal
  (o) ethyl cyclopentyl ketone
  (p) *p*-methoxybenzaldehyde

7.4 In 1985, the U. S. State Department accused the Russians (the KGB) of dusting ordinary objects like doorknobs and steering wheels with an invisible (and possibly carcinogenic) powder to track the movements and contacts of Americans in Moscow. This compound is 5-(4-nitrophenyl)-2,4-pentadien-1-al. The charges were adamantly denied by Russia. Draw the structural formula of the compound.

7.5 As we shall see in Chapter 11, 2,3-dihydroxypropanal and 1,3-dihydroxyacetone are important carbohydrates. Draw their structural formulas.

7.6 Glutaraldehyde (pentanedial) is a germicide that is replacing formaldehyde as a sterilizing agent. It is less irritating to the eyes, nose, and skin. Draw the structural formula of glutaraldehyde.

7.7 Write complete balanced equations for these oxidations.
  (a) 1-propanol by chromic oxide in pyridine, methylene chloride, and HCl
  (b) 3-methylcyclohexanol by potassium permanganate and potassium hydroxide. (In alkaline solutions permanganate is reduced to manganese dioxide.)

7.8 Account for the fact that the oxidation of primary alcohols usually gives poorer yields of aldehydes than the oxidation of secondary alcohols gives of ketones.

7.9 Give the structures of the alcohols that could be oxidized to the following aldehydes or ketones.
  (a) 4-methylcyclohexanone
  (b) 2,2-dimethylpropanal
  (c) 3-bromopentanal
  (d) 2-pentanone
  (e) benzophenone
  (f) o-methylbenzaldehyde

7.10 Arrange the following compounds in order of decreasing boiling points.

  (a) $CH_3CH_2CH_2C\!\!\begin{array}{c}{}^{\displaystyle O}\\{}_{\displaystyle H}\end{array}$ $CH_3CH_2CH_2CH_2OH$

  $CH_3CH_2OCH_2CH_3$

  (b) $CH_3CHOHCH_3$ $CH_3OCH_2CH_3$ $CH_3\overset{\displaystyle O}{\overset{\|}{C}}CH_3$

  (c) $CH_3C\!\!\begin{array}{c}{}^{\displaystyle O}\\{}_{\displaystyle H}\end{array}$ $CH_3CH_2\overset{\displaystyle O}{\overset{\|}{C}}CH_3$ (cyclohexanone)

7.11 Arrange the following compounds in order of decreasing solubility in water.

  (a) $CH_3\!-\!C\!\!\begin{array}{c}{}^{\displaystyle O}\\{}_{\displaystyle H}\end{array}$ $CH_3CH_2CH_2CH_2OH$

  $CH_3CH_2\overset{\displaystyle O}{\overset{\|}{C}}CH_2CH_3$

  (b) $CH_3CH_2\overset{\displaystyle O}{\overset{\|}{C}}CH_3$ $H\!-\!C\!\!\begin{array}{c}{}^{\displaystyle O}\\{}_{\displaystyle H}\end{array}$ (benzaldehyde)

  (c) CH_3OCH_3 $CH_3CHOHCH_2CH_3$

  $CH_3\overset{\displaystyle O}{\overset{\|}{C}}CH_2CH_2CH_3$

7.12 o-Hyroxybenzaldehyde and p-hydroxybenzaldehyde can each form intermolecular hydrogen bonds between the carbonyl group and the hydroxyl group.
  (a) Which one of these compounds can also form intramolecular hydrogen bonds?
  (b) Which compound has the higher boiling point? Why?

7.13 Both alcohols and ketones have lower boiling points than alcohols of comparable molecular weight, yet they have approximately the same solubility in water. Explain.

7.14 Alkenes and carbonyl compounds both undergo addition to the double bond.
  (a) Which type of double bond is more polar?
  (b) What type of reagent makes the initial attack on the carbon–carbon double bond (see Section 3.4-b)?
  (c) What type of reagent makes the initial attack on the carbonyl double bond?
  (d) To which atom of the carbonyl group does the attacking agent add?

7.15 Complete the following equations, giving only the organic products (if any).

  (a) $H\!-\!C\!\!\begin{array}{c}{}^{\displaystyle O}\\{}_{\displaystyle H}\end{array} + H_2O \rightleftharpoons$

  (b) $CH_3CH_2C\!\!\begin{array}{c}{}^{\displaystyle O}\\{}_{\displaystyle H}\end{array} + CH_3CH_2OH \overset{H^+}{\rightleftharpoons}$

  (c) $CH_3\overset{\displaystyle O}{\overset{\|}{C}}CH_3 + CH_3OH \overset{H^+}{\rightleftharpoons}$

  (d) $CH_3C\!\!\begin{array}{c}{}^{\displaystyle O}\\{}_{\displaystyle H}\end{array} + 2\,CH_3CH_2CH_2OH \overset{H^+}{\longrightarrow}$

  (e) $CH_3\overset{\displaystyle O}{\overset{\|}{C}}CH_2CH_2CH_3 + 2\,CH_3CH_2OH \overset{H^+}{\longrightarrow}$

  (f) (phenyl)$\!-\!C\!\!\begin{array}{c}{}^{\displaystyle O}\\{}_{\displaystyle H}\end{array} + HCN \longrightarrow$

  (g) $CH_3\overset{\displaystyle O}{\overset{\|}{C}}CH_2CH_3 + HCN \longrightarrow ? \overset{H^+}{\underset{HOH}{\longrightarrow}}$

  (h) $CH_3CH_2C\!\!\begin{array}{c}{}^{\displaystyle O}\\{}_{\displaystyle H}\end{array} \overset{OH^-}{\longrightarrow} ? \overset{\Delta}{\longrightarrow}$

  (i) $CH_3CH_2\overset{\displaystyle O}{\overset{\|}{C}}CH_2CH_3 \overset{OH^-}{\longrightarrow} ? \overset{\Delta}{\longrightarrow}$

  (j) (cyclohexanone) $\overset{OH^-}{\longrightarrow} ? \overset{\Delta}{\longrightarrow}$

(k)

(l) $(CH_3)_3C-\underset{H}{\overset{O}{C}} + CH_3-\overset{O}{C}-\text{(phenyl)} \xrightarrow{OH^-}$

(m) $CH_3CH_2\overset{O}{C}_H \xrightarrow[H^+]{K_2Cr_2O_7}$

(n) $CH_3CH_2CH_2\overset{O}{C}CH_3 \xrightarrow[H^+]{KMnO_4}$

(o) $CH_3CH_2CH_2\overset{O}{C}_H \xrightarrow[Ni]{H_2}$

(p) (cyclopentanone) $=O \xrightarrow[Ni]{2H_2}$

(q) (phenyl)$-CH=CHCH_2-\overset{O}{C}-CH_3 \xrightarrow[\text{ether}]{LiAlH_4} \xrightarrow[H_2O]{H^+}$

(r) (cyclopentene)$=O \xrightarrow[CH_3CH_2OH]{NaBH_4}$

7.16 Mixed aldol condensations that use two different aliphatic aldehydes are not preparatively useful because they yield a mixture of too many products. Write formulas for all addition products obtained from the base catalyzed reaction of acetaldehyde and propanal.

7.17 Formaldehyde, benzaldehyde, and benzophenone do not exist in the enol form. Explain.

7.18 Draw all the possible enol forms for each of the following ketones.
(a) cyclobutanone
(b) butanone
(c) 2-methylcyclopentanone
(d) 2-pentanone

7.19 Draw the resonance hybrids for the enolate ion derived from (a) propanal and (b) 3-pentanone.

7.20 Draw the enol form(s) of the following dicarbonyl compounds. Also show the intramolecular hydrogen bond that is formed in each of the enol tautomers.
(a) $CH_3\overset{O}{C}CH_2\overset{O}{C}OCH_3$
(b) $CH_3\overset{O}{C}CH(CH_3)\overset{O}{C}CH_3$

(c)

(d)

7.21 Write two half-reactions for the following oxidation–reduction reaction.

(phenyl)$-\overset{O}{C}-H + 2\,Ag(NH_3)_2{}^+\,OH^- \longrightarrow$

(phenyl)$-\overset{O}{C}-O^-NH_4{}^+ + 2\,Ag + 2\,NH_3 + NH_4OH$

7.22 Write equations for the preparation of the following.
(a) 2-propen-1-ol from 2-propenal
(b) acetic acid from acetaldehyde
(c) 2-hexanone from the appropriate alcohol
(d) cyclohexylmethanol from cyclohexylmethanal
(e) acetone diethyl ketal
(f) 3-hydroxy-2-methylpentanal, then 2-methyl-2-pentenal, then 2-methyl-2-penten-1-ol from propanal

7.23 Define the following terms and give a suitable example for each.
(a) carbonyl group   (b) hemiacetal
(c) hemiketal   (d) acetal
(e) ketal   (f) cyanohydrin
(g) aldol condensation (h) α-hydrogens
(i) enolization   (j) enol form
(k) keto form   (l) enolate ion
(m) tautomerism   (n) tautomers

7.24 What is the effective chemical reagent in (a) Benedict's solution, (b) Fehling's solution, and (c) Tollens' reagent?

7.25 Devise simple chemical tests that would enable you to distinguish each compound from the others in each of the following.
(a) 2-butanol, 2-methyl-1-propanol, and 2-methyl-2-propanol
(b) 1-butanol, butanal, and butanone
(c) propene and propanal
(d) benzaldehyde and cyclohexanone

7.26 Give an equation for the commercial preparation and give one use for each of the following compounds.
(a) formaldehyde   (b) acetaldehyde
(c) acetone   (d) chloral hydrate

7.27 Compound A, whose molecular formula is C_5H_8O, decolorizes bromine water and gives a positive Benedict's test. Oxidation of A gave compound B ($C_5H_8O_2$), which gave $(CH_3)_2CHCH_2COOH$ upon catalytic reduction with hydrogen. What are the possible formulas for A and B?

8

Acids and Esters

Organic acids contain the *functional group* $\overset{\displaystyle O}{\overset{\displaystyle \|}{-\text{C}}}-\textbf{OH.}$ This particular grouping of atoms is called the **carboxyl** group (derived from a combination of *carb*onyl and hydr*oxyl*), and thus organic acids are generally referred to as **carboxylic acids.** The general formula for the carboxylic acids is written **RCOOH,** where **R** may be either an aliphatic or an aromatic group.

Carboxylic acid derivatives are formally obtained by substituting another atom or group of atoms for the hydroxyl group. Replacement by a halogen atom, an acyloxy group, an alkoxy group, and an amino group gives rise to *acyl halides, acid anhydrides, esters,* and *amides,* respectively.

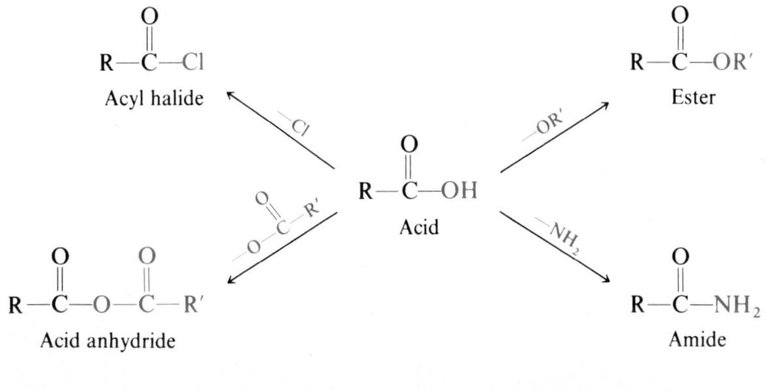

$$
\begin{array}{ccc}
\overset{\displaystyle O}{\overset{\displaystyle \|}{\text{R}-\text{C}}}-\text{Cl} & & \overset{\displaystyle O}{\overset{\displaystyle \|}{\text{R}-\text{C}}}-\text{OR}' \\
\text{Acyl halide} & & \text{Ester}
\end{array}
$$

$$
\overset{\displaystyle O}{\overset{\displaystyle \|}{\text{R}-\text{C}}}-\text{OH}
$$

Acid

$$
\begin{array}{ccc}
\overset{\displaystyle O}{\overset{\displaystyle \|}{\text{R}-\text{C}}}-\text{O}-\overset{\displaystyle O}{\overset{\displaystyle \|}{\text{C}}}-\text{R}' & & \overset{\displaystyle O}{\overset{\displaystyle \|}{\text{R}-\text{C}}}-\text{NH}_2 \\
\text{Acid anhydride} & & \text{Amide}
\end{array}
$$

Acids

8.1 Nomenclature of Carboxylic Acids

The acids most frequently encountered are known by their common names. Many of these are based upon Latin and Greek names and are related to the source of the acid. When using common names, substituted acids are named by locating the

TABLE 8.1 Nomenclature of Some Aliphatic Carboxylic Acids

Formula	Common Name	IUPAC Name
HCOOH	Formic acid	Methanoic acid
CH_3COOH	Acetic acid	Ethanoic acid
CH_3CH_2COOH	Propionic acid	Propanoic acid
$CH_3(CH_2)_2COOH$	Butyric acid	Butanoic acid
$CH_3(CH_2)_3COOH$	Valeric acid	Pentanoic acid
$CH_3(CH_2)_4COOH$	Caproic acid	Hexanoic acid
$CH_3(CH_2)_6COOH$	Caprylic acid	Octanoic acid
$CH_3(CH_2)_8COOH$	Capric acid	Decanoic acid
$CH_3(CH_2)_{14}COOH$	Palmitic acid	Hexadecanoic acid
$CH_3(CH_2)_{16}COOH$	Stearic acid	Octadecanoic acid
CH_3 | CH_3CH—COOH	Isobutyric acid (α-Methylpropionic acid)	2-Methylpropanoic acid
Cl | Cl—C—COOH | Cl	Trichloroacetic acid	2,2,2-Trichloroethanoic acid
CH_3CHCH_2—COOH | OH	β-Hydroxybutyric acid	3-Hydroxybutanoic acid
HOOC—CH_2—CH—COOH | OH	Malic acid	2-Hydroxybutanedioic acid
CH_3CH=CH—COOH	Crotonic acid	2-Butenoic acid
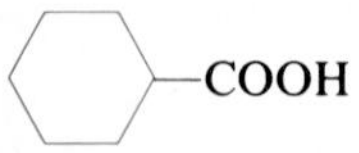 ⬡—COOH	Cyclohexanecarboxylic acid	Cyclohexanecarboxylic acid
CH_3 | $(CH_3)_2CHCH_2$—⬡—CHCOOH	Ibuprofen (in Motrin, Nuprin, and Advil)	2-(p-Isobutylphenyl)propanoic acid

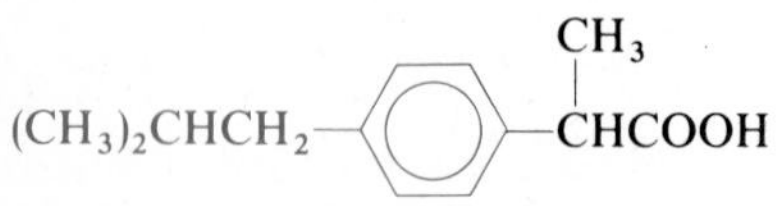

position of the substituent group by means of the Greek letters α, β, γ, δ, etc. These letters refer to the position of the carbon atom in relation to the carboxyl carbon as illustrated.

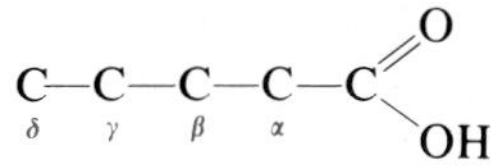

In the IUPAC system, the parent hydrocarbon is taken to be the one that corresponds to the longest continuous chain containing the carboxyl group. The -*e* ending of the parent alkane is replaced by the suffix -*oic,* and the word *acid* follows. As in the case of the aldehydes, the carboxyl carbon is understood to be carbon number one. The IUPAC system uses only numbers, not Greek letters. Table 8.1 contains the names and formulas of some of the most frequently encountered carboxylic acids.

Derivatives of the aliphatic dicarboxylic acids are very important in biological systems. These acids are almost always referred to by their common names; a mnemonic for remembering these names is given in Table 8.2.

The frequently encountered aromatic acids are known by their common names; others are named as derivatives of the parent acid, benzoic acid. If the aromatic group is not bonded directly to the carboxyl group, then the aromatic group is named as a substituent.

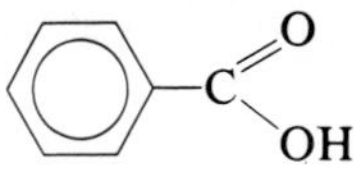

Benzoic acid
(Phenylmethanoic acid)

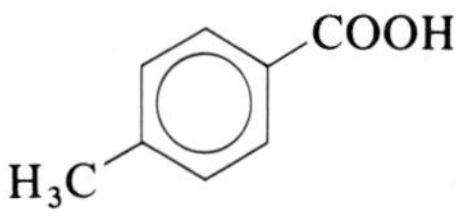

4-Methylbenzoic acid
(*p*-Methylbenzoic acid)

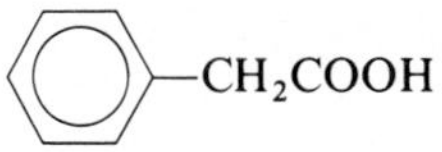

Phenylacetic acid
(Phenylethanoic acid)

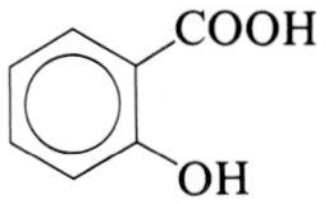

Salicylic acid
(*o*-Hydroxybenzoic acid)

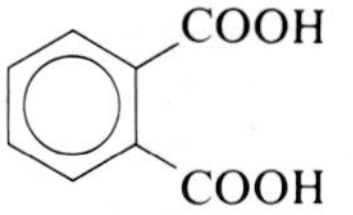

Phthalic acid
(Benzene-1,2-dicarboxylic acid

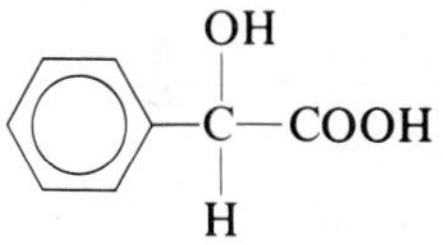

Mandelic acid
(2-Hydroxy-2-phenylethanoic acid)

TABLE 8.2 Aliphatic Dicarboxylic Acids

Formula	Common Name	Mnemonic
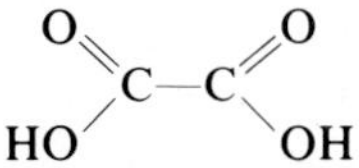	Oxalic acid	Oh
$HOOC-CH_2-COOH$	Malonic acid	My
$HOOC-(CH_2)_2-COOH$	Succinic acid	Such
$HOOC-(CH_2)_3-COOH$	Glutaric acid	Good
$HOOC-(CH_2)_4-COOH$	Adipic acid	Apple
$HOOC-(CH_2)_5-COOH$	Pimelic acid	Pie

8.2 *Preparation of Carboxylic Acids*

a. Oxidation

Carboxylic acids are the final oxidation products of aldehydes and/or primary alcohols. The acid produced contains the same number of carbon atoms as did the precursor aldehyde or alcohol.

General Equation

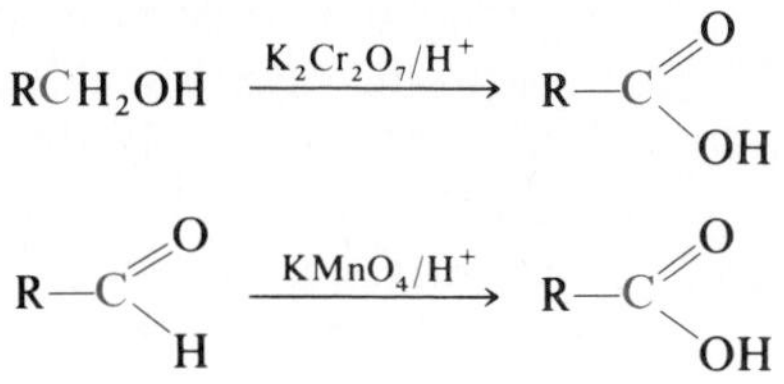

$$RCH_2OH \xrightarrow{K_2Cr_2O_7/H^+} R-C\underset{OH}{\overset{O}{\lessgtr}}$$

$$R-C\underset{H}{\overset{O}{\lessgtr}} \xrightarrow{KMnO_4/H^+} R-C\underset{OH}{\overset{O}{\lessgtr}}$$

Specific Equation

$$3\,CH_3CH_2OH + 2\,K_2Cr_2O_7 + 8\,H_2SO_4 \longrightarrow 3\,CH_3COOH + 11\,H_2O + 2\,K_2SO_4 + 2\,Cr_2(SO_4)_3$$

Ethanol   (orange)   Acetic acid   (green)

The preceding reaction forms the basis of the roadside breath test to detect drunken drivers. The Breathalyzer contains a solution of sulfuric acid and potassium dichromate (orange). If a suspect's breath causes the color to change to green (Cr^{3+}), this is an indication that his/her blood contains more than the legal level of alcohol (0.1%). Further tests are then carried out at the police station to confirm the suspicion.

Our bodies accomplish this same oxidation of alcohols to acids whenever we drink alcoholic beverages. The liver contains enzymes that convert the ethyl alcohol to acetic acid. Acetic acid is utilized to provide energy (Section 16.8), or it is converted into fat (Section 17.2). Excess alcohol that is not oxidized in the liver continues to circulate in the blood and eventually causes intoxication.

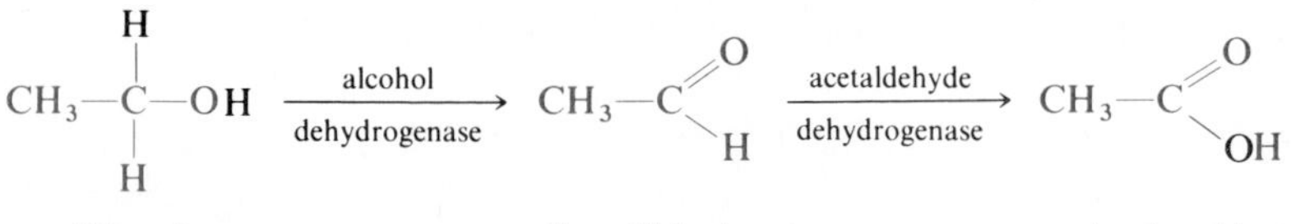

$$CH_3-\overset{\displaystyle H}{\underset{\displaystyle H}{C}}-OH \xrightarrow[\text{dehydrogenase}]{\text{alcohol}} CH_3-C\underset{H}{\overset{O}{\lessgtr}} \xrightarrow[\text{dehydrogenase}]{\text{acetaldehyde}} CH_3-C\underset{OH}{\overset{O}{\lessgtr}}$$

Ethanol   Acetaldehyde   Acetic acid

Recall that in Section 6.5, we stated that methanol and ethylene glycol are poisonous to the body. It is not these compounds themselves, but rather their oxidation products, that cause the toxic effects. 1,2-Propylene glycol, on the other hand, is harmless because its oxidation product is pyruvic acid, which is a normal intermediate of carbohydrate metabolism.

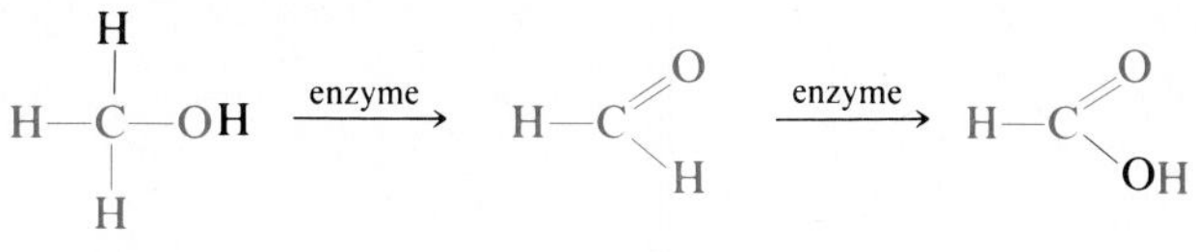

Methanol → Formaldehyde → Formic acid

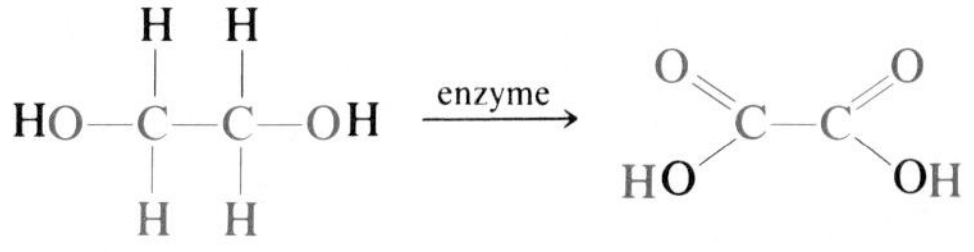

Ethylene glycol → Oxalic acid

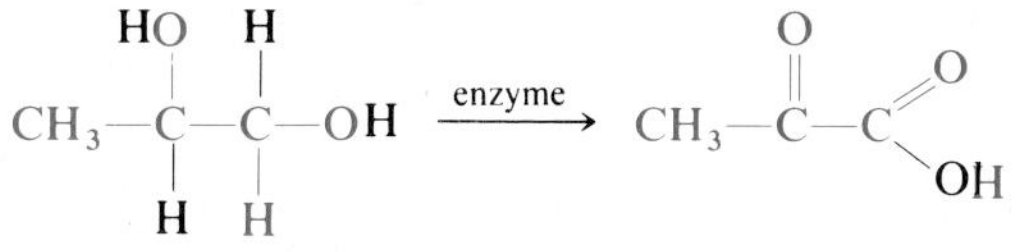

1,2-Propylene glycol → Pyruvic acid

Aromatic acids are prepared by direct oxidation of alkylbenzenes. The length of the carbon chain is not a factor. Benzoic acid[1] is the only acid formed; the remaining carbon atoms of the chain are oxidized to carbon dioxide and water.

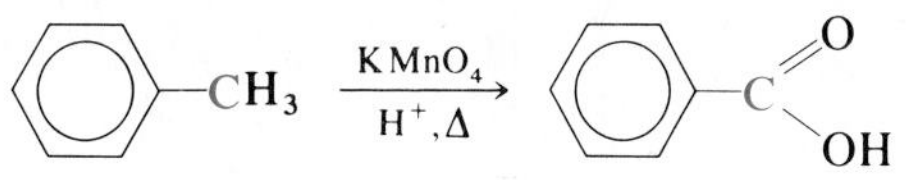

Toluene → Benzoic acid

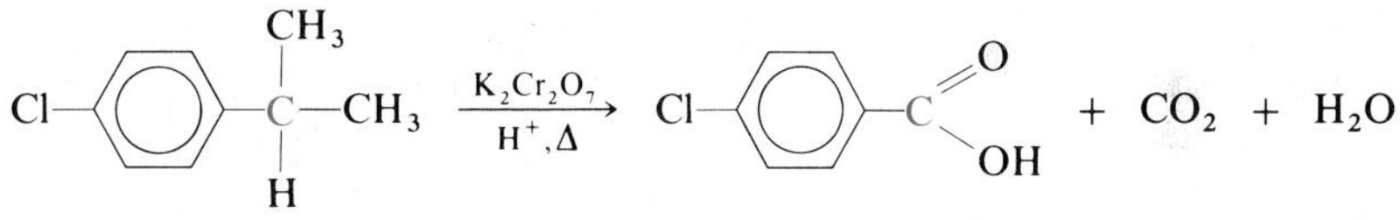

p-Chloroisopropylbenzene → p-Chlorobenzoic acid $+ CO_2 + H_2O$

b. From Alkyl Halides

This preparation works best for primary alkyl halides, and it involves a two-step reaction sequence. The alkyl halide is first converted to its corresponding nitrile in a

---

[1] Recall that in Section 4.4 we said that benzene is highly poisonous to humans, whereas toluene is less toxic. Recall also that animals are unable to synthesize the benzene ring, nor because of its stability, can they break it down. Benzene and toluene readily enter the body through the lungs and the skin. They are transported to the liver where toluene can be metabolized to benzoic acid by oxidative enzymes. The benzoic acid is then conjugated with the amino acid glycine (H_2NCH_2COOH) to form a water-soluble compound, hippuric acid ($C_6H_5CONHCH_2COOH$). Hippuric acid is transported to the kidney and eliminated in the urine. Benzene cannot be converted to a water-soluble metabolite for excretion. Instead, it accumulates in the liver (and other body tissues, e.g., bone marrow) where benzene exerts its toxic effects.

simple substitution reaction.

$$\underset{\text{Ethyl bromide}}{CH_3CH_2Br} + KCN \longrightarrow \underset{\substack{\text{Ethyl cyanide}\\ \text{(Propionitrile)}^2}}{CH_3CH_2CN} + KBr$$

In a second step, the nitrile is hydrolyzed to the acid or to its salt. A strong acid or base is the catalyst. Notice that the reaction sequence yields an acid containing one more carbon atom than the initial alkyl halide.

$$CH_3CH_2C{\equiv}N \xrightarrow[\text{acid hydrolysis}]{+\,2HOH\,+\,HCl} CH_3CH_2C{\overset{O}{\lessgtr}}_{OH} + NH_4Cl$$

$$CH_3CH_2C{\equiv}N \xrightarrow[\text{basic hydrolysis}]{+\,HOH\,+\,NaOH} CH_3CH_2C{\overset{O}{\lessgtr}}_{O^-\,Na^+} + NH_3$$

8.3 *Physical Properties of Carboxylic Acids*

The first nine members of the carboxylic acid series are colorless liquids having very disagreeable odors. The odor of vinegar is that of acetic acid; the odor of rancid butter is primarily that of butyric acid. Caproic acid (Latin, *caper,* goat) is present in the hair and the secretions of goats. The acids from C_5 to C_{10} all have "goaty" odors (odor of Limburger cheese). These acids are produced by the action of skin bacteria on perspiration oils; hence the odor of unaired locker rooms. The acids above C_{10} are waxlike solids, and because of their low volatility are practically odorless. As was the case with the alcohols, the lower acid homologs are soluble in water, but the solubility decreases sharply after butyric acid as the hydrocarbon character of the molecule increases. The aromatic acids are odorless solids and are sparingly soluble in water. All the acids are soluble in organic solvents such as alcohol, benzene, carbon tetrachloride, and ether.

Table 8.3 contains physical constants for the first ten members of the aliphatic carboxylic acid family. Notice that the melting points show no regular increase with increasing molecular weight. The boiling points of the acids are abnormally high, and increase in approximately 20°C increments. They are higher than the boiling points of alcohols or alkyl halides of comparable molecular weights. This low volatility is attributed to the strong association of acid molecules in the liquid state. In fact, molecular weight measurements indicate that the acids exist primarily as

[2] By naming the products as nitriles rather than cyanides, the relationship to the acids is stressed. Nitriles are named according to the acids that they yield upon hydrolysis. Thus CH_3CN is acetonitrile and CH_3CH_2CN is propionitrile. These alkyl cyanides should not be confused with alkyl isocyanates, R—N=C=O (e.g., CH_3NCO is methyl isocyanate). The inadvertent release of methyl isocyanate gas was the cause of a major chemical disaster in Bhopal, India, in 1984.

TABLE 8.3 Physical Constants of Carboxylic Acids

Formula	Name of Acid	MP (°C)	BP (°C)	Solubility (g/100 g of water)	K_a (25°C)
HCOOH	Formic	8	101	Miscible	1.8×10^{-4}
CH_3COOH	Acetic	17	118	Miscible	1.8×10^{-5}
CH_3CH_2COOH	Propionic	−22	141	Miscible	
$CH_3(CH_2)_2COOH$	Butyric	−5	163	Miscible	
$CH_3(CH_2)_3COOH$	Valeric	−35	187	5	
$CH_3(CH_2)_4COOH$	Caproic	−3	205	1	1.5×10^{-5}
$CH_3(CH_2)_5COOH$	Enanthic	−8	224	0.24	
$CH_3(CH_2)_6COOH$	Caprylic	16	238	0.07	
$CH_3(CH_2)_7COOH$	Pelargonic	14	254	0.03	
$CH_3(CH_2)_8COOH$	Capric	31	268	0.02	

dimers, or double molecules. For example, acetic acid is believed to have the following hydrogen-bonded structure.

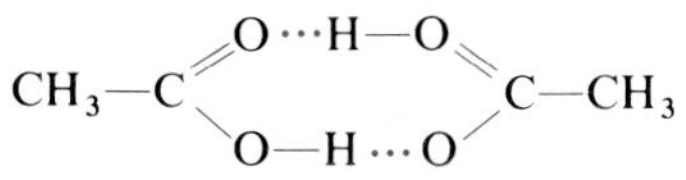

$$CH_3-C \overset{\displaystyle O \cdots H-O}{\underset{\displaystyle O-H \cdots O}{}} C-CH_3$$

8.4 *Acidity, Hydrogen Ion Concentration, and pH*

Aqueous solutions of organic acids turn litmus red, neutralize bases, have a sour taste, and conduct an electric current. In this respect, carboxylic acids are functionally, although not structurally, similar to the common inorganic acids such as hydrochloric, nitric, and sulfuric acids. However, the two types of acids are quite different with respect to a property which we call *acid strength*. HCl, HNO_3, H_2SO_4, HBr, and HI are referred to as strong acids. That is to say, they are completely dissociated in aqueous solution.[3]

$$HCl \longrightarrow H^+ + Cl$$

Carboxylic acids, on the other hand, are only very slightly dissociated (less than 5%) in aqueous solution. The following equilibrium condition lies far to the left.

$$RCOOH \rightleftharpoons H^+ + RCOO$$

In this respect, then, the organic acids are most similar to the weak inorganic acids such as HF ($K_a = 3.5 \times 10^{-4}$) and HNO_2 ($K_a = 4.6 \times 10^{-4}$).

[3]Whenever the solvent is not specified, it is to be understood that the dissociation occurs in an aqueous solution. Therefore, the entire equation should be $HCl + H_2O \rightarrow H_3O^+ + Cl^-$. In this book we follow the convention of omitting water from the dissociation equation and of writing the symbol H^+ to represent the hydronium ion.

One method of comparing relative acid strengths is to compare the values of the acid dissociation constant (K_a) for each particular acid under consideration. Acid dissociation constants are obtained from the mass law expression involving the equilibrium between the undissociated acid, hydrogen ions, and anions. For any acid HA, which dissociates in water, the following equations can be written.

$$HA \rightleftharpoons H^+ + A^-$$

$$K_a = \frac{[H^+][A^-]}{[HA]}$$

It is clear that the highly dissociated acids will have large values for K_a because their dissociation produces high concentrations of hydrogen ion and low concentrations of undissociated acid. Strong acids have K_a values greater than 10. Conversely, weak acids, which are only slightly dissociated, have low concentrations of hydrogen ion and high concentrations of undissociated acid. From the K_a values given in Table 8.3 we know that formic acid is a stronger acid than acetic acid, which in turn is stronger than the other higher-membered acids, which are all of comparable acidity.

At this point, a worthwhile exercise is the calculation of the hydrogen ion concentration of an acid of a given concentration. If we calculate the hydrogen ion concentrations of equimolar solutions of formic and acetic acids, we should expect to find that the solution of formic acid will have the higher concentration of hydrogen ions.

Example 8.1 Calculate the hydrogen ion concentration present (a) in a 0.10 M solution of formic acid and (b) in a 0.10 M solution of acetic acid.

Solution

(a)

$$HCOOH \rightleftharpoons H^+ + HCOO^-$$

$$K_a = \frac{[H^+][HCOO^-]}{[HCOOH]}$$

Letting x represent the hydrogen ion concentration, we obtain

$$1.8 \times 10^{-4} = \frac{(x)(x)}{0.10 - x}$$

Assume x is negligible compared to 0.10, then

$$1.8 \times 10^{-4} = \frac{x^2}{0.10}$$

$$x^2 = 1.8 \times 10^{-5}$$

$$x = 4.2 \times 10^{-3} M$$

Therefore, the hydrogen ion concentration in a solution of 0.10 M formic acid $= 4.2 \times 10^{-3}$ M. This represents a dissociation of

$$\frac{(4.2 \times 10^{-3})(100)}{0.10}$$

or 4.2%, which means that less than five molecules of formic acid out of every hundred are dissociated at any given time.

(b) For acetic acid, we have

$$CH_3COOH \rightleftharpoons H^+ + CH_3COO^-$$

$$K_a = \frac{[H^+][CH_3COO^-]}{[CH_3COOH]}$$

Letting y represent the hydrogen ion concentration, we obtain

$$1.8 \times 10^{-5} = \frac{(y)(y)}{0.10 - y}$$

$$y = 1.3 \times 10^{-3}$$

The hydrogen ion concentration of a 0.10 M acetic acid solution is, as predicted, less than that for formic acid. The percent dissociation for acetic acid is

$$\frac{(1.3 \times 10^{-3})(100)}{0.10} = 1.3\%$$

A knowledge of the hydrogen ion concentration present in all aqueous reaction mixtures is of utmost importance to the chemist. The student will recall that pure water contains a hydrogen ion concentration of 1.0×10^{-7} M. Water is said to be a neutral substance because it also contains an equal concentration of hydroxide ions (1.0×10^{-7} M). Accordingly, an **acid** is often defined as *a substance that increases the hydrogen ion concentration of water beyond 1.0×10^{-7} M*, and a **base** is defined as *a substance that reduces the hydrogen ion concentration of water below 1.0×10^{-7} M*.

It often becomes cumbersome to work with powers of ten, and so for the sake of convenience, another term has been introduced to specify the hydrogen ion concentration of a given aqueous solution. The term is **pH**, and it is defined as the *negative logarithm of the hydrogen ion concentration.*

$$pH = -\log[H^+] = \log\frac{1}{[H^+]}$$

Therefore, a hydrogen ion concentration of 1.0×10^{-7} M corresponds to a pH of 7.0. Values of pH between 0 and 14 provide a convenient scale for the comparison of

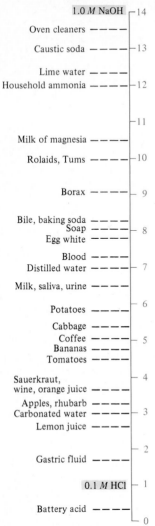

FIGURE 8.1 pH values of some common substances.

relative acidities. All neutral solutions have a pH of 7: an acid solution is one whose pH is less than 7; a basic solution has a pH greater than 7. Figure 8.1 lists pH values for some common substances. Note that most ingestible substances have pH values below 7.

It is often necessary in both organic chemistry and biochemistry to convert a given hydrogen ion concentration to its equivalent pH value. Many students have difficulty with this calculation because of a lack of familiarity with logarithms. The following examples illustrate a short method for treating common pH problems. The method derives directly from the longer, conventional method.

Example 8.2 Calculate the pH of a solution whose hydrogen ion concentration is 2.7×10^{-6} M.

Solution

Long Method: $\quad pH = -\log [H^+] = \log \dfrac{1}{[H^+]}$

$$pH = -\log (2.7 \times 10^{-6})$$
$$= -\log 2.7 + (-\log 10^{-6})$$
$$= -0.43 + 6 = 5.57$$

Short Method: For any hydrogen concentration that is expressed in the form

$$[H^+] = a.b \times 10^{-z}$$
$$\text{then the } pH = z - \log a.b$$

Thus,

(1)   if $[H^+] = 2.7 \times 10^{-6}$
     then $pH = 6 - \log 2.7$
          $= 6 - 0.43 = 5.57$

(2)   if $[H^+] = 8.3 \times 10^{-8}$
     then $pH = 8 - \log 8.3$
          $= 8 - 0.92 = 7.08$

(3)   if $[H^+] = 2.0 \times 10^{-3}$
     then $pH = 3 - \log 2.0$
          $= 3 - 0.30 = 2.70$

(4)   if $[H^+] = 1.0 \times 10^{-5}$
     then $pH = 5 - \log 1.0$
          $= 5 - 0 = 5.0$

For all cases such as the last, where $[H^+] = 1.0 \times 10^{-n}$, the pH will always be equal to the exponent since the logarithm of 1.0 is zero. These particular problems are the easiest and can be done just by inspection. For example,

$$\text{if } [H^+] = 1.0 \times 10^{-12}$$
$$\text{then pH} = 12.0$$

Conversely,

$$\text{if pH} = 9.0$$
$$\text{then } [H^+] = 1.0 \times 10^{-9}$$

8.5 *Buffers*

In a large number of chemical reactions (especially those carried out in aqueous solution), it is important that the pH remain relatively constant at a fixed value. Practically all biological and physiological processes occur most readily at some unique pH value. The pH of human blood, for example, varies only slightly between the values of 7.35 and 7.45, and for any given person, the pH of venous and arterial blood differs by only 0.02 pH unit despite the production of numerous acids and bases in the cells. If the pH were to fall outside this range, the blood would lose its ability to transport oxygen from the lungs to the cells. *Any solution that maintains a near constancy of pH is referred to as a* buffer *or a* buffer solution. Such a solution contains a combination of two chemical compounds that work as a pair in resisting marked changes in pH. Solutions containing a weak acid and a salt of the weak acid, or a weak base and a salt of the weak base, serve as excellent buffer systems.

The buffering phenomenon is best understood by considering the processes that occur when a strong acid or a strong base is added to a buffer solution. We shall limit our discussion[4] to the buffer pair consisting of the weak acid RCOOH and its salt, RCOO⁻ Na⁺.

When a strong acid is added to a buffer system, the added hydrogen ions immediately combine with the excess anions, RCOO⁻, present in the solution, forming the weakly dissociated

[4] This generalized discussion will apply to the organic acids as well as to the weak inorganic acids. Of particular importance to blood chemistry are the bicarbonate buffer pair (H_2CO_3/HCO_3^-) and the phosphate buffer pair ($H_2PO_4^-/HPO_4^{2-}$).

acid RCOOH. The hydrogen ion concentration of the resulting solution is therefore not appreciably altered from that of the original solution.

$$RCOO^- Na^+ \longrightarrow RCOO^- + Na^+$$
$$+$$
$$HCl \longrightarrow H^+ + Cl^-$$
$$\downarrow$$
$$RCOOH$$

If a strong base is added to a buffer system, the hydroxyl ions are neutralized by the hydrogen ions produced by the dissociation of the acid. As hydrogen ions are removed from the solution, more acid molecules will dissociate in order to reestablish equilibrium concentrations. This dissociation of the acid tends to maintain a nearly constant hydrogen ion concentration.

$$RCOOH \rightleftharpoons RCOO^- + H^+$$
$$+$$
$$NaOH \longrightarrow Na^+ + OH^-$$
$$\downarrow$$
$$HOH$$

The effect, then, of the addition of an acid or a base to a buffer, is to increase the concentration of the weak acid or the anion, respectively. However, since these species are present in excessive amounts in the buffer, the increase in their concentrations is insignificant, and there is little change in the pH of the solution. A buffer will continue to function until the concentration of added acid or base exceeds the concentration of the weak acid or anion present in the system.

By starting with the mass action expression for the dissociation of a weak acid, we can easily derive an equation that permits the calculation of the pH of a solution containing any buffer pair.

$$K_a = \frac{[H^+][RCOO^-]}{[RCOOH]}$$

1. Rearrange.

$$[H^+] = K_a \frac{[RCOOH]}{[RCOO^-]}$$

2. Take the logarithm of both sides of the equation.

$$\log[H^+] = \log K_a + \log \frac{[RCOOH]}{[RCOO^-]}$$

3. Multiply both sides of the equation by -1.

$$-\log[H^+] = -\log K_a + -\log \frac{[RCOOH]}{[RCOO^-]}$$

Since $-\log[H^+] = pH$, and $-\log K_a = pK_a$, and

$$-\log \frac{[RCOOH]}{[RCOO^-]} = +\log \frac{[RCOO^-]}{[RCOOH]}$$

we have

$$pH = pK_a + \log \frac{[RCOO^-]}{[RCOOH]}$$

This expression is referred to as the **Henderson–Hasselbalch** equation.

Since pK_a is the negative logarithm of the dissociation constant of the acid, it, too, is a constant, and thus the pH of any particular buffer system varies with the ratio of the concentration of the anion (provided by the salt) to that of the acid. If the weak acid and its anion are present in equal concentrations, the hydrogen ion concentration of the buffered solution will be equal to the dissociation constant of the weak acid (that is, $pH = pK_a$). Why? Reference to a table of dissociation constants for weak acids allows us to choose a system that will buffer at almost any point within the entire pH scale. A few common buffer systems are indicated in Table 8.4.

Using the values in Table 8.4 we can calculate the concentrations of bicarbonate ion relative to carbonic acid and of monohydrogen phosphate relative to dihydrogen phosphate required to maintain the pH of blood plasma at 7.4. For the bicarbonate pair,

$$pH = pK_a + \log \frac{[HCO_3^-]}{[H_2CO_3]}$$

$$7.4 = 6.37 + \log \frac{[HCO_3^-]}{[H_2CO_3]}$$

TABLE 8.4   Buffer Systems

Weak Acid	Anion	K_a	Buffer pH $= pK_a$ when $\frac{[anion]}{[weak\ acid]} = 1$
HSO_4^-	SO_4^{2-}	1.2×10^{-2}	1.92
H_3PO_4	$H_2PO_4^-$	7.5×10^{-3}	2.12
HNO_2	NO_2^-	4.6×10^{-4}	3.37
HF	F^-	3.5×10^{-4}	3.45
HCOOH	$HCOO^-$	1.8×10^{-4}	3.75
$CH_3CHOHCOOH$	$CH_3CHOHCOO^-$	1.4×10^{-4}	3.85
C_6H_5COOH	$C_6H_5COO^-$	6.5×10^{-5}	4.19
CH_3COOH	CH_3COO^-	1.8×10^{-5}	4.74
H_2CO_3	HCO_3^-	4.3×10^{-7}	6.37
HSO_3^-	SO_3^{2-}	1.0×10^{-7}	7.00
$H_2PO_4^-$	HPO_4^{2-}	6.3×10^{-8}	7.20
H_3BO_3	$B(OH)_4^-$	7.3×10^{-10}	9.14
HCN	CN^-	4.9×10^{-10}	9.31
HCO_3^-	CO_3^{2-}	5.6×10^{-11}	10.25
HPO_4^{2-}	PO_4^{3-}	2.2×10^{-13}	12.67

Chapter 8 · Acids and Esters

$$\log \frac{[\text{HCO}_3{}^-]}{[\text{H}_2\text{CO}_3]} = 1.03$$

$$\therefore \frac{[\text{HCO}_3{}^-]}{[\text{H}_2\text{CO}_3]} \approx 10$$

The concentration of bicarbonate ion in the blood is ten times that of carbonic acid. The disproportionate concentrations of the two components of the system would seem to indicate that this particular buffer pair has an extremely narrow buffering capacity. However, as we shall see in Section 19.8, the respiratory removal of carbon dioxide keeps this buffer ratio nearly constant. Similarly, it can be determined from the Henderson–Hasselbalch equation that at a pH of 7.4 the concentration of $\text{HPO}_4{}^{2-}$ in the blood is 1.6 times that of $\text{H}_2\text{PO}_4{}^-$. This calculation is left to the student in Exercise 8.14.

8.6 *Effect of Structure on Acidity*

In Section 8.4 we mentioned that most organic acids are only slightly dissociated in solution and hence they are classified as weak acids. We ought now to ask: Why are carboxylic acids much more acidic (by a factor of about 10^{11}) than alcohols? The reason usually given is that the carboxylate anion formed upon dissociation is stabilized by resonance (two equivalent resonance structures) relative to the original acid molecule. Thus the negative charge is spread over the three-atom **O—C—O** grouping. In the alkoxide ion, RO^-, the negative charge is not delocalized, but is concentrated on the single oxygen atom. Although the tendency of organic acids to dissociate is slight, the stabilization of the anion through resonance facilitates the ionization process.

Resonance-stabilized anion $K_a \approx 10^{-5}$

$$\text{R—OH} \rightleftharpoons \text{H}^+ + \text{R—O}^- \qquad K_a \approx 10^{-16}$$

If resonance were the only factor that accounted for the acidity, we would expect all carboxylic acids to be of comparable strength. We know that this is not so. Table 8.5 lists some of the acids and their respective dissociation constants. Notice that a wide range of values is possible. Carboxylic acids that contain strong electron-attracting groups (like halogens) on the α-carbon are far more acidic than the unsubstituted acid. The methyl group, on the other hand, is an electron-releasing group. This accounts for the tenfold decrease in acidity in going from formic acid to acetic acid. Electron-attracting groups facilitate ionization of the carboxylic acid by withdrawing electrons from the carboxylate group, weakening the oxygen–hydrogen bond, and thus aiding in the release of the proton. These groups also help to stabilize the acid anion, by delocalizing the negative charge from the carboxylate group to adjacent atoms. Recall (Section 3.4-b) that inductive effects are additive, and the greater the number of electron-withdrawing groups on the α-carbon, the stronger the acid. (Likewise, the more numerous the electron-releasing groups, the weaker the acid.) Inductive effects diminish rapidly with distance; thus a halogen on a β-carbon is not nearly as effective as one on an α-carbon (compare α- and β-chlorobutyric acids in Table 8.5). In the following equations the arrowheads indicate the inductive effect of withdrawing electrons from (stabilizing), or releasing electrons to (destabilizing), the carboxylate anion.

TABLE 8.5 Factors Influencing Acid Strength of Carboxylic Acids

Factor	Formula	Name of Acid	K_a
Electronegativity of substituent	CH_2FCOOH	Fluoroacetic	260×10^{-5}
	$CH_2BrCOOH$	Bromoacetic	130×10^{-5}
	CH_2ICOOH	Iodoacetic	75×10^{-5}
	CH_3COOH	Acetic	1.8×10^{-5}
Number of electron-withdrawing groups on α-carbon	CCl_3COOH	Trichloroacetic	200×10^{-3}
	$CHCl_2COOH$	Dichloroacetic	33×10^{-3}
	$CH_2ClCOOH$	Chloroacetic	1.4×10^{-3}
Distance from carboxyl group	$CH_3CH_2CHClCOOH$	α-Chlorobutyric	140×10^{-5}
	$CH_3CHClCH_2COOH$	β-Chlorobutyric	8.9×10^{-5}
	$CH_2ClCH_2CH_2COOH$	γ-Chlorobutyric	3.0×10^{-5}
	$CH_3CH_2CH_2COOH$	Butyric	1.5×10^{-5}

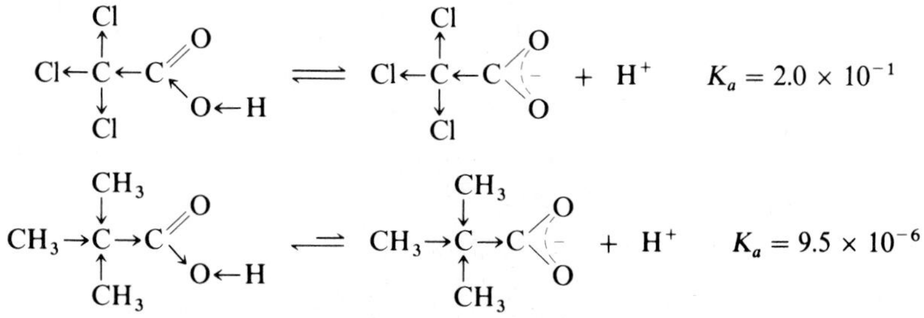

$$Cl \leftarrow \overset{\overset{\displaystyle Cl}{\uparrow}}{\underset{\underset{\displaystyle Cl}{\downarrow}}{C}} \leftarrow C \overset{O}{\underset{O \leftarrow H}{\diagup}} \rightleftharpoons Cl \leftarrow \overset{\overset{\displaystyle Cl}{\uparrow}}{\underset{\underset{\displaystyle Cl}{\downarrow}}{C}} \leftarrow C \overset{O}{\underset{O}{\diagup}}{}^{-} + \ H^+ \qquad K_a = 2.0 \times 10^{-1}$$

$$CH_3 \rightarrow \overset{\overset{\displaystyle CH_3}{\downarrow}}{\underset{\underset{\displaystyle CH_3}{\uparrow}}{C}} \rightarrow C \overset{O}{\underset{O \leftarrow H}{\diagup}} \rightleftharpoons CH_3 \rightarrow \overset{\overset{\displaystyle CH_3}{\downarrow}}{\underset{\underset{\displaystyle CH_3}{\uparrow}}{C}} \rightarrow C \overset{O}{\underset{O}{\diagup}}{}^{-} + \ H^+ \qquad K_a = 9.5 \times 10^{-6}$$

Note also that the phenyl group is an electron-attracting group, so that we find benzoic acid to be a slightly stronger acid than acetic acid but weaker than formic acid (see Exercise 8.15-j).

8.7 Chemical Properties of Carboxylic Acids

All of the reactions of the carboxylic acids can be considered to be substitution reactions. The nature of the other reacting species will determine whether substitution of the hydroxyl hydrogen or the hydroxyl group will occur.

a. Substitution of Acidic Hydrogen—Neutralization

Characteristic of all acids is the ability to react with bases to form salts.

$$\underset{\text{Acid}}{HCl} + \underset{\text{Base}}{NaOH} \longrightarrow \underset{\text{Salt}}{NaCl} + \underset{\text{Water}}{HOH}$$

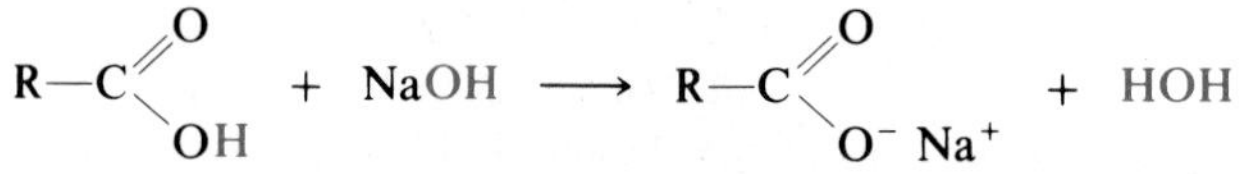

$$R-C\overset{O}{\underset{OH}{\diagup}} + NaOH \longrightarrow R-C\overset{O}{\underset{O^- Na^+}{\diagup}} + HOH$$

Salts so formed are quite different from their parent acids. They are water-soluble, ionic solids and are completely dissociated in solution. They are insoluble in non-polar solvents such as ether, benzene, and carbon tetrachloride. Use is made of this property in separating carboxylic acids from nonpolar organic compounds. For example, the addition of a sodium hydroxide solution to an ether extract containing hexanoic acid will result in the formation of a salt (sodium hexanoate) that is soluble in the aqueous phase. After separation, hydrochloric acid is added to neutralize the solution, and insoluble hexanoic acid then separates and can be recovered.

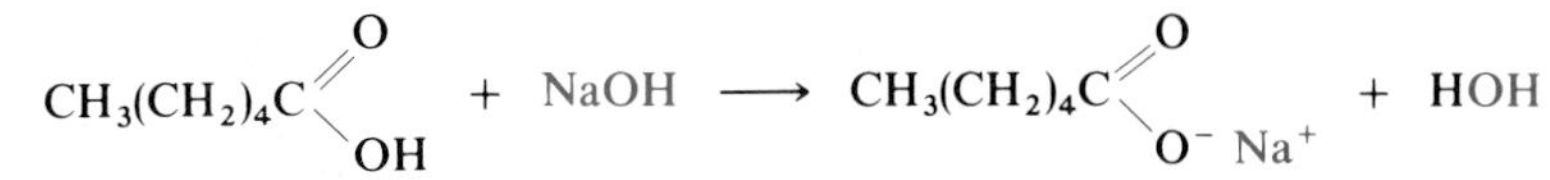

$$CH_3(CH_2)_4C\overset{O}{\underset{OH}{\big<}} + NaOH \longrightarrow CH_3(CH_2)_4C\overset{O}{\underset{O^-\,Na^+}{\big<}} + HOH$$

Hexanoic acid   Sodium hexanoate
(soluble)

$$CH_3(CH_2)_4C\overset{O}{\underset{O^-\,Na^+}{\big<}} + HCl \longrightarrow CH_3(CH_2)_4C\overset{O}{\underset{OH}{\big<}} + Na^+\,Cl^-$$

(insoluble)

All carboxylic acids, whether or not they are water soluble, liberate carbon dioxide from a solution of the weak base sodium bicarbonate ($NaHCO_3$). This reaction may be used to distinguish the carboxylic acids from all other classes of organic compounds. (Recall that the phenols are much weaker acids. They can react with sodium hydroxide but *not* with sodium bicarbonate.)

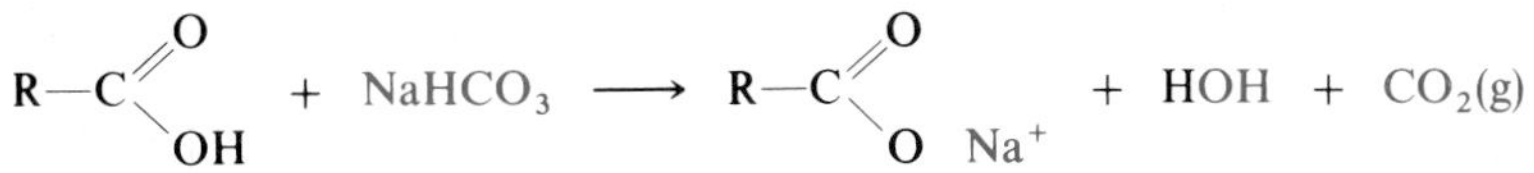

$$R-C\overset{O}{\underset{OH}{\big<}} + NaHCO_3 \longrightarrow R-C\overset{O}{\underset{O\,Na^+}{\big<}} + HOH + CO_2(g)$$

Organic salts are named in the same manner as the inorganic salts. The name of the cation is followed by the name of the organic anion. The name of the anion is obtained by dropping the -*ic* ending of the acid name and replacing it with the suffix -*ate*. This applies whether we are using common names or IUPAC names. Following are some examples.

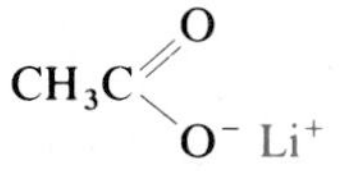

$$CH_3C\overset{O}{\underset{O^-\,Li^+}{\big<}} \qquad CH_3CH_2CH_2C\overset{O}{\underset{O^-\,K^+}{\big<}} \qquad \bigcirc\!\!-C\overset{O}{\underset{O^-\,Na^+}{\big<}}$$

Lithium acetate   Potassium butyrate   Sodium benzoate
(Lithium ethanoate) 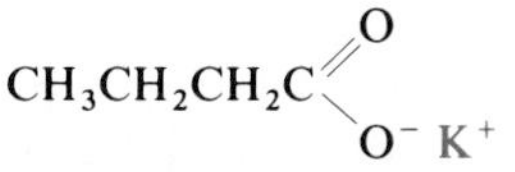   (Potassium butanoate)

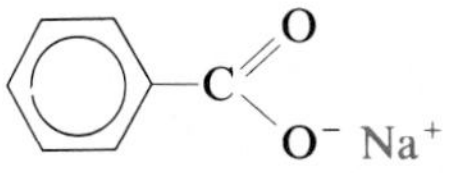

The sodium or potassium salts of long-chain carboxylic acids are called soaps. We discuss the chemistry of soaps in considerable detail in Section 12.8.

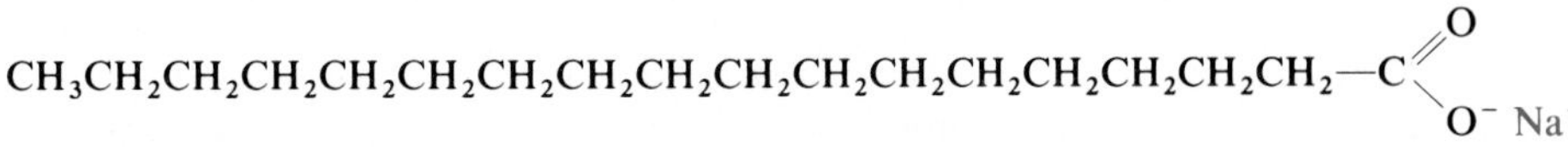

$$CH_3CH_2CH_2CH_2CH_2CH_2CH_2CH_2CH_2CH_2CH_2CH_2CH_2CH_2CH_2CH_2CH_2-C\overset{O}{\underset{O^-\,Na^+}{\big<}}$$

Sodium stearate (a soap)

Other salts that have commercial significance are those used as preservatives. They act to prevent spoilage by inhibiting the growth of bacteria and fungi. Calcium and sodium propionate are added to processed cheese and bakery goods; sodium benzoate is used as a preservative in cider, jellies, pickles, and syrups; and sodium and potassium sorbate are added to fruit juices, sauerkraut, soft drinks, and wine. (Look for these salts on ingredient labels the next time you are shopping.)

$$\left(CH_3CH_2C \underset{O^-}{\overset{O}{\diagup}} \right)_2 Ca^{2+} \qquad\qquad CH_3CH{=}CHCH{=}CHC \underset{O^- \; K^+}{\overset{O}{\diagup}}$$

Calcium propionate   Potassium sorbate

b. Substitution of the Hydroxyl Group

1. Formation of Acyl Halides An **acyl** group $\left(R{-}\overset{\overset{\displaystyle O}{\|}}{C}{-} \right)$ is *a carboxylic acid minus its hydroxyl group.* The hydroxyl group of an acid may be replaced by a halide ion. The phosphorous halides and thionyl chloride are the reagents commonly employed in this reaction. Thionyl chloride, a low boiling liquid, is a particularly good reagent for the preparation of acyl chlorides. Any excess reagent can be separated out by distillation and the by-products of the reaction are both gases and are easily removed.

$$R{-}C \underset{OH}{\overset{O}{\diagup}} \; + \; SOCl_2 \; \longrightarrow \; R{-}C \underset{Cl}{\overset{O}{\diagup}} \; + \; SO_2(g) + HCl(g)$$

Acyl bromides and iodides can be prepared by similar procedures, but acyl chlorides are made more easily and they are more stable. Acyl halides are the most reactive of the carboxylic acid derivatives. This increased reactivity is due to the presence of the electronegative halogen atom directly bonded to the carbonyl carbon. The inductive effect of the halogen is to withdraw electrons from the carbon atom making the carbon more susceptible to attack by nucleophilic reagents. The chief function of acyl halides is to serve as intermediates in the preparation of many other organic compounds. The name of the acyl group is obtained from the parent acid by dropping the *-ic* ending and adding *-yl.* The acetyl group is an extremely important functional group in both organic chemistry and biochemistry.

$$CH_3{-}C \overset{O}{\diagup} \qquad CH_3{-}C \underset{Cl}{\overset{O}{\diagup}} \qquad CH_3{-}C \underset{OPO_3{}^{2-}}{\overset{O}{\diagup}} \qquad \bigcirc\hspace{-1.8em}\bigcirc{-}C \underset{Cl}{\overset{O}{\diagup}}$$

Acetyl group   Acetyl chloride   Acetyl phosphate   Benzoyl chloride

2. Acid Anhydrides Inorganic anhydrides are obtained by the elimination of a molecule of water from the corresponding acid or base.

$$H_2SO_4 \longrightarrow SO_3 + H_2O$$

$$Ca(OH)_2 \longrightarrow CaO + H_2O$$

Structurally organic acid anhydrides are formally derivable by the removal of water between two molecules of acid. (Cyclic anhydrides arise from the intramolecular dehydration of a dicarboxylic acid.) In practice, the noncyclic anhydrides are more easily prepared by the reaction of an acyl chloride and the sodium salt of an acid.

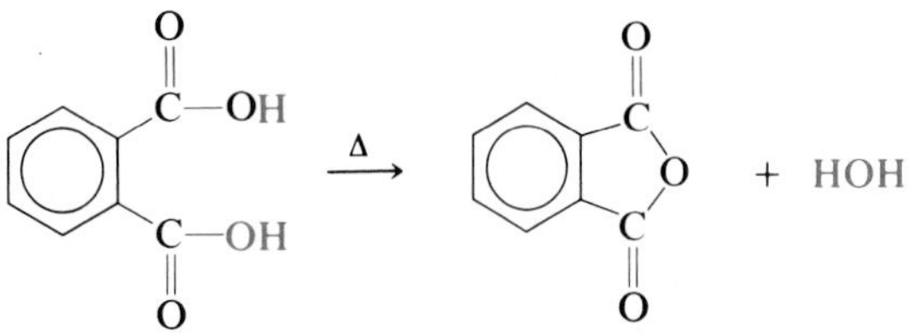

$$\underset{R-\overset{\displaystyle O}{\overset{\|}{C}}-OH}{} + \underset{HO-\overset{\displaystyle O}{\overset{\|}{C}}-R}{} \longrightarrow \underset{R-\overset{\displaystyle O}{\overset{\|}{C}}-O-\overset{\displaystyle O}{\overset{\|}{C}}-R}{} + HOH$$

Phthalic acid $\xrightarrow{\Delta}$ Phthalic anhydride   + HOH

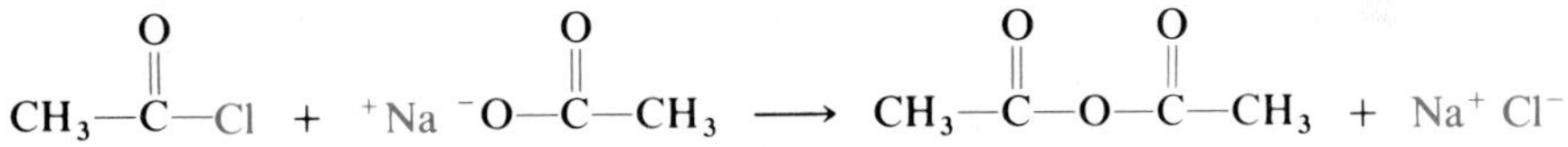

$$\underset{\text{Acetyl chloride}}{CH_3-\overset{\displaystyle O}{\overset{\|}{C}}-Cl} + \underset{\text{Sodium acetate}}{^+Na\ ^-O-\overset{\displaystyle O}{\overset{\|}{C}}-CH_3} \longrightarrow \underset{\text{Acetic anhydride}}{CH_3-\overset{\displaystyle O}{\overset{\|}{C}}-O-\overset{\displaystyle O}{\overset{\|}{C}}-CH_3} + Na^+\ Cl^-$$

Both acetyl chloride and acetic anhydride are known as *acetylating agents* because they will react with many substances that have a replaceable hydrogen to form acetyl derivatives. The following reactions with water (hydrolysis), alcohol (alcoholysis), and ammonia (ammonolysis) illustrate the synthetic usefulness of these compounds.

Hydrolysis—preparation of acids.   Acyl halides and acid anhydrides are readily converted back to their parent acids by reaction with water.

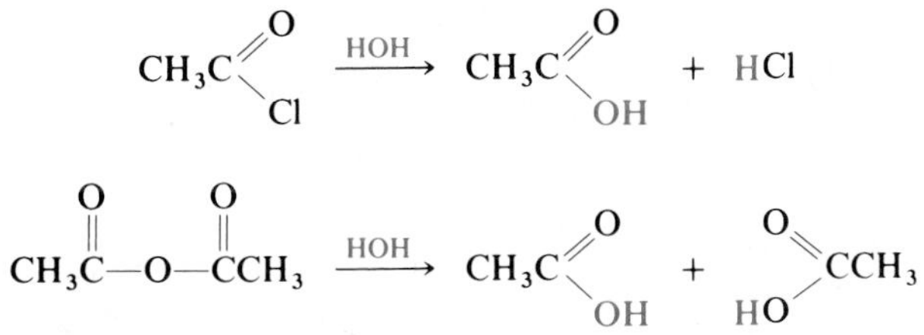

$$CH_3C\underset{Cl}{\overset{O}{\diagup}} \xrightarrow{HOH} CH_3C\underset{OH}{\overset{O}{\diagup}} + HCl$$

$$CH_3\overset{\displaystyle O}{\overset{\|}{C}}-O-\overset{\displaystyle O}{\overset{\|}{C}}CH_3 \xrightarrow{HOH} CH_3C\underset{OH}{\overset{O}{\diagup}} + \underset{HO}{\overset{O}{\diagdown}}CCH_3$$

Alcoholysis—preparation of esters.   Reactions between alcohols and acyl halides or acid anhydrides occur readily, and they are irreversible. In comparison, when carboxylic acids combine with alcohols to form esters, the reaction is reversible (see Section 8.10).

$$CH_3C\underset{Cl}{\overset{O}{\diagup}} \xrightarrow[\text{pyridine}]{CH_3CH_2OH} CH_3C\underset{OCH_2CH_3}{\overset{O}{\diagup}} + HCl$$

Ester

$$CH_3\overset{O}{\overset{\|}{C}}-O-\overset{O}{\overset{\|}{C}}CH_3 \xrightarrow{CH_3CH_2OH} CH_3C\underset{OCH_2CH_3}{\overset{O}{\diagup}} + \underset{HO}{\overset{O}{\diagdown}}CCH_3$$

Ammonolysis — preparation of amides (see Section 9.2)

$$CH_3C\underset{Cl}{\overset{O}{\diagup}} \xrightarrow{2\,HNH_2} CH_3C\underset{NH_2}{\overset{O}{\diagup}} + NH_4{}^+\ Cl$$

Amide

$$CH_3\overset{O}{\overset{\|}{C}}-O-\overset{O}{\overset{\|}{C}}CH_3 \xrightarrow{2\,HNH_2} CH_3C\underset{NH_2}{\overset{O}{\diagup}} + \underset{{}^+NH_4\quad O}{\overset{O}{\diagdown}}CCH_3$$

3. **Formation of Esters (Esterification)** When an acid and an alcohol are heated in the presence of an acid catalyst, a condensation reaction occurs that produces an ester and water. The esterification reaction is described in considerable detail in Section 8.10.

General Equation

$$R-C\underset{OH}{\overset{O}{\diagup}} + ROH \underset{}{\overset{H^+}{\rightleftharpoons}} R-C\underset{OR}{\overset{O}{\diagup}} + HOH$$

Ester

Specific Equation

$$CH_3-C\underset{OH}{\overset{O}{\diagup}} + CH_3CH_2OH \overset{H^+}{\rightleftharpoons} CH_3-C\underset{OCH_2CH_3}{\overset{O}{\diagup}} + HOH$$

Ethyl acetate

8.8 *Important Carboxylic Acids*

a. Formic Acid

Formic acid (Latin, *formica*, ant) is a principal component of the secretions of bees and ants. It is responsible for the blistering of the skin that follows a bee or an ant sting. Formic acid may be prepared commercially by heating carbon monoxide with sodium hydroxide. The resulting salt, sodium formate, is treated with sulfuric acid and formic acid is liberated. It is used commercially to remove hair from animal hides in leather making.

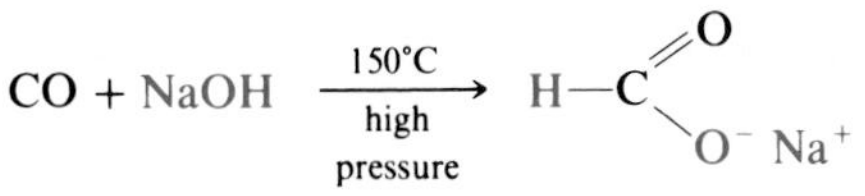

$$CO + NaOH \xrightarrow[\substack{\text{high} \\ \text{pressure}}]{150°C} H-C\underset{O^- Na^+}{\overset{O}{\diagup\!\!\!\backslash}}$$

Sodium formate

$$H-C\underset{O^- Na^+}{\overset{O}{\diagup\!\!\!\backslash}} + H_2SO_4 \longrightarrow H-C\underset{OH}{\overset{O}{\diagup\!\!\!\backslash}} + NaHSO_4$$

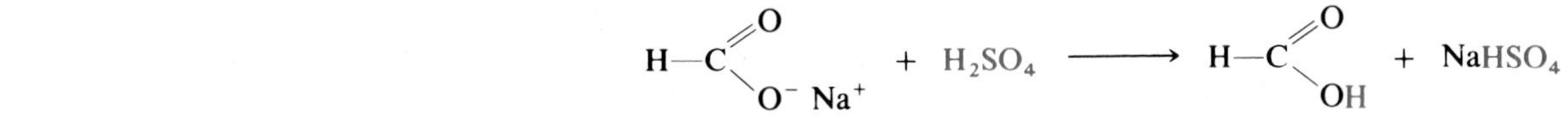

Unlike all the other carboxylic acids, formic acid possesses an aldehydic group as well as a carboxyl group.

$$H-C\underset{OH}{\overset{O}{\diagup\!\!\!\backslash}} \qquad H-C\underset{OH}{\overset{O}{\diagup\!\!\!\backslash}}$$

Carboxyl group   Aldehydic group

It is therefore not surprising to find that formic acid is capable of undergoing reactions that are not possible for the higher homologs of the acid series. Unlike most acids, it is a powerful reducing agent. It is easily oxidized to carbon dioxide and water, and thus gives a positive test with Tollens' reagent. (It will also decolorize potassium permanganate solution, but will not reduce Fehling's or Benedict's solution.)

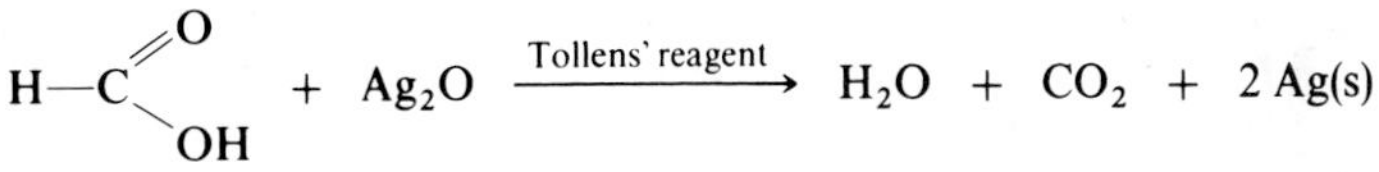

$$H-C\underset{OH}{\overset{O}{\diagup\!\!\!\backslash}} + Ag_2O \xrightarrow{\text{Tollens' reagent}} H_2O + CO_2 + 2 Ag(s)$$

Furthermore, unlike all the other acids, formic acid can be both dehydrogenated and dehydrated. The latter reaction is employed in chemical laboratories to generate small amounts of pure carbon monoxide.

Dehydrogenation (Decarboxylation)

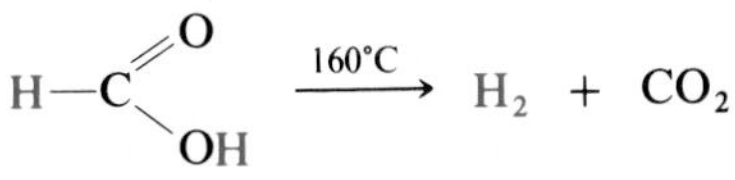

$$H-C\underset{OH}{\overset{O}{\diagup\!\!\!\backslash}} \xrightarrow{160°C} H_2 + CO_2$$

Dehydration

$$H-C\underset{OH}{\overset{O}{\diagup\!\!\!\backslash}} \xrightarrow{H_2SO_4} H_2O + CO$$

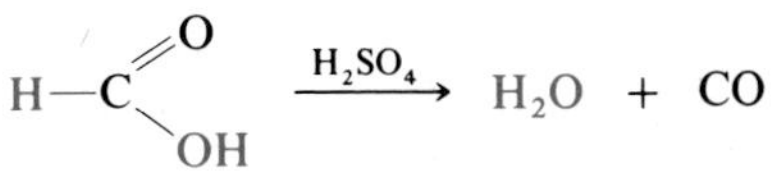

Another unique characteristic of formic acid is its inability to form a stable acyl chloride. All attempts to prepare formyl chloride yield only hydrogen chloride and carbon monoxide.

$$ H-C{\overset{O}{\underset{OH}{}}} \xrightarrow[\text{or SOCl}_2]{\text{PCl}_3, \text{PCl}_5,} \left[H-C{\overset{O}{\underset{Cl}{}}} \right] \longrightarrow HCl + CO $$

Unstable

b. Acetic Acid

Acetic acid (Latin, *acetum,* vinegar) is by far the most important of the monocarboxylic acids. It is one of the earliest known organic compounds. Vinegar is a 4–5% solution of acetic acid in water. It is the acetic acid that imparts to the water the characteristic sour taste of vinegar (and "sour wine"). Pure acetic acid freezes at a temperature of 17°C (63°F). At normal room temperature, acetic acid is a colorless liquid. However, in poorly heated laboratories, it may begin to crystallize, forming a glass-like solid. Hence the pure acid is often called *glacial* acetic acid.

Commercially acetic acid is made by the oxidation of acetaldehyde involving direct air or oxygen in the presence of an acetate salt (cobalt, copper, or manganese). Acetic acid is resistant to further oxidation and is frequently employed as a solvent for the oxidation of alkenes and alcohols. It is also used in the production of synthetic polymers such as cellulose acetate (acetate rayon) and polyvinyl acetate.

c. Butyric Acid

The name of this acid is also indicative of its chief source (Latin, *butyrum,* butter); butyric acid is liberated from butter left standing in the air. As previously mentioned, the disagreeable odor associated with rancid butter is that of butyric acid.

d. Oxalic Acid

Oxalic acid is found as the monopotassium salt ($HOOC—COO^- K^+$) in the leaves and roots of many plants, especially spinach and rhubarb. It also occurs in human and animal urine, and the calcium salt is a major constituent of kidney stones.

Like formic acid, and unlike all the other dicarboxylic acids, oxalic acid is readily oxidized to carbon dioxide and water by potassium permanganate. It is used in the quantitative analysis laboratory to standardize permanganate solutions.

$$ 5\,H_2C_2O_4 + 2\,KMnO_4 + 3\,H_2SO_4 \longrightarrow 2\,MnSO_4 + K_2SO_4 + 10\,CO_2 + 8\,H_2O $$

Oxalic acid is the strongest of the dicarboxylic acids. Because of its high acidity, it is irritating to the skin and mucous membranes and is highly poisonous if ingested. (The cooking process decomposes the acid in foods.) It is used as a bleaching agent and as a remover of rust and ink stains (by complexing with the metal ions).

e. Lactic Acid

Lactic acid forms in sour milk as a result of the action of bacteria (*Lactobacillus*) upon lactose (milk sugar).

$$C_{12}H_{22}O_{11} \ + \ H_2O \ \xrightarrow[\text{in bacteria}]{\text{enzymes}} \ 4\,CH_3CHOHCOOH$$

Lactose   Lactic acid

Commercially it is synthesized from acetaldehyde and hydrogen cyanide followed by hydrolysis of the cyanohydrin (page 125). It may also be prepared by the bacterial fermentation of cane sugar or corn starch. The presence of lactic acid in certain sour foods (such as sauerkraut) prevents bacterial spoilage because most bacteria are unable to survive in an acidic environment. It is as strong an acid as formic acid. As we shall see in Section 16.4, lactic acid is an important compound biochemically. Some of the energy necessary for muscular activity is supplied by the breakdown of carbohydrates to lactic acid. The accumulation of lactic acid in muscle cells causes muscle fatigue. It is primarily used as a preservative in foods and in pharmaceutical products.

f. Tartaric Acid

Tartaric acid (HOOC—CHOH—CHOH—COOH) occurs in grapes and other fruits. It is obtained as a by-product of the wine industry. When the wine is aged in the casks, a white salt crystallizes out. This salt is potassium hydrogen tartrate (known as cream of tartar) and it is widely used in the manufacture of baking powder. Recall that Rochelle salt, used in Fehling's solution, is the potassium sodium salt of tartaric acid ($^{+}Na\ ^{-}OOC$—CHOH—CHOH—$COO^{-}\ K^{+}$). The free acid and its salts are used in medicine, dyeing, and in some foods and soft drinks. We refer to tartaric acid in Section 10.4 in connection with its optical activity.

g. Citric Acid

Citric acid (HOOC—CH_2—COH(COOH)—CH_2—COOH) is one of the most widely distributed naturally occurring acids. As its name implies, it is particularly abundant in citrus fruits (lemon juice contains 5–8%). The acid is prepared commercially by the fermentation of glucose by a certain mold. This tricarboxylic acid is a normal constituent of blood serum and it is present in all living cells that obtain their energy from the aerobic metabolism of carbohydrates (see Section 16.8). Citric acid is widely used in the food industry and in the preparation of beverages because of its solubility in water and mildly sour taste. It forms a variety of salts, some of which are used in pharmaceutical preparations. For example, sodium citrate is an anticoagulant (prevents blood from clotting—see Section 19.5), and magnesium citrate is a cathartic (stimulates evacuation of the bowels).

h. Salicylic Acid

Salicylic acid, $C_6H_4(OH)COOH$, was originally extracted from the willow and other plants. It is a white, odorless solid that may discolor gradually in sunlight. Early Greeks and Romans (~400 B.C.) chewed on willow bark to relieve pain and reduce fever. Salicyclic acid is a pain reliever, but it has an unpleasant taste and it is very irritating to the stomach lining. A more powerful germicide than phenol, salicylic acid is commonly used in shampoos and ointments for topical treatment of dandruff, acne, and skin diseases. It causes the dead, outer layer of skin to fall off, but it does not destroy the healthy tissue. About 60% of the salicylic acid produced in the United States is used in the manufacture of aspirin. See Section 8.13 for a discussion of the important esters of salicylic acid.

Esters

Esters are perhaps the most important class of derivatives of the carboxylic acids. The general formula for an ester is **RCOOR′**, where R may be a hydrogen, an alkyl group, or an aryl group, and where R′ may be alkyl or aryl, but *not* hydrogen.

Esters are widely distributed in nature. Their occurrence is particularly important in fats and vegetable oils, which are esters of long-chain fatty acids and glycerol (Chapter 12). Esters of phosphoric acid are of the utmost importance to life, and they are enumerated in Section 8.14.

TABLE 8.6 Nomenclature of Esters

Formula	Common Name	IUPAC Name
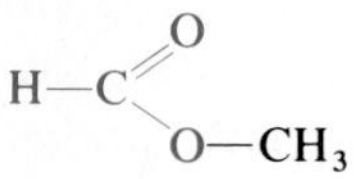 $H-C(=O)O-CH_3$	Methyl formate	Methyl methanoate
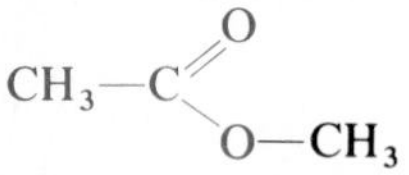 $CH_3-C(=O)O-CH_3$	Methyl acetate	Methyl ethanoate
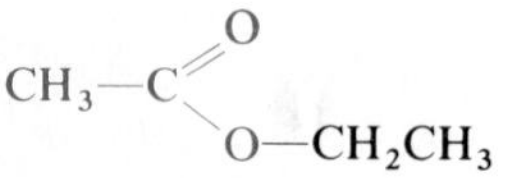 $CH_3-C(=O)O-CH_2CH_3$	Ethyl acetate	Ethyl ethanoate
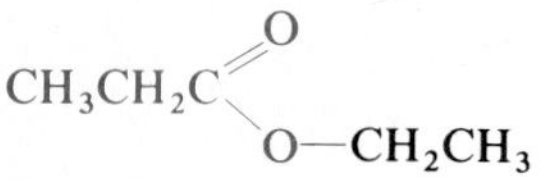 $CH_3CH_2C(=O)O-CH_2CH_3$	Ethyl propionate	Ethyl propanoate
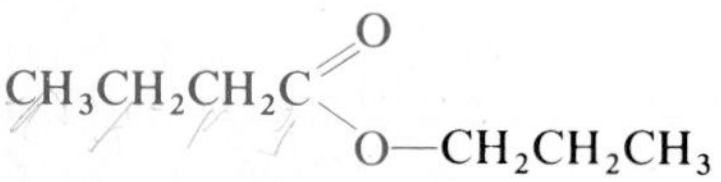 $CH_3CH_2CH_2C(=O)O-CH_2CH_2CH_3$	*n*-Propyl butyrate	Propyl butanoate
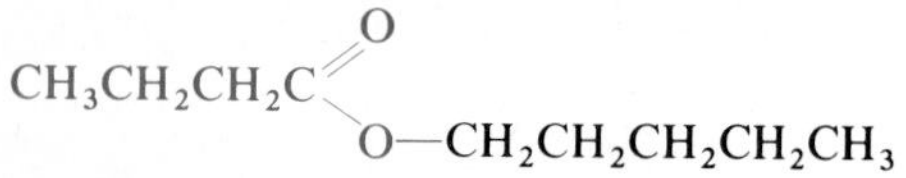 $CH_3CH_2CH_2C(=O)O-CH_2CH_2CH_2CH_2CH_3$	*n*-Pentyl butyrate (Amyl butyrate	Pentyl butanoate
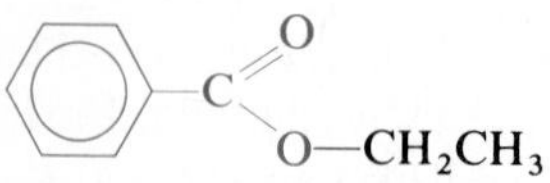 $C_6H_5-C(=O)O-CH_2CH_3$	Ethyl benzoate	Ethyl benzoate
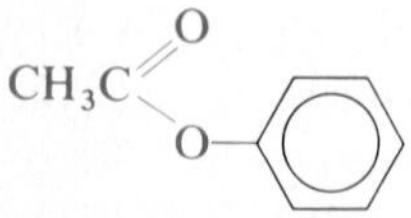 $CH_3C(=O)O-C_6H_5$	Phenyl acetate	Phenyl ethanoate

8.9 Nomenclature of Esters

Esters are named in a manner analogous to that used for naming organic salts. The group name of the alkyl or aryl portion is given first and is followed by the name of the acid portion. In both common and IUPAC nomenclature, the *-ic* ending of the parent acid is replaced by the suffix *-ate*. Table 8.6 illustrates the nomenclature of several esters.

8.10 Preparation of Esters

The preparation of esters from acyl chlorides and acid anhydrides via alcoholysis has already been discussed (Section 8.7-b). Mention has also been made of the esterification reaction between an acid and an alcohol. We now examine this very important reaction in greater detail.

Recall that the equation for the reaction between acetic acid and ethyl alcohol is

$$CH_3-C{\overset{O}{\underset{OH}{}}} \;+\; CH_3CH_2OH \;\underset{\Delta}{\overset{H^+}{\rightleftharpoons}}\; CH_3-C{\overset{O}{\underset{OCH_2CH_3}{}}} \;+\; HOH$$

Notice the presence of the double arrows, indicating the reversibility of the reaction. When equilibrium is achieved, the reaction mixture contains approximately 66% of the ester and 34% of both the acid and the alcohol. An approximate value for the equilibrium constant for the esterification reaction can be calculated from the law of chemical equilibrium.

$$K = \frac{[CH_3COOCH_2CH_3][HOH]}{[CH_3COOH][CH_3CH_2OH]}$$

$$= \frac{(0.66)(0.66)}{(0.34)(0.34)} \approx 4$$

Experimental data reveal that, for any esterification reaction, the composition of the equilibrium mixture is governed by the value of the equilibrium constant regardless of the relative amounts of initial alcohol and acid. Consideration of the law of chemical equilibrium and Le Châtelier's principle allows predetermination of reaction conditions that produce a maximum yield of the desired ester. It is evident that the use of an excess of either the alcohol or the acid (whichever is less expensive) produces an increase in the amount of the ester formed. In addition, if the water is removed from the reaction mixture as soon as it is formed, the equilibrium is shifted continuously to the right to conform to the law of chemical equilibrium. This can be accomplished if the acid, the alcohol, and the ester all boil at temperatures above 100°C. The water may then be distilled away from the reaction mixture as it is formed. A second method involves the addition of some inert substance, such as benzene, which forms an azeotropic mixture (boiling at 65°C) with water. In this instance, the water may be distilled off at a relatively low temperature. A third method involves the use of a dehydrating reagent that is inert toward the reaction but will absorb the water. Zinc chloride is sometimes used for this purpose.

In general, the reaction between an acid and an alcohol is extremely slow, and a catalyst must be employed to speed it up. Sulfuric acid is commonly used because it is both an acid and a good dehydrating agent. Thus it serves both to catalytically increase the rate of reaction and to shift the equilibrium to the right by reacting with water to form hydronium ions, H_3O^+, thus increasing the yield of the ester.

Theoretically, the esterification reaction can be considered as occurring by either of two pathways. The condensation of acid and alcohol could occur by the elimination of water through the loss of the hydrogen of the acid and the hydroxyl of the alcohol, or vice versa. The following equations illustrate the two possibilities. Note that both alternatives yield the *same* ester.

$$
RC\!\!\diagdown\!\!{}^{O}_{O\,H} \;+\; HO\;R' \;\underset{}{\overset{H^+}{\rightleftharpoons}}\; RC\!\!\diagdown\!\!{}^{O}_{OR'} \;+\; HOH
$$

$$
RC\!\!\diagdown\!\!{}^{O}_{OH} \;+\; H\;OR' \;\underset{}{\overset{H^+}{\rightleftharpoons}}\; RC\!\!\diagdown\!\!{}^{O}_{OR'} \;+\; HOH
$$

It is known that (for most esterifications) the reactions proceeds according to the second equation. This knowledge was obtained by the use of isotopic labeling, a useful technique for the understanding of chemical mechanisms. Elucidation of this mechanism was accomplished by an esterification involving methyl alcohol that contained a heavy isotope of oxygen (^{18}O). When the alcohol was allowed to react with acetic acid, it was found that the resulting methyl acetate contained all of the isotopic oxygen; the water contained none of the labeled oxygen.

$$
CH_3C\!\!\diagdown\!\!{}^{O}_{OH} \;+\; CH_3{}^{18}O\;H \;\underset{}{\overset{H^+}{\rightleftharpoons}}\; CH_3C\!\!\diagdown\!\!{}^{O}_{{}^{18}OCH_3} \;+\; HOH
$$

Similar experiments have been performed with a wide variety of acids and alcohols. In every case it was found that the hydroxyl group is lost from the acid and that the hydrogen is lost from the alcohol. It should be reemphasized that the ester formed in either case would be identical. We are concerned here only with a better understanding of the pathway of the reaction.

A commercially important esterification reaction occurs between a dicarboxylic acid and a dialcohol. Such a reaction yields an ester that contains a free carboxyl group and a free alcohol group on either end. Hence further condensation reactions can occur to produce polyesters (polymers). The most significant polyester, Dacron, is made from terephthalic acid and ethylene glycol monomers. Dacron polyester is found in permanent press garments, carpets, tires, and many other products. Dacron is inert and is used in surgery (in the form of a mesh) to repair or replace diseased sections of blood vessels.

$$HO-\overset{\overset{\displaystyle O}{\|}}{C}-\bigcirc-\overset{\overset{\displaystyle O}{\|}}{C}-OH \ + \ HOCH_2CH_2OH \ \longrightarrow \ -O\left[\overset{\overset{\displaystyle O}{\|}}{C}-\bigcirc-\overset{\overset{\displaystyle O}{\|}}{C}-OCH_2CH_2O\right]_n\overset{\overset{\displaystyle O}{\|}}{C}- \ + \ n\,HOH$$

| Terephthalic acid | Ethylene glycol | Dacron |

8.11 Physical Properties of Esters

Unlike the carboxylic acids from which they are derived, the esters have very pleasant odors. In fact, the specific aromas of many flowers and fruits are due to the presence of esters. Esters are utilized in the manufacture of perfumes and as flavoring agents in the confectionery and soft drink industries. (A mixture of nine esters is utilized to produce an artificial raspberry flavor.) Vapors of esters are nontoxic unless inhaled in large concentrations. (There is, therefore, danger in the fads of glue sniffing and smoking banana peels. The toxic ester in each case is amyl acetate.) Esters find their most important use as industrial solvents. Ethyl acetate is commonly used to extract organic solutes from an aqueous solution. It is also used as a fingernail polish remover and is the major constituent of paint removers. Most esters are colorless liquids, insoluble in and lighter than water. They are neutral substances and usually have lower boiling points and melting points than the acids or alcohols of comparable molecular weight. Since they do not contain hydroxylic hydrogens, there is no possibility of intermolecular hydrogen bonding. Table 8.7 lists the physical properties of some common esters.

TABLE 8.7 Physical Properties of Esters

Formula	Name	MW	MP (°C)	BP (°C)	Aroma
$HCOOCH_3$	Methyl formate	60	-99	32	
$HCOOCH_2CH_3$	Ethyl formate	74	-80	54	Rum
CH_3COOCH_3	Methyl acetate	74	-98	57	
$CH_3COOCH_2CH_3$	Ethyl acetate	88	-84	77	
$CH_3CH_2COOCH_3$	Methyl propionate	88	-88	80	
$CH_3CH_2COOCH_2CH_3$	Ethyl propionate	102	-74	99	
$CH_3CH_2CH_2COOCH_3$	Methyl butyrate	102	-85	102	Apple
$CH_3CH_2CH_2COOCH_2CH_3$	Ethyl butyrate	116	-101	121	Pineapple
$CH_3COO(CH_2)_4CH_3$	Amyl acetate	130	-71	148	Banana
$CH_3COOCH_2CH_2CH(CH_3)_2$	Isoamyl acetate	130	-79	142	Pear
$CH_3COOCH_2C_6H_5$	Benzyl acetate	150	-51	215	Jasmine
$CH_3CH_2CH_2COO(CH_2)_4CH_3$	Amyl butyrate	158	-73	185	Apricot
$CH_3COO(CH_2)_7CH_3$	Octyl acetate	172	-39	210	Orange

8.12 Chemical Properties of Esters

The most significant reaction of the esters is their hydrolysis (Greek, *hydro,* water; *lysis,* loosening). **Hydrolysis,** then, is *the splitting of molecules by the addition of water.* Most organic and biochemical compounds react very slowly, if at all, with water. The chemist employs acids or bases to facilitate the hydrolysis reactions in the laboratory. Living organisms utilize enzymes to catalyze these same reactions. In the laboratory, the process may take several hours; in living systems, it is complete in a fraction of a second.

a. Acid Hydrolysis of Esters

The acid hydrolysis of esters is simply the reverse of the esterification reaction (Section 8.10). The ester is refluxed with a large excess of water containing a strong acid catalyst. However, the equilibrium is unfavorable for ester hydrolysis ($K \sim \frac{1}{4}$) and so the reaction never goes to completion.

General Equation

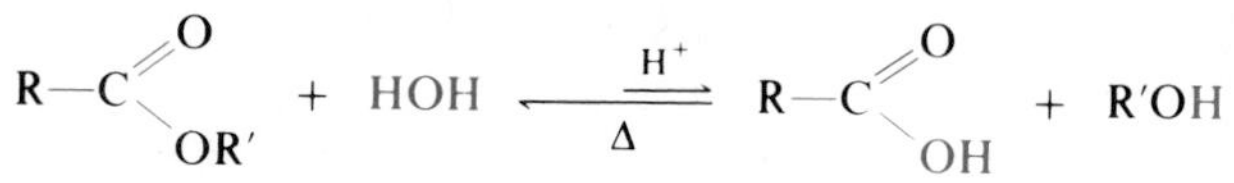

$$R-C\overset{O}{\underset{OR'}{\diagdown}} + HOH \underset{\Delta}{\overset{H^+}{\rightleftharpoons}} R-C\overset{O}{\underset{OH}{\diagdown}} + R'OH$$

Specific Equation

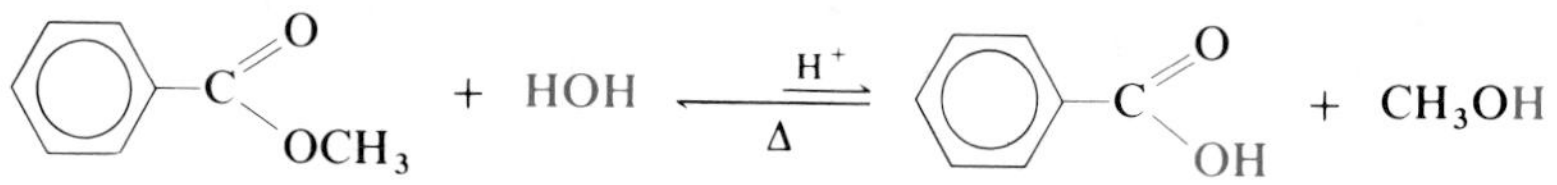

Methyl benzoate + HOH $\underset{\Delta}{\overset{H^+}{\rightleftharpoons}}$ Benzoic acid + CH$_3$OH

b. Alkaline Hydrolysis of Esters (Saponification)

When a base (such as sodium or potassium hydroxide) is used to hydrolyze an ester, the reaction goes to completion since the carboxylic acid is removed from the equilibrium by its conversion to a salt. Organic salts do not react with alcohols, so the reaction is essentially irreversible. Accordingly, ester hydrolysis is usually carried out in basic solution. Ethanol is the most common solvent for the reaction since most esters are insoluble in water. Both the ester and the base are soluble in alcohol, and this offers a homogeneous medium. Because soaps are prepared by the alkaline hydrolysis of fats and oils, the term **saponification** (Latin, *sapon,* soap; *facere,* to make) *is used to describe the alkaline hydrolysis of all esters* (see also Section 12.5-a). Note that in the saponification reaction, the base is a reactant and thus is not a catalyst.

General Equation

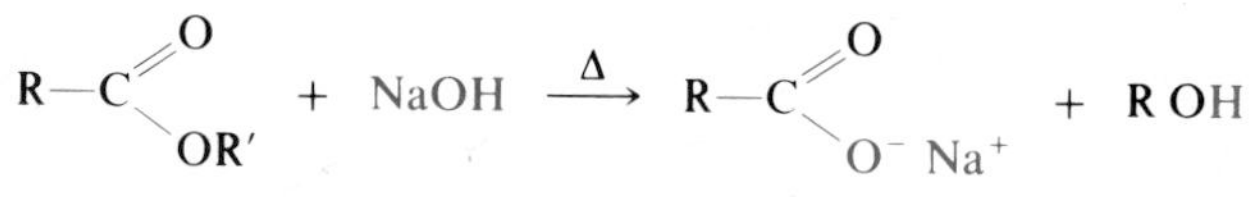

$$R-C\overset{O}{\underset{OR'}{\diagdown}} + NaOH \overset{\Delta}{\longrightarrow} R-C\overset{O}{\underset{O^- Na^+}{\diagdown}} + R\,OH$$

Specific Equation

$$CH_3-C\overset{O}{\underset{OCH_2CH_3}{\diagdown}} + NaOH \overset{\Delta}{\longrightarrow} CH_3-C\overset{O}{\underset{O^- Na^+}{\diagdown}} + CH_3CH_2OH$$

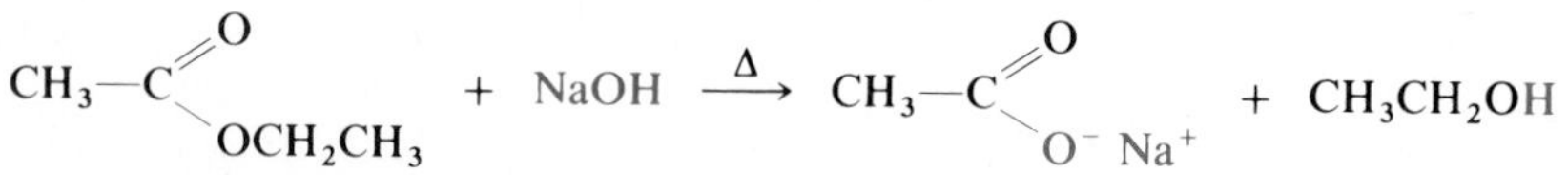

Ethyl acetate   Sodium acetate   Ethanol

c. Reduction of Esters (Formation of Alcohols)

The catalytic hydrogenation of esters (and carboxylic acids) requires unusually high temperatures and pressures. Therefore, in the laboratory, reducing agents such as lithium aluminum hydride and lithium borohydride are utilized. When esters are reduced, two alcohols are formed. The best reagent for the reduction of carboxylic acids is diborane (B_2H_6).

General Equation

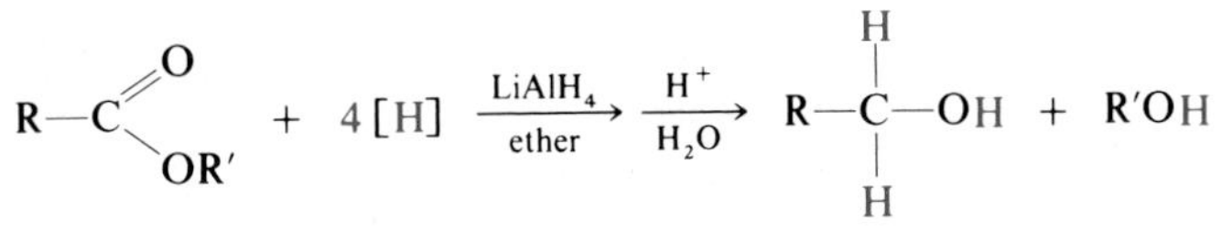

$$R-\underset{\underset{OR'}{|}}{\overset{\overset{O}{\|}}{C}} + 4[H] \xrightarrow[\text{ether}]{\text{LiAlH}_4} \xrightarrow[\text{H}_2\text{O}]{\text{H}^+} R-\underset{\underset{H}{|}}{\overset{\overset{H}{|}}{C}}-OH + R'OH$$

Specific Equations

$$CH_3CH_2CH_2\underset{\underset{OCH_3}{|}}{\overset{\overset{O}{\|}}{C}} + 4[H] \xrightarrow[\text{ether}]{\text{LiBH}_4} \xrightarrow[\text{H}_2\text{O}]{\text{H}^+} CH_3CH_2CH_2CH_2OH + CH_3OH$$

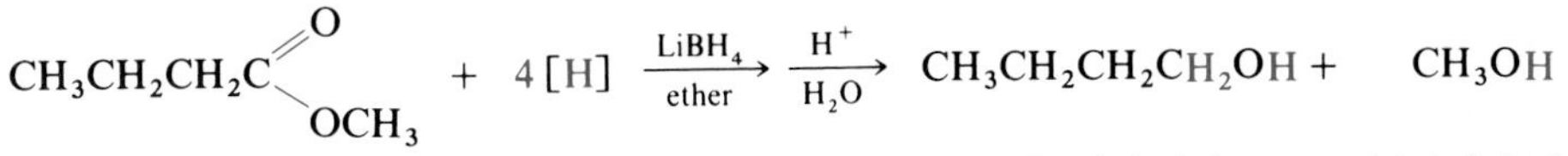

Methyl butyrate · n-Butyl alcohol · Methyl alcohol

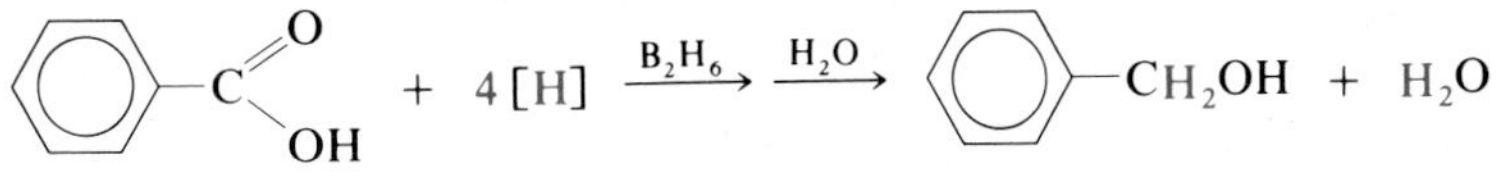

Benzoic acid + 4[H] $\xrightarrow{\text{B}_2\text{H}_6}$ $\xrightarrow{\text{H}_2\text{O}}$ Benzyl alcohol + H_2O

8.13 Esters of Salicylic Acid

Salicylic acid is a bifunctional aromatic compound containing both a carboxyl group and a hydroxyl group. Thus, it may function as an acid or as an alcohol in an esterification reaction.

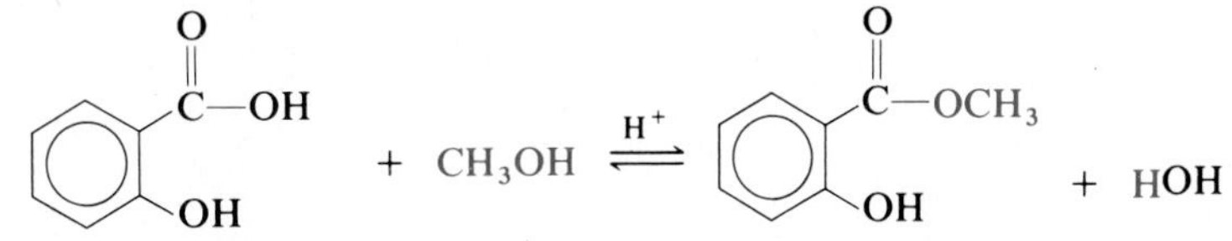

o-Hydroxybenzoic acid (Salicylic acid) + Methanol $\rightleftharpoons$ Methyl salicylate (Oil of wintergreen) + HOH

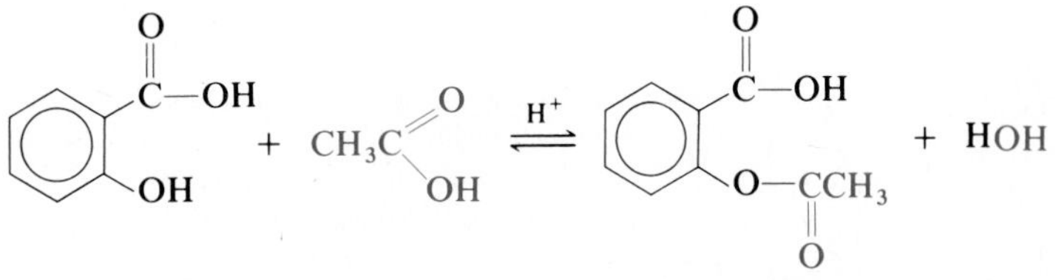

Acetic acid · Acetylsalicylic acid (Aspirin)

Acyl chloride of salicylic acid + Phenol ⇌ Phenyl salicylate (Salol) + HCl

Methyl salicylate is an oil found in numerous plants and has a fragrance associated with wintergreen. Commercially, it is used in perfumes and for flavoring candy. It finds widespread use as the pain-relieving ingredient in liniments such as Ben-Gay. When rubbed on the skin, this ester has the unusual ability of penetrating the surface. Hydrolysis then occurs, liberating salicylic acid, which relieves the soreness.

Acetylsalicylic acid was first synthesized in 1853 from the reaction of acetyl chloride and sodium salicylate. The white solid is sparingly soluble in water (1 g/300 mL) and melts at 136°C. It is undoubtedly the most widely used drug in the world. Aspirin acts as a fever reducer (*antipyretic*), a pain depresser (*analgesic*), and an *anti-inflammatory* agent (reduces swelling in injured tissues and rheumatic joints). However, if taken in large quantities, it is also an effective poison (30–40 g constitutes a lethal dose)! By law, each aspirin tablet must contain 0.32 g (5 grains) of aspirin mixed with starch or some other inert binder to hold the tablet together. Tablets intended for use by children contain only one fourth (1.25 grains) as much aspirin. More than 40 million lb of aspirin is produced annually in the United States (by the reaction of acetic anhydride and salicylic acid). Of this amount, about half is converted into aspirin tablets, and the rest is sold in the form of combination pain relievers and cold remedies.

We are beginning to understand how aspirin accomplishes its extraordinary effects. It was found that aspirin acetylates, and thus inhibits, an enzyme (*cyclooxygenase*) necessary for the synthesis of prostaglandins (see Section 12.15). Among their many functions, prostaglandins are involved in inflammation (see footnote 10, page 293), increased blood pressure, and the contraction of smooth muscle. Elevated concentrations of prostaglandins appear to activate pain receptors in the tissues, making the tissues more sensitive to any pain stimulus. Generally speaking, prostaglandins enhance inflammatory effects, whereas aspirin diminishes them.

Certain hazards are associated with the use of aspirin. Among these are a slight deterioration of the stomach lining (anyone with an ulcer should avoid taking aspirin), some intestinal bleeding,[5] a prolongation of labor, and the development of

---

[5]Aspirin can inhibit the normal blood-clotting mechanism, causing prolongation of bleeding time and hemorrhaging (see footnote 5, page 500). Therefore it should not be taken after surgery. However, by interfering with the clotting mechanism, aspirin may be effective in reducing the incidence of second heart attacks. Aspirin appears to interfere with the formation of blood clots that block coronary arteries. Researchers have found that daily doses (one tablet) of aspirin reduced the risk of fatal and nonfatal heart attacks by 20%.

Reye's syndrome.[6] A small percentage of the population (about 1 in 10,000) is allergic to aspirin. Susceptible individuals experience skin rashes, asthmatic attacks, and even loss of consciousness. These people must be careful to avoid aspirin alone or in any combination. The most satisfactory alternatives available are *acetaminophen* (Tylenol, Tempra, Liquiprin—see Table 9.1) and *ibuprofen* (nonprescription Advil and Nuprin, Motrin by prescription—Table 8.1). Ibuprofen, like aspirin, is a prostaglandin inhibitor (see page 298), whereas acetaminophen is an analgesic and an antipyretic, but it does not relieve inflammation of muscles and joints as do aspirin and ibuprofen.

Salol is the phenyl ester of salicylic acid and is most easily prepared from the acyl chloride rather than the free acid. Salol is a widely used intestinal antiseptic. It is not hydrolyzed by acids and therefore passes through the stomach unchanged. In the alkaline medium of the intestines, hydrolysis occurs to yield phenol and the salicylate ion. Salol is also employed as a coating for some medicinal pills in order to permit them to pass through the stomach intact, but to disintegrate in the intestines.

8.14   *Esters of Phosphoric Acid*

Phosphate or pyrophosphate esters are present in every plant and animal cell. They are biochemical intermediates in the transformation of food into usable energy (in the form of ATP). Organic phosphates are also important structural constituents of phospholipids (Section 12.9), nucleic acids (Section 15.1), coenzymes (Table 14.2), and insecticides (Section 8.15).

Pyrophosphoric acid ($H_4P_2O_7$) is formed by heating phosphoric acid above 210°C. A molecule of water is eliminated from two molecules of phosphoric acid; another phosphate can be added in a similar fashion to form a triphosphate (tripolyphosphoric acid). Sodium salts of tripolyphosphoric acid are utilized as builders (sequestering agents) in some detergents to bind to the metal ions that are present in hard water. Sodium tripolyphosphate ($Na_5P_3O_{10}$), in which the five acidic hydrogens of tripolyphosphoric acid are replaced by sodium ions, is the most commonly used phosphate compound in commercial detergents (see Section 12.8-b).

Phosphoric acid   Pyrophosphoric acid   Tripolyphosphoric acid

[6]The publicity about Reye's syndrome has been one of the factors responsible for the decline in sales of aspirin in recent years. Reye's syndrome is a brain disease that also causes fatty degeneration of organs such as the liver. It can occur during recovery from chicken pox or flu, with a child vomiting and then becoming lethargic, confused, irritable, or aggressive. Coma, brain damage, and death sometimes result. Labels on aspirin bottles contain the warning that a physician should be consulted before administering the medicine to children, including teenagers, with chicken pox or flu.

Replacement of one or more of the hydrogen atoms of phosphoric and pyrophosphoric acid by organic groups yields phosphate esters. Usually the name of the ester specifies the number of hydrogens that have been replaced by organic groups.

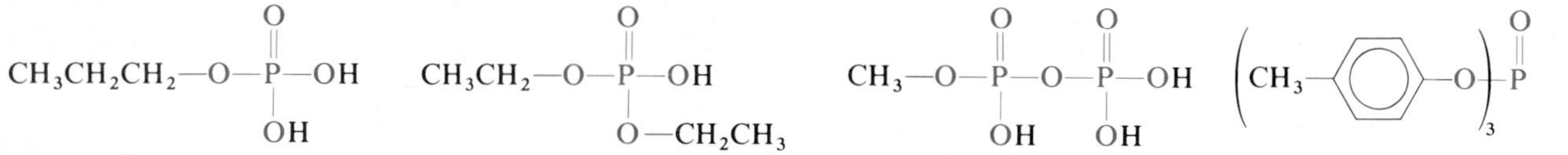

$$CH_3CH_2CH_2-O-\overset{\overset{O}{\|}}{\underset{\underset{OH}{|}}{P}}-OH \qquad CH_3CH_2-O-\overset{\overset{O}{\|}}{\underset{\underset{O-CH_2CH_3}{|}}{P}}-OH \qquad CH_3-O-\overset{\overset{O}{\|}}{\underset{\underset{OH}{|}}{P}}-O-\overset{\overset{O}{\|}}{\underset{\underset{OH}{|}}{P}}-OH \qquad \left(CH_3-\bigcirc-O\right)_3-\overset{\overset{O}{\|}}{P}$$

n-Propyl phosphate   Diethyl phosphate   Methyl pyrophosphate   Tricresyl phosphate (TCP) (component of high octane fuels)

Alkyl esters of phosphoric and pyrophosphoric acid are prepared in a similar fashion to those of carboxylic acids. One of the best methods involves the reaction of an alcohol with the acyl chloride of phosphoric acid ($POCl_3$).

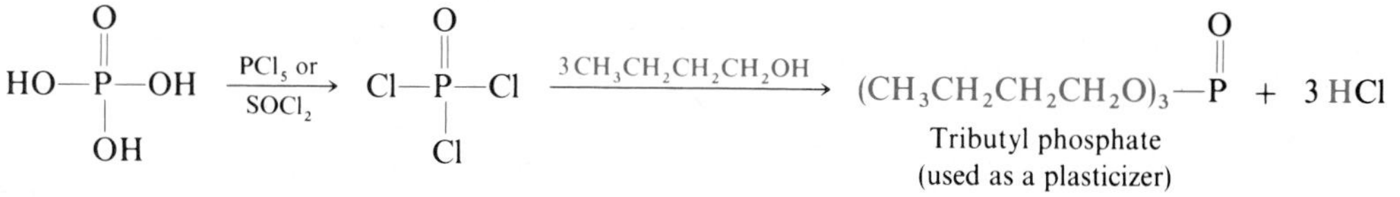

$$HO-\overset{\overset{O}{\|}}{\underset{\underset{OH}{|}}{P}}-OH \xrightarrow{\underset{SOCl_2}{PCl_5 \ or}} Cl-\overset{\overset{O}{\|}}{\underset{\underset{Cl}{|}}{P}}-Cl \xrightarrow{3\,CH_3CH_2CH_2CH_2OH} (CH_3CH_2CH_2CH_2O)_3-\overset{\overset{O}{\|}}{P} + 3\,HCl$$

Tributyl phosphate (used as a plasticizer)

A wide variety of phosphate esters occur naturally and are compounds of central importance in metabolism. The majority of the substances obtained from our food must be converted to phosphate esters before they can be utilized by the cells. Many of them are formed by phosphorylation reactions—the interaction of alcohols with anhydrides of phosphoric acid.

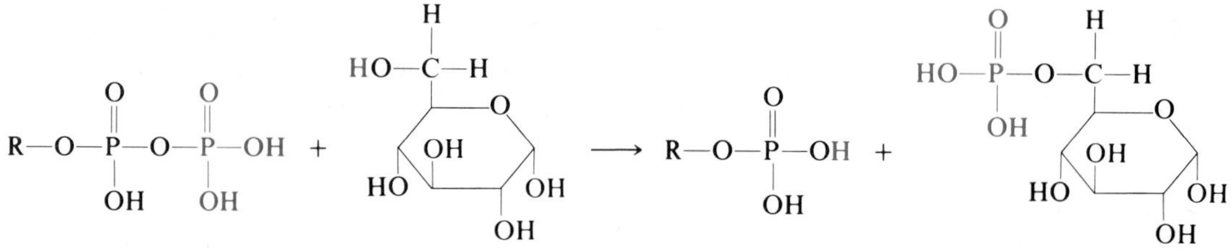

A polyphosphate   Glucose   Glucose 6-phosphate

Figure 8.2 presents only a few of the phosphate esters found in living cells. Notice that at physiological pH (≈ 7), the phosphate groups are ionized.

Probably the most important phosphate compound is **adenosine triphosphate (ATP).** Adenosine triphosphate was first isolated from skeletal muscle tissue and has since been shown to occur in all types of plant and animal cells. The concentration of ATP in the cell varies from 0.5 to 2.5 mg/mL of cell fluid. ATP is a nucleoside triphosphate composed of adenine, ribose, and three phosphate groups (Figure 8.3). As the names indicate, adenosine diphosphate (ADP) contains two phosphate groups, and adenosine monophosphate (AMP) contains one phosphate group (Figure 8.4).

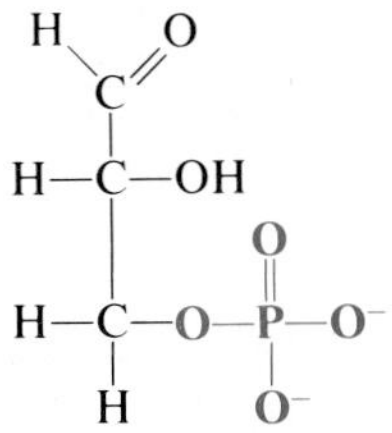

FIGURE 8.2 Some phosphate compounds of biological importance.

Glyceraldehyde 3-phosphate

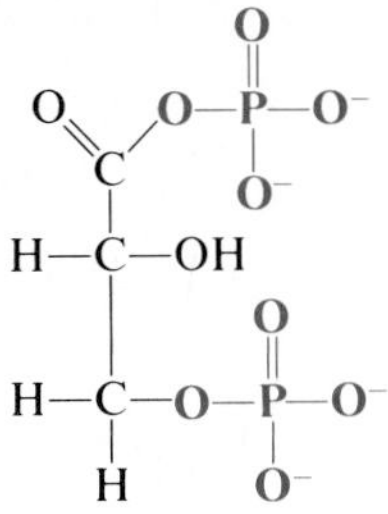

1,3-Diphosphoglyceric acid

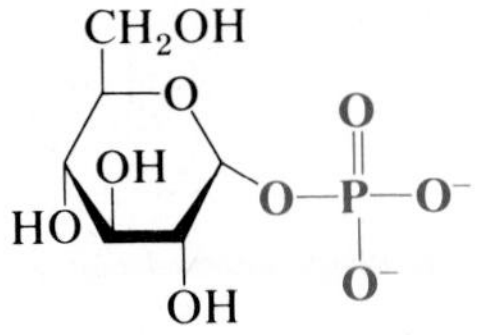

Glucose 1-phosphate

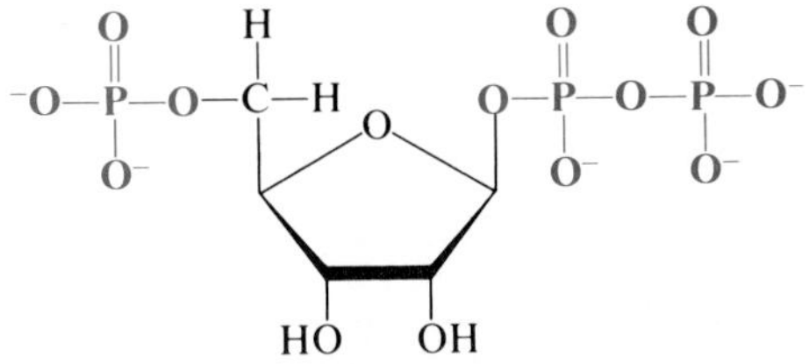

Phosphoribosyl pyrophosphate

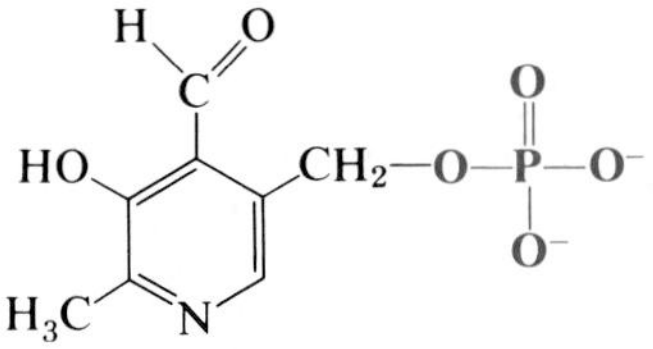

Pyridoxal phosphate

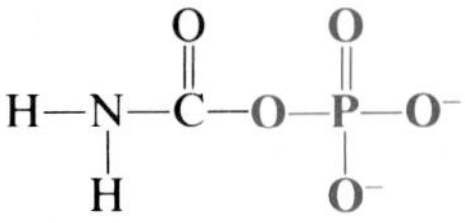

Carbamyl phosphate

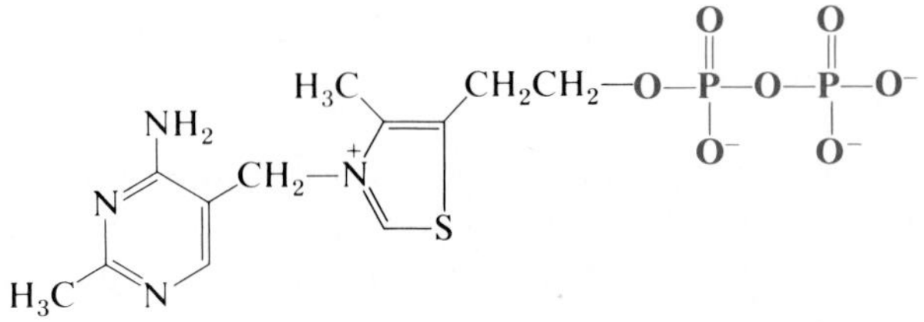

Thiamine pyrophosphate

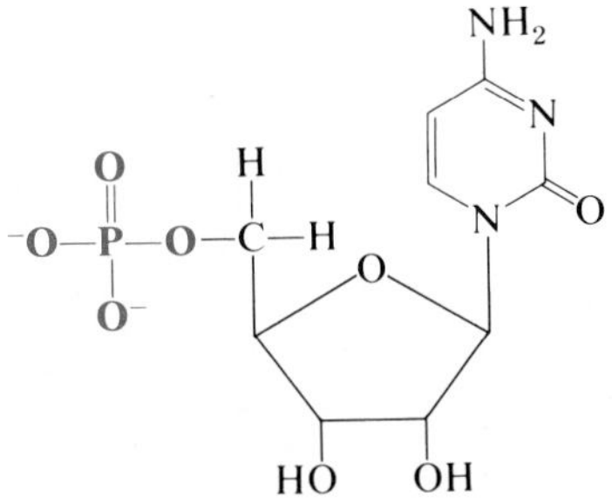

Cytidine nucleotide

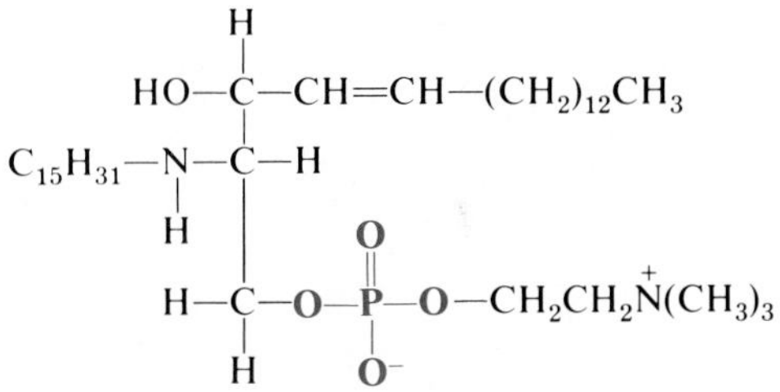

Sphingomyelin

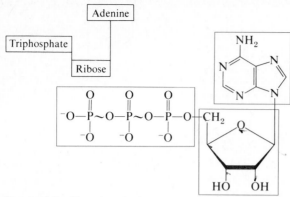

FIGURE 8.3 The chemical structure of adenosine triphosphate (ATP).

The most significant feature of the ATP molecule is the presence of the phosphoric acid anhydride, or pyrophosphate, linkage.

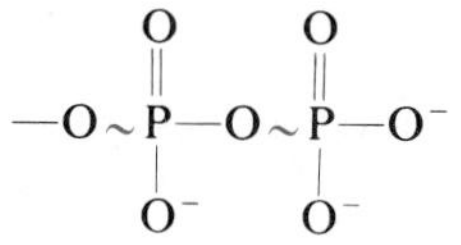

$$-O\sim\overset{\overset{\displaystyle O}{\|}}{\underset{\underset{\displaystyle O^-}{|}}{P}}-O\sim\overset{\overset{\displaystyle O}{\|}}{\underset{\underset{\displaystyle O^-}{|}}{P}}-O^-$$

ATP is often called an *energy-rich compound;* its pyrophosphate bonds are referred to as high-energy bonds and are sometimes symbolized by a squiggle bond (∼). A **high-energy bond** is one that *releases a relatively large amount of energy* (>7000 cal/mole) *when it is hydrolyzed.* In this case, one of the driving forces for the reaction is to relieve the electron–electron repulsions associated with the negatively charged phosphate groups. It should be noted, however, that there is nothing special about the bonds themselves. The symbol ∼ is just a device to focus attention on a portion of the molecule that undergoes reaction. Energy-rich compounds, then, are substances having particular structural features that yield high energies of

FIGURE 8.4 The relationships of ATP, ADP, and AMP.

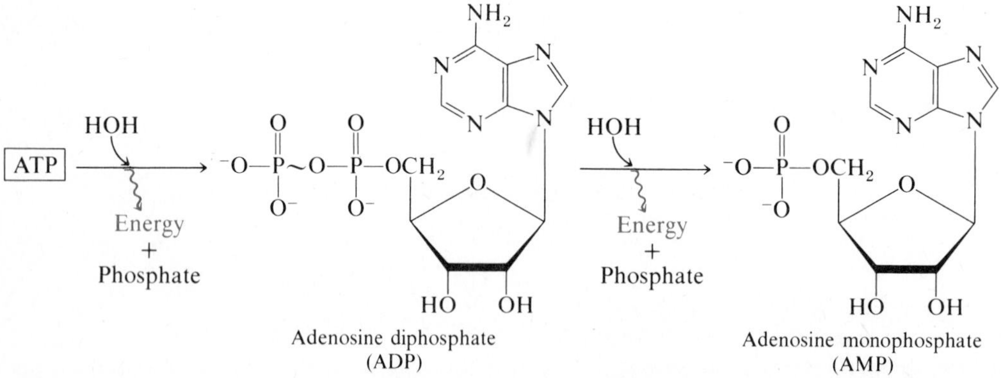

Adenosine diphosphate
(ADP)

Adenosine monophosphate
(AMP)

TABLE 8.8 Standard Free Energies of Hydrolysis of Some Phosphate Compounds

Type	Example	$\Delta G'$ (cal/mole)	See Page
Enol phosphate 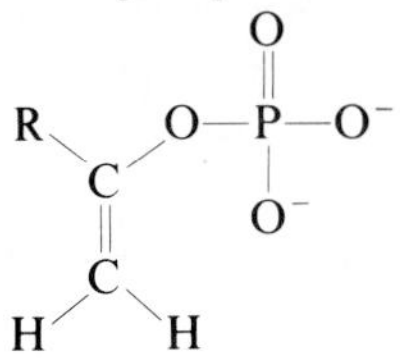	Phosphoenolpyruvic acid	−14,800	435
Acyl phosphate 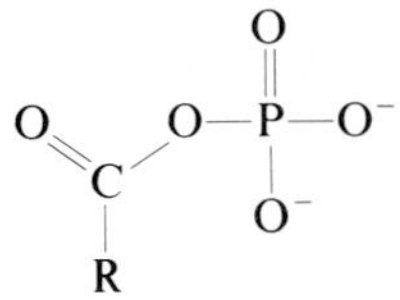	1,3-Diphosphoglyceric acid	−11,800	434
	Acetyl phosphate	−10,300	
Guanidine phosphate 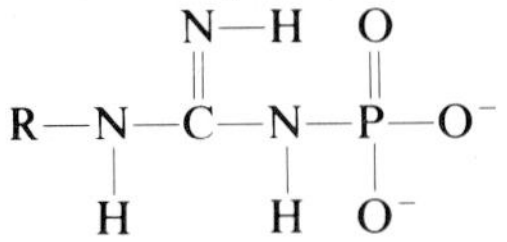	Creatine phosphate	−10,300	441
	Arginine phosphate	−7,700	441
Pyrophosphate 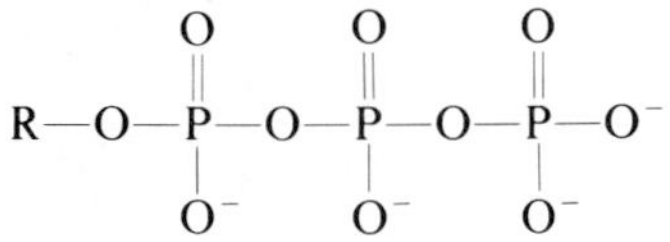	ATP $\longrightarrow$ AMP + PP$_i$	−7,700	399
	ATP $\longrightarrow$ ADP + P$_i$	−7,300	431
	ADP $\longrightarrow$ AMP + P$_i$	−7,300	
	PP$_i$ $\longrightarrow$ 2 P$_i$	−6,500	465
Sugar phosphate 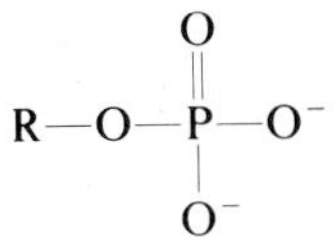	Glucose 1-phosphate	−5,000	430
	Fructose 6-phosphate	−3,800	432
	AMP $\longrightarrow$ Adenosine + P$_i$	−3,400	
	Glucose 6-phosphate	−3,300	431
	Glycerol 3-phosphate	−2,200	470

Structures shown in the Type column:

Enol phosphate:
$$R-C(=CH_2)-O-P(=O)(O^-)(O^-)$$

Acyl phosphate:
$$O=C(R)-O-P(=O)(O^-)(O^-)$$

Guanidine phosphate:
$$R-N(H)-C(=N-H)-N(H)-P(=O)(O^-)(O^-)$$

Pyrophosphate:
$$R-O-P(=O)(O^-)-O-P(=O)(O^-)-O-P(=O)(O^-)-O^-$$

Sugar phosphate:
$$R-O-P(=O)(O^-)(O^-)$$

hydrolysis, and for this reason they are able to supply energy for energy-requiring biochemical processes.

$$\text{ATP} \underset{}{\overset{H_2O}{\rightleftharpoons}} \text{ADP} + P_i{}^7 + 7300\ \text{cal/mole}^8$$

[7] P$_i$ is the symbol for the inorganic phosphate anions $H_2PO_4{}^-$ and $HPO_4{}^{2-}$ that are present in the intra- and extracellular fluids. About 1.2 g of phosphate is needed in the daily diet to replace the amount excreted in the urine.

[8] The values in the literature for the energy released when ATP is hydrolyzed vary somewhat. This situation is due in part to the fact that reaction conditions (concentration, temperature, pH) have not always been the same in different laboratories and in part to the difficulties in obtaining exact values for the equilibrium constants. Most texts and research articles report values between −7000 and −8000 cal/mole for the free energy of hydrolysis of ATP at 25°C and a pH of 7.0. We shall use a value of −7300 throughout this text. However, this is only an approximate value since the concentrations of cellular reactants are not molar, as required for energy calculations.

The important feature of this biochemical reaction is its reversibility. The hydrolysis of ATP releases energy; its synthesis requires energy. In a typical cell, an ATP molecule is consumed ("turned over") within 1 minute after its formation. Thus ATP is produced by those processes that supply energy to an organism (absorption of radiant energy of the sun in green plants and breakdown of foodstuffs in animals), and ATP is hydrolyzed by those processes that require energy (syntheses of carbohydrates, lipids, proteins; transmission of nerve impulses; muscle contraction; etc.). This coupling of the synthesis of ATP to processes that release energy, and of the hydrolysis of ATP to processes that require energy, is one of the striking characteristics of living matter.

Although ATP is the principal medium of energy exchange in biological systems, it is not the only high-energy compound. There are several other phosphate esters that are utilized to provide energy for certain energy-requiring reactions. Table 8.8 lists a number of phosphate compounds and reference is given to subsequent pages in the text where the use of these compounds is illustrated. Notice that the free energy of hydrolysis of ATP is approximately midway between those of the high-energy and the low-energy phosphate compounds.

8.15 Organophosphates and Nerve Transmission

Certain phosphate esters, in stark contrast to the above-mentioned metabolic intermediates, exert powerful inhibitory effects upon the transmission of the nerve impulse. The first such organophosphates were developed in Germany in the late 1930s in the course of research on antipersonnel nerve gases. An example of a nerve gas developed for warfare is diisopropyl fluorophosphate, DFP (Figure 8.5).

Nerve cells (neurons) interact with other nerve cells, and with muscles and glands, at junctions called *synapses* (Figure 8.6). Nerve impulses are transported across synapses by small molecules known as **neurotransmitters.** Neurons utilize a number of different molecules as neurotransmitters. The two

FIGURE 8.5 Some poisonous phosphate esters.

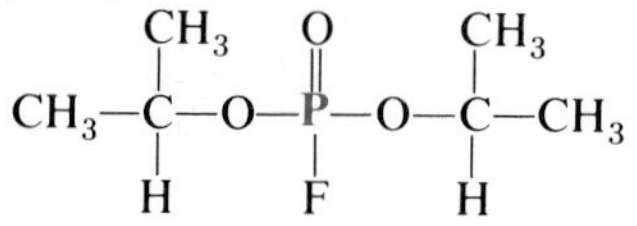

DFP
(Diisopropyl fluorophosphate)

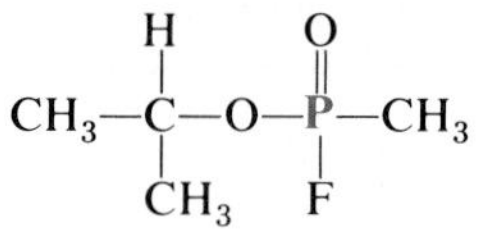

Sarin
(Nerve gas)

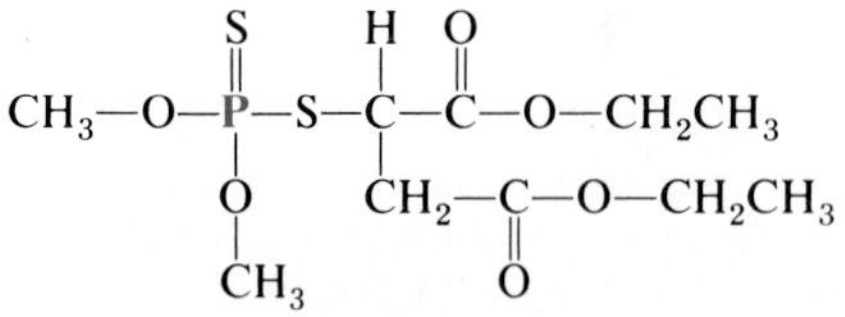

Malathion
(Dimethyl dithiophosphate ester
of diethyl mercaptosuccinate)

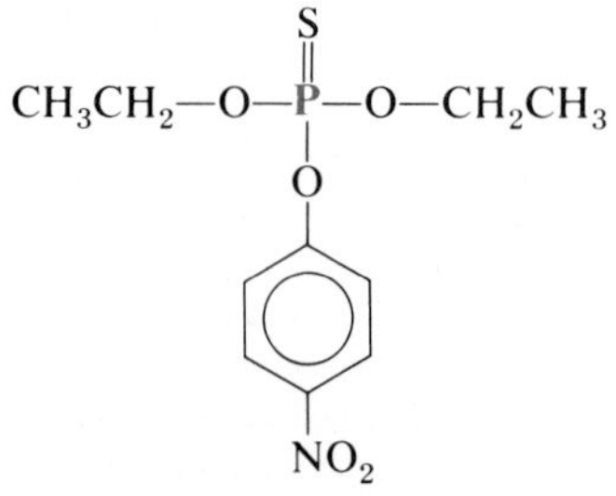

Parathion
(Diethyl *p*-nitrophenyl thiophosphate)

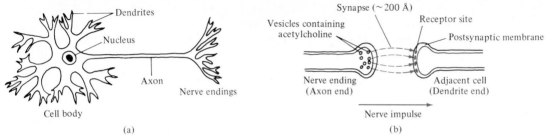

FIGURE 8.6 (a) A neuron. (b) Transmission of nerve impulses between neurons.

major neurotransmitters are *acetylcholine* and *norepinephrine* (see Section 9.12-d).

Acetylcholine is synthesized from acetyl-CoA and choline and is stored in special vesicles at the axon ends of neurons. The arrival of a nerve impulse leads to the release of acetylcholine into the synapses.[9] The acetylcholine molecules then diffuse across the synapse, where they combine with specific receptor protein molecules embedded in the postsynaptic membrane of the adjacent neuron (or muscle, or gland). Binding of acetylcholine to the receptors causes a change in membrane permeability of the receiving neuron (or muscle, or gland). Sodium ions move into the cell, potassium ions move out, and this ion flux causes the signal (the action potential) to be sent along the entire neuron until it reaches another synapse. (The so-called sodium pump is a special protein complex that requires energy to pass the sodium and potassium ions back across the membrane to reestablish and maintain the concentration gradients.)

Once the impulse has been passed on, the acetylcholine must be immediately deactivated so that the receptor molecules can receive the next stimulus. The deactivation occurs by the hydrolysis of acetylcholine to choline and acetic acid, through the catalytic activity of the enzyme *acetylcholinesterase* (which is located in the synaptic cleft). This enzyme is characterized by an extremely high turnover number. It is estimated that this enzyme-catalyzed hydrolysis reaction occurs in 40 μsec (40×10^{-6} sec). This speed is essential because nerve fibers can transmit 1000 impulses per second as long as the postsynaptic membrane is continually available to receive new acetylcholine molecules.[10]

The organophosphate nerve gases affect all biological systems

[9] Botulism toxin, one of the most poisonous substances known, and certain snake venoms act by inhibiting the release of acetylcholine from the axon. Thus these toxic substances effectively block nerve transmissions that utilize acetylcholine as their neurotransmitter.

[10] Curare, used for centuries on the arrows of South American Indians, causes skeletal muscle paralysis. Curare exerts its effects by competing with acetylcholine for the receptor sites on the postsynaptic muscle cell membranes. Thus it blocks the receptor sites and prevents the message from being transmitted from nerve to muscle.

$$CH_3-\overset{\overset{\displaystyle O}{\|}}{C}-OCH_2CH_2-\overset{\overset{\displaystyle CH_3}{|}}{\underset{\underset{\displaystyle CH_3}{|}}{N^+}}-CH_3 \;+\; HOH \;\xrightarrow{\text{acetylcholin-esterase}}$$

Acetylcholine

$$CH_3-\overset{\overset{\displaystyle O}{\|}}{C}-OH \;+\; HOCH_2CH_2-\overset{\overset{\displaystyle CH_3}{|}}{\underset{\underset{\displaystyle CH_3}{|}}{N^+}}-CH_3$$

Acetic acid   Choline

in a similar manner. They inhibit *acetylcholinesterase*. If the enzyme is inhibited, acetylcholine is not broken down. Instead it accumulates on the receptor sites, causing a continuous stimulation of the motor nerves and the glands and muscles that the nerves control.[11] The movements of the entire body become

[11] The nerve cells, muscle cells, or other receptor cells that are sensitive to acetylcholine are said to be *cholinergic*. Pharmacologists have developed drugs that affect cholinergic nerves by either enhancing or blocking the action of acetylcholine on the receptor cells. Succinylcholine, an inhibitor of acetylcholine, is used to produce muscular relaxation in surgical procedures. On the other hand, an insufficient secretion of acetylcholine by motor neurons results in the condition called *myasthenia gravis*. The person suffers from muscular weakness and may have trouble contracting the muscles associated with breathing, chewing, eye movements, and speaking. This condition can be treated by administering neostigmine, which inhibits *acetylcholinesterase*. If the enzyme is inhibited, enough acetylcholine may accumulate in the neuromuscular synapse to stimulate muscle contraction.

$$(CH_3)_3\overset{+}{N}CH_2CH_2O\overset{\overset{\displaystyle O}{\|}}{C}CH_2CH_2\overset{\overset{\displaystyle O}{\|}}{C}OCH_2CH_2\overset{+}{N}(CH_3)_3$$

Succinylcholine

$$(CH_3)_3\overset{+}{N}\;\text{—}\!\!\!\bigcirc\!\!\!\text{—}\;O-\overset{\overset{\displaystyle O}{\|}}{C}-N(CH_3)_2$$

Neostigmine

uncoordinated; tremors, muscular spasms, convulsions, paralysis, and death quickly follow. For this reason, the organic phosphate insecticides are among the most poisonous chemicals in the world. The structures of two of the common phosphorus-containing insecticides are given in Figure 8.5.

Parathion is one of the most powerful and toxic organophosphate insecticides. (Many people involved in the manufacture and application of parathion are killed each year as a direct result of exposure to it.) Its only saving grace, albeit a small one, is that it decomposes rather rapidly. Unlike the chlorinated hydrocarbons, parathion will disappear from the environment within 15–20 days through biological degradation. (By contrast, at least half of the DDT may still be present five years after application.)

Since the banning of DDT, **malathion** has become one of the principal insecticides for use against household and garden insects. It is the least toxic of the organic phosphates to the higher animals because of the presence of an enzyme in the liver that quickly decomposes it. (Recently, however, it was discovered that malathion is highly toxic to fish. Care must be taken with its use near streams and lakes.) Malathion is widely employed to control fleas in dog kennels and to protect dairy and other livestock. The organophosphate compounds are more expensive insecticides, and because of their rapid biodegradation, they have to be applied more frequently.

Exercises

8.1 Draw and name all the isomeric acids and esters having the molecular formula $C_4H_8O_2$.

8.2 Name the following compounds.

(a) $(CH_3)_2CHCH_2C\overset{O}{\underset{OH}{}}$

(b) $CH_3C\overset{O}{\underset{OCH_3}{}}$

(c) $(CH_3)_3CCH(CH_3)CH_2C\overset{O}{\underset{OH}{}}$

(d) $CH_3CH_2CH_2CH_2C\overset{O}{\underset{OCH_2CH_2CH_3}{}}$

(e) $CH_2OHCH_2CH_2C\overset{O}{\underset{OH}{}}$

(f) $CH_3CH_2C\overset{O}{\underset{O^- Na^+}{}}$

(g) $CH_3CH{=}CHCH_2CH_2C\overset{O}{\underset{OH}{}}$

(h) $(CH_3)_2CHCCl_2CH_2CH_2C\overset{O}{\underset{OH}{}}$

(i) 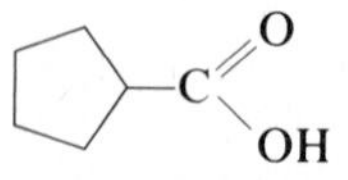 cyclopentyl$-C\overset{O}{\underset{OH}{}}$

(j) 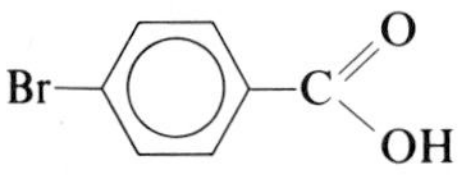 $Br-$⟨benzene ring⟩$-C\overset{O}{\underset{OH}{}}$

(k) $(CH_3O)_3P$

(l) $CH_3(CH_2)_4C\overset{O}{\underset{O^- K^+}{}}$

(m) $CH_3CHOHCH_2CHIC\overset{O}{\underset{OH}{}}$

(n) $\overset{O}{\underset{HO}{}}C(CH_2)_2C\overset{O}{\underset{OH}{}}$

(o) $H-C\overset{O}{\underset{OCH(CH_3)_2}{}}$

(p) $CH_3C\overset{O}{\underset{Cl}{}}$

(q) $(CH_3CH_2)_2CHC\overset{O}{\underset{OCH_2CH_3}{}}$

(r) $CH_3CH(OCH_3)CH(CH_2CH_3)CH_2C\overset{O}{\underset{OH}{}}$

(s) 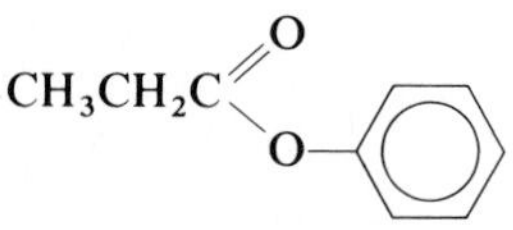 $CH_3CH_2C\overset{O}{\underset{O-}{}}$⟨benzene ring⟩

(t) 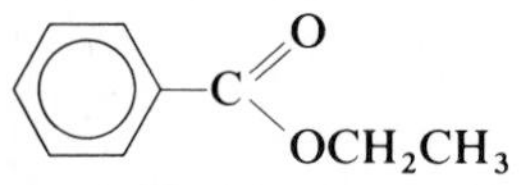 ⟨benzene ring⟩$-C\overset{O}{\underset{OCH_2CH_3}{}}$

172

8.3 Write structural formulas for the following compounds.
  (a) 2,2-dimethylpropanoic acid
  (b) 4-chloro-2-methylpentanoic acid
  (c) oxalic acid
  (d) ethyl acetate
  (e) acetic anhydride
  (f) α-iodobutyric acid
  (g) calcium acetate
  (h) *m*-bromobenzoic acid
  (i) isopropyl benzoate
  (j) phenyl 5-methylhexanoate
  (k) potassium propanoate
  (l) methyl 3,3-dimethylbutanoate
  (m) ethyl phosphate
  (n) isopropyl butanoate
  (o) 4-bromo-3-pentenoic acid
  (p) methyl lactate
  (q) *p*-nitrobenzoic acid
  (r) salicylic acid
  (s) 3-phenylpropanoic acid
  (t) β-ethoxypropionic acid

8.4 Write equations showing the preparation of butyric acid from each of the following.
  (a) 1-butanol   (b) butanal
  (c) 1-bromopropane   (d) ethyl butanoate
  (e) sodium butyrate

8.5 Offer an explanation for the following.
  (a) Acetic acid is more acidic than ethyl alcohol.
  (b) Trichloroacetic acid is a stronger acid than acetic acid.
  (c) Methyl acetate has a lower boiling point (57°C) than either methyl alcohol (65°C) or formic acid (101°C), even though it has a higher molecular weight.
  (d) Formic acid melts about 30°C higher than propionic acid, even though its molecular weight is considerably lower.
  (e) Butyric acid is soluble in water, whereas its isomer ethyl acetate is insoluble.
  (f) A solution containing equal concentrations of benzoic acid and sodium benzoate resists changes in pH.
  (g) Sodium benzoate is soluble in water, whereas benzoic acid is insoluble.
  (h) In the carboxyl group there are two different carbon–oxygen bond distances, whereas in the carboxylate anion there is only one carbon–oxygen bond distance.
  (i) The alkaline hydrolysis of ester is irreversible, whereas the acidic hydrolysis of esters is reversible.
  (j) Salicylic acid is the starting material for the synthesis of both aspirin and oil of wintergreen.
  (k) Succinylcholine causes muscle relaxation, whereas neostigmine causes muscle contraction.

8.6 Arrange each set of compounds in order of decreasing solubility in water.

(a) C_6H_5–C(=O)–OH CH_3CH_2C(=O)–OH
    $CH_3CH_2CH_2CH_2C$(=O)–OH

(b) CH_3CH_2C(=O)–OH $CH_3CH_2CHOHCH_3$
    CH_3C(=O)–$OCH_2CH_2CH_3$

(c) cyclohexyl–C(=O)–OH C_6H_5–C(=O)–$O^- Na^+$
    $CH_3CH_2CH_2CH_2C$(=O)–OH

8.7 Arrange each set of compounds in Exercise 8.6 in order of decreasing boiling point.

8.8 Write an equation to illustrate what occurs when an organic acid is placed in water. What determines the strength of an acid?

8.9 A 0.01 *M* solution of the weak organic acid, RCOOH, dissociates to the extent of 3.4%. What is the dissociation constant of the acid?

8.10 Calculate the concentration of H^+ and OH^- in a 1.0 *M* solution of acetic acid. What is the pH of the solution?

8.11 What is the hydrogen ion concentration of a 0.01 *M* solution of chloroacetic acid ($K_a = 1.4 \times 10^{-3}$)? What is the pH of this solution?

8.12 A solution containing 1.0 *M* sodium acetate and 1.0 *M* acetic acid, buffers at a pH about 4. Show by equations what happens when acids or bases are added to this buffer solution.

8.13 Compare in a general way the pH of a 0.1 *M* acetic acid solution with that of a 0.1 *M* acetic acid solution that is also 0.1 *M* with sodium acetate. Account for these differences.

8.14 At a pH of 7.4, what is the ratio of HPO_4^{2-} to $H_2PO_4^-$? (Use the Henderson–Hasselbalch equation.)

8.15 Arrange each of the following groups of compounds in order of increasing acid strength.
  (a) butanoic acid, 2-chlorobutanoic acid, 3-chlorobutanoic acid, 4-chlorobutanoic acid.
  (b) pentanoic acid, 2-fluoropentanoic acid, 2-chloropentanoic acid, trimethylacetic acid
  (c) propanoic acid, 2,2-dichloropropanoic acid, 2-chloropropanoic acid, acetic acid.
  (d) acetic acid, dichloroacetic acid, bromoacetic acid, 3-chlorobutanoic acid.

(e) water, trichloroacetic acid, butyric acid, α-chloro-butyric acid

(f) acetic acid, formic acid, methyl alcohol, fluoroacetic acid

(g) 2,2-dimethylbutanoic acid, butanoic acid, 2-methylbutanoic acid, 2-bromobutanoic acid

(h) cyclohexanecarboxylic acid, benzoic acid, phenol, cyclohexanol

(i) acetoacetic acid ($pK_a = 3.58$), lactic acid ($pK_a = 3.85$), benzoic acid ($pK_a = 4.20$), pyruvic acid ($pK_a = 2.50$)

(j) phenylacetic acid ($pK_a = 4.31$), acetic acid ($pK_a = 4.74$), benzoic acid ($pK_a = 4.20$), formic acid ($pK_a = 3.68$)

8.16 Complete the following equations, giving only the organic products (if any).

(a) $CH_3CH_2CH_2OH + K_2Cr_2O_7 \xrightarrow{H^+}$

(b) $+ KMnO_4 \xrightarrow{H^+}$

(c) $-CH_2OH + Na_2Cr_2O_7 \xrightarrow{H^+}$

(d) $H_3C-$$-CH_3 + KMnO_4 \xrightarrow[\Delta]{H^+}$

(e) $+ K_2Cr_2O_7 \xrightarrow{H^+}$

(f) $CH_3CH_2Br + KCN \longrightarrow ? \xrightarrow[HOH]{H^+}$

(g) $CH_3C\overset{O}{\underset{OH}{\diagdown}} + HCl \longrightarrow$

(h) $(CH_3)_3CCH_2C\overset{O}{\underset{OH}{\diagdown}} + NaOH \longrightarrow$

(i) $H_3C-$$-C\overset{O}{\underset{O^- K^+}{\diagdown}} + HCl \longrightarrow$

(j) $CH_3CH=CHC\overset{O}{\underset{OH}{\diagdown}} + NaHCO_3 \longrightarrow$

(k) $-CH_2CH_2CH_2C\overset{O}{\underset{OH}{\diagdown}} + SOCl_2 \longrightarrow$

(l) $CH_3-\overset{O}{\overset{\|}{C}}-O-\overset{O}{\overset{\|}{C}}-CH_3 + HOH \longrightarrow$

(m) $-C\overset{O}{\underset{OH}{\diagdown}} + CH_3OH \underset{\Delta}{\overset{H^+}{\rightleftharpoons}}$

(n) $(CH_3)_2CHC\overset{O}{\underset{O-}{\diagdown}}$—$-CH_3 + KOH \xrightarrow{\Delta}$

(o) $CH_3CHBrC\overset{O}{\underset{OCH(CH_3)_2}{\diagdown}} \underset{HOH}{\overset{H^+}{\rightleftharpoons}}$

(p) $CH_3C\overset{O}{\underset{OCH_2CH_3}{\diagdown}} \xrightarrow[ether]{LiAlH_4} \xrightarrow[HOH]{H^+}$

(q) $Br-$$-C\overset{O}{\underset{OH}{\diagdown}} \xrightarrow{B_2H_6} \xrightarrow{HOH}$

(r) $-C\overset{O}{\underset{OH}{\diagdown}}$ with OH $+ CH_3OH \overset{H^+}{\rightleftharpoons}$

(s) $-C\overset{O}{\underset{OH}{\diagdown}}$ with OH $+ CH_3C\overset{O}{\underset{Cl}{\diagdown}} \longrightarrow$

(t) $CH_3-C\overset{O}{\underset{OCH_2CH_2N^+(CH_3)_3}{\diagdown}} \xrightarrow[enzyme]{HOH}$

8.17 Identify the acid or ester in each of the following.
(a) a substance with a wintergreen odor
(b) a well-known analgesic
(c) the sting of red ants
(d) the sour taste and sharp odor of vinegar
(e) the odor of oranges
(f) the odor of bananas
(g) the odor of rancid butter
(h) spinach and rhubarb
(i) citrus fruits
(j) sour milk and sauerkraut
(k) grapes
(l) willow bark
(m) monomer used in the production of Dacron
(n) component of high octane fuels
(o) high-energy compound found in all living cells
(p) a neurotransmitter

8.18 Assume that the equilibrium constant at 25°C for the esterification reaction between RCOOH and ROH is 3.5. Tell how the equilibrium will be affected by each of the following.
(a) rise in temperature

(b) addition of a small amount of sulfuric acid

(c) removal of water

(d) addition of sodium bicarbonate

8.19 Complete the following equations. Be sure to include the label in one of the products.

(a) $\text{C}_6\text{H}_5\text{-C}(\!=\!O)\text{-OH} + CH_3CH_2CH_2{}^{18}OH \underset{\Delta}{\overset{H^+}{\rightleftharpoons}}$

$(CH_3)_2CHC(\!=\!O){}^{18}OH + (CH_3)_2CHCH_2OH \underset{\Delta}{\overset{H^+}{\rightleftharpoons}}$

(c) $\text{C}_6\text{H}_5\text{-}CH_2C(\!=\!O){}^{18}OH + NaOH \longrightarrow$

(d) $CH_3-C(\!=\!O)-O^{14}CH_3 + HOH \underset{\Delta}{\overset{H^+}{\rightleftharpoons}}$

(e) $\text{C}_6\text{H}_5\text{-}{}^{14}C(\!=\!O)\text{-O-}\text{C}_6\text{H}_5 + NaOH \overset{\Delta}{\longrightarrow}$

(f) $\text{C}_6\text{H}_5\text{-C}(\!=\!{}^{18}O)\text{-}OCH_3 \xrightarrow[\text{ether}]{LiAlH_4} \xrightarrow[HOH]{H^+}$

(g) $(CH_3)_2CHCH_2-{}^{18}O\text{-C(}\!=\!O)\text{-}H \xrightarrow[HOH]{H^+}$

(h) $CH_3CH_2-O\text{-C(}\!=\!O)\text{-}{}^{14}CH_2CH_3 + NaOH \longrightarrow$

8.20 The following compounds contain a carbon–oxygen–carbon linkage. Compare their relative reactivity toward water.

(a) ethyl ether

(b) acetic anhydride

(c) ethyl acetate

8.21 Devise simple tests that would enable you to distinguish between the members of the following pairs of compounds. Do not use the litmus test.

(a) methyl butanoate and valeric acid

(b) propanoic acid and propanal

(c) butanoic acid and 1-butanol

(d) pentanoic acid and 1-pentene

(e) benzoic acid and phenol

8.22 Illustrate the following terms with a suitable example or a chemical equation.

(a) buffer solution

(b) neutralization

(c) a soap

(d) acid anhydride

(e) a nitrile

(f) an alkyl isocyanate

(g) aspirin

(h) ATP

(i) pyrophosphate

(j) glacial acetic acid

(k) cream of tartar

(l) esterification

(m) hydrolysis

(n) saponification

(o) ammonolysis

(p) synapse

(q) neurotransmitter

(r) acetylcholinesterase

(s) organophosphate pesticide

(t) organophosphate nerve gas

8.23 A compound, $C_7H_{14}O_2$, was treated with potassium hydroxide to yield ethanol and a potassium salt. Write four possible structural formulas of the original compound.

8.24 An ester, $C_6H_{12}O_2$, was hydrolyzed in aqueous acid to yield an acid (Y) and an alcohol (Z). Oxidation of the alcohol with potassium permanganate resulted in the identical acid (Y). What is the structural formula of the ester?

8.25 5.1 g of a monocarboxylic acid was required to neutralize 125 mL of a 0.4 M NaOH solution. Write all possible structural formulas for the acid.

8.26 If 3.0 g of acetic acid reacted with excess methanol, how many grams of methyl acetate could be formed?

8.27 How many milliliters of a 0.10 M barium hydroxide solution would be required to neutralize 0.50 g of dichloroacetic acid?

9

Amides and Amines

The compounds studied in the preceding chapters can be considered as derivatives of the alkanes and of water. In this chapter, we deal with the organic nitrogen-containing compounds related to ammonia. *Replacement of one or more hydrogen atoms of ammonia with alkyl* (or aryl) *groups yields* **amines,** RNH_2, R_2NH, and R_3N (or $ArNH_2$, Ar_2NH, Ar_3N). *When a hydrogen atom of ammonia is replaced instead by an acyl group,* $RCO-$, *the resultant compound is an* **amide,** $RCONH_2$.

Amides

Amides are also considered to be derivatives of carboxylic acids. Replacement of the hydroxyl group by the $-NH_2$ group (the amino group), the $-NHR$ group, or the $-NR_2$ group gives rise to an unsubstituted, a monosubstituted, or a disubstituted amide, respectively.

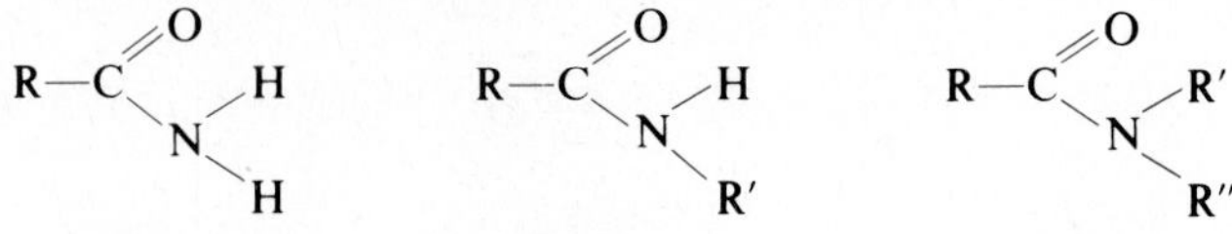

Unsubstituted amide	Monosubstituted amide	Disubstituted amide

The functional group of the amides is the $-\overset{\overset{\displaystyle O}{\|}}{C}-N\diagup$ grouping. The *carbonyl carbon–nitrogen bond* is referred to as the **amide linkage.** This bond is very stable and is found in the repeating units of protein molecules (Chapter 13), in nylon (Section 9.2), and in many other industrial polymers.

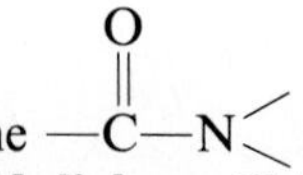

9.1 Nomenclature of Amides

Amides are named as derivatives of organic acids. The *-ic* ending of the common name or the *-oic* ending of the IUPAC name is replaced with the suffix *-amide*. Alkyl or aryl substituents on the nitrogen atom are denoted by prefixing the name of the amide by *N-*, followed by the name of the substituent group. Table 9.1 gives some examples of amide nomenclature.

9.2 Preparation of Amides

The unsubstituted amides are commonly prepared by the addition of ammonia to the carboxylic acid derivatives (acyl chlorides or acid anhydrides). The route through the acyl chloride is usually considered to be the most convenient. Each mole of acyl chloride requires 2 moles of ammonia; the second mole neutralizes the liberated hydrogen chloride.

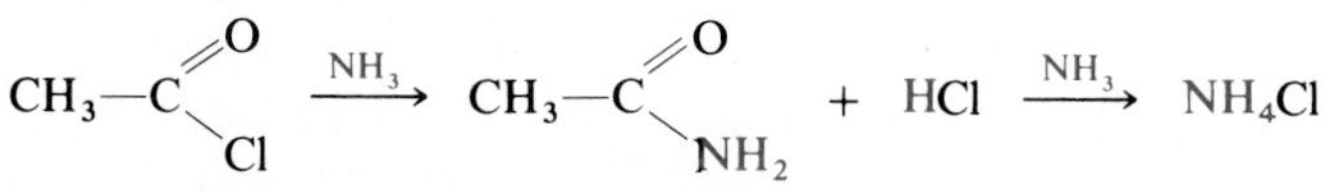

Acetyl chloride   Acetamide

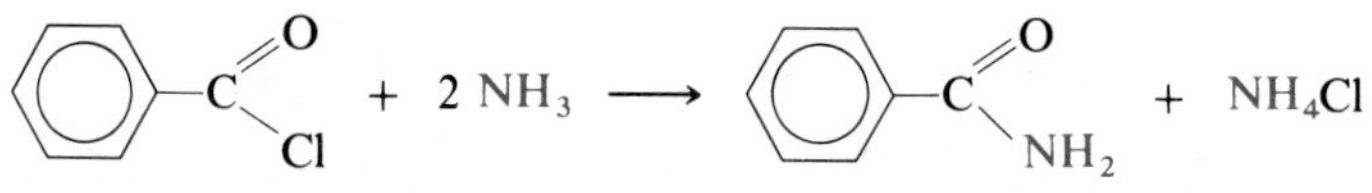

Benzoyl chloride   Benzamide

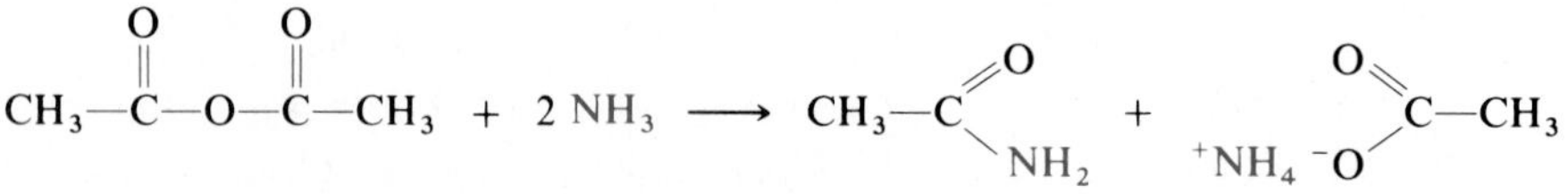

Acetic anhydride   Acetamide   Ammonium acetate

TABLE 9.1 Nomenclature of Amides

Formula	Common Name	IUPAC Name
$H-\overset{\displaystyle O}{\underset{\displaystyle NH_2}{C}}$	Formamide	Methanamide
$CH_3-\overset{\displaystyle O}{\underset{\displaystyle NH_2}{C}}$	Acetamide	Ethanamide
$CH_3CH_2-\overset{\displaystyle O}{\underset{\displaystyle NH_2}{C}}$	Propionamide	Propanamide
$CH_3CH_2CH_2-\overset{\displaystyle O}{\underset{\displaystyle NH_2}{C}}$	Butyramide	Butanamide
$CH_3\overset{\displaystyle CH_3}{\underset{\displaystyle}{CH}}-\overset{\displaystyle O}{\underset{\displaystyle NH_2}{C}}$	Isobutyramide	2-Methylpropanamide
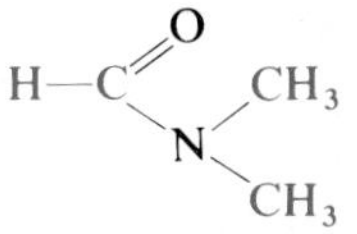 $H-\overset{\displaystyle O}{\underset{\displaystyle N}{C}}\overset{CH_3}{\underset{CH_3}{}}$	N,N-Dimethylformamide (DMF—important organic solvent)	N,N-Dimethylmethanamide
Benzamide formula	Benzamide	Benzenecarboxamide
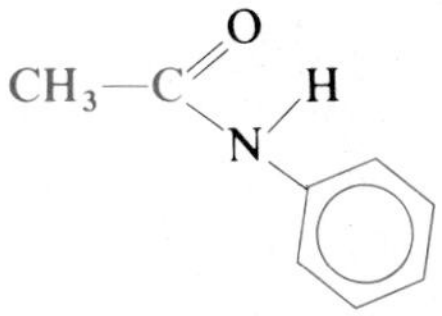 $CH_3-\overset{\displaystyle O}{\underset{\displaystyle N}{C}}\overset{H}{}$ (phenyl)	Acetanilide (N-Phenylacetamide)	N-Phenylethanamide
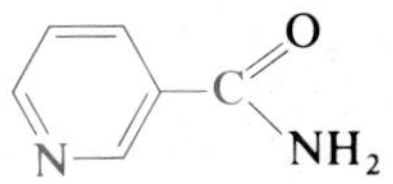 Nicotinamide formula	Nicotinamide (Niacin, a B vitamin)	Pyridine-3-carboxamide
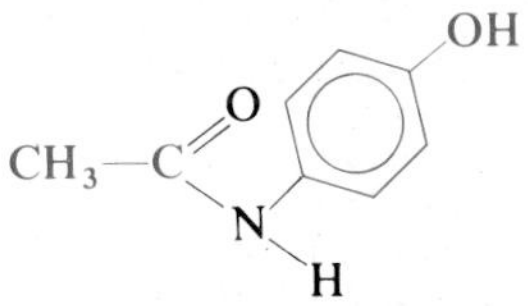 Acetaminophen formula	Acetaminophen (pain killer—major ingredient in Tylenol, Tempra, and Liquiprin)	N-p-Hydroxyphenylethanamide

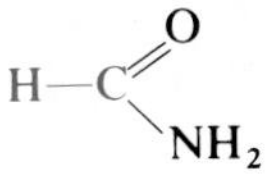

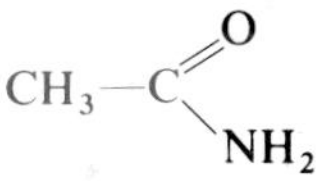

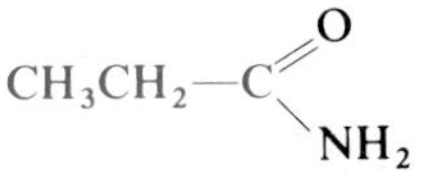

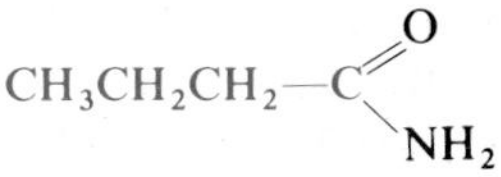

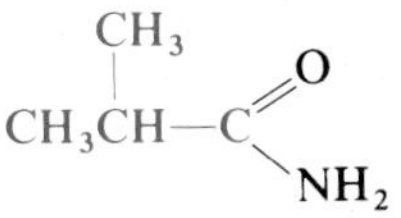

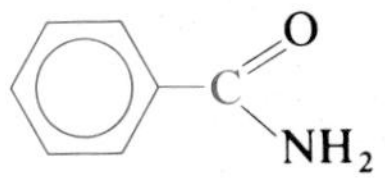

Formation of amides by addition of ammonia to the free acid will occur, but the reaction is very slow at room temperature. The ammonium salt of the acid is formed first; then water can be split out if the reaction temperature is maintained above

100°C. The second step is reversible; the equilibrium favors salt formation. Continuous removal of the water shifts the equilibrium to the right.

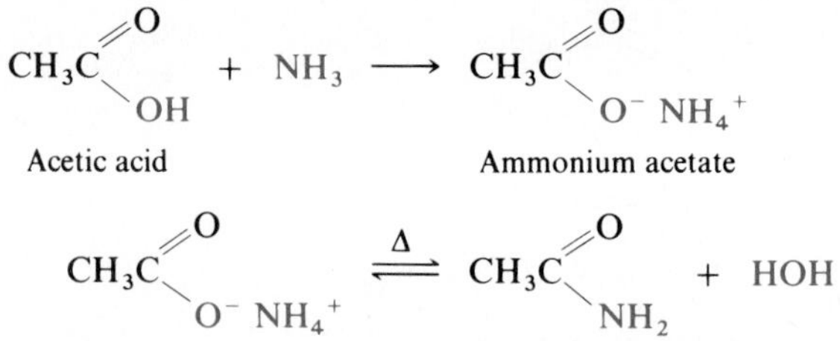

$$CH_3C \overset{O}{\underset{OH}{<}} + NH_3 \longrightarrow CH_3C \overset{O}{\underset{O^- NH_4^+}{<}}$$

Acetic acid   Ammonium acetate

$$CH_3C \overset{O}{\underset{O^- NH_4^+}{<}} \overset{\Delta}{\rightleftharpoons} CH_3C \overset{O}{\underset{NH_2}{<}} + HOH$$

On page 160, we discussed the condensation polymerization reaction to form polyesters. An even more important reaction is one that yields polyamides. Again, two difunctional monomers are employed, usually adipic acid and hexamethylenediamine. The monomers condense by splitting out water to form a new product that is still difunctional and thus can react continuously to yield a polymer.

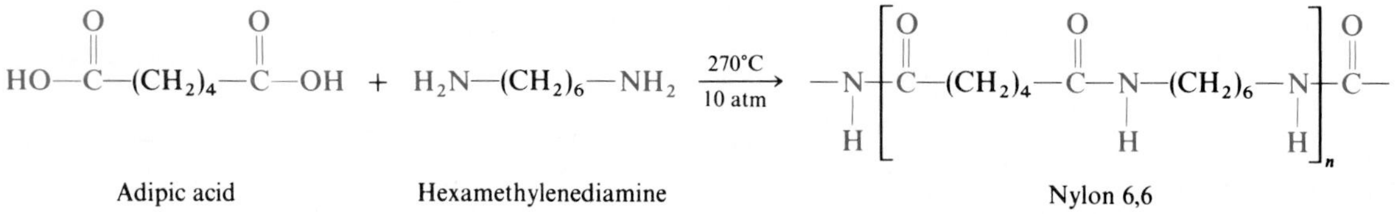

$$HO-\overset{O}{\overset{\|}{C}}-(CH_2)_4-\overset{O}{\overset{\|}{C}}-OH + H_2N-(CH_2)_6-NH_2 \xrightarrow[\text{10 atm}]{270°C} -N \begin{bmatrix} \overset{O}{\overset{\|}{C}}-(CH_2)_4-\overset{O}{\overset{\|}{C}}-N-(CH_2)_6-N \end{bmatrix}_n \overset{O}{\overset{\|}{C}}-$$

Adipic acid   Hexamethylenediamine   Nylon 6,6

Nylon is the collective name of several different synthetic polyamide fibers. Nylon 6,6 is the most common; the 6,6 designation refers to the number of carbon atoms in each of the monomers. It is possible to make other nylons by varying the number of carbons in either the dicarboxylic acid or the diamine. Nylon is a remarkable polymer. It is stable in dilute acids or bases, has a high melting point (260°C), and is very strong. Nylon is one of the most widely used synthetic fibers, for example, in rope, sails, carpets, clothing, tires, brushes, parachutes, and so on. It also can be molded into blocks for use in electrical equipment, gears, bearings, and valves.

9.3   Physical Properties of Amides

With the exception of formamide, which is a liquid, all unsubstituted amides are solids (Table 9.2). Most amides are colorless and odorless. The lower members of the series are soluble in both water and alcohol, water solubility decreases as molecular

TABLE 9.2   Physical Constants of Some Unsubstituted Amides

Formula	Name	MP (°C)	BP (°C)	
$HCONH_2$	Formamide	2	193	
CH_3CONH_2	Acetamide	82	222	Soluble in water
$CH_3CH_2CONH_2$	Propionamide	81	213	
$CH_3CH_2CH_2CONH_2$	Butyramide	115	216	
$C_6H_5CONH_2$	Benzamide	132	290	Insoluble

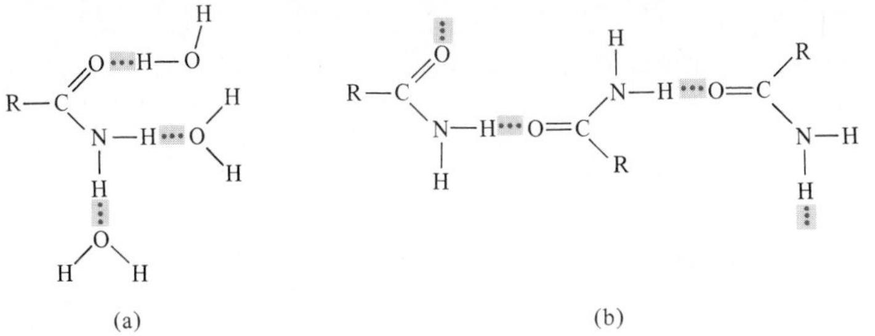

FIGURE 9.1 (a) Hydrogen bonding of amides with water molecules.
(b) Intermolecular hydrogen bonding in amides.

weight increases. The amide group is polar and, unlike the amines, amides are neutral molecules. The unshared electron pair is not localized on the nitrogen atom but is delocalized by resonance onto the oxygen atom of the carbonyl group. The dipolar ion structure restricts free rotation about the carbon–nitrogen bond. This geometric restriction has important consequences for protein structure (see Section 13.3).

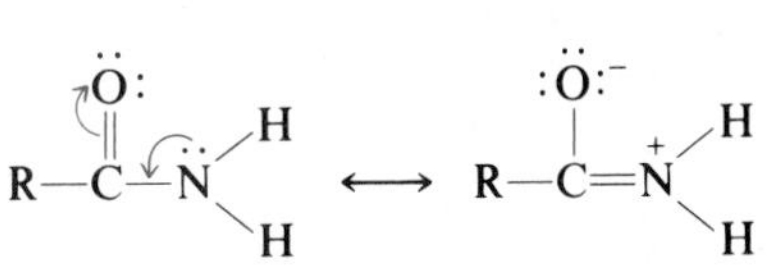

The amides have abnormally high boiling points and melting points. This phenomenon, as well as the water solubility of the amides, is a result of the polar nature of the amide group and the formation of hydrogen bonds (Figure 9.1). Thus electrostatic forces and hydrogen bonding combine to account for the very strong intermolecular attractions found in the amides. Note, however, that disubstituted amides have no hydrogens bonded to nitrogen and thus are incapable of hydrogen bonding. N,N-Dimethylacetamide has a melting point of $-20°C$, which is about $100°C$ lower than the melting point of acetamide.

9.4 Chemical Properties of Amides

The most important reaction of the amides is their hydrolysis. Amide hydrolysis reactions are strictly analogous to the hydrolysis of proteins, which is discussed in Chapter 13. The hydrolysis may be effected by heating the amide in either acidic or basic solution.

General Equation

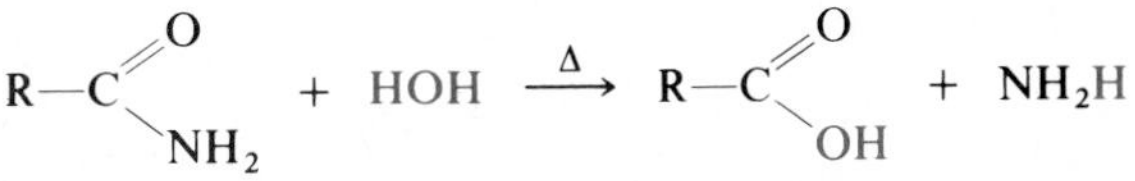

a. Acid Hydrolysis

Hydrolysis of unsubstituted amides in acid solution produces the free organic acid and an ammonium salt. Monosubstituted and disubstituted amides are hydrolyzed to their corresponding acid and amine salt.

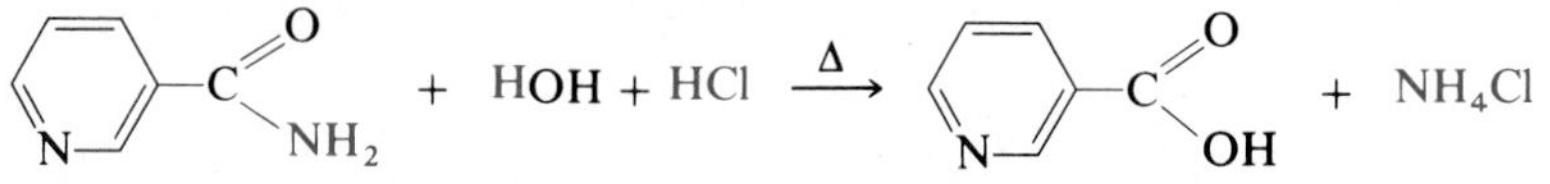

Nicotinamide → Nicotinic acid + NH_4Cl

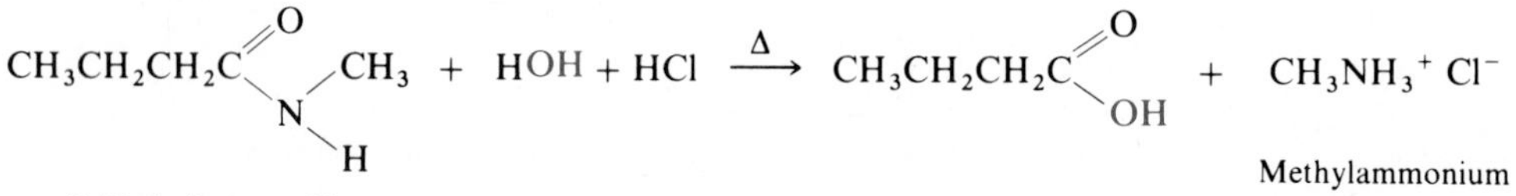

N-Methylbutyramide → Butyric acid + Methylammonium chloride

b. Basic Hydrolysis

Hydrolysis of unsubstituted amides in basic solution produces a salt of the organic acid and ammonia. The strong odor of ammonia signifies its formation as a reaction product.

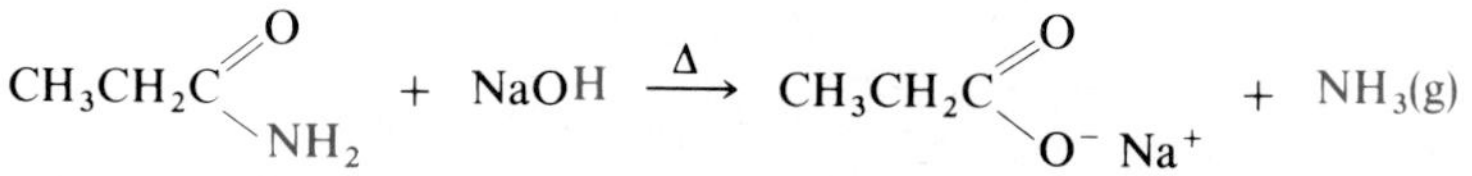

Propionamide + NaOH → Sodium propionate + $NH_3(g)$

9.5 Urea

Urea is the diamide of carbonic acid ($HO-\overset{\overset{O}{\|}}{C}-OH$). Although the acid itself is unstable, many of its derivatives are known and are of great importance. You are familiar with salts of carbonic

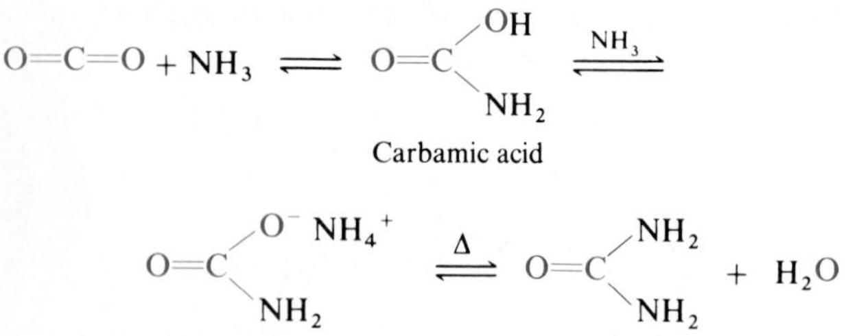

Carbamic acid

Ammonium carbamate   Urea

acid such as sodium bicarbonate (baking soda) and calcium carbonate (limestone). Recall that Wöhler's synthesis of urea from inorganic raw materials (lead cyanate and ammonium hydroxide) was instrumental in the demise of the vital force theory and in the birth of organic chemistry as a separate science (page 1).

Urea is an odorless, white solid that is the end product of protein metabolism in mammals (see Section 18.5). The normal adult excretes about 28–30 g of urea daily in urine. Industrially, urea is prepared by heating carbon dioxide with ammonia at high pressure and temperature.

Several reactions are unique to urea, but we mention only one that is of clinical significance. The action of sodium hypobromite on urea yields nitrogen gas. A measurement of the volume of nitrogen released affords a quantitative determination of urea in various body fluids.

$$NH_2-\overset{\displaystyle O}{\overset{\|}{C}}-NH_2 \ + \ 3\,NaOBr + 2\,NaOH \ \longrightarrow$$

$$N_2(g) + Na_2CO_3 + 3\,H_2O + 3\,NaBr$$

The chief commercial uses of urea are as a fertilizer (to add nitrogen to the soil), as a livestock protein supplement, and as a starting material in the production of barbiturates and urea–formaldehyde plastics. The condensation of urea with various substituted malonic esters yields important cyclic amides. Barbituric acid is synthesized by the reaction of urea with diethyl malonate in the presence of sodium ethoxide. It is the parent compound of the barbiturates, which are used as anesthetics, hypnotics, and sedatives (see Section 9.12-b).

Amines

9.6 · Classification and Nomenclature of Amines

It will be recalled that the amines are the alkyl and aryl derivatives of ammonia. Amines are classified as *primary, secondary,* or *tertiary* according to the number of alkyl or aryl groups bonded to the nitrogen (or the number of hydrogen atoms of ammonia that have been replaced).

$$\overset{\displaystyle H}{\overset{\|}{H-N-H}} \qquad \overset{\displaystyle H}{\overset{\|}{R-N-H}} \qquad \overset{\displaystyle R'}{\overset{\|}{R-N-H}} \qquad \overset{\displaystyle R'}{\overset{\|}{R-N-R''}}$$

| Ammonia | A primary amine | A secondary amine | A tertiary amine |

The simple aliphatic amines are commonly named by specifying the name(s) of the alkyl group(s), in alphabetical order, and adding the suffix -*amine* (Table 9.3). (Note that the name is written as one word in a manner analogous to the naming of the amides and in contrast to the naming of the alcohols.) The prefixes *sec-* and *tert-*, when part of the name of an amine, have no bearing on the classification of that amine. They refer instead to the nature of the carbon atom to which the nitrogen is attached; for example, *sec*-butylamine (a primary amine) and *tert*-butylamine (a primary amine) signify that the amino group is bonded to a secondary carbon and a tertiary carbon, respectively (see Table 9.3). When identical alkyl substituents occur in the same amine, the prefixes *di-* and *tri-* are employed. In the IUPAC system amines are named as derivatives of a parent hydrocarbon. The *amino group* ($-NH_2$) and any other substituents on the carbon chain are located by a number, whereas substituents bonded to nitrogen are identified by a capital *N*. Amine salts are named as derivatives of the ammonium ion. Aniline ($C_6H_5NH_2$) is the most important aromatic amine, and certain compounds are commonly named as derivatives of aniline.

9.7 · Preparation of Primary Amines

Several nitrogen-containing compounds can be easily reduced to the corresponding primary amine, either by the use of hydrogen and a catalyst or by chemical reducing agents. The nitrogen compounds most frequently employed are the amides, nitro

TABLE 9.3 Nomenclature of Amines

Amine Type	Formula	Name
Primary	CH_3NH_2	Methylamine (Aminomethane)
Primary	$CH_3CH_2NH_2$	Ethylamine (Aminoethane)
Secondary	$CH_3-\overset{\overset{\displaystyle H}{\mid}}{N}-CH_3$	Dimethylamine
Secondary	$CH_3-\overset{\overset{\displaystyle H}{\mid}}{N}-CH_2CH_3$	Ethylmethylamine
Tertiary	$CH_3-\overset{\overset{\displaystyle CH_3}{\mid}}{N}-CH_3$	Trimethylamine
Tertiary	$CH_3-\overset{\overset{\displaystyle CH_2CH_3}{\mid}}{N}-CH_2CH_2CH_3$	Ethylmethylpropylamine
Primary	$CH_3CH_2\overset{\overset{\displaystyle NH_2}{\mid}}{C}HCH_3$	*sec*-Butylamine (2-Aminobutane)
Primary	$CH_3-\overset{\overset{\displaystyle CH_3}{\mid}}{\underset{\underset{\displaystyle CH_3}{\mid}}{C}}-NH_2$	*tert*-Butylamine (2-Amino-2-methylpropane)
Primary	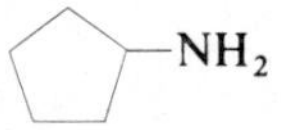 ⬠—NH_2	Cyclopentylamine (Aminocyclopentane)
Primary	$H_2N-CH_2CH_2CH_2CH_2CH_2CH_2-NH_2$	Hexamethylenediamine (1,6-Diaminohexane)
Salt	$\left[CH_3-\overset{\overset{\displaystyle CH_3}{\mid}}{\underset{\underset{\displaystyle CH_3}{\mid}}{N}}-CH_3\right]^+ Cl$	Tetramethylammonium chloride
Primary	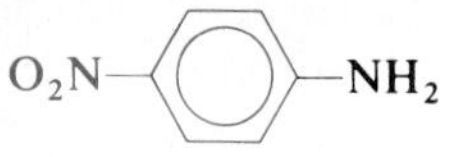 O_2N—⬡—NH_2	*p*-Nitroaniline (4-Nitrobenzenamine)
Secondary	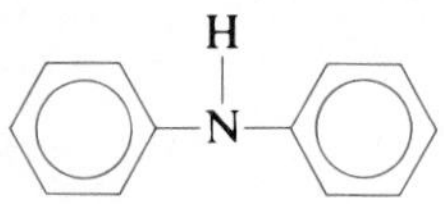 ⬡—$\overset{\overset{\displaystyle H}{\mid}}{N}$—⬡	*N*-Phenylaniline (Diphenylamine)
Tertiary	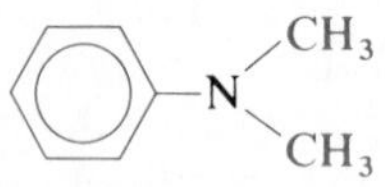 ⬡—$N\overset{\displaystyle CH_3}{\underset{\displaystyle CH_3}{<}}$	*N,N*-Dimethylaniline (Dimethylphenylamine)
Primary	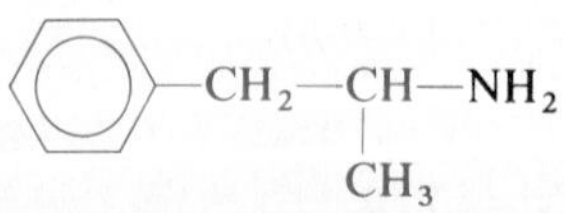 ⬡—$CH_2-\overset{\overset{\displaystyle }{}}{\underset{\underset{\displaystyle CH_3}{\mid}}{C}}H-NH_2$	Benzedrine (2-Amino-1-phenylpropane) (Amphetamine, ingredient in "pep pills")

compounds, and nitriles, all of which are themselves easily prepared from readily accessible organic compounds.

a. Reduction of Amides

Amides are reduced by lithium aluminum hydride (LiAlH$_4$) to yield amines. This reduction reaction is unusual because the carbonyl group is reduced directly to a methylene group (—CH$_2$—) rather than to an alcohol. Since amides are readily prepared from acyl halides and acid anhydrides, this is perhaps the most general and best laboratory method for preparing amines. Note that secondary or tertiary amines may also be synthesized, depending upon the structure of the amide used.

General Equation

$$R-C{\overset{O}{\underset{NH_2}{}}} + 4[H] \xrightarrow[\text{ether}]{\text{LiAlH}_4} RCH_2NH_2 + H_2O$$

Specific Equations

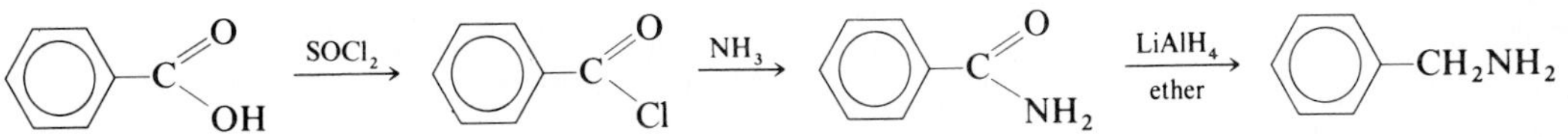

Benzoic acid $\xrightarrow{\text{SOCl}_2}$ Benzoyl chloride $\xrightarrow{\text{NH}_3}$ Benzamide $\xrightarrow[\text{ether}]{\text{LiAlH}_4}$ Benzylamine (—CH$_2$NH$_2$)

$$CH_3CH_2CH_2C{\overset{O}{\underset{\underset{CH_3}{N}}{}}}CH_3 \xrightarrow[\text{ether}]{\text{LiAlH}_4} CH_3CH_2CH_2CH_2-N{\overset{CH_3}{\underset{CH_3}{}}}$$

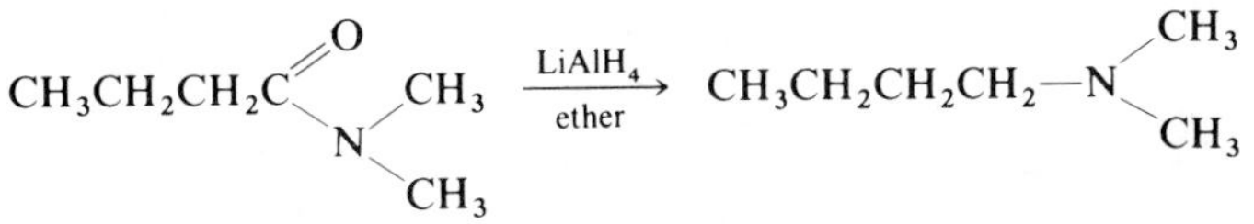

N,N-Dimethylbutanamide   Butyldimethylamine

b. Reduction of Nitro Compounds

Reduction of aromatic nitro compounds with tin or iron and hydrochloric acid is a particularly useful method for the formation of primary aromatic amines. Sodium hydroxide is added to convert the amine salt to the amine. Nitroalkanes, because they are less readily available, are seldom used for the preparation of aliphatic amines.

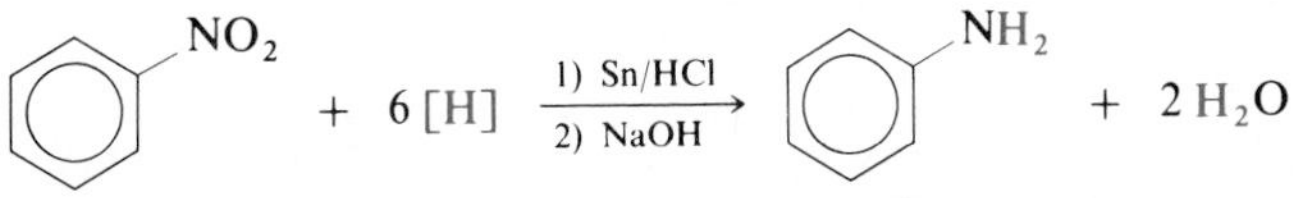

$$\text{NO}_2 \text{ (benzene)} + 6[H] \xrightarrow[\text{2) NaOH}]{\text{1) Sn/HCl}} \text{NH}_2 \text{ (benzene)} + 2H_2O$$

Nitrobenzene   Aniline

c. Reduction of Nitriles (Cyanides)

Nitriles are reduced by catalytic hydrogenation or by lithium aluminum hydride to give primary amines. Since nitriles are readily available from alcohols or alkyl halides (Section 5.4-b), the procedure can be used to synthesize a wide variety of primary amines. The resulting amine contains one more carbon than the initial alcohol or alkyl halide.

General Equation

$$R-C\equiv N + 4\,[H] \xrightarrow[\text{or } H_2/Ni]{\text{LiAlH}_4/\text{ether}} R-\overset{\displaystyle H}{\underset{\displaystyle H}{\overset{\displaystyle |}{\underset{\displaystyle |}{C}}}}-\overset{\displaystyle |}{\underset{\displaystyle H}{N}}-H$$

Specific Equation

$$CH_3OH \xrightarrow{PBr_3} \underset{\substack{\text{Methyl}\\\text{bromide}}}{CH_3Br} \xrightarrow{KCN} \underset{\substack{\text{Methyl cyanide}\\\text{(Acetonitrile)}}}{CH_3CN} \xrightarrow{H_2/Ni} \underset{\text{Ethylamine}}{CH_3CH_2NH_2}$$

$\underset{\text{Methanol}}{}$

9.8 *Physical Properties of Amines*

The lower members of the amine series resemble ammonia. They are colorless gases, soluble in water, and have pronounced odors that are somewhat similar to ammonia but are less pungent and more fish-like. The characteristic odor of fish is attributed to the presence of amines in the body fluid of the fish (for example, dimethylamine and trimethylamine are constituents of herring brine).

Some of the decomposition products found in decaying flesh are diamino-alkanes. They have foul odors, and their names attest either to their odor or to their source (for example, $NH_2CH_2CH_2CH_2CH_2NH_2$ is putrescine and

TABLE 9.4 Physical Properties of the Amines

Formula	Name	MP (°C)	BP (°C)	K_b
NH_3	Ammonia	−78	−33	1.8×10^{-5}
CH_3NH_2	Methylamine	−94	−6	3.7×10^{-4}
$CH_3CH_2NH_2$	Ethylamine	−81	17	6.4×10^{-4}
$(CH_3)_2NH$	Dimethylamine	−93	7	5.4×10^{-4}
$(CH_3)_3N$	Trimethylamine	−117	3	6.4×10^{-5}
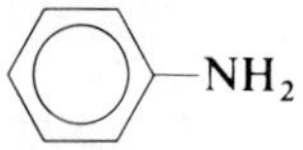 ⬡—NH₂	Aniline	−6	184	4.3×10^{-10}
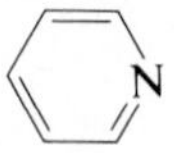 ⬡N	Pyridine	−42	115	1.8×10^{-9}
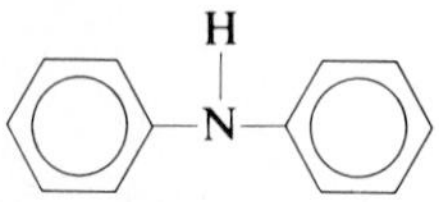 ⬡—N(H)—⬡	Diphenylamine	54	302	6.2×10^{-14}

$NH_2CH_2CH_2CH_2CH_2CH_2NH_2$ is cadaverine). They arise from the decarboxylation of ornithine and lysine, respectively, amino acids found in animal cells (see Section 18.3-c).

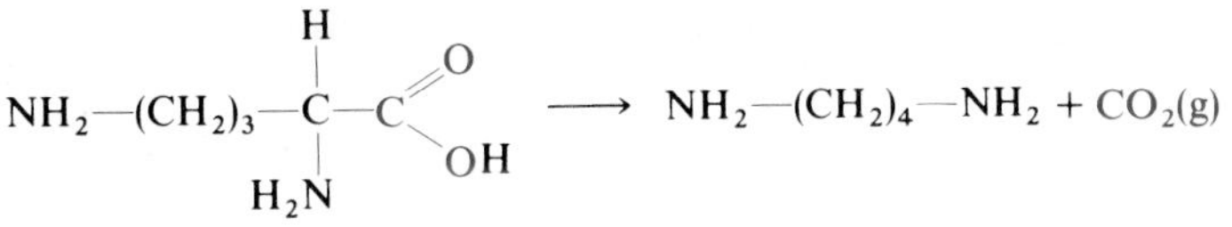

$$NH_2-(CH_2)_3-\underset{\underset{H_2N}{|}}{\overset{\overset{H}{|}}{C}}-C\overset{O}{\underset{OH}{}} \longrightarrow NH_2-(CH_2)_4-NH_2 + CO_2(g)$$

Ornithine   Putrescine

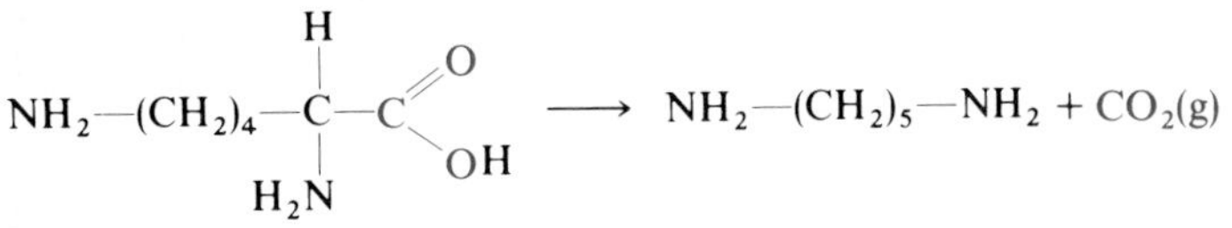

$$NH_2-(CH_2)_4-\underset{\underset{H_2N}{|}}{\overset{\overset{H}{|}}{C}}-C\overset{O}{\underset{OH}{}} \longrightarrow NH_2-(CH_2)_5-NH_2 + CO_2(g)$$

Lysine   Cadaverine

(a)

(b)

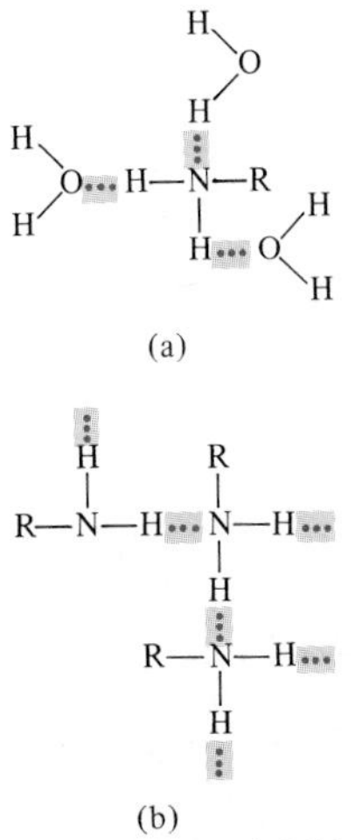

FIGURE 9.2 (a) Hydrogen bonding of amine with water molecules. (b) Intermolecular hydrogen bonding in amines.

Primary amines containing three to eleven carbon atoms are liquids; the higher homologs are solids. Dimethylamine is the only gaseous secondary amine and trimethylamine is the only gaseous tertiary amine. Amines with small **R**— groups are soluble in water because of hydrogen bonding (Figure 9.2-a). As the **R**— groups become more bulky, hydrogen bonding by water molecules becomes less effective, and water solubility decreases. Aromatic amines are generally much less soluble in water than are aliphatic amines. Aniline and most substituted anilines are toxic and should be handled cautiously. Chronic exposure causes tumors and nervous disorders. There is some slight intermolecular association among primary and secondary amine molecules due to hydrogen bonding (Figure 9.2-b); however, since nitrogen is less electronegative than oxygen, the bonding is less pronounced than in the alcohols. For example, compare the boiling points of CH_3NH_2 ($-6°C$) and CH_3OH ($65°C$). Tertiary amines are polar molecules by virtue of the unshared pair of electrons on nitrogen, but because they lack a hydrogen bonded to nitrogen, they are incapable of intermolecular hydrogen bonding. Consequently, tertiary amines have lower boiling points and melting points than isomeric primary and secondary amines. Table 9.4 lists the physical characteristics of some of the common amines.

9.9   Basicity of Amines

Whereas carboxylic acids are organic acids, amines are the organic bases. In Section 8.4, we defined an acid as a proton donor. In the same manner, therefore, a base is defined as a proton acceptor. This definition may not appear, at first, to be useful in the case of the strong bases such as sodium or potassium hydroxide. The complete dissociation of such a base is usually written as

$$NaOH \longrightarrow Na^+ + OH^-$$
$$KOH \longrightarrow K^+ + OH^-$$

Once again, the convention here is the omission of the solvent water molecules from the equation. However, the water molecules are usually included in the equations describing the ionization of weak bases. Ammonia is probably the most familiar example of a weak base; a solution of ammonia in water is commonly referred to as ammonium hydroxide.

$$H-\overset{\cdot\cdot}{N}-H \ + \ H-O \ \rightleftharpoons \ NH_4OH$$
$$\quad\ |\qquad\qquad\ |$$
$$\quad\ H\qquad\qquad\ H$$

To facilitate writing the mass law expression for this equilibrium, some texts consider ammonium hydroxide as a molecular species and represent its dissociation as follows.

$$NH_4OH \ \rightleftharpoons \ NH_4^+ \ + \ OH^-$$

Strictly speaking, this treatment is incorrect. No one has, as yet, been able to isolate discrete molecules of NH_4OH, and it is therefore believed that molecules of NH_4OH do not enjoy independent existence in solution. A more correct representation of the process is

$$H-\overset{\cdot\cdot}{N}-H \ + \ H-O \ \rightleftharpoons \ H-\overset{\overset{\textstyle H}{\cdot\cdot\,+}}{N}-H \ + \ OH^-$$
$$\quad\ |\qquad\qquad\ |\qquad\qquad\quad\ |$$
$$\quad\ H\qquad\qquad\ H\qquad\qquad\quad\ H$$

This treatment more clearly illustrates the proton-accepting function of a base, which is a direct result of the availability of an unshared pair of electrons on the nitrogen atom. The mass action expression for the preceding equation is written as

$$\frac{[NH_4^+][OH^-]}{[NH_3]}$$

(A molecule of water is omitted from the denominator since water is the solvent and its concentration remains essentially constant.)

When equilibrium is attained, the mass action expression is equal to a constant, K_b, the **basicity constant.**

$$K_b = 1.8 \times 10^{-5} = \frac{[NH_4^+][OH^-]}{[NH_3]}$$

Every weak base has its own characteristic basicity constant. The larger the value of K_b, the stronger the base (that is, the greater the tendency to *accept* a proton). Recall that K_a, the acidity constant, is a measure of the ability of acids to *donate* protons.

According to Table 9.4, the three classes of aliphatic amines have K_b values between 10^{-4} and 10^{-5}, and thus are slightly stronger bases than ammonia.

$$CH_3NH_2 + HOH \rightleftharpoons CH_3NH_3^+ + OH^-$$

$$K_b = 3.7 \times 10^{-4} = \frac{[CH_3NH_3^+][OH^-]}{[CH_3NH_2]}$$

The pH of an amine solution is calculated in a manner analogous to the calculation of the pH of an acid solution, which has previously been described (page 145). In this case, we first determine the pOH (the negative logarithm of the hydroxide ion concentration) and then subtract it from 14 to obtain the desired pH. Since

$$K_w = [H^+][OH^-] = 1.0 \times 10^{-14},$$

$$\log[H^+] + \log[OH^-] = \log 1.0 \times 10^{-14}$$

$$pH + pOH = 14$$

Example 9.1 What is the pH of a 0.30 M solution of trimethylamine ($K_b = 6.4 \times 10^{-5}$)?

$$(CH_3)_3N + HOH \rightleftharpoons (CH_3)_3NH^+ + OH^-$$

$$K_b = 6.4 \times 10^{-5} = \frac{[(CH_3)_3NH^+][OH^-]}{[(CH_3)_3N]}$$

Solution: Solving for $[OH^-]$ in a similar manner to that explained in Section 8.4, we have

$$6.4 \times 10^{-5} = \frac{x^2}{0.30}$$

$$x^2 = 1.9 \times 10^{-5}$$

$$x = [OH^-] = 4.4 \times 10^{-3} \ M$$

$$pOH = 3 - \log 4.4 = 2.4$$

$$pH = 14 - 2.4 = 11.6$$

Amines constitute the most important class of organic bases. By proper choice of aliphatic, aromatic, or heterocyclic amine (Section 9.11), a broad spectrum of base strengths can be obtained. The degree of basicity of amines is dependent on the availability of the unshared electron pair (the more available the electrons, the stronger the base). An increase in the electron density about the nitrogen atom makes the electron pair more available; conversely, a decrease in electron density makes these electrons less available. The fact that the aliphatic amines are stronger bases than ammonia is explainable in terms of the inductive effect of the alkyl groups. Recall that alkyl groups are electron-releasing and thus they tend to increase the electron density at the nitrogen atom and also to disperse the positive charge on the cation that is formed.

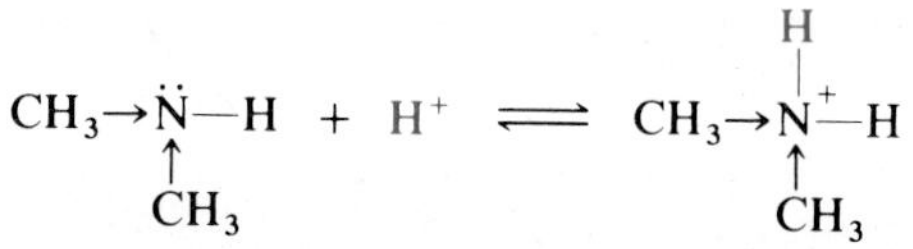

$$CH_3 \rightarrow \overset{\underset{\displaystyle CH_3}{\uparrow}}{\ddot{N}}-H \ + \ H^+ \ \rightleftharpoons \ CH_3 \rightarrow \overset{\underset{\displaystyle CH_3}{\uparrow}}{\overset{\displaystyle H}{\underset{}{N^+}}}-H$$

We find that tertiary amines are usually more basic than ammonia, yet less basic than secondary or primary amines. From this observation, we must conclude that steric considerations outweigh inductive effects. The crowding of bulky groups about the nitrogen reduces the accessibility of the electron pair. In other words, the R groups hinder the approach of hydrogen ions (and other electrophiles).

Aromatic amines in which the nitrogen is directly bonded to the aromatic ring are weaker bases than the aliphatic amines. This is attributed to the delocalization of the electron pair by resonance. The net effect of this resonance interaction is to render the lone pair less accessible for bonding to a proton.

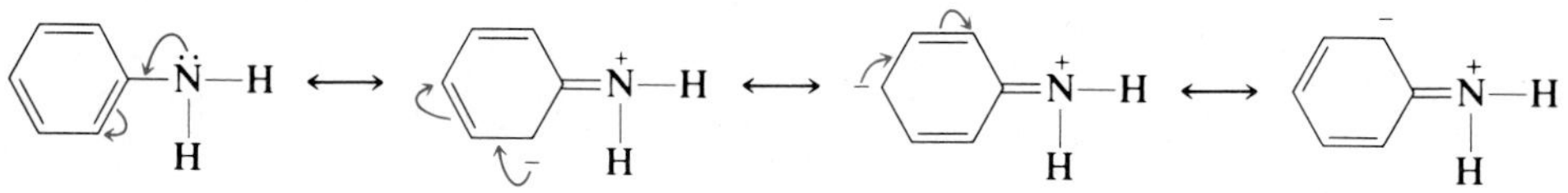

9.10 *Chemical Properties of Amines*

a. Salt Formation

The most characteristic property of the amines is their ability to react with acids to form amine salts. These salts are similar to ammonium salts. They are formed by treating the amine with a strong acid such as hydrochloric, hydrobromic, or sulfuric acid, or with the weaker carboxylic acids.

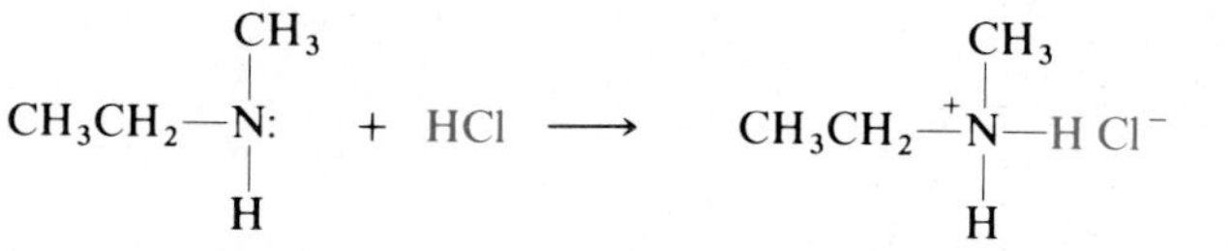

$$CH_3CH_2\overset{\underset{\displaystyle H}{|}}{\underset{}{\overset{\displaystyle CH_3}{|}}{N}}: \ + \ HCl \ \longrightarrow \ CH_3CH_2\overset{\underset{\displaystyle H}{|}}{\underset{}{\overset{\displaystyle CH_3}{|}}{\overset{+}{N}}}-H \ Cl^-$$

Ethylmethylamine   Ethylmethylammonium chloride

$$CH_3-\ddot{N}H_2 \ + \ CH_3COOH \ \longrightarrow \ CH_3-NH_3^+ \ CH_3COO^-$$
Methylamine   Acetic acid   Methylammonium acetate

Since amine salts are ionic compounds, they are soluble in water even though their corresponding amines may be water insoluble. As we shall see in Section 9.12, the majority of drugs contain an amine group. To facilitate injection in aqueous solution and/or for ease of transport through the blood, pharmaceutical companies often convert an insoluble amine into the more soluble amine salt. Examples are morphine sulfate, chlorpromazine (Thorazine) hydrochloride, and dextropropoxyphene (Darvon) hydrochloride.

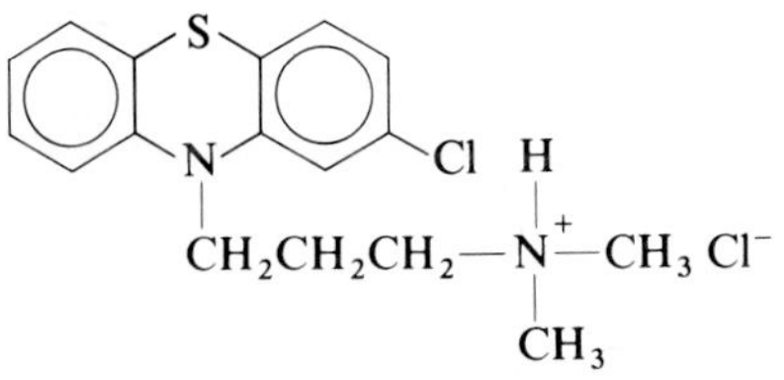

Chlorpromazine hydrochloride
(Thorazine hydrochloride)

We shall see in Chapter 13 that an important characteristic of both amino acids and proteins is their ability to combine with protons (from acids) to form salts.

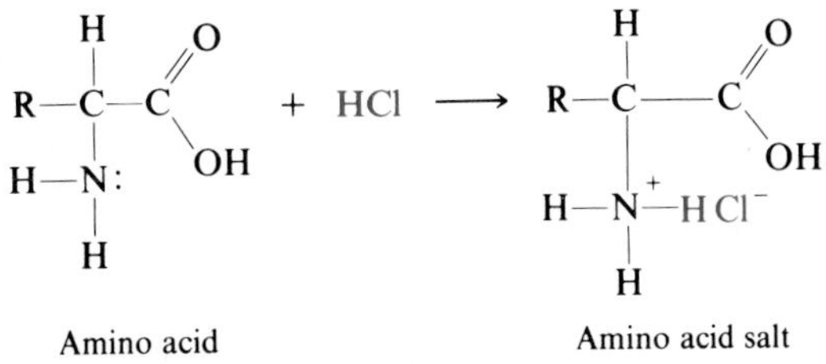

Amino acid   Amino acid salt

b. Alkylation

Like ammonia, amines react readily with alkyl halides to form more highly substituted amines. This reaction is employed to substitute alkyl groups for the hydrogens of primary and secondary amines. Unfortunately, the reaction has severe limitations from a synthetic standpoint. Like the chain reaction for the successive chlorination of the alkanes (Section 2.4), the reaction is difficult to control. The amines that are products of the initial reaction may react further to yield a mixture[1] of primary, secondary, and tertiary amines and substances known as *quaternary ammonium salts*[2] (four organic groups are bonded to nitrogen). The mechanism of the reaction is analogous to the displacement of halide ion from alkyl halides by hydroxide in the preparation of alcohols (Section 6.2-b). Quaternary ammonium salts, like amine salts, are ionic compounds. They are high-melting solids and are generally soluble in water.

[1] In the commercial synthesis of amines, separation of the mixture is accomplished by fractional distillation.
[2] Some quaternary ammonium salts that contain one large alkyl group are useful detergents and bactericides (e.g., trimethylhexadecylammonium chloride—see page 282). We have previously discussed the neurotransmitter acetylcholine, $[CH_3COOCH_2CH_2N(CH_3)_3]^+ OH^-$, and we shall mention the important quaternary ammonium salt choline, $[CH_2OHCH_2N(CH_3)_3]^+ OH^-$, in connection with the structure of phospholipids (Section 12.9).

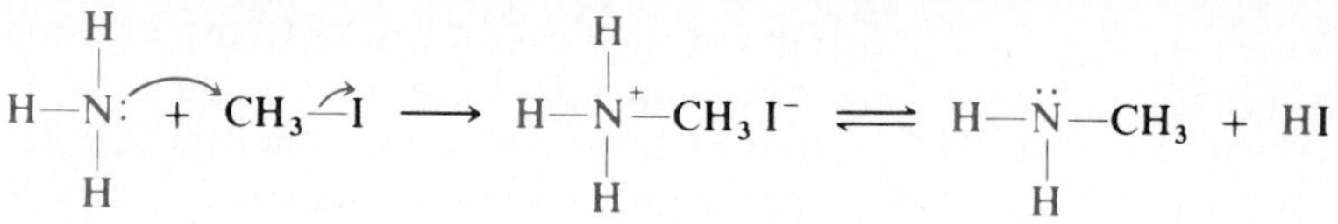

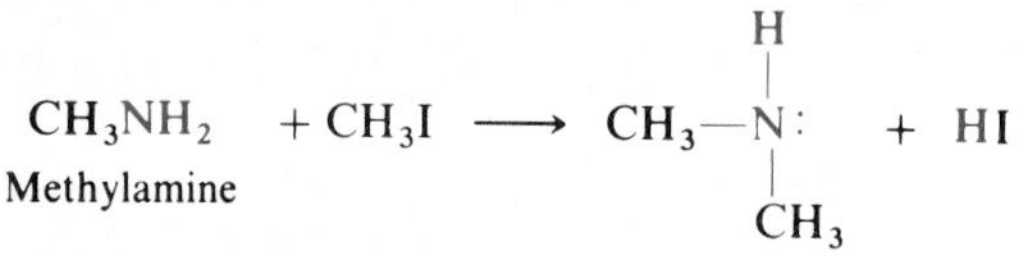

$$H-\overset{\overset{\displaystyle H}{|}}{\underset{\underset{\displaystyle H}{|}}{N}}: + \overset{\curvearrowright}{CH_3}\overset{\displaystyle \nearrow}{I} \longrightarrow H-\overset{\overset{\displaystyle H}{|}}{\underset{\underset{\displaystyle H}{|}}{\overset{+}{N}}}-CH_3 \; I^- \rightleftharpoons H-\overset{\overset{\displaystyle H}{|}}{\underset{\underset{\displaystyle H}{|}}{\ddot{N}}}-CH_3 + HI$$

$$CH_3NH_2 \quad + CH_3I \longrightarrow \quad CH_3-\overset{\overset{\displaystyle H}{|}}{\underset{\underset{\displaystyle CH_3}{|}}{N}}: \quad + \; HI$$
Methylamine

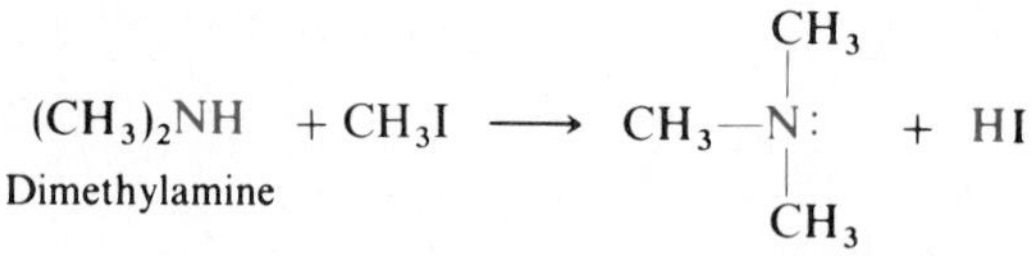

$$(CH_3)_2NH \quad + CH_3I \longrightarrow \quad CH_3-\overset{\overset{\displaystyle CH_3}{|}}{\underset{\underset{\displaystyle CH_3}{|}}{N}}: \quad + \; HI$$
Dimethylamine

$$(CH_3)_3N \quad + CH_3I \longrightarrow \quad CH_3-\overset{\overset{\displaystyle CH_3}{|}}{\underset{\underset{\displaystyle CH_3}{|}}{\overset{+}{N}}}-CH_3 \; I^-$$

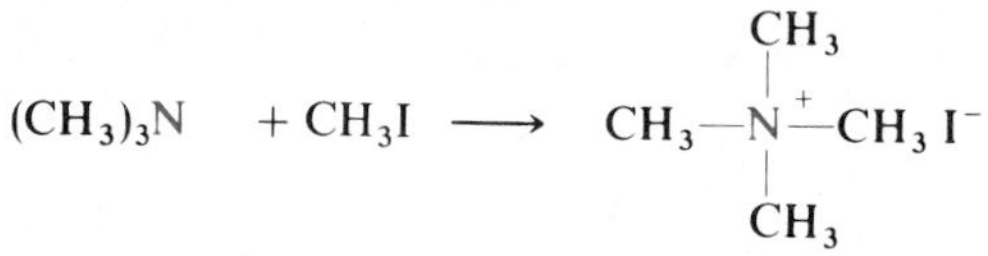

Trimethylamine   Tetramethylammonium
iodide

c. Acylation (Formation of Amides)

The reaction between a primary or secondary amine and an acyl chloride or acid anhydride yields an amide in which the acyl group has been substituted for a hydrogen bonded to the amine nitrogen. Tertiary amines do not undergo this reaction.

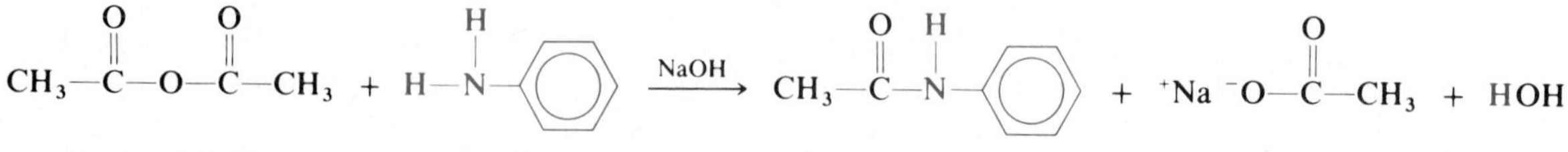

$$CH_3-\overset{\overset{\displaystyle O}{\|}}{C}-O-\overset{\overset{\displaystyle O}{\|}}{C}-CH_3 + H-\overset{\overset{\displaystyle H}{|}}{N}-\bigcirc \xrightarrow{\text{NaOH}} CH_3-\overset{\overset{\displaystyle O}{\|}}{C}-\overset{\overset{\displaystyle H}{|}}{N}-\bigcirc + \; ^+Na \; ^-O-\overset{\overset{\displaystyle O}{\|}}{C}-CH_3 + HOH$$

Acetic anhydride   Aniline   Acetanilide

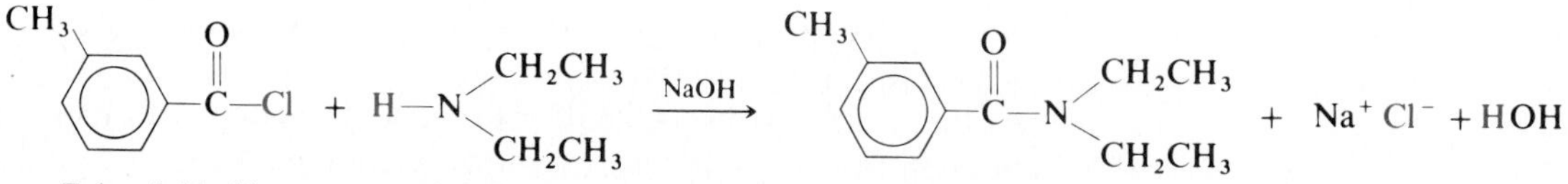

$$CH_3-\bigcirc-\overset{\overset{\displaystyle O}{\|}}{C}-Cl + H-N\overset{\displaystyle CH_2CH_3}{\underset{\displaystyle CH_2CH_3}{}} \xrightarrow{\text{NaOH}} CH_3-\bigcirc-\overset{\overset{\displaystyle O}{\|}}{C}-N\overset{\displaystyle CH_2CH_3}{\underset{\displaystyle CH_2CH_3}{}} + \; Na^+ \; Cl^- + HOH$$

m-Toluoyl chloride   Diethylamine   N,N-Diethyl-m-toluamide

The product N,N-diethyl-m-toluamide, also known as m-delphene, is the active ingredient in the popular insect repellent *Off!*. It is interesting that the two other isomers (o- and p-delphene) are ineffective as insect repellents. This remarkable specificity-of-structure requirement was mentioned previously in connection with pesticides (Section 5.6).

d. Nitrosation of Amines

In contrast to nitric acid, nitrous acid (HNO_2) does not react like a typical inorganic acid with respect to amino compounds. That is, the majority of amines are not converted to their corresponding amine salts. Rather, the reaction products depend on whether the amine is primary, secondary, or tertiary, and whether it is aliphatic or aromatic.

Aromatic primary amines react with nitrous acid to form *arenediazonium salts,* which are important intermediates in the synthesis of a wide variety of commercially important aromatic compounds. Nitrous acid is unstable and so it is generated in the reaction mixture (*in situ*) by the combination of a nitrite salt and a strong acid.

$$\text{C}_6\text{H}_5{-}\text{NH}_2 \xrightarrow[\text{0--5°C}]{\text{NaNO}_2/\text{HCl}} \text{C}_6\text{H}_5{-}\overset{+}{\text{N}}{\equiv}\text{N Cl}^- \ + \ 2\,\text{H}_2\text{O}$$

A diazonium salt

When secondary amines react with nitrous acid, the products are known as *N*-nitroso compounds or *nitrosoamines* (the —N=O group is the *nitroso group*).

General Equation

$$\underset{\overset{|}{\text{R}'}}{\text{R}{-}\text{N}{-}\text{H}} \ + \ \text{HO}{-}\text{N}{=}\text{O} \ \xrightarrow[\text{HCl}]{\text{NaNO}_2} \ \underset{\overset{|}{\text{R}'}}{\text{R}{-}\text{N}{-}\text{N}{=}\text{O}} \ + \ \text{HOH}$$

A nitrosoamine

Specific Equation

$$\underset{\overset{|}{\text{CH}_3}}{\text{CH}_3{-}\text{N}{-}\text{H}} \ + \ \text{HONO} \ \xrightarrow[\text{HCl}]{\text{NaNO}_2} \ \underset{\overset{|}{\text{CH}_3}}{\text{CH}_3{-}\text{N}{-}\text{N}{=}\text{O}} \ + \ \text{HOH}$$

Dimethylamine   Nitrosodimethylamine

The nitrosoamines are not soluble in the aqueous medium, and they generally separate out as yellow oils. These nitroso compounds are known to be potent animal carcinogens, and there has been concern about the use of nitrites and nitrates as color enhancers and food preservatives (to retard spoilage and to prevent botulism). It is possible that bacteria in the stomach reduce nitrates to nitrites and that, in the presence of hydrochloric acid in the stomach, the nitrites are converted into nitrous acid. The nitrous acid then could react with certain secondary amines in food proteins or with proteins in our body to form nitrosoamines. It may be that, like the polycyclic aromatic compounds (page 60), the carcinogens are not the nitrosoamines themselves but rather some metabolite of the nitrosoamines.

In 1978 the Department of Agriculture required meat firms to reduce the sodium nitrite added to preserve bacon from 200 to 120 ppm. The amount of nitrite added to ham and frankfurters was also reduced, and other preservatives are being used. The meat industry points out, however, that there is no evidence to link the ingestion of nitrites and the increased incidence of stomach and intestinal cancer. They say that the amount of nitrite that occurs naturally in foods far surpasses the amount that is used as an additive. Nitrosoamines have also been found in alcoholic beverages, cosmetics, and pesticides.

TABLE 9.5 Nitrogen Heterocycles

Formula	Name	Occurrence	Example
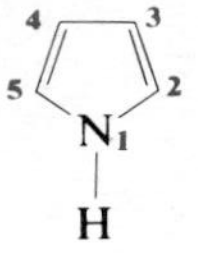	Pyrrole	Chlorophyll Hemoglobin Vitamin B_{12} Cytochromes	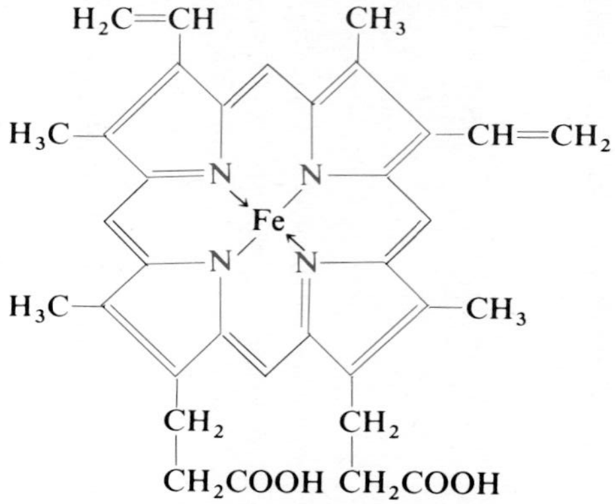 Heme portion of the hemoglobin molecule
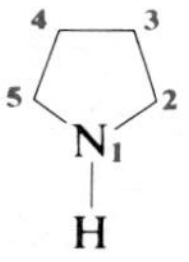	Pyrrolidine	Proline Hydroxyproline $\}$ amino acids Nicotine	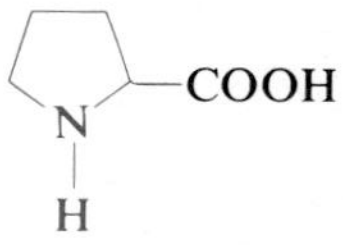 Proline
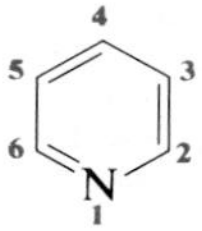	Pyridine	Nicotine Niacin Pyridoxine $\}$ vitamins NAD^+ Pyridoxal phosphate $\}$ coenzymes	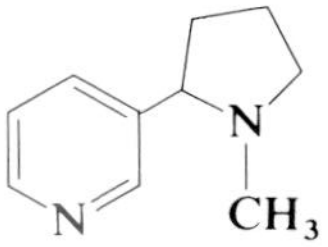 Nicotine
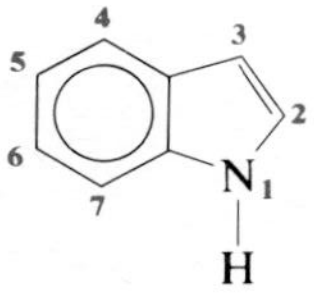	Indole	Tryptophan amino acid Serotonin Lysergic acid $\}$ indole Strychnine $\}$ alkaloids Reserpine	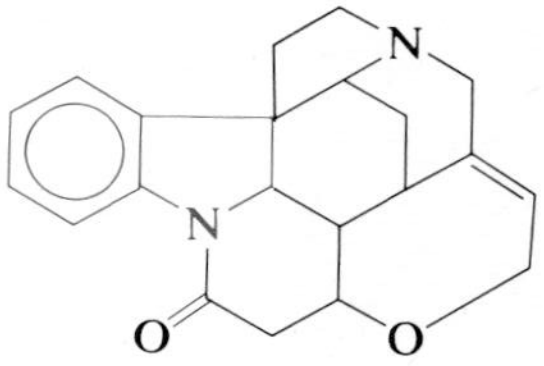 Strychnine

194

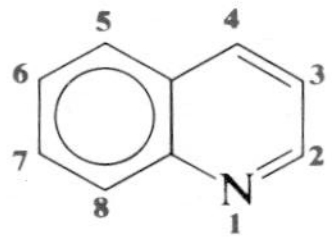

CH=CH₂ → $CH=CH_2$

CH₂CH(NH₂)COOH → $CH_2CH(NH_2)COOH$

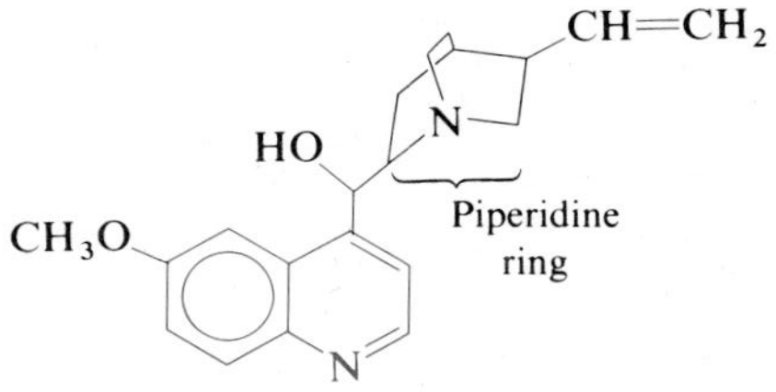

Piperidine ring

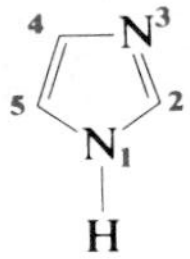

HO

CH_3O

Quinine

Histidine

NH_2

CH_3 CH_2CH_2OH

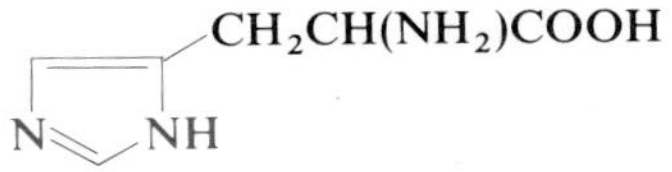

Cl^- CH_3

N^+ S

CH_2

CH_3

Vitamin B₁
(Thiamine)

NH_2

OH   OH

O

CH_2

O—P—O

O⁻

O—P—O⁻

O⁻

O—P—O⁻

O⁻

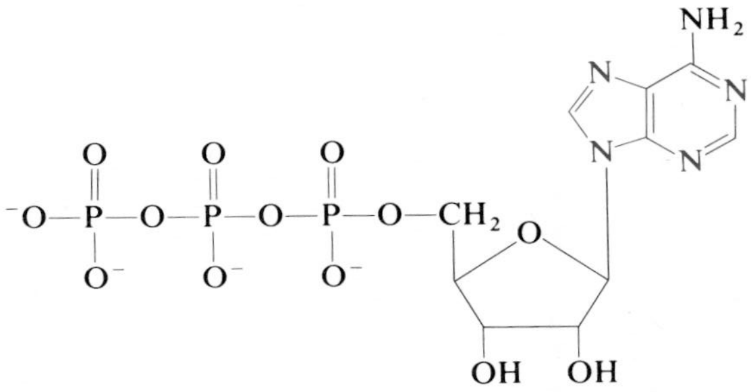

Adenosine triphosphate (ATP)

Quinoline	Quinine	Quinine
		Curare
Imidazole	Histidine	Histidine amino acid
Pyrimidine	Vitamin B₁	
	Cytosine	
	Uracil	} in nucleic acids
	Thymine	
Purine	Adenine	} in nucleic acids
	Guanine	
	NAD⁺, FAD	} coenzymes
	Coenzyme A	
	ATP, ADP, AMP	
	Caffeine	
	Dramamine	

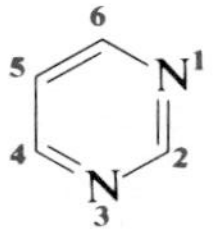

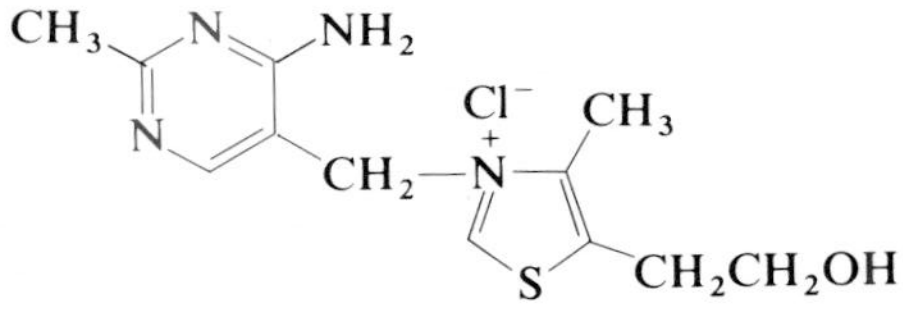

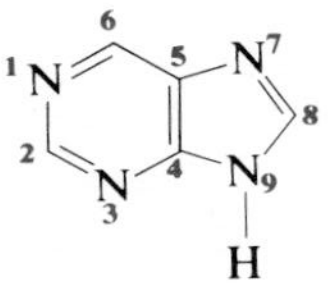

Quinoline

Imidazole

Pyrimidine

Purine

9.11 Nitrogen Heterocycles

All of the cyclic compounds that have been mentioned thus far have had only one element, carbon, in the ring. There is a large class of *organic compounds that contain two or more different elements in the same ring structure.* Such cyclic compounds are called **heterocyclic** (Greek, *heteros,* other). The majority of heterocycles are those in which the hetero atom is nitrogen, oxygen, or sulfur. Nitrogen-containing ring compounds comprise a large segment of organic chemistry and are widely distributed in all living systems. The parent compounds of this class possess relatively simple structures. However, their derivatives, many of which exhibit physiological activity, are often quite complex. The common names and structures of the biologically significant nitrogen heterocycles are presented in Table 9.5.

Many nitrogen heterocycles have aromatic properties similar to benzene. Pyridine and pyrrole are the chief examples of such compounds and they serve as representative systems about which we can make some useful generalizations. In Section 9.12 we discuss the important role that the heterocyclic compounds play in the chemistry of drugs.

a. Pyridine

Pyridine is the direct nitrogen analog of benzene. Like benzene, pyridine is an aromatic molecule. It is one of the most abundant heterocyclic compounds and is obtained commercially from coal tar. Pyridine is the parent of a series of compounds that is important in agricultural, industrial, and medicinal chemistry. It has an obnoxious odor and is soluble in water as well as in most organic solvents. The unshared electron pair on nitrogen is not involved in the ring electron system; therefore, these electrons are available to combine with hydrogen ions. Thus, pyridine is a weak base and forms salts with acids. It is widely used in organic chemistry as a water-soluble base.

$K_b = 1.8 \times 10^{-9}$

Pyridinium
ion

Pyridinium
chloride

The nitrogen can be alkylated by primary alkyl halides to yield quaternary ammonium salts.

N-Methylpyridinium
iodide

The pyridine ring, like the benzene ring, is resistant to oxidation. This is demonstrated by the oxidation of nicotine to produce nicotinic acid.

Nicotine

Nicotinic acid
(Niacin)

Nicotine is an alkaloid present in tobacco. It sometimes is used as an agricultural insecticide, but it is also highly toxic to humans. Niacin, one of the essential B vitamins, is synthesized by plant and animal cells from the amino acid tryptophan. Its corresponding amide, nicotinamide, is a component of the coenzyme NAD$^+$ (see Table 14.2).

Partial reduction of pyridine yields dihydropyridine, reduction occurring at the 4-position. This reaction can be reversed, and it is the basis of the very important oxidation–reduction reactions that take place in living systems, catalyzed by the coenzyme NAD$^+$ (see Section 16.9). Complete reduction of pyridine results in the formation of piperidine.

Dihydropyridine

Piperidine

b. Pyrrole

Pyrrole is a colorless liquid that darkens on exposure to air. If pure, it has an odor similar to that of chloroform. Pyrrole is soluble in alcohol, benzene, and ether, but is only sparingly soluble in water. Pyrrole, like pyridine, is an aromatic heterocycle. However, unlike pyridine, the unshared pair of electrons is part of the aromatic system. Consequently, these electrons are not available to bond to a hydrogen ion and pyrrole is an example of an amine that does not behave like a base.

c. Imidazole

Imidazole is the name of the five-membered heterocycle that contains two nitrogen atoms. The imidazole ring is found in the amino acid histidine (Table 9.5), and it is responsible for many of the properties of that amino acid. Unlike pyrrole, imidazole is a weak base. The second nitrogen (N-3) in the ring does have an unshared pair of electrons that can bind to hydrogen ions. Imidazole is about 100 times more basic than pyridine because the resulting imidazolium ion is stabilized by resonance.

$$K_b = 10^{-7}$$

d. Indole

Indole contains a pyrrole ring fused to a benzene ring. It is found in the amino acid tryptophan and in skatole. The latter is a metabolite of tryptophan that is eliminated in the feces and is partly responsible for the odor of feces. The drug LSD (page 206) and the neurotransmitter serotonin (page 206) also contain the indole group.

Indole   Skatole

Tryptophan

9.12   Psychoactive Drugs

According to the broadest definition, a **drug** *is any chemical substance that affects an individual in such a way as to bring about physiological, emotional, or behavioral change.* In recent years the use of drugs, on a worldwide basis, has reached alarming proportions. Of particular concern are the so-called "pop" or "mind" drugs—the drugs that affect the brain and spinal cord. These organs make up the central nervous system (CNS), the control center not only of the body but also of everything we call the mind—that is, emotion, sensation, thought. Almost all drugs exhibit more than one type of response or activity, depending upon the dosage and the individual. The World Health Organization has classified six different types of psychoactive drugs according to the patterns of their action and the major responses they bring about. These are opiates, barbiturates, cocaine, cannabis (marijuana), amphetamines, and hallucinogens (LSD). Table 9.6 indicates some of the basic criteria by which part of this classification was made. We also discuss the tranquilizers and "designer drugs" because of their widespread use and abuse.

a. Alkaloids of Opium

An **alkaloid** *is any nitrogen-containing compound that is obtained from plants and that has physiological activity* (nicotine, atropine, quinine, etc.). The opium alkaloids, sometimes referred to as *opiates* or *opioids,* originally meant a drug obtained from the opium poppy (*Papaver somniferum*—not the common garden-variety poppy). But today the term includes all the synthetic compounds that have *morphine-like* activity. That is, when administered in medicinal doses, they *relieve pain* (analgesics) and they *induce sleep* (hypnotics).

Morphine, which has been used extensively for many years, is still the most important sedative-analgesic. Morphine remains the standard against which new analgesics are measured. Although morphine can be synthesized in the laboratory, it is still obtained from opium. When pure, morphine is an odorless, white crystalline solid, having a bitter taste, and it is insoluble in water. (Recall, we said that morphine is converted into its salt in order to increase its water solubility (Section 9.10).

The dried juice from the unripened seed pod of the poppy plant is commercial opium. Opium is a complex mixture

TABLE 9.6 Some Characteristics of Certain Types of Drugs

Type of Drug	Basic Action	Psychological (Psychic) Dependence	Physical Dependence	Development of Tolerance
Opiate	Depressant	Yes, strong	Yes	Yes
Barbiturate	Depressant	Yes	Yes	Yes
Cocaine	Stimulant	Yes, strong	Yes	Yes
Cannabis (marijuana)	Stimulant	Yes, moderate to strong	No	Yes
Amphetamine	Stimulant	Yes, variable	Yes	Yes
Hallucinogen (LSD)	Stimulant	Yes	No	Yes

containing more than 20 different compounds. Morphine, which constitutes about 10% of opium, was first isolated in 1804, followed thereafter by codeine (methylmorphine—0.5% of opium) and other important medical agents, such as papaverine and noscapine, which are anticonvulsants. Heroin (diacetylmorphine), a semisynthetic derivative of morphine, was first prepared in 1874 by heating morphine with acetic acid (acetylation). Since then, a variety of synthetic drugs with morphine-like physiological activities (such as Demerol, Darvon, methadone) have been produced and have found widespread application

FIGURE 9.3 The chemical structures of some analgesic drugs.

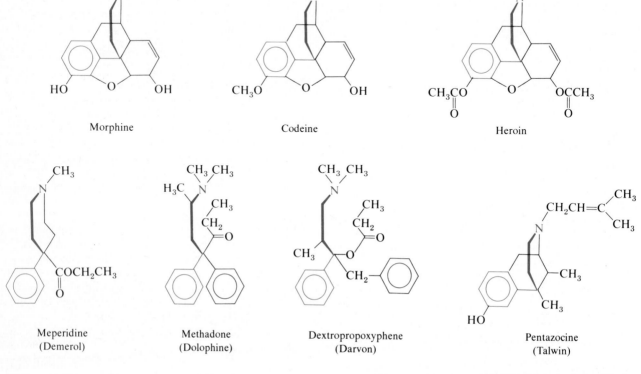

Morphine

Codeine

Heroin

Meperidine (Demerol)

Methadone (Dolophine)

Dextropropoxyphene (Darvon)

Pentazocine (Talwin)

(Figure 9.3). These drugs are all classified, in a legal sense, as **narcotics.** That is, they *all produce physical addiction* (withdrawal symptoms), *and their use is regulated by federal laws.*

The depressant action of the opiates is exerted entirely on the central nervous system (CNS). Among the principal effects are analgesia, sedation (freedom from anxiety), hypnosis (drowsiness and lethargy), and euphoria (a sense of well-being and contentment—see page 205 for one of the theories to explain euphoria). The combination of elevation in mood and relief of apprehension is considered the major reason for drug abuse. As time passes, the euphoria wears off and the user becomes apathetic and gradually falls asleep. The opiates also cause pupillary constriction, and they depress the cough center (codeine being the most effective, thus its use in cough medicines). In the brain stem these drugs will depress the respiratory center. Death from an overdose results from respiratory failure.

Morphine is available in ampules of 1 mL or larger or in tablets of various sizes for preparing an injectable solution. The normal dosage range is 5–15 mg, and administration is intravenously and subcutaneously (under the skin). The chief medical function of morphine is as an analgesic for the relief of postoperative pain, cardiac pain, the pain of terminal cancer, and in childbirth. The problem, however, is that addiction liability seems to parallel analgesic activity. Many of the synthetic compounds are more potent pain killers than morphine, but they are either dangerously addicting or hallucinogenic. Morphine also causes constriction of the smooth muscles such as those found in the intestines and can therefore be used in the treatment of dysentery. Paregoric, formerly given to children suffering from diarrhea, is a mixture of compounds and contains some morphine.

Heroin is currently the most commonly abused narcotic. (It is estimated that there are 200,000 heroin addicts in New York City alone.) It is five to ten times more potent than morphine (because it can more easily pass into the brain), and because of the greater risk of addiction, heroin has been outlawed in the United States (except for research purposes). The major characteristics of heroin addiction are the development of both a tolerance to the drug and a physical dependence on it. **Tolerance** is a *term used to signify the body's ability to adapt to a drug.* If a tolerance is developed, increased quantities of the drug are required to produce the same effect as the original dose. Eventually the abuser may require doses that would be lethal to a nonuser.

Physical dependence *refers to an altered physiological state produced by repeated taking of a drug, so that, on discontinuing use of the drug, certain withdrawal symptoms occur.*[3] If the heroin user is without the drug for 10–12 hours, he experiences vomiting, diarrhea, pains and tremors, restlessness, and mental disturbances. Opiate dependency can be passed from mother to fetus. A child born to an addict must spend the first days of its life withdrawing from the drug or death may result. Treatment can remove an addict's physical dependence upon a drug, but it is very difficult to abstain from further use of the drug because of the **psychological dependency** (*an uncontrollable desire for the drug*) that is produced.

The principal method of treating heroin addiction in this country is by administering minimal doses of methadone (Figure 9.3). Methadone was synthesized in Germany in 1943, when opiate analgesics were not available because of the war. It was first called Dolophine after Adolf Hitler. The drug was subsequently used in the United States for chronic pain and as a cough suppressant. In the late 1940s, it was used for the withdrawal of heroin addicts.

Although methadone is also addictive, patients in the various maintenance programs have shown increased productivity and decreased antisocial behavior. Unlike the other narcotic drugs, methadone does not normally induce euphoria and it eliminates the heroin hunger and withdrawal pains that would otherwise accompany giving up heroin. Moreover, at appropriate dosage levels, methadone blocks the pleasurable sensations derived from heroin, thus further reducing heroin-seeking behavior. Another advantage of methadone is that it can be taken orally so that the dangers of infectious hepatitis and AIDS[4] are eliminated. The usual oral dose is 40 mg/day; it is only effective for 24–36 hours, so methadone must be taken every day. Critics have voiced opposition to the methadone treatment. They question whether this approach achieves any lasting benefit for the addict, and they insist that methadone's long-term safety and effectiveness have not been adequately demonstrated. Withdrawal from methadone is generally regarded as being at least as difficult as withdrawal from heroin. If taken in excess by a person not tolerant of the drug, it can be lethal.

Note that the analgesic drugs shown in Figure 9.3 have similar structural characteristics. This suggests that there is a specificity for molecules of a particular shape. It has been discovered that the central nervous system contains receptor sites for naturally occurring pain-relieving substances (see below). It is now

---

[3] *Withdrawal symptoms* (abstinence syndrome) are the characteristic pattern of reactions and behavior that occurs when one abruptly stops taking a drug after the body has developed a physical dependence on the drug. The withdrawal symptoms vary with the drug, the dosage, the length of time the drug has been used, and with the individual taking the drug.

[4] These diseases are caused by viruses that can be transmitted by using syringes and/or needles of infected persons. Heroin addicts' use of needles contaminated with acquired immune deficiency syndrome (AIDS) is the second most common means of spreading the disease. In 1985, New York City considered legalizing the sale of hypodermic needles and syringes to reduce the spread of AIDS.

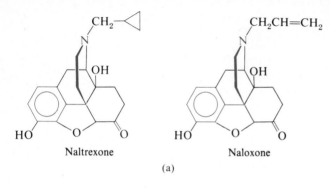

FIGURE 9.4 The narcotic antagonists (a) and the antagonist analgesics (b) are structurally similar to the morphine narcotics.

believed that the opiates produce their effects by binding to these so-called "opiate-receptor" sites, and thereby blocking the release of certain neurotransmitters that transmit the pain message across the synapse between nerve cells. The pain is still present, but the individual is insensitive to it because the message never gets to the brain.

During the past 50 years, a great deal of research has been devoted to the synthesis of the ideal analgesic–one that is effective, yet nonaddicting and free of side effects such as respiratory depression and hallucinogenic activity. A few of the synthetic analgesics that have been developed are less effective than morphine as a pain killer, but they are noneuphoric. Demerol has the advantage that it can be taken orally rather than intravenously, and it is given to individuals who suffer from severe nausea when treated with morphine. However, its major disadvantage is that of addiction. Pentazocine (Talwin) has been proclaimed to have very low addiction liability, and it is not subject to narcotic controls. It is an especially useful drug in the control of chronic pain. Darvon may be considered to be a derivative of methadone. It is claimed to be a nonaddictive analgesic without the gastrointestinal side effects of the opiates.

Pentazocine and Darvon are approximately equivalent to codeine as analgesics. Clinical evidence indicates, however, that they are also addicting in some individuals.

Still other compounds have been developed to treat acute morphine poisoning and to replace methadone in the treatment of heroin addicts. Chief among these are naltrexone and naloxone (Figure 9.4-a). They are termed *narcotic antagonists* because they block the euphoric effects of the opiates. Antagonists exert their effect by preferentially occupying the body's opiate receptor sites. (Naloxone has about 25 times the affinity for the opiate receptors that heroin has.) If a large enough dose of an antagonist is ingested before the addict uses heroin, the heroin will produce no euphoria, no sedation, and no analgesia because the opiate receptors are occupied by the antagonist. The antagonists do not produce tolerance nor do they cause physical or psychological dependence.

Further research on the narcotic antagonists through modification of existing structures has yielded new drugs that are termed *antagonist analgesics*. These substances are as potent as morphine in the relief of pain, and since they block the euphoric effects, they seem minimally likely to cause addiction. Two of these compounds, available to physicians in injectable form, are nalbuphine and butorphanol (Figure 9.4-b).

In 1975, scientists culminated many years of exhaustive research efforts by isolating two pentapeptides from the brains of pigs and calves. These peptides,[5] called *enkephalins* (from the Greek word for head), were remarkable analgesics and were similar to molecules previously discovered in the pituitary glands of camels, animals that are extremely insensitive to pain. It is likely that the enkephalins are some of the molecules that the body uses to counteract pain (Figure 9.5). The enkephalins may inhibit the release of an excitatory neurotransmitter (maybe substance P—see Exercise 13.38) from sensory neurons, especially those carrying information about pain.

Tyr-Gly-Gly-Phe-Met	Tyr-Gly-Gly-Phe-Leu
Methionine enkephalin	Leucine enkephalin

In 1976, several longer polypeptides were discovered in the pituitary gland. As analgesics, these peptides were more potent than the enkephalins and about as potent as morphine. These endogenous pain killers were termed *endorphins* (the morphine within). It is now known that there are several different endorphins designated as α, β, and γ. Furthermore, the endorphins and the enkephalins are structurally related in that the first five amino acids of β-endorphins are those present in leucine

[5] As we shall learn in Section 13.3, *peptides* are relatively short chains of amino acids. Notice that four of the five amino acids are identical in both of the enkephalins. See Table 13.1 for the formulas of the amino acids; also see Exercise 13.21.

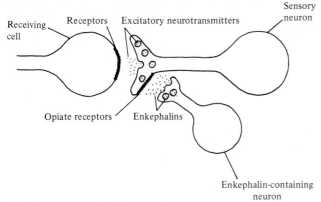

FIGURE 9.5 A proposed mechanism of opiate activity. The enkephalins bind to the "opiate" receptors and prevent the release of neurotransmitters that convey the pain message to the brain. [From S. H. Snyder, *Chem. Eng. News*, Nov. 28, 1977, p. 26.]

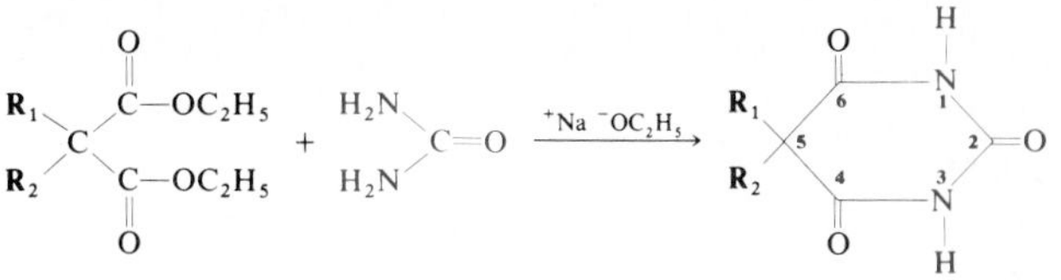

$$R_1 \underset{R_2}{\overset{}{C}} \begin{matrix} C-OC_2H_5 \\ \\ C-OC_2H_5 \end{matrix} \;+\; \begin{matrix} H_2N \\ \\ H_2N \end{matrix} C=O \;\xrightarrow{\;^+Na\ ^-OC_2H_5\;}\; R_1 \underset{R_2}{\overset{}{\longleftrightarrow}} \text{(barbiturate ring)}$$

R_1	R_2	Name	Trade Name
C_2H_5-	C_2H_5-	Barbital	Veronal
C_2H_5-	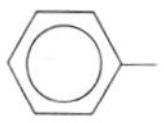 (phenyl)	Phenobarbital	Luminal
C_2H_5-	$CH_3CHCH_2CH_2-$ (CH_3)	Amobarbital "Blue heaven"	Amytal
C_2H_5-	$CH_3CH_2CH_2CH-$ (CH_3)	Pentobarbital "Yellow jackets"	Nembutal
$CH_2=CHCH_2-$	$CH_3CH_2CH_2CH-$ (CH_3)	Secobarbital "Red devils"	Seconal

FIGURE 9.6 The preparation and structures of some barbiturates.

enkephalin. There is no evidence, however, that enkephalins arise directly from endorphins.

Endorphins are another example of neurotransmitters, and they are located throughout the central nervous system and peripheral nervous system. They help mediate phenomena as diverse as pain relief, sexual behavior, hormonal balance, and possibly schizophrenia. Further research on enkephalins and endorphins may lead to the synthesis of more effective pain killers with fewer side effects (including addiction). Unlike morphine, these endogenous chemicals are quickly degraded after binding to the receptors. They do not accumulate in the synapse to induce the tolerance seen in morphine addicts. The body's release of endorphins in response to certain stimuli may be the explanation for the medical mysteries of acupuncture, electrical stimulation analgesia, Indian fakir's walking over hot coals, and people losing limbs yet not complaining of pain. It also may explain the "high" reported by distance runners.

b. Barbiturates

The barbiturates belong to a class of synthetic drugs that have *sedative* (soothing, calming) and *hypnotic* effects when administered in therapeutic doses. They were first prepared in 1903 by heating urea with suitably substituted diethyl malonates. Figure 9.6 presents a synopsis of the preparation and structure of some of the well-known barbiturates. (To increase their water solubility, these drugs are formulated as salts.) Curiously, in the United States, the names of the barbiturates end in -al even though they are ketones rather than aldehydes. The British spelling uses the suffix -one.

Barbituric acid itself lacks central nervous system depressant activity, but the presence of alkyl or aryl substituents at position

5 gives pharmacologically active compounds. The nature of the substituents plays an important role in establishing the pharmacological action of a particular barbiturate. Generally, the more highly lipid soluble a barbiturate is made by modifying the substituents, the more rapid is the onset of its action and the shorter is the duration of its pharmacological effects. Seconal and Nembutal are short-acting barbiturates (3–4 hours duration). They are used most often as sleeping pills and as preanesthetic sedatives, which are given to patients 30–60 minutes prior to surgery. Amytal is intermediate-acting (6–8 hours duration), whereas Luminal is a long-acting barbiturate (10–12 hours duration). Once used widely as a tranquilizer, it is now employed chiefly as an anticonvulsant agent in the treatment of epilepsy and other brain disorders. The mechanism of Luminal's action is unknown, but it is thought to suppress the spread of abnormal electrical discharges given off by the brain during a seizure.

Barbiturates, in the terminology of the drug culture, are "downers." The principal action of these drugs is on the central nervous system. It is believed that they depress the activity of the brain cells by interfering with oxygen consumption and the use and storage of energy. An average therapeutic dose to counteract insomnia or anxiety is 10–20 mg. An individual who repeatedly uses a larger amount develops a tolerance, and soon many times the original dose becomes necessary. Habitual doses in excess of 800 mg/day depress the brain cells and effect a general depression of the central nervous system.

Of all the sedatives, the barbiturates are the most abused. (Ten billion barbiturate tablets are manufactured each year in the United States.) It has definitely been established that barbiturates are habit-forming drugs. The symptoms of barbiturate addiction are similar to those of chronic alcoholism. A small dose can make the user rather sociable, relaxed, and good-humored, but his alertness and ability to react are decreased. Larger amounts result in blurred speech, mental sluggishness, confusion, loss of emotional control, and belligerence, and may induce a deep sleep or coma. Without immediate attention, the user may never wake.

Barbiturate dependency is becoming a more serious problem than opiate dependency. One reason is that the withdrawal symptoms are usually far more dangerous than those resulting from withdrawal of the opiates. Also, as tolerance to morphine develops, the user's body is able to handle what would be considered a lethal dosage for the nonuser without serious effects. However, with the barbiturates, even though a tolerance is developed, the lethal dose remains essentially constant. Thus, in a relatively short time, the barbiturate addict's margin of safety decreases as his drug intake increases. Barbiturates kill more people (mostly through deliberate overdoses) in the United States than any other drug, and they account for more than one fifth of all the poisonings reported. Since alcohol enhances the action of barbiturates, a particularly lethal combination is "booze" and sleeping pills.

As with the morphine narcotics, the addictive nature of barbiturates has led to the development of nonaddictive agents to treat insomnia. Sominex and Nytol contain pyrilamine (Figure 9.7), which is structurally similar to the antihistamines (see Figure 18.5). Dalmane is a safe sedative and muscle relaxant that can be obtained by prescription. Dalmane may induce sleep indirectly by relieving anxieties. Notice that it, like the tranquilizers Valium and Librium (see Figure 9.10), contains a benzodiazepine nucleus. In the past few years, the therapeutic use of the barbiturates as sedative–hypnotic drugs has declined (except for a few specialized cases). They have largely been replaced by the benzodiazepines because (1) barbiturates lack specificity of effect in the central nervous system, (2) they are not as effective as the benzodiazepines, (3) tolerance to barbiturates occurs more frequently, (4) the liability for abuse of barbiturates is greater, and (5) barbiturates interact with a considerable number of other drugs.

One sleep-inducing substitute for the barbiturates is not safe, although it was originally approved by the FDA. The drug is methaqualone, which is known on the street as Quaalude. Major reasons for the popularity of Quaalude were the mistaken notions that it had aphrodisiac activity and that it induced a "high" without the drowsiness caused by the barbiturates. Methaqualone is as dangerous and as addictive as the barbiturates. Severe central nervous system depression occurs when Quaalude is taken in combination with alcohol or other depressants. Withdrawal from methaqualone has resulted in grand mal convulsions and even death when the detoxification program was not properly supervised.

c. Marijuana

Marijuana, colloquially known as "pot," "grass," "reefer," or "joint," is found in the flowering tops of the female hemp plant (*Cannabis sativa*). (The longer male plant is used for fiber or hemp.) Hashish (*charas*) is a purer and more concentrated form of marijuana. The major active constituent in marijuana is believed to be a derivative of tetrahydrocannabinol (Δ^9-THC) that is obtained from the resinous exudate of hemp plants. The potency of marijuana is related to its THC content. Typical marijuana available in the United States contains 0.5–2% THC, imported pot contains about 2–5% THC, and the THC content of hashish is normally in the range of 5–10%. It has been suggested that Δ^9-THC is rapidly metabolized to 11-hydroxy-Δ^9-tetrahydrocannabinol, and it is this latter compound that is responsible for the majority of the pharmacological effects. This metabolic product of THC persists in the bloodstream for more than 72 hours, so that a frequent user of marijuana can get high on a smaller dose than that needed by the occasional user. By

FIGURE 9.7 Nonbarbiturate sleep-inducing drugs.

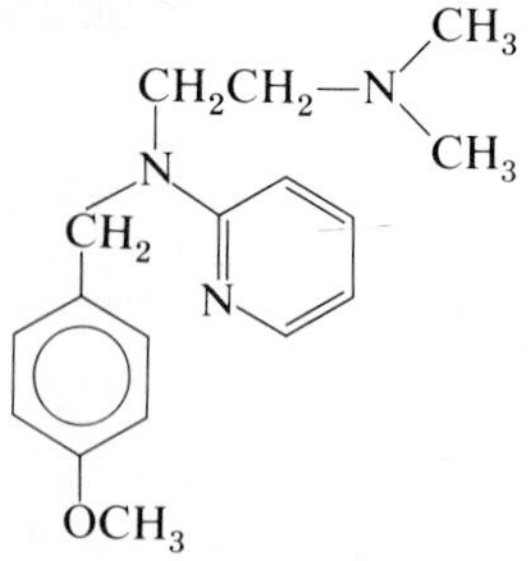

Pyrilamine

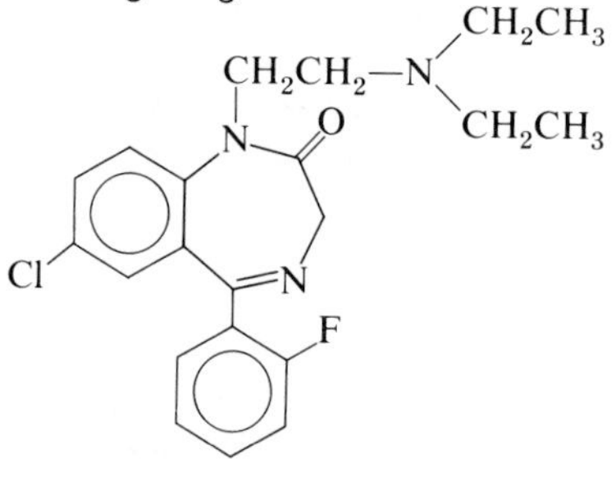

Flurazepam
(Dalmane)

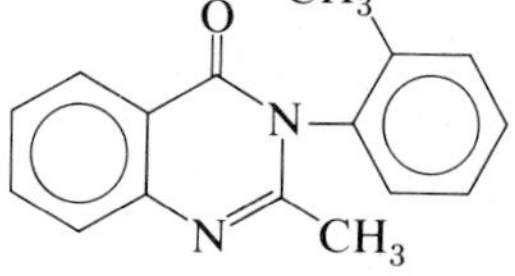

Methaqualone
(Quaalude)

experimenting with synthetic THC derivatives of known purity, scientists hope to understand the pharmacological activity of marijuana.

CH$_3$

Δ^9-Tetrahydrocannabinol

CH$_3$

11-Hydroxy-Δ^9-tetrahydrocannabinol

Marijuana acts on both the central nervous system and the cardiovascular system. When it is smoked, the effects of the drug can be felt within a few minutes and may last 2–3 hours. If marijuana is eaten with food, the onset of drug action usually occurs within 30 minutes to 1 hour and the effects may persist from 3 to 5 hours. Among the physiological effects of the drug are some or all of the following: increase in the pulse rate, bloodshot eyes, and a voracious appetite. The subjective effects (getting "high" or "stoned") include lethargy and hilarity, often at the same time. The user may experience an intensification of visual and auditory stimuli; time, space, and touch are usually distorted.

A great deal of misinformation, both pro and con, has developed about the smoking of pot. A publication by the National Clearinghouse for Mental Health Information deals with some of the fallacies associated with marijuana use.

Fable: Marijuana is a narcotic.
Fact: Marijuana is not a narcotic except by statute.

Fable: Marijuana is addictive.
Fact: Marijuana does not cause physical addiction, since tolerance to its effects and symptoms on sudden withdrawal do not occur. It can produce habituation (psychological dependence), resulting in restlessness, anxiety, irritability, insomnia, or other symptoms.

Fable: Marijuana causes violence and crime.
Fact: Persons under the influence of marijuana tend to be passive. Sometimes a crime is committed by a person while under the influence of marijuana. The personality of the user is as important as the type of drug in determining whether chemical substances lead to criminal or violent behavior.

Fable: Marijuana leads to increase in sexual activity.
Fact: Marijuana has no aphrodisiac property.

Fable: Marijuana is harmless.
Fact: Instances of acute panic, depression, and psychotic states are known, although they are infrequent. Certain kinds of individuals can also become overinvolved in marijuana use and center their lives around it. We do not know the effects of long-term use, although there is evidence that long-term users are typically apathetic and sluggish both physically and mentally.

(Note: A report on "Marijuana and Health" issued by the Department of Health, Education and Welfare expresses concern that marijuana's greatest danger involves operation of motor vehicles rather than the more widely publicized biological damage.)

Fable: Occasional use of marijuana is less harmful than occasional use of alcohol.
Fact: We do not know. Research on the effects of various amounts of each drug for various periods is under way.

Fable: Marijuana use leads to heroin.
Fact: We know nothing in the nature of marijuana that predisposes to heroin abuse. It is estimated that less than 5% of chronic users of marijuana will progress to experiment with heroin. (One study shows that more heroin addicts started out on alcohol than on marijuana.)

Fable: Marijuana enhances creativity.
Fact: Marijuana might bring fantasies of enhanced creativity but they are illusory, as are "instant insights" reported by marijuana users.

Fable: More severe penalties will solve the marijuana problem.
Fact: Marijuana use has increased enormously in spite of severely punitive laws.

In recent years, as research on the effects of marijuana has continued, the following observations have been reported for regular, long-term users.

1. The individuals are apathetic and sluggish, both mentally and physically.
2. Marijuana suppresses the body's immune response, which could increase the susceptibility to disease.
3. There is decreased fertility in males that is apparently attributable to a decrease of the male hormone, testosterone.
4. There are adverse effects on the bronchial tract and the lungs. This is probably caused by the greater amount of tar in marijuana cigarettes compared to ordinary cigarettes.
5. There is impairment of short-term memory (i.e., tasks such as learning and remembering new information or remembering and following a sequence of directions).

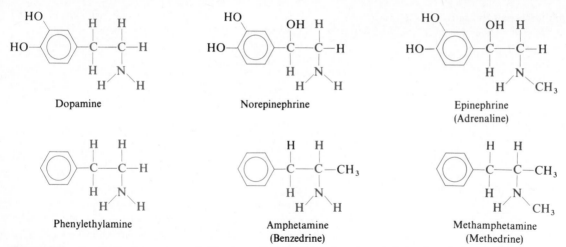

FIGURE 9.8 Amphetamines are synthetic drugs that possess the phenylethylamine skeleton. They are structurally similar to epinephrine and norepinephrine, and therefore they can mimic the actions of the catecholamine neurotransmitters.

Finally, it should be mentioned that THC is particularly effective in reducing the pressure of the optic fluid; hence it is helpful in relieving the symptoms of glaucoma. It has also been used to control the nausea and vomiting suffered by cancer patients receiving chemotherapy. In 1985 the FDA approved the marketing of a new drug that consists of a synthetic form of THC. This drug will be available to those patients receiving chemotherapy who fail to respond to other antinausea drugs.

d. Amphetamines

The amphetamines ("uppers" or "pep pills") represent a group of synthetic drugs that stimulate the central nervous system. The parent compound, amphetamine is not a single compound, but is a mixture of two isomers (see Exercise 10.27) that has been marketed under the name Benzedrine. It was synthesized in 1927 as a drug to simulate the actions of epinephrine (adrenaline). The latter is a hormone produced by the adrenal medulla; it plays a major role in making glucose rapidly available to tissues that require immediate energy (see Section 16.3). The amphetamines are often referred to as *sympathomimetic amines* since they mimic the action of epinephrine and norepinephrine on the central nervous system (Figure 9.8). Norepinephrine is released from storage sites at the end of the sympathetic nerves,[6] and its stimulating effects are thought to be associated primarily with excitatory responses (contraction) of smooth muscles. Increased

stimulation of the sympathetic nerves results in an elevation of the pulse rate and blood pressure, an increased rate of respiration, and a loss of appetite (anorexia). The neurotransmitters epinephrine, norepinephrine, and dopamine are collectively called the *catecholamines*. They are synthesized in the nerve cells from the amino acid tyrosine.

Medically, the stimulating effects of the amphetamines make them useful for the treatment of a variety of conditions, including mild depression, narcolepsy (an overwhelming compulsion to sleep), certain behavioral disorders (hyperkinetics) in children, and obesity. (Most physicians do not recommend the use of amphetamines as appetite suppressants for weight reduction because as soon as the drug is withdrawn, the patient's former appetite will return unless a new dietary pattern has been established.) Ordinary therapeutic doses of 10–30 mg/day provide a feeling of well-being, a decrease in fatigue, and increased alertness. Overtired businessmen and truck drivers, students cramming for exams, and athletes who need to "get up" for a game are frequent misusers of these drugs.

Prolonged abuse of amphetamines can have ruinous effects upon a person's physical and mental health. Once ingested (or injected), they are not easily broken down, and by passing across the blood–brain barrier they tend to concentrate in the brain and cerebrospinal fluid. Extensive use of higher than normal doses can lead to long periods of sleeplessness, loss of weight, and severe paranoia. When excited, the person is liable to become dangerously violent and may hurt or kill somebody.

The most widely abused amphetamine has been methamphetamine ("speed"). A user generally begins by taking the drug orally, but then progresses to intravenous injections ("shooting speed") to attain the desired effects. A tolerance is rapidly

---

[6]The sympathetic nerves are a division of the autonomic nervous system. This system has the specialized task of regulating body functions over which an individual has no control such as circulation, digestion, heart rate, and respiration. Sympathetic nerves are also concerned with emotions.

developed, and soon enormous amounts are required. With continued injections the "speed-freak" or "methhead" can stay awake for days. This constant wakefulness, which is accompanied with incessant babbling and tremendous nervous energy, is called a "run" or "speed-binge." Following a three or four day run, his body becomes exhausted and he falls into a coma-like sleep for 12–18 hours, usually with the aid of barbiturates or heroin. Upon wakening, depression sets in and the entire procedure is begun once again. As is the case with heroin users, death from serum hepatitis (due to insanitary syringe needles) is quite common.

e. Cocaine

Cocaine ("coke") occurs in the leaves of the coca plant (*Erythroxylon coca*) that grows in Bolivia and Peru. For centuries these leaves have been chewed for their stimulating effect by the people of the region. When cocaine is smuggled into the United States, it is in the form of glistening white crystals (hence the slang name "snow").

The major pharmacological activity—and hence medicinal use—of cocaine is as a local anesthetic. It is applied topically (to skin and mucous membranes) because it is too toxic to be injected. Cocaine is occasionally used in ear, nose, or throat operations because of its ability to block the initiation or conduction of nerve impulses. In general, however, the use of cocaine as a local anesthetic has been replaced by other drugs (e.g., Lidocaine) that are less addicting and less toxic.

Cocaine exerts its effects by blocking the uptake (reabsorption) of norepinephrine from the synaptic cleft. Unlike acetylcholine, which is deactivated enzymatically (by *acetylcholinesterase*—see page 171), the major deactivation pathway for norepinephrine is reabsorption into the presynaptic membrane. If this reabsorption is blocked, then the sending (presynaptic) neuron will have insufficient norepinephrine to be released in response to new impulses. This explains the anesthetic activity of cocaine. The stimulating effect (constriction of blood vessels, increased blood pressure, increased heart rate) is due to the increased amounts of norepinephrine that remain in the synapses because of the blockage in the reabsorption mechanism.

Until recently, explanations for the euphoric effects of cocaine (and other drugs as well) were not available. There is some evidence to suggest that the opiates produce euphoria by stimulating the release of the neurotransmitter dopamine in brain cells.[7] This is the *dopamine hypothesis of euphoria.*

[7]As we shall see in Section 18.3-c, the amount of dopamine secreted in brain cells is vital to normal well-being. A deficiency of dopamine causes Parkinson's disease and an excess of dopamine, or increased sensitivity of dopamine receptors in the brain, may possibly be a cause of schizophrenia.

According to this theory, the experience of pleasure is due to the stimulation of the pleasure center in the midbrain. The receptors of the pleasure center bind to dopamine, and the more dopamine secreted by nerves adjacent to the pleasure center, the greater the pleasure. The opiates bind to certain receptors in the brain cells that cause a greater release of dopamine. Cocaine blocks the reabsorption of dopamine; therefore there is a greater amount of dopamine remaining in the synapse. Both classes of drugs serve to increase the amount of dopamine in the synapse, hence increasing the pleasurable sensations.

Cocaine is metabolized by the body (in the liver) very rapidly. Therefore, one of the major problems with the use of cocaine is that the "high" only lasts for about 15–20 minutes. To recapture the same feeling, the user will take the drug repeatedly every 30–40 minutes. If taken by inhalation ("snorted"), cocaine takes about 3–5 minutes to reach the brain; if it is injected (as its hydrochloride salt), it gets to the brain in 14 seconds, and if *free basing* (inhaling cocaine vapors through a glass pipe), it gets there in 6 seconds. (Recently, a smokable form of cocaine, called "crack" from the crackling noise it makes when smoked, has become an increasingly popular drug of abuse.)

Cocaine, like amphetamines, causes physical and psychological dependence. Tolerance develops rapidly. There are reports of users taking 3 g of cocaine at the end of a day of constant use, where 1.5 g would have been lethal at the beginning of the day. Repeated use or overdoses often lead to hallucinations, paranoia, psychosis, and even death due to respiratory failure or heart attack. Although the greatest notoriety has been achieved by Hollywood celebrities and sports personnel, the pervasive use of cocaine among all segments of society has reached alarming proportions. Ten percent of the American population has tried cocaine, and about 6 million people abuse it, despite having to pay $100 or more for a single gram.

Cocaine

f. Hallucinogens

The hallucinogens are drugs that stimulate sensory perceptions (such as hearing strange voices and seeing alarming sights) that have no basis in physical reality. They are also referred to as *psychotomimetic drugs,* because in some cases their effects seem to mimic psychosis, and *psychedelic drugs,* because of the visions produced. (The term psychedelic has come to be synonymous with hallucinogenic). Among the more abused

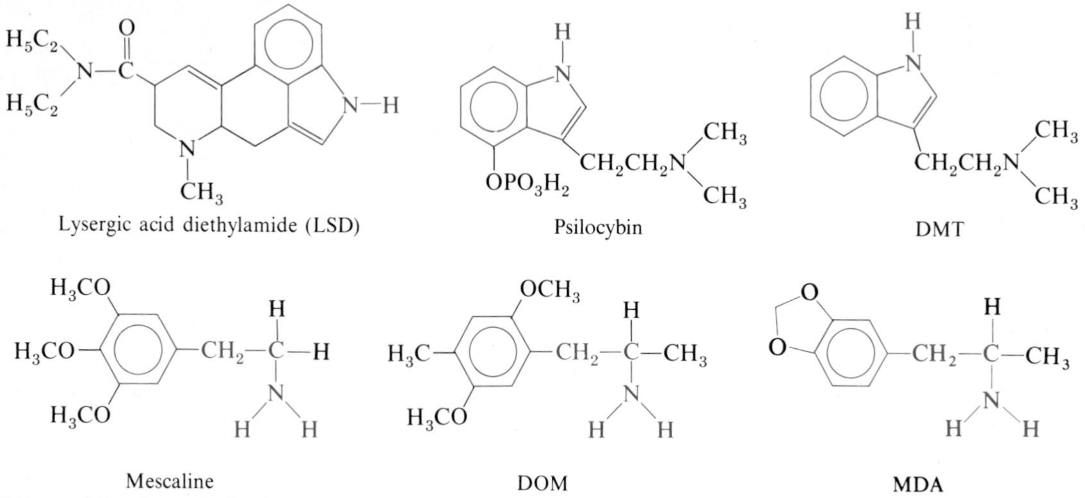

FIGURE 9.9 Some hallucinogenic drugs.

hallucinogens are LSD (lysergic acid diethylamide), mescaline (3,4,5-trimethoxyphenylethylamine), psilocybin (4-phosphoryl-N,N-dimethyltryptamine), DOM (2,5-dimethoxy-4-methyl-amphetamine), DMT (N,N-dimethyltryptamine), and MDA (3,4-methylenedioxyamphetamine). These drugs can be divided into two broad categories: indole derivatives and phenylethylamine derivatives (Figure 9.9). This latter group is collectively known as *psychotomimetic amphetamines* because they combine the stimulant effects of amphetamines and the psychedelic effects of drugs like mescaline.

The plants from which the major hallucinogens are obtained have been known for millenniums. Mescaline comes from the peyote cactus; psilocybin from certain species of mushrooms; and lysergic acid is derived from the parasitic fungus ergot, which grows on rye and wheat. The diethylamide of lysergic acid is then chemically synthesized in the laboratory. LSD is a white, crystalline powder that is odorless, tasteless, and readily soluble in water, It is, by far, the most powerful of the hallucinogens. In order to achieve the equivalent effects of 0.1 mg of LSD (an average "trip" dose), the user requires 5 mg of DOM, 12 mg of psilocybin, and 300 mg of mescaline.

Oral doses of LSD as low as 20–25 μg (0.020–0.025 mg) are capable of producing psychological responses in sensitive individuals. The effects of the drug on the central nervous system are usually apparent within 1 hour following ingestion and most trips last from 8 to 12 hours. Tolerance to LSD is rapidly developed. The basic physiological effects are those typical of a mild stimulation of the sympathetic nervous system. These include elevated body temperature, dilated pupils, increased blood glucose, elevated blood pressure, and increased heart rate. The major subjective effect is a change in visual perception.

Many users feel a new awareness of the physical beauty of the world, particularly of visual harmonies, colors, the play of light, and the exquisiteness of detail.

Many attempts have been made to explain and localize the action of the hallucinogens. The structural resemblance of mescaline and DOM to dopamine and norepinephrine (Figure 9.8) suggests a possible link between the drugs and the neurotransmitters. LSD (and the other substituted tryptamines) are structurally similar to serotonin, a neurotransmitter that is found in neurons in the upper brainstem and throughout the central nervous system. Serotonin is an inhibitory neurotransmitter; its effects may be to modulate our responses to various stimuli involving such behavior as aggressiveness, motor activity, mood, and sexual desire. It also regulates our sleep/wake patterns. One theory is that LSD blocks the release of serotonin and/or competes with serotonin for the receptor sites on the postsynaptic membranes.

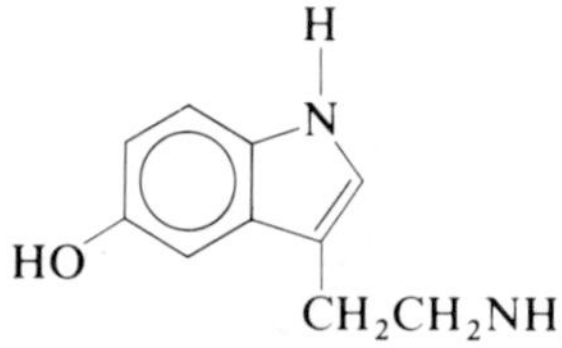

Serotonin
(5-Hydroxytryptamine, 5-HT)

There have been conflicting reports as to the teratogenic effects of LSD. (**Teratogenesis** refers to the *production by drugs of physical defects in offspring while in the uterus.*) Some investigators have reported an unusually high incidence of genetic damage to the chromosomes of the white blood cells of infants born to mothers who used LSD during their pregnancy. But other studies showed no such evidence, and these latter investigators conclude that pure LSD, ingested in moderate doses, does not produce chromosome damage detectable by available methods. Moreover, despite all claims to the contrary, LSD does not enhance creativity. In light of these contradictory findings, it is obvious that a great deal of further research is required.

Another hallucinogenic drug that has been greatly abused is phencyclidine (PCP, "angel dust" or "crystal"), which was first developed in the 1950s as a general anesthetic. Its use was discontinued in humans because the patients experienced delirium when they emerged from anesthesia. However, it is still used by veterinarians as a potent animal tranquilizer and is marketed for this purpose under the trade name Sernylan.

Phencyclidine (PCP)
[1-(1-Phenylcyclohexyl)piperidine]

The legitimate use of PCP makes it available for diversion to the illicit market. Because it also can be easily made in underground laboratories, PCP is obtained readily and cheaply and thus has become one of the most destructive drugs in the United States. It is next in popularity to pot and alcohol among teenagers. In addition, the analysis of confiscated street samples purported to be LSD, mescaline, psilocybin, and marijuana has shown them to contain a high percentage of PCP. Although originally ingested orally, PCP now is commonly smoked or snorted. Intravenous use is less frequent, but has been reported.

From a pharmacological viewpoint, the compound is of interest because it possesses a variety of actions. It can both stimulate and depress the central nervous system; it is hallucinogenic and an analgesic. There are few drugs that seem to induce so wide a range of effects. The immediate danger is that PCP is an extremely toxic substance, particularly when mixed with alcohol (it triggers violent, psychotic behavior). PCP is considered more dangerous than heroin because medically nothing is known about its effects, nor are there drugs available to counteract it. It is very easy to overdose on PCP to the point of convulsions.

g. Tranquilizers

Tranquilizers are drugs that are taken (1) to modify psychotic behavior without inducing sleep or (2) to reduce anxiety, excitement, and restlessness. Tranquilizing drugs fall into two broad categories, depending upon which of these effects is desired (Figure 9.10).

The major (strong) tranquilizers are used in the treatment of severe mental disorders. Chlorpromazine (Thorazine) and reserpine were the first and best known of the major tranquilizers. Reserpine is extracted from the snakeroot plant *Rauwolfia serpentina*. Reserpine reduces blood pressure, induces calm, lowers aggression, and aids sleep.

Chlorpromazine was synthesized in France in 1950, and its use in psychiatric patients in 1952 marked the beginnings of modern psychopharmacology. It reduces muscle activity and produces sedation, causing the psychotic patient to become less agitated and restless. Chlorpromazine has been found to be extremely effective against the symptoms of schizophrenia and manic depression. Recall (footnote 7, page 205) that the condition of schizophrenia has been linked to the excessive production and release of dopamine in brain cells. The hypothesis is that chlorpromazine and the other antipsychotic drugs relieve the symptoms of schizophrenia and mania by blocking the postsynaptic dopamine receptors. That is, some of the switches are shut down; there is less firing of brain neurons. These drugs do not cure the patients; they do enable them to leave institutions and live an almost normal life, just as long as they continue to take the drug. It has been estimated that at least 50 million patients worldwide have received some form of chlorpromazine therapy; 95% of all schizophrenics no longer need hospitalization.

The minor (mild) tranquilizers are used to relieve the tensions that develop in "normal" individuals. Sales of these drugs are enormous. Diazepam (Valium)[8] and chlordiazepoxide (Librium) are among the most frequently prescribed drugs in the world (about $500 million is spent annually). A third large-selling minor tranquilizer is meprobamate, which is marketed under the trade names Equanil or Miltown. Alprazolam (Xanax) is a recently marketed benzodiazepine derivative that has been used in the treatment of panic attacks.

Notice (Figure 9.10) that Valium and Librium have the same complex benzodiazepine ring structure. The benzodiazepines are good at reducing anxiety, and they have anticonvulsant activity. Clinically, Librium is used in the treatment of neuroses and withdrawal symptoms of acute alcoholism, whereas Valium is used to relieve muscle spasms and to improve sleep patterns. Both of these compounds can lead to drug dependence and withdrawal symptoms. Individuals who received excessive doses of Valium over an extended period of time experienced confusion, temporary memory loss, and withdrawal symptoms similar

[8]In 1985, the FDA approved the marketing of Valium in its generic form, diazepam. The use of the generic version was expected to decrease the cost of prescriptions by 30–40%. At the time, Valium was the fourth largest selling prescription drug in the United States.

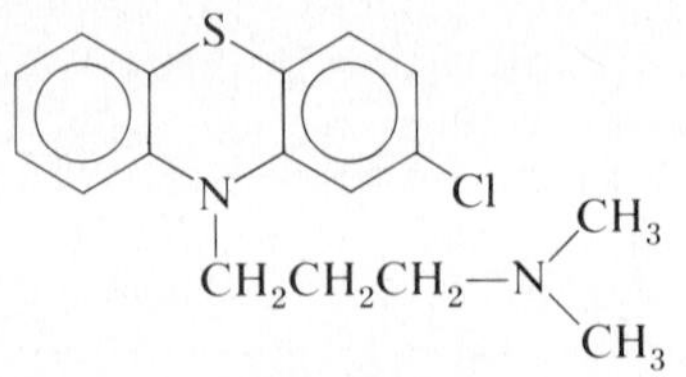

FIGURE 9.10 Some of the major (a) and minor (b) tranquilizers.

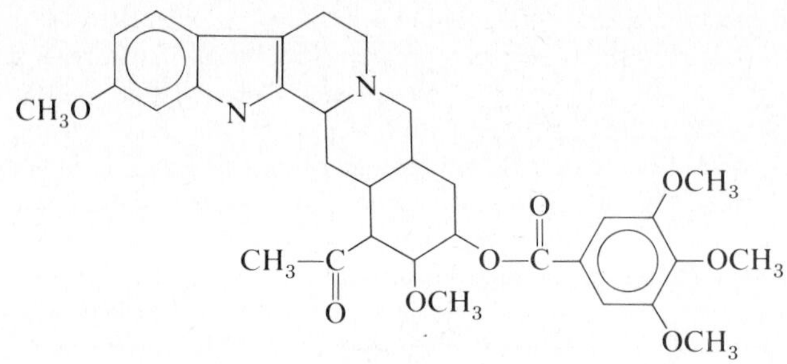

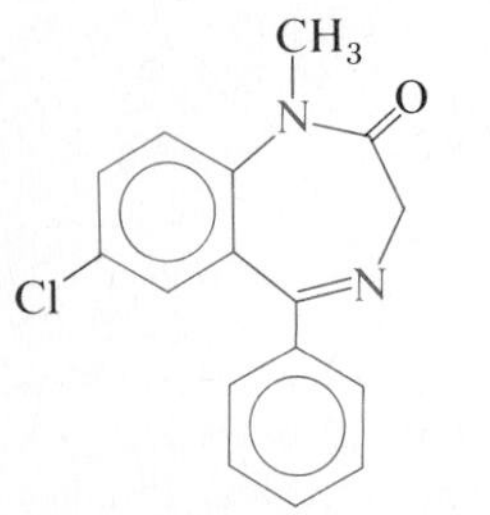

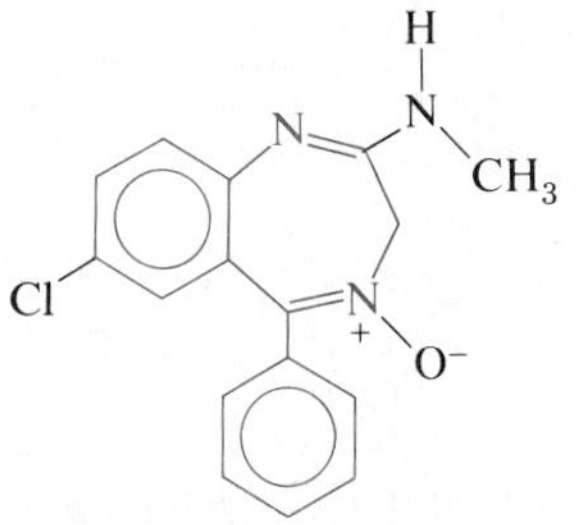

in character to symptoms associated with barbiturates and alcohol.

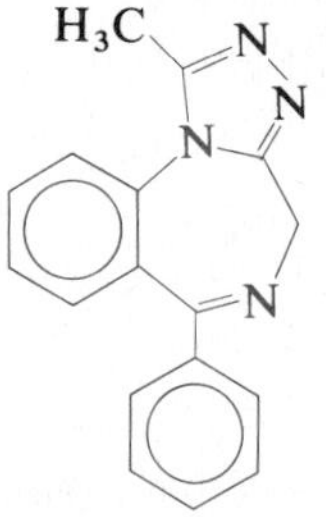

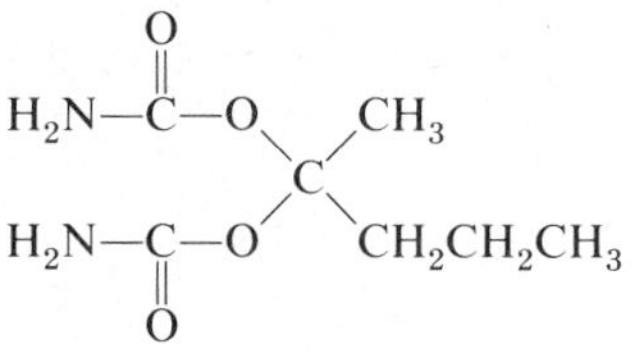

The mode of action of the benzodiazepines is unknown. However, the recent discovery of brain receptors for this type of compound is similar to the discovery of the opiate receptors. Does the brain also possess natural tranquilizers? It was found that these receptors are linked closely to receptors for the neurotransmitter GABA (γ-aminobutyric acid—see Figure 13.1). GABA, like serotonin, is an inhibitory neurotransmitter, and one theory suggests that the benzodiazepines potentiate the inhibitory effects of GABA.

Finally, there are many over-the-counter drugs (Compoz, Cope, Vanquish, etc.) that claim to be tranquilizers. These products usually contain aspirin and an antihistamine (page 482), which causes drowsiness. This latter effect is often equated with the tranquilizing effect of Valium or Librium, but the physiological activities are certainly not the same. Automobile driving should be avoided when taking any of these drugs.

h. Designer Drugs

So-called designer drugs are analogs of compounds that have some proven pharmacological activity. These compounds first surfaced in California in the early 1980s. They are synthesized in clandestine laboratories and then sold on the street as heroin. (PCP, which had been synthesized in the 1950s, can be considered the forerunner of the designer drugs.) Like PCP, the designer drugs do not fit into any one drug category; rather, they possess a wide diversity of pharmacological effects.

The term, designer drugs, was first used to describe analogs of fentanyl (Figure 9.11). Fentanyl is a powerful narcotic that is marketed under the trade name Sublimaze. It is a short-acting anesthetic, and it has been used in about 70% of all surgical procedures in the United States (since the early 1970s). It is just as addictive as heroin, and its euphoric effects are similar to those of heroin, although of shorter duration. By making slight modifications in the structure of fentanyl, the underground chemists have developed more potent, longer lasting drugs that give a greater feeling of euphoria. For example, α-methyl-fentanyl ("China White") is purported to be 20–40 times more potent as a euphoric drug than heroin. 3-Methylfentanyl is currently the most common fentanyl analog on the street. It is an extremely potent narcotic, and its duration of action and euphoriant effects are indistinguishable from those of heroin.

Chapter 9 · Amides and Amines

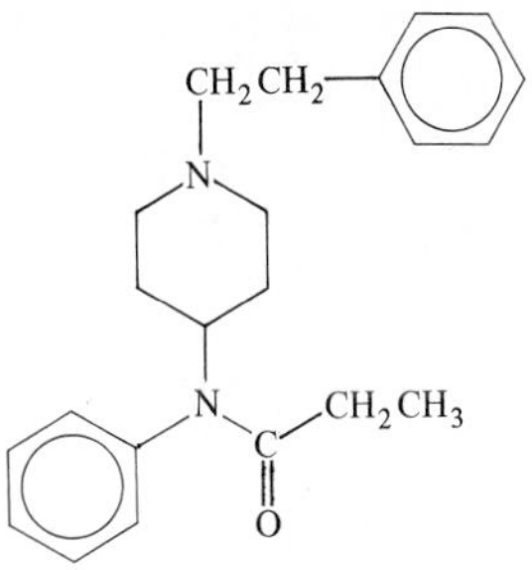

FIGURE 9.11   Designer drugs.

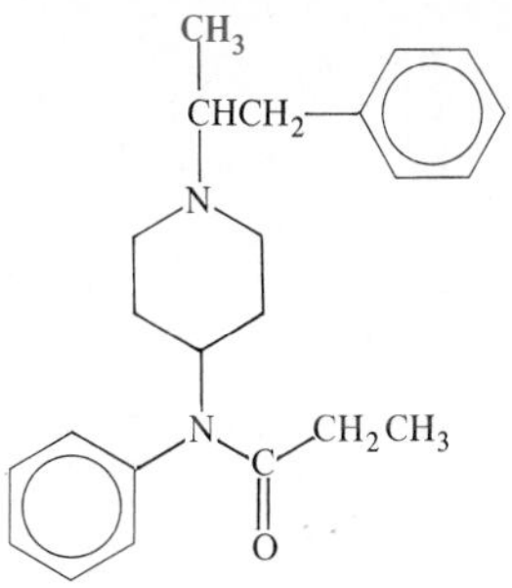

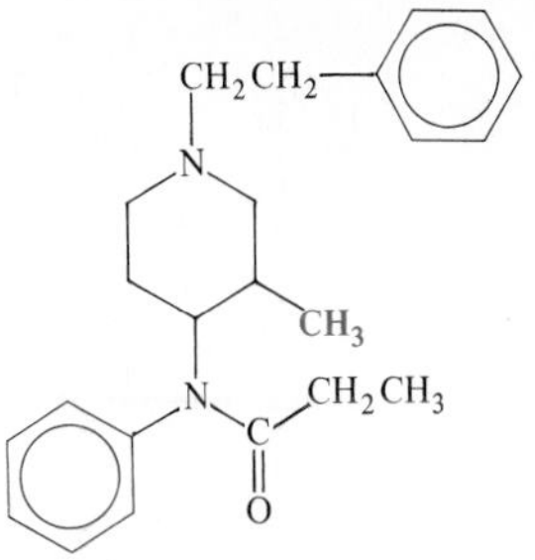

MDMA (3,4-methylenedioxymethamphetamine), also called the "Yuppie" drug, "Ecstasy," and "Adam," has been classified as a designer drug because it appeared on the drug scene at about the same time as the other drugs. It was first synthesized in 1914 as an appetite suppressant but never marketed. The interesting aspect of this compound is that although it bears a close structural similarity to both methamphetamine (Figure 9.8) and to MDA (Figure 9.9), it bears little pharmacological relationship to those drugs. Unlike MDA, MDMA appears to have almost no hallucinogenic properties, nor does it have the stimulating effects of methamphetamine. Instead, it seems to enhance open communication between people and to ease psychic trauma. Some psychiatrists have been using MDMA since the 1970s in counseling sessions as an adjunct to psychotherapy. They report that under clinical conditions, MDMA has few negative side effects and can act to ease psychic trauma and increase communicativeness.

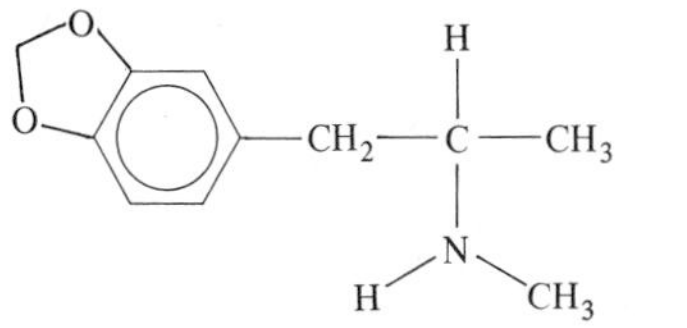

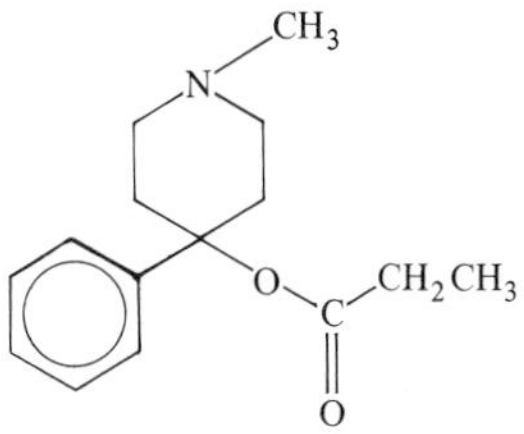

To recreational users, MDMA is a pleasant way of raising one's consciousness without experiencing the hallucinating properties of LSD. However, a report of the National Institute on Drug Abuse (NIDA) calls MDMA "a nationwide problem as well as a serious health threat." The report states that MDMA users experience problems similar to those associated with the use of amphetamines and cocaine. The Drug Enforcement Administration (DEA) outlawed the use of MDMA because research suggested it may cause permanent brain damage to its users (mostly college students and young professionals).

A third class of designer drugs is analogs of meperidine (Demerol—Figure 9.3). The principal analog that has appeared on the street is MPPP (1-methyl-4-phenyl-4-propionoxypiperidine). MPPP is about three times as potent as morphine and 25 times as potent as Demerol. An impurity in the synthesis of MPPP has caused several cases of irreversible Parkinson's disease among the addicts who used it. The lack of quality control in the production of these designer drugs is a major problem. The drugs are all contaminated with impurities, and the only animal testing is done on the humans who buy the drugs. It is therefore no wonder that designer drugs are often quite lethal. Over 100 drugs overdose deaths in California have been attributed to the use of designer drugs.

Exercises

9.1   Draw and name all the isomeric amides having the molecular formula C_4H_9NO.

9.2   Draw and name all the isomeric amines having the molecular formula $C_4H_{11}N$.

9.3   Name the following compounds.

(a) $CH_3CH_2CH(CH_3)C$
$$\overset{O}{\underset{NH_2}{\diagup}}$$

(b) CH_3NH_2

(c) 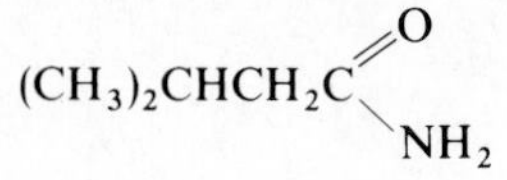 $(CH_3)_2CHCH_2C\overset{O}{\underset{NH_2}{\big<}}$

(d) $(CH_3CH_2)_2NH$

(e) 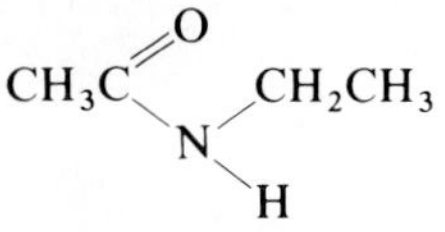 $CH_3C\overset{O}{\underset{\underset{H}{N}}{\big<}}CH_2CH_3$

(f) $CH_3CH(NH_2)CH_2CH_2CH_3$

(g) 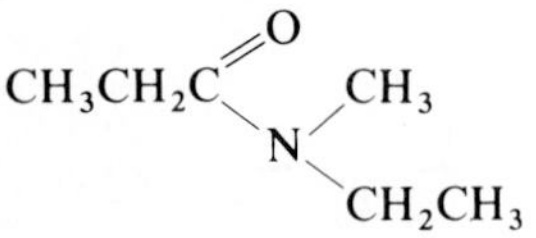 $CH_3CH_2C\overset{O}{\underset{\underset{CH_2CH_3}{N}}{\big<}}CH_3$

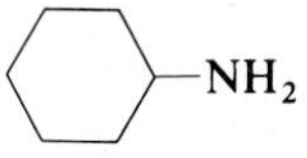

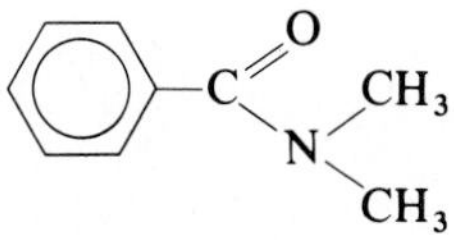

(h) $H_2NCH_2CH_2CH_2NH_2$

(i)  $CH_3CHOHCH_2C\overset{O}{\underset{NH_2}{\big<}}$

(j) $(CH_3CH_2)_4N^+\ I^-$

(k)

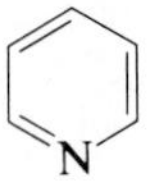

(l) 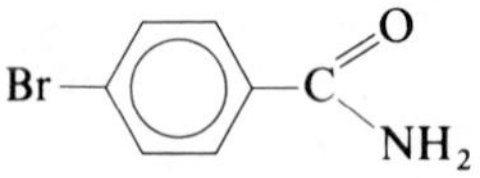 $Br-\underset{}{\bigcirc}-C\overset{O}{\underset{NH_2}{\big<}}$

(m) $\bigcirc-NH_2$

(n) $\bigcirc-C\overset{O}{\underset{\underset{CH_3}{N}}{\big<}}CH_3$

(o) $CH_3CH_2-\bigcirc-NH_2$

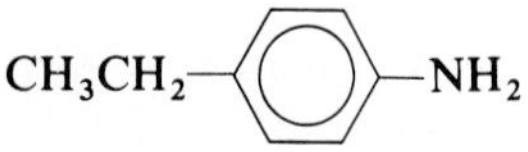

(p) 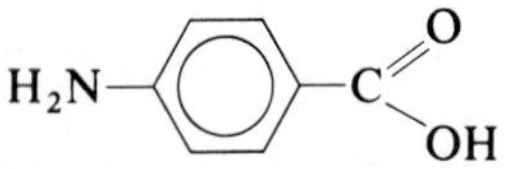 $H_2N-\bigcirc-C\overset{O}{\underset{OH}{\big<}}$

(q) 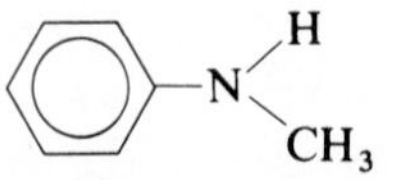 $\bigcirc-\overset{H}{\underset{CH_3}{N}}$

(r)

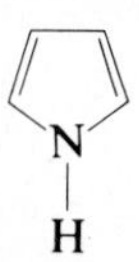

9.4 Write structural formulas for the following compounds.
  (a) acetamide
  (b) benzamide
  (c) β-hydroxybutyramide
  (d) N,N-diethylethanamide
  (e) 2-methyl-4-hexenamide
  (f) N-ethylbenzamide
  (g) N-methyl-3-phenylpropanamide
  (h) N-phenyl-2-aminoethanamide
  (i) trimethylamine
  (j) 3-amino-2-pentanol
  (k) 1,4-diaminobutane
  (l) benzyltrimethylammonium bromide
  (m) cyclopropylmethylamine
  (n) N,N-diethylaniline
  (o) 3,5-dibromoaniline
  (p) diphenylamine
  (q) imidazole
  (r) indole

9.5 Explain and illustrate the meaning of the terms primary, secondary, and tertiary when applied to amines.

9.6 Write equations for two preparations of acetamide.

9.7 Write equations for two preparations of ethylamine.

9.8 Contrast the physical properties of amides with those of amines.

9.9 Ethylamine and ethyl alcohol have comparable molecular weights, yet ethyl alcohol has a much higher boiling point. Explain.

9.10 The boiling point of propylamine is 48°C, whereas the boiling point of trimethylamine is 3°C. Explain the considerable difference between the boiling points of these two isomers.

9.11 Arrange the compounds in each set in order of decreasing solubility in water.

(a) 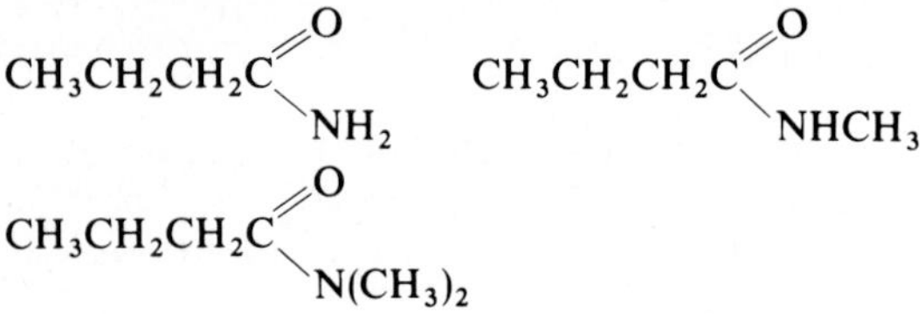 $CH_3CH_2CH_2C\overset{O}{\underset{NH_2}{\big<}}$ $CH_3CH_2CH_2C\overset{O}{\underset{NHCH_3}{\big<}}$

$CH_3CH_2CH_2C\overset{O}{\underset{N(CH_3)_2}{\big<}}$

(b) 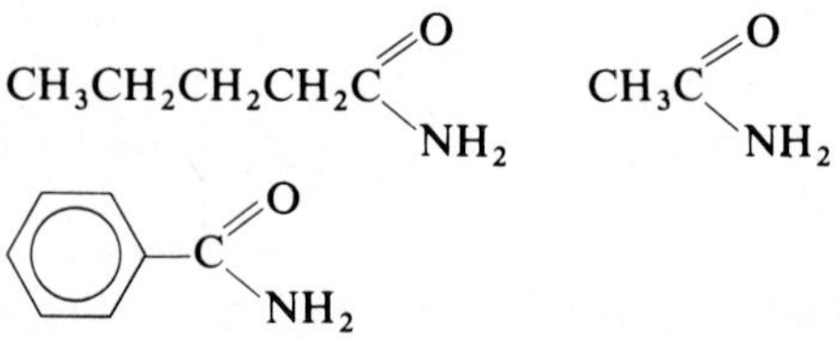 $CH_3CH_2CH_2CH_2C\overset{O}{\underset{NH_2}{\big<}}$ $CH_3C\overset{O}{\underset{NH_2}{\big<}}$

$\bigcirc-C\overset{O}{\underset{NH_2}{\big<}}$

(c) $CH_3CH_2NH_2$ 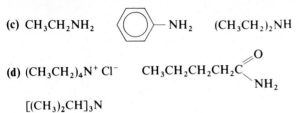 NH_2 $(CH_3CH_2)_2NH$

(d) $(CH_3CH_2)_4N^+\ Cl^-$ $CH_3CH_2CH_2CH_2C\overset{\displaystyle O}{\underset{\displaystyle NH_2}{\diagup\diagdown}}$

$[(CH_3)_2CH]_3N$

9.12 Arrange the compounds in each set in order of decreasing boiling point.

(a) $CH_3CH_2CH_2C\overset{\displaystyle O}{\underset{\displaystyle NH_2}{\diagup\diagdown}}$ $CH_3C\overset{\displaystyle O}{\underset{\displaystyle NH_2}{\diagup\diagdown}}$

$CH_3-C\overset{\displaystyle O}{\underset{\displaystyle N(CH_3)_2}{\diagup\diagdown}}$

(b) $CH_3C\overset{\displaystyle O}{\underset{\displaystyle NHCH_3}{\diagup\diagdown}}$ $H_2NCH_2CH_2NH_2$

$CH_3CH_2CH_2NH_2$

(c) $CH_3CH_2CH_2NH_2$ $(CH_3)_3N$ $CH_3CH_2NHCH_3$

(d) NH_2 $\text{⬡}-C\overset{\displaystyle O}{\underset{\displaystyle NH_2}{\diagup\diagdown}}$

$CH_3CH_2CH_2CH_2NH_2$

9.13 N,N-Dimethylformamide is a very polar molecule that is soluble in water even though it has no nitrogen–hydrogen bonds. Explain.

9.14 Write an equation to represent what occurs when dimethylamine is added to water.

9.15 Suggest a reason why pyridine is soluble in water but benzene is not.

9.16 Arrange the compounds in each set in order of decreasing basicity.

(a) $CH_3CH_2C\overset{\displaystyle O}{\underset{\displaystyle NH_2}{\diagup\diagdown}}$ $CH_3CH_2CH_2NH_2$ $(CH_3)_3N$

(b) NH_3 $CH_3C\overset{\displaystyle O}{\underset{\displaystyle NH_2}{\diagup\diagdown}}$ $\text{⬡}-NH_2$

(c) $(CH_3)_2NH$ $(CH_3CH_2)_2NH$ $(CH_3CH_2)_3N$

(d) $\text{⬡}-NH_2$ ⬡(N) $\text{⬡}-\underset{\displaystyle H}{N}-\text{⬠}$

(e) $\text{⬠}\underset{\displaystyle H}{N}$ ⬡(N) $\text{⬠}\underset{\displaystyle H}{N}(N)$

(f) $CH_3\overset{\displaystyle Cl}{\underset{}{C}}HNH_2$ $CH_3CH_2NH_2$ $CH_3CH_2NH_3^+\ Cl^-$

(g) $\text{⬡}NH_2$ $\text{⬡}CH_2NH_2$ $\text{⬡}-C\overset{\displaystyle O}{\underset{\displaystyle NH_2}{\diagup\diagdown}}$

9.17 Amines are basic, amides are neutral, yet phthalimide is weakly acidic. Explain.

Phthalimide

9.18 Complete the following equations, giving only the organic products (if any).

(a) $\text{⬡}-C\overset{\displaystyle O}{\underset{\displaystyle Cl}{\diagup\diagdown}}$ + NH_3 ⟶

(b) $H-C\overset{\displaystyle O}{\underset{\displaystyle NHCH_3}{\diagup\diagdown}}$ + HCl + HOH ⟶

(c) $CH_3(CH_2)_5NH_2$ + HBr ⟶

(d) $CH_3C\overset{\displaystyle O}{\underset{\displaystyle N(CH_2CH_3)_2}{\diagup\diagdown}}$ + $NaOH$ ⟶

(e) $H_2NCH_2CH_2CH_2CH_2NH_2$ + H_2SO_4 ⟶

(f) CH_3NH_2 + $3\ CH_3CH_2Br$ ⟶

(g) $\text{⬡}-CH_2C\overset{\displaystyle O}{\underset{\displaystyle NH_2}{\diagup\diagdown}}$ $\xrightarrow[\text{ether}]{LiAlH_4}$

(h) $CH_3NHCH(CH_3)_2$ $\xrightarrow[\text{HCl}]{NaNO_2}$

(i) $H_2N-\text{⬡}-NO_2$ $\xrightarrow[\text{(2) NaOH}]{\text{(1) Sn/HCl}}$

(j) $CH_3\overset{\displaystyle O}{\underset{}{C}}-O-\overset{\displaystyle O}{\underset{}{C}}-CH_3$ + $(CH_3CH_2)_2NH$ ⟶

(k) $\text{⬠}-CN$ $\xrightarrow[\text{ether}]{LiAlH_4}$

(l) ⬡(N) + CH_3I ⟶

9.19 Define, explain, or given an example of each of the following terms.

(a) monosubstituted amide
(b) amide linkage
(c) nylon 6,6
(d) urea
(e) amino group
(f) cadaverine
(g) quaternary ammonium salt
(h) nitrosation
(i) nitrosoamine
(j) nitrogen heterocycle
(k) drug
(l) alkaloid
(m) analgesic
(n) sedative
(o) narcotic
(p) tolerance
(q) euphoria
(r) enkephalins
(s) endorphins
(t) barbiturates
(u) hypnotic
(v) pot
(w) amphetamines
(x) catecholamines
(y) physical dependence
(z) psychological dependence
(aa) hallucinogens
(bb) serotonin
(cc) teratogen
(dd) tranquilizers
(ee) benzodiazepine
(ff) designer drugs

9.20 Match each of the following with one of the numbers below.

(a) constitutes 10% of opium 15
(b) narcotic used in some cough syrups 4
(c) the most commonly abused narcotic
(d) used in the treatment of heroin addicts 9
(e) used in the treatment of epilepsy 7

(f) Quaalude 11
(g) THC 8
(h) Benzedrine 1
(i) speed 10
(j) obtained from the leaves of coca plant 3
(k) obtained from ergot 6
(l) obtained from the peyote cactus 17
(m) obtained from certain mushrooms 16
(n) angel dust 7
(o) a major tranquilizer 2
(p) a minor tranquilizer 18
(q) China white 2
(r) Ecstasy 13

1. amphetamine	10. methamphetamine
2. chlorpromazine	11. methaqualone
3. cocaine	12. α-methylfentanyl
4. codeine	13. MDMA
5. heroin	14. mescaline
6. LSD	15. morphine
7. Luminal	16. psilocybin
8. marijuana	17. phencyclidine
9. methadone	18. Valium

9.21 MPPP is often referred to as the "reverse ester" of meperidine. Examine both structures (Figures 9.3 and 9.11) and tell what is meant by this expression.

9.22 Dextromethorphan (Romilar) is a synthetic compound that has replaced codeine as a cough suppressant in many cough medicines. It is not an analgesic and it is not addictive, yet its antitussive potency is the equal of codeine. Structurally it is identical to codeine except there is no hydroxyl group and the heterocyclic oxygen is replaced by a CH_2 group. Draw the structure of dextromethorphan.

9.23 How much 0.15 M hydrochloric acid (in milliliters) is required to neutralize 0.25 g of n-butylamine?

9.24 The dissociation constant for a primary amine, RNH_2, is 1.0×10^{-6}. Calculate the hydroxide ion concentration and the pH of a 0.30 M RNH_2 solution.

9.25 The pK_b of pyridine at 25°C is 8.7. Calculate the pH of a 0.10 M solution of pyridine in water.

10

Stereoisomerism

Isomerism has been defined as the phenomenon whereby two or more *different* compounds are represented by *identical* molecular formulas. Isomeric molecules have different physical and chemical properties, and these differences are attributed to the existence of different structural formulas.

Two types of **structural isomers** have been mentioned—positional isomers and functional group isomers. *Positional isomers* result from the presence of an atom or a group of atoms at different positions on the carbon chain. These isomers were discussed in the chapters on alkanes, alkyl halides, and alcohols; examples are listed in Table 10.1.

TABLE 10.1 Examples of Positional Isomers

ALKANES	ALKYL HALIDES	ALCOHOLS
$CH_3CH_2CH_2CH_3$	CH_3CHCl_2	$CH_3CH_2CH_2OH$
n-Butane	1,1-Dichloroethane	1-Propanol
$CH_3CH(CH_3)_2$	CH_2ClCH_2Cl	$CH_3CHOHCH_3$
Isobutane	1,2-Dichloroethane	2-Propanol

TABLE 10.2 Examples of Functional Group Isomers

ALCOHOLS AND ETHERS

CH_3CH_2OH CH_3OCH_3

Ethanol   Dimethyl ether

ALDEHYDES AND KETONES

$$CH_3CH_2\overset{\displaystyle O}{\overset{\|}{C}}H \qquad CH_3\overset{\displaystyle O}{\overset{\|}{C}}CH_3$$

Propanal   Acetone

ACIDS AND ESTERS

$$CH_3\overset{\displaystyle O}{\overset{/}{\underset{\backslash}{C}}}{}_{OH} \qquad H{-}\overset{\displaystyle O}{\overset{/}{\underset{\backslash}{C}}}{}_{OCH_3}$$

Acetic acid   Methyl formate

Two molecules that have the same molecular formula but contain different functional groups are *functional group isomers*. Examples have been given for functional group isomers of alcohols and ethers, aldehydes and ketones, and acids and esters (Table 10.2).

The second major category of isomerism is **stereoisomerism,** or space isomerism. Unlike structural isomers, stereoisomers have the identical order of atoms and identical functional groups. They differ only with respect to the spatial arrangement of atoms or groups of atoms within the molecule. Therefore, **stereoisomers** are *isomers that have the same structural formulas, but differ in the arrangement of atoms in three-dimensional space.* They may differ in configuration, in conformation, or in both configuration and conformation.

The term **configuration** refers to the unique arrangement of atoms that characterizes a particular compound. *Stereoisomers that can be interconverted only by the breaking and remaking of one or more bonds* are called **configurational isomers.** *Stereoisomers that are easily interconverted by rotation about a bond* are called **conformational isomers** (recall footnote 6, page 29). In our discussion of stereoisomerism, we will refer exclusively to configurational isomers.

10.1 *Plane-Polarized Light*

Ordinary light may be described as exhibiting an electromagnetic wave motion. As a light wave moves along in one direction, electromagnetic vibrations occur perpendicular to that direction (Figure 10.1-a). If we ignore the electrical and magnetic components, we can simplify the situation and say that ordinary light consists of a multitude of light waves that can vibrate in all directions in the plane perpendicular to the direction of travel (Figure 10.1-b).

If a beam of light is treated in such a manner as to allow only those waves traveling in one plane to be transmitted, the ensuing light beam is said to be **plane-polarized** (that is, *vibrating in a single plane*—Figure 10.1-c). Ordinary light is converted to plane-polarized light by passing through certain materials such as Polaroid (a plastic substance that is used in sunglass lenses) or calcite (a particular crystalline form of calcium carbonate). Both materials transmit only those light waves that are vibrating in a single plane and deny passage to those waves vibrating in other planes.

FIGURE 10.1 (a) The waves in a beam of light vibrate in all directions in the plane perpendicular to the direction of propagation. (b) A cross-sectional view of a beam of monochromatic light vibrating in all planes. (c) A cross-sectional view of a beam of plane-polarized light.

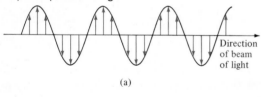

        (a)   (b)   (c)

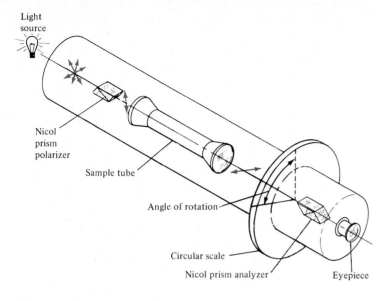

Light source

Nicol prism polarizer

Sample tube

Angle of rotation

Circular scale

Nicol prism analyzer

Eyepiece

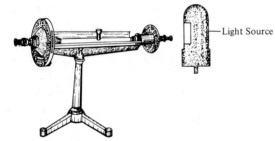

Light Source

FIGURE 10.2 A polarimeter.

A **polarimeter** is *an instrument that detects and measures the effects of various substances upon plane-polarized light.* A diagram of a polarimeter is given in Figure 10.2. The polarimeter consists of a monochromatic light source (that is, single-wavelength light generated from a sodium or mercury lamp), two Nicol prisms (made of calcite), a tube to contain the sample,[1] and an eyepiece or phototube. These are arranged so that the light passes through the first prism, the polarizer, and is converted to plane-polarized light. The polarized light then passes into the sample tube. If the sample rotates the polarized light a certain number of degrees either to the right or to the left, it is said to be **optically active.** When the light leaves the sample tube, it will still be vibrating in one plane, but the plane will be at a different angle from the original plane. The rotated light next strikes the second, or analyzer, prism. The light will not pass through the analyzer prism until the prism is rotated the same

[1]The sample may be a pure gas, a pure liquid, a pure crystalline solid, or a solute dissolved in an appropriate solvent.

number of degrees in the same direction as the rotated light. A round dial, marked off in degrees, is attached to the analyzer. This permits recording of the direction and number of degrees through which the analyzer must be rotated in order to align it properly with the polarized light. The *angle between the original and final planes of polarization* is known as the **optical rotation.**

The optical rotation is proportional to the number of optically active molecules in the path of the light beam. Therefore, it is proportional to the length of the sample tube and the amount of sample. Chemists have established a set of standard conditions to provide for comparison of all optically active substances no matter what the concentration or phase of the substance. The **specific rotation,** $[\alpha]$, is defined as *the amount of rotation caused by 1 g of substance per cubic centimeter in a sample tube 1 dm long.* The specific rotation is as important and just as characteristic a property of a compound as its melting point, boiling point, or density. For example, the specific rotation of an aqueous solution of sucrose is $[\alpha]_D^{20} = +66.5°$. Since specific rotation is affected by wavelength and temperature, these parameters should always be specified (as a subscript and superscript, respectively).

$$[\alpha]_D^t = \frac{\alpha}{(l)(c)}$$

where $[\alpha]$ = specific rotation
$\quad$ D = sodium line of spectrum (5893 Å)
$\quad$ t = temperature (°C)
$\quad$ α = observed rotation
$\quad$ l = length of tube in decimeters
$\quad$ c = concentration (g/mL) of a solution or density of a pure liquid

10.2 *Structure and Optical Activity*

Optical activity has been defined as *the ability of a substance to rotate the plane of polarized light.* Conversely, if a substance does not deflect plane-polarized light, it is optically inactive. If light is rotated to the right (clockwise), the substance is **dextrorotatory** (Latin, *dexter,* right); substances that rotate light to the left (counterclockwise) are **levorotatory** (Latin, *laevus,* left) (Figure 10.3). To denote the direction of rotation, a positive sign (+) is given to dextrorotatory substances and a negative sign (−) to levorotatory substances.[2] Sucrose is said to be dextrorotatory since it rotates plane-polarized light 66.5° in the clockwise direction, and it is designated as (+)-sucrose.

So far, the properties of optically active compounds have been described, and certain terms have been defined for use in dealing with them. Certain fundamental questions have yet to be answered.

1. Why do some compounds exhibit optical activity?
2. What are the spatial arrangements of these compounds?

[2] In the earlier chemical literature, the symbol *d* was used to designate dextrorotatory substances and *l* was used for levorotatory substances.

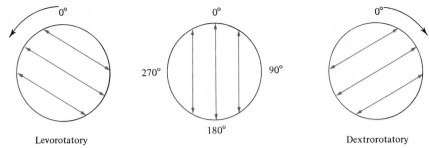

FIGURE 10.3 Direction of rotation of analyzer.

270° 90°

0° 0° 0°

180°

Levorotatory   Dextrorotatory

3. How do they differ from one another and from compounds that are not optically active?

The background to the discovery and explanation of optically active isomers is one of the most interesting sagas in the history of chemistry. The student is directed to a more comprehensive organic chemistry textbook or, for full enlightenment, to the original papers of Pasteur (1848), of Wislicenus (1873), and of van't Hoff and Le Bel (both 1874). The key to their explanation of optical activity in organic compounds was the *tetrahedral carbon atom.*

Recall that the configuration of the methane molecule is tetrahedral. If any or all hydrogen atoms are substituted by other atoms or groups of atoms, the tetrahedral arrangement about the central carbon atom is still retained (Figure 10.4). If

FIGURE 10.4 Tetrahedral configuration of methane and a trisubstituted compound.

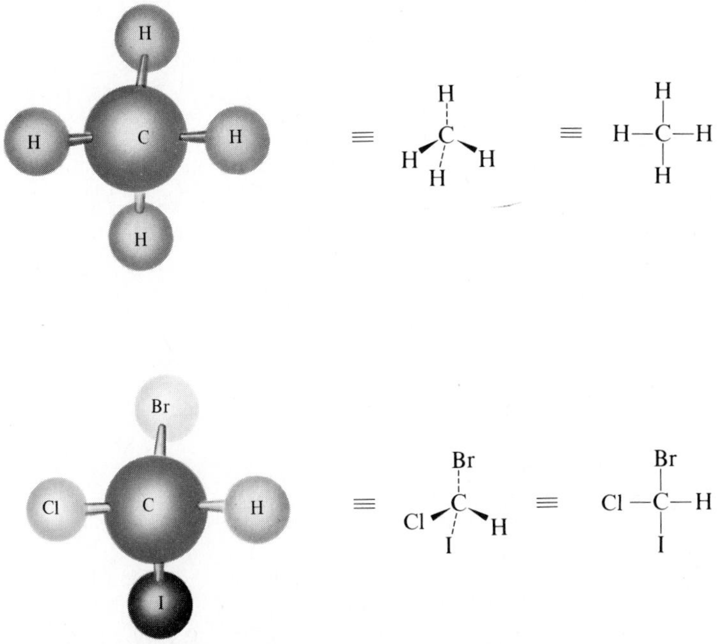

additional carbon atoms are added and if they all contain single bonds, the configuration about each of the carbon atoms will be tetrahedral.

Consider the two generalized molecules shown in Figure 10.5. Compound I contains two identical substituents and two different substituents bonded to the central carbon atom, and compound II contains four dissimilar substituents. Compound I is a symmetrical molecule; that is, a plane of symmetry passes through b, d, and the central carbon atom. Compound II is a nonsymmetrical molecule; there is no plane of symmetry that can be drawn through the molecule.

To understand better why compound I is classified as a symmetric molecule and compound II as a nonsymmetric molecule, we can place both compounds before a mirror and attempt to impose the mirror images upon the original molecules (Figure 10.6). A symmetric molecule is superimposable on its mirror image, whereas a nonsymmetric molecule cannot be superimposed upon its mirror image. The bonds may be twisted and turned, but none of them may be broken. A beginning student

FIGURE 10.5 Compound I contains two similar and two dissimilar substituents. Compound II contains four dissimilar substituents.

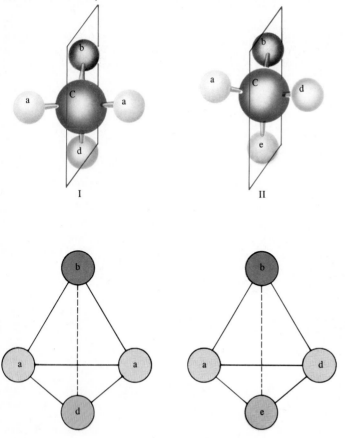

I II

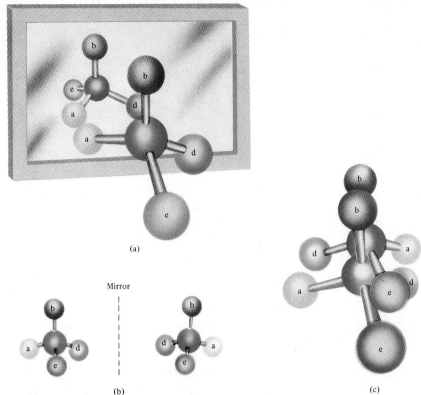

FIGURE 10.6 (a) and (b) Mirror images of a chiral molecule. (c) A chiral molecule cannot be superimposed on its mirror image.

who lacks experience in thinking in three-dimensional terms will fully understand symmetry concepts only after working with molecular models.

A useful analogy can be drawn between nonsuperimposable mirror-image compounds and your right and left hands (Figure 10.7). Regardless of how you twist and turn them, you cannot superimpose your right hand upon your left, or vice versa. This is so because, although your right and left hands are nearly perfect mirror images of each other, they are *not* identical. This difference becomes immediately apparent if you try to place your right hand in a left-hand glove. The general property of "handedness" is called **chirality** (Greek, *cheir*, hand). *An object that is not superimposable upon its mirror image* is said to be **chiral**.[3] An object that is superimposable upon (that is, identical to) its mirror image is **achiral.**

A **chiral carbon** is defined as *a carbon atom that is bonded to four different groups.* If a compound contains only one chiral atom, the molecule is always chiral. However, as we shall see in Section 10.4, sometimes a molecule may be achiral if it contains more than one chiral atom.

[3]The word *chiral* (pronounced KYE-ral) is now widely accepted and has displaced the earlier terms dissymmetric and asymmetric.

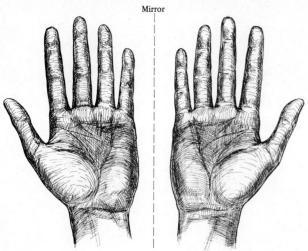

FIGURE 10.7 The left and right hands are nonsuperimposable mirror images.

Molecules that are nonsuperimposable (nonidentical) mirror images are called **enantiomers** (Greek, *enantios,* opposite). Enantiomers have identical physical and chemical properties, and they rotate plane-polarized light the same number of degrees—*but in opposite directions.* For example, two forms of lactic acid, $CH_3CHOHCOOH$, are known to exist in nature. One form, isolated from muscle tissue, is identical in all respects to another form that is isolated from yeast, except that the former is dextrorotatory ($[\alpha]_D^{20} = +2.3°$) and the latter is levorotatory ($[\alpha]_D^{20} = -2.3°$). The second carbon atom is the only one that is bonded to four different groups, and it is therefore the one that is responsible for the chirality of the molecule.

It is preferable to draw perspective formulas when representing enantiomers since these formulas convey the three-dimensionality of the molecules. This is especially important because the very nature of stereoisomerism depends upon the three-dimensional spatial arrangement of the atoms in a molecule. However, when we deal with most organic and biochemical molecules, writing perspective formulas is a cumbersome task. A convention was therefore established by Emil Fischer (see page 244) to represent enantiomers meaningfully by the use of their structural formulas. It was arbitrarily decided that substituents written at either side of the chiral carbon project forward toward the reader, whereas those positioned above and below the carbon extend backward into the page. The chiral carbon itself lies in the plane of the paper. The enantiomers of lactic acid are illustrated in Figure 10.8.

Thus far we have mentioned those compounds that contain only one chiral carbon atom. For such compounds there always exist two isomers that are nonsuperimposable mirror images (enantiomers), a dextrorotatory form and a levorotatory form. A convenient method of drawing the enantiomer of an optically active compound is to maintain the position of two of the substituents (usually the larger ones) about the chiral center and invert the positions of the other two.

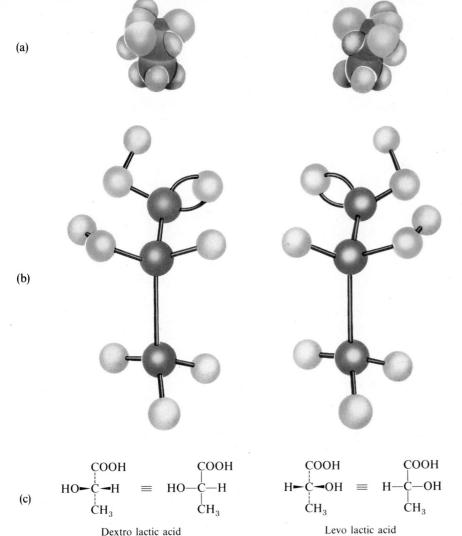

(a)

(b)

(c)

$$\underset{\substack{|\\ \mathrm{CH_3}}}{\overset{\substack{\mathrm{COOH}\\|}}{\mathrm{HO{-}C{\rightarrow}H}}} \equiv \underset{\substack{|\\ \mathrm{CH_3}}}{\overset{\substack{\mathrm{COOH}\\|}}{\mathrm{HO{-}C{-}H}}}$$

Dextro lactic acid

$$\underset{\substack{|\\ \mathrm{CH_3}}}{\overset{\substack{\mathrm{COOH}\\|}}{\mathrm{H{\leftarrow}C{-}OH}}} \equiv \underset{\substack{|\\ \mathrm{CH_3}}}{\overset{\substack{\mathrm{COOH}\\|}}{\mathrm{H{-}C{-}OH}}}$$

Levo lactic acid

FIGURE 10.8 Enantiomers of lactic acid. (a) Space-filling models; (b) ball-and-stick models; (c) structural formulas.

Example 10.1 2-Methyl-1-butanol exists in two optically active forms. The specific rotation of one is +5.756° and the other −5.756°. Draw structural formulas of the enantiomers.

Solution: The structural formula for one enantiomer is written and the chiral center is identified. (To allow for valid comparisons, the convention is to draw the carbon chain vertically.) The other enantiomer is then generated by interchanging the positions of two of the groups about the chiral center while maintaining the positions of the other two groups.

$$CH_3$$
$$|$$
$$CH_2$$
$$|$$
$$H-C-CH_3$$
$$|$$
$$CH_2OH$$

I

$$CH_3$$
$$|$$
$$CH_2$$
$$|$$
$$H_3C-C-H$$
$$|$$
$$CH_2OH$$

II

It is not possible to tell which isomer is dextrorotatory and which is levorotatory merely by inspection of the structural formulas. The distinction can only be made by measuring the optical rotation of each compound in a polarimeter.

A word of caution is required with this shorthand method of representing the three-dimensional structures of an enantiomeric pair of isomers. The unwary student must not be misled into thinking that he can superimpose the two formulas (I and II) simply by flipping one over on top of the other. Identical two-dimensional formulas would then result. But molecules are three-dimensional; the substituents positioned above and below the flipped molecule would project *forward*, and those positioned on either side would project to the *rear*, thus the molecules would *not* coincide. When molecules are compared to see whether or not they are enantiomers,

FIGURE 10.9 Enantiomers are nonsuperimposable mirror images.

"Mirror, mirror on the wall, who is the enantiomerest of them all?"

convention dictates that two-dimensional structural formulas may only be rotated 180° *in the plane* of the paper; they may never be flipped over or lifted out of the plane (Figure 10.9). Convince yourself of this by working with molecular models!

Table 10.3 lists some optically active and inactive compounds from several classes of organic compounds. A worthwhile exercise is the drawing of the enantiomers for each of the optically active compounds listed.

10.3 *Diastereomers*

Molecules that contain two or more chiral carbons can exist in more than two stereoisomeric forms. The first chemist to realize this was van't Hoff (first Nobel Prize in Chemistry, 1901), who formulated a statement (**van't Hoff's rule**) allowing the prediction of the total number of possible stereoisomers for a molecule with more than one chiral center. *The maximum number of different configurations is 2^n, where n is the number of chiral carbon atoms.* The rule is best illustrated with a specific example.

Example 10.2 Draw all the possible stereoisomers of 2-methyl-1,3-butanediol.

Solution
1. The structural formula is first drawn, and the number of chiral carbon atoms is noted. By use of the formula 2^n, the number of possible stereoisomers is calculated.

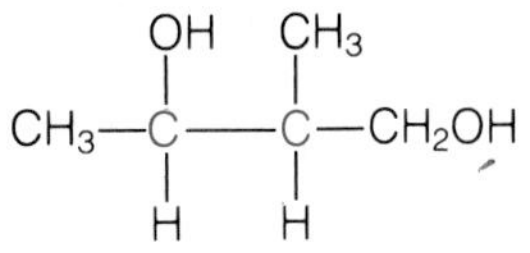

$$CH_3-\underset{\underset{H}{|}}{\overset{\overset{OH}{|}}{C}}-\underset{\underset{H}{|}}{\overset{\overset{CH_3}{|}}{C}}-CH_2OH$$

There are two chiral carbons and therefore there are 2^2 or 4 possible stereoisomers.

2. Since two configurations are possible for each chiral carbon, there will be two sets of mirror images.

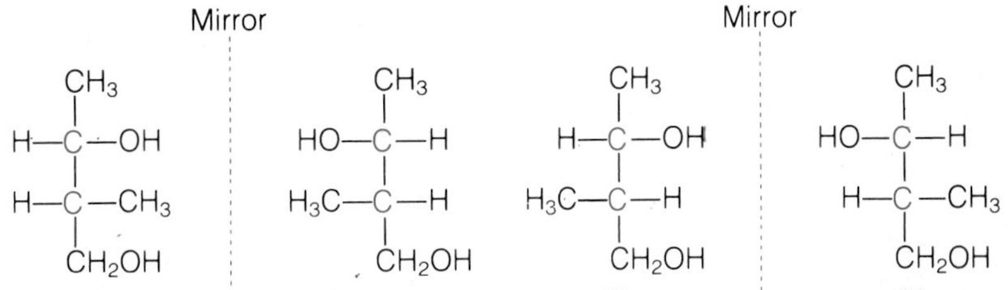

Structures I and II, and III and IV represent pairs of enantiomers[4] since they are nonsuperimposable mirror images. They have identical melting points, boiling

[4]The number of enantiomeric pairs of stereoisomers is always equal to $2^n/2$.

TABLE 10.3 Optically Active and Inactive Organic Compounds

Family	Optically Active	Optically Inactive
Alkane	$CH_3-\overset{\displaystyle H}{\underset{\displaystyle CH_2CH_3}{C}}-CH_2CH_2CH_3$	$CH_3-\overset{\displaystyle H}{\underset{\displaystyle CH_3}{C}}-CH_2CH_2CH_2CH_3$
Alkyl halide	$CH_3-\overset{\displaystyle H}{\underset{\displaystyle Cl}{C}}-CH_2CH_3$	$H-\overset{\displaystyle H}{\underset{\displaystyle Cl}{C}}-CH_2CH_2CH_3$
Alcohol	$CH_3CH_2-\overset{\displaystyle OH}{\underset{\displaystyle CH_3}{C}}-H$	$CH_3CH_2CH_2-\overset{\displaystyle H}{\underset{\displaystyle H}{C}}-OH$
Ether	$CH_3CH_2-\overset{\displaystyle H}{\underset{\displaystyle CH_3}{C}}-OCH_3$	$CH_3CH_2O-\overset{\displaystyle CH_3}{\underset{\displaystyle H}{C}}-CH_3$
Aldehyde	$CH_3CH_2-\overset{\displaystyle CH_3}{\underset{\displaystyle H}{C}}-C\overset{\displaystyle O}{\underset{\displaystyle H}{\big\backslash\!\!/}}$	$CH_3-\overset{\displaystyle CH_3}{\underset{\displaystyle CH_3}{C}}-C\overset{\displaystyle O}{\underset{\displaystyle H}{\big\backslash\!\!/}}$
Ketone	$CH_3\overset{\displaystyle O}{C}-\overset{\displaystyle H}{\underset{\displaystyle CH_3}{C}}-CH_2CH_3$	$CH_3\overset{\displaystyle O}{C}-\overset{\displaystyle H}{\underset{\displaystyle H}{C}}-CH_2CH_2CH_3$
Amine	$CH_3CH_2-\overset{\displaystyle H}{\underset{\displaystyle CH_3}{C}}-NH_2$	$CH_3CH_2CH_2-\overset{\displaystyle H}{\underset{\displaystyle H}{C}}-NH_2$
Hydroxy acid	$CH_3-\overset{\displaystyle H}{\underset{\displaystyle OH}{C}}-CH_2C\overset{\displaystyle O}{\underset{\displaystyle OH}{\big\backslash\!\!/}}$	$H-\overset{\displaystyle H}{\underset{\displaystyle OH}{C}}-CH_2CH_2C\overset{\displaystyle O}{\underset{\displaystyle OH}{\big\backslash\!\!/}}$
Amino acid	$CH_3-\overset{\displaystyle H}{\underset{\displaystyle NH_2}{C}}-C\overset{\displaystyle O}{\underset{\displaystyle OH}{\big\backslash\!\!/}}$	$H-\overset{\displaystyle H}{\underset{\displaystyle NH_2}{C}}-CH_2C\overset{\displaystyle O}{\underset{\displaystyle OH}{\big\backslash\!\!/}}$
Ester	$CH_3CH_2-\overset{\displaystyle H}{\underset{\displaystyle CH_3}{C}}-C\overset{\displaystyle O}{\underset{\displaystyle OCH_3}{\big\backslash\!\!/}}$	$CH_3-\overset{\displaystyle H}{\underset{\displaystyle CH_3}{C}}-C\overset{\displaystyle O}{\underset{\displaystyle OCH_2CH_3}{\big\backslash\!\!/}}$
Amide	$C_6H_5-\overset{\displaystyle CH_3}{\underset{\displaystyle H}{C}}-C\overset{\displaystyle O}{\underset{\displaystyle NH_2}{\big\backslash\!\!/}}$	$CH_3-C_6H_4-\overset{\displaystyle H}{\underset{\displaystyle H}{C}}-C\overset{\displaystyle O}{\underset{\displaystyle NH_2}{\big\backslash\!\!/}}$

points, densities, solubilities, and so on. They rotate plane-polarized light to the same extent, but in opposite directions. This is not true of structures I and III, I and IV, II and III, or II and IV. They are all pairs of stereoisomers, but they are not mirror images. Such pairs of stereoisomers are diastereomers. **Diastereomers,** then, are *stereoisomers that are not enantiomers.* Diastereomers do not have identical physical properties, nor do they rotate plane-polarized light to the same extent. They will have the same type of chemical properties, but the rate at which they react may be different.

10.4   *Racemic Mixtures and Meso Compounds*

We have mentioned that enantiomers rotate plane-polarized light the same number of degrees in different directions. One would predict, therefore, that a mixture containing equal amounts of dextrorotatory and levorotatory isomers would be optically inactive since the rotation caused by the molecules of one form would be exactly canceled by the rotation caused by the molecules of the other form. Experimental data bear out this prediction. A mixture containing equal amounts of a pair of enantiomers is called a **racemic mixture** or a **racemate;** it is symbolized by the notation ($\pm$). There are just as many molecules rotating the plane-polarized light to the right as to the left; hence the net rotation observed in the polarimeter is zero.

Another phenomenon of stereoisomerism results when a molecule contains more than one *similar* chiral carbon atom. Similar chiral carbon atoms are those in which the four unlike groups bonded to one chiral carbon atom are identical to the four bonded to the other chiral carbon atom. Tartaric acid, HOOC—CHOH—CHOH—COOH, is the classic example of this phenomenon. The molecule contains two chiral carbon atoms, and four stereoisomers might be predicted.

$$
\begin{array}{cccc}
\text{COOH} & \text{COOH} & \text{COOH} & \text{COOH} \\
\text{H—C—OH} & \text{HO—C—H} & \text{H—C—OH} & \text{HO—C—H} \\
\text{HO—C—H} & \text{H—C—OH} & \text{H—C—OH} & \text{HO—C—H} \\
\text{COOH} & \text{COOH} & \text{COOH} & \text{COOH} \\
\text{I} & \text{II} & \text{III} & \text{IV} \\
\end{array}
$$

Enantiomers		Identical	

Structures I and II are indeed nonsuperimposable mirror images, and thus are both optically active. However, although structures III and IV are mirror images of each other, they are *superimposable* mirror images. This can be seen easily by rotating either structure 180° in the plane of the paper. Both structures are identical, and this one compound is an optically inactive diastereomer of both compound I and compound II. Furthermore, observe that if the bond between carbon-2 and carbon-3 of compounds III and IV is bisected by a mirror, the top half of the molecule is the mirror image of the bottom half. No similar plane of symmetry can be drawn

TABLE 10.4 Physical Properties of Tartaric Acid Isomers

	MP (°C)	Density (g/cm³)	Solubility (g/100 g H₂O)	$[\alpha]_D^{20}$ water
(+)-Tartaric acid (dextro)	170	1.76	139	$+15°$
(−)-Tartaric acid (levo)	170	1.76	139	$-15°$
meso-Tartaric acid	147	1.67	125	0

through structures I and II.

```
        COOH
         |
   HO—C—H
- - - - - -|- - - - - - - Mirror
   HO—C—H
         |
        COOH
```

The existence of a plane of symmetry (or a center of symmetry) indicates that a molecule is achiral and cannot exist in an enantiomeric form. The achiral structure is referred to as *meso*-tartaric acid.[5] It is optically inactive because of *internal compensation.* Any effect that the upper half of the molecule will have upon plane-polarized light will be exactly opposed to the effect of the lower half. As expected, the physical and chemical properties of *meso*-tartaric acid are different from those of the optically active dextro and levo forms. Table 10.4 lists some properties of the isomeric tartaric acids.

10.5 *The R,S Convention for Absolute Configuration*

A problem arises in naming stereoisomers of compounds that contain more than one chiral carbon atom. That is, there is no evident correlation between the sign of optical rotation and the absolute configuration of the molecule. It therefore became necessary to devise a convention for the assignment of a specific designation to a given configuration. Such a system, referred to as the **R,S convention,** has been introduced by R. S. Cahn, C. Ingold, and V. Prelog. This system, which has been adopted by organic chemists, is not yet widely used in biochemistry, where the D- and L-notations still prevail (see Section 11.4).

[5] A *meso* compound is superimposable on its mirror image even though it contains chiral carbon atoms. For this reason it is incorrect to say that all molecules that contain chiral carbon atoms are chiral and thus optically active. Another example of a *meso* compound is ribitol, whose structural formula is

```
        CH₂OH
         |
    H—C—OH
         |
- - - H—C—OH - - - -Plane of symmetry
         |
    H—C—OH
         |
        CH₂OH
```

The two configurations for a chiral molecule are designated R (Latin, *rectus*, right) and S (Latin, *sinister*, left) and are governed by a set of rules that establish a priority sequence for the four substituents.

1. Assign an order of priority, a, b, c, or d, to the atoms bonded directly to the chiral carbon with the atom of *highest atomic number* getting the highest priority such that $a > b > c > d$ (that is, for the halogens the priority sequence is $I > Br > Cl > F$; other priorities are equally obvious, for example, $Cl > O > N > C$). H is the lowest priority.
2. Once priorities are assigned, visualize the orientation of the molecule in such a manner that it approximates the steering wheel of a car. The bond joining the chiral carbon atom to the group of lowest priority is considered to be the steering column. The other three groups are considered to be the spokes of the steering wheel. Trace a path from a to b to c. If the direction of this path is clockwise, then the molecule is assigned the R configuration. If the path follows a counterclockwise direction, the molecule is assigned the S configuration.

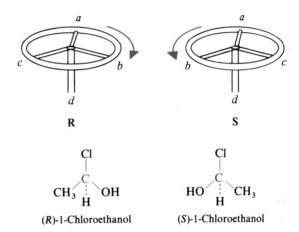

(R)-1-Chloroethanol   (S)-1-Chloroethanol

3. When two or more groups on the chiral carbon atom contain the same atoms, priority is given to the group with the highest atomic number in the second atom from the chiral carbon. For example, the compound 2-aminobutane contains two carbons bonded to the chiral carbon atom. However, the C of the $-CH_2CH_3$ group is bonded to a carbon, whereas the C of the $-CH_3$ group is bonded to a hydrogen. The priority order for 2-aminobutane is $NH_2 > CH_2CH_3 > CH_3$.

(R)-2-Aminobutane   (S)-2-Aminobutane

10.6   *Geometric Isomerism (Cis-Trans Isomerism)*

It has been mentioned that molecules having a cyclic structure or a carbon–carbon double bond have certain restrictions placed upon them. In Table 3.1 three isomers of butene, C_4H_8, were identified.

$$\underset{\substack{\text{1-Butene}\\ \text{I}}}{CH_3CH_2-\overset{\displaystyle H}{\overset{|}{C}}=\overset{\displaystyle H}{\overset{|}{C}}-H} \qquad \underset{\substack{\text{2-Butene}\\ \text{II}}}{CH_3-\overset{\displaystyle H}{\overset{|}{C}}=\overset{\displaystyle H}{\overset{|}{C}}-CH_3} \qquad \underset{\substack{\text{2-Methylpropene}\\ \text{III}}}{CH_3-\overset{\displaystyle H_3C}{\overset{|}{C}}=\overset{\displaystyle H}{\overset{|}{C}}-H}$$

Experimental evidence has shown, however, that there are four different butene molecules, all having distinctly different physical properties. The fourth isomer also has the structure that we have designated as 2-butene, and therefore these two isomeric 2-butenes, although they are structural isomers of compounds I and III, are not structural isomers of one another. Our knowledge of stereoisomerism leads us to the conjecture that these two molecules might differ in the spatial configuration of their atoms. Because they have different physical properties, however, they cannot be enantiomers; a different type of configurational explanation must be sought to account for this phenomenon. The explanation is based upon the geometric arrangement of the carbon–carbon double bond.

Recall that the two carbon atoms of a C=C double bond and the four atoms attached to them are all in the same plane and that rotation around the double bond is prevented. (This is in sharp contrast to the free rotation enjoyed by carbon atoms that are linked to one another by single bonds.) Construction of ball-and-stick models indicates that there are two possible ways to arrange the atoms of 2-butene that are in keeping with its structural formula (Figure 10.10). These three-dimensional ball and stick models are more simply represented as follows.

$$\underset{\substack{\text{IIa}\\ \textit{cis}\text{-2-Butene}\\ \text{mp } -139°C\\ \text{bp }\quad 4°C}}{\overset{\displaystyle H}{\underset{\displaystyle CH_3}{}}\!\!\!\diagdown\!\!\overset{}{\underset{}{C=C}}\!\!\diagup\!\!\!\overset{\displaystyle H}{\underset{\displaystyle CH_3}{}}} \qquad\qquad \underset{\substack{\text{IIb}\\ \textit{trans}\text{-2-Butene}\\ \text{mp } -106°C\\ \text{bp }\quad 1°C}}{\overset{\displaystyle CH_3}{\underset{\displaystyle H}{}}\!\!\!\diagdown\!\!\overset{}{\underset{}{C=C}}\!\!\diagup\!\!\!\overset{\displaystyle H}{\underset{\displaystyle CH_3}{}}}$$

In structure IIa, both methyl groups lie on the same side of the molecule, and this compound is the *cis* isomer (Latin, *cis*, on this side). The methyl groups of structure IIb are on opposite sides of the molecule; it is the *trans* isomer (Latin, *trans,* across). Because of the restriction on free rotation about the double bond, the structures are clearly nonsuperimposable and hence not identical. *cis*-2-Butene and *trans*-2-butene are geometric isomers of each other.

Geometric isomers *are compounds that have different configurations because of the presence of a rigid structure in the molecule.* Geometric isomers are diastereomers because they are stereoisomers that are not enantiomers. For alkenes there are *only*

228

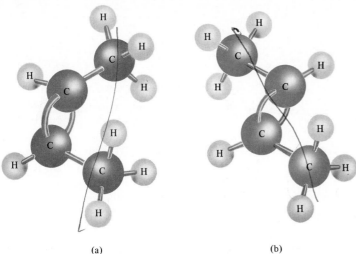

(a)   (b)

FIGURE 10.10   (a) *cis*-2-Butene. (b) *trans*-2-Butene.

two geometric isomers that correspond to each double bond (*cis* and *trans*). Because there are no geometric isomers of 1-butene or 2-methyl-1-propene, it is evident that the presence of a double bond is not the only criterion (nor is it a necessary one) for the occurrence of geometric isomerism. Geometric isomerism is not possible when there are identical substituents on either of the double-bonded carbons.

$$\underset{H}{\overset{H}{>}}C=C\underset{CH_2CH_3}{\overset{H}{<}} \quad \text{is identical to} \quad \underset{H}{\overset{H}{>}}C=C\underset{H}{\overset{CH_2CH_3}{<}}$$

$$\underset{CH_3}{\overset{CH_3}{>}}C=C\underset{H}{\overset{H}{<}} \quad \text{is identical to} \quad \underset{H}{\overset{H}{>}}C=C\underset{CH_3}{\overset{CH_3}{<}}$$

In general, geometric isomerism occurs in the alkenes when each carbon atom of a double bond bears two different substituents. When two identical (or nearly identical) substituents are on the same side of the double bond, the compound is the *cis* isomer; the *trans* isomer is the one in which similar groups are on opposite sides of the double bond. Following are some examples of geometric isomers.

$$\underset{H}{\overset{Cl}{>}}C=C\underset{H}{\overset{Cl}{<}}$$
cis-1,2-Dichloroethene
mp $-80°C$
bp   60°C

$$\underset{H}{\overset{Cl}{>}}C=C\underset{Cl}{\overset{H}{<}}$$
trans-1,2-Dichloroethene
mp $-50°C$
bp   48°C

$$\underset{CH_3}{\overset{H}{>}}C=C\underset{CH_3}{\overset{CH_2CH_3}{<}}$$
cis-3-Methyl-2-pentene

$$\underset{H}{\overset{CH_3}{>}}C=C\underset{CH_3}{\overset{CH_2CH_3}{<}}$$
trans-3-Methyl-2-pentene

The *cis-trans* system for naming geometric isomers often leads to confusion. For example, is the isomer of 2-bromo-1-iodopropene *cis* or *trans*?

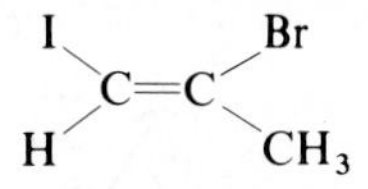

2-Bromo-1-iodopropene

To resolve this kind of ambiguity the **E,Z nomenclature** was devised. In this system the two atoms or groups of atoms bonded to each carbon of the double bond are assigned priorities as is done in naming enantiomers by the *R,S* convention (Section 10.5). The group of higher priority on one carbon is then compared with the group of higher priority on the other carbon. If the two groups of higher priority are on the same side of the molecule, the compound is the *Z* isomer (from the German *zusammen*, together). If the two groups of higher priority are on opposite sides of the molecule, the compound is the *E* isomer (from the German *entgegen*, in opposition to).

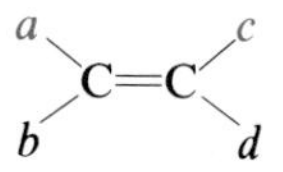

Z isomer   E isomer

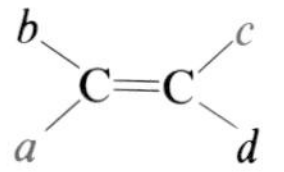

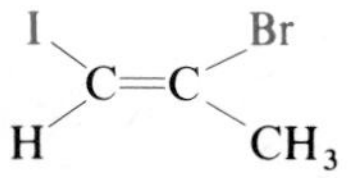

(Z)-2-Bromo-1-iodopropene 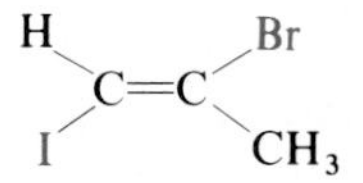   (E)-2-Bromo-1-iodopropene 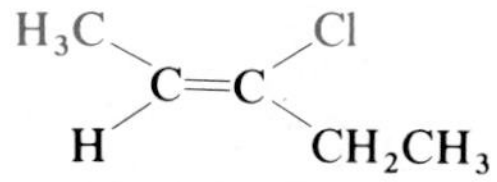   (Z)-3-Chloro-2-pentene 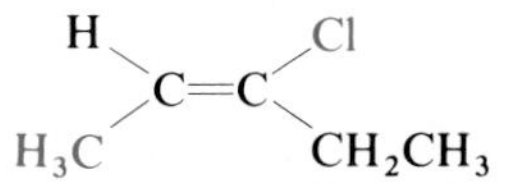   (E)-3-Chloro-2-pentene

Maleic and fumaric acids are classic examples of geometric isomers that have widely different chemical and physical properties (Figure 10.11). Because of the proximity of its carboxyl groups, maleic acid readily loses water to form an anhydride upon gentle heating.

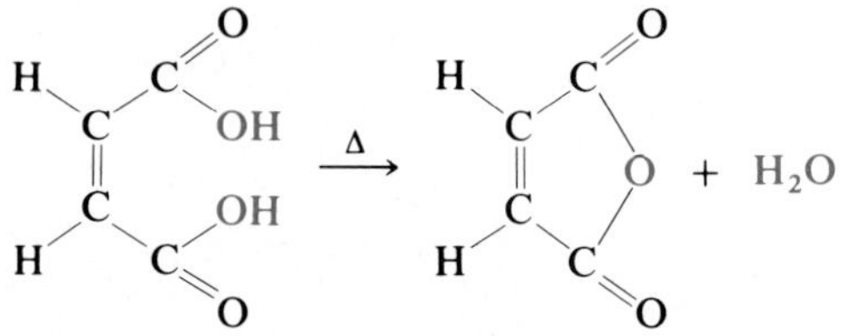

Maleic acid   Maleic anhydride

Fumaric acid is incapable of anhydride formation under the same reaction conditions. If it is heated to high temperatures ($\sim 300°C$), fumaric acid rearranges to form maleic acid, which then loses water to form the anhydride.

The nature of the bonding in the cycloalkanes also imposes geometric restraints on the substituents bonded to the ring carbon atoms. Cyclopropane is a planar molecule, whereas cyclobutane and cyclopentane have slightly puckered rings. Recall that rings of six or more carbons are decidedly nonplanar (Section 2.8). Common to all ring structures, however, is the inability of substituents to rotate

230

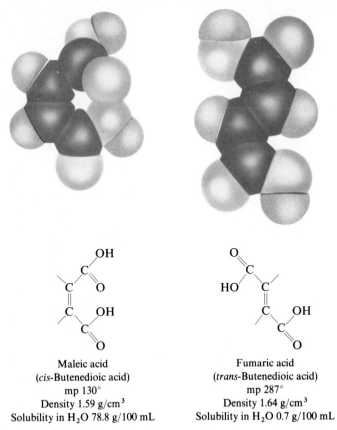

Maleic acid
(*cis*-Butenedioic acid)
mp 130°
Density 1.59 g/cm³
Solubility in H₂O 78.8 g/100 mL

Fumaric acid
(*trans*-Butenedioic acid)
mp 287°
Density 1.64 g/cm³
Solubility in H₂O 0.7 g/100 mL

Figure 10.11 Space-filling models, structural formulas, and properties of maleic acid and fumaric acid.

about any of the ring carbon–carbon bonds. Therefore, substituents can either be on the same side of the ring (*cis*) or on opposite sides of the ring (*trans*). For our purposes here, we represent all cycloalkanes as planar structures, but we clearly indicate the positions of the substituents, either above or below the plane of the ring.

cis-1,2-Dibromocyclopropane

trans-1,2-Dimethylcyclobutane

cis-1-Chloro-3-iodocyclopentane

trans-4-Ethylcyclohexanol

10.7 *Biochemical Significance*

Molecular configurations are of the utmost importance in biochemistry. The example of the two enantiomers of lactic acid has already been cited. The dextrorotatory form is isolated from muscle tissue and the levorotatory isomer is found in yeast and some bacteria. Laboratory synthesis of lactic acid from either acetaldehyde or pyruvic acid produces a racemic mixture of ($\pm$)-lactic acid; it is impossible to synthesize chemically either of the chiral forms. This is invariably the case. Whenever achiral reagents are utilized in a chemical synthesis, the resultant products will always be achiral, even though one or more chiral centers may have been created in the new compounds. (A detailed explanation of this interesting phenomenon of organic synthesis may be found in a comprehensive text on organic chemistry.)

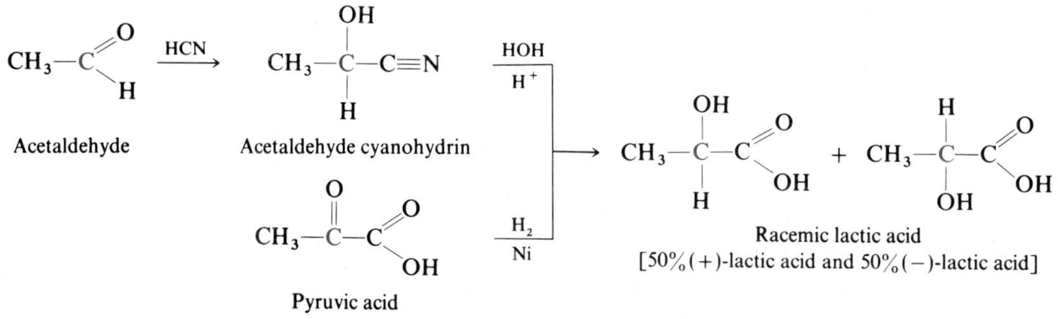

The obvious question, then, is how do muscle cells synthesize only ($+$)-lactic acid, and yeast only the ($-$)-isomer? The explanation here arises many times during the study of biochemistry—enzymatic control is the answer.

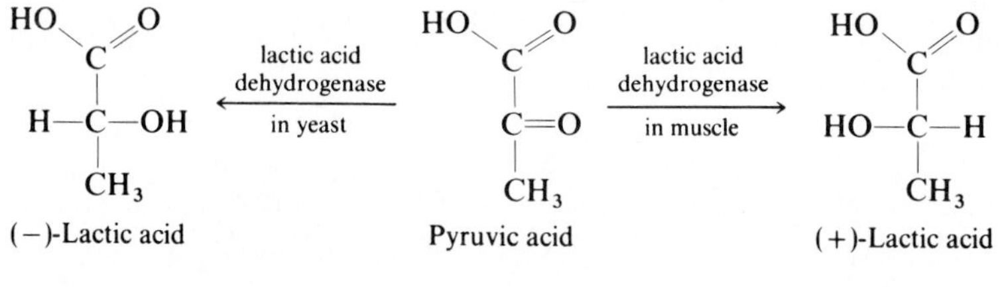

Enzymes are biological catalysts that are themselves chiral organic compounds. Similarly, almost every organic compound that occurs in living organisms is one enantiomer of a pair. Foods and medicines must have the proper molecular configurations if they are to be beneficial to the organism. For example, the popular meat-flavoring agent Accent is levorotatory monosodium glutamate.[6] The

---

[6]L-Monosodium glutamate (MSG) does not itself impart any taste, but rather it somehow enhances the flavor of foods that it combines with. Some people are allergic to MSG, and display symptoms such as dizziness, hot flashes, numbness, sweating, and swelling of the hands and feet. Since Chinese restaurants use large amounts of MSG, the occurrence of these symptoms has become known as *Chinese restaurant syndrome.*

dextrorotary form of this salt would not enhance the flavor of meat because our taste buds could not recognize it (Figure 10.12). Similarly, the natural form of epinephrine is levorotatory. It has a physiological activity about 15–20 times greater than that of its dextrorotatory enantiomer.

Because of the dangers of pesticides (Section 5.6), scientists have been searching for alternate methods to control insects. One method that has made a promising start involves the use of chemicals known as pheromones. **Pheromones** *are chemicals that are used for communication between members of the same species of insects.* Insects emit pheromones for a variety of purposes, such as sending an alarm, social regulation, attracting a mate, trail marking, and territorial marking.

The most important pheromone for insect control is the sex attractant. The females of many insect species depend upon an attractant to lure males for mating. (In a few insect species, including the boll weevil, the male emits the pheromone.) These chemicals are remarkably powerful; a few drops can attract males within a range of 2 miles. Traps baited with the sex attractant, and also containing an insecticide, can be used to lure all the males (of that species) to their deaths. One such compound is trimedlure, which has been found to be strongly attractive to the male Mediterranean fruit fly. Trimedlure has eight possible stereoisomers, and they differ considerably in attraction for the insect. The fly is most strongly drawn to the isomer in which the methyl and ester groups are *trans* to each other.

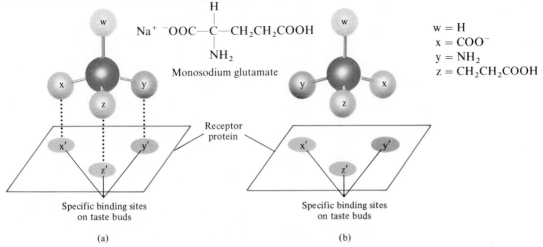

Methyl and ester group are *cis*.   Methyl and ester group are *trans*.

FIGURE 10.12 (a) The levorotatory stereoisomer of monosodium glutamate forms a precise fit with the bonding sites on the receptor protein of our taste buds. (b) The dextrorotatory isomer does not fit and therefore cannot bind to the receptor sites.

$$Na^+ \ {}^-OOC-\overset{\overset{\displaystyle H}{|}}{\underset{\underset{\displaystyle NH_2}{|}}{C}}-CH_2CH_2COOH$$

Monosodium glutamate

$w = H$
$x = COO^-$
$y = NH_2$
$z = CH_2CH_2COOH$

Receptor protein

Specific binding sites on taste buds

(a)

Specific binding sites on taste buds

(b)

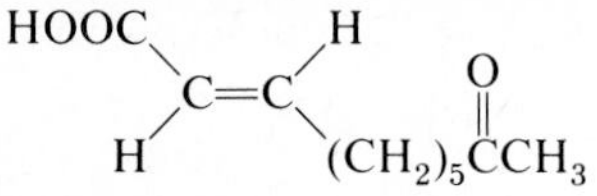

9-Oxo-*trans*-2-decenoic acid
(a)

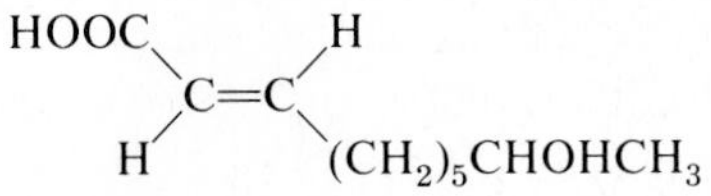

9-Hydroxy-*trans*-2-decenoic acid
(b)

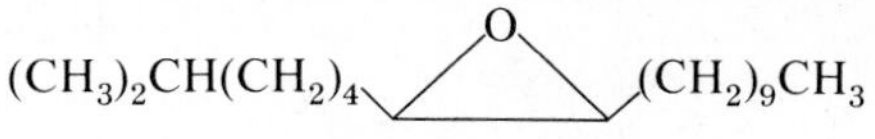

cis-7,8-Epoxy-2-methyloctadecane
(Disparlure)
(c)

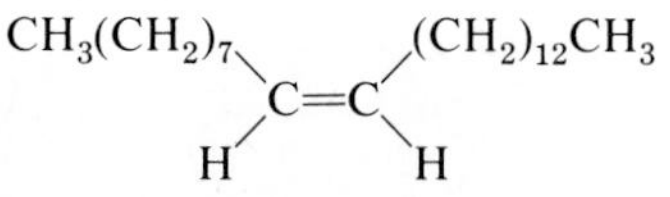

cis-9-Tricosene
(Muscalure)
(d)

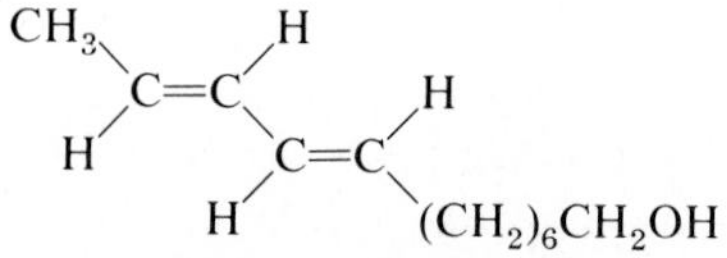

trans-8-*trans*-10-Dodecadien-1-ol
(e)

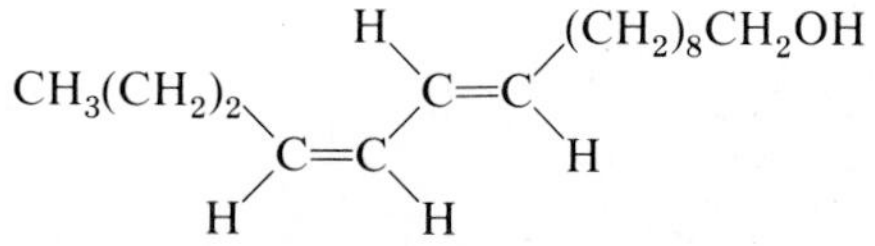

trans-10-*cis*-12-Hexadecadien-1-ol
(Bombykol)
(f)

FIGURE 10.13 Sex attractants of some female insects. (a) and (b) Queen honeybee, (c) gypsy moth, (d) common house fly, (e) codling moth, (f) silkworm moth.

Sex attractants of more than 30 insects have been identified. In most cases, just one of the possible stereoisomers is physiologically active (Figure 10.13).

The very subtle differences in structural configurations of organic molecules are of primary importance to life. We shall hold in abeyance an explanation of the stereoselectivity of enzymes until we first examine the compositions of the three major classes of biochemical compounds—the carbohydrates, the lipids, and the proteins. It is necessary to observe strictly the proper configurational formulas of these compounds. If enzymes can recognize such subtle differences of shape and structure, so must we.

10.8 *Molecules to See and to Smell*

When light strikes the retina of the eye, a complex series of reactions is initiated by a *cis–trans* isomerization. This reaction is termed a **photochemical isomerization** (photoisomerization) because the energy of light causes the geometric change to occur. The only function of light in vision is to alter the shape of the absorbing molecule, 11-*cis*-retinal, to the *trans* configuration (Figure 10.14). This primary event in vision occurs within a few picoseconds (psec, 10^{-12} sec). Proteins in the eye are altered by this single photochemical act, and some energy is released that triggers a nerve impulse (Figure 10.15). The impulse is transmitted via the optic nerve to the brain. Then an enzyme converts *trans*-retinal back to *cis*-retinal so it can bind to opsin to await the next exposure to light. Notice that *trans*-retinal is analogous to vitamin A in all respects except that it contains a carbonyl group, whereas the vitamin contains a primary alcohol group. Vitamin A is the reduced form of *trans*-retinal (or *trans*-retinal is the oxidized form of vitamin A), and the oxidation–reduction interconversions are essential to the chemical events of vision (Figure 10.16).

TABLE 10.5 Primary Odors

Primary Odor	Chemical Example	Familiar Substance
Camphoraceous	Camphor	Moth repellent
Musky	Pentadecanolactone	Angelica root oil
Floral	3-Methyl-1-phenyl-3-pentanol	Roses
Peppermint	Menthone	Mint candy
Ethereal	Ethylene dichloride	Drycleaning fluid
Pungent	Acetic acid	Vinegar
Putrid	Butyl mercaptan	Rotten eggs

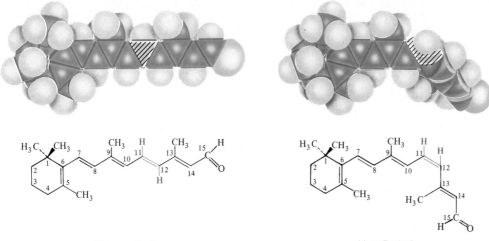

All-*trans* retinal   11-*cis*-Retinal

FIGURE 10.14 Fundamental molecule of vision is retinal ($C_{20}H_{28}O$), also known as retinene, which combines with proteins called opsins to form visual pigments. Because the nine-member carbon chain in retinal contains an alternating sequence of single and double bonds, it can assume a variety of bent forms. Two isomers of retinal are depicted here. In the space-filling models (*above*) carbon atoms are dark except carbon-11, which is shown hatched; hydrogen atoms are light. The large atom attached to carbon-15 is oxygen. When tightly bound to opsin, retinal is in the bent and twisted form known as 11-*cis*. When struck by light, it straightens out into the all-*trans* configuration. This simple photochemical event provides the basis for vision. [From "Molecular Isomers in Vision" by Ruth Hubbard and Allen Kropf. Copyright © 1967 by Scientific American, Inc. All rights reserved.]

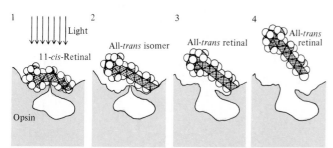

FIGURE 10.15 Molecular events in vision can be inferred from the known changes in the configuration of 11-*cis*-retinal after the absorption of light. In these schematic diagrams the twisted isomer is shown attached to its binding site on the protein molecule of opsin (1). After absorbing light the 11-*cis*-retinal straightens into the all-*trans* isomer (2). Presumably a change in the shape of opsin (3) facilitates the release of all-*trans* retinal (4). The configuration of the binding site in opsin is not yet known. [From "Molecular Isomers in Vision" by Ruth Hubbard and Allen Kropf. Copyright © 1967 by Scientific American, Inc. All rights reserved.]

Vitamin A

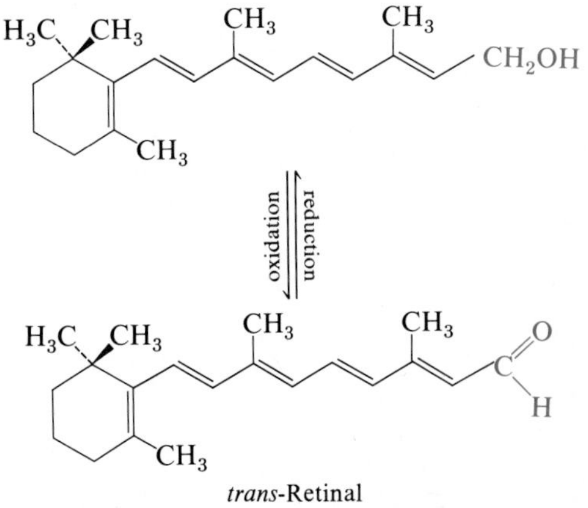

trans-Retinal

FIGURE 10.16 Vitamin A and *trans*-retinal are interconvertible via oxidation–reduction reactions.

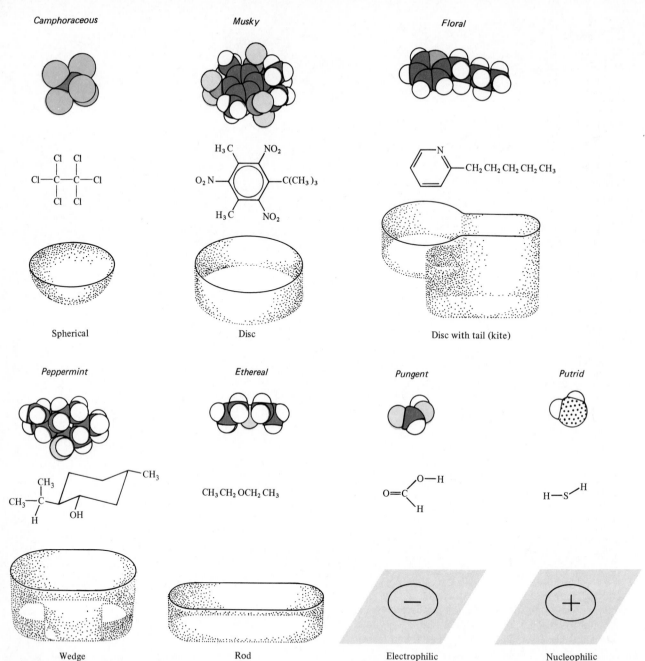

FIGURE 10.17 Olfactory receptor sites are shown for each of the primary odors together with molecules representative of each odor. The first five respectively are hexachloroethane, xylene musk, α-amylpyridine, (−)-menthol, and diethyl ether. Pungent (formic acid) and putrid (hydrogen sulfide) molecules fit on the receptor sites because of polarity rather than shape. [From ''The Stereochemical Theory of Odor'' by J. Amoore, J. Johnston, Jr., and M. Rubin. Copyright © 1964 by Scientific American, Inc. All rights reserved.]

The current theory of olfaction was first postulated in 1949 by R. W. Moncrieff. He proposed that the olfactory system is composed of a limited number of receptor cells, each one representing a distinct primary odor. Furthermore, all odorous molecules produce their effects by fitting closely on the receptor sites of these cells. (This hypothesis is essentially the same as the lock-and-key theory that we discuss in connection with enzymes in Section 14.5.) In essence, then, the major factor in determining the odor of a substance is the three-dimensional geometric shape of its molecules (Figures 10.17 and 10.18).

In 1952 John Amoore identified seven primary odors (Table 10.5). Every known odor can be made from these seven by mixing them in various proportions (in the same way that all the colors can be made from the three primary colors—red, yellow, and blue). In addition, Amoore described the size, shape, and chemical affinities of the seven different kinds of receptor sites that would recognize each of the primary odors. These receptor sites must necessarily have a distinctive shape, size, and chemical affinity so as to accept only those molecules with the correct geometric configuration or polarity. The shape, size, and polarity of the molecules representing the seven primary odors are illustrated in Figure 10.17. If the molecules of a substance can fit into more than one receptor site, then the brain will perceive a more complex signal and the resultant odor will be a combination of the primary sites occupied by the molecules. The

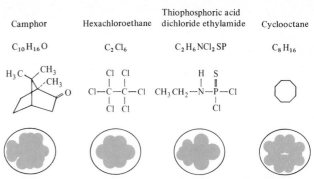

FIGURE 10.18 Unrelated chemicals with camphor-like odors show no resemblance in structural formulas. Yet, because the size and shape of their molecules are similar, they all fit the bowl-shaped receptor for camphoraceous molecules. [From "The Stereochemical Theory of Odor" by J. Amoore, J. Johnston, Jr., and M. Rubin. Copyright © 1964 by Scientific American, Inc. All rights reserved.]

olfactory receptors are similar to those of taste (the taste buds—Section 11.2). These two senses function together, and they aid in our food selection, since many foods are smelled at the same time that they are tasted.

Exercises

10.1 Define or explain the meaning of each of the following.
  (a) stereoisomers
  (b) configurational isomers
  (c) conformational isomers
  (d) polarimeter
  (e) plane-polarized light
  (f) optically active
  (g) optical rotation
  (h) specific rotation
  (i) levorotatory
  (j) (+) isomer
  (k) enantiomers
  (l) diastereomers
  (m) chiral molecule
  (n) racemic mixture
  (o) geometric isomers
  (p) meso compound
  (q) pheromones
  (r) photochemical isomerization

10.2 What requirement must be met if a compound is to exhibit optical activity?

10.3 Which of the heptane isomers are optically active? (Refer to Exercise 2.1.)

10.4 Draw all the enantiomers for compounds having the molecular formulas **(a)** C_4H_9Cl and **(b)** $C_4H_8Br_2$.

10.5 Write formulas for pairs of isomers that represent the following.
  (a) positional isomers
  (b) functional group isomers
  (c) conformational isomers
  (d) configurational isomers
  (e) enantiomers
  (f) diastereomers
  (g) noncyclic geometric isomers
  (h) cyclic geometric isomers

10.6 In what ways do enantiomers resemble each other? How do they differ?

10.7 How does a racemic mixture differ from a meso compound? In what ways are they similar?

10.8 How many chiral carbon atoms are there in each of the following compounds?
  (a) $CH_3CHOHCH(CH_3)CH_2CH_3$
  (b) $CH_3CH=CHCH_2CH=CHBr$
  (c) $CH_3CHClCH(CH_3)C\overset{\displaystyle O}{\underset{\displaystyle OCH(CH_3)CH_2CH_3}{\Big\backslash}}$

(d) $(CH_3)_2CHCHOHCH(CH_3)_2$

(e) $CH_3CH=CHCHBrCH=CH_2$

(f) $CH_3CH_2-N \overset{\displaystyle CH(CH_3)CH_2CH_3}{\underset{\displaystyle CH_3}{}}$

10.9 Which of these compounds contains a plane of symmetry?

(a) 2-propanol

(b) 2-butanol

(c) *cis*-1,2-dichlorocyclopentane

(d) *trans*-1,4-dichlorocyclohexane

(e) butyric acid

(f) 2,3-diiodobutyric acid

(g) *cis*-1,3-dibromocyclopentane

(h) *trans*-1,3-dibromocyclopentane

(i)

$$
\begin{array}{c}
CH_3 \\
H-C-OH \\
HO-C-H \\
H-C-OH \\
CH_3
\end{array}
$$

(j)

$$
\begin{array}{c}
H \quad O \\
C \\
H-C-OH \\
H-C-OH \\
CH_2OH
\end{array}
$$

10.10 Which of the following molecules are chiral? For those that are chiral, draw all of the configurational isomers.

(a) $CH_3CHBrCH_2CH_3$

(b) CH_3CH_2CHBrI

(c) ⬡$-CH_3$

(d) $CH_3CH=CHCH_2CH_3$

(e) $CH_3CH_2CH(NH_2)COOH$

(f) ⬡$-CHOHC \overset{\displaystyle O}{\underset{\displaystyle H}{}}$

(g) $(CH_3)_2CHCH_2CH_2OH$

(h) $CH_3CH_2\overset{\displaystyle O}{\overset{\displaystyle \|}{C}}CHClCH_3$

(i) ⬠$-CH_3$

(j) $CH_3CH=CHBr$

(k) ⬡$-CHBrCH_3$

(l) $CH_3CHICH_2CHBrCH_3$

(m) ◇$-CHOH-CHOH-$◇

(n) $CH_3CHBrCH_2CH_2CHBrCH_3$

(o) $CH_3CHOHCH(CH_3)CHOHCH_3$

(p) $CH_3CH_2CH(SH)CH_2CH_3$

10.11 Draw configurational formulas for each of the following. Indicate which are (1) optically active, (2) optically inactive, (3) enantiomers, (4) diastereomers, and (5) *meso* forms.

(a) 3-bromo-1-butene

(b) 1-phenyl-1-ethanol

(c) 1,2-diiodopentane

(d) 2-methyl-1-butanol

(e) 3-methyl-2-butanol

(f) 2,3-dichlorobutanedioic acid

(g) 3,4-dimethyl-3,4-hexanediol

(h) 2,4-dibromopentane

(i) 2,3-dichlorobutanoic acid

(j) 2,4-dibromo-3-iodopentane

(k) 3,4-dihydroxypentanal

(l) 3,4-dihydroxy-2-pentanone

10.12 Write formulas showing the configurations of all the possible isomers of ribitol. Indicate which of these structures are enantiomers, which diastereomers, and which *meso* forms.

10.13 Draw the configurational isomers for the remaining three stereoisomers of 3-phenyl-2-butanol. Given the following properties for the stereoisomer shown, predict as many properties as you can for the other three. Include a statement as to why you were not able to predict some of the properties.

$$
\begin{array}{c}
CH_3 \\
H-C-⬡ \\
HO-C-H \\
CH_3
\end{array}
$$

bp (25 mmHg) = 118°C

$[\alpha]_D^{25} = +30.9°$

10.14 Write the formulas of the geometric isomers for each of the following compounds. Label them *cis* and *trans* (or *Z* and *E*). If there are no geometric isomers, write None.

(a) 2-hydroxy-2-pentene

(b) 3-hexene

(c) 4-methoxy-2-pentyne

(d) 1,1-dibromo-1-butene

(e) 2-butenoic acid

(f) 4-methyl-2-pentene

(g) 2,3-dimethyl-2-pentene

(h) 1,1-dibromo-2-ethylcyclopropane

(i) 1,2-dibromocyclohexane

(j) 2-chlorocyclohexanol
(k) 1-bromo-3-chlorocyclobutane
(l) 1,2,3-trimethylcyclopropane

10.15 1,2-Dimethylcyclobutane exists as *cis* and *trans* isomers. One isomer is chiral, the other is achiral. Draw mirror images for both sets of geometric isomers, and identify which isomers are identical and which isomers are enantiomers. Are *cis*- and *trans*-1,2-dimethylcyclobutanes examples of diastereomers?

10.16 For 1,2-dibromocyclopropane, there are three stereoisomers. Draw structures for the three stereoisomers.

10.17 Draw the other geometric isomers for all of the pheromones given in Figure 10.13.

10.18 Urushiol is an unsaturated phenolic compound that is the active agent in poison ivy and poison sumac. Draw all of the geometric isomers for urushiol.

$$HO \quad OH$$

$$-(CH_2)_7CH=CHCH_2CH=CHCH_2CH_2CH_3$$

10.19 If the four bonds of carbon were directed toward the corners of a square, how many isomers of CH_2BrCl would exist? Draw them.

10.20 Draw and name all the geometric isomers (both noncyclic and cyclic) corresponding to the molecular formula C_5H_{10}.

10.21 Only one of the isomers of pentanal is optically active. What is its formula? Draw the two enantiomers of this compound and designate each as either R or S.

10.22 Give the formula of the smallest noncyclic alkane that could be optically active.

10.23 An alcohol has the molecular formula $C_4H_{10}O$.
(a) Write all the possible structural formulas for the alcohol.

(b) If the alcohol can be separated into two optically active forms, which of the formulas in (a) is correct?
(c) Assign an R or S designation to each of the enantiomers in (b).

10.24 Draw the enantiomers of each of the following compounds. Assign an R or S designation to each.
(a) 2-butanethiol
(b) 2-methoxypropanoic acid
(c) 2,3-dihydroxypropanal
(d) 2-bromobutanoic acid

10.25 The concentration of an optically active compound in water was 12 g/L. A 10 mL aliquot has an observed rotation of $+1.4°$ in a 10 cm polarimeter tube. Calculate the specific rotation of this compound.

10.26 2.00 g of an optically active compound, dissolved in 10.0 mL of ethanol, rotated a beam of plane-polarized light 25° to the left in a 1 dm polarimeter tube. What is its specific rotation?

10.27 Benzedrine is a racemic mixture of (+)- and (−)-amphetamines. The pure dextrorotatory enantiomer, Dexedrine, has a much greater physiological activity than either the racemic mixture or the pure levorotatory form. Draw the structural formulas for the enantiomers of amphetamine (see Figure 9.8).

10.28 Lysergic acid diethylamide (LSD) contains two chiral carbon atoms (see Figure 9.9). Identify which carbon atoms are chiral and then draw structural formulas of the four stereoisomers of LSD. (Only one of these four isomers, (+)-lysergic acid diethylamide, is physiologically active.)

10.29 Examine the labels of some common household products (detergents, foods, drugs, sprays, cosmetics). Write structural formulas for the chemical compounds contained in these products.

11

Carbohydrates

Carbohydrate literally means *hydrate of carbon*. This name is derived from the investigations of early chemists who found that when they heated sugars for a long period of time in an open test tube, they obtained a black residue (carbon), and droplets of water condensed on the sides of the tube. Furthermore, chemical analysis of sugars and other carbohydrates indicated that they contained only carbon, hydrogen, and oxygen, and many of them were found to have the general formula $C_x(H_2O)_y$.[1]

It is now known that some carbohydrates contain nitrogen and sulfur in addition to carbon, hydrogen, and oxygen. They definitely are not hydrated compounds, as are many inorganic salts (for example, $CuSO_4 \cdot 5H_2O$). Today the name **carbohydrate** is used to designate the *large class of compounds that are polyhydroxy aldehydes, polyhydroxy ketones, or substances that yield such compounds upon acid hydrolysis.*

Carbohydrates are the major constituents of most plants, comprising from 60 to

[1] Most carbohydrates do conform to the formula $C_x(H_2O)_y$. However, this formula is not unique to the carbohydrates, nor is it a necessary criterion for membership in the carbohydrate class. For example, formaldehyde (CH_2O), acetic acid ($C_2H_4O_2$), and lactic acid ($C_3H_6O_3$) are certainly not carbohydrates, but the compounds rhamnose ($C_6H_{12}O_5$) and deoxyribose ($C_5H_{10}O_4$) are carbohydrates.

90% of their dry mass (mostly as cellulose). In contrast, most animal tissue contains a comparatively small amount of carbohydrate (e.g., less than 1% in humans). Plants utilize carbohydrates both as a source of energy and as supporting tissue in the same manner that proteins are used by animals. Plants are able to synthesize their own carbohydrates from the carbon dioxide of the air and from water taken from the soil. Animals are incapable of this synthesis, and therefore are dependent upon the plant kingdom as a source of these vital compounds. Humans, the special animals, utilize carbohydrates not only for their food (about 60–65% by mass of the average diet) but also for their clothing (cotton, linen, rayon), shelter (wood), fuel (wood), and paper (wood).

11.1 Classification of Carbohydrates

Carbohydrates are classified according to their acid hydrolysis products. Three major categories are recognized.

1. The monosaccharides, or *simple sugars, cannot be broken down into smaller molecules by hydrolysis.*
2. The disaccharides *yield two monosaccharide molecules upon hydrolysis.*
3. The polysaccharides *yield many monosaccharide molecules upon hydrolysis.*

For the most part, the mono- and disaccharides are sweet-tasting, white, crystalline solids that are readily soluble in water and insoluble in nonpolar organic solvents. Polysaccharides are frequently tasteless, insoluble, amorphous compounds with exceedingly high molecular weights.

11.2 Sweeteners

A sweet taste is one of the four primary taste sensations that can be distinguished by the taste buds on the surface of the tongue. The others are sour, salty, and bitter. Although sweetness is commonly associated with most mono- and disaccharides, it is not a specific property of carbohydrates. Many sugars are sweet to varying degrees, but several organic compounds have been synthesized that are far superior as sweetening agents. Sucaryl sodium, for example, is about 30 times as sweet as sucrose, whereas saccharin is about 400 times as sweet as sucrose. These synthetic compounds have no caloric value, and therefore they are useful for those persons (e.g., diabetics) who must minimize their carbohydrate intake.

Sucaryl sodium and the analogous calcium salt, sucaryl calcium are referred to collectively as the **cyclamates** (i.e., salts of cyclamic acid). The cyclamates were first marketed in 1950 and their production soon became a multimillion dollar business. Then in 1969 experimental findings showed that large doses (50 times the daily maximum recommended for human consumption) of cyclamates induced cancer in rat bladders. Citing the Delaney Clause, the Food and Drug Administration (FDA) ordered all beverages and other food products containing cyclamates off the market. (The Delaney Clause, a 1958 amendment to the Food, Drug, and Cosmetic Act of 1938, requires removal from the market of any food additive that has been shown to cause cancer in laboratory animals.) It should be stated, however, that there was never any connection between cyclamates and cancer or any other disease or disorder in humans. Many subsequent studies completed since 1969 by the FDA, the National Cancer Institute, and researchers in other countries have indicated that cyclamates are not carcinogenic. In 1985 a committee of the National Academy of Sciences said that "cyclamates do not appear to cause cancer, but they may promote tumor development." The committee called for ad-

Sucaryl sodium

Saccharin

ditional animal testing to determine whether or not cyclamates can enhance the effects of carcinogenic chemicals.

During the period 1969–1977 **saccharin** was the only artificial sweetener permitted in the United States, although it too had been proclaimed by some to be carcinogenic. In 1977 the FDA disclosed that a three-year study carried out in Canada showed that rats fed a diet containing 5% saccharin developed bladder cancers. As a result, the FDA, again citing the Delaney Clause, proposed a ban on the sale of saccharin in the United States. This proposed ban was highly controversial. Opponents of the ban estimated that an individual would have to consume 800 12-ounce cans of diet soda each day for seven years to obtain the same amount of saccharin that was fed to the rats. Because of public (and diet food industry) opposition to the ban, Congress in November 1977 prohibited the FDA from acting against saccharin for 18 months and authorized a study on its safety by the National Academy of Sciences (NAS). One year later a committee of the NAS unequivocally affirmed the earlier FDA decision to ban saccharin. The panel concluded that (1) saccharin is a carcinogen in animals, although of low potency, (2) it is a potential human carcinogen, (3) the compound itself, not impurities from manufacturing, is responsible for carcinogenic activity, and (4) there is no evidence that saccharin offers any meaningful health benefits. (However, more recent epidemiological studies have found no link between human bladder cancer and saccharin.) The saccharin controversy has not been settled, and saccharin remains on the market. The original moratorium has been extended four times, with the research period now scheduled to run until May 1, 1987. Products containing saccharin are labeled with a warning about its effects upon laboratory animals. In 1985 the American Medical Association reported that "recent studies provide no evidence of increased risk of bladder cancer among users of artificial sweeteners, including saccharin, and because there is no ideal alternative sweetener, saccharin should continue to be available as a food additive. However, the AMA is not implying that it condones the use of saccharin." The AMA urged careful monitoring for possible adverse health effects in all users, as well as a continued search for an ideal sweetener.

In 1981, after eight years of extensive testing (and controversy regarding its safety), the FDA approved the use of another low calorie sweetener, **aspartame** (L-aspartyl-L-phenylalanine methyl ester—see Exercise 13.19). This white crystalline compound is about 180 times sweeter than sucrose and does not leave the bitter aftertaste often associated with saccharin. The FDA approved aspartame for use in more than 70 products, including soft drinks (Nutrasweet), cereals, gelatins, and chewing gum, and as tablets to be used as sugar substitutes. It is interesting that the two constituent amino acids are not sweet. L-Aspartic acid has a flat taste, whereas L-phenylalanine is bitter. Aspartame is used as a sweetener for a wide variety of foods because it can blend well with other food flavors. In the body, aspartame is hydrolyzed to aspartic acid, phenylalanine, and methanol. The small amount of methanol does not seem to be a problem, but the release of phenylalanine is a matter of concern to those on low phenylalanine diets (see Section 15.6). A report was released in the early 1980s that pregnant women who consume aspartame may have babies with permanent brain damage, and there was a later report of similar damage to infants who ingested it during the six months following birth. In 1985, the American Medical Association completed its investigations and concluded that aspartame is safe, and only people who are sensitive to phenylalanine need regulate their intake. (It should be noted that the shelf-life of aspartame is limited because at high temperatures the molecule breaks down and loses its sweetness.)

11.3 *Monosaccharides—General Terminology*

The general names for the monosaccharides are obtained in a manner analogous to the naming of organic compounds by the IUPAC system. The number of carbon atoms in the molecule is denoted by the appropriate prefix; the suffix -*ose* is the generic designation for any sugar. For example, the terms triose, tetrose, pentose, and hexose signify three-, four-, five-, and six-carbon monosaccharides, respectively. In addition, those *monosaccharides that contain an aldehyde group* are called **aldoses;** *those containing a ketone group* are **ketoses.** By combining these terms, both the type of carbonyl group and the number of carbon atoms in the molecule are easily expressed. Thus, monosaccharides are generally referred to as aldotetroses, aldopentoses, ketopentoses, ketoheptoses, etc. Glucose and fructose are specific examples of an aldose and a ketose.

Glucose
(An aldohexose)

Fructose
(A ketohexose)

11.4 Stereochemistry

The simplest monosaccharides are the trioses, glyceraldehyde and dihydroxy-acetone. Glyceraldehyde possesses a chiral carbon atom and thus may exist in two optically active forms. Dihydroxyacetone does not contain a chiral center. Figure 11.1 shows condensed ball-and-stick models of the two forms of glyceraldehyde and the structural formula of dihydroxyacetone. The two glyceraldehyde isomers are clearly mirror images. Except for the direction in which they rotate plane-polarized light, they have identical physical properties. One form has a specific rotation of $+8.7°$, the other a rotation of $-8.7°$. The great German chemist Emil Fischer (Nobel Prize, 1902) initiated the convention of projecting the formulas on a two-dimensional plane so that the aldehyde group is written at the top, with the hydrogen and hydroxyl written to the right and left. (Formulas of chiral molecules represented in this manner are referred to as Fischer projections, Fischer models, or Fischer configurations.) Arbitrarily, Fischer then decided that the formula of glyceraldehyde in which the hydroxyl group is positioned to the right of the chiral carbon atom represents the dextrorotatory isomer. He assigned the letter D as its prefix. The levorotatory isomer, in which the —OH group is positioned to the left of the chiral carbon atom, was accordingly assigned the letter L as its prefix.[2]

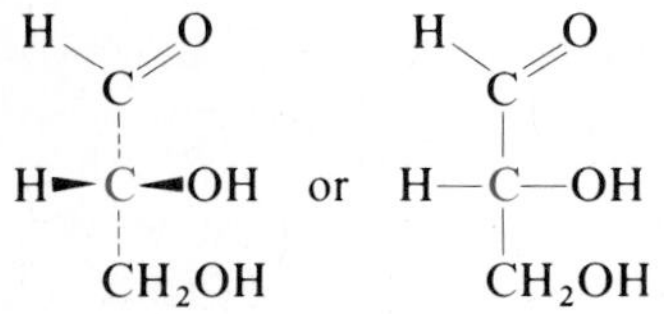

The two forms of glyceraldehyde are especially important because the more complex sugars may be considered to be derived from them. They serve as a reference point for designating and drawing all other monosaccharides. Sugars whose Fischer projections terminate in the same configuration as D-glyceraldehyde are designated as **D-sugars;** those derived from L-glyceraldehyde are designated as **L-sugars.** The convention is illustrated by a consideration of the following four stereoisomeric aldotetroses.

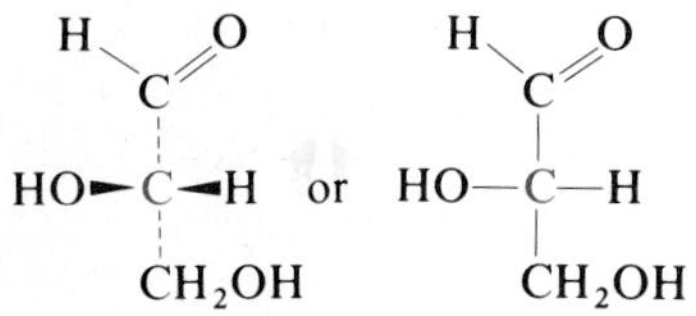

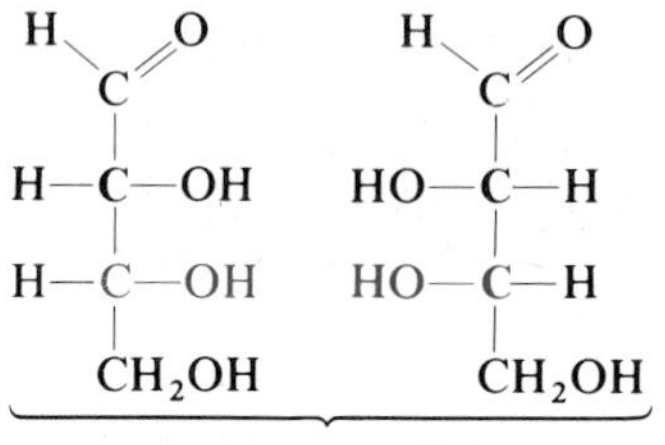

D-Glyceraldehyde

L-Glyceraldehyde

Mirror images

D-Erythrose   L-Erythrose

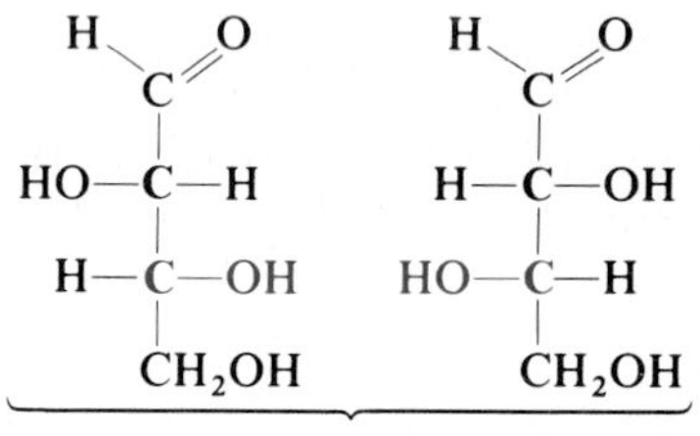

Mirror images

D-Threose   L-Threose

Notice that D- and L-erythrose and D- and L-threose are pairs of enantiomers. However, D-threose and L-threose are diastereomers of D-erythrose and L-erythrose. Thus, aside from having the same molecular formulas, the erythroses and the threoses are completely different sugars, having different chemical and physical properties.

[2]Fischer's arbitrary assignment proved to be correct. In 1951 chemists, with the aid of x-ray crystallography, determined the absolute configurations of the glyceraldehyde enantiomers and found that the *D*-isomer was indeed dextrorotatory.

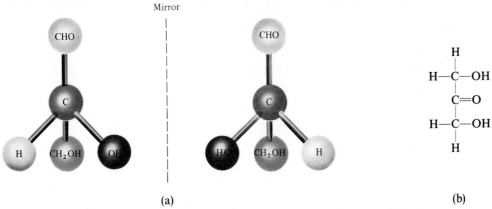

FIGURE 11.1 (a) Enantiomers of glyceraldehyde. (b) Structural formula of dihydroxyacetone.

The letters D and L are often very misleading for the beginning student. It must be emphasized that these prefixes serve only to signify the absolute configuration of a molecule. *A* D-**sugar** is *one that has the same configuration about the* **penultimate**[3] *carbon atom as does* D-*glyceraldehyde*. The letters do not in any way refer to the optical rotation of the molecule. D-Glyceraldehyde just happens to be dextrorotatory. It is not to be expected, however, that all compounds derived from it will have the same optical rotation, since such compounds will have additional chiral carbon atoms. In fact, D-erythrose and D-threose are both found to be levorotatory. The direction of rotation of plane-polarized light is a specific property of each optically active molecule. It is not at all dependent upon the configuration about the penultimate carbon. The symbols plus ($+$) and minus ($-$) are used to denote the optical rotation of the monosaccharides; ($+$) indicates a clockwise, or dextrorotatory, rotation, and ($-$) indicates a counterclockwise, or levorotatory, rotation. The correct designations for the isomers of glyceraldehyde, erythrose, and threose, that indicate both their configuration and their optical rotation, are given as follows: D-($+$)-glyceraldehyde, L-($-$)-glyceraldehyde, D-($-$)-erythrose, L-($+$)-erythrose, D-($-$)-threose, and L-($+$)-threose.

Figure 11.2 shows all the possible stereoisomeric aldopentoses and their complete names. Notice that the aldopentoses contain three chiral carbon atoms and that there are four enantiomeric pairs. Aldohexoses (Figure 11.3) contain four chiral carbon atoms, and thus there are 8 enantiomeric pairs, or 16 isomers. Fortunately for students of biochemistry, only 3 of the 16 isomers are commonly found in nature, D-($+$)-glucose, D-($+$)-mannose, and D-($+$)-galactose. (All 16 isomers have been prepared synthetically.) No one of these three stereoisomers is a mirror image of either of the others, so all three are diastereomers of one another. Furthermore, D-($+$)-glucose differs from D-($+$)-mannose and D-($+$)-galactose only in the

---

[3]The penultimate (next to last) carbon has been chosen, by convention, to be the reference carbon atom. It is the chiral carbon atom that is farthest from the aldehyde or ketone group.

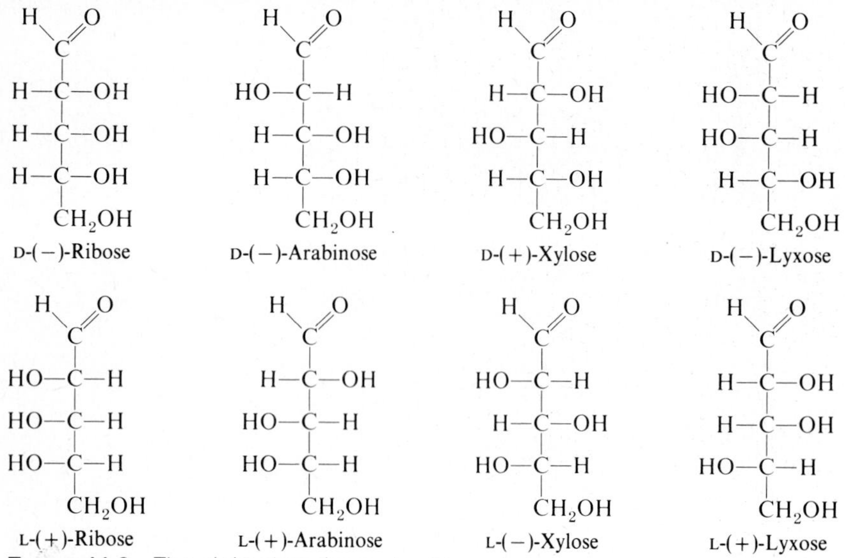

FIGURE 11.2 The eight stereoisomeric aldopentoses.

D-(−)-Ribose   D-(−)-Arabinose   D-(+)-Xylose   D-(−)-Lyxose

L-(+)-Ribose   L-(+)-Arabinose   L-(−)-Xylose   L-(+)-Lyxose

FIGURE 11.3 Configurations of the D-aldohexoses.

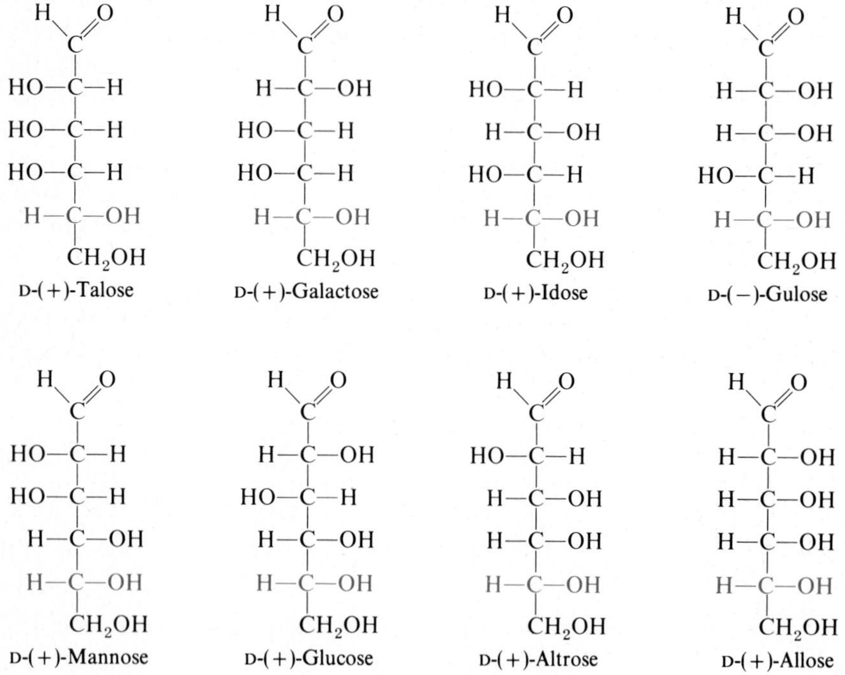

D-(+)-Talose   D-(+)-Galactose   D-(+)-Idose   D-(−)-Gulose

D-(+)-Mannose   D-(+)-Glucose   D-(+)-Altrose   D-(+)-Allose

246

configuration about one chiral carbon atom, carbon-2 and carbon-4, respectively. D-(+)-Mannose and D-(+)-galactose are said to be epimers of D-(+)-glucose. **Epimers** are *a pair of diastereomers that differ only in the configuration about a single carbon atom.* No epimeric relationship exists between D-(+)-mannose and D-(+)-galactose. Finally, it should be mentioned that the aldohexoses are structural isomers of the only naturally occurring ketohexose, D-(−)-fructose.

```
        ┌── Epimers ──┐ ┌── Epimers ──┐

        O    H         O    H         O    H
         \\  /          \\  /          \\  /         CH₂OH
          C              C              C            |
          |              |              |            C=O
      H—C—OH         H—C—OH         HO—C—H        HO—C—H
     HO—C—H         HO—C—H         HO—C—H         H—C—OH
     HO—C—H          H—C—OH         H—C—OH         H—C—OH
      H—C—OH          H—C—OH         H—C—OH         H—C—OH
         CH₂OH           CH₂OH          CH₂OH          CH₂OH

   D-(+)-Galactose   D-(+)-Glucose   D-(+)-Mannose   D-(−)-Fructose
```

11.5   Cyclic Forms of Sugars

Thus far we have considered the sugars only as open-chain structures containing a carbonyl group. Studies of the chemical and physical properties of these sugars indicate that other forms predominate, both in solution and in the solid state. The most convincing evidence was supplied by Charles Tanret (1895), a French pharmacist who first isolated two crystalline forms of D-glucose, each having distinctly different specific rotations. We consider this phenomenon further in Section 11.6.

The existence of two forms of glucose can best be explained by considering the configuration of the molecule as represented by Figure 11.4. Observe that the hydroxyl group on carbon-5 lies in proximity to the aldehyde group. In Section 7.5-a, we mentioned that alcohols react intermolecularly with aldehydes to form unstable compounds called hemiacetals.

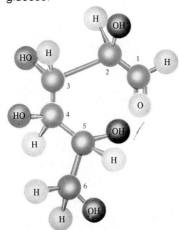

FIGURE 11.4 A ball-and-stick model of the open chain form of glucose.

$$R—\overset{\displaystyle O}{\underset{\displaystyle H}{C}} \;+\; R'OH \;\rightleftharpoons\; R—\overset{\displaystyle OH}{\underset{\displaystyle OR'}{C}}—H$$

Aldehyde   Alcohol   Hemiacetal

Recall that molecules containing both a hydroxyl group and a carbonyl group can react intramolecularly to form stable cyclic compounds. Because of the geometry of the glucose molecule, the hydroxyl group on carbon-5 can react intramolecularly with the carbonyl group to form a stable cyclic hemiacetal.

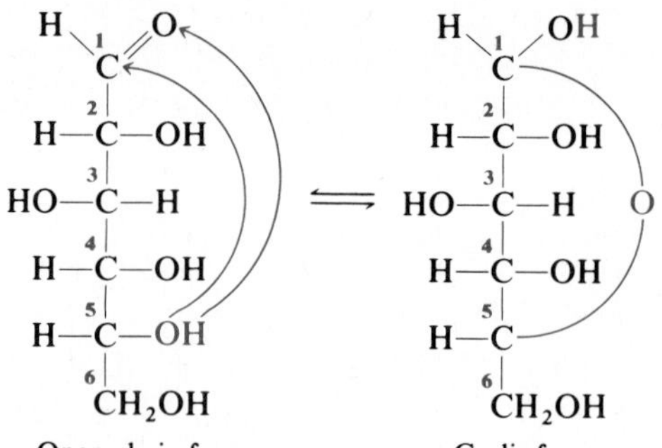

Open-chain form   Cyclic form

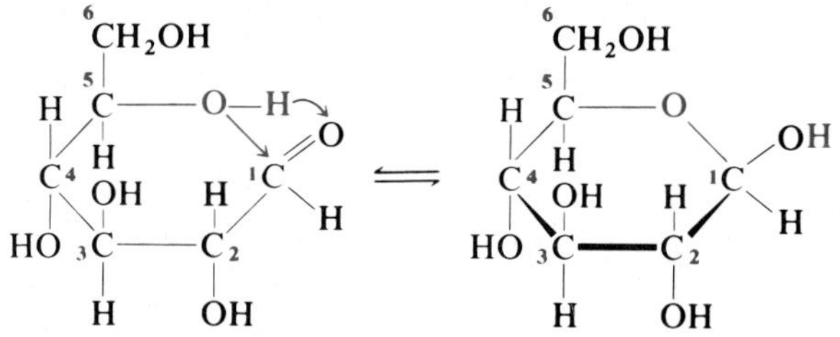

As a result of the cyclization reaction, two new problems of stereochemistry and nomenclature confront us.

1. A six-membered ring containing five carbon atoms and one oxygen atom has been formed. Since the reaction only involved an intramolecular transfer of atoms rather than the loss or gain of atoms, the new cyclic compound is a structural isomer of the open-chain compound.
2. Examination of the cyclic structure reveals that carbon-1 has now become chiral, so that we would expect to find two stereoisomers of the ring compound. Additional terminology will be needed to distinguish among the three different structures of D-glucose—one open-chain form and two cyclic forms.

As we learn in the next section, a solution of glucose actually exists as an equilibrium mixture of the three forms. In this book, the convention first suggested by an English chemist, Sir W. N. Haworth, for representing the formulas of the cyclic forms is used. The molecules are drawn as planar hexagonal slabs with darkened edges toward the viewer. Ring carbon atoms and the hydrogen atoms directly attached to them are not shown. The disposition of the hydroxyl groups, positioned either above or below the plane of the ring, is sufficient to define the correct configuration of the molecule. Any group of atoms written to the right in the Fischer projection appears below the plane of the ring, and any group written to the left appears above the plane. The formulas for the three forms of D-glucose are shown.

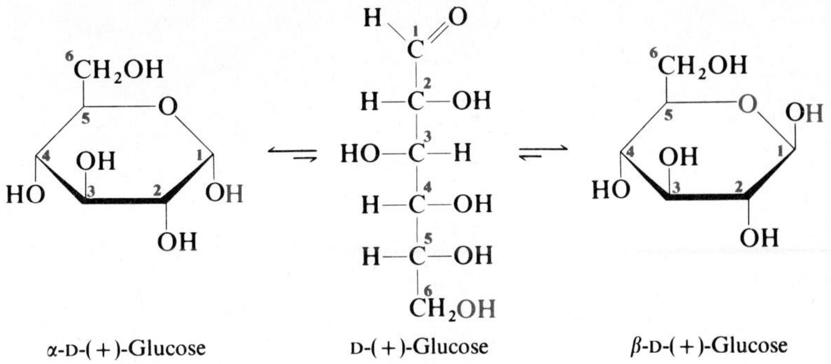

α-D-(+)-Glucose   D-(+)-Glucose   β-D-(+)-Glucose

The two ring structures differ only in the relative configurations of the hydrogen and hydroxyl group with respect to the hemiacetal carbon (the former aldehyde carbon). Thus they are not enantiomers, but are related as epimeric diastereomers and have different chemical and physical properties. Such isomers are called **anomers,** and the hemiacetal carbon is referred to as the **anomeric carbon.** By convention, the two anomers are commonly differentiated by the Greek letters α and β. In the D-series of sugars, the α-anomer is the one in which the hydroxyl group is positioned below the plane of the ring (in the form shown above). If the hydroxyl group on the anomeric carbon is positioned above the plane of the ring, it is the β-anomer. For the aldohexoses, the β-anomer is the one that has the —OH group at carbon-1 on the same side of the ring as the —CH₂OH group at carbon-5.

Intramolecular hemiacetal formation is not unique to glucose. It occurs in galactose, mannose, and the naturally occurring aldopentoses and aldoheptoses. Fructose and other ketose sugars form intramolecular hemiketals. Figure 11.5 illustrates the equilibrium between the three forms of D-galactose, D-mannose, and D-fructose. Notice that galactose and mannose, like glucose, form a six-membered cyclic structure. Fructose can also exist in this form, but is most commonly found in nature existing as a five-membered ring. Thus we are faced with another problem of nomenclature; we must have a way of distinguishing between sugars that form five-membered cyclic structures and those whose ring compounds contain six atoms. *Sugars with six-membered rings* are called **pyranoses** because they are related to the heterocyclic compound pyran; *those with five-membered rings* are **furanoses** and are related to the heterocyclic compound furan. By reference to the pyranose or furanose form of a monosaccharide one can unambiguously designate the proper ring form of a sugar. For example, the following formulas both represent β-D-fructose. (Notice that carbon-2 is the anomeric carbon in fructose.)

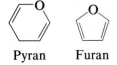

Pyran   Furan

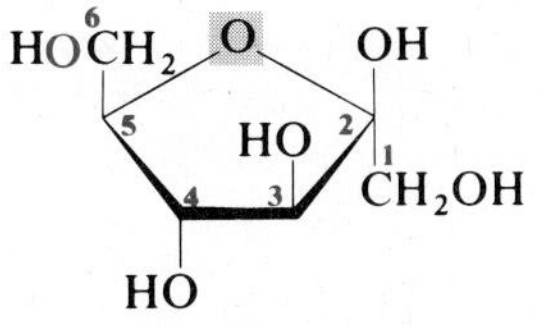

β-D-Fructose
(furanose form)

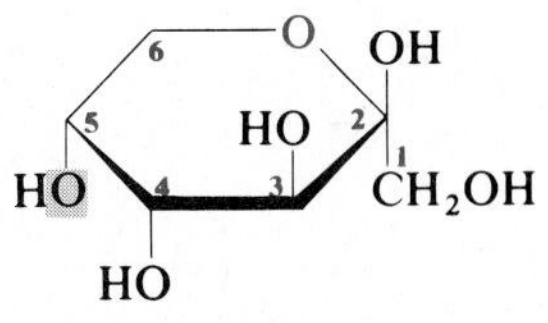

β-D-Fructose
(pyranose form)

Only by the use of the terms β-D-fructofuranose and β-D-fructopyranose can we differentiate between them.

<h1>11.6 | Mutarotation</h1>

We have seen that α-D-(+)-glucose and β-D-(+)-glucose (α-D-glucopyranose and β-D-glucopyranose) are diastereomers. Thus, they have different physical properties and can be separated by crystallization. If D-glucose is dissolved in water or in a 70% ethyl alcohol solution and is allowed to crystallize (by evaporation of the solvent or by lowering the temperature of the solution), the α-form of the sugar is obtained. Its melting point is 146°C and the specific rotation is +112°. If glucose is crystallized from acetic acid or pyridine, the β-isomer is obtained. This form has a melting point of 150°C and a specific rotation of +19°. Both α- and β-glucoses are stable in the

FIGURE 11.5   Mutarotation of D-galactose, D-mannose, and D-fructose.

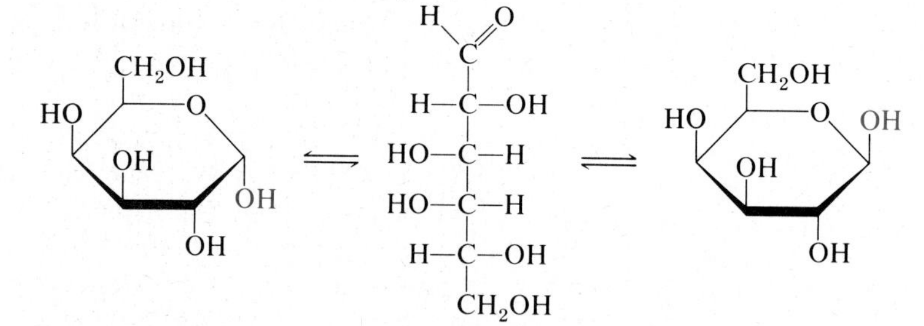

α-D-(+)-Galactose   D-(+)-Galactose   β-D-(+)-Galactose

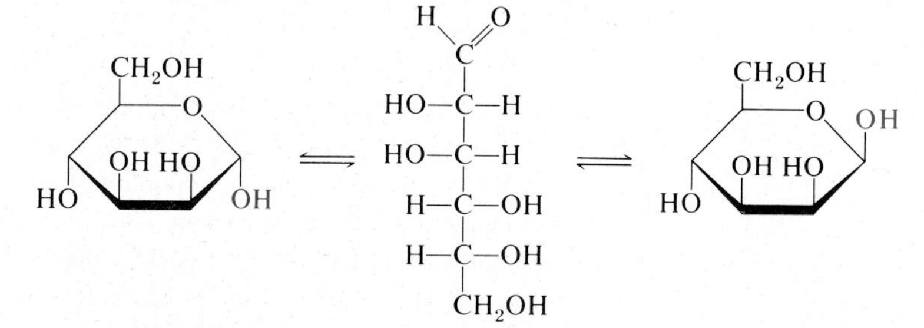

α-D-(+)-Mannose   D-(+)-Mannose   β-D-(+)-Mannose

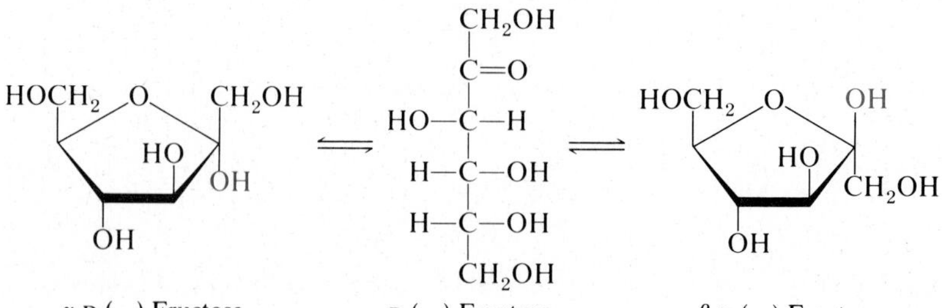

α-D-(−)-Fructose   D-(−)-Fructose   β-D-(−)-Fructose

crystalline form. When either of these isomers is dissolved in water, the specific rotation of each gradually changes with time and approaches a final equilibrium value of $+53°$. A quick calculation reveals that this value is not the average of the rotations of α- and β-glucoses. Instead of a 50–50 mixture of the two forms, the equilibrium condition must be such that one of the two isomers predominates. The percentages of α- and β-glucoses in an equilibrium mixture may be calculated by the use of simultaneous equations.

Let x = percent of β-isomer expressed as a decimal
Let y = percent of α-isomer expressed as a decimal

Therefore

$$x(19°) + y(112°) = 53°$$

(The percentage of the β-isomer multiplied by its specific rotation plus the percentage of the α-isomer multiplied by its specific rotation must equal the specific rotation of the final solution.)

and

$$x + y = 1.00$$

(Since both x and y are percentages expressed as a decimal, their sum must be unity.)

Solving for x and y in the two equations, we obtain

$$y = 0.366 = 37\%$$

$$x = 1.00 - y = 0.634 = 63\%$$

Thus, an equilibrium mixture of glucose contains 63% of the β-isomer and 37% of the α-form. This *gradual change of specific rotation,* which continues until equilibrium is established, is known as **mutarotation.** It is a phenomenon that is

TABLE 11.1 Specific Rotations of Some Mono- and Disaccharides

Sugar	Specific Rotation (degrees)		
	α	β	Equilibrium Mixture
D-Glucose	+112	+19	+53
D-Fructose	−21	−133	−92
D-Galactose	+151	−53	+80
D-Mannose	+29	−16	+14
D-Lactose	+84	+34	+54
D-Maltose	+173	+112	+130

common to all monosaccharides (and to some disaccharides) that can exist in α- and β-cyclic structures. There is sufficient evidence to believe that all monosaccharides that undergo mutarotation pass through the open-chain form, and a trace of this form is therefore present in the equilibrium mixture (less than 0.1%). It is for this reason that the monosaccharides are said to have a *potential carbonyl group*. Table 11.1 lists specific rotation values for the common sugars.

11.7 *Important Monosaccharides*

a. Trioses

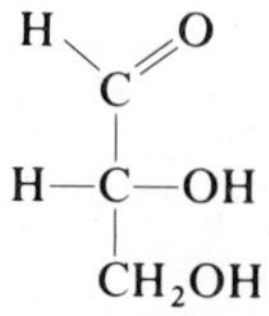

Glyceraldehyde 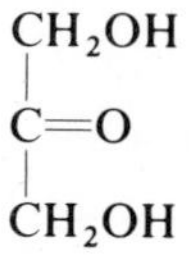   Dihydroxyacetone

Both of these compounds are found in plant and animal cells and play an important role in carbohydrate metabolism (Chapter 16).

b. Pentoses

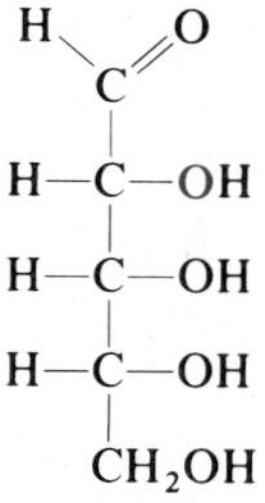

D-(−)-Ribose

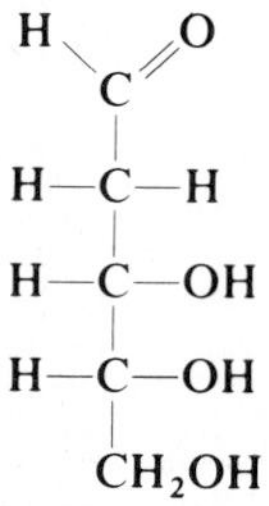

(−)-2-Deoxy-D-ribose
(The prefix 2 is usually omitted from the name of the compound)

Ribose and deoxyribose are commonly found in nature. They are integral components of nucleic acids (Chapter 15) in which they possess β-furanose structures. Ribose is also an intermediate in the pathway of carbohydrate metabolism and is a constituent of ATP as well as several of the coenzymes (see Table 14.2)

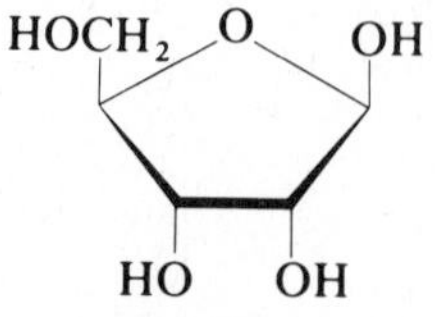

β-D-Ribose 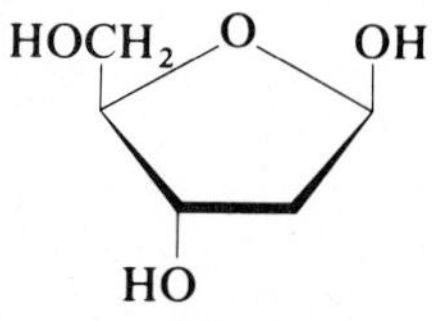   β-Deoxy-D-ribose

c. Hexoses

1. Glucose D-Glucose is the most abundant sugar found in nature. It is commonly found in fruits, especially in ripe grapes, and for this reason it is often referred to as *grape sugar*. It is also known as *dextrose*, a name that derives from the fact that the predominant natural form of the sugar is dextrorotatory.

    Most of the carbohydrates taken in by the body are eventually converted to glucose in a series of metabolic pathways. Since glucose requires no digestion, it can be given intravenously (as a 5% solution) to patients who are unable to take food orally. Glucose is the circulating carbohydrate of animals; the blood contains about 0.08% glucose and normal urine may contain anywhere from a trace to 0.2% glucose.

    Commercially, glucose is made by the hydrolysis of starch. In the United States, cornstarch is used in the process; in Europe, the starch is obtained from potatoes. Glucose is only 70% as sweet as table sugar (sucrose), but it has the same caloric value.

2. Galactose D-Galactose is formed by the hydrolysis of lactose, a disaccharide composed of a glucose unit and a galactose unit. It does not occur in nature in the uncombined state. The galactose needed by the human body for the synthesis of lactose (in the mammary glands) is obtained by the conversion of D-glucose into D-galactose. In addition, galactose is an important constituent of the glycolipids (Section 12.10) that occur in the brain and in the myelin sheath of nerve cells. The genetic disease *galactosemia* results from the absence of an enzyme that converts galactose to glucose (see Section 15.6).

3. Fructose D-Fructose is the only naturally occurring ketohexose. It exists in two structural forms, depending upon whether it is in the free state or combined. In the free state, fructose exists predominantly as a pyranose structure, whereas in the combined form (as in sucrose, inulin, and several phosphate esters), it exists as a furanose. Fructose (Latin, *fructus,* fruit) is also referred to as *levulose* because it has a specific rotation that is strongly levorotatory ($-92°$). It is the sweetest sugar (1.7 times sweeter than sucrose), and it is found, together with glucose and sucrose, in honey (which contains 40% fructose) and sweet fruits. Fructose is the major energy source for spermatozoa and is formed from glucose in the prostate gland.

11.8 *The Disaccharides*

Most naturally occurring carbohydrates contain more than one monosaccharide unit. The manner in which two monosaccharide molecules are joined together is of particular interest. We have seen that the cyclic structures of the monosaccharides arise from intramolecular hemiacetal (or, in the case of fructose, hemiketal) formation. Recall that hemiacetals can react with a molecule of an alcohol to form a stable compound known as an acetal (Section 7.5-a). If either α- or β-glucose is treated with methanol in the presence of an acid catalyst, a mixture of two acetals,

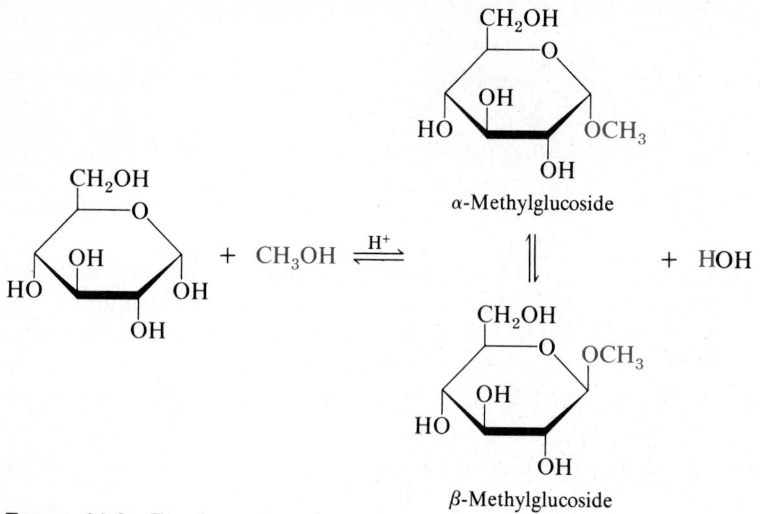

α-Methylglucoside

β-Methylglucoside

FIGURE 11.6 The formation of an equilibrium mixture of α- and β-methylglucosides.

known as *glucosides,* is formed (Figure 11.6). The more general term **glycoside** is used to refer to *the acetal that is formed when any carbohydrate reacts with a hydroxy compound.* Specific compounds may be named according to the sugars from which they are derived: hence, glucoside from glucose, galactoside from galactose, etc. *The carbon–oxygen–carbon linkage that joins the two components of the acetal* is called the **glycosidic linkage.** We see in the next sections that the biologically significant disaccharides and polysaccharides are composed of monosaccharide units that are joined by glycosidic linkages. The disaccharides all have the same molecular formula, $C_{12}H_{22}O_{11}$, and are therefore structural isomers of each other. They differ with respect to their constituent monosaccharide units and the type of glycosidic linkage connecting them.

a. Sucrose

Sucrose is known as beet sugar, cane sugar, table sugar, or simply as sugar. It is probably the largest selling pure organic compound in the world. As its names imply, sucrose is obtained from sugar canes and sugar beets (whose juices contain 14–20% of the sugar) by evaporation of the water and recrystallization. The dark brown liquid that remains after crystallization of the sugar is sold as molasses. Brown sugar is obtained when the recrystallization process is not entirely completed. The average American ingests about 100 lb of sucrose every year.

A molecule of sucrose may be envisioned to result from the combination of one molecule of α-D-glucopyranose and one molecule of β-D-fructofuranose; a molecule of water is eliminated in the process. The unique feature that characterizes the sucrose molecule is its glucosidic linkage; it involves the hydroxyl group on carbon-1 of α-D-glucose and the hydroxyl group on carbon-2 of β-D-fructose. By convention, sugars are read from left to right (or top to bottom). This connecting linkage is therefore an **α-1,2-glucosidic linkage.** This bonding bestows certain properties upon sucrose that are quite different from those of the other disaccharides.

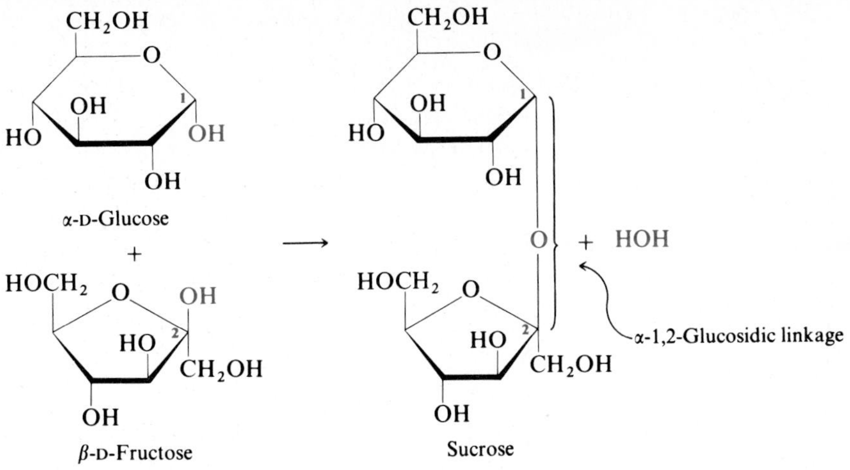

α-D-Glucose

+

β-D-Fructose

⟶

Sucrose

α-1,2-Glucosidic linkage

+ HOH

Sucrose, unlike the other disaccharides, is incapable of mutarotation. Thus, it exists in only one form both in the solid state and in solution. The presence of the 1,2-glucosidic linkage makes it impossible for sucrose to exist in the α- or β-configuration or in the open-chain form. This is a direct result of the fact that the anomeric carbon of the glucose moiety and the anomeric carbon of the fructose moiety have both been tied up in the formation of the 1,2 (head-to-head) linkage. As long as the sucrose molecule remains intact, it cannot uncyclize to form the open-chain structure. Sucrose, therefore, does not undergo reactions that are typical of aldehydes and ketones, and it is said to be a *nonreducing* sugar (see Section 11.10-b).

The human body is unable to utilize sucrose or any other disaccharide directly because such molecules are too large to pass through cell membranes. Therefore, the disaccharide must first be broken down by hydrolysis into its two constituent monosaccharide units. In the body, this hydrolysis reaction is catalyzed by enzymes. The same hydrolysis reaction may be carried out in a test tube with dilute acid as a catalyst, but the reaction rate is much slower. The equation for the hydrolysis reaction is the reverse of the one given for the formation of sucrose.

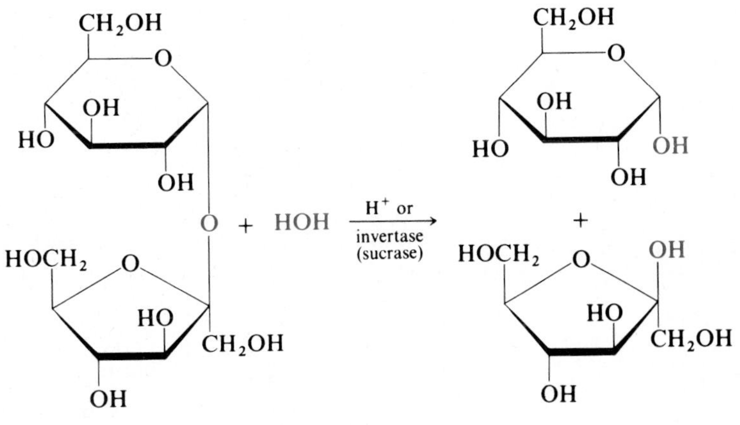

H^+ or invertase (sucrase)

+ HOH

An interesting feature of this catalytic reaction involves the specific rotation of the hydrolyzate (the products of the hydrolysis reaction). Sucrose is dextrorotatory, $[\alpha]_D^{20} = +66.5°$, but upon hydrolysis the sign of the specific rotation changes (inverts) from positive to negative, $[\alpha]_D^{20} = -19.5°$. The product of the reaction is an equimolar mixture of D-glucose and D-fructose, both of which undergo mutarotation. The hydrolyzate actually contains an equilibrium mixture of α- and β-glucoses and α- and β-fructoses. The specific rotation of the hydrolyzate is the sum of the specific rotations of each of the components, multiplied by their concentrations in grams per cubic centimeter (g/cm^3).

$$\text{Sucrose} \xrightarrow[\text{enzyme}]{\text{acid or}} \text{Glucose} + \text{Fructose}$$

$$\underbrace{\quad\quad 0.5 \text{ g/cm}^3 \quad\quad 0.5 \text{ g/cm}^3}_{\text{Invert sugar}}$$

Sucrose 1.0 g/cm³

$$[\alpha]_D^{20} = 0.5(+53°) + 0.5(-92°)$$
$$= -19.5°$$

The term **inversion** is often applied to the hydrolysis of sucrose since it is the only case in which the hydrolysis of a disaccharide effects a change in the sign of the specific rotation. *An equimolar mixture of D-glucose and D-fructose is called* **invert sugar.** The enzyme that catalyzes the hydrolysis is often called *invertase* instead of *sucrase,* its systematic name.

The inversion reaction has several practical applications. Since sucrose can exist in only one molecular configuration, it is one of the most readily crystallizable sugars. Invert sugar has a much greater tendency to remain in solution. In the manufacture of jelly and candy and in the canning of fruit, crystallization of the sugar is undesirable; therefore, conditions leading to the hydrolysis of sucrose are employed in these processes. Since fructose is sweeter than sucrose, the hydrolysis adds to the sweetening effect. Bees carry out this reaction during their production of honey.

The widespread use of sucrose has generated much adverse publicity. Various health magazines have reported that excess sugar causes cancer, heart disease, migraine headaches, hyperkinetic children, obesity, and tooth decay. Only the latter two claims have been substantiated. As we shall see in Chapter 17, carbohydrate is converted to fat when the caloric intake exceeds the body's requirements. Sucrose does cause tooth decay by serving as part of the plaque that sticks to a tooth. The bacteria contained within the plaque use sucrose both as an adhesive and as a food source. We shall see in Chapter 16 that bacteria can decompose sugar to lactic acid, and this acid corrodes the teeth and leads to the destruction of the gums. The best way to remove plaque is by daily flossing, and the amount formed can be decreased by reducing the quantity of sucrose that is ingested.

b. Maltose

Maltose does not occur in the free state in nature to an appreciable extent. It occurs in animals as the principal sugar formed by the enzymatic (*ptyalin*) hydrolysis of starch. It is fairly abundant in germinating grain where it is formed by the enzymatic (*diastase*) breakdown of starch. In the manufacture of beer, maltose is liberated by the action of malt (germinating barley) on starch, and for this reason it is often referred to as *malt sugar.* Maltose is about 30% as sweet as sucrose.

Maltose is a reducing sugar, and it exhibits mutarotation. An equilibrium mixture of the maltose isomers is highly dextrorotatory, $[\alpha]_D^{20} = +136°$ (Figure 11.7). When maltose is hydrolyzed, either enzymatically or by means of an acid catalyst, two molecules of D-glucose are produced. The formula of maltose must therefore incorporate two glucose molecules in such a way that one free anomeric hydroxyl group exists. The glucose units in maltose are joined in a *head-to-tail* fashion through an α-linkage from carbon-1 of one glucose molecule to carbon-4 of the second glucose molecule (that is, an **α-1,4-glucosidic linkage**).

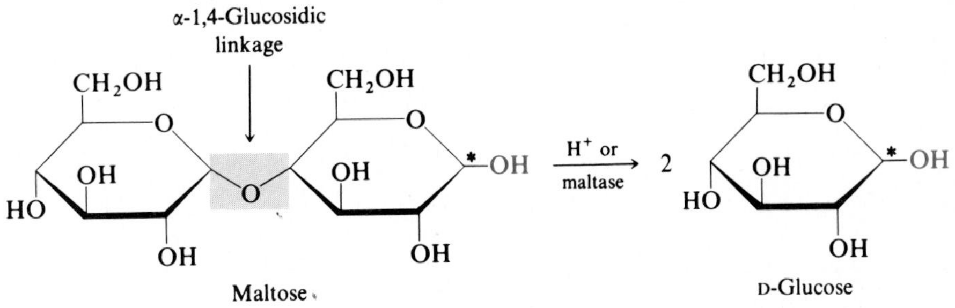

(*We use this convention for writing the hydroxyl group on the anomeric carbon when we do not wish to specify either the α-anomer or the β-anomer.)

Another disaccharide, **cellobiose,** is a stereoisomer of maltose. It is also composed of two glucose units, but in this case the two sugar moieties are joined by a **β-1,4-glucosidic linkage.** Like maltose, cellobiose is a reducing sugar and undergoes mutarotation. The enzyme *maltase* is stereospecific; it is capable only of hydrolyzing α-1,4-glucosidic linkages and for this reason does not act upon cellobiose. Cellobiose

FIGURE 11.7 Equilibrium mixture of maltose isomers.

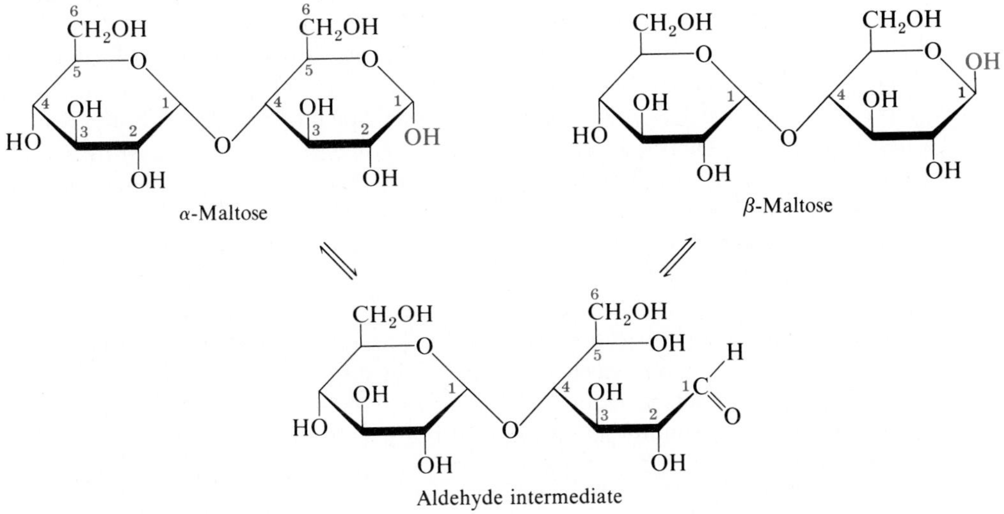

α-Maltose

β-Maltose

Aldehyde intermediate

is obtained by the partial hydrolysis of cellulose. It may be further hydrolyzed to yield two molecules of D-glucose by the action of the enzyme *cellobiase* (which is specific for β-glucosidic linkages).

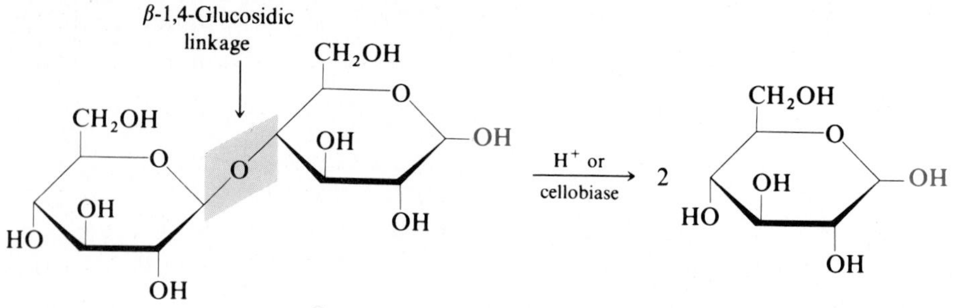

Cellobiose   D-Glucose

c. Lactose

Lactose is known as *milk sugar* because it occurs in the milk of humans, cows, and other mammals. Human milk contains about 7.5% lactose, whereas cow's milk, which is not as sweet, contains about 4.5% lactose. Unlike most carbohydrates, which are plant products, lactose is one of the few carbohydrates associated exclusively with the animal kingdom. (The biosynthesis of lactose is confined to the mammary tissue.) It is produced commercially from whey, which is obtained as a by-product in the manufacture of cheese. Lactose is one of the lowest ranking sugars in terms of sweetness (about one sixth as sweet as sucrose). It is a reducing sugar and exhibits mutarotation. An equilibrium mixture of lactose has a specific rotation of $+54°$. The α-form of the sugar is of commercial importance as an infant food and in the production of penicillin. Drug dealers use lactose to "cut" their heroin and thus increase their profits.

Lactose is composed of one molecule of D-galactose and one molecule of D-glucose joined by a **β-1,4-galactosidic bond.** The two monosaccharides are obtained from lactose by acid hydrolysis or by catalytic action of the enzyme *lactase*. Lactose makes up about 40% of an infant's diet during the first year of life. Infants and small children contain one form of *lactase* within their small intestine. However, an adult form of the enzyme is less active and is missing in about 70% of the world's adult population (especially Africans and Orientals). These people suffer from **lactose intolerance** and are unable to digest the sugar found in milk. For some people, the inability to synthesize sufficient enzyme increases with age. Many adult Americans suffer from a degree of lactose intolerance. Some of the unhydrolyzed lactose passes into the intestine, and its presence tends to draw water from the interstitial fluid into the intestinal lumen by osmosis. At the same time, intestinal bacteria may act on the lactose to produce organic acids and gases. The intake of water, plus the bacterial decay products, leads to the abdominal distention, cramps, and diarrhea that are symptoms of the condition. Symptoms disappear completely if milk is excluded from the diet.

Yeasts can metabolize sucrose and maltose, but not lactose because they do not contain the *lactase* enzyme. Certain bacteria can metabolize lactose, and they form lactic acid as one of the products. As we shall see (page 339), this is responsible for the "souring" of milk.

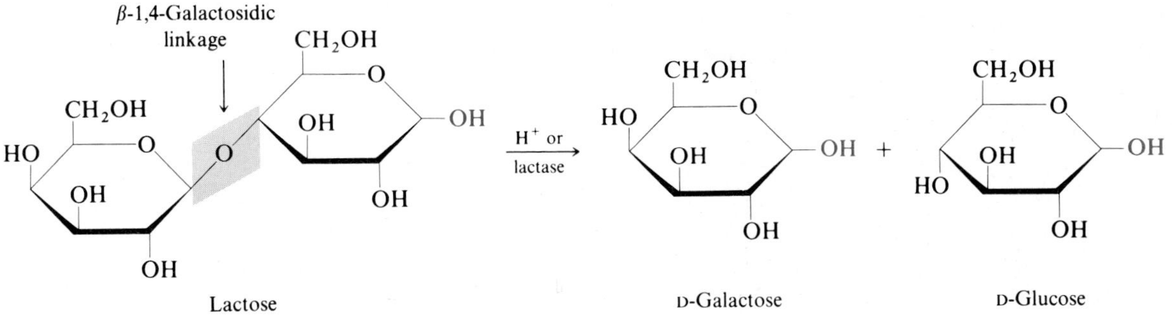

11.9 *The Polysaccharides*

The polysaccharides are the most abundant of the carbohydrates found in nature. They serve as reserve food substances and as structural components of plant cells. Polysaccharides are high molecular weight (25,000–15,000,000) polymers of monosaccharides joined together by glycosidic linkages. Biochemically, the three most significant polysaccharides are starch, glycogen, and cellulose. They are also referred to as *homopolymers* since each of them yields only one type of monosaccharide (D-glucose) upon complete hydrolysis. (Inulin, another homopolymer, is found in the tubers of the Jerusalem artichoke and is composed entirely of fructofuranose units.) *Heteropolymers* may contain sugar acids, amino sugars, or noncarbohydrate substances. They are very common in nature (gums, pectins, hyaluronic acid), but are not discussed in this text. The polysaccharides are nonreducing carbohydrates, are not sweet tasting (probably because of their limited solubility), and do not undergo mutarotation.

a. Starch

Starch is the most important source of carbohydrates in the human diet. (It accounts for more than 50% of our carbohydrate intake.) Starch occurs in plants in the form of granules. Starch granules are particularly abundant in the seeds, where they serve as a storage form of carbohydrate. The breakdown of starch to glucose nourishes the plant during periods of reduced photosynthetic activity. We often think of potatoes as a "starchy" food, yet other plants contain a much greater percentage of starch (e.g., potatoes 15%, wheat 55%, corn 65%, and rice 75%).

Starch is a mixture of two polymers, **amylose** and **amylopectin,** which can be separated from each other by physical and/or chemical methods. Natural starches consist of about 20–30% amylose and 70–80% amylopectin. Amylose is a straight-chain polysaccharide composed entirely of D-glucose units joined by an

α-1,4-glucosidic linkage, as in maltose. Thus, amylose might be thought of as either polymaltose or polyglucose.[4]

Repeating unit; $n = 100-1000$

Amylose

Amylopectin is a branched-chain polysaccharide composed of glucose units that are linked primarily by **α-1,4-glucosidic bonds,** but have occasional **α-1,6-glucosidic linkages,** which are responsible for the branching. It has been estimated that there are over 1000 glucose units in amylopectin, and that branching occurs about once every 25–30 units (Figure 11.8-b). The helical structure of amylopectin is disrupted by the branching of the chain, so instead of the deep blue color amylose gives with iodine, amylopectin produces a less intense red-brown color.

The complete hydrolysis of starch (amylose and amylopectin) yields, in three successive stages, dextrins, maltose, and glucose. Dextrins are glucose polysaccharides of intermediate size. The shine and stiffness imparted to clothing by starch are due to the presence of dextrins formed when the clothing is ironed. Because of their characteristic stickiness upon wetting, dextrins are used as adhesives on stamps, envelopes, and labels, and as pastes and mucilages. Since dextrins are more easily digested than starch, they are extensively used in the commercial preparation of infant foods (Dextrimaltose). A dried mixture of dextrins, maltose, and milk is used in the preparation of malted milk. Starch can be hydrolyzed by heating in the presence of dilute acid. In the human body, it is degraded sequentially by several enzymes known collectively as *amylase.*

$$\text{Starch} \xrightarrow[\text{amylase}]{\text{H}^+, \Delta \text{ or}} \text{Dextrins} \xrightarrow[\text{amylase}]{\text{H}^+, \Delta \text{ or}} \text{Maltose} \xrightarrow[\text{maltase}]{\text{H}^+, \Delta \text{ or}} \text{Glucose}$$

b. Glycogen

Glycogen, often called *animal starch,* is the reserve carbohydrate of animals. Practically all mammalian cells contain some glycogen for the purpose of storing carbohydrate. However, it is especially abundant in the liver, 4–8% (per weight of tissue), and in skeletal muscle cells, 0.5–1.0%.[5] When fasting or during periods

---

[4]Experimental evidence indicates that the molecule is not a straight chain of glucose units; rather, it is coiled like a spring with six glucose monomers per turn (Figure 11.8-a). When coiled in this fashion, amylose has just enough room in its core to accommodate an iodine molecule. The characteristic blue-violet color that starch gives when treated with iodine is due to the formation of the amylose–iodine complex.

[5]These percentages may be misleading. Although the percent of glycogen is higher in the liver, the much greater mass of skeletal muscle stores a greater *total* amount of glycogen. About 70% of the total glycogen in the body is stored in muscle cells.

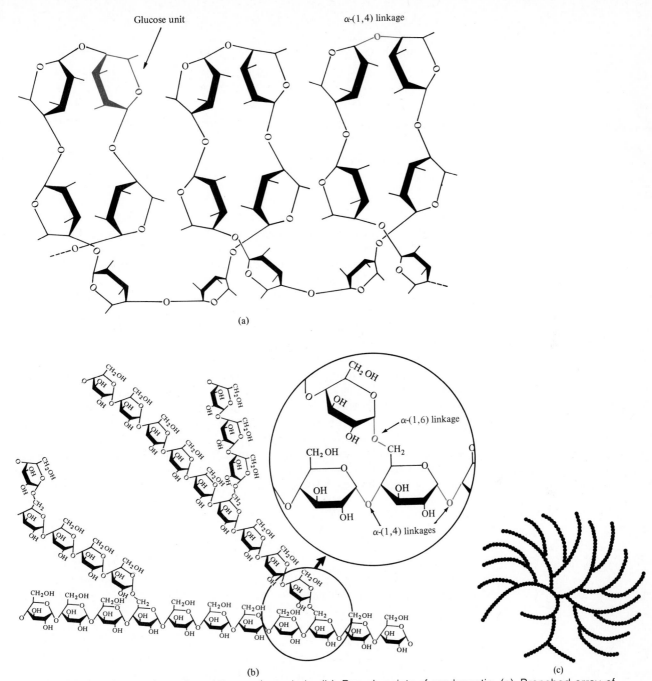

FIGURE 11.8 (a) The conformation of the amylose chain. (b) Branch points of amylopectin. (c) Branched array of glucose units in amylopectin or glycogen.

Glucose unit

α-(1,4) linkage

CH₂OH

α-(1,6) linkage

CH₂OH

CH₂

α-(1,4) linkages

(a)

(b)

(c)

of starvation, animals draw upon these glycogen reserves to obtain the glucose needed to maintain a state of metabolic balance.

In terms of structure, glycogen is quite similar to amylopectin, but it is more highly branched and its branches are shorter (8–12 glucose units in length). When treated with iodine, glycogen gives a red-brown color. Glycogen can be broken down into its D-glucose subunits by acid hydrolysis or by means of the same enzymes that attack starch. In animals, the enzyme *phosphorylase* catalyzes the breakdown of glycogen into phosphate esters of glucose (see Section 16.4).

c. Cellulose

Cellulose is a fibrous carbohydrate found in all plants; it is the structural component of plant cell walls. Since the earth is covered with vegetation, cellulose is the most abundant of all carbohydrates, accounting for over 50% of all the carbon found in the vegetable kingdom. Cotton fiber and filter paper are almost entirely cellulose (about 95%), wood is about 50% cellulose, and the dry weight of leaves is about 10–20% cellulose. From an industrial and economic standpoint, cellulose is the most important of all the carbohydrates. The largest use of cellulose is in the manufacture of paper and paper products. Although there is increasing use of synthetic fibers, rayon (made from cellulose) and cotton account for over 70% of textile production.

Complete hydrolysis (40% aqueous HCl) of cellulose yields only D-glucose, whereas partial hydrolysis of cellulose yields the disaccharide cellobiose. Thus cellulose must be composed of chains of D-glucose units (about 2000–3000) joined by **β-1,4-glucosidic linkages.** The chains are almost exclusively linear, unlike those of starch, which are highly branched. The linear nature of the cellulose chains allows a great deal of hydrogen bonding between hydroxyl groups on adjacent chains. As a result, the chains are closely packed into fibers, and there is little interaction with water or with any other solvent. Cotton and wood, for example, are completely insoluble in water and of considerable mechanical strength. Since there is no helical structure, cellulose does not bind to iodine to give a colored product.

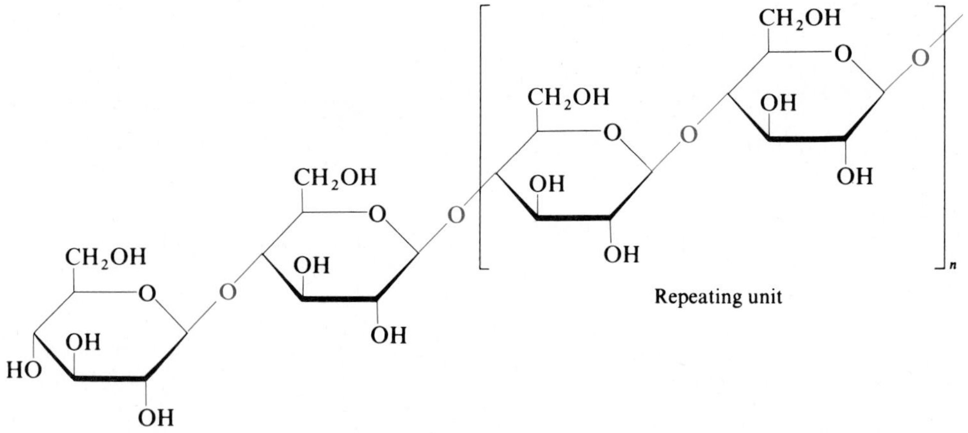

Cellulose

Cellulose yields D-glucose upon complete acid hydrolysis, yet humans (and all vertebrates) cannot utilize cellulose as a source of glucose. Our digestive juices lack the enzymes that hydrolyze β-glucosidic linkages. Carnivorous animals need cellulose in the diet to provide bulk[6] for the feces, thus preventing constipation. Many microorganisms and herbivorous animals (cow, horse, sheep) can digest cellulose. These higher animals can do so only because they contain microorganisms in their digestive tracts whose enzyme (*cellulase*) catalyzes cellulose hydrolysis. Termites also contain *cellulase*-secreting microorganisms and thus can subsist on a wood diet. This, once again, emphasizes the extreme stereospecificity of biochemical processes.

11.10 *Tests for Carbohydrates*

The chemical tests employed to detect carbohydrates and to distinguish among them may be divided into two broad categories: (1) tests based on the production of furfural derivatives and (2) tests based on the reducing property of sugars.

a. Tests Based on the Production of Furfural and Its Derivatives

When a monosaccharide is treated with strong acid, dehydration of the monosaccharide occurs. If the monosaccharide is a pentose, the dehydrated product is *furfural;* a hexose gives rise to *hydroxymethylfurfural.* (Trioses and tetroses are incapable of undergoing this reaction since they do not possess the requisite minimum of five carbon atoms.) Under vigorous conditions, all carbohydrates containing five or more carbons can be made to react; under milder conditions only certain compounds undergo the dehydration reaction, thus affording us with more specific tests.

A pentose   Furfural

[6] A topic that has aroused considerable interest among nutritionists as well as the general public is the subject of fiber (roughage) in the diet. **Dietary fiber** consists of cellulose and related substances (such as lignin) that make up the cell walls of plants. These substances are highly hydrophilic and thus aid in the retention of water in the intestinal contents, producing a softer stool. Some scientists believe that a low fiber diet (about 20 g/day) is associated with the increased incidence of colon cancer, diverticulosis, coronary heart disease, atherosclerosis, gallstones, and hemorrhoids. Other scientists dispute some of these claims. They point out that a high fiber diet (about 100 g/day) is a largely vegetarian diet, and thus individuals consume relatively little meat and saturated fat (see Section 12.17).

A recent proposal is that a high fiber diet is useful in the treatment of diabetes (see page 425). It has long been known that diabetics ought to avoid ingestion of rapidly digested sugars such as glucose and sucrose and eat only carbohydrates that are slowly digested (e.g., starch). The presence of fiber in the diet reduces the rate of absorption of glucose, and therefore the peak blood sugar concentration is lowered.

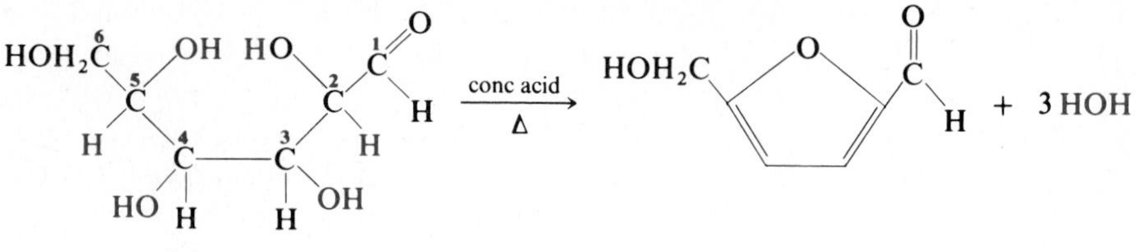

A hexose   Hydroxymethylfurfural

In the presence of concentrated acid, various phenolic compounds (e.g., α-naphthol, orcinol, resorcinol) will condense with the furfural or the hydroxymethylfurfural to form colored dyes. The formation of these colored compounds is a positive test for the presence of carbohydrates (as described below).

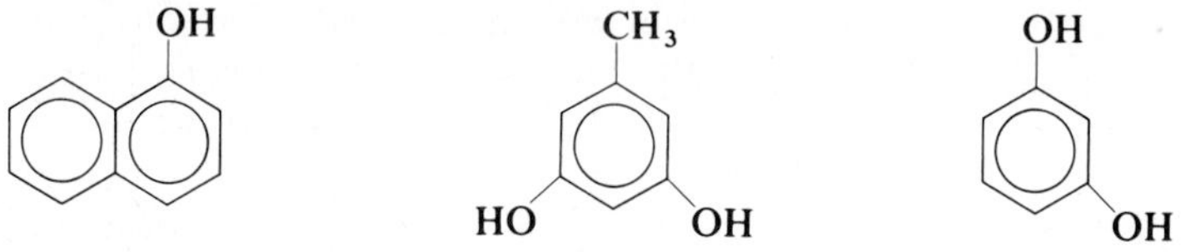

α-Naphthol   Orcinol   Resorcinol
(α-Hydroxynaphthalene)   (3,5-Dihydroxytoluene)   (m-Hydroxyphenol)

1. **Molisch Test** The Molisch test is the most general test for the presence of carbohydrates because it gives a positive test (indicated by the appearance of a purple ring) with all carbohydrates larger than tetroses. Concentrated H_2SO_4 will first hydrolyze all di- and polysaccharides to monosaccharides, the monosaccharides then form furfural or hydroxymethylfurfural, and these compounds condense with α-naphthol to form a purple condensation product.

2. **Bial's Orcinol Test** Bial's orcinol test is important in the determination of pentoses and the pentose-containing subunits of nucleic acids. The reaction is *not* specific for pentoses, however, since other compounds such as trioses and certain heptoses produce identical bright blue condensation products with orcinol. It is therefore important that this test be applied to reasonably pure samples. Bial's test is useful in differentiating between hexoses and pentoses because orcinol condenses with furfural (from pentoses) to form a blue-green compound and with hydroxymethylfurfural (from hexoses) to form a yellow-brown product.

3. **Seliwanoff's Test** This test makes use of the fact that a hot hydrochloric acid solution dehydrates the ketohexoses to form hydroxymethylfurfural much faster than it acts upon the corresponding aldohexoses. Therefore, the test distinguishes aldohexoses from ketohexoses based on their differential rates of reaction. During a given time interval (usually chosen to be 60 sec), fructose and those di- and polysaccharides that are hydrolyzed to fructose by hydrochloric acid will react with the resorcinol in Seliwanoff's reagent to yield compounds having a deep red color. Aldohexoses yield substances that are just slightly pink.

b. Tests Based on the Reducing Property of Sugars

Because sugars are polyhydroxy aldehydes and ketones, they are usually able to undergo all of the normal reactions of the alcohols and of the carbonyl compounds. All monosaccharides and some disaccharides have the ability to reduce an alkaline solution of cupric ion. Any carbohydrate capable of this reduction without first undergoing hydrolysis is said to be a **reducing sugar.** It is not possible to write a complete oxidation–reduction equation for the process because a variety of oxidized products are formed. (This reaction has been adopted as a simple and rapid diagnostic test for the presence of glucose in the blood or in the urine.)

$$\text{Reducing sugar} \ + \ Cu^{2+} \ \xrightarrow{\ OH^-\ } \ Cu_2O_{(s)} \ + \ \text{Oxidized products of the sugar}$$

The oxidation of sugars is a process that is best understood in terms of chemical equilibria. Recall that in solution only a very small fraction of most mono- and disaccharides exists in the open-chain, free-aldehyde form (Section 11.6). The cupric ions react with this small amount of free aldehyde, and in so doing, serve to continuously remove the open-chain form from the equilibrium mixture. This causes a continuous shifting of the equilibrium *away* from the cyclic forms *toward* the open-chain form. Gradually, all the sugar in the ring forms is converted to the open-chain form and is subsequently oxidized. In Section 11.8-a it was mentioned that sucrose is not a reducing sugar because it does not contain a free anomeric hydroxyl group. That is, the anomeric hydroxyl groups of both glucose and fructose are lost in the formation of the glycosidic bond. Lactose and maltose are reducing sugars because each possesses a free anomeric hydroxyl group.

The alkaline reaction conditions facilitate the attainment of a tautomeric equilibrium between the enol and keto forms of the open-chain structure (Section 7.5-b). Enolization may result in the formation of glucose from fructose or vice versa. This fact is vital to an understanding of why fructose is a reducing sugar even though it contains a ketonic (rather than an aldehydic) carbonyl group. The following interconversion explains the phenomenon and is discussed again in Chapter 16.

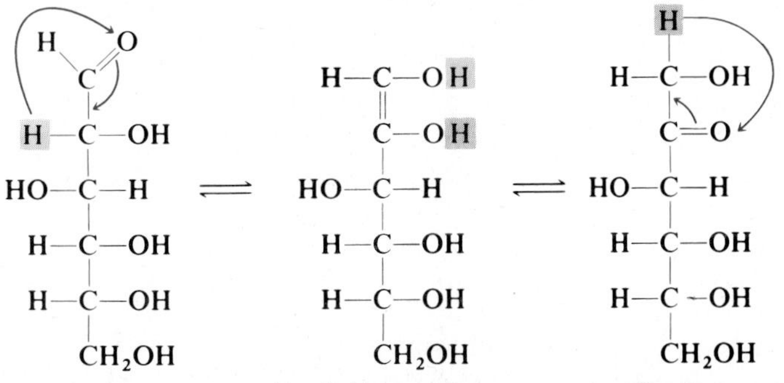

D-Glucose   Enediol intermediate   D-Fructose

1. Benedict's Test and Fehling's Test (See Section 7.5-c.) These tests are performed under mildly alkaline conditions. They are extremely sensitive tests for the presence

of carbohydrates, but are by no means specific for any particular sugar. Aldehydes, 2-ketoses, and many other compounds will give a positive test with these reagents. The formation of a red cuprous oxide precipitate is the criterion of a positive test. Clinitest tablets, which are used in clinical laboratories to test for sugar in the urine, contain cupric ions and are based on the same principles as the Benedict test. A green color indicates very little sugar, whereas a brick-red color indicates sugar in excess of 2 g/100 mL of urine.

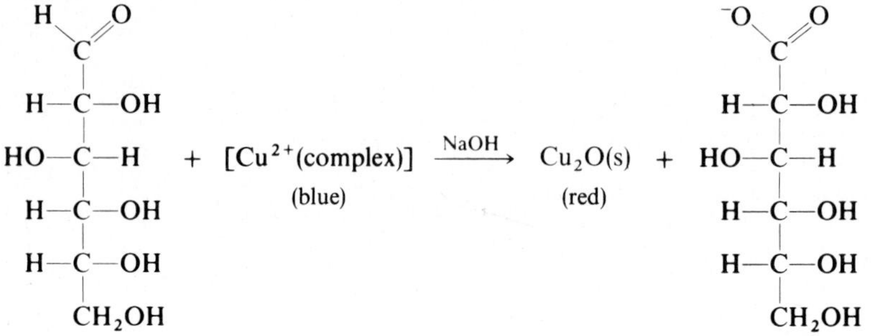

$$\text{D-Glucose} + [Cu^{2+}(\text{complex})] \xrightarrow{\text{NaOH}} Cu_2O(s) + \text{D-Gluconate}$$

D-Gluconate (undergoes further oxidation)

2. **Barfoed's Test** The Barfoed test is used to distinguish between reducing mono-saccharides and reducing disaccharides. A differential rate of reaction with cupric acetate in acetic acid is the basis of the test. Within a given time interval (usually 10 minutes), only monosaccharides will reduce the cupric ion, again forming the red insoluble cuprous oxide. Under acid conditions, cupric ion is a weaker oxidizing agent and is therefore only capable of oxidizing the monosaccharides within the assigned time. If heating is prolonged, the disaccharides may be hydrolyzed by the acid, and the resulting monosaccharides will give a positive test.

Exercises

11.1 Define, explain, or draw a formula for each of the following.
- (a) carbohydrate
- (b) monosaccharide
- (c) disaccharide
- (d) polysaccharide
- (e) hexose
- (f) aldotetrose
- (g) ketopentose
- (h) mutarotation
- (i) β-D-glucopyranose
- (j) glycosidic linkage
- (k) β-maltose
- (l) α-cellobiose
- (m) α-lactose
- (n) lactose intolerance
- (o) dextrins
- (p) reducing sugar

11.2 Distinguish between the terms in each of the following pairs.
- (a) aldose and ketose
- (b) saccharin and aspartame
- (c) D-sugar and L-sugar
- (d) penultimate carbon and anomeric carbon
- (e) anomer and epimer
- (f) epimer and enantiomer
- (g) furanose sugar and pyranose sugar
- (h) dextrose and levulose
- (i) glucoside and glycoside
- (j) sucrose and invert sugar

11.3 Draw the structural formulas for three compounds that have the empirical formula $C_x(H_2O)_y$, but are not carbohydrates.

11.4 Are all the monosaccharides and the disaccharides soluble in water? Explain.

11.5 Draw all the ketose stereoisomers of erythrulose (a ketotetrose). Label them D- or L-.

11.6 Draw structural formulas for the open-chain and cyclic forms of the four naturally occurring hexoses. Of these four hexoses, which is least significant for humans?

11.7 Draw all the possible stereoisomeric ketohexoses.

11.8 Why are (+)-glucose and (−)-fructose both classified as D-sugars?

11.9 L-Sorbose is the C-5 epimer of D-fructose. It is used in the commercial synthesis of vitamin C. Draw the structure of L-sorbose.

11.10 What is the relationship between each pair of compounds? Select your answers from the following terms: enantiomers, epimers, anomers, diastereomers, none of these. (Consult Figures 11.2 and 11.3.)
  (a) α-D-glucose and α-D-galactose
  (b) β-D-glucose and β-D-fructose
  (c) β-D-xylose and α-D-ribose
  (d) β-D-glucose and β-L-glucose
  (e) D-(+)-mannose and D-(+)-talose
  (f) D-(−)-arabinose and D-(+)-xylose
  (g) D-(+)-idose and D-(−)-gulose
  (h) α-D-fructose and β-D-fructose
  (i) D-(−)-lyxose and L-(+)-lyxose
  (j) α-D-altrose and α-D-allose

11.11 How can it be shown that a solution of α-D-glucose exhibits mutarotation? Do any of the isomeric aldotetroses exhibit mutarotation? Explain.

11.12 How does ribose differ from deoxyribose? Write a cyclic formula for each.

11.13 When ribose is dissolved in water, both furanose and pyranose forms are present at equilibrium. Draw the two pyranose anomers.

11.14 L-Sugars occur in nature, but they are not nearly as abundant as D-sugars. L-Fucose (6-deoxy-L-galactose) and L-rhamnose (6-deoxy-L-mannose) are constituents of some bacterial cell walls. Draw structures for L-fucose and L-rhamnose.

11.15 Draw formulas for each of the following.
  (a) glycolaldehyde ($C_2H_4O_2$)
  (b) L-(−)-glucose
  (c) L-(+)-fructose
  (d) the enantiomer of D-mannose
  (e) a ketopentose in the furanose form
  (f) a ketohexose in the pyranose form
  (g) an aldoheptose in the pyranose form
  (h) β-D-galactofuranose
  (i) α-D-idose
  (j) β-L-gulose
  (k) α-L-arabinofuranose
  (l) α-D-xylopyranose
  (m) α-methyl-D-fructoside
  (n) β-D-mannopyranosyl-α-D-fructofuranoside
  (o) two mannose molecules joined by an α-1,4-glycosidic bond
  (p) a trisaccharide composed of three different hexose units

11.16 Give the common name of each of the following systematically named sugars.
  (a) 4-(β-D-galactopyranosyl)-D-glucose
  (b) 4-(α-D-glucopyranosyl)-D-glucose
  (c) 4-(β-D-glucopyranosyl)-D-glucose
  (d) α-D-glucopyranosyl-β-D-fructofuranoside

11.17 Write an equation to show the mixture of glucosides that results from the reaction of methanol (plus acid) and α-D-ribofuranose.

11.18 Gentiobiose is a disaccharide composed of two glucose units joined by a β-1,6-glucosidic bond. Draw the structure of gentiobiose.

11.19 Raffinose is a trisaccharide (found in sugar beets) containing D-galactose, D-glucose, and D-fructose. The enzyme α-*galactase* catalyzes the hydrolysis of raffinose to galactose and sucrose. Draw the structure of raffinose. (The bond from galactose to the glucose moiety is α-1,6.)

11.20 Explain the structural similarities and differences between each of the following.
  (a) maltose and cellobiose
  (b) starch and cellulose
  (c) amylose and amylopectin
  (d) amylopectin and glycogen

11.21 Explain why starch is digestible by humans, but cellulose is not.

11.22 Gun cotton is cellulose trinitrate, a fluffy white material that is used to make smokeless powder. Draw a partial structure of cellulose trinitrate.

11.23 The shells of lobsters and crabs and the hard body coverings of many insects are composed of a polysaccharide called chitin. What is the formula of the D-glucosamine (the monosaccharide) that is obtained by the complete hydrolysis of chitin?

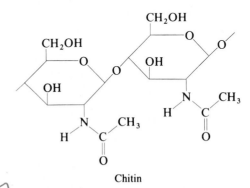

Chitin

11.24 What purposes do starch and cellulose serve in plants? What purpose is served by glycogen in animals?

11.25 What structural characteristics are necessary if a disaccharide is to be a reducing sugar? Draw the structure of a hypothetical nonreducing disaccharide composed of two aldopentoses.

11.26 Trehalose is a nonreducing disaccharide (found in mushrooms) that yields glucose upon hydrolysis. Its systematic name is α-D-glucopyranosyl-α-D-glucopyranoside. Draw the structural formula of trehalose.

11.27 D-Glucose can be oxidized at C-1 to form D-gluconic acid, at C-6 to yield D-glucuronic acid, and at both C-1 and C-6 to yield D-glucaric acid. Draw structures of these three oxidation products of glucose.

11.28 The carbonyl groups of aldoses and ketoses can be reduced to the corresponding alcohols (recall Section 7.5-d). The reduction of glucose and mannose with $NaBH_4$ yields D-sorbitol and D-mannitol, respectively. Draw their structures. When D-fructose is reduced with $NaBH_4$, the product is a mixture of D-sorbitol and D-mannitol. Explain.

11.29 Ketones cannot be oxidized by mild oxidizing agents, yet fructose is oxidized by Benedict's reagent. Explain.

11.30 Dilute alkali catalyzes the interconversion of fructose and mannose and of dihydroxyacetone and glyceraldehyde. Write equations to show these interconversions.

11.31 Which of the following will give a positive Benedict's test?

(a) L-galactose   (b) levulose
(c) D-mannose   (d) malt sugar
(e) cane sugar   (f) invert sugar

(g) milk sugar   (h) cellobiose
(i) inulin   (j) starch
(k) cellulose   (l) glycogen

11.32 Give the hydrolysis products of the compounds in Exercise 11.31 where possible.

11.33 What is the general test for a carbohydrate? Explain.

11.34 What test would you use to distinguish between the members of each of the following pairs?

(a) glucose and starch
(b) glucose and fructose
(c) sucrose and invert sugar
(d) mannose and maltose
(e) galactose and ribose
(f) fructose and erythrose

11.35 Write equations for each of the following.

(a) hydrolysis of sucrose
(b) stepwise hydrolysis of starch
(c) oxidation of glucose with Benedict's reagent
(d) enolization of D-fructose to D-glucose
(e) reaction of α-D-glucose with H_3PO_4 to produce α-D-glucose 6-phosphate

11.36 α-D-Fructose has a specific rotation of $-21.0°$, and the specific rotation of β-d-fructose is $-133.0°$. An equilibrium mixture of the two forms exhibits a specific rotation of $-92.0°$. Calculate the percentages of α- and β-fructoses present in the equilibrium mixture.

12

Lipids

The lipids are a heterogeneous group of organic compounds that are important constituents of plant and animal tissues. They are arbitrarily classed together according to their solubility in organic solvents such as benzene, ether, chloroform, carbon tetrachloride (the so-called fat solvents) and their insolubility in water. Their solubility properties are a function of their alkane-like structures. Edible lipids constitute approximately 30% of the diet of the average American, and they serve as a starting material for the production of many important commodities such as soap products. Within the past two decades, the role of lipids in the diet has received a great deal of attention because of the apparent connection of saturated fats and blood cholesterol with arterial disease. In addition, the National Cancer Institute recommends that we should reduce our consumption of fats because of the correlation between high fat intake and some forms of cancer. As we shall learn in Chapter 17, the lipids are the most important energy storage compounds in the animal kingdom. In contrast, you will recall, plants store most of their energy in the form of carbohydrates, primarily as starch. In addition, lipids provide insulation for the vital organs, protecting them from mechanical shock and maintaining optimum body temperature. Lipids that contain both polar and nonpolar groups are an integral component of cell membrane structure and, as such, are associated with transportation across cellular membranes.

12.1 Classification of Lipids

Unlike polysaccharides and proteins, lipids are not polymers—they lack a repeating monomeric unit. Lipids have been classified in several different ways. We have chosen to classify them on the basis of whether they are *saponifiable* or *nonsaponifiable*. (Recall from Section 8.12-b that saponification is the alkaline hydrolysis of esters to yield salts of acids.) The saponifiable lipids are further subdivided according to their hydrolysis products (Table 12.1).

TABLE 12.1 Classification of Lipids

Saponifiable Lipids	Hydrolysis Products	Nonsaponifiable Lipids
Fats and oils	Fatty acids and glycerol	Steroids
Waxes	Fatty acids and long-chain alcohols	Terpenes
		Prostaglandins
Phosphoglycerides	Fatty acids, glycerol, phosphoric acid, and a polar alcohol group	Fat-soluble vitamins
Sphingolipids	Fatty acids, sphingosine, phosphoric acid, and a polar alcohol group	
Glycolipids	Fatty acids, sphingosine *or* glycerol, and one or more monosaccharides	

12.2 Fats and Oils

Fats and oils[1] are the most abundant lipids found in nature. Both types of compounds are called **triacylglycerols**[2] because they are *esters* composed of *three fatty acids* joined to *glycerol*, a *trihydroxy alcohol*.

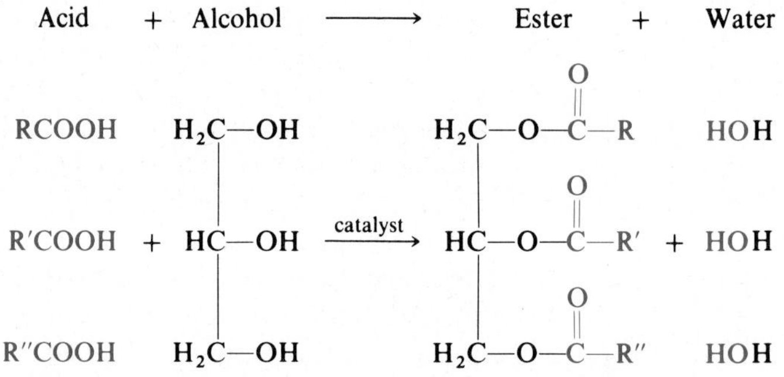

$$\text{Acid} \quad + \quad \text{Alcohol} \longrightarrow \text{Ester} \quad + \quad \text{Water}$$

$$
\begin{array}{lll}
\text{RCOOH} & \text{H}_2\text{C}-\text{OH} & \text{H}_2\text{C}-\text{O}-\overset{\displaystyle \overset{O}{\|}}{\text{C}}-\text{R} \qquad \text{HOH}\\[2mm]
\text{R'COOH} & + \;\; \text{HC}-\text{OH} \;\; \xrightarrow{\;\text{catalyst}\;} & \text{HC}-\text{O}-\overset{\displaystyle \overset{O}{\|}}{\text{C}}-\text{R'} \; + \; \text{HOH}\\[2mm]
\text{R''COOH} & \text{H}_2\text{C}-\text{OH} & \text{H}_2\text{C}-\text{O}-\overset{\displaystyle \overset{O}{\|}}{\text{C}}-\text{R''} \qquad \text{HOH}\\
\text{Three} & \text{Glycerol} & \text{A triacylglycerol}\\
\text{fatty acids} & &
\end{array}
$$

[1] The student should not confuse the term *oil*, used here to refer to a particular group of lipids, with the hydrocarbon petroleum oils.

[2] Formerly these compounds were called *triglycerides*. An international nomenclature commission has recommended that this chemically inaccurate term no longer be used.

Further classification of triacylglycerols is made on the basis of their physical states at room temperature. In general, a lipid is called a **fat** if it is a *solid at 25°C,* and an **oil** if it is a *liquid at the same temperature.* (These differences in melting points reflect differences in the degree of unsaturation of the constituent fatty acids.) Furthermore, lipids obtained from animal sources are usually solids, whereas oils are generally of plant origin. Therefore we commonly speak of *animal fats* and *vegetable oils.*

12.3 *Fatty Acids*

Since fats and oils both contain glycerol, the differences between them must be due to differences in their fatty acid components. It is therefore customary to describe triacylglycerols in terms of their fatty acids.

Fatty acids are long-chain, monocarboxylic acids; more than 70 fatty acids have been identified in nature. They may be either saturated or unsaturated, but they are invariably straight chains rather than branched or cyclic.[3] Furthermore, fatty acids that are obtained from lipids almost always contain an even number of carbon atoms.[4] The most abundant fatty acids are given in Table 12.2. Fatty acids are usually called by their common names. They are derived from Greek or Latin words that indicate the source of the compound.

TABLE 12.2 Common Fatty Acids

Name	MP (°C)	Condensed Structural Formula	Abbreviated[a] Formula
Lauric	44	$CH_3(CH_2)_{10}COOH$	$C_{11}H_{23}COOH$
Myristic	54	$CH_3(CH_2)_{12}COOH$	$C_{13}H_{27}COOH$
Palmitic	63	$CH_3(CH_2)_{14}COOH$	$C_{15}H_{31}COOH$
Stearic	70	$CH_3(CH_2)_{16}COOH$	$C_{17}H_{35}COOH$
Palmitoleic	−1	$CH_3(CH_2)_5CH{=}CH(CH_2)_7COOH$	$C_{15}H_{29}COOH$
Oleic	13	$CH_3(CH_2)_7CH{=}CH(CH_2)_7COOH$	$C_{17}H_{33}COOH$
Linoleic	−5	$CH_3(CH_2)_3(CH_2CH{=}CH)_2(CH_2)_7COOH$	$C_{17}H_{31}COOH$
Linolenic	−11	$CH_3(CH_2CH{=}CH)_3(CH_2)_7COOH$	$C_{17}H_{29}COOH$
Arachidonic	−50	$CH_3(CH_2)_3(CH_2CH{=}CH)_4(CH_2)_3COOH$	$C_{19}H_{31}COOH$

Essential f.a.

[a]Saturated fatty acids have the general formula $C_nH_{2n+1}COOH$; unsaturated fatty acids are of the form $C_nH_{2n-1}COOH$, $C_nH_{2n-3}COOH$, $C_nH_{2n-5}COOH$, and so on.

[3]Two fatty acids are exceptions to this statement. Both malvalic acid and sterculic acid contain the very highly strained cyclopropene ring.

$CH_3(CH_2)_7$ $(CH_2)_6COOH$
Malvalic acid

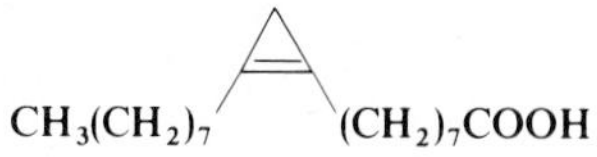

$CH_3(CH_2)_7$ $(CH_2)_7COOH$
Sterculic acid

[4]The reason for this is that the hydrocarbon chain of a given fatty acid is biosynthesized two carbon units at a time.

Notice that stearic acid, a saturated fatty acid, contains the same number of carbon atoms as three of the unsaturated acids listed in the table, yet it has a much higher melting point. It is generally the case that the greater the degree of unsaturation of a fatty acid, the lower its melting point. The explanation for this generalized statement is based upon the particular stereochemistry of the unsaturated fatty acids found in lipids. *The configuration about their double bonds is almost always **cis** rather than **trans**.*

The carbon chains of the saturated fatty acids are extended in a linear, zigzag fashion (Figure 12.1-a). Recall (Section 2.3) that unbranched, saturated hydrocarbons can be closely packed together and therefore experience stronger van der Waals attractions. (Thus, they will have higher melting points.) The presence of a *trans* double bond does not distort the linearity of the zigzag chain (Figure 12.1-b). *Cis* double bonds, however, place a severe bend in the chain (Figure 12.1-c, d), resulting in a looser packing of molecules, weaker intermolecular attractions, and a lowering of the melting point. In general, then, we can say that *fats contain a greater proportion of saturated fatty acids, whereas oils contain a greater percentage of unsaturated fatty acids.*

If all three hydroxyl groups of the glycerol molecule are esterified with the same fatty acid, the resulting ester is called a **simple triacylglycerol.** Although some simple triacylglycerols have been synthesized in the laboratory, they rarely occur in nature. All of the triacylglycerols obtained from naturally occurring fats and oils contain two or three different fatty acid components and are thus termed **mixed triacylglycerols.**

FIGURE 12.1 Extended chains of fatty acids: saturated (a), *trans*-unsaturated (b), *cis*-unsaturated (c), and *cis, cis*-unsaturated (d).

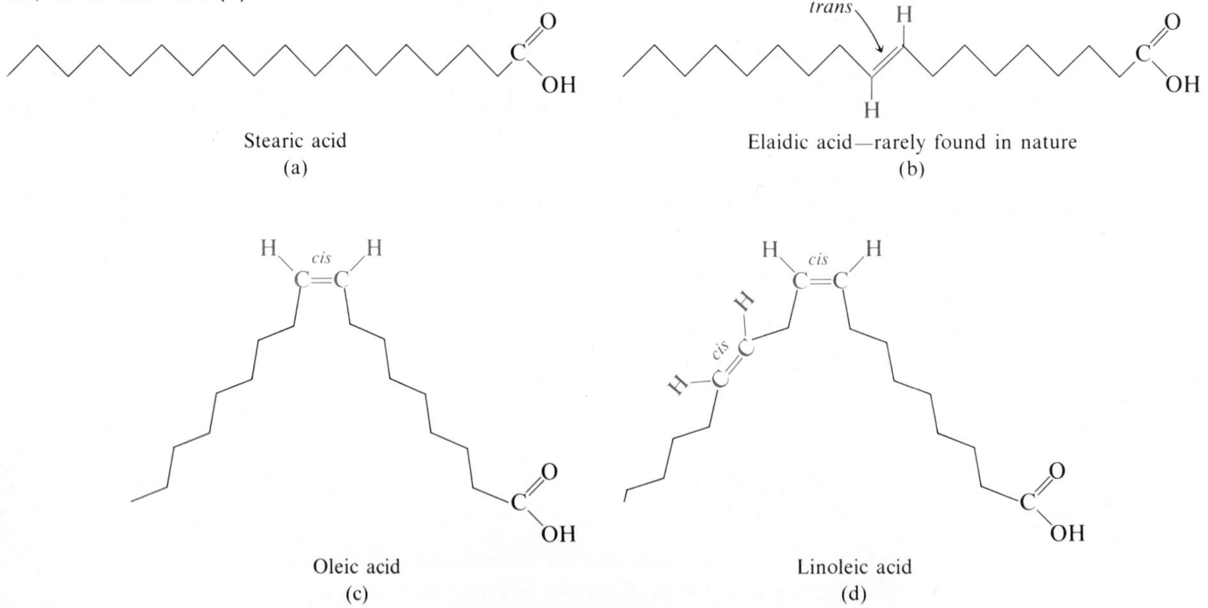

Stearic acid
(a)

Elaidic acid—rarely found in nature
(b)

Oleic acid
(c)

Linoleic acid
(d)

$$
\begin{array}{c}
\text{C}_{17}\text{H}_{35}-\overset{\displaystyle O}{\overset{\|}{\text{C}}}-\text{O}-\overset{\displaystyle H}{\overset{|}{\text{C}}}-\text{H} \\
\text{C}_{17}\text{H}_{35}-\overset{\displaystyle O}{\overset{\|}{\text{C}}}-\text{O}-\overset{|}{\text{C}}-\text{H} \\
\text{C}_{17}\text{H}_{35}-\overset{\displaystyle O}{\overset{\|}{\text{C}}}-\text{O}-\overset{|}{\underset{\displaystyle H}{\text{C}}}-\text{H}
\end{array}
\qquad
\begin{array}{c}
\text{C}_{11}\text{H}_{23}-\overset{\displaystyle O}{\overset{\|}{\text{C}}}-\text{O}-\overset{\displaystyle H}{\overset{|}{\text{C}}}-\text{H} \\
\text{C}_{15}\text{H}_{31}-\overset{\displaystyle O}{\overset{\|}{\text{C}}}-\text{O}-\overset{|}{\text{C}}-\text{H} \\
\text{C}_{17}\text{H}_{33}-\overset{\displaystyle O}{\overset{\|}{\text{C}}}-\text{O}-\overset{|}{\underset{\displaystyle H}{\text{C}}}-\text{H}
\end{array}
\qquad
\begin{array}{c}
\text{C}_{17}\text{H}_{31}-\overset{\displaystyle O}{\overset{\|}{\text{C}}}-\text{O}-\overset{\displaystyle H}{\overset{|}{\text{C}}}-\text{H} \\
\text{C}_{17}\text{H}_{31}-\overset{\displaystyle O}{\overset{\|}{\text{C}}}-\text{O}-\overset{|}{\text{C}}-\text{H} \\
\text{C}_{17}\text{H}_{31}-\overset{\displaystyle O}{\overset{\|}{\text{C}}}-\text{O}-\overset{|}{\underset{\displaystyle H}{\text{C}}}-\text{H}
\end{array}
$$

Glyceryl stearate
(Tristearin)
(a simple triacylglycerol)
mp 71°C

Glyceryl lauropalmitooleate
(a mixed triacylglycerol)

Glyceryl linoleate
(Trilinolein) 3 of *Linoleic*
(a simple triacylglycerol)
mp 9°C

No single formula can be written to represent the naturally occurring fats and oils since they are highly complex mixtures of molecules in which many different fatty acids are represented. Table 12.3 shows the fatty acid composition of some common fats and oils. Notice that a fairly wide range of values occurs. The range is wide because the composition of lipids is variable and depends upon the plant or animal species involved as well as dietetic and climatic factors. For example, lard from corn-fed hogs is more highly saturated than lard from peanut-fed hogs. Linseed oil obtained from cold climates is more unsaturated than linseed oil from warm

TABLE 12.3 Fatty Acid Components of Some Common Fats and Oils

	Component Acids (%)[a]							Typical Iodine Number
	Lauric (C_{12})	Myristic (C_{14})	Palmitic (C_{16})	Stearic (C_{18})	Oleic (C_{18})	Linoleic (C_{18})	Linolenic (C_{18})	
Fats								
Butter	1–4	8–13	25–32	8–13	22–29	2–4		36
Tallow (beef)		2–3	24–32	20–25	37–43	2–3		50
Lard (hog)		1–2	25–30	12–16	40–50	3–8		59
Edible Oils								
Coconut oil	44–50	13–18	7–10	1–4	5–8	1–3		10[b]
Olive oil	0–1	0–2	7–20	2–3	53–86	4–22		81
Peanut oil		0–1	6–11	3–6	40–65	17–38		93
Cottonseed oil		0–3	17–23	1–3	23–44	34–55		106
Corn oil		1–2	8–12	2–5	29–49	34–56		123
Soybean oil		0–1	6–10	2–5	20–30	50–60	2–10	130
Safflower oil			6–7	2–3	12–14	75–80	0–2	145
Nonedible Oil								
Linseed oil		0–1	5–9	4–7	9–29	8–29	45–67	179

[a]Totals less than 100% indicate the presence of lower or higher acids in small amounts.
[b]Coconut oil is a highly saturated oil, hence the very low iodine number. It contains an unusually high percentage (53–70%) of the low-melting C_8, C_{10}, and C_{12} saturated fatty acids. Coconut oil is a liquid in the warmer, tropical climates, but at room temperature in the temperate zone, it is a solid.

climates. Palmitic acid is the most abundant of the saturated fatty acids, and oleic acid is the most abundant unsaturated fatty acid. Unsaturated fatty acids predominate over saturated ones for most plants and animals.

12.4 Physical Properties of Triacylglycerols

As previously mentioned, the triacylglycerols may be either liquids or noncrystalline solids at room temperature. Contrary to popular belief, *pure* fats and oils are colorless, odorless, and tasteless. The characteristic colors, odors, and flavors associated with these lipids are imparted to them by foreign substances that have been absorbed by the lipids and are soluble in them. For example, the yellow color of butter is due to the presence of the pigment carotene; the taste of butter is a result of two compounds, diacetyl ($CH_3COCOCH_3$) and 3-hydroxy-2-butanone ($CH_3COCHOHCH_3$), that are produced by bacteria in the ripening of the cream. Fats and oils are lighter than water, having densities of about 0.8 g/cm^3. They are poor conductors of heat and electricity and therefore serve as excellent insulators for the body.

12.5 Chemical Properties of Triacylglycerols

a. Saponification

Triacylglycerols may be hydrolyzed by several procedures, the most common of which utilizes alkali or enzymes called *lipases*. Alkaline hydrolysis is termed **saponification** because one of the products of the hydrolysis is a soap, generally sodium or potassium salts of fatty acids (Section 8.12). Notice that, stoichiometrically, complete saponification of 1 mole of triacylglycerol requires 3 moles of base.

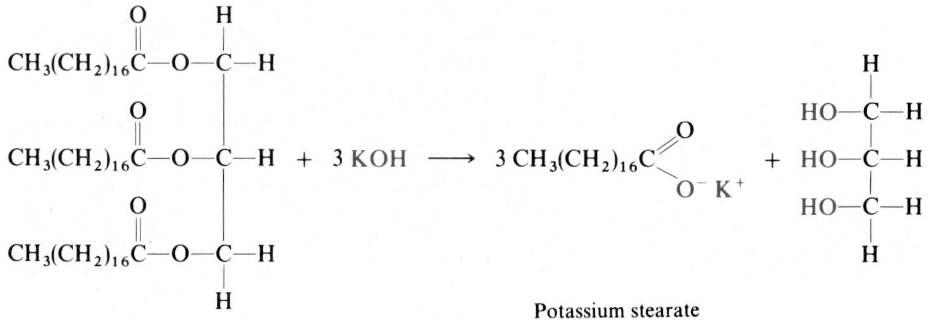

Tristearin

Potassium stearate
(a soap)

b. Halogenation

Unsaturated fatty acids, whether they are free or combined as esters in fats and oils, react with halogens by addition at the double bond(s). The reaction (halogenation) results in the decolorization of the halogen solution (Section 3.4-a). Since the extent of addition by a fat or oil is proportional to the number of double bonds in the fatty acid moieties, the amount of halogen absorbed by a lipid can be used as an index of the degree of unsaturation. The index value is called the **iodine number** and is defined as the *number of grams of iodine (or iodine equivalent) that will add to 100 g of fat or oil.* This value is influenced by a number of factors, such as percentage of unsaturated fatty acids in the triacylglycerol molecule and the degree of unsaturation of each fatty acid. A high iodine number indicates a high degree of unsaturation. Natural fats that have a preponderance of saturated fatty acids have iodine numbers of about 10–50; those that contain an abundance of polyunsaturated fatty acids have iodine numbers of 120–150.

$$
\begin{array}{l}
CH_3(CH_2)_7CH=CH(CH_2)_7\overset{\overset{\displaystyle O}{\|}}{C}-O-\overset{\overset{\displaystyle H}{|}}{C}-H \\[4pt]
CH_3(CH_2)_7CH=CH(CH_2)_7\overset{\overset{\displaystyle O}{\|}}{C}-O-\overset{|}{C}-H \quad +\ 3\,I_2 \\[4pt]
CH_3(CH_2)_7CH=CH(CH_2)_7\overset{\overset{\displaystyle O}{\|}}{C}-O-\overset{\underset{\displaystyle H}{|}}{C}-H
\end{array}
$$

Triolein

$$
\longrightarrow
\begin{array}{l}
CH_3(CH_2)_7\overset{\overset{\displaystyle I}{|}}{C}H-\overset{\overset{\displaystyle I}{|}}{C}H(CH_2)_7\overset{\overset{\displaystyle O}{\|}}{C}-O-\overset{\overset{\displaystyle H}{|}}{C}-H \\[4pt]
CH_3(CH_2)_7\overset{\overset{\displaystyle I}{|}}{C}H-\overset{\overset{\displaystyle I}{|}}{C}H(CH_2)_7\overset{\overset{\displaystyle O}{\|}}{C}-O-\overset{|}{C}-H \\[4pt]
CH_3(CH_2)_7\overset{\overset{\displaystyle I}{|}}{C}H-\overset{\overset{\displaystyle I}{|}}{C}H(CH_2)_7\overset{\overset{\displaystyle O}{\|}}{C}-O-\overset{\underset{\displaystyle H}{|}}{C}-H
\end{array}
$$

The preceding equation indicates the addition of molecular iodine. In actual practice, however, the reagents used are the interhalogens iodine monochloride (IC1) or iodine monobromide (IBr), both of which are more reactive than iodine alone. A weighed sample of lipid is treated with an excess of the iodine reagent. After the reaction is complete, the unused halogen is determined by titration with a standard solution of sodium thiosulfate.

c. Hydrogenation

A large-scale commercial industry has been developed for the purpose of transforming vegetable oils into edible fats. The chemistry of this conversion process is essentially identical to the catalytic hydrogenation reaction that has been described for the alkenes in Section 3.4-a. *The process of converting oils to fats by means of hydrogenation* is sometimes referred to as **hardening.** One method consists of bubbling hydrogen gas under pressure (25 lb/in.^{2}) into a tank of hot oil (200°C) containing a finely dispersed nickel catalyst. An example is the conversion of triolein to tristearin.

$$
\begin{array}{l}
CH_3(CH_2)_7-CH=CH-(CH_2)_7-\overset{\overset{\displaystyle O}{\|}}{C}-O-\overset{\overset{\displaystyle H}{|}}{C}-H \\[4pt]
CH_3(CH_2)_7-CH=CH-(CH_2)_7-\overset{\overset{\displaystyle O}{\|}}{C}-O-\overset{|}{C}-H \quad +\ 3\,H_2 \\[4pt]
CH_3(CH_2)_7-CH=CH-(CH_2)_7-\overset{\overset{\displaystyle O}{\|}}{C}-O-\overset{\underset{\displaystyle H}{|}}{C}-H
\end{array}
$$

Triolein

$$
\xrightarrow[\Delta]{Ni}
\begin{array}{l}
CH_3(CH_2)_{16}-\overset{\overset{\displaystyle O}{\|}}{C}-O-\overset{\overset{\displaystyle H}{|}}{C}-H \\[4pt]
CH_3(CH_2)_{16}-\overset{\overset{\displaystyle O}{\|}}{C}-O-\overset{|}{C}-H \\[4pt]
CH_3(CH_2)_{16}-\overset{\overset{\displaystyle O}{\|}}{C}-O-\overset{\underset{\displaystyle H}{|}}{C}-H
\end{array}
$$

Tristearin

The equation represents the complete saturation of an unsaturated lipid. In the actual hardening process, the extent of hydrogenation is controlled so as to maintain a certain number of unsaturated linkages. (If all the bonds are hydrogenated, the product would become hard and brittle like tallow.) If reaction conditions are properly controlled, it is possible to prepare a fat with a desirable physical consistency (soft and pliable). In this manner, inexpensive and abundant vegetable oils (cottonseed, corn, soybean) are converted into oleomargarine and cooking fats (Spry, Crisco, etc.—Figure 12.2). In the preparation of margarine, the partially hydrogenated oils are mixed with water, salt, and nonfat dry milk. Flavoring agents, coloring agents, and vitamins A and D are added to approximate butter. (Preservatives and antioxidants are also added.) The peanut oil in peanut butter has been partially hydrogenated to prevent the oil from separating out. Today, because of the possible connection between saturated fats and arterial disease (see Section 12.12-a), many people are cooking with the vegetable oils (especially safflower seed oil) rather than with the hydrogenated products.

If the hydrogenation of an oil is allowed to continue for a long period of time, glycerol and long-chain alcohols are formed. This reaction is analogous to the previously mentioned reduction of esters (Section 8.12-c). These long-chain alcohols are employed in the manufacture of synthetic detergents (see Section 12.8-b).

Figure 12.2 Hydrogenation of an oil. The beaker on the left contains a clear vegetable oil before hydrogenation; on the right is the same oil hardened by hydrogenation.

Chapter 12 · Lipids

$$\begin{array}{l} CH_3(CH_2)_{10}\overset{\displaystyle O}{\overset{\displaystyle \|}{C}}-O-\overset{\displaystyle H}{\underset{\displaystyle |}{C}}-H \\[2mm] CH_3(CH_2)_{10}\overset{\displaystyle O}{\overset{\displaystyle \|}{C}}-O-\overset{\displaystyle |}{\underset{\displaystyle |}{C}}-H \quad + \quad 6\,H_2 \quad \xrightarrow{\text{catalyst}} \quad 3\,CH_3(CH_2)_{10}CH_2OH \quad + \\[2mm] CH_3(CH_2)_{10}\overset{\displaystyle O}{\overset{\displaystyle \|}{C}}-O-\underset{\displaystyle H}{\overset{\displaystyle |}{C}}-H \end{array}$$

$$\begin{array}{l} HO-\overset{\displaystyle H}{\underset{\displaystyle |}{C}}-H \\[1mm] HO-\overset{\displaystyle |}{\underset{\displaystyle |}{C}}-H \\[1mm] HO-\underset{\displaystyle H}{\overset{\displaystyle |}{C}}-H \end{array}$$

Trilaurin (abundant in coconut oil)	Lauryl alcohol	Glycerol

12.6 Special Features of Triacylglycerols

a. Rancidity

The term **rancid** is applied to *any fat or oil that develops a disagreeable odor.* Two principal chemical reactions are responsible for causing rancidity—*hydrolysis* and *oxidation.*

Butter is particularly susceptible to hydrolytic rancidity because it contains many of the lower molecular weight acids (butyric, caproic), all of which have offensive odors. Under moist and warm conditions, hydrolysis of the ester linkages occurs, liberating the volatile acids. Microorganisms present in the air furnish the enzymes (*lipases*) that catalyze the process. Hydrolytic rancidity can easily be prevented by storing butter covered in a refrigerator.

Oxidative rancidity occurs in triacylglycerols containing polyunsaturated fatty acids. The reaction is quite complex, but it is believed that the first step involves the formation of a free radical, followed by production of hydroperoxides. Further oxidation reactions occur in which bonds are cleaved and the short-chain, offensive-smelling carboxylic acids are liberated.

$$\begin{array}{c} \overset{\displaystyle H}{\underset{}{|}}\ \overset{\displaystyle H}{\underset{}{|}}\ \overset{\displaystyle H}{\underset{}{|}}\ \overset{\displaystyle H}{\underset{}{|}}\ \overset{\displaystyle H}{\underset{}{|}} \\ -C=C-C-C=C- \\ \underset{\displaystyle H}{|} \end{array} \quad \xrightarrow{O_2} \quad \begin{array}{c} \overset{\displaystyle H}{\underset{}{|}}\ \overset{\displaystyle H}{\underset{}{|}}\ \overset{\displaystyle H}{\underset{}{|}}\ \overset{\displaystyle H}{\underset{}{|}}\ \overset{\displaystyle H}{\underset{}{|}} \\ -C=C-C-C=C- \\ \underset{\displaystyle O-O-H}{|} \end{array}$$

A hydroperoxide

Rancidity is a major concern of the food industry, and chemists involved in this area are continually seeking new and better substances to act as *antioxidants.* Such compounds are added in very small amounts (0.01–0.001%) to suppress rancidity. They have a greater affinity for oxygen than the lipid to which they are added and thus function by preferentially depleting the supply of adsorbed oxygen. We mentioned the synthetic antioxidant BHT in Section 2.5; two of the naturally occurring antioxidants are tocopherol (vitamin E) and ascorbic acid (vitamin C) (see Section 12.16).

b. Drying Oils

A **drying oil** is *any substance that causes a paint or varnish to develop a hard, protective coating.* It is the susceptibility of highly unsaturated oils to react with oxygen that accounts for their usefulness in the paint industry. Linseed oil is especially reactive and is most commonly used. The term *drying* may be a misnomer because it implies that the protective coating is formed by the evaporation of the solvent. Instead, the drying process involves an oxidation followed by a polymerization reaction that results in the formation of a vast interlocking network of triacylglycerols joined by peroxide bridges. These oxidation–polymerization reactions are catalyzed by metal ions (manganese, cobalt), and salts of these metals are included in paint tò hasten the drying process. *Oil paints* are suspensions of very finely divided pigments in linseed oil.

The reactions involved in the drying process are exothermic, and heat is therefore given off to the surroundings. If the surroundings happen to be a poorly ventilated container, this heat will raise the oil to its kindling temperature and cause *spontaneous combustion.* Thus paint or oil rags should never be bundled together and stored in wood or cardboard containers.

12.7 *Waxes*

A **wax** is *an ester of a long-chain alcohol* (usually monohydroxy) *and a fatty acid.* (Household paraffin wax, which is a mixture of high molecular weight hydrocarbons, has wax-like properties but is not a wax.) The acids and alcohols normally found in waxes have chains of the order of 12–34 carbon atoms in length. Waxes are low-melting solids that are widely distributed in nature and are found in both plants and animals. They are not as easily hydrolyzed as the triacylglycerols and therefore are useful as protective coatings. Plant waxes are found on the surfaces of leaves, stems, flowers, and fruits and serve to protect the plant from dehydration and from invasion by harmful microorganisms. (You can polish an apple to a high luster because of the waxes present in its skin.) Carnauba wax, largely myricyl cerotate ($C_{25}H_{51}COOC_{30}H_{61}$), is obtained from the leaves of certain Brazilian palm trees and is used as a floor and automobile wax and as a coating on carbon paper and mimeograph stencils.

Animal waxes also serve as protective coatings. They are found on the surface of feathers, skin, and hair and help to keep these surfaces pliable and water repellent. The waxy coating on the feathers of water birds (ducks, gulls) helps them to stay afloat. If this wax is dissolved as a result of the bird swimming in an oil slick, the feathers become wet and heavy; the bird cannot maintain its buoyancy and will drown. Earwax protects the delicate lining of the inner ear. Many marine species store waxes as fuels. Spermaceti wax, mainly cetyl palmitate ($C_{15}H_{31}COOC_{16}H_{33}$), is found in the head cavities and the blubber of the sperm whale. Spermaceti wax crystallizes in heavy white flakes when whale oil is exposed to air and chilled. It is used primarily in ointments, in cosmetics, and in the manufacture of candles. Lanolin, obtained by washing sheep's wool, is a mixture of fatty acid esters of the steroids lanosterol and agnosterol. It finds widespread medical applications as a base for creams, ointments, and salves.

Insects also secrete waxes. Beeswax, which is mostly myricyl palmitate ($C_{15}H_{31}COOC_{30}H_{61}$), is secreted by the wax glands of the bee and is used to construct the combs in which bees store their honey. Beeswax is obtained by heating the honeycombs in water and skimming off the wax that floats on the surface. It is the wax most often used in candles, cosmetics, wax paper, and medicinals.

12.8 Soaps and Synthetic Detergents

It has already been mentioned that soaps are alkali metal salts of long-chain fatty acids. Alkaline hydrolysis of a fat or oil produces a soap and glycerol.[5] The older method of soap production consisted of treating molten tallow (the fat of cattle and sheep) with a slight excess of alkali in large open kettles. The mixture was heated and steam was bubbled through it. After the saponification process was completed, the soap was precipitated by the addition of sodium chloride and then filtered and washed several times with water. It was then reprecipitated from the aqueous solution by the addition of more sodium chloride. The glycerol was recovered from the aqueous wash solutions.

Today most soaps are prepared by a continuous process wherein triacylglycerols (frequently tallow and/or coconut oil) are hydrolyzed by water under high pressures and temperatures (700 lb/in.² and 200°C). Sodium carbonate is used instead of the more expensive sodium hydroxide to neutralize the fatty acids.

$$
\begin{array}{l}
CH_2OOC(CH_2)_nCH_3 \\
| \\
CHOOC(CH_2)_nCH_3 \\
| \\
CH_2OOC(CH_2)_nCH_3
\end{array}
\xrightarrow[\substack{heat, \\ pressure}]{H_2O}
\text{Glycerol} + 3\ CH_3(CH_2)_nCOOH
\xrightarrow{Na_2CO_3}
3\ CH_3(CH_2)_nCOO^-\ Na^+
$$

Fatty acid   Sodium salt of a fatty acid (a soap)

The crude soap is used as industrial soap without further processing. Pumice or sand may be added to produce scouring soap. Other ingredients, such as dyes, perfumes, and antiseptics, are added to produce colored soaps, fragrant soaps, and deodorizing soaps, respectively. If air is blown through molten soap, a floating soap is produced. Such a soap is not necessarily purer than other soaps; it merely contains more air. Ordinary soap is a mixture of the sodium salts of various fatty acids. Potassium soaps (soft soap) are more expensive but produce a softer lather and are more soluble. They are used in liquid soaps, shampoos, and shaving creams. Tincture of green soap is an alcoholic solution of a potassium soap that is commonly used in hospitals.

a. The Cleansing Action of Soaps

A soap molecule can be considered to be composed of a large nonpolar hydrocarbon portion (*hydrophobic*—repelled by water) and a carboxylate salt end (*hydrophilic*—water soluble).

[5] Recall that the glycerol obtained as a by-product in the manufacture of soaps is used to make the explosive nitroglycerin (Section 6.5-e). During World Wars I and II, people saved excess cooking fats and oils and turned them in for reclamation of the glycerol.

$$\overset{\text{Hydrophobic}}{\text{(Lipophilic)}} \qquad\qquad\qquad\qquad\qquad\qquad\qquad\qquad \overset{\text{Hydrophilic}}{\text{(Lipophobic)}}$$

$$CH_3CH_2CH_2CH_2CH_2CH_2CH_2CH_2CH_2CH_2CH_2CH_2CH_2CH_2CH_2CH_2CH_2-C \overset{\displaystyle O}{\underset{\displaystyle O^- \ Na^+}{\diagdown}}$$

Sodium stearate

When soap is added to water, the hydrophilic ends of the molecules are attracted to the water and dissolve in it, but the hydrophobic ends are repelled by the water molecules. Consequently, a thin film (suds) forms on the surface of the water, drastically lowering its surface tension and increasing its "wetting" capacity (Figure 12.3). The soap solution can spread out more and penetrate fabrics, rather than just beading up on the surface. Soil can then be more readily loosened and removed. (You can observe this reduced surface tension by placing a drop of water on the back of one hand and a drop of soap solution on the other hand.)

When the soap solution is brought into contact with grease or oil (most dirt is held to clothes by a thin film of grease or oil), soap molecules become reoriented. The hydrophobic portions dissolve in the grease or the oil, and the hydrophilic ends remain dissolved in the aqueous phase. Mechanical action, such as scrubbing, causes the oil or grease to disperse into tiny droplets, and soap molecules arrange themselves around the surface of the globules. Oil or grease droplets surrounded by soap molecules are examples of micelles. (**Micelles** *are aggregations of molecules that contain both polar and nonpolar groups.*) Because the polar carboxylate ends of the soap molecules project outward, the surface of each drop is negatively charged; therefore the drops repel one another and do not coalesce (Figure 12.4). The entire micelle becomes water soluble and is able to be washed away by a stream of water. The cleansing process involves both the lowering of the surface tension of water and **emulsification** (*the conversion of large, water-insoluble fat globs into a suspension of smaller fat globules*). As we shall see in Section 17.1, the bile salts serve as emulsifying agents in the digestion of dietary lipids.

One major disadvantage involved in the use of soap results from the presence of certain metal ions in hard water. Calcium and magnesium ions form precipitates

FIGURE 12.3 (a) The formation of a thin film (a monolayer) on the surface of water. (b) A schematic representation. The surface of the liquid changes from arrays of strongly associated water molecules to nonassociated, nonpolar hydrocarbon chains.

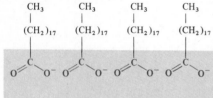

(a)

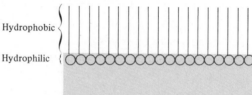

(b)

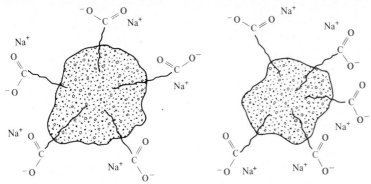

FIGURE 12.4 Two micelles will not coalesce because of the repulsions between their surrounding carboxylate groups.

with the carboxylate ions of fatty acids. An equation representing the action of hard water upon soap is

$$2\ CH_3(CH_2)_{16}C\diagdown\substack{O \\ O^-\ Na^+}\ +\ Ca^{2+}\ \longrightarrow\ \left(CH_3(CH_2)_{16}C\diagdown\substack{O \\ O^-}\right)_2 Ca^{2+}\ +\ 2\ Na^+$$

Sodium stearate
(soluble)

Calcium stearate
(insoluble)

These soap precipitates form the "bathtub ring" and the white insoluble curd found at the bottom of washing machines. The various methods used for softening hard water all involve the removal of the calcium and magnesium ions.

A second disadvantage of soaps is that they cannot be used in any region where the wash water is acidic. In the presence of excess hydrogen ions, soaps are converted back to fatty acids. The fatty acids are much less soluble in water than their salts, and they precipitate out of the water, forming a scum. (Sodium stearate is soluble in water, whereas the solubility of stearic acid is about 0.0003 g/100 mL of water.)

b. Synthetic Detergents

The term **detergent** is a rather general one meaning *any cleansing agent*. Soaps would fall under such a broad definition. However, the popular use of the word generally refers to synthetic detergents, also called *syndets*. Syndets have the desirable property of not forming precipitates with the ions of hard water. There are close to a thousand synthetic detergents commercially available in the United States, and worldwide production exceeds 25 million tons. They may be classified as anionic, cationic, or nonionic as determined by their water-soluble portions.

1. Anionic Detergents Anionic detergents contain a negatively charged ion incorporated at one end of the long hydrocarbon portion. The major anionic detergents are sulfates of long-chain alcohols (linear alkyl sulfates, LAS) or sulfonate salts of aromatic hydrocarbons (alkylbenzenesulfonates, ABS).

$$CH_3(CH_2)_{11}O-\overset{\overset{\displaystyle O}{\|}}{\underset{\underset{\displaystyle O}{\|}}{S}}-O^-\ Na^+$$

Sodium lauryl sulfate
(in shampoos)

Hydrophobic ⎯⎯⎯ Hydrophilic

$$CH_3(CH_2)_{11}-\hexagon-\overset{\overset{\displaystyle O}{\|}}{\underset{\underset{\displaystyle O}{\|}}{S}}-O^-\ Na^+$$

Sodium dodecylbenzenesulfonate
(in laundry detergents)

2. **Cationic Detergents** Cationic detergents are sometimes referred to as *invert soap* because their water-soluble end carries a positive, rather than a negative, charge. In addition to being good cleansing agents, they possess germicidal properties and are widely used in hair rinses and mouthwashes and for surgical scrubs and preoperative cleansing of patients in hospitals.

$$\hexagon-CH_2-\overset{\overset{\displaystyle CH_3}{|}}{\underset{\underset{\displaystyle C_8H_{17}}{|}}{\overset{+}{N}}}-CH_3\ Cl^-$$

Benzyldimethyloctylammonium
chloride

Hydrophilic

$$CH_3(CH_2)_{15}-\overset{\overset{\displaystyle CH_3}{|}}{\underset{\underset{\displaystyle CH_3}{|}}{\overset{+}{N}}}-CH_3\ Cl^-$$

Trimethylhexadecylammonium
chloride

3. **Nonionic Detergents** Nonionic detergents contain polar covalent structures that provide the required water solubility. They tend to be low sudsing and are used extensively in liquid soaps and dishwashing liquids, and on all occasions that call for the absence of inorganic ions.

Hydrophobic ⎯⎯⎯ Hydrophilic

$$CH_3(CH_2)_{14}\overset{\overset{\displaystyle O}{\|}}{C}-OCH_2-\overset{\overset{\displaystyle CH_2OH}{|}}{\underset{\underset{\displaystyle CH_2OH}{|}}{C}}-CH_2OH$$

Pentaerythrityl palmitate

4. Environmental Effects The large-scale use of synthetic detergents during the 20 years following World War II created a serious disposal problem. Soaps, which contain straight-chain alkyl groups, can be *removed from waste water through degradation by microorganisms in the soil* (septic tanks) or in sewage treatment plants; soaps are therefore said to be **biodegradable.** Many of the synthetic detergents could not be removed in this manner. The metabolism of microorganisms is adapted to the straight-chain alkyl groups found in soaps and natural fats, but could not break down the highly branched analogs used in the early syndets. The synthetic detergents continued to foam and make suds,[6] which clogged waste disposal plants, killed fish and wildlife by polluting streams, and even managed to make their way into city drinking water. Since 1966, all United States companies have been using straight-chain hydrocarbons in the

[6]It is interesting to note that many effective detergents do not foam in water. Experiments have proved that the degree of sudsing has very little to do with the efficiency of a detergent. However, the consumer has come to associate sudsing with cleaning ability, so manufacturers often add sudsing agents to their products.

production of syndets. Although this involves a greater expense, it seems to have alleviated some of the pollution problem.

A second environmental problem caused by detergents has not been solved. In their search for more effective cleansing agents, manufacturers have added "builders" to their detergents. Builders have little detergent effectiveness alone; their functions are (1) to tie up metal ions, (2) to prevent soil from redepositing on clothes, (3) to maintain a proper level of alkalinity in the wash water, and (4) to add bulk to detergent formulations (i.e., to fill up the detergent box so that consumers believe they are purchasing a full box of detergent). Although a number of inorganic compounds (carbonates, bicarbonates, borates, silicates) have been used as builders, the phosphates are the most effective. When phosphates were first added to detergents, a typical detergent might have contained as much as 50% phosphate. Approximately half of the phosphate content of domestic sewage was contributed by detergents, the remainder being derived from human wastes.

As later chapters will show, phosphate is a nutrient required for plant (and animal) growth. It has thus been implicated as the principal cause of **eutrophication** of lakes and rivers. (Eutrophic derives from the Greek *eu,* well, and *trophos,* pertaining to nourishment.) Eutrophic lakes and rivers contain an over-abundance of plants, especially algae (algae blooms), and decreased levels of oxygen. Only certain species of fish (carp, crappie, perch, bullhead) can live in such an environment. Moreover, the algae can produce chemicals that cause unpleasant tastes and smells in water and that, in some cases, are toxic to animals.

As a result of the environmental impact of eutrophication, states and local communities have passed laws that either ban phosphate detergents or drastically reduce the amount of phosphates. One method of reducing the phosphate concentration in natural waters is to substitute other builders for phosphates in detergents. One such compound developed was the sodium salt of nitrilotriacetic acid, NTA.

$$^+Na\ ^-O-\overset{\overset{\displaystyle O}{\|}}{P}-O-\overset{\overset{\displaystyle O^-\ Na^+}{\|}}{P}-O-\overset{\overset{\displaystyle O}{\|}}{P}-O^-\ Na^+$$

Sodium tripolyphosphate, $Na_5P_3O_{10}$
(most common phosphate builder)

$$^+Na\ ^-OOCCH_2-\overset{\overset{\displaystyle}{\underset{\underset{\displaystyle CH_2COO^-\ Na^+}{|}}{N}}}{}-CH_2COO^-\ Na^+$$

Sodium nitrilotriacetate
(NTA)

In the late 1960s several detergents were marketed in Sweden using NTA as a builder. Preliminary testing uncovered no hazards to human or animal life, and NTA did not act as a significant nutrient source for algae. Then, in 1970, tests performed in the United States showed that NTA, when administered in high doses to rats and mice, in combination with cadmium and mercury (two metals commonly found in waste water), caused a significant increase in birth defects in the test animals. NTA therefore appeared to be teratogenic in combination with certain metals and was considered to be a hazardous chemical. The detergent industry removed NTA detergents from the United States market.

The Environmental Protection Agency subsequently re-evaluated the environmental and health risks from NTA and concluded "that the projected levels of exposure from the use of NTA in laundry detergents are generally low and therefore that the associated risks also would be low." Hence, the EPA will allow the use of laundry detergents containing NTA, but they have asked the manufacturers of NTA to limit occupational exposure to NTA and to refrain from using NTA in consumer products such as shampoos and foods, in which there would be direct skin or oral contact. It is expected that detergents containing NTA will be marketed in areas of the United States where phosphate-containing detergents are forbidden.

12.9 *Phospholipids*

A **phospholipid** (also called a phosphatide) is *any lipid that contains a phosphate group.* Therefore, both phosphoglycerides and sphingolipids are subclasses of phospholipids. Phosphoglycerides are derived from glycerol, whereas sphingolipids contain sphingosine as their backbone.

Phospholipids are found in all living organisms. They are particularly abundant in the biological membranes that surround individual cells and in the membranes surrounding certain organelles within the cell. Figure 12.5 illustrates the three major cell types found in living organisms. Phospholipids are the most polar of lipids, and

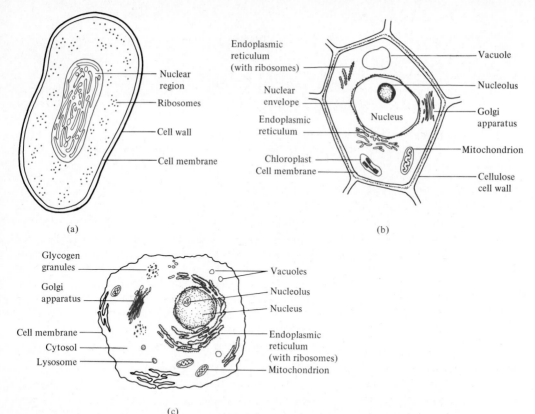

Figure 12.5 The structural features of living cells. (a) The bacterium *E. coli* is an example of a prokaryote. They are simple cells that are packed with ribosomes and are surrounded by both a cell membrane and a polysaccharide cell wall. They lack the internal membranes that are characteristic of the eukaryotes as exemplified by (b) a typical plant cell and (c) a typical animal cell.

they contain both hydrophilic and hydrophobic groups. This characteristic property is utilized in the structure of membranes (see Section 12.11).

Phospholipids also play other vital roles within the organism. They take part in fat metabolism by promoting the transport of lipids in the bloodstream as lipoprotein complexes (see Section 12.17). They are involved in oxidative phosphorylation, in secretory processes, and in the transport of certain molecules across cell membranes.

a. Phosphoglycerides

The parent compound of the phosphoglycerides is *phosphatidic acid*. Two of the hydroxyl groups of glycerol are esterified to two fatty acid chains, and the third hydroxyl group is esterified to phosphoric acid. Frequently, one of the fatty acids is saturated and the other is unsaturated.

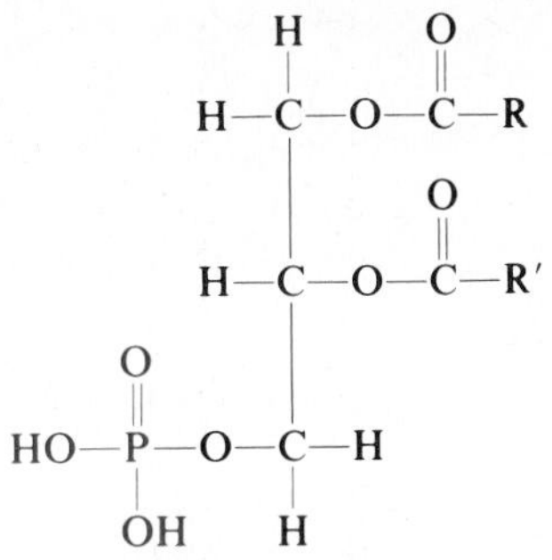

$$\begin{array}{c}
\text{H} \quad\quad \text{O} \\
\quad\quad\quad\quad\quad\quad \| \\
\text{H}-\text{C}-\text{O}-\text{C}-\text{R} \\
\quad\quad\quad\quad\quad\quad \text{O} \\
\quad\quad\quad\quad\quad\quad \| \\
\text{H}-\text{C}-\text{O}-\text{C}-\text{R}' \\
\text{O} \\
\| \\
\text{HO}-\text{P}-\text{O}-\text{C}-\text{H} \\
| \quad\quad\quad | \\
\text{OH} \quad\quad \text{H}
\end{array}$$

Phosphatidic acid
(Diacylglycerol 3-phosphoric acid)

Only small amounts of phosphatidic acid occur in the cell. It serves as the precursor for the biosynthesis of the phosphoglycerides. There are many different phosphoglycerides, depending upon which fatty acids and which alcohol groups they contain. The most abundant phosphoglycerides in higher plants and animals are lecithin (phosphatidylcholine) and the cephalins (phosphatidylethanolamine and phosphatidylserine). These compounds are major components of most plant and animal cell membranes.

Lecithin is probably the most common of the phosphoglycerides. It contains the quaternary ammonium salt *choline*, $HOCH_2CH_2N^+(CH_3)_3$, joined to a phosphoric acid residue by means of an ester linkage. At a pH of 7, the nitrogen in choline carries a positive charge and the phosphate a negative charge so that at physiological pH values, lecithin exists as an **internal salt** or **zwitterion.** (We discuss zwitterions at greater length in the next chapter.) The structure and hydrolysis products of lecithin are shown in the following equation.

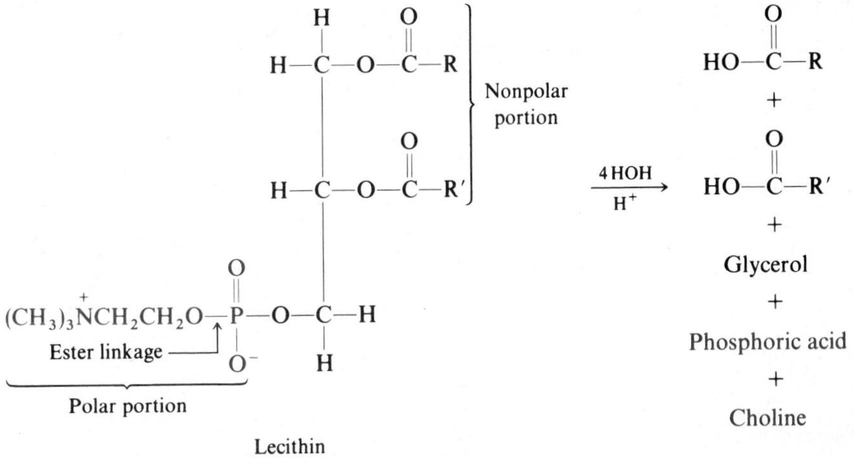

Lecithin
(Phosphatidylcholine)

Pure lecithin is an odorless, tasteless, waxy, white solid that quickly darkens when exposed to air (because of the peroxidation of the unsaturated fatty acid portion). In contrast to fats and oils, it is colloidally dispersed in water and is insoluble in anhydrous acetone. It is therefore possible to separate lecithin from an ether extract by the addition of acetone. Lecithin is especially abundant in egg yolk and soybeans. Commercially, lecithin is obtained almost entirely from soybeans, and it is used to stabilize emulsions such as margarine and mayonnaise.

The term **cephalin** is derived from the location of its chief occurrence in the body, namely, the head and spinal tissue (Greek, *kephalikos,* head). Cephalins are also essential in the blood-clotting mechanism. The chief difference between the cephalins and lecithin lies in the component linked to the phosphate group (i.e., ethanolamine and serine instead of choline).

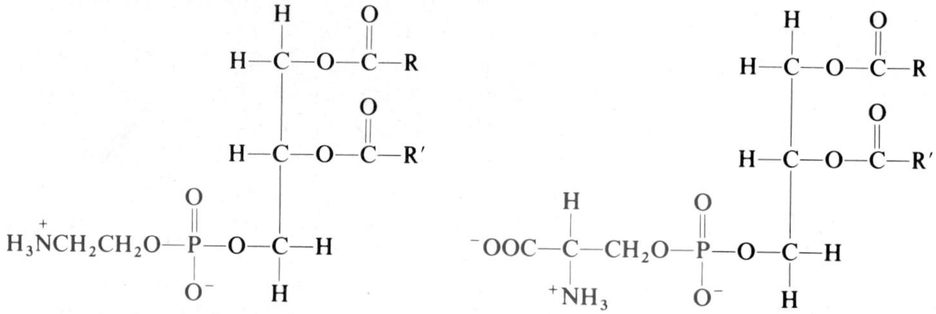

Phosphatidylethanolamine

Phosphatidylserine

The cephalins

b. Sphingolipids

Sphingolipids occur in the membranes of both plant and animal cells, with only a minor amount found in fat deposits. The long-chain unsaturated amino alcohol *sphingosine* is the major backbone of the sphingolipids found in mammals. Also present are fatty acids, phosphate, and a polar alcohol component. The most abundant sphingolipids in animals are the **sphingomyelins,** which contain choline or ethanolamine as the polar groups. Sphingomyelins are particularly abundant in the myelin sheath surrounding nerve fibers. They, like the phosphoglycerides, exist as zwitterions at a pH of 7.

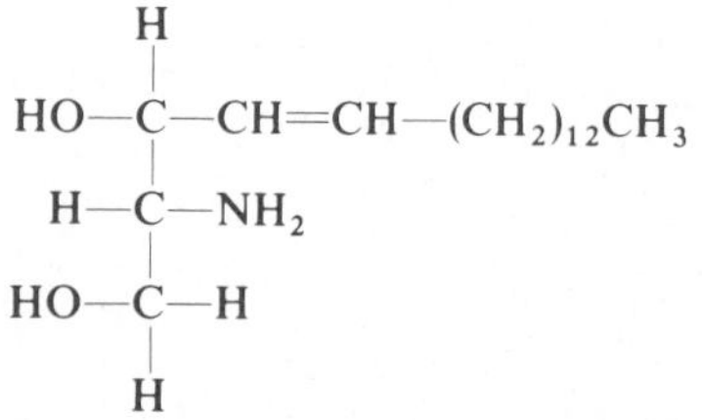

Sphingosine

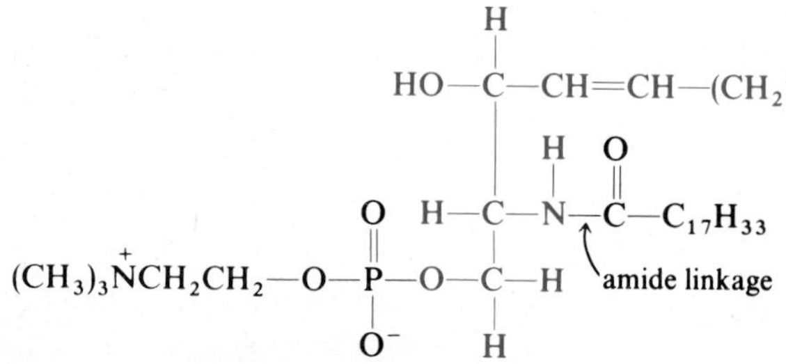

A sphingomyelin

12.10 *Glycolipids*

Several groups of compounds contain both lipid and carbohydrate structures. Those that are water soluble are termed *liposaccharides* and are considered as derived carbohydrates. Those that retain solubility in nonpolar organic solvents are classed as **glycolipids.** Glycolipids generally constitute about 5% of the lipid molecules in the outer surfaces of cell membranes where they function as distinct cell-surface markers to distinguish one tissue from another and in the detection of foreign cells by the immune system. In bacteria and plant cells almost all glycolipids contain glycerol as the backbone, whereas in animal cells they are usually derived from sphingosine. The latter are termed *glycosphingolipids,* and these membrane molecules have become the subject of intense research by cancer immunologists (see *Scientific American,* May 1986).

The simplest glycosphingolipids are the **cerebrosides,** which contain only one sugar group, either galactose or glucose, attached by an acetal linkage at carbon-1 of sphingosine. Galactocerebroside is particularly abundant in the myelin sheath that

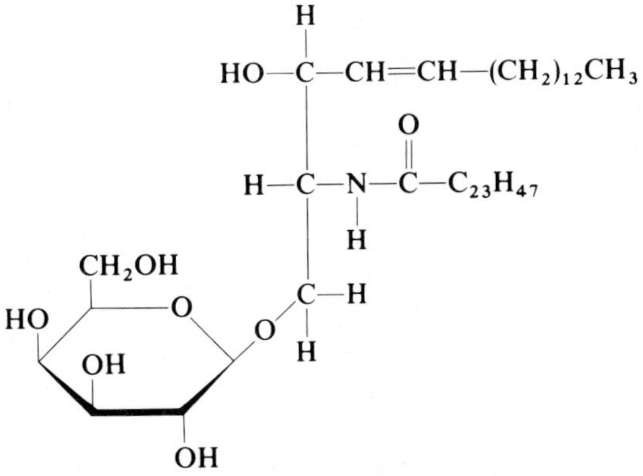

A cerebroside
(Galactocerebroside)

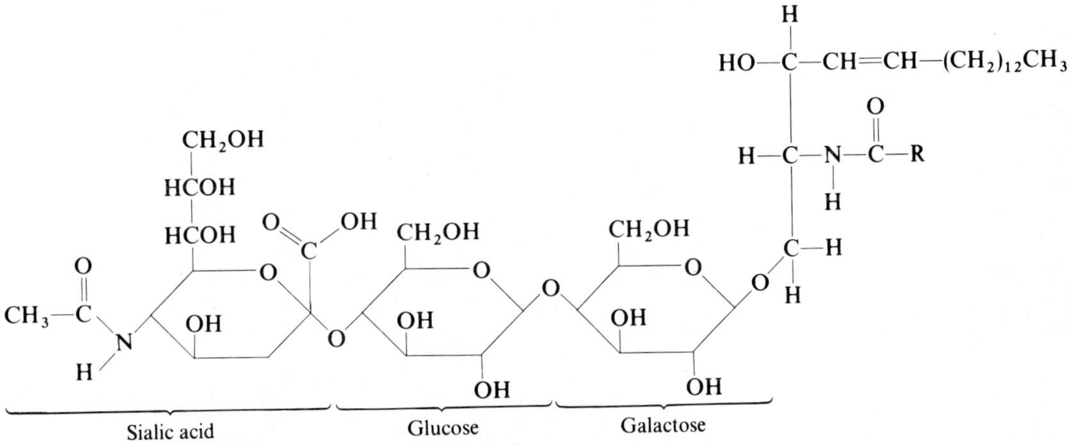

A ganglioside

insulates nerves. It is believed to play a principal role in the membrane wrapping process that is unique to myelination.

Gangliosides are the most complex class of glycosphingolipids. They usually contain a branched chain of three to eight monosaccharides and/or substituted sugars. Gangliosides are most prevalent in the outer membrane of nerve cells, although they are found in smaller quantities in most cell types. There is considerable variation in their sugar components, and about 130 varieties of glycosphingolipids have been identified. It is the sequence of sugars that most often determines cell-to-cell recognition and communication (e.g., blood group antigens).

12.11 *Cell Membranes*

Biological membranes were once viewed as inert barriers that just served to contain the cell and the organelles (nucleus, mitochondria, Golgi bodies, etc.) within the cell. We now recognize them to be intrinsically involved in the transport of materials and as receptors of external stimuli. Cell membranes consist of *a continuous double layer* (a **bilayer**) of lipid molecules in which various proteins are embedded (Figure 12.6). Most cell membranes are composed of about 50% (by mass) lipid and 50% protein, although large variations from these percentages can exist in certain cells.[7] The membranes are referred to as **semipermeable** because *only certain materials are allowed to pass from one side of the membrane to the other.*

The three major classes of lipid molecules in the membrane bilayer are phospholipids, glycolipids (about 5% and confined to the outer layer), and cholesterol (see Section 12.12-a). Roughly equal numbers of phospholipid and cholesterol molecules are present in the cell membranes of eukaryotic cells. As shown in Figure 12.6, the lipid bilayer consists of two rows of phospholipid molecules arranged tail to tail. The hydrophobic tails (the fatty acid portions) are directed toward each other and can interact by hydrophobic and van der Waals forces.[8] At the same time they are isolated from the aqueous environment that exists within and outside the cells. The polar portions of the phospholipids and glycolipids, therefore, project from the inner and outer surfaces of the membrane and interact with water molecules.

If membranes were comprised only of lipids, then they would act as a barrier against the passage of ions or polar molecules. The passage of polar species across the membrane is facilitated by proteins that either are attached to the surfaces (*peripheral proteins*) or are partly or fully embedded in the bilayer (*integral proteins*).

---

[7] The percentage of each component of the membrane is related to the function of the particular tissue. For example, the myelin sheath of nerve cells can contain up to 80% lipid because the major function of the membrane is insulation and protection. The inner mitochondrial membrane, on the other hand, is unique in having a large protein component (75–80%). The proteins (enzymes) within the membrane play an integral role in the energy conversion function of the mitochondria (see Section 16.9). It is important to emphasize that these are percents by mass. Lipid molecules are much smaller than proteins; hence there are always more lipid than protein molecules. In an average membrane there are 50 lipid molecules for one protein molecule.

[8] This interaction is relatively weak because of the presence of unsaturated (*cis* branching) fatty acids. As a result, the lipid portion is not rigid but is quite fluid and allows movement within the membrane. Cholesterol is also a key moderator of membrane fluidity. It breaks up the hydrophobic and van der Waals interactions of the fatty acid chains by fitting between these chains.

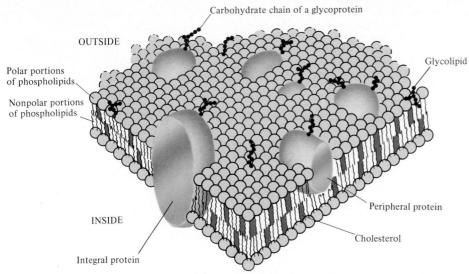

Carbohydrate chain of a glycoprotein

OUTSIDE

Polar portions
of phospholipids

Nonpolar portions
of phospholipids

Glycolipid

INSIDE

Peripheral protein

Cholesterol

Integral protein

FIGURE 12.6 Schematic diagram of the fluid-mosaic model of membrane structure. The cell membrane is a phospholipid bilayer in which cholesterol and various kinds of proteins are embedded. Attached to the membrane proteins are one (or several) short polysaccharide chains.

These proteins give each kind of membrane its distinctive identity and help to implement its specialized functions. For example, proteins serve to transport ions and polar molecules through the "sea" of lipids, whereas nonpolar compounds can diffuse through the "sea." The membrane proteins also serve as receptor sites to bind messenger molecules (e.g., hormones and neurotransmitters) that can influence the activity of the cell without entering into the cell. The message is transmitted to the inside of the cell via a change in configuration of the receptor protein (see Section 16.3).

12.12 *Steroids*

Perhydrocyclopentanophenanthrene

Over 30 different steroidal compounds have been found in nature. They occur in plant and animal tissues, yeasts, and molds, but not in bacteria, and may exist free or combined with fatty acids or carbohydrates. All steroids have a perhydro-cyclopentanophenanthrene ring system, which consists of a completely saturated phenanthrene moiety fused to a cyclopentane ring. The rings are designated by capital letters, and the carbon atoms are numbered as shown.

Although the parent ring structure is that of a cycloalkane, several steroids contain one, two, or three double bonds, and many of them possess one or more hydroxyl groups. (These steroids are often termed *sterols*, signifying the presence of the hydroxyl group.) Most steroids contain methyl groups at carbon-10 and carbon-13 and a side chain at carbon-17. Since the steroid skeleton is a rigid cyclic system, any substituents will be situated either above or below the plane of the molecule. This admits the possibility of geometric isomerism. In addition, many of the ring carbon atoms are chiral, and there are numerous diastereomers. Since the steroids are essentially high molecular weight hydrocarbons, they are soluble only in the fat solvents. They differ from the other lipids in that they do *not* undergo saponification. The steroids are one of the most versatile classes of compounds found in nature; they commonly function as hormones and other regulators of physiological processes. Very small amounts of steroids and slight variations in structure or in the nature of substituent groups effect profound changes in biological activity.

a. Cholesterol

Cholesterol rarely occurs in plants, but it is the best known and most abundant (about 240 g) steroid in the human body. About one half of the total body cholesterol is present in cell membranes interspersed among the phospholipid molecules (recall Figure 12.6). Much of the cholesterol in the body is converted into cholic acid, which is used in the formation of bile salts. Cholesterol is also an important precursor in the biosynthesis of the sex hormones, adrenal hormones, and vitamin D. Excess cholesterol that is not utilized by the body is released from the liver and transported by the blood to the gallbladder. Normally, it stays in solution and is secreted into the intestine (in the bile) to be eliminated. Sometimes the cholesterol precipitates in the gallbladder, producing gallstones. Its name is derived from this source (Greek, *chole,* bile; *stereos,* solid).

Cholesterol

Most meats and foods derived from animal products such as eggs, butter, cheese, and cream are particularly rich in cholesterol. The average American consumes about 600 mg of cholesterol each day. The liver is the major site of cholesterol biosynthesis, although other tissues (intestines, adrenals, gonads) are also involved. The human body synthesizes about 1 g of cholesterol each day; all 27 carbon atoms are derived from acetyl-CoA molecules. The plasma cholesterol level controls the synthesis of cholesterol by the liver. When the cholesterol level in the blood exceeds 150 mg/100 mL, the rate of cholesterol biosynthesis is halved. Hence, if cholesterol is present in the diet, there is a feedback mechanism that supresses its biosynthesis in the liver. However, this is not a one-to-one ratio. The reduction in biosynthesis does not equal the amount of cholesterol ingested. Fasting also inhibits the biosynthesis of cholesterol because of the limited availability of acetyl-CoA. Conversely, diets high in carbohydrate or fat tend to accelerate cholesterol biosynthesis because they increase the amount of acetyl-CoA in the liver. The lipids of fish and poultry contain relatively more unsaturated fatty acids than the lipids of beef, lamb, and pork. Fish (fish oils in particular) and poultry are recommended for people who wish to lower their serum cholesterol levels. (It has been recommended that people do not exceed 300 mg of cholesterol per day in their diets.)

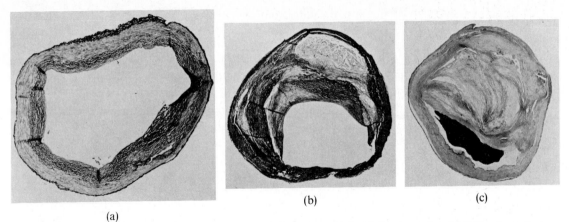

FIGURE 12.7 Comparison of normal artery (a) with those containing fatty deposits. (b) Fatty deposits in vessel wall; (c) plugged artery with fatty deposits and clot. [Reproduced with permission. © American Heart Association.]

Cholesterol has received much attention because of the suspected correlation between the cholesterol level in the blood and certain types of heart disease (see Section 12.17). The cholesterol content of blood varies considerably with age, diet, and sex. Young adults average about 170 mg of cholesterol per 100 mL of blood, whereas males at age 55 may have 250 mg/100 mL or higher (because the rate of cholesterol metabolism decreases with age). Females tend to have lower blood cholesterol levels than males.

About 85% of deaths due to cardiovascular disease are directly linked to arteriosclerosis. This degenerative condition is characterized by the loss of elasticity (hardening) of the arteries, which produces degenerative heart disease, stroke, and other arterial diseases. **Atherosclerosis** (a form of arteriosclerosis) *results from the deposition of fatty substance, especially cholesterol.* When cholesterol precipitates out of the blood, it accumulates on the inside of the arterial wall (Figure 12.7). Such deposits are called *plaque,* and the resulting constriction reduces blood flow, diminishes oxygen supply, and leads to high blood pressure. The danger of a blood clot increases as the plaque builds up. (Note that high blood pressure is also aggravated by other factors such as lack of exercise, obesity, heredity, stress, and smoking.) Investigators are attempting to establish that reducing the blood cholesterol level by minimizing lipid intake through dietary restrictions, or by administering anticholesterol drugs (such as cholestyramine), will help to prevent heart disease. Cardiovascular disease accounts for more than half of the deaths each year in the United States (see Section 12.17).

b. Bile Salts

Bile is a yellowish green liquid with a pH range of 7.8–8.6. In addition to water, it contains bile salts, bile pigments (from the degradation of hemoglobin molecules— page 333), and cholesterol. Bile salts are the most important constituents of bile, and their major function is to aid in the digestion of dietary lipids. The bile salts are sodium salts of amide-like combinations of bile acids and glycine or the rare amino acid taurine ($H_2NCH_2CH_2SO_3^-$). They are synthesized from cholesterol in the liver, stored in the gallbladder, and then secreted in bile into the small intestine. (About 500 mL of bile is secreted each day.) Bile salts are highly effective detergents because they contain both hydrophobic and hydrophilic groups. Thus, they function by acting as emulsifying agents; that is, they break down large fat globules into smaller ones and keep these smaller globules suspended in the aqueous digestive environment (see Section 17.1). The *lipases* can then hydrolyze the fat molecules more effectively. Bile salts also aid in the absorption of fatty acids, cholesterol, and the fat-soluble vitamins by forming complexes (micelles) that can diffuse into the intestinal epithelial cells.

Cholic acid
(a bile acid)

Sodium glycocholate
(a bile salt)

Glycine

c. Adrenocortical Hormones

Hormones are *a heterogeneous group of molecules that are synthesized by specific tissues (endocrine glands).* They are secreted by these glands and transported by the blood to their target sites (heart, liver, muscle, kidney, etc.), where they cause an alteration of activity within these organs. That is, hormones can affect (1) the permeability of cell membranes, (2) the rate of enzymatic reactions, or (3) the rate of synthesis of certain proteins. Some hormones, such as epinephrine and thyroxine are derivatives of amino acids, others are small peptides (page 322) or proteins (Section 16.3) that are polar molecules and cannot readily pass through cell membranes to enter cells. Recall (Section 12.11) that these hormones bind to receptor molecules on the outer surfaces of their target cells.

Steroid hormones (e.g., adrenocortical hormones, sex hormones), on the other hand, are lipids. They are soluble in the lipid components of the cell membrane and can easily diffuse into cells. Inside the cell they combine with specific receptor molecules in the cytosol. They may influence enzymatic reactions directly, or the steroid–receptor complex can enter the nucleus. In the nucleus, steroids bind to specific sites on DNA where they increase the rate of synthesis of mRNA (see Section 15.4-a), thus increasing the rate of biosynthesis of cellular enzymes. Since the primary effect of steroid hormones is on gene expression (the regulation of protein synthesis), their effects are much slower than other hormones (i.e., hours rather than minutes).

The outer part, or cortex, of the adrenal[9] glands utilizes cholesterol to produce a

FIGURE 12.8 The anatomical location and composition of the adrenal gland.

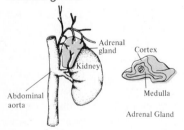

Abdominal aorta

Adrenal gland

Kidney

Cortex

Medulla

Adrenal Gland

[9]The term *adrenal* comes from the gland's location in the body, *ad*jacent to the *renal* (kidney—Figure 12.8).

mixture of many steroids (about 40) that are essential to life. These steroids constitute a family of hormones of which **aldosterone** and **cortisol** are the major representatives. Aldosterone is called a *mineralocorticoid*. This name alludes to its function in regulating the exchange of sodium, potassium, and hydrogen ions. Although aldosterone acts on most cells in the body, it is particularly effective in enhancing the rate of reabsorption of sodium ions in the kidney tubule and increasing the secretion of potassium ions and/or hydrogen ions by the tubule. Since the concentration of sodium ions is the major factor in water retention in the tissues, aldosterone also promotes water retention (controls tissue swelling) and reduces urine output. It thus supplements the action of vasopressin (see page 323).

Cortisol (Hydrocortisone)   Aldosterone   Cortisone   Prednisolone

Cortisol and its keto derivative, cortisone, are called *glucocorticoids*. These hormones regulate a number of key metabolic reactions (e.g., they increase glucose production and mobilize fatty acids and amino acids). They also inhibit the inflammatory response.[10] Glucocorticoids are used for immunosuppression after transplant operations and in the treatment of autoimmunity, severe skin allergies, and rheumatoid arthritis. The hormones or their analogs are injected, taken orally, or applied directly to the site of inflammation.

Pharmaceutical companies have been able to obtain large quantities of synthetic cortisone by means of an involved process utilizing cholic acid isolated from the bile juices of cattle. Prolonged use of cortisone can have serious side effects including high blood pressure, wasting of muscles, and resorption of bone. Cortisone has been supplemented with a synthetic analog, prednisolone, which is effective in much smaller doses, thereby greatly reducing the side effects. Prednisolone has been used in combination with cyclosporin in the treatment of autoimmunity, and for immunosuppression after kidney and liver transplant operations.

[10]**Inflammation** is *a tissue response to injury or stress* (e.g., chemical irritants, exposure to radiation or to extreme temperature). The blood vessels in the surrounding area become dilated to bring extra blood to the affected area. (This accounts for the "redness" associated with inflammation.) Dilated blood vessels are more permeable, and there is a tendency for fluid to leave the blood and enter the damaged tissue, causing swelling. As we shall see in Section 12.15, prostaglandins contribute to the inflammatory response by (1) promoting vasodilation of the blood vessels, (2) increasing the permeability of the capillaries, and (3) stimulating the pain receptors. It may be that the glucocorticoids are effective anti-inflammatory agents because they inhibit an enzyme (*phospholipase*) that is necessary for the synthesis of arachidonic acid, a prostaglandin precursor.

d. Sex Hormones and Anabolic Steroids

The sex hormones are a class of steroids that are secreted by the gonads (ovaries or testes), the placenta, and the adrenal glands. The primary male sex hormones (called *androgens*) such as **testosterone** and **androstenedione** are produced in the testes (and in lesser amounts by the adrenal cortex and the ovary). They control the primary sexual characteristics of males, that is, the development of the male genital organs and the continued production of sperm. Androgens are also responsible for the development of the secondary male characteristics such as facial hair, deep voice, and muscle strength.

Two sex hormones are of particular importance in females. **Progesterone** effects the changes that occur in the uterus during the rhythmic reproductive cycle (see Section 12.13). The **estrogens** are mainly responsible for the development of female secondary sexual characteristics such as breast development and increased deposition of adipose tissue in the breast, buttocks, and thighs. Progesterone is the precursor for the synthesis of the androgens, and the estrogens are synthesized from the androgens. Estrone is derived from androstenedione, whereas estradiol (the major estrogen hormone) is formed from testosterone. Thus, both males and females secrete androgens and estrogens. The difference between the sexes is the amount of secreted hormones, not the total absence of one or the other group. Notice that the male and female hormones exhibit only very slight structural differences, yet their physiological effects differ enormously.

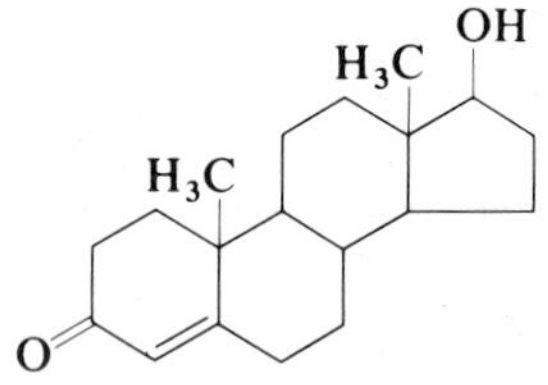

Testosterone

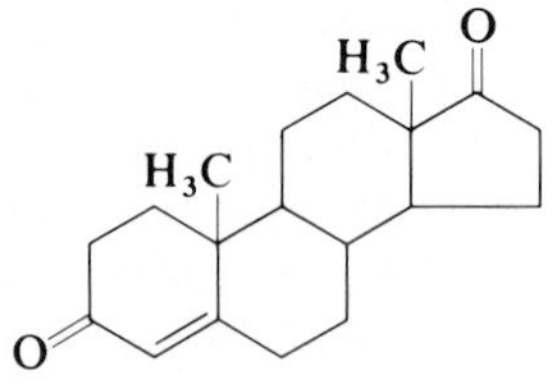

Androstenedione

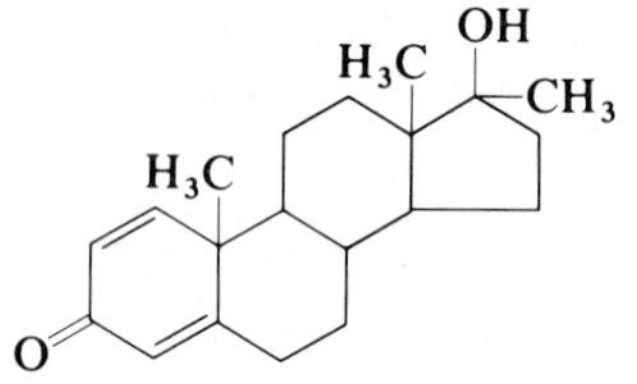

Methandienone
(Dianabol)

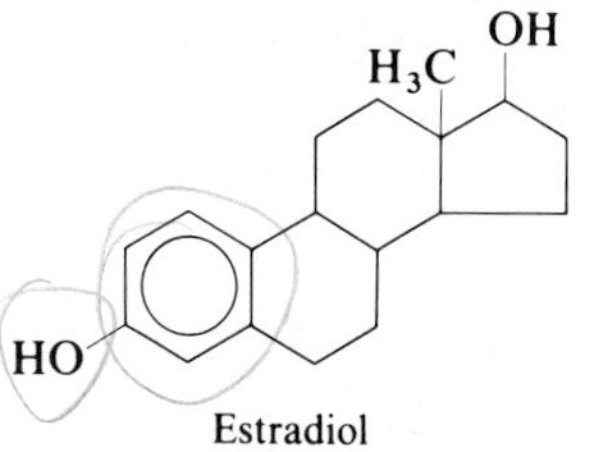

Estradiol

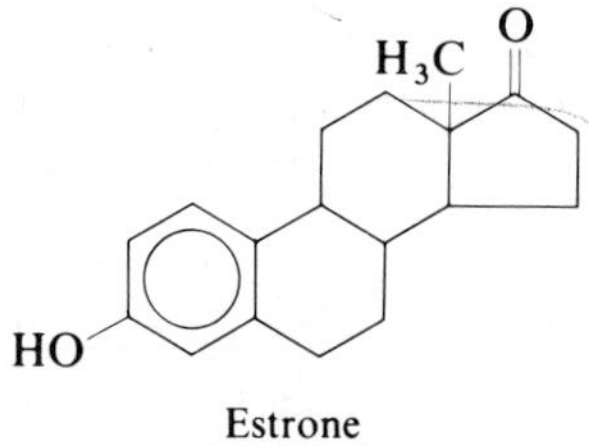

Estrone

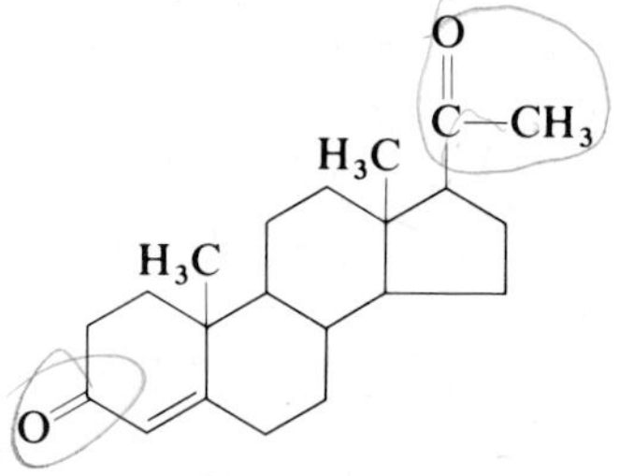

Progesterone

Estrogens

There has been a great deal of controversy generated by the widespread use of steroids among athletes. The drugs in question are synthetic androgens that stimulate protein synthesis (especially in skeletal muscle cells) without affecting the sex glands. We mentioned that testosterone is responsible for an increase in the amount of muscle cells (and more aggressive behavior), as well as its virilizing activity. However, testosterone is not very active if taken orally because it is metabolized in the liver. The incorporation of a methyl group at carbon-17 prevents this metabolism. Introduction of a second double bond in ring A produces a compound, methandienone (Dianabol), that has anabolic activity (stimulation of protein synthesis) but little of the virilizing effects of testosterone.

Many male and female athletes take more than 100 mg of steroids each day to increase their muscle bulk and muscle strength. Some reports indicate that under conditions of regular training, anabolic steroids do increase muscle strength, increase aggressiveness, and decrease fatigue. However, the same reports mention the adverse side effects that constitute a serious health hazard with long-term use of steroids. These include kidney, liver, and heart disease, cancer, sterility, impotence, and increased risk of diabetes. In females, there are increased masculinization and termination of menstruation. Sports officials can detect the presence of synthetic steroids in the body by monitoring the urine and testing for degradation products. Anabolic steroids are marketed for use in the treatment of senile debility, anorexia, and anemia and during convalescence.

Synthetic derivatives of the female sex hormones have attracted widespread attention. When taken regularly, these drugs effectively function to prevent ovulation. The oral contraceptives are usually mixtures of two compounds that are analogs of progesterone and estradiol. For example, Enovid is a combination of norethynodrel and mestranol, whereas Ortho-Novum contains norethindrone and ethinyl estradiol. Most of the combination pills sold in the United States contain 1 mg of the progesterone analog and less than 0.03 mg of the estrogen analog.

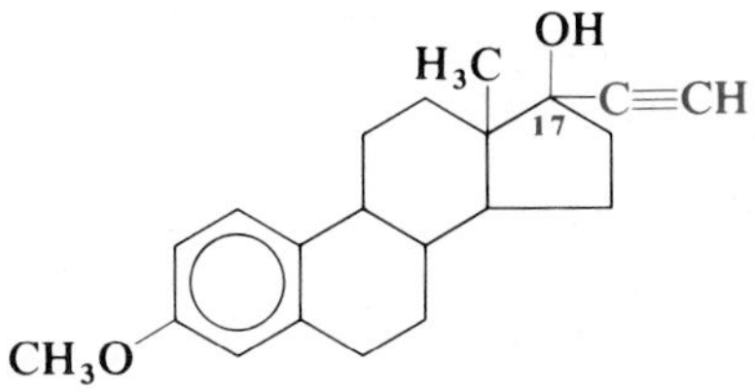

Mestranol

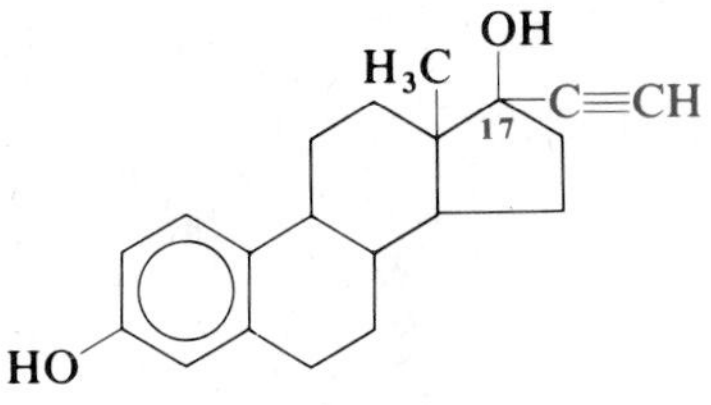

Ethinyl estradiol

(analogs of the estrogens)

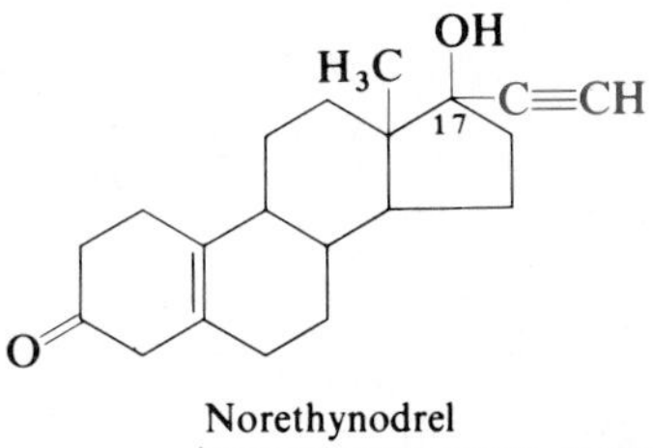

Norethynodrel

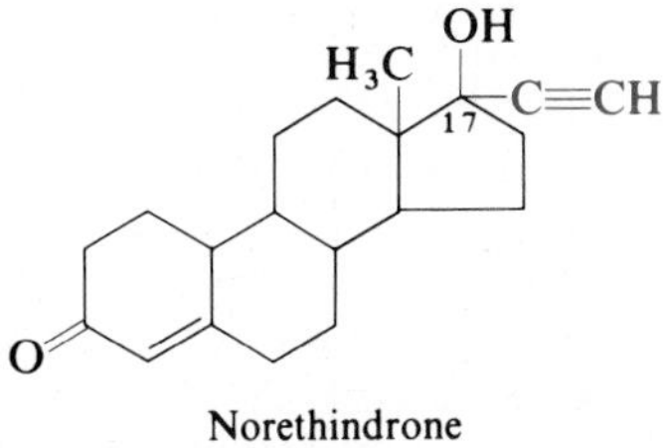

Norethindrone

(analogs of progesterone)

The prevention of ovulation is also effected by administration of progesterone and estradiol, but these hormones must be injected into the body for maximal results.

Only a slight structural difference (incorporation of the acetylenic group at carbon-17) confers upon the synthetic compounds the ability to be taken orally and to function in the same manner as the steroids produced by the body.

12.13   *Oral Contraceptives*

Control of fertility continues to be of paramount concern throughout the world, even though the population growth rate has begun to decline in some countries. This is due partly to government sanctions (e.g., China), partly to education, and partly to the extensive use of oral contraceptives. Several billion pills are produced annually, and these are consumed by more than 50 million women all over the world. (Oral contraceptives are second only to sterilization as a birth control method in the United States.)

There have been reports of increased incidence of breast and cervical cancers, thromboembolisms (blood clots), hypertension (high blood pressure), depression and psychiatric disturbances, benign liver tumors, gallbladder disease, urinary tract infections, and other risks among women who have taken the pill. In most cases, the increased risk of adverse side effects has been linked to the estrogen component in the pill. For this reason, the estrogen content in most products was decreased. There are now "minipills" that contain only progesterone analogs (called *progestins*). Minipills have to be taken every day, unlike the combination pill. They are less effective in suppressing ovulation, but they produce fewer side effects. A 1975 report by the Rockefeller Foundation concluded that the pill is a "highly effective and generally safe method of birth control—but not for all women." The major points of the report, based on a review of oral contraceptive research from medical centers around the world, are

1. Oral contraceptives are highly effective and generally safe for most women.
2. The risk of developing serious illness as a consequence of taking the pill is small.
3. The number of deaths associated with the oral contraceptives is of a very low order of magnitude.
4. The pill's long-term side effects must continue to be monitored closely to safeguard users.
5. There is no evidence connecting use of oral contraceptives with human cancers of the breast, uterus, or ovary.
6. The increased risk of death from thromboembolism, established in early studies as approximately 3 per 100,000 women per year, is confirmed by more recent research. Increased risk of venous thrombosis of the legs and of cerebral thrombosis has also been established.
7. Smokers, women over 40, and women with a history of

blood clots are at a higher risk of experiencing clot formation.
8. Women who use oral contraceptives have a greater risk than nonusers of having gallbladder disease requiring surgery. The increased risk may first appear within one year of use and may double after four or five years.

In order to understand how oral contraceptives work, it is first necessary to understand the normal physiology of the female menstrual cycle (Figure 12.9). This cycle is characterized by rhythmic monthly changes in the rates of secretion of the sex hormones (estrogens and progesterone) and corresponding changes in the sex organs themselves. Menstruation marks the terminal events of the normal ovarian cycle. The cycle begins when the pituitary gland releases *follicle stimulating hormone*

FIGURE 12.9   Hormone interactions in the female reproductive system.

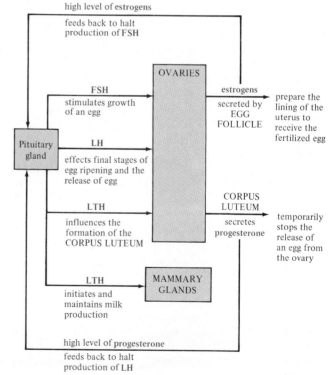

high level of estrogens

feeds back to halt production of FSH

Pituitary gland

FSH — stimulates growth of an egg

LH — effects final stages of egg ripening and the release of egg

LTH — influences the formation of the CORPUS LUTEUM

LTH — initiates and maintains milk production

OVARIES

EGG FOLLICLE

estrogens secreted by EGG FOLLICLE — prepare the lining of the uterus to receive the fertilized egg

CORPUS LUTEUM

secretes progesterone — temporarily stops the release of an egg from the ovary

MAMMARY GLANDS

high level of progesterone

feeds back to halt production of LH

(FSH) into the bloodstream. FSH is primarily responsible for stimulating the growth of one of the eggs in the ovary. Under the influence of FSH, the follicle, or sack surrounding the egg, secretes the estrogen hormones. The estrogens function to maintain secondary sexual traits and to prepare the lining of the uterus to receive the fertilized egg. As the follicle ages, the secretion of estrogens increases and this rise in the estrogen level of the body has a negative feedback effect on the pituitary gland. As the amount of estrogens increases, the output of FSH decreases. At this stage another hormone, *luteinizing hormone* (LH), is secreted by the pituitary gland. LH oversees the final stages of egg maturation and ensures that the egg is ejected from the ovary (ovulation).

Following ovulation, a third hormone, *luteotrophic hormone* (LTH), is secreted by the pituitary gland. LTH influences the formation of a tissue called the *corpus luteum* (yellow tissue). As it grows in size, the corpus luteum releases estrogens and progesterone, which temporarily stops the release of another egg from the ovary. Again there is negative feedback control whereby increased amounts of estrogens and progesterone result in a diminution of LH secretion by the pituitary gland. Further egg maturation is thus prevented until the fate of the ovulated egg has been decided.

If the egg is not fertilized, the life of the corpus luteum is short and it disintegrates. When this occurs, many of the small blood vessels that are present in the uterus rupture and bleed. This small amount of blood, along with the fragments of glands and the mucin from these glands, forms the menstrual discharge. Also, if fertilization does not occur, there is a sharp decrease in the amounts of estrogens and progesterone. As a result, the pituitary gland secretes more FSH and LH and the cycle begins anew.

When an embryo is implanted in the uterus, its cells secrete the hormone HCG[11] (*human chorionic gonadatropin*). This hormone causes the corpus luteum to be maintained and to continue secreting estrogens and progesterone. Thus, the uterine wall continues to grow and develop and the menstrual cycles are inhibited. In the last six months of pregnancy, the secretion of HCG by the implanted embryo declines because the estrogens and progesterone are produced in large quantities by the placenta. The high level of these hormones maintains the uterine wall, stimulates the further development of the mammary glands, and prevents the pituitary from secreting FSH and LH. In this way no additional mature eggs are released during pregnancy.

The two synthetic analogs of the female sex hormones that are in the combination pill deceive the female body into thinking it is pregnant and acting accordingly. Finding an abundant supply of the sex hormones in the blood, the pituitary gland shuts down its supply of FSH and LH. The egg follicles are not developed nor are eggs released into the fallopian tube. The synthetic hormones also enhance the contraceptive effect by increasing the viscosity of the mucus around the cervix, and thus setting up a barrier that is difficult for the sperm to penetrate.

12.14 *Estrogen Replacement Therapy*

In most women, menstrual cycles continue to occur into the late forties at which time they become increasingly irregular and finally cease altogether. This period is called *menopause*. Follicles no longer mature, ovulation does not occur, and the plasma estrogen concentration level sharply decreases.

As a result of lower estrogen levels, the female secondary sexual characteristics are modified. In addition, there is no longer the feedback control by estrogens over the secretion of FSH and LH. These hormones are released continuously by the pituitary gland, and they seem to be responsible for some of the unpleasant sensations that menopausal women experience (e.g., hot flashes, irritability, anxiety, and fatigue). These discomforts can be overcome by the administration of daily doses of estrogens or estrogen substitutes. The drugs also seem to reduce the risk of brittle bones (osteoporosis) and may reduce the risk of heart disease and stroke.[12]

It is estimated that more than 20% of the women in the United States have taken estrogen therapy, many for the treatment of menopause and/or to retain their youthfulness. However, one study by the University of Pennsylvania School of Medicine concludes that women who use estrogens for five years or more are 15 times more likely to develop uterine cancer than nonusers of estrogen.[13]

---

[11] Pregnancy can usually be diagnosed about 10 days after the first missed menstrual period by a test that detects the presence of HCG in a woman's urine.

[12] The results of two studies released in 1985 reached opposite conclusions about whether estrogen treatments for menopause increase the risk of heart and blood vessel disease. One Harvard Medical School study found that postmenopausal hormones substantially protect women against nonfatal heart attacks and fatal heart disease. But a study by the Framingham Heart Study found that similar uses of estrogens substantially increase risks. It may be that the differences between the two conclusions are a result of the slight differences in the ages of the women studied. The Framingham study examined women who tended to be older. Additional investigations are in progress.

[13] A subsequent study of more than 1200 menopausal women confirmed that the use of estrogen therapy increases the risk of contracting cancer and that the cancer spreads to tissue surrounding the uterus and to distant parts of the body. The study also found that the risks did not decline even 10 years after the therapy was discontinued.

Estrogens such as estrone, estradiol, ethinyl estradiol and mestranol are used in estrogen replacement therapy, particularly in the menopausal and postmenopausal female. Other compounds that are structurally similar to the estrogens are also used. One such compound is *diethylstilbestrol* (DES). Other medical applications of DES are suppression of lactation, treatment of breast and prostate carcinoma, treatment of acne, and postcoital contraception. It had also been used to stimulate female fertility. However, DES was found to cause mammary tumors in some strains of mice. Additionally, daughters born to women who took DES during their pregnancy developed rare forms of vaginal cancer (clear cell adenocarcinoma) when they were in their early twenties. The FDA has ordered that no estrogens may be prescribed during pregnancy. The FDA also removed from the market DES tablets that were used as "morning after pills." The use of DES as a postcoital contraceptive is lawful only following rape or incest. (Notice that the *trans* form of DES structurally approximates the estrogen steroids.)

Diethylstilbestrol (DES)

12.15 *Prostaglandins*

Prostaglandins are *hormone-like substances* that were originally isolated from semen found in the prostate gland. In the mature male, the prostate gland secretes about 0.1 mg/day of prostaglandins. However, they are biosynthesized by most mammalian tissue, and they affect almost all organs in the body.

Prostaglandins are a family of unsaturated fatty acids, each containing 20 carbon atoms and having the same basic skeleton as prostanoic acid (Figure 12.10-a). The major classes are PGA, PGB, PGE and PGF followed by a subscript that denotes the number of double bonds outside the five-carbon ring. Prostaglandins are not stored as such in cells. Rather they are synthesized on demand from arachidonic acid, a 20-carbon, polyunsaturated fatty acid that is released from phospholipids within cell membranes (by the action of the enzyme *phospholipase*). Arachidonic acid is then converted into an endoperoxide by an enzyme complex called *prostaglandin cyclooxygenase*. The endoperoxide intermediate can then be transformed into the prostaglandins or into two related groups of compounds, the thromboxanes and the prostacyclins. This sequence

of reactions is sometimes referred to as the *arachidonic cascade* (Figure 12.10).

The prostaglandins are among the most potent biological substances known. Slight structural changes are responsible for quite distinct biological effects; however, all prostaglandins exhibit some ability to induce smooth muscle contraction, to lower blood pressure, and to contribute to the inflammatory response (recall footnote 10, page 293). We have mentioned (Section 12.12-c) that certain steroid drugs (cortisol and its synthetic analogs) exert their anti-inflammatory action by inhibiting the release of arachidonic acid from membrane phospholipids (i.e., they may inhibit *phospholipase*). On the other hand, aspirin and the other nonsteroidal anti-inflammatory agents (e.g., indomethacin—Indocin; ibuprofen—Motrin, Advil, Nuprin), obstruct the synthesis of prostaglandins by inhibiting the *cyclooxygenase* enzyme, which converts arachidonic acid to endoperoxides.

The prostaglandins are believed to have myriad potential therapeutic uses. These include regulating menstruation and fertility, preventing conception, inducing labor, decreasing gastric secretions, controlling blood pressure, sensitizing the pain receptors, maintaining body temperature, and relieving asthma and nasal congestion. The two compound groups related to prostaglandins have opposing effects. *Thromboxanes* induce blood clotting by stimulating blood platelet aggregation, whereas *prostacyclins* prevent clotting by inhibiting platelet aggregation. Recall (page 164) that one of the side effects of aspirin is a prolonged bleeding time. This is due to the inhibition of platelet aggregation by blocking thromboxane biosynthesis.

The major clinical use of prostaglandins and their analogs concerns the induction of abortion. Their mechanism is uncertain, but is different from that of the steroids. The prostaglandins cause regression of the corpus luteum, uterine contractions, and abortion of the embryo. Because they induce abortion, prostaglandins would only have to be taken once a month or only if a menstrual period were missed. Obviously, a great deal of controversy will be generated if and when these compounds become commercially available.[14] Practically every major pharmaceutical company now has active prostaglandin research programs under way to develop new syntheses and to discover new natural sources of prostaglandins. A rich source of prostaglandins exists in a sea plant (the sea whip or sea fan), which grows in coral reefs off the coast of Florida. Chemists have developed several totally synthetic sequences for the preparation of prostaglandins. S. K. Bergstrom, B. I. Samuelsson,

---

[14] As therapeutic agents, prostaglandins must be administered by either intravenous or intrauterine injection. They cannot be taken orally because they are rapidly degraded in the digestive tract. A major goal of researchers is to develop prostaglandin analogs that are orally effective.

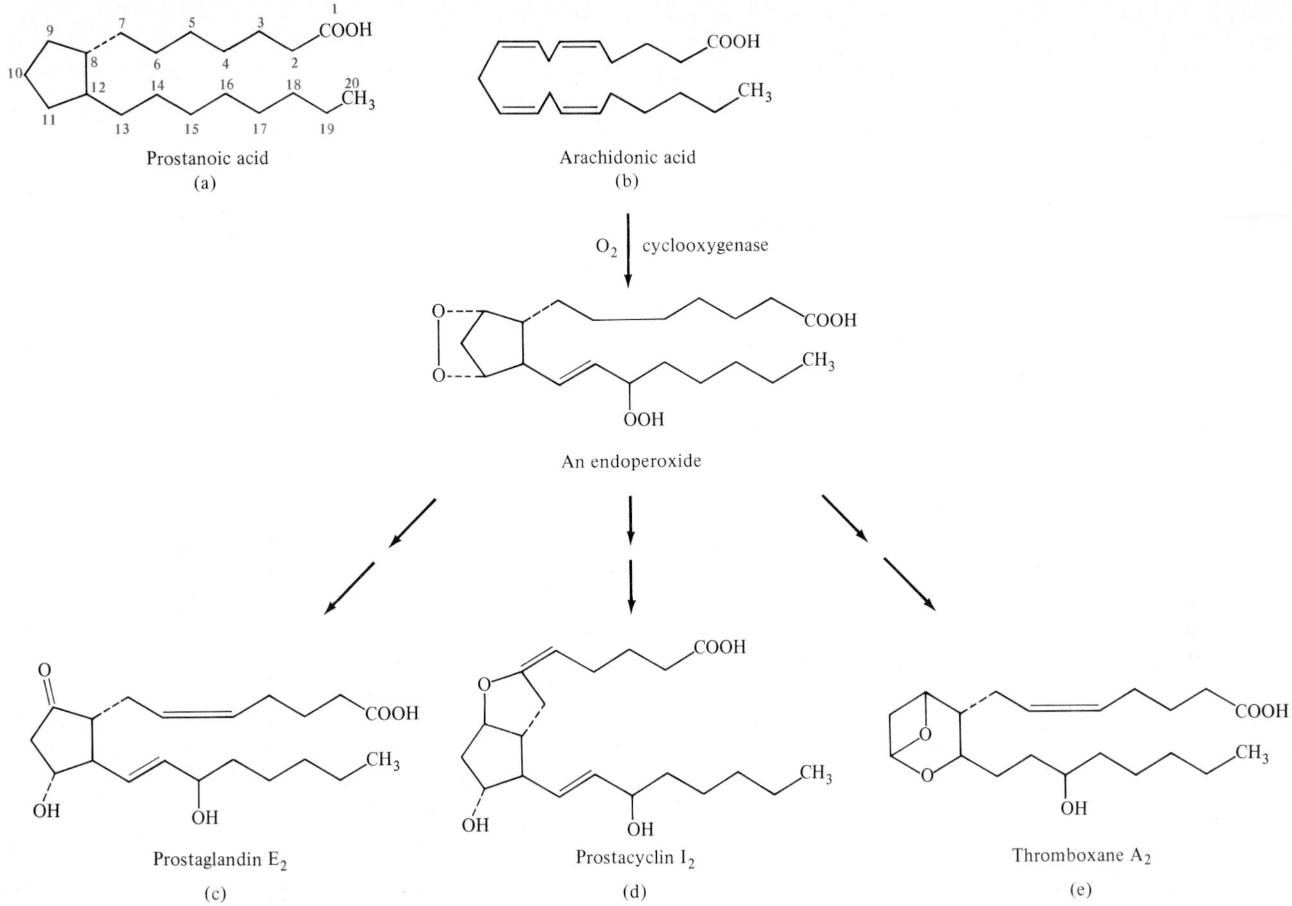

FIGURE 12.10 The arachidonic acid cascade. (a) Prostanoic acid is the parent compound of the prostaglandins. (b) Arachidonic acid is released from membrane phospholipids and is the precursor for the synthesis of the prostaglandins (c) and two related groups of compounds, prostacyclins (d) and thromboxanes (e).

and J. R. Vane shared the 1982 Nobel Prize in Physiology and Medicine for their work on prostaglandins.

12.16 *The Fat-Soluble Vitamins and Vitamin C*[15]

In 1912, the Polish biochemist Casimir Funk proposed the name vitamin (originally spelled vitamine) to describe a vital amine (thiamine) obtained from rice husks that relieved the paralytic effects of beriberi. **Vitamins** are *organic compounds that cannot be synthesized by an organism, but nevertheless are essential for the maintenance of normal metabolism and therefore must be included in the diet.* Since organisms differ in their synthetic abilities, a substance that is a vitamin for one species may not be so for another. Over the years, scientists have isolated 13 vitamins needed by humans. Researchers today are in general agreement that no more vitamins are likely to be found. One reason is that about 40 years have elapsed since the last one (vitamin B_{12}) was discovered, despite an intensive search since then for others. Another reason is that many people have lived for years without exhibiting signs of vitamin deficiency on intravenous solutions containing only the known vitamins and other nutrients.

[15]This section has been taken, in part, from Howard J. Sanders, "Nutrition and Health," *Chemical and Engineering News*, Vol. 57, pp. 27–46 (1979).

Vitamins often are classified as either water soluble (vitamin C and the B complex vitamins) or fat soluble (vitamins A, D, E, and K). Most water-soluble vitamins act as coenzymes or are required for the synthesis of coenzymes (see Section 14.4). The fat-soluble vitamins have more varied functions. In general, the fat-soluble vitamins are obtained from fish, liver, dairy products, green vegetables, and vegetable oils. All the fat-soluble vitamins and vitamin C contain double bonds and/or a phenolic group. Consequently, these vitamins are susceptible to attack by oxidizing agents or the oxygen in the air. Fat-soluble vitamins, if taken in high doses, can accumulate in hazardously large amounts (hypervitaminosis) because they are stored in body fat. People who consume too much vitamin D, for example, can develop bone pain, bone-like deposits in the kidneys, and mental retardation. Water-soluble vitamins, on the other hand, generally are rapidly excreted in the urine. Thus, even when taken in relatively large amounts, they usually are not toxic.

Minimum required doses of the vitamins have been set by examining the levels below which deficiency diseases occur.[16] (General warning signs of vitamin deficiency are slow healing of wounds, tiredness, and frequent illness.) There is no agreement, however, on the optimum levels of dietary vitamins. Vitamins have been the subject of more fads and more misrepresentations than any other group of nutrients. Claims have been made that, among other things, vitamins cure cancer, arthritis, and mental illness; increase sexual potency; prevent colds; and overcome muscular weakness. It is small wonder that the public is baffled by such claims, which many nutritionists reject.

a. Vitamin A

Vitamin A is a compound that occurs only in the animal kingdom. It was first isolated from halibut oil and is also present in cod liver oil and in butter. However, the plant pigment β-carotene (Section 3.5) is a precursor substance (a provitamin) that can be converted to vitamin A by animals, and thus most green and yellow vegetables (carrots, lettuce, spinach, yams) are a good source of the vitamin. Some foods are fortified by the addition of vitamin A obtained by extraction from fish liver oils or made synthetically. The recommended daily allowance of vitamin A is 0.7 mg. If vitamin A is present in excess, it is stored

---

[16]The RDA (recommended daily allowance) is a government estimate of the amount of vitamins that the average healthy person should eat daily to maintain good nutrition and health. A committee of the National Academy of Sciences is presently revising the RDAs, but the new figures have not yet been released. The RDA should not be confused with the MDR (the minimum daily requirement). The MDR is the smallest amount of a nutrient that if ingested will prevent a nutritional deficiency. The RDA significantly exceeds the MDR for each nutrient.

in the liver. Adult livers can store enough vitamin A to last for several months. On the other hand, the liver of infants and children does not store much of the vitamin. Consequently, infants and children are more likely to develop vitamin A deficiencies if their diets are inadequate.

In Section 10.8, we discussed the well-known role of vitamin A in vision (see Figure 10.16 for the structure of vitamin A). It also stimulates fluid secretion by the epithelial cells of the eye. Thus, if the dietary supply of vitamin A is inadequate, the cornea of the eye becomes dried, or keratinized. The cornea then becomes extremely vulnerable, and even the slightest nick or scratch may cause it to perforate, which leads to blindness. Blindness in this case is brought about by a vitamin A deficiency disease called *keratomalacia*. The disease is the major cause of blindness in young children in most of the developing countries, and it is estimated to affect hundreds of thousands of children throughout the world.

In addition, vitamin A is important to the growth and maintenance of epithelial tissue. In the past, some physicians used large doses of vitamin A for treating acne. Not only were the doses potentially harmful (painful joints, loss of hair), but they were not really effective. However, it has been found that a synthetic derivative of vitamin A (13-*cis*-retinoic acid) appears promising as a drug for treating severe acne.

Population studies suggest that vitamin A may be protective against cancers of the lungs, bladder, mouth, esophagus, larynx, breast, and cervix. A 1984 study of lung-cancer patients and healthy volunteers at the Cancer Center of Hawaii found that men who consumed a diet with low levels of vitamin A had twice the lung cancer rates of those with high levels of the vitamin.

b. Vitamin D

Vitamin D is the name given to a closely related group of compounds that are effective in preventing rickets, a childhood disease in which the organic matrix of new bone is not mineralized. (The bones are soft and deformed because of faulty calcium deposition.) In adults, vitamin D deficiency can produce the bone disease *osteomalacia* (increased tendency for bone fractures to occur). The two most important members of the group are vitamin D_2 (ergocalciferol) and vitamin D_3 (cholecalciferol). (Vitamin D_1, originally called vitamin D, was found to be a mixture of vitamins D_2 and D_3.) The D vitamins are produced by ultraviolet radiation of cholesterol derivatives. Vitamin D_2 is derived from ergosterol, a steroid found in yeast, ergot, and other fungi. Vitamin D_3 is formed in the skin of animals by the action of sunlight on 7-dehydrocholesterol. Notice that these compounds differ only in the structure of the side chain at carbon-17, and in both cases, the B ring of the steroid skeleton has been cleaved.

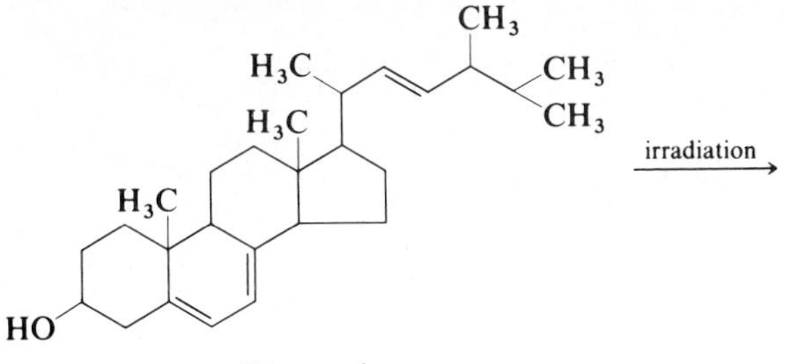

CH₃ structures on left column:

CH_3

H_3C

CH_3
CH_3

H_3C

H_3C

HO

irradiation →

Ergosterol

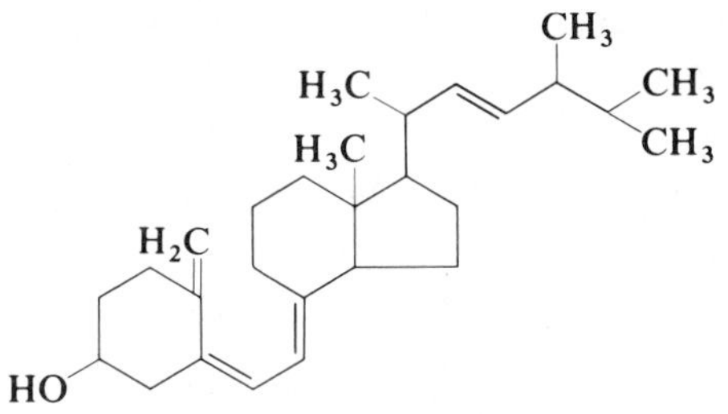

CH_3

H_3C
CH_3
CH_3

H_3C

H_2C

HO

Vitamin D₂
(Ergocalciferol)

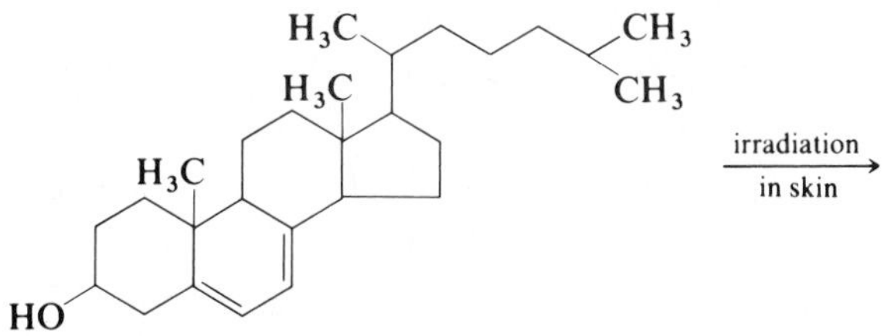

H_3C
CH_3
CH_3

H_3C

H_3C

HO

irradiation
in skin →

7-Dehydrocholesterol

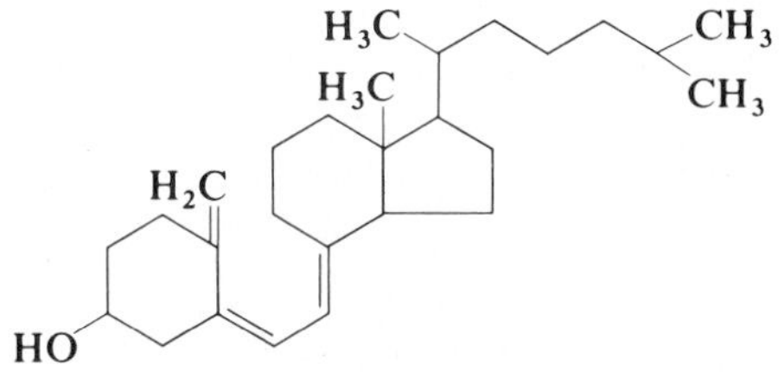

H_3C
CH_3
CH_3

H_3C

H_2C

HO

Vitamin D₃
(Cholecalciferol)

Most foods contain little or no vitamin D. The best natural sources of the vitamin are fish liver oils and egg yolks. Irradiated ergosterol (from yeast) is added to milk (10 μg per quart) and margarine as a supplemental source of vitamin D. Individuals with a reasonable proportion of their skin exposed to sunlight rarely suffer from a vitamin D deficiency. It is recommended that children receive about 20 μg of vitamin D in their daily diet. For adults, exposure to sunlight for 30 minutes satisfies the daily requirements of the vitamin. The vitamin is activated in the skin (from 7-dehydrocholesterol) by exposure to solar radiation and is then transported in the body for utilization or storage in the liver. In the liver, cholecalciferol is hydroxylated to form 25-hydroxycholecalciferol, which is then transported to the kidneys where another hydroxylation reaction produces 1,25-dihydroxycholecalciferol (Figure 12.11). This dihydroxy compound is the active form of the vitamin, and it behaves in a hormone-like manner. Its principal function is to promote the synthesis of a calcium-binding protein in the intestinal mucosal cells. This protein is necessary for the absorption of calcium from the intestine into the bloodstream, so that the concentration of calcium ion in the blood reaches a level high enough to permit proper bone mineralization. (Thus vitamin D plays an important role in the control of the plasma calcium concentration.) The ingestion of excessive amounts of vitamin D is hazardous and produces abnormalities of bones and teeth, as well as calcification of the lungs and kidneys.

c. Vitamin E

Eight naturally occurring compounds possessing vitamin E activity have been isolated and identified. α-Tocopherol is the most abundant and biologically active form in animal tissue. Vitamin E is widely distributed in many foods, and occurs in the highest concentration in wheat germ and vegetable oils. Deficiencies of this vitamin in humans occur almost exclusively in premature babies who are fed improper infant formulas. In infants, a shortage of vitamin E can cause anemia, produced by the accelerated destruction of red blood cells. (The RDA of vitamin E is 5–10 mg.)

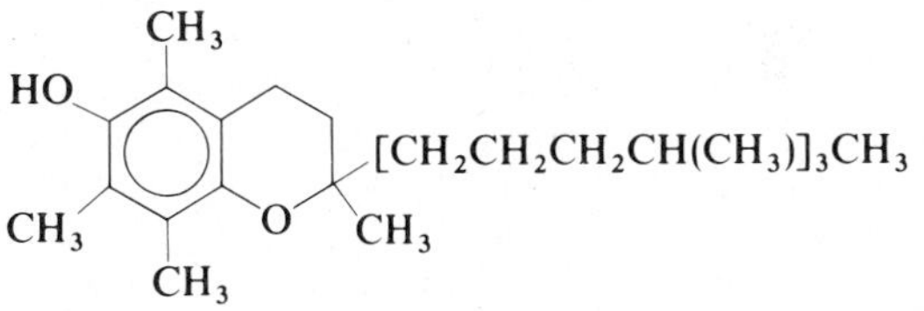

CH_3
HO
CH_3
CH_3 O CH_3
CH_3
$[CH_2CH_2CH_2CH(CH_3)]_3CH_3$

α-Tocopherol
(Vitamin E)

Vitamin E is one of the most vigorously promoted and highly controversial vitamins. Among the sensational claims made for vitamin E are that it increases virility and sexual endurance in humans. Some of the claims for the vitamin have resulted, at least in part, from the erroneous extrapolation to humans of tests carried out in laboratory animals. Some years ago, scientists found that vitamin E is among the factors required to prevent sterility in male rats and to permit normal pregnancy in female rats. From these findings, some people concluded that

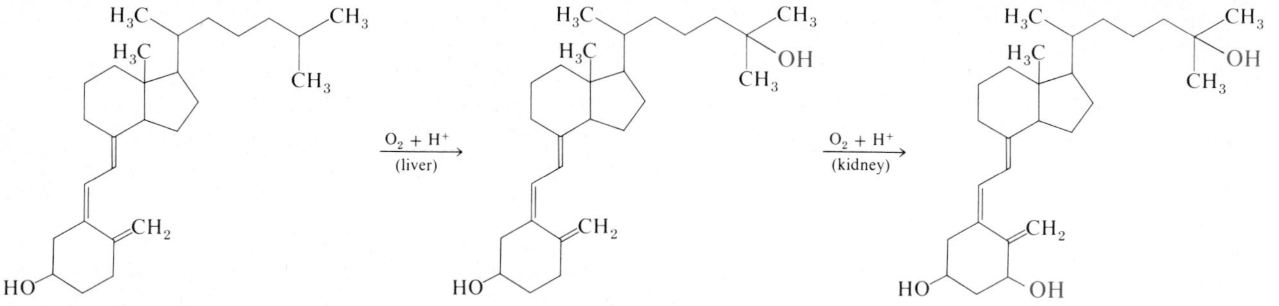

FIGURE 12.11 Vitamin D must acquire two hydroxyl groups to be active.

vitamin E in humans is effective in preventing male impotence or sterility and in promoting successful pregnancy. This view, however, is not supported by scientific or medical evidence gathered over the past 50 years.

The exact mechanism whereby vitamin E functions in the body is not known. It probably acts as an antioxidant controlling oxidation–reduction reactions in a variety of tissues, particularly in protecting the unsaturated fatty acids in cellular membranes from oxidative attack. Some investigators have reported that, when taken in large amounts, vitamin E may help to protect the lungs from oxidative damage caused by ozone and other air pollutants (Section 2.5). On the other hand, vitamin E can accumulate in the body and cause problems. Some people who take large daily doses develop nausea, intestinal disorders, and blurred vision.

d. Vitamin K

In 1935, Henrik Dam (Nobel Prize, 1943) in Copenhagen discovered vitamin K, which he showed could combat hemorrhaging in chicks. Later, other researchers found that this vitamin is stored in limited amounts in the liver, where it functions in the synthesis of prothrombin, a protein vital to the blood-clotting process (see Section 19.5). (The vitamin got its name from the Danish word *koagulation*.) It is found in green, leafy vegetables, and vitamin K deficiencies in humans are rare because intestinal bacteria synthesize the body's requirements. Prolonged treatment with antibiotics has the adverse effect of killing these bacteria, and the body's supply of vitamin K is temporarily reduced.

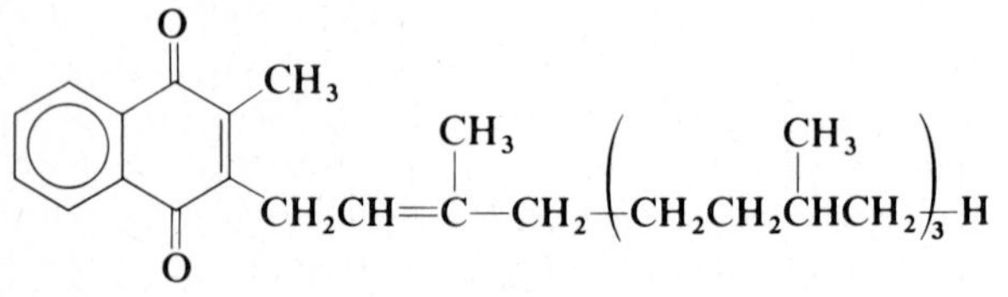

Vitamin K

e. Vitamin C

Although vitamin C is a water-soluble vitamin, it is included here (rather than in Section 14.4) because it does not function as a coenzyme in the manner of the B vitamins. Like vitamin E, the role of vitamin C in nutrition is still a subject of controversy. Whereas vitamin E protects the lipid portion of cells, vitamin C (a highly polar compound) serves as an antioxidant in the aqueous regions. Vitamin C participates in several biological oxidation reactions, such as the hydroxylation of proline and lysine moieties in collagen. It reacts with oxygen and/or oxidizing agents to form dehydroascorbic acid.

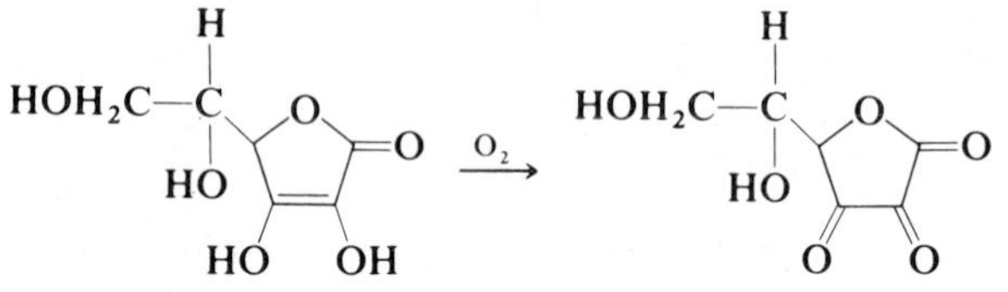

Ascorbic acid
(Vitamin C)

Dehydroascorbic acid

In 1747, James Lind discovered that citrus fruit was effective in treating sailors suffering from scurvy, a weakening of the collagenous tissues. The symptoms are swollen gums, loose teeth, sore joints, bleeding under the skin, and slow healing of wounds. It was not until 1932, however, that the vitamin was isolated from citrus fruit. Vitamin C was the first dietary component to be recognized as essential for preventing a human disease. It is widely distributed in plants[17] and animals.

[17] As is well known, vitamin C is particularly abundant in vegetables and citrus fruits, but for the vitamin to be useful, the food must be reasonably fresh because ascorbic acid is slowly oxidized by air. Vitamin C is one of the least stable vitamins. It can be destroyed by heat, light, and alkali as well as oxidizing agents. Therefore, cooking vegetables destroys an appreciable amount of vitamin C activity. It even deteriorates when kept for long periods in the refrigerator in well-capped bottles.

However, humans, other primates, guinea pigs, and some bats, birds, and fish lack an enzyme for its biosynthesis.

The controversy surrounding vitamin C received its impetus in 1970, when Linus C. Pauling published his best-selling book, *Vitamin C and the Common Cold*. He stated that vitamin C in doses ranging from 1 to 5 g a day could prevent colds and that as much as 15 g a day could cure a cold. Scientists investigating Pauling's claims have obtained conflicting results. (The RDA of vitamin C for adults is 60 mg.)

Some scientists report that taking 1.5 g of vitamin C every half hour over a period of 2 hours can cure a cold. Others concede that this may be true for a small group of the population. However, as one researcher puts it: "For the remainder of the population, vitamin C is relatively or completely ineffective in curing the common cold, since large-scale controlled studies have been quite unimpressive."

There have also been claims that vitamin C can prevent cancer and that it is useful in the therapy and management of cancer patients. Its role in cancer/cold prevention is related to its supposed stimulation of the immune system and its activity as an antioxidant in suppressing the damage of free radicals and in blocking the formation of nitrosamines. There is as yet no concrete evidence to substantiate or to invalidate these claims.

Since vitamin C is water soluble, excessive amounts are excreted rather than stored in the body. However, some scientists point to the possible dangers of taking massive doses of this vitamin. Such doses, for example, may raise the uric acid level in body fluids and thus cause gout in people predisposed to this disease (see Section 18.5). Also, it is known that the ingestion of 5 g of ascorbic acid by the normal adult human will cause diarrhea.

12.17 Cholesterol and Cardiovascular Disease[18]

In the past 20 years, few subjects in the nutrition field have attracted as much public attention as cholesterol. Today, everything from margarine and vegetable oils to egg substitutes and meat analogs is advertised on the basis that it contains little or no cholesterol. Cholesterol is believed to be a primary factor in the development of atherosclerosis, coronary heart disease, and stroke. As the leading cause of death in the United States, heart attack and stroke together take about 840,000 lives a year.

The disorder underlying both heart attack and stroke is atherosclerosis, characterized by the buildup of deposits (plaques) on the inner surfaces of arteries (recall Figure 12.7). If a blood clot forms in such a constricted artery leading to the heart or brain and causes a complete stoppage of blood flow, a heart attack or stroke occurs almost instantly. In atherosclerotic plaques, the lipids in highest concentration are cholesterol and cholesterol esters. Present in lower concentrations are two other types of lipids—phospholipids and triacylglycerols.

Scientists generally agree that elevated cholesterol levels in the blood, as well as high blood pressure and cigarette smoking, are associated in humans with an increased risk of heart attack. A long-term investigation by the NIH has shown that, among men aged 30–49, the incidence of coronary heart disease was five times greater if their cholesterol level was 260 mg/100 mL of serum or more, compared to the men whose level was 200 mg/100 mL of serum or less.

What has not been entirely proved, say many scientists, is that a lowering of serum cholesterol levels in humans by the use of diet or drugs will reduce the incidence of coronary heart disease. In 1980, the National Research Council's Food and Nutrition Board released a report entitled "Toward Healthful Diets," which stated that there is no need for the average person to cut down on the amount of cholesterol in the diet. The authors found no evidence that reducing serum cholesterol levels by dietary changes could prevent coronary heart disease. The report also noted that the body synthesizes its own cholesterol and that only 10–50% of the cholesterol coming into the body through the diet is absorbed.

The conclusions in this report were in direct conflict with studies made by the Departments of Agriculture and of Health and Human Services and many other groups that had all recommended that cholesterol intake be reduced. Many scientists believe that research in humans already has demonstrated that a lowered cholesterol level does reduce the risk of coronary heart disease. The gathering of proof to support this contention is exceedingly difficult because large numbers of people must be studied over long periods of time. Useful data can be obtained only by a slow, laborious process, partly because only a small percentage of any group will die of coronary heart disease in any given year. Predictably, the conflicting opinions have aroused much anger and controversy among food and nutrition scientists and total confusion among the public at large.

Several alternative theories have been proposed to explain the cause of atherosclerosis. The most recent one suggests that defects in the lipid-transporting system are responsible for a buildup of lipids in the blood, which eventually triggers plaque formation. Because lipids such as cholesterol are not soluble in water, they cannot be transported in the blood (an aqueous medium) unless complexed with water-soluble proteins (lipoproteins).[19] Lipoproteins generally are classified according to

---

[18] This section has been taken, in part, from Howard J. Sanders, "Nutrition and Health," *Chemical and Engineering News,* Vol. 57, pp. 27–46 (1979).

[19] Plasma lipoproteins are generally spherical particles with a surface that consists largely of phospholipids, free cholesterol, and protein and a core that contains mostly triacylglycerols and cholesterol esters.

TABLE 12.4 Composition of Lipoprotein Isolated from Normal Subjects

Lipoprotein Class	Density Range (g/mL)	Protein	Triacyl-glycerol	Cholesterol Free	Cholesterol Ester	Phospho-lipid
				Composition (wt. %)		
Chylomicrons	<0.94	1–2	85–95	1–3	2–4	3–6
VLDL	0.94–1.006	6–10	50–65	4–8	16–22	15–20
LDL	1.006–1.063	18–22	4–8	6–8	45–50	18–24
HDL	1.063–1.21	45–55	2–7	3–5	15–20	26–32

their density and composition. There are four broad categories: *chylomicrons* (density of less than 0.94 g/mL and made in the intestine), *very low density lipoproteins* (density of 0.94–1.006 and made mainly in the liver), *low density lipoproteins* (1.006–1.063), and *high density lipoproteins* (1.063–1.21). The density of lipoproteins is determined by the relative content of protein and lipid. Since lipids are less dense than proteins, lipoproteins containing a greater amount of lipid are less dense than those containing a greater proportion of protein. The chylomicrons contain up to 99% by weight of lipids, and the very low density lipoproteins (VLDLs) contain up to about 90% lipids. The low density lipoproteins (LDLs) contain about 80% lipids, and the high density lipoproteins (HDLs) only about 50%. The protein component makes up the remainder of the lipoprotein molecule (e.g., HDLs are composed of 50% lipid and 50% protein—Table 12.4).

In research on cholesterol and its role in heart disease, the types of lipoproteins that have received the greatest attention in recent years have been the LDLs and HDLs. The reason is that they almost always contain a higher percentage of cholesterol than do the chylomicrons or the VLDLs, which serve to transport the triacylglycerols. One of the most fascinating discoveries in this field is that cholesterol that is bound to HDLs reduces a person's risk of developing coronary heart disease. On the other hand, cholesterol that is bound to LDLs increases that risk. Notice in Table 12.4 that LDLs are the major cholesterol-carrying lipoproteins, whereas chylomicrons and VLDLs are the major carriers of triacylclcerols in the plasma.

Research evidence indicated that atherosclerosis and coronary heart disease are associated with elevated levels of serum LDLs, rather than with serum lipoproteins in general. It was also reported that the level of serum LDLs was a better predictor of coronary heart disease risk than was the level of serum cholesterol. (The normal concentration of LDLs in humans is about 120 mg/100 mL of serum). Persons who, because of hereditary factors, have high levels of LDLs in their blood, have a higher incidence of coronary heart disease.

Most of the serum cholesterol is transported as an LDL–cholesterol complex, which delivers the cholesterol directly to

cells that need it. Low density lipoproteins contain about 55% cholesterol, whereas high density lipoproteins contain only about 25% cholesterol. LDLs are believed to promote coronary heart disease by first penetrating the coronary artery wall, where they are broken down enzymatically to cholesterol, cholesterol esters, and protein. The cholesterol and cholesterol esters are then deposited in the artery wall, becoming major parts of the atherosclerotic plaque.

How do HDLs reduce the risk of developing coronary heart disease? No one knows for sure, but two theories have been suggested.

1. HDLs competitively inhibit the uptake of LDLs by cells by occupying, and thus blocking, the LDL receptor sites on the cell membranes.[20]
2. One role of HDLs is to transport excess cholesterol *from* the cells to the liver. Therefore HDLs aid in removing cholesterol from the smooth muscle cells of the arterial wall and transporting it to the liver where it can be metabolized.

Assuming that HDL helps to protect the body against coronary heart disease, what can be done to increase serum levels of this lipoprotein? One way is to be born a female. Women have a higher average HDL value in the blood than men (55 mg/100 mL versus 45 mg/100 mL), and this may be a reason that women have a lower rate of heart disease then men. Another way is by sustained exercise (e.g., aerobic training). It was reported that marathon runners had a decidedly higher

[20]LDL-receptor proteins on the cell membranes are vital to the uptake of cholesterol from the blood. If these receptors are absent or deficient, the inherited condition called *familial hypercholesterolemia* results. Cholesterol levels in the blood rise dramatically and the excess cholesterol that can't enter cells is deposited in certain tissues, particularly skin, tendons, and in arterial plaques. Most individuals who are homozygous for the disorder have almost no LDL receptors and they die of coronary artery disease in childhood. Somewhat less than one percent of the U. S. population is heterozygous; they have about half the number of receptors, and often develop atherosclerosis by age 50.

average HDL level (~65 mg/100 mL) than did a group of sedentary men. Still another way is to lose weight. It was shown that HDL levels can be increased by losing weight. Finally, we should mention a method that has received quite a bit of publicity. It was reported that HDL levels in the blood can be increased (to about 80–100 mg/100 mL) by drinking alcohol *in moderation*. The amount of alcohol used in this study was equivalent to about three 12-ounce bottles of beer a day.

To sum up—a reduction of low density lipoprotein levels is believed by many scientists to be desirable. This may be accomplished by eating a diet low in cholesterol and low in saturated fat. And some physicians recommend that overweight people go on reducing diets to achieve their recommended weight.

Heart disease kills more Americans than all other diseases combined. Every year, approximately 1.5 million people suffer heart attacks. Almost half of them will die. Of those who survive, many are left with permanent damage, and a significant number have subsequent attacks. A heart attack usually does not happen suddenly. The body has an early warning system. The American Heart Association compiled a list of early warning signals.

1. One of the first signs is pressure or pain in the middle of the chest. That's where the heart is, not on the left as many believe.
2. This pain can get worse and spread through the whole chest as well as down the left arm.
3. The pain may also spread to both arms, shoulders, neck or jaw. A sensation of pressure, fullness, or squeezing may occur in the abdomen and is often mistaken for indigestion.
4. Pain may occur in any one or a combination of these areas at the same time. It could go away and return later. Many times, sweating, nausea, vomiting, or shortness of breath may come with the pain.

At the first sign of any of these symptoms, go to the nearest hospital emergency room at once! Get there as fast as possible. Have someone take you or dial 0 for an operator, say there is a heart attack victim at (location). In the hospital, insist it is an emergency; you have chest pain (or other symptoms) and may be having a heart attack.

Exercises

12.1 Define, explain, or draw formulas for each of the following.
   (a) triacylglycerol
   (b) triolein
   (c) palmitic acid
   (d) linolenic acid
   (e) lipases
   (f) iodine number
   (g) hardening
   (h) antioxidant
   (i) spontaneous combustion
   (j) micelle
   (k) nonionic detergent
   (l) biodegradable
   (m) builders
   (n) eutrophication
   (o) phosphatidic acid
   (p) steroid
   (q) inflammation
   (r) anabolic steroids
   (s) oral contraceptives
   (t) estrogen therapy
   (u) prostaglandins
   (v) RDA

12.2 Distinguish between the terms in each of the following pairs.
   (a) fat and oil
   (b) fat and wax
   (c) saponifiable lipid and nonsaponifiable lipid
   (d) *cis* double bond and *trans* double bond
   (e) simple triacylglycerol and mixed triacylglycerol
   (f) butter and margarine
   (g) oxidative rancidity and hydrolytic rancidity
   (h) hydrophobic and hydrophilic
   (i) saponification and emulsification
   (j) rancid oil and drying oil
   (k) hard water and soft water
   (l) hard soap and soft soap
   (m) soaps and syndets
   (n) anionic detergent and cationic detergent
   (o) prokaryotes and eukaryotes
   (p) phosphoglycerides and sphingolipids
   (q) lecithin and cephalins
   (r) peripheral proteins and integral proteins
   (s) arteriosclerosis and atherosclerosis
   (t) bile and bile salts
   (u) hormones and vitamins
   (v) androgens and estrogens
   (w) thromboxanes and prostacyclins
   (x) keratomalacia and osteomalacia
   (y) ergocalciferol and cholecalciferol
   (z) LDLs and HDLs

12.3 What functions does fat serve in the body?

12.4 Contrast the physical properties of fats and fatty acids. In what solvents are fats and oil soluble?

12.5 Why do unsaturated fatty acids have lower melting points than saturated fatty acids?

12.6 The melting point of elaidic acid (see Figure 12.1) is 52°C. How does this compare with the melting points of stearic acid and oleic acid? Explain. Would you expect the melting point of *trans*-hexadecenoic acid, $CH_3(CH_2)_5CH=CH(CH_2)_7COOH$, to be lower or higher than elaidic acid? Why?

12.7 Name each of the following compounds.

(a) 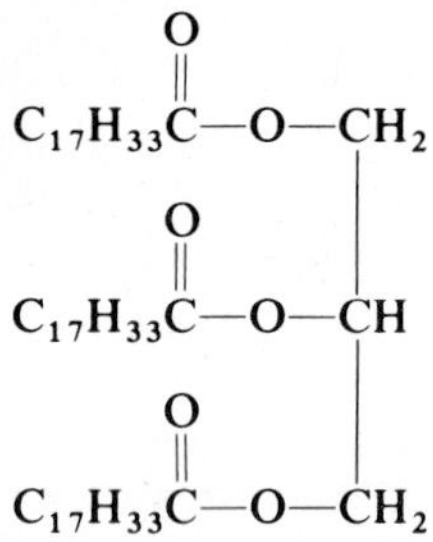
$$\text{C}_{17}\text{H}_{33}\overset{\displaystyle O}{\overset{\|}{\text{C}}}-\text{O}-\text{CH}_2$$
$$\text{C}_{17}\text{H}_{33}\overset{\displaystyle O}{\overset{\|}{\text{C}}}-\text{O}-\text{CH}$$
$$\text{C}_{17}\text{H}_{33}\overset{\displaystyle O}{\overset{\|}{\text{C}}}-\text{O}-\text{CH}_2$$

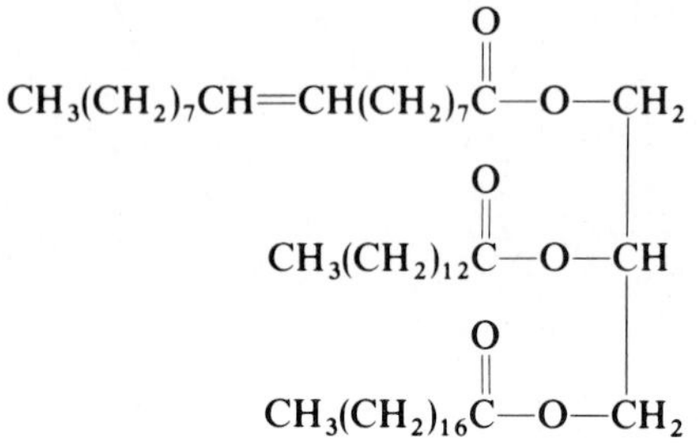

(b)
$$\text{CH}_3(\text{CH}_2)_7\text{CH}{=}\text{CH}(\text{CH}_2)_7\overset{\displaystyle O}{\overset{\|}{\text{C}}}-\text{O}-\text{CH}_2$$
$$\text{CH}_3(\text{CH}_2)_{12}\overset{\displaystyle O}{\overset{\|}{\text{C}}}-\text{O}-\text{CH}$$
$$\text{CH}_3(\text{CH}_2)_{16}\overset{\displaystyle O}{\overset{\|}{\text{C}}}-\text{O}-\text{CH}_2$$

(c)
$$\text{C}_{17}\text{H}_{35}\text{C}\overset{\displaystyle O}{\underset{\text{O}^-\,\text{Na}^+}{\diagdown}}$$

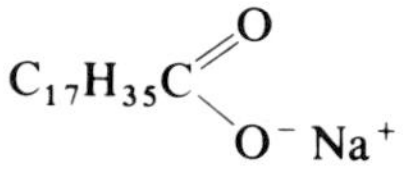

(d) 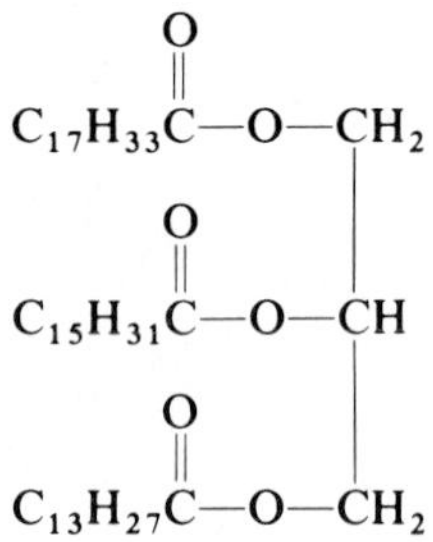
$$\text{C}_{17}\text{H}_{33}\overset{\displaystyle O}{\overset{\|}{\text{C}}}-\text{O}-\text{CH}_2$$
$$\text{C}_{15}\text{H}_{31}\overset{\displaystyle O}{\overset{\|}{\text{C}}}-\text{O}-\text{CH}$$
$$\text{C}_{13}\text{H}_{27}\overset{\displaystyle O}{\overset{\|}{\text{C}}}-\text{O}-\text{CH}_2$$

(e) 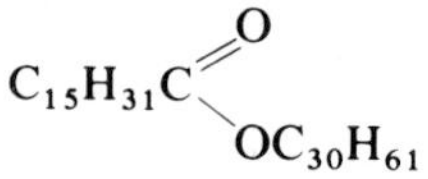
$$\text{C}_{15}\text{H}_{31}\text{C}\overset{\displaystyle O}{\underset{\text{OC}_{30}\text{H}_{61}}{\diagdown}}$$

(f) 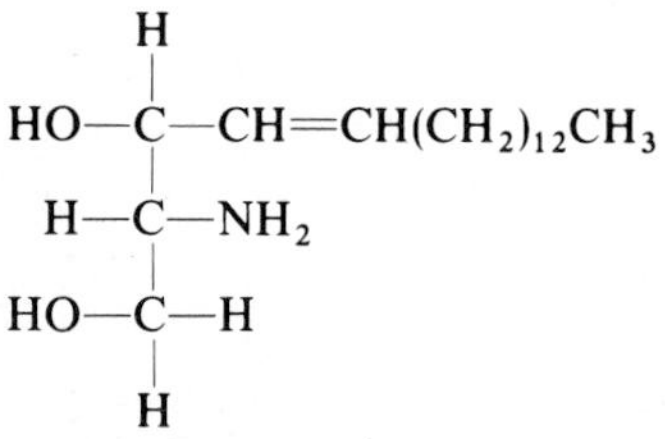
$$\text{HO}-\overset{\displaystyle H}{\overset{|}{\text{C}}}-\text{CH}{=}\text{CH}(\text{CH}_2)_{12}\text{CH}_3$$
$$\text{H}-\overset{|}{\text{C}}-\text{NH}_2$$
$$\text{HO}-\overset{|}{\underset{|}{\text{C}}}-\text{H}$$
$$H$$

12.8 How many mixed triacylglycerols are possible by combining stearic acid and oleic acid with glycerol?

12.9 Write structural formulas for the following.
(a) glyceryl palmitate
(b) cetyl stearate
(c) a triacylglycerol likely to be found in cottonseed oil
(d) a highly unsaturated oil

(e) sodium palmityl sulfate
(f) phosphatidic acid

12.10 Phosphatidylinositol is another phosphoglyceride that is found in cell membranes. Given the structure of inositol, draw a structural formula for phosphatidylinositol. (Linkage occurs at the colored hydroxyl group.)

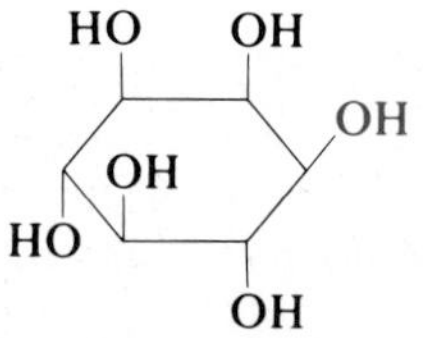

Inositol

12.11 Which of the compounds in Exercise 12.7 would react with hydrogen gas? How many moles of hydrogen would react per mole of lipid?

12.12 What products are formed (a) when a fat is hydrolyzed and (b) when a fat is saponified?

12.13 (a) What disagreeable odor is formed when butter becomes rancid?
(b) How is it formed?
(c) How may rancidity be prevented?

12.14 Can waxes be converted into soaps? Explain.

12.15 Name four waxes obtained from plants and animals, and give one use for each.

12.16 Briefly describe how soaps clean.

12.17 (a) What structural features are necessary for a compound to be a good detergent?
(b) What advantages do detergents have over ordinary soap?
(c) What are the disadvantages of detergents?
(d) Which type of detergent most resembles a soap?

12.18 Write an equation for each of the following reactions.
(a) hydrogenation of triolein
(b) halogenation of triolein
(c) saponification of tristearin
(d) acid hydrolysis of trilinolein
(e) sodium stearate and magnesium ions
(f) potassium oleate and hydrochloric acid
(g) phosphatidic acid and ethanolamine
(h) acid hydrolysis of phosphatidylserine

12.19 Why are phospholipids referred to as polar lipids?

12.20 Start with one molecule each of glycerol, palmitic acid, oleic acid, phosphoric acid, and choline and construct a typical lecithin molecule. Circle all of the ester bonds.

12.21 (a) What type of bond joins the fatty acid to sphingosine in cerebrosides?
(b) What type of bond joins the sugar unit to sphingosine in cerebrosides?

12.22 (a) What are the three types of lipids found in cell membranes?

(b) What are the functions of the proteins found in cell membranes?

12.23 (a) Draw the formula of the steroid nucleus and number the carbon atoms.

(b) Identify, by number, which of the carbon atoms of cholesterol are chiral carbons.

12.24 Why are the steroids classified as nonsaponifiable lipids?

12.25 Write equations for a possible reaction between cholesterol and **(a)** Br_2 and **(b)** acetic acid.

12.26 What functional groups occur in each of the following steroids?

(a) cholesterol   **(b)** cholic acid
(c) cortisone   **(d)** aldosterone
(e) estradiol   **(f)** progesterone

12.27 What are the differences in biological function of the mineralocorticoids and the glucocorticoids?

12.28 What is a major structural difference between the hormones progesterone and estrogen and the synthetic analogs found in oral contraceptives?

12.29 (a) Which fatty acid is the precursor of the prostaglandins?

(b) What are some potential therapeutic uses of prostaglandins?

12.30 How does aspirin function to relieve pain, reduce fever, and reduce inflammation?

12.31 Give one function and one deficiency disease associated with each of the following.

(a) vitamin A   **(b)** vitamin D
(c) vitamin E   **(d)** vitamin C

12.32 What is the role of cholesterol in heart disease?

12.33 Examine the labels on margarines and shortenings and list the oils used in the various brands.

12.34 Examine several different detergents, and note the active ingredient in each. What other additives are present?

13

Proteins

The class of compounds known as proteins is an essential constituent of all cells. Proteins are the most abundant organic molecules found in living cells. On the average, more than 50% of the cellular content of plants and animals is protein. The Dutch chemist G. J. Mulder (1838) is credited as being one of the first scientists to recognize their importance. He used the term *protein* (Greek, *proteios*, first) to describe these vital compounds.

> There is present in plants and animals a substance which ... is without a doubt the most important of the known components of living matter, and without it, life would be impossible. This substance has been named protein.

It is now apparent that the name was well chosen. Proteins are the major structural components of animal tissue, just as cellulose provides for the structure of the plants. Proteins are components of skin, hair, wool, feathers, nails, horns, hoofs, muscles, tendons, connecting tissue, and supporting tissue such as cartilage. In addition, proteins are involved in communication and transport (membrane proteins), defense (antibodies), metabolic regulation (hormones), biochemical catalysis (enzymes), and oxygen transport (hemoglobin). Proteins are utilized in the building of new tissue and in the maintenance of tissue that is already developed.

Whereas the carbohydrates and lipids are used primarily as energy sources, the primary function of the proteins is body building and maintenance. Lipids and carbohydrates are stored by the body as energy reserves, but proteins are not stored to any appreciable extent. It is possible for humans to survive for a short period of time on a diet consisting of protein, vitamins, and minerals; we could not survive over the same period of time on a protein-free diet containing lipids, carbohydrates, vitamins, and minerals.

Proteins are giant polymeric molecules (linear polymers of amino acids) that vary greatly in molecular dimensions. Their molecular weights may range from several thousand to several million daltons.[1] In addition to carbon, hydrogen, and oxygen, all proteins contain nitrogen, and many contain sulfur, phosphorus, and traces of other elements. The composition of most proteins is remarkably constant at about 51% carbon, 7% hydrogen, 23% oxygen, 16% nitrogen, 1–3% sulfur, and less than 1% phosphorus.

13.1 *Amino Acids*

An α-amino acid

Proline

Hydroxyproline

Proteins may be defined as *high molecular weight compounds consisting largely or entirely of chains of amino acids*. In this respect, proteins may be considered to be polymers analogous to the polysaccharides. However, 20 *different* structural monomeric units are commonly found in proteins, and these are the amino acids. The proteins in all living species, from bacteria to humans, are constructed from the basic set of 20 amino acids. Several other amino acids (e.g., hydroxyproline), which occur to some extent in certain proteins, are all derivatives of the common amino acids and are modified *after* incorporation into the polypeptide chain. With the exception of proline (which contains an alpha secondary nitrogen atom), the amino acids that are the building blocks of proteins are characterized by a primary amino group alpha to the carboxyl group.

Each individual amino acid has unique characteristics as a result of the size, shape, solubility, and ionization properties of the different **R** groups. (The **R** group is referred to as the *amino acid side chain*.) As we shall see in Section 13.5, the side chains of amino acids exert a profound effect on the conformation and the biological activity of proteins.

Amino acids may be classified in several ways. We choose to group them, according to the polarity of their side chains, into four different classes: (1) nonpolar, (2) polar and neutral, (3) acidic, and (4) basic. The structures of the common amino acids, their three-letter abbreviations, and certain of their distinctive features are given in Table 13.1. The amino acids are known exclusively by their common names, as the IUPAC names are too cumbersome. For example, asparagine was the first amino acid to be isolated (1806), and was given its name because it was obtained from protein found in asparagus juice. Glycine, the major amino acid found in

---

[1] Many biologists and biochemists use the term *dalton* as a unit of mass. A dalton is equivalent to the mass of one hydrogen atom $(1.67 \times 10^{-24}$ g). The molecular weight of a molecule expressed in daltons is numerically equivalent to its molecular weight expressed in units of grams per mole. Thus a protein whose molecular weight is 10,000 daltons has a molecular weight of 10,000 g/mole.

TABLE 13.1 Naturally Occurring Amino Acids

Name	Abbreviation	Structural Formula	MW (daltons)	Distinctive Features
1. Amino Acids with a Nonpolar R— Group				
Alanine	Ala	(structure) *hydrophobic*	89	The least hydrophobic member of this class because of its small **R** group (methyl).
Valine	Val	(structure) *hydrophobic*	117	
Leucine	Leu	(structure) *hydrophobic*	131	Most animals cannot synthesize branched-chain amino acids. They are, therefore, essential in the diet.
Isoleucine	Ile	(structure)	131	
Phenylalanine	Phe	(structure)	165	
Tryptophan	Trp	(structure)	204	A heterocyclic amino acid (a derivative of indole).

(Table continues)

TABLE 13.1 (continued)

Name	Abbreviation	Structural Formula	MW (daltons)	Distinctive Features
Methionine	Met	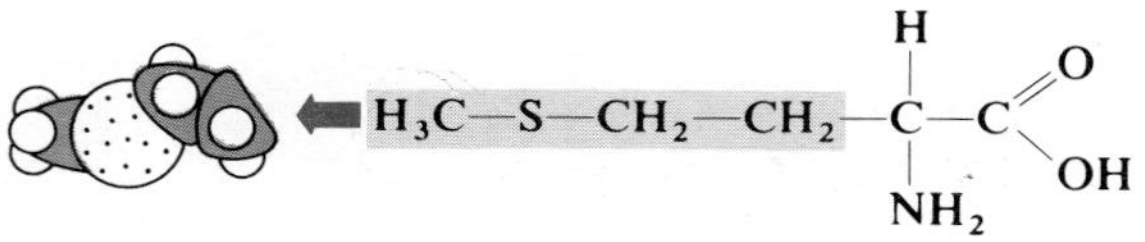 $H_3C-S-CH_2-CH_2$ ... C with H, COOH, NH₂	149	Contains a sulfur atom in the nonpolar side chain and is important as a donor of methyl groups.
Proline	Pro	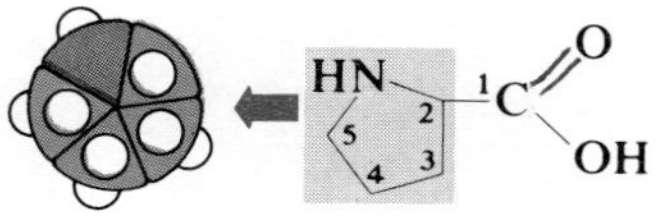 (ring structure) HN, C, COOH	115	Contains a secondary amino group rather than a primary amino group and so is referred to as an *α-imino acid*. A major constituent of the structural protein collagen. Hydroxylation of proline yields 4-hydroxyproline (Hypro), which is also abundant in collagen.

2. Amino Acids with a Polar But Neutral R— Group

Glycine	Gly	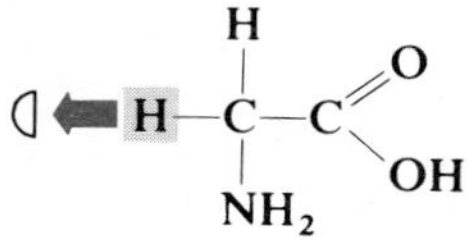 H—C—COOH, NH₂, H *(handwritten: hydrogen bonds)*	75	The only amino acid lacking a chiral carbon. Sometimes classified as a nonpolar amino acid, but its single hydrogen **R** group is too small to influence the polarity of the molecule.
Serine	Ser	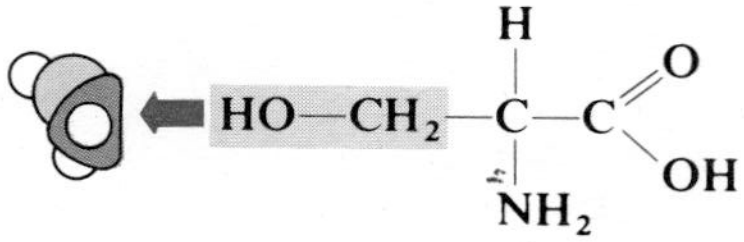 HO—CH₂—C—COOH, NH₂, H *(handwritten: hydrophilic OH + hydrogen bonds)*	105	Occurs at the active site of many enzymes. The hydroxyl group may take part in the usual alcoholic reactions such as ester formation.
Threonine	Thr	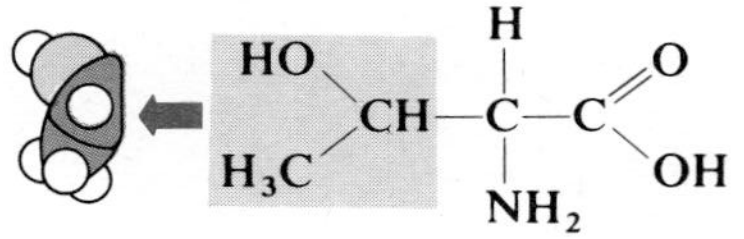 HO—CH—C—COOH, H₃C, NH₂, H *(handwritten: philic)*	119	Named for its similarity to the sugar threose (contains two chiral carbons).

Cysteine	Cys	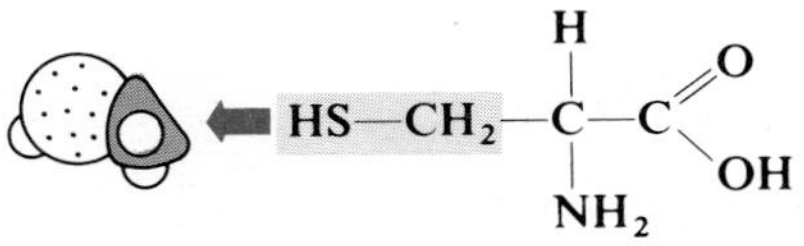	121	Often occurs in proteins in its oxidized form, cystine (see page 109). The disulfide bond of cystine serves in many proteins as a cross-link between loops of a single chain or between two separate polypeptide chains. These disulfide bonds are indicated by heavy cross-links in the figures of proteins in this chapter.
Tyrosine	Tyr	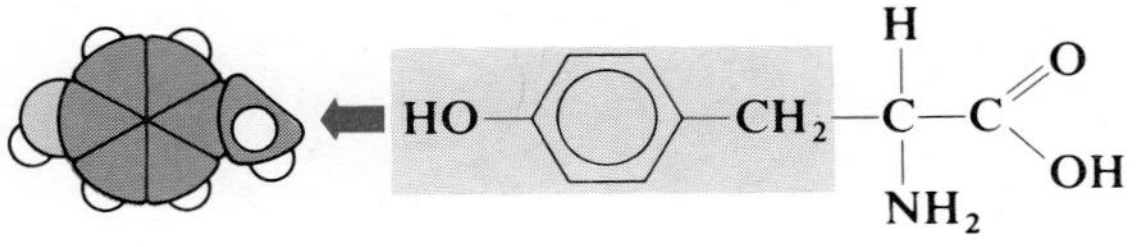	181	The *p*-hydroxy derivative of phenylalanine. The phenolic group is weakly acidic. It loses its proton at pH values above 9.
Asparagine	Asn	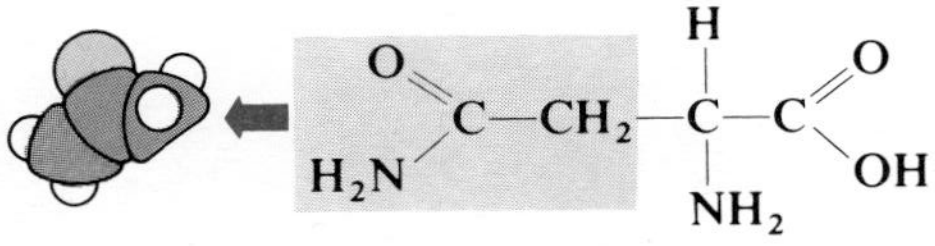	132	The amide of aspartic acid
Glutamine	Gln	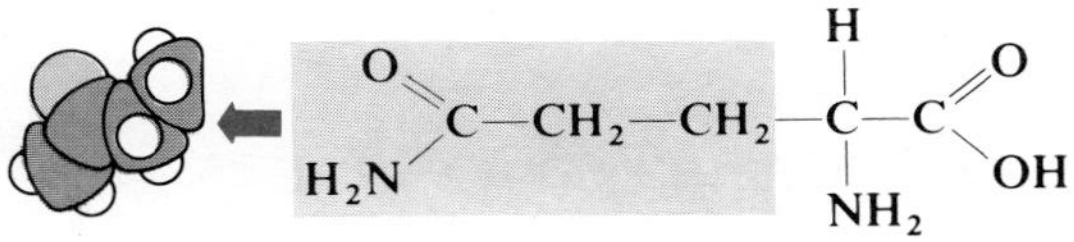	146	The amide of glutamic acid

(Table continues)

TABLE 13.1 *(continued)*

Name	Abbreviation	Structural Formula	MW (daltons)	Distinctive Features
3. Acidic Amino Acids				
Aspartic acid	Asp	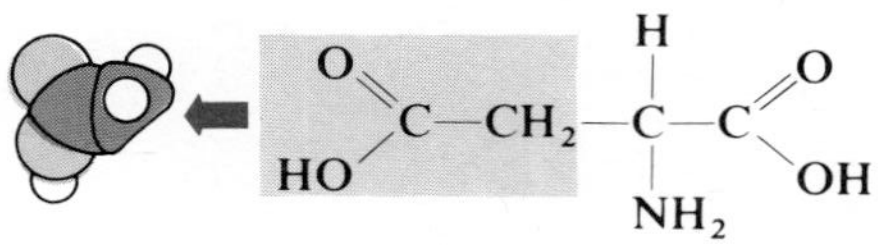	133	The second carboxyl group is ionized. These amino acids are therefore negatively charged at physiological pH.
Glutamic acid	Glu	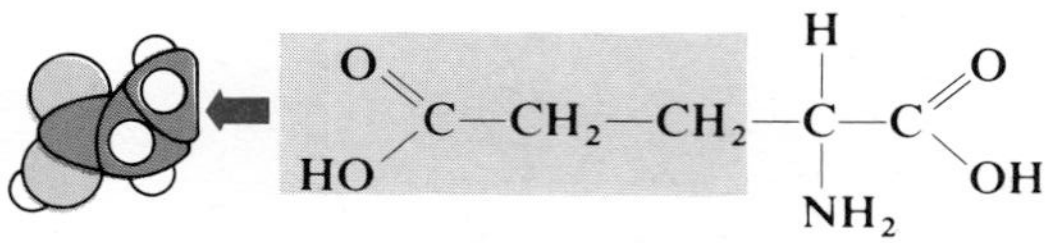	147	
4. Basic Amino Acids				
Lysine	Lys	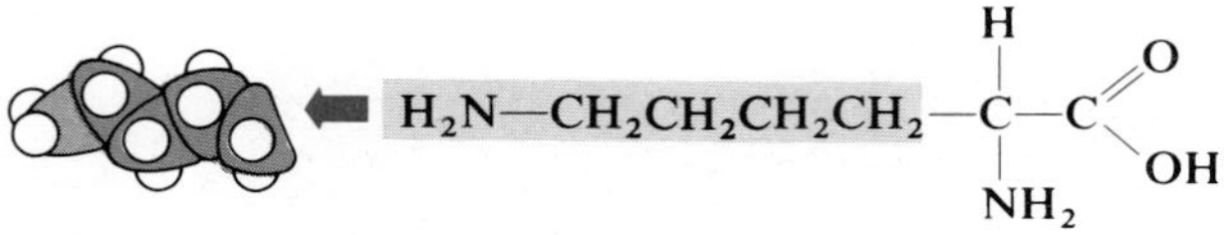	146	The ε-amino group of lysine is protonated and thus positively charged at physiological pH.
Arginine	Arg	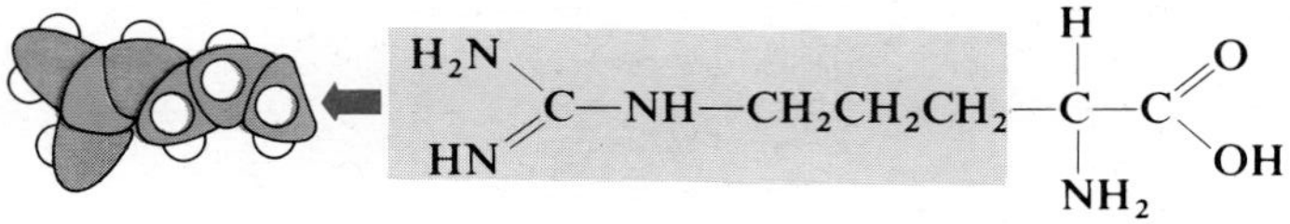	174	Almost as strong a base as NaOH because the guanidyl cation is stabilized by resonance. 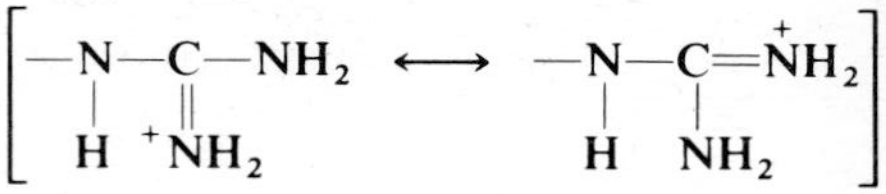
Histidine	His	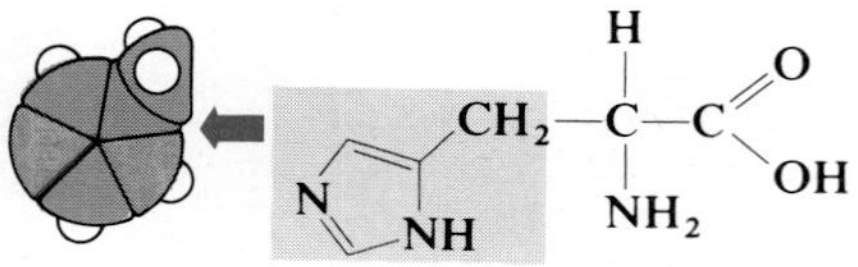	154	The only amino acid whose **R** group has a pK_a (6.0) near physiological pH and whose isoelectric pH (7.6) is near physiological pH. Thus the imidazole ring can be charged (+) or uncharged in the physiological pH range.

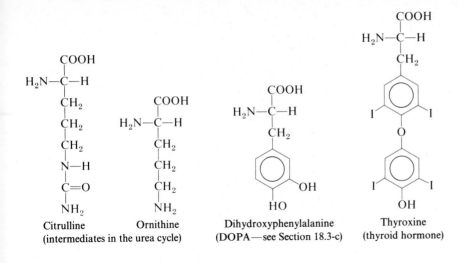

Citrulline
(intermediates in the urea cycle)

Ornithine

Dihydroxyphenylalanine
(DOPA—see Section 18.3-c)

Thyroxine
(thyroid hormone)

Homocysteine
(intermediate in the
synthesis of methionine)

Homoserine
(intermediate in the
synthesis of threonine)

β-Alanine
(component of
coenzyme A)

γ-Aminobutyric acid
(GABA—inhibitory
neurotransmitter)

FIGURE 13.1 Some biologically important nonprotein amino acids.

gelatin, received its name because of its sweet taste (Greek, *glykys*, sweet). More than 150 other amino acids occur in nature (particularly in the plant kingdom), but they do not appear in proteins (Figure 13.1). These amino acids perform important biological functions (for example, as intermediates in metabolic pathways) either as single molecules or combined in molecules of relatively small size.

13.2 *General Properties of Amino Acids*

a. Configuration

Notice, in Table 13.1, that glycine is the only amino acid whose α-carbon atom is not chiral. Therefore, with the exception of glycine, all the amino acids are optically active and may exist in either the D- or L-enantiomeric form (Figure 13.2). Once again, the reference compound for the assignment of configuration is glyceraldehyde. (The amino group of the amino acid takes the place of the hydroxyl group of glyceraldehyde.)

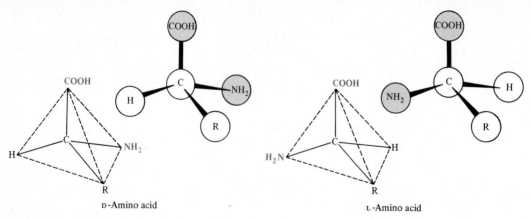

FIGURE 13.2 Enantiomers of an amino acid.

L-(−)-Glyceraldehyde   L-Amino acid   D-Amino acid

It is interesting to note that the naturally occurring sugars belong to the D-series, whereas nearly all known plant and animal proteins are composed entirely of L-amino acids.[2] Recall that the letters D and L refer only to a specific configuration, not to the direction of optical rotation. For example, in a neutral solution L-alanine is dextrorotatory, whereas L-serine is levorotatory. The optical rotation of any amino acid is very much dependent upon the pH of the solution.

b. Dipolar Ion Structure

The amino acids are colorless, nonvolatile, crystalline solids, melting with decomposition at temperatures above 200°C. Certain amino acids—glycine, alanine, proline, threonine, lysine, and arginine—are quite soluble in water; the others are sparingly soluble in varying degrees. All amino acids are insoluble in nonpolar organic solvents. Their properties diverge widely from those of their unsubstituted carboxylic acid analogs. Organic acids of comparable molecular weight are liquids or low melting solids that are soluble in organic solvents but have limited solubility in water. In fact, the properties of the amino acids are more similar to those of inorganic salts than to those of amines or organic acids.

The salt-like character of the amino acids is more readily accounted for if we assign a dipolar ion (also called **inner salt** or **zwitterion**) structure to amino acids in the solid state and in neutral solution. Since amino acids contain both acidic

[2]Certain bacteria are known to possess some D-amino acids as components of their cell walls. *Streptococcus faecalis* has a requirement for D-alanine; *Staphylococcus aureus* requires D-glutamic acid. Several antibiotics (e.g., actinomycin D and the gramicidins) contain varying amounts of D-leucine, D-phenylalanine, and D-valine (see Figure 14.16 for the structure of gramicidin S).

(—COOH) and basic (—NH$_2$) groups within the same molecule, we may postulate an intramolecular neutralization reaction, leading to salt formation.

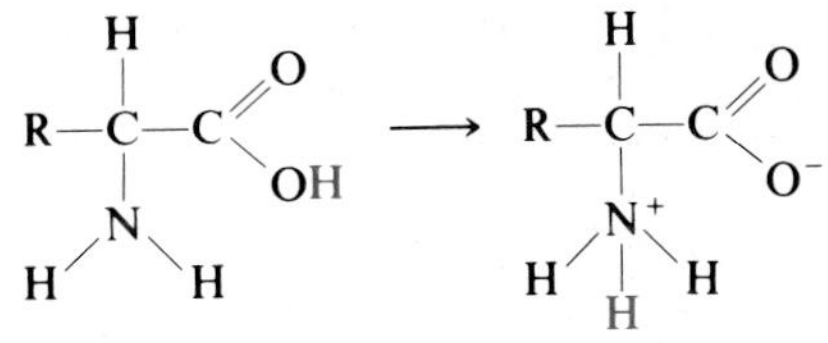

Zwitterion form
of an amino acid
(a dipolar ion)

Conductivity measurements confirm that at certain pH values, all amino acids exist in the neutral zwitterion form. When placed in an electric field, the dipolar ions migrate neither to the anode nor to the cathode, thus indicating a net charge of zero. *The pH at which an amino acid exists in solution as a zwitterion* is called the **isoelectric pH** (that is, the pH at which the amino acid is electrically neutral and has no tendency to migrate toward either electrode). Each amino acid has its own characteristic isoelectric pH. The isoelectric pH of the neutral amino acids ranges from 5.5 to 6.5, that of the acidic amino acids ranges from 3.0 to 3.2, and the isoelectric pH of the basic amino acids occurs between 7.6 and 10.8. Table 13.2 gives the isoelectric pHs for some representative amino acids.

Dipolar ions are **amphoteric** substances; they are *capable of donating or accepting protons* and thus may behave either as acids or as bases. When an acid is added to a solution of an amino acid that is at its isoelectric pH, hydrogen ions are accepted; the species becomes positively charged and will migrate to the negative electrode. Addition of a base causes the zwitterion to donate protons, thus becoming negatively charged; the species will now migrate toward the positive electrode. Figure 13.3 illustrates the acid–base behavior of the neutral amino acids.

For the neutral amino acids the pK_a of the α-carboxyl group (—COOH) is approximately 2–3, and the pK_a of the α-amino group (—NH$_3{}^+$) is approximately 9–10. The isoelectric pH can be determined for these neutral amino acids by simply

TABLE 13.2 Isoelectric pHs of Some Representative Amino Acids

Amino Acid	Type of Amino Acid	Isoelectric pH
Alanine	Neutral, nonpolar	6.0
Valine	Neutral, nonpolar	6.0
Serine	Neutral, polar	5.7
Threonine	Neutral, polar	6.5
Aspartic acid	Acidic	3.0
Glutamic acid	Acidic	3.2
Histidine	Basic	7.6
Lysine	Basic	9.7
Arginine	Basic	10.8

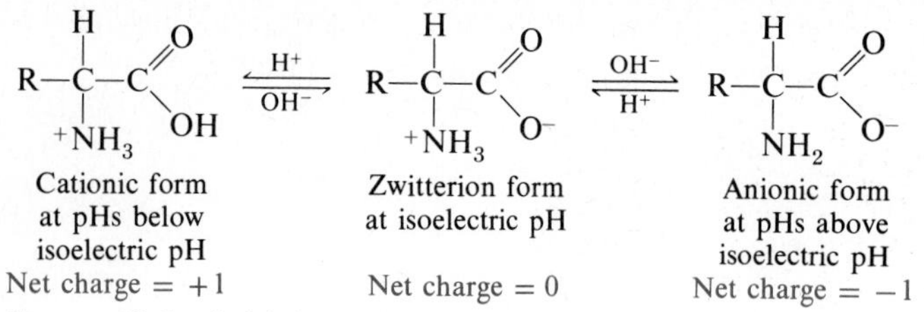

FIGURE 13.3 Acid–base behavior of neutral amino acids.

averaging the pK_a values for the two ionizable groups. Thus for alanine

$$pK_a(-COOH) = 2.3$$
$$pK_a(-NH_3^+) = 9.7$$
$$\overline{ 12.0}$$

$$\text{isoelectric pH} = \frac{12.0}{2} = 6.0$$

We shall see shortly that the structure and properties of proteins depend largely upon the nature of the R groups of their constituent amino acids. Of particular importance are those amino acids that contain either acidic or basic side chains. Each of these groups has a characteristic pK_a; it is determined by titrating the amino acid with standard solutions of acid (such as HCl) and base (such as NaOH). The pK_a values of the acidic and basic amino acids are listed in Table 13.3. Figure 13.4 depicts the different species that exist in solution for aspartic acid and lysine. Again, we can determine the isoelectric pH for these amino acids by averaging two pK_a values. For the acidic amino acids, the isoelectric pH is the average of the pK_as of the two carboxyl groups, whereas the isoelectric pH for the basic amino acids is the average of the pK_as of the two amino groups.

FIGURE 13.4 Structures and charges for aspartic acid (a) and lysine (b) in solutions of varying hydrogen ion concentration.

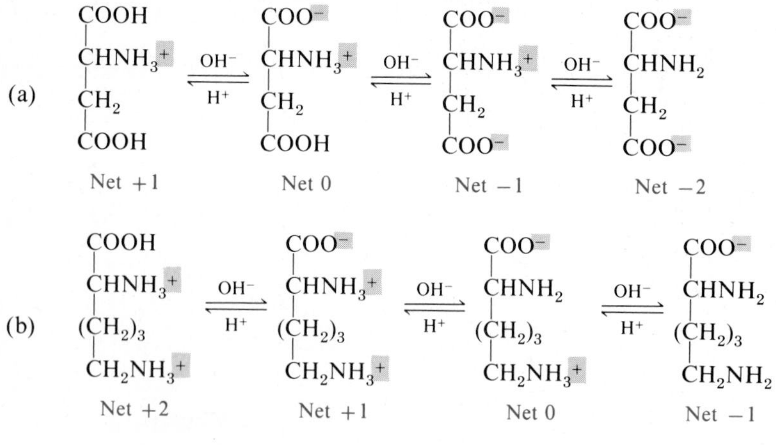

318

TABLE 13.3 pK_a Values for the Acidic and Basic Amino Acids

Amino Acid	pK_a (—COOH)	pK_a (—$\overset{+}{N}H_3$)	pK_a (R group)
Aspartic acid	2.1	9.8	3.9
Glutamic acid	2.2	9.7	4.3
Arginine	2.2	9.0	12.5
Lysine	2.2	9.0	10.5
Histidine	1.8	9.2	6.0

13.3   *Primary Structure of Proteins — Peptides*

Partial hydrolysis of proteins with dilute hydrochloric acid yields smaller molecules known as **peptides.** In 1902, Emil Fischer postulated that these peptides *contain amino acids joined together by amide linkages between the amino group of one molecule and the carboxyl group of another* (the union is accompanied by the elimination of a molecule of water). In protein chemistry, such an amide linkage is given the name **peptide bond** to indicate its relevance to the combination of amino acids. Peptides containing two, three, and four amino acids are called dipeptides, tripeptides, and tetrapeptides, respectively. The term **polypeptide** is applied to relatively small molecules containing 5–35 amino acids, and the term **protein** is arbitrarily used for longer polypeptides (that is, those having molecular weights above 5000 daltons). However, this is not meant to imply that a protein is composed of only one polypeptide chain. The enzymes *lysozyme* (Figure 13.5) and *ribonuclease* (Figure 13.6) contain 129 and 124 amino acids, respectively, in only one polypeptide chain, but the hormone insulin (Figure 13.7) contains two polypeptide chains (one chain contains 21 amino acids, the other 30) joined by disulfide linkages, and the hemoglobin molecule contains four amino acid chains. The **primary structure** of a protein refers to *the number and sequence of the amino acids in its polypeptide chain(s).*

By convention, we represent the structure of peptides beginning with the amino acid whose amino group is free (the so-called *N-terminal end*). The other end, therefore, contains a free carboxyl group and is referred to as the *C-terminal end.* Each amino acid in the peptide, with the exception of the C-terminal amino acid, is named as an acyl group in which the suffix *-ine* is replaced by *-yl.* (In naming the individual units it is not necessary to include the prefix L since it is understood that the amino acids have the L-configuration.) Figure 13.8 illustrates the bonding and the naming of some peptides.

There are several naturally occurring peptides that possess significant biological activity. We have previously mentioned the enkephalins (page 200), pain-relieving peptides that are released by the brain. The hormones *oxytocin* and *vasopressin* are nonapeptides produced by the pituitary gland (Figure 13.9). Notice that seven of the nine amino acids are identical in both peptides, yet their physiological effects are markedly different. Oxytocin stimulates lactation and causes the contraction of smooth muscles in the uterine wall. It is often administered at childbirth for the induction of labor.

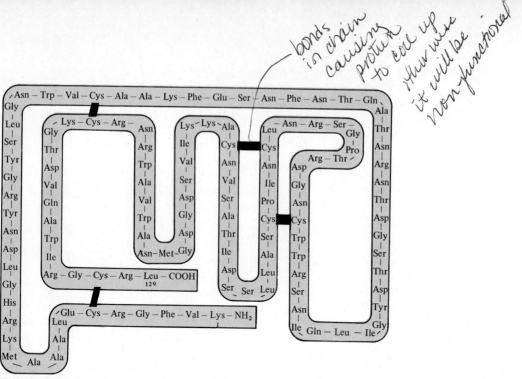

bonds in chain causing protein to coil up thus wise it will be non functional

FIGURE 13.5 The primary structure of lysozyme. Lysozyme is a protein enzyme that is found in many tissues and secretions of the human body, in plants, and in the whites of eggs. Its function is to destroy invading bacteria by catalyzing the cleavage of the polysaccharide chains that form part of the bacterial cell wall. Without a rigid cell wall, the bacterial cell bursts because of the sudden influx of water.

FIGURE 13.6 The primary structure of bovine ribonuclease. Ribonuclease is a protein enzyme that catalyzes the hydrolysis of ribonucleic acid (RNA).

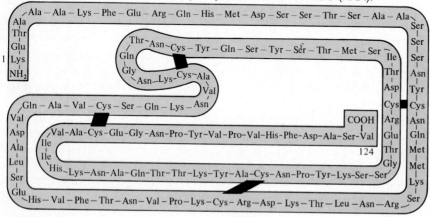

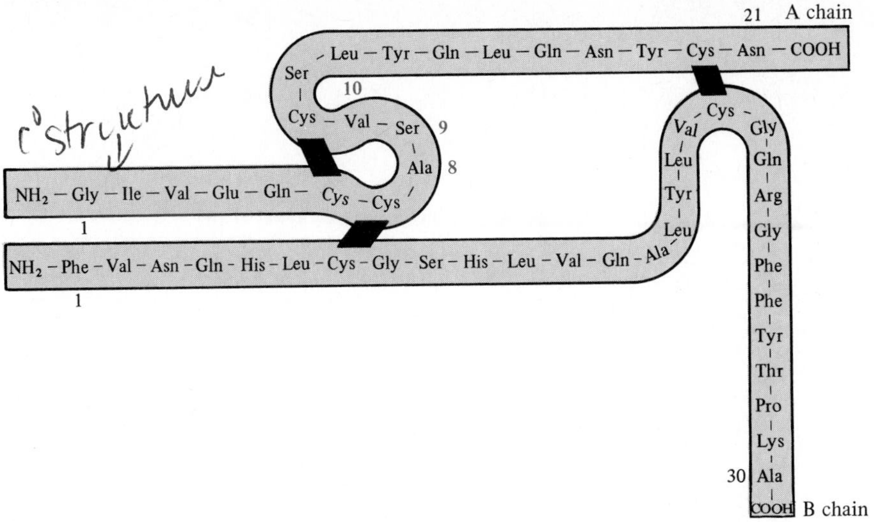
21 A chain
Leu – Tyr – Gln – Leu – Gln – Asn – Tyr – Cys – Asn – COOH
Ser
10
Cys – Val – Ser 9
Ala 8
NH₂ – Gly – Ile – Val – Glu – Gln – Cys – Cys
1
NH₂ – Phe – Val – Asn – Gln – His – Leu – Cys – Gly – Ser – His – Leu – Val – Gln – Ala
1
Cys
Val
Leu
Tyr
Leu
Gly
Gln
Arg
Gly
Phe
Phe
Tyr
Thr
Pro
Lys
30 Ala
COOH B chain

Species	Amino Acid Position #8	#9	#10
Cow	Ala	Ser	Val
Sheep	Ala	Gly	Val
Horse	Thr	Gly	Ile
Human	Thr	Ser	Ile
Pig	Thr	Ser	Ile
Whale	Thr	Ser	Ile
Dog	Thr	Ser	Ile

FIGURE 13.7 The primary structure of bovine insulin. Insulin is a protein hormone produced in the pancreas. It is essential for the regulation of carbohydrate metabolism. Insulin from other mammalian species has the identical structure except for amino acid positions 8, 9, and 10 of the A chain, which differ as shown. In addition, human insulin differs from all the others at position 30 of the B chain where a threonine replaces the alanine.

FIGURE 13.8 (a) Formation of a dipeptide. (b) A tripeptide. (c) A tetrapeptide.

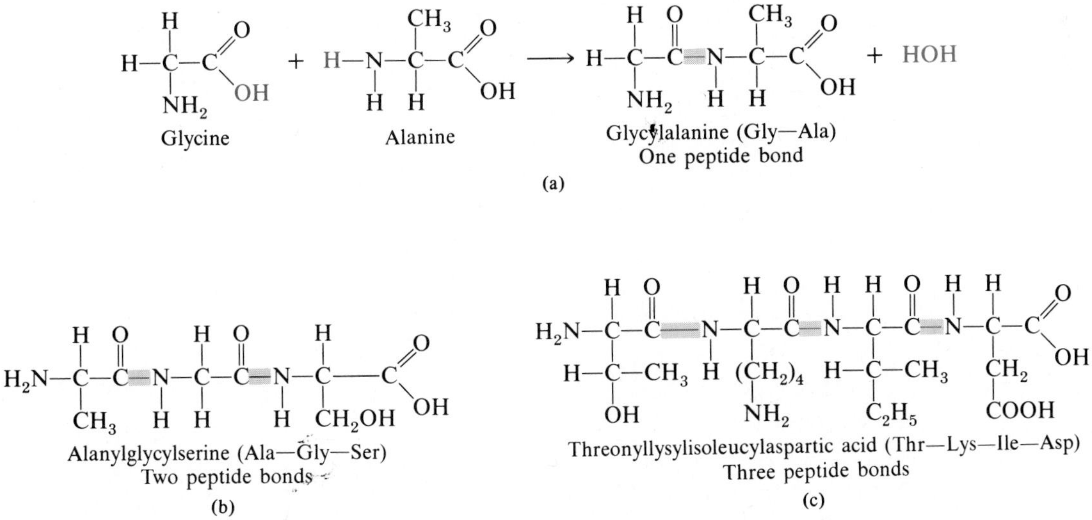
Glycine
Alanine
+ HOH
Glycylalanine (Gly—Ala)
One peptide bond
(a)
Alanylglycylserine (Ala—Gly—Ser)
Two peptide bonds
(b)
Threonyllysylisoleucylaspartic acid (Thr—Lys—Ile—Asp)
Three peptide bonds
(c)

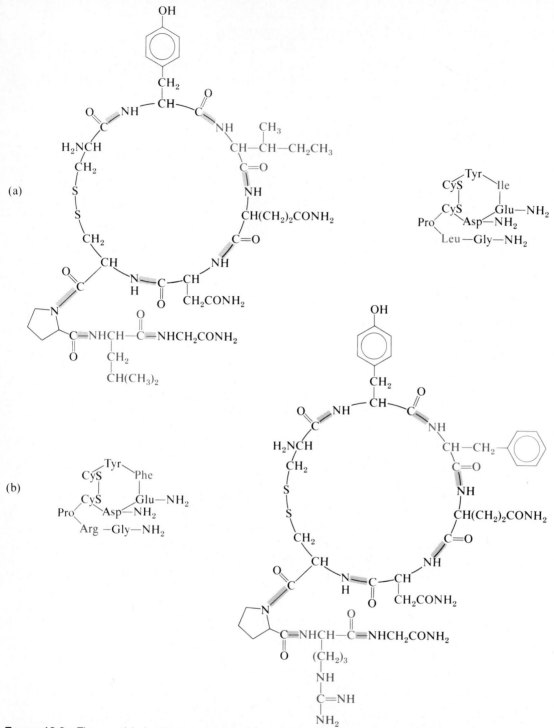

FIGURE 13.9 The peptide hormones oxytocin (a) and vasopressin (b). Vasopressin differs in structure from oxytocin only in the replacement of isoleucine by phenylalanine and leucine by arginine. Notice that the C-terminal glycine and the carboxyl groups in the aspartic acid and glutamic acid side chains are present as amides.

Vasopressin is called the *antidiuretic*[3] *hormone* (ADH) because it acts on the kidneys to reduce the amount of water excreted. Thus, the major function of vasopressin is to increase water reabsorption in the kidney. After drinking alcohol, many people excrete more urine than can be accounted for by the volume of water in their drinks. It is thought that alcohol inhibits the secretion of vasopressin, and therefore, the kidney does not reabsorb as much water. A deficiency of vasopressin, or an inability of the kidney to respond to vasopressin, results in *diabetes insipidus,* in which too much urine is excreted (> 10 L/day). This disease (which is treated by administering vasopressin) should not be confused with diabetes mellitus (see Section 16.3). In addition, vasopressin stimulates the contractions of muscles in the walls of blood vessels and thus increases the blood pressure. It has been used to overcome low blood pressure caused by shock following surgery.

One can begin to appreciate the enormous complexity of protein molecules from a consideration of these simple peptides. For example, if you take three different amino acids, A, B, and C, and use each one only once in forming a tripeptide, six possible isomers can be formed: ABC, ACB, BAC, BCA, CAB, and CBA. If there are x different amino acids in a polypeptide, each one appearing only once, then the number of different possible structures is x factorial.

$$x(x - 1)(x - 2)\cdots$$

Application of this formula to a peptide containing ten different amino acids reveals that there are over 3.6 million different possible structures.

Now consider further that in most proteins certain amino acids appear not once, but several times. A typical protein contains anywhere from 100 to 300 amino acids, and these units are joined together in a definite, characteristic sequence. Such a protein may have an enormous number of structural isomers based solely upon differences in amino acid sequence. It is estimated that a typical animal cell contains about 9000 different proteins and the human body contains over 100,000 different protein molecules, all of which are characterized by different sequential arrangements of the 20 fundamental building blocks. The best analogy of the relationship between the 20 amino acids and the vast number of protein molecules they constitute is drawn between the letters of the alphabet and the number of words they form. All the words in the English language are composed of different sequential combinations of just 26 letters.

The truly remarkable phenomenon of living systems is that with so many possibilities for protein synthesis, generation after generation of cells produce identical protein molecules. This ability of the cell to produce the same proteins as its parent is vital to life because it is the sequence of amino acids within a protein that bestows biological function upon the molecule. In other words, if a protein molecule is to fulfill its physiological function, its amino acids must be ordered in a particular definite sequence. The substitution of even one amino acid in a large protein

[3] *Diuretics* are substances that increase the volume of urine, and therefore any substance that reduces the volume of urine is an antidiuretic. Diuretics are often given to people with high blood pressure (see Section 19.4) to cause the loss of water and sodium ions, both of which contribute to the elevated blood pressure.

molecule may completely alter its biological function. A classic example of this phenomenon occurs in the disease sickle cell anemia.[4] The hemoglobin of the afflicted individual contains valine in place of glutamic acid. As we shall learn in Section 13.6, hemoglobin consists of 574 amino acids that are arranged in four polypeptide chains. The substitution of valine for glutamic acid at position 6 occurs only in two of the identical chains (i.e., only 2 amino acids out of 574 are affected). This seemingly insignificant alteration in the molecule causes a drastic change in the properties of the molecule.

Normal hemoglobin: Val-His-Leu-Thr-Pro-Glu-Glu-Lys...
$1234567$8

Sickle cell hemoglobin: Val-His-Leu-Thr-Pro-Val-Glu-Lys

Two formidable research problems facing scientists today are to determine (1) the primary structure of proteins and (2) the relationship between amino acid sequence and protein function. The pioneer work on the elucidation of the primary structure of proteins was performed by Frederick Sanger (1953) at Cambridge, England. Sanger received the Nobel Prize for Chemistry in 1958 for his work on the structure of proteins, especially that of insulin. He chose to study insulin because of its small size (MW = 5733 daltons) and because it was readily available in crystalline form from a variety of sources.[5] By employing methods of controlled hydrolysis of the protein to recognizable peptides, Sanger succeeded in determining the amino acid sequence in the two polypeptide chains of insulin. (It is beyond the scope of this book to elaborate on the techniques used in the analysis of protein structure. For detailed accounts of these methods, you are directed to more advanced texts.)

[4]This is an inherited disease that, if left untreated, may be fatal (due to infection, blood clots, cardiac or renal failure). The change from a polar amino acid (Glu) to a nonpolar one (Val) reduces the overall charge on the hemoglobin molecule. If the altered hemoglobin molecule is fully oxygenated, there is no problem and it remains in solution. However, if the level of oxygenation decreases (e.g., at high altitudes or during vigorous physical exercise), the less soluble, deoxygenated hemoglobin molecules clump together, forming long insoluble fibers, and force the red blood cell to assume a crescent or sickled shape. The abnormal red blood cells become trapped in the capillaries and impair circulation. The resultant blockage of blood further decreases the oxygen supply to the affected areas of the body and increases the sickling of additional red blood cells. The abnormal cells are subsequently destroyed by the spleen, and this leads to anemia (and its subsequent tiredness and other permanent damage). In the United States about 10% (more than 2 million) of the black population are genetic carriers of the disease (heterozygotes), and 0.25% have the symptoms of sickle cell anemia (homozygotes). Genetic screening tests can determine whether prospective parents carry the sickle cell trait. Sickle cell anemia is the most prevalent of the genetic diseases (see Section 15.6).

[5]An interesting fact about living systems is that different ones contain, in many cases, almost identical protein molecules. Insulin is a striking example of this phenomenon (see Figure 13.7). Not only are the primary structures of a variety of mammalian insulin molecules similar, but they also have the same biochemical properties. This is fortunate for some diabetics who develop an allergy to a particular type of insulin; they can often be treated with insulin from another species that does not produce the allergic reaction. In this instance, alteration of some of the amino acids does not cause the protein to have an altered physiological effect, whereas in hemoglobin, oxytocin, and vasopressin it does. This explanation of the special physiological functions of proteins in terms of their structure will become apparent when we consider the concept of the *active site* in relation to enzymes in Section 14.5. (As we shall see in Section 15.8, human insulin is being produced by recombinant DNA technology and this is diminishing the allergy problem.)

TABLE 13.4 Some Proteins Whose Sequence of Amino Acids is Known

Protein	Function	Number of Amino Acid Residues
Enzymes		
Ribonuclease	Hydrolyze RNA	124
Lysozyme	Cleaves bacterial cell walls	129
Papain	Digests proteins	212
Trypsinogen	Digests proteins	229
Chymotrypsinogen	Digests proteins	245
Carbonic anhydrase (human)	Hydration of CO_2	260
Subtilisin (a bacterial protease)	Digests bacterial proteins	274
Carboxypeptidase A (bovine)	Digests proteins	307
Alcohol dehydrogenase (horse)	Oxidizes alcohol	374
Others		
Nisin	Antibiotic	29
Insulin	Variety of metabolic functions	51
Trypsin inhibitor (bovine pancreas)	Inhibits trypsin	58
Cytochrome c (human, horse, pig, rabbit, chicken)	Electron transport	104
Cytochrome c (yeast)	Electron transport	108
Hemoglobin (human)	Oxygen transport	
α-chain		141
β-chain		146
Calmodulin (bovine)	Calcium transport	148
Myoglobin	Oxygen storage	153
Tobacco mosaic virus protein subunit	Virus protein	158
Myelin (bovine)	Protects nerve cells	170
Myelin (human)	Protects nerve cells	172
Human growth hormone	Necessary for normal growth	191
α-Casein (bovine)	Milk protein	199
Human serum albumin	Blood protein	584
Immunoglobulin, IgG	Antibody	1320

Sanger's extraordinary achievement opened the door to the elucidation of the structures of many additional proteins. Today, the primary structures are known for over 3000 polypeptides, and partial sequences are known for many others (Table 13.4).

13.4 *Secondary Structure of Proteins*

The primary structure describes only the sequence of amino acids in the protein chain but tells nothing about the shape of the molecule. Since most of the bonds in protein molecules are single bonds, one might assume that there is completely free rotation and that the molecules can assume an infinite number of shapes. However, it is known from a variety of physical measurements that each protein occurs in nature in a particular three-dimensional conformation. *The fixed conformation of the polypeptide backbone* is referred to as the secondary structure of a protein.

Two major considerations are involved in the secondary structure of proteins. The first involves the manner in which the protein chain is folded and bent; the second involves the nature of the bonds that stabilize this structure.

Based upon x-ray studies, Linus Pauling (Nobel Prize for Chemistry, 1954, and for Peace, 1962) and Robert Corey postulated that some proteins have a spiral shape (that is, they are shaped like a helix). This shape is best visualized as a spring coiled about an imaginary cylinder. The spring may be either *left-* or *right-handed*. The spiral structure assumed by proteins (which, you will recall, contain only L-amino acids) is invariably found to be right-handed; i.e., the coils turn in a clockwise fashion around the axis (Figure 13.10). *The spiral, or helix, is stabilized by hydrogen bond formation between the amide hydrogen of one peptide bond and the carbonyl oxygen above it which is located on the next turn of the helix.* This **intrachain** hydrogen-bonded structure is designated as α-*helical*. X-ray data indicate that the helix makes one turn for every 3.6 amino acids, and that the side chains of these amino acids project outward from the coiled backbone. The α-keratins, found in hair and wool, are exclusively α-helical in conformation.

Not all proteins assume a helical conformation. Some proteins, such as gamma globulin, chymotrypsin, and cytochrome c, have little or no helical structure. Other proteins, such as hemoglobin and myoglobin, are helical in certain regions of the polypeptide chain; the remaining portions assume random conformations. The polypeptide chains of structural proteins such as silk fibroin and certain enzymes such as *carboxypeptidase A* and *lysozyme* are aligned side by side in a sheet-like arrangement. In these proteins, segments of the polypeptide chains lie next to one

FIGURE 13.10 Alpha helical conformation of proteins: right-handed (a) and left-handed (b).

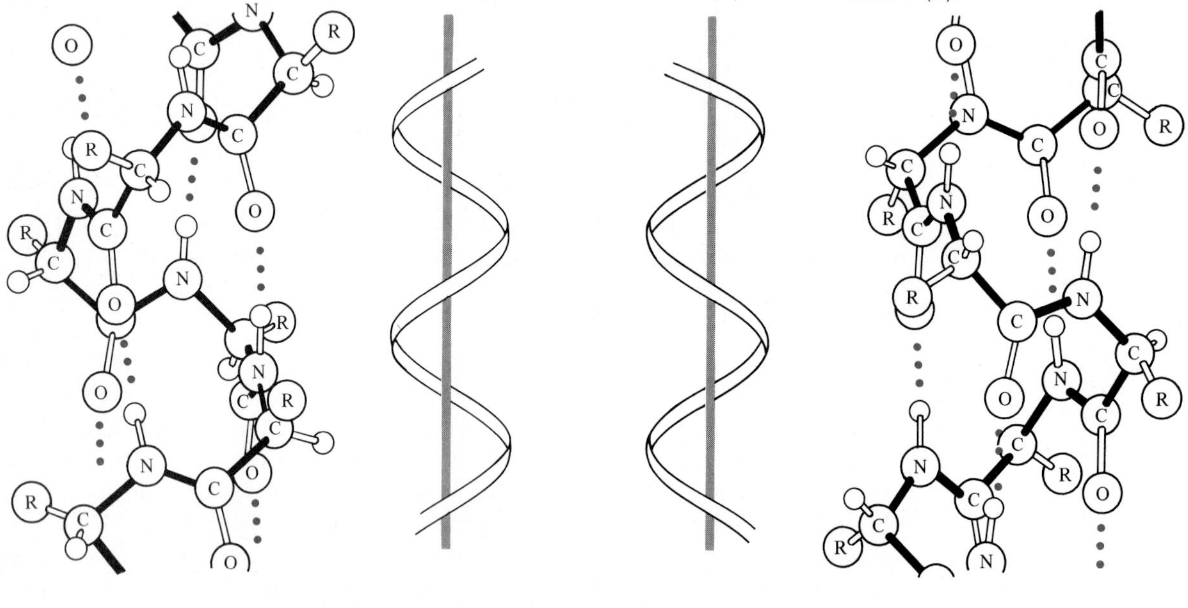

(a)   (b)

another and run either parallel or antiparallel, with hydrogen bonding connecting the adjacent strands (Figure 13.11). This structural arrangement is designated as the β-pleated sheet conformation,[6] and it occurs when two extended polypeptide chains (or two separate regions on the same chain) are aligned side by side.

The physical characteristics of wool and silk are a result of their structural conformations. Wool is very flexible and extensible. It can stretch to twice its normal length without breaking, and the fiber will return to its original state upon release of tension. The stretching process involves breaking hydrogen bonds along turns of the α-helix (covalent bonds remain intact). The disulfide bonds between helices, together

FIGURE 13.11 Extended polypeptide chain (a) and schematic diagram (b) of the β-pleated sheet conformation of silk fibroin. The peptide bonds lie in the plane of the pleated sheet; the side chains lie above or below the sheet and alternate along the chain. The polypeptide chains are held together by interchain hydrogen bonds.

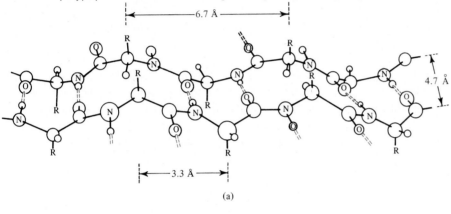

(a)

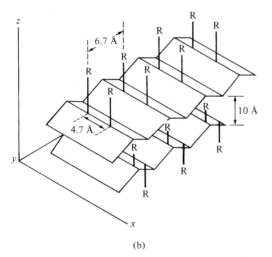

(b)

[6]The designation β was used by Pauling and Corey because it was the second protein conformation that they discovered, α-helix having been the first.

with re-formed hydrogen bonds, provide the forces that operate to restore the helix when tension is released. Silk, on the other hand, is already stretched out in the β-sheet conformation. The sheets are held together very tightly by hydrogen bonds. As a result, silk fiber is strong, but it is resistant to stretching.

13.5 *Tertiary Structure of Proteins*

The picture of a protein that we have thus far constructed is that of a loosely coiled spring or an extended sheet that lacks definite geometric arrangement. Intrachain hydrogen bonds are adequate for maintaining the conformations of proteins in the crystalline state, but they are not sufficiently strong to stabilize these conformations in aqueous solution. Recall that water is a good hydrogen bonding substance, and therefore water molecules can successfully compete for the hydrogen bonding sites on the protein backbone. This would cause disruption of the internal bonds, and the protein would assume a random conformation simultaneous with the disruption. Experimental evidence has shown, however, that the conformation of the protein is *not* destroyed when the molecule is dissolved in water. We must therefore conclude that there are other forces involved in compressing the long spiral chains and extended sheets into the definite geometric structures characteristic of each different protein. *This unique three-dimensional shape, which results from the precise folding and bending of the protein backbone,* is referred to as the **tertiary structure** of the protein (Figure 13.12). The tertiary structure of a protein is intimately involved with the proper biochemical functioning of that protein, as we shall see in the next chapter on enzymes.

The linkages responsible for the tertiary structure of a protein are a function of the nature of the amino acid side chains within the molecule. Globular proteins (see

FIGURE 13.12 The tertiary structure of a protein. This tube or "sausage" model shows the coiled protein backbone and the approximate volume occupied by the protein.

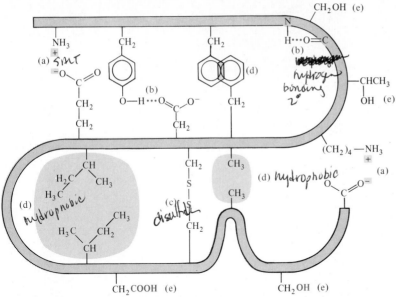

FIGURE 13.13 Bonds that stabilize the tertiary structure of proteins. (a) Salt linkages, (b) hydrogen bonds, (c) disulfide linkages, (d) hydrophobic interactions, (e) polar group interactions with water.

Section 13.7) are extremely compact, almost spherical in shape. Such proteins have their nonpolar side chains directed toward the interior of the molecule (the hydrophobic or nonaqueous region) and their polar side chains project outward from the surface of the molecule toward the aqueous environment. The resulting picture is very similar to that of a micelle, which was discussed in connection with the properties of soap (Section 12.8). Some of the linkages that contribute to the tertiary structure of proteins are shown in Figure 13.13. Table 13.5 gives an indication of the relative strengths of interactions involving the noncovalent bonds found in proteins.

a. Salt Linkages

Salt linkages (ionic bonds) result from electrostatic interactions between positively and negatively charged groups on the side chains of the basic and acidic amino acids. For example, the mutual attraction between an aspartic acid carboxylate ion and a lysine ammonium ion helps to maintain a particular folded area of the protein.

Aspartic acid   Lysine

TABLE 13.5 Noncovalent Bonds and Interactions in Polypeptides[a]

Example	Type of Bond	Approximate Stabilization Energy (kcal/mole)
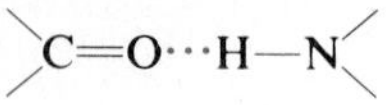 $C=O\cdots H-N$	Hydrogen bond between peptides	2–5
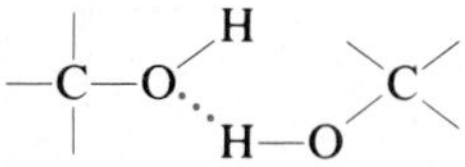 $-C-O\cdots H-O-C$ (with H)	Hydrogen bond between neutral groups	2–5
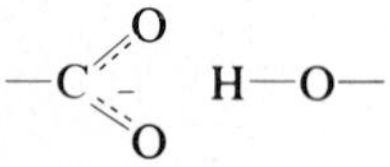 $-C(=O)(O)\cdots H-O-$	Hydrogen bond between neutral and charged groups	2–5
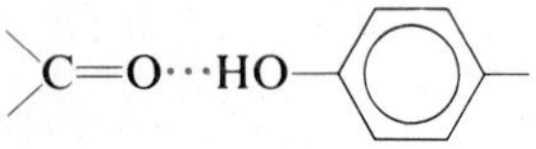 $C=O\cdots HO-\bigcirc-$	Hydrogen bond between peptide and **R** group	2–5
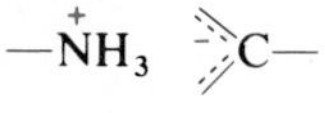 $-\overset{+}{N}H_3 \quad \overset{-}{}C-$	Salt linkage *or* ionic bond between charged groups (strongly dependent on distance)	<10
$-CH_3 \ CH_3-$	Hydrophobic interaction	0.3
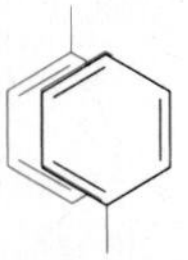 (aromatic rings)	Hydrophobic interaction—stacking of aromatic rings	1.5
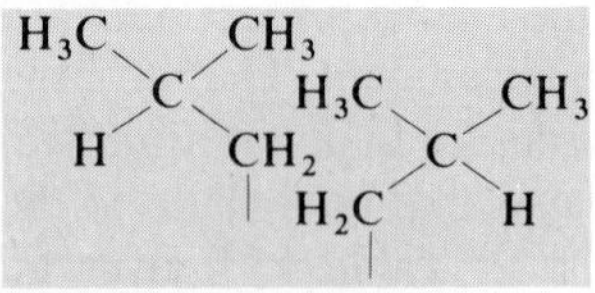 $H_3C \ CH_3 / H_3C \ CH_3$ (branched)	Hydrophobic interaction	1.5
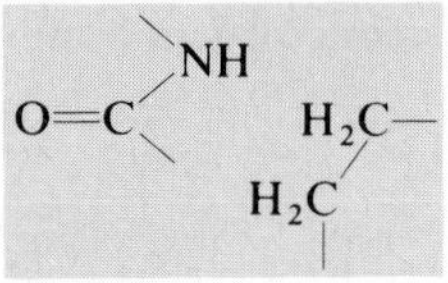 $O=C(NH)\cdots H_2C-, H_2C$	Hydrophobic interaction between **R** group (or part of **R** group, as in lysine) and the peptide bond. This interaction usually involves the hydrophobic **R** groups of the amino acid contributing the carboxyl group to the bond.	$0.3/CH_2$
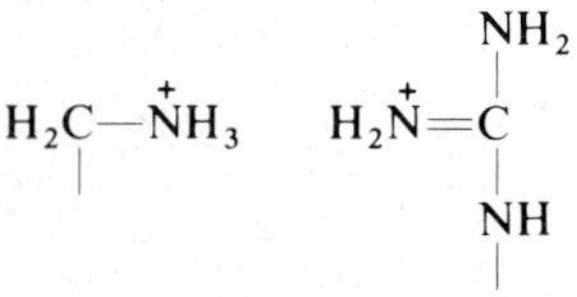 $H_2C-\overset{+}{N}H_3 \quad H_2\overset{+}{N}=C(NH_2)(NH)$	Repulsive interactions between similarly charged groups (strongly dependent on distance)	< −5

[a]From Robert Barker, *Organic Chemistry of Biological Compounds,* 1971, p. 103. Reprinted by permission of Prentice-Hall, Inc., Englewood Cliffs, NJ.

b. Hydrogen Bonding

Hydrogen bonds are formed principally between the side chains of the polar amino acids and between a carboxyl oxygen and a hydrogen donor group. The hydrogen-bonding capabilities of the terminal amino group of lysine and the terminal carboxyl groups of aspartic acid and glutamic acid are pH dependent. These groups can serve as both hydrogen-bond acceptors and hydrogen-bond donors only over a certain range of pH. Hydrogen bonds (as well as salt linkages) are extremely important in the interaction of proteins with other molecules.

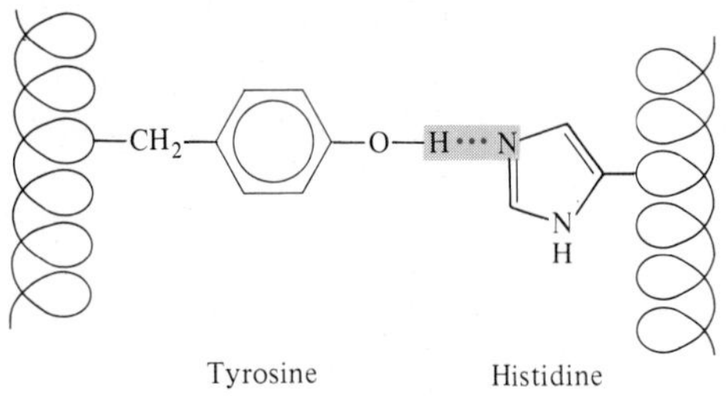

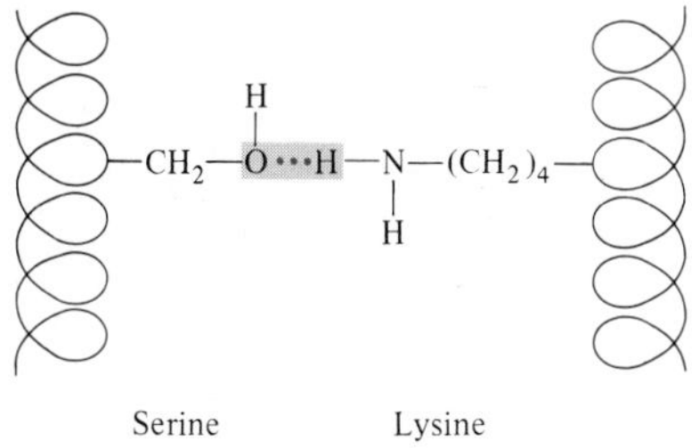

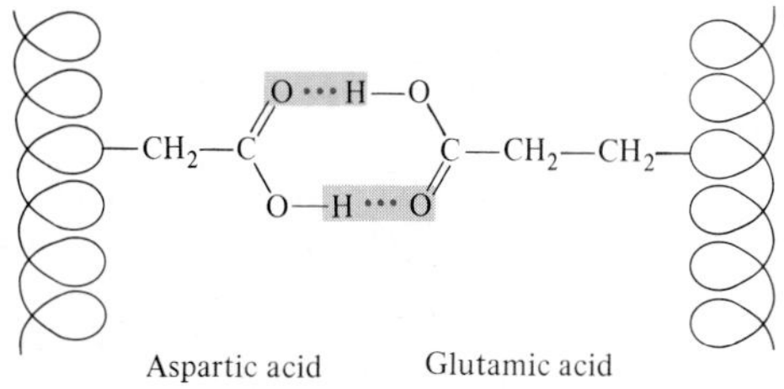

Tyrosine   Histidine

Serine   Lysine

Aspartic acid   Glutamic acid

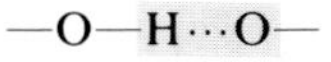

—O—H···O—
Strong H-bond

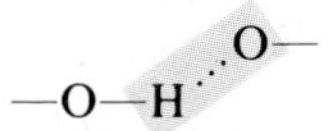

—O—H···O—
Weak H-bond

A significant feature of hydrogen bonds is that they are highly directional. The strongest hydrogen bond results when the hydrogen donor and the acceptor atom are colinear. If the acceptor atom is at an angle to the covalently bonded hydrogen atom, the hydrogen bond is much weaker.

c. Disulfide Linkages

Two cysteine residues may come in proximity as the protein molecule folds. The disulfide linkage results from the subsequent oxidation of the highly reactive sulfhydryl (—SH) groups to form cystine (Section 6.16).

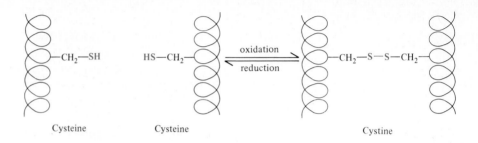

Cysteine   Cysteine   Cystine

This disulfide bridge is the second most important covalent interaction involved in protein structure. (Recall that the peptide bond is the most important covalent interaction in the structure of proteins.) Intrachain disulfide linkages are frequently found in proteins as a general aid to the stabilization of the tertiary structure. Note, however, that one or more of these bonds may join one portion of a polypeptide chain covalently to another, thus interfering with the helical structure. Interchain disulfide bonds are important forces that link two separate polypeptide chains. Such linkages are clearly indicated in the protein structures given in Figures 13.5, 13.6, and 13.7.

d. Hydrophobic Interactions

The noncovalent hydrophobic forces are considered to be the most significant in stabilizing the conformation of a polypeptide chain. It is not because they are so strong, but rather because there are so many of them. The majority of the nonpolar amino acid groups cluster together at the interior of the protein, and the strength of all their hydrophobic interactions is considerable.

Phenylalanine   Phenylalanine   Valine   Leucine

13.6   Quaternary Structure of Proteins

Proteins that contain two or more polypeptide chains are known as *oligomeric* proteins, and their individual chains are called subunits. *The manner in which the subunits fit together* in oligomeric proteins is referred to as the **quaternary structure** of the protein. The quaternary structure is stabilized by the same kinds of forces (ionic bonds, hydrogen bonds, interchain disulfide bonds, and hydrophobic bonds) that are involved in maintaining the tertiary structure.

J. C. Kendrew and M. F. Perutz (Nobel Prize winners in 1962) were able, through the use of x-ray diffraction studies, to elucidate completely the primary, secondary, and tertiary structures of myoglobin, a protein consisting of 153 amino acids

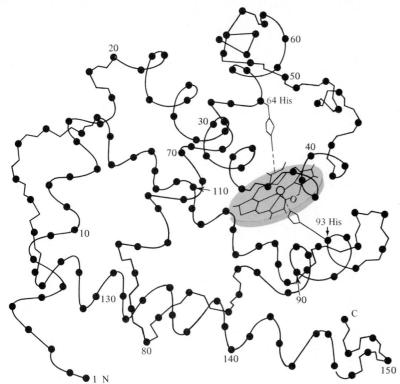

FIGURE 13.14 The conformation of myoglobin deduced from x-ray diffraction studies. The disk shape represents the heme group.

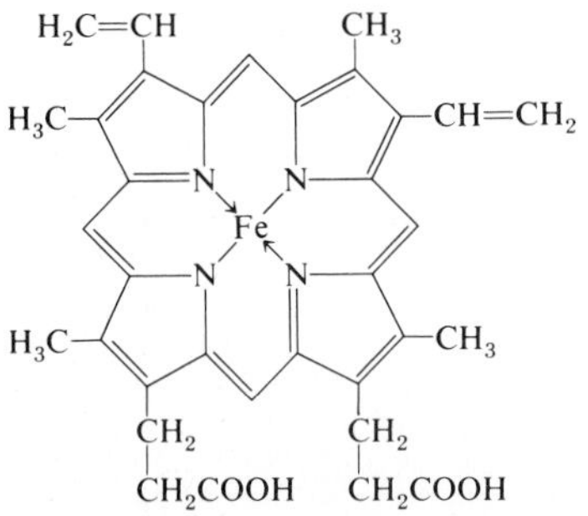

In the figure: 20, 60, 50, 64 His, 30, 40, 70, 110, 93 His, 10, 90, 130, C, 80, 140, 150, 1 N

$H_2C=CH$ CH_3

H_3C $CH=CH_2$

N   N

Fe

N   N

H_3C CH_3

CH_2 CH_2

CH_2COOH CH_2COOH

FIGURE 13.15 Heme is a complex organometallic compound that is present in both myoglobin and hemoglobin.

arranged in a single chain (Figure 13.14). Myoglobin can combine reversibly with molecular oxygen, and it functions to store oxygen in muscle cells. It is particularly abundant in marine animals, such as whales, seals, and porpoises, enabling them to remain under water for prolonged periods. (In humans, myoglobin is found mainly in heart muscle.) The primary structure of myoglobin results from the formation of the single polypeptide chain. The secondary structure involves the coiling of this chain into an α-helix (about 70% of the protein strand has a spiral conformation). The tertiary structure results from the nonuniform folding of the chain to form a stable compact structure. The polar side chains are on the outside of the molecule, and almost all of the nonpolar ones are on the inside. The heme group (Figure 13.15) is located in a crevice of the polypeptide chain.

The structure of the hemoglobin molecule was also deduced by Kendrew and Perutz. It consists of four separate polypeptide chains or subunits—two identical α-chains (141 amino acids) and two identical β-chains (146 amino acids). Each of the four chains is very similar in structure to the single polypeptide chain of myoglobin (Figure 13.16). Since each subunit contains a heme group, one hemoglobin molecule can bind four molecules of oxygen. The four hemoglobin subunits are held

FIGURE 13.16 The quaternary structure of hemoglobin. The heme molecules are shown within the folds of the four polypeptide chains. Hemoglobin has an almost spherical shape as a result of the fitting together of the four subunits.

together by noncovalent surface interactions between the polar side chains (and probably hydrophobic interactions as well).

Myoglobin has no quaternary structure (since it is composed of a single polypeptide chain), whereas the protein coats of several viruses are composed almost entirely of polypeptide subunits arranged in a highly ordered conformation

FIGURE 13.17 Schematic diagrams of the structures of several proteins.

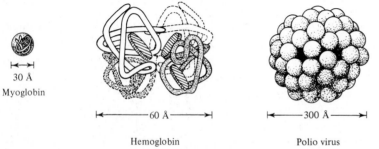
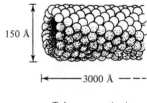

30 Å
Myoglobin

|←——— 60 Å ———→|

Hemoglobin

|←—— 300 Å ——→|

Polio virus

150 Å

|←—— 3000 Å ——— - - -|

Tobacco mosaic virus

(a)   (b)

FIGURE 13.18 Schematics of two fibrous proteins. (a) The α-keratins consist of three polypeptide chains in right-handed α-helical coils. (b) The protein collagen is a rigid cable made from three left-handed polypeptide chains. The cable is about 3000 Å long and 14 Å thick. The intertwined chains are maintained by interchain hydrogen bonding.

(Figure 13.17). The polio virus contains 130 polypeptide chains. The tobacco mosaic virus contains a grand total of about 345,000 amino acids arranged in 2130 individual polypeptide chains that are assembled around a central core of nucleic acids (see Section 15.7). The average molecular weight of each polypeptide chain is about 18,000 daltons.

Collagen, which is the most abundant of all proteins in higher vertebrates (about one third of the total protein in our bodies), has a distinctive shape because of its unique amino acid composition. Although collagens of different species differ somewhat in amino acid sequence, most collagen molecules contain about 30% glycine, 12% proline, and 9% hydroxyproline. When proline and hydroxyproline form peptide bonds, they lose the single hydrogen atom bonded to the ring nitrogen (see Table 13.1). Thus, they are unable to form the intrachain hydrogen bonds that maintain the α-helical conformation of most other proteins. The quaternary structure of α-keratin,[7] for example, consists of three right-handed α-helical polypeptide chains coiled around one another. The three-stranded rope is maintained primarily by interchain *disulfide bonds*. In contrast, collagen is a left-handed *triple helix* of polypeptide chains (each chain contains about 1000 amino acids—Figure 13.18). The three chains are wrapped around one another in cable fashion and are cross-linked by interchain *hydrogen bonding*. In tendons, the collagen fibers are arranged in parallel bundles to yield structures that have nearly the tensile strength of steel wire, but little or no capacity to stretch. In bone and teeth, the collagen cables form the matrix upon which is built the network of calcium salts. When collagen is placed in boiling water, it is converted into the soluble protein *gelatin*.

13.7 *Classification of Proteins*

Because of the great complexity of protein molecules, it is not possible to classify the proteins systematically in the same way as the carbohydrates and lipids are categorized, that is, on the basis of structural similarities. An older convention classified proteins under either of two descriptive headings. Stringy, elongated proteins such as collagen, silk fibroin, and keratin were termed **fibrous;** spherical, compact proteins such as egg albumin, casein, and most enzymes and antibodies were termed **globular.** Fibrous proteins tend to be insoluble in water and in most other solvents, whereas globular proteins are soluble in water and in solutions of salt and water.

Presently, there are two alternative methods for classifying proteins. One method classifies them according to composition as either simple or conjugated. A **simple protein** yields exclusively α-amino acids upon hydrolysis. A **conjugated protein** yields a nonproteinaceous material, called a **prosthetic group,** upon hydrolysis in addition

---

[7]There are two classes of keratins. The α-keratins (such as hair and wool) have the α-helix conformation, and they contain many cysteine amino acids, which provide interchain disulfide bonds. The β-keratins (such as silk fibroin) exist in the pleated-sheet conformation; they do not contain cysteine and therefore cannot form disulfide bonds.

to the α-amino acids. The simple proteins are further subdivided according to their solubility in various solvents. Subdivision of the conjugated proteins rests upon the nature of the prosthetic group. The second method of classification groups the proteins according to their function. Both systems are outlined here.

a. Classification of Proteins According to Composition

1. Simple Proteins **Albumins** constitute the most important and the most common group of simple proteins. They are present in egg white (egg albumin) and in blood (serum albumin). They are soluble in water and in dilute salt solutions. Serum albumins serve as blood buffers, and they are important transport vehicles for the fatty acids.

    Globulins are insoluble in water, but are soluble in dilute salt solutions. Some globulins function as antibodies, whereas others are involved in the transport of lipid molecules.

    Histones are basic proteins, since they contain a high proportion of basic amino acids (lysine and/or arginine). Histones are found associated with nucleic acids in the nucleoproteins of the cell. They are soluble in water and insoluble in dilute ammonium hydroxide.

    Scleroproteins (albuminoids) have structural and protective functions and are characterized by their insolubility in water and in most other solvents. Examples are keratin, collagen, and elastin (elastic fibers of connective tissue).

2. Conjugated Proteins **Phosphoproteins** are converted to phosphoric acid and amino acids upon complete hydrolysis. Casein (in milk) and vitellin (in egg yolk) are important members of this class.

    Glycoproteins contain carbohydrates or carbohydrate derivatives as their prosthetic group covalently bonded to the protein chain. Heparin, which prevents the clotting of blood, and mucin, a constituent of saliva, are glycoproteins. Glycoproteins are located at the external surface of cell membranes and play an important role in cell–cell recognition and other membrane functions.

    Chromoproteins consist of a pigmented prosthetic group combined with a simple protein. Examples are hemoglobin and myoglobin, both of which possess the iron-containing pigment heme (see Figure 13.15).

    Nucleoproteins are complex substances that occur abundantly in the nuclei of plant and animal cells. The prosthetic groups are complex polymers of high molecular weight and are called nucleic acids (DNA and RNA—see Chapter 15).

    Lipoproteins consist of triacylglycerols, cholesterol esters, and phospholipids attached to protein molecules. Most, if not all, of the lipid in mammalian blood is transported in the form of lipoprotein complexes (recall Section 12.17). The electron transport system in the mitochondria contains large amounts of lipoprotein. Lipoproteins are also found in egg yolk, cell nuclei, ribosomes, and the myelin sheath of nerves.

b. Classification of Proteins According to Function

1. Structural Proteins More than half of the total protein of the mammalian body is collagen, found in cartilage, tendons, blood vessels, skin, teeth, and bone. Elastin is

found in most connective tissue along with collagen. It is particularly abundant in ligaments, in the walls of blood vessels, and in the necks of grazing animals. As its name implies, elastins return to their original forms after they are stretched.

2. Contractile Proteins Both striated and smooth muscle are composed chiefly of proteins. Examples are myosin and actin isolated from skeletal muscle.

3. Protective Proteins Examples are the keratins of hair, nails, wool, scales, feathers, hooves, and horns. Leather is almost pure keratin.

4. Enzymes Enzymes represent the largest class of proteins. Over 2000 different kinds of enzymes are known. These biological catalysts are vitally important to all living systems and are discussed in detail in the next chapter.

5. Hormones Several hormones have already been mentioned in connection with steroids. However, many hormones, such as insulin and growth hormone, are proteins; also the cell membrane receptors that bind to hormones and neurotransmitters are proteins.

6. Antibodies The body produces antibodies to destroy any foreign materials (antigens) released into the body by an infectious agent. Immunoglobulins are antibodes (see Section 19.2).

7. Blood Proteins The albumins, globulins, and fibrinogen are the three major protein constituents of blood (see Section 19.1-a).

8. Storage Proteins Casein is a food reserve in mammalian milk. Ovalbumin is the protein reservoir in egg white.

9. Transport Proteins Hemoglobin transports oxygen within the red blood cells, and myoglobin transports oxygen in muscle cells. The cytochromes are proteins that transport electrons.

13.8 *Electrochemical Properties of Proteins*

When amino acids combine to form the polypeptide chain(s) of protein molecules, the majority of their amino and carboxyl groups are tied up in peptide bond formation. However, the side chains of aspartic and glutamic acids, and those of lysine, arginine, and histidine all retain their acidic and basic groups. In proteins, just as in the free amino acids, these groups exist in solution as charged species, such as $-COO^-$ and $-NH_3^+$. Accordingly, proteins are also amphoteric substances. Since all proteins contain some of the acidic and basic amino acids, positive and negative charges are found throughout the molecule.

When the net charge on a protein is zero, that is, when the number of negatively charged groups is equal to the number of positively charged groups, the protein will not migrate in an electric field. The pH value at which this charge cancellation

TABLE 13.6 Isoelectric pH of Various Proteins

Protein	Isoelectric pH
Pepsin	< 1.1
Pepsinogen	3.7
Casein	4.6
Egg albumin	4.7
Serum albumin	4.8
Urease	5.0
Insulin	5.3
Fibrinogen	5.5
Catalase	5.6
Hemoglobin	6.8
Myoglobin	7.0
Ribonuclease	9.5
Cytochrome c	10.6
Lysozyme	11.0

occurs is the isoelectric pH of the protein under consideration. The isoelectric pH is characteristic of a given protein; it is dependent on the number, kind, and arrangement of the acidic and basic groups within the molecule. Those proteins having a high proportion of basic amino acids usually have relatively high isoelectric pHs, and those with a preponderance of acidic amino acids have relatively low isoelectric pHs. Table 13.6 lists the isoelectric pHs of several proteins.

Because of the presence of ionized groups in their structure, proteins behave as either cations or anions, depending upon the pH of the solution. This effect is exploited in the electrophoresis of proteins (Figure 13.19). The process of *electrophoresis*[8] is a very powerful tool that is used to separate and identify specific proteins in a mixture of proteins by subjecting them to an electric field. The protein mixture is applied on a solid support, such as a strip of cellulose acetate, that is soaked in a buffer solution at a certain pH. A current is then applied and the proteins migrate toward the oppositely charged electrodes. Those proteins having the greatest number of negative charges will migrate the most rapidly toward the positive electrode, and vice versa. A dye (such as ninhydrin) is used to make the separated protein spots visible. This technique is used on blood samples in hospital laboratories to assess certain diseases by detecting the relative concentrations of the plasma proteins.

FIGURE 13.19 Electrophoresis of a mixture of proteins. Protein 2 has not migrated from the point of application. Its isoelectric pH is equal to the pH of the buffer. Protein 1 has migrated toward the negative electrode, and must have an isoelectric pH that is higher than the buffer pH. Protein 3 has migrated toward the positive electrode; its isoelectric pH is lower than the buffer pH.

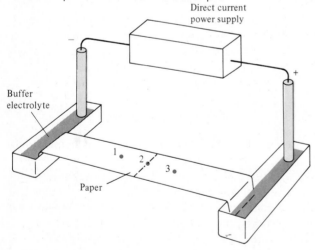

[8]Electrophoresis is also used in the laboratory to separate mixtures of amino acids.

The solubility of proteins in water is greatly dependent on pH. As a general rule, proteins are least soluble at their isoelectric pHs. At pH values below its isoelectric pH, a protein bears a net positive charge; at pH values above its isoelectric pH, it bears a net negative charge. In either case, the net electrostatic effect is repulsion between adjacent protein molecules thus preventing coalescence. At the isoelectric pH, there is no *net* charge on the molecule. Individual molecules now have a greater tendency to approach one another because of the electrostatic attractions between the oppositely charged groups. They tend to clump together (coagulate) and precipitate out of solution.

The phosphoprotein casein is the major protein component of milk, and it will precipitate from the milk in the form of white curds at its isoelectric pH of 4.6. The souring of milk results from the production of lactic acid by bacteria. The lactic acid lowers the pH value of milk from its normal value of about 6.6 to that of the isoelectric pH of casein. Casein is used in the manufacture of cheese. It can be obtained either by adding acid to milk or by bacterial action.

13.9 Denaturation of Proteins

Denaturation may be defined as *any change that alters the unique three-dimensional conformation of a protein molecule without causing a concomitant cleavage of the peptide bonds* (Figure 13.20). Accompanying the disruption of the secondary, tertiary, and quaternary structures of proteins are dramatic changes in the physical and functional nature of the molecules (e.g., decreased solubility and loss of biological activity). A wide variety of reagents and conditions can effect protein denaturation—some of them are outlined here.

a. Heat and Ultraviolet Radiation

Heat and ultraviolet radiation supply kinetic energy to protein molecules, causing their atoms to vibrate more rapidly and thus disrupting the relatively weak *hydrogen bonds* and *hydrophobic bonds*. Most proteins are denatured when heated above 50°C, and this results in coagulation of the protein. The most common example is the frying or boiling of an egg. Heat and ultraviolet radiation are employed in sterilization techniques since they denature the enzymes in bacteria and, in so doing, destroy them. Denatured proteins are usually easier to chew and easier for enzymes to digest; hence we cook our protein-containing food.

b. Treatment with Organic Compounds (Ethyl Alcohol, Formaldehyde, Urea, Rubbing Alcohol)

These reagents are capable of forming intermolecular hydrogen bonds with protein molecules and so disrupt the intramolecular *hydrogen bonding* within the molecule. A 70% alcohol solution is used as a disinfectant in cleansing the skin prior to an injection. The alcohol functions to denature the protein (enzymes in particular) of any bacteria present in the area of the injection. A 70% alcohol solution effectively penetrates the bacterial cell wall, whereas 95% alcohol coagulates proteins at the surface, forming a crust that prevents the alcohol from entering into the cell.

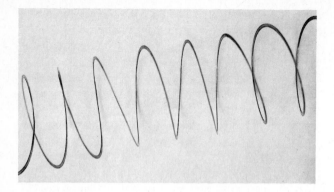

(a)

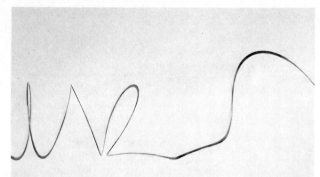

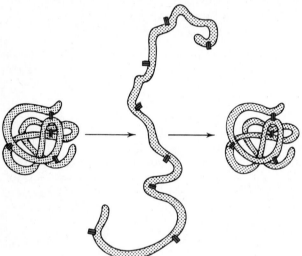

= Areas of forces (—S—S—, hydrogen bonding, ionic, etc.) stabilizing conformation

(b)

FIGURE 13.20 Denaturation of a protein. (a) Irreversible denaturation. The coiled spring represents the helical structure of a protein; when the elastic limit of the helix is exceeded, the shape is irreversibly altered. (b) If the denaturing treatment is mild (e.g., a dilute urea solution), denaturation can be reversed. The unfolded polypeptide chain spontaneously re-forms into the original conformation when the denaturant is removed.

c. Acids or Bases

Acidic and basic reagents disrupt *salt linkages* by altering the state of ionization of the carboxyl and amino groups. Often the molecules of the denatured protein interact with each other, clumping together and causing the protein to precipitate (coagulate). If a coagulated protein is permitted to remain in prolonged contact with an acid or a base, cleavage of the peptide linkages will eventually occur. When proteins are hydrolyzed to their constituent amino acids, denaturation usually precedes hydrolysis.

d. Salts of Heavy Metal Ions (Hg^{2+}, Ag^+, Pb^{2+})

The cations of these heavy metals form very strong bonds with the carboxylate anions of the acidic amino acids and with the sulfhydryl groups of the sulfur-containing amino acids. Therefore they disrupt *salt linkages* and *disulfide linkages* and cause the protein to precipitate out of solution as insoluble metal–protein salts. This property makes some of the heavy metal salts suitable for use as antiseptics. For example, a 1% solution of silver nitrate (also called lunar caustic), which is used to prevent gonorrhea infections in the eyes of newborn infants, and mercuric chloride, another antiseptic, act to precipitate the proteins in infectious bacteria. Most heavy metal salts are toxic when taken internally because they precipitate the proteins of all the cells with which they come in contact.[9] Substances high in protein, such as egg whites and milk, are used as antidotes for heavy metal poisoning because their proteins readily combine with the heavy metal ions to form insoluble solids. The resulting insoluble matter is removed from the stomach by the use of an emetic, thus preventing the digestive juices from destroying the protein and once again liberating the poisonous heavy metal ions.

e. Alkaloid Reagents (Picric Acid, Tannic Acid)

These reagents are so named because they were originally used to study the structure of the alkaloids (morphine, cocaine, quinine). They function in a manner analogous to the heavy metal cations, but the picrate and tannate anions combine with the positively charged amino groups to disrupt the *salt linkages*. In the manufacture of leather, tannic acid is used to precipitate the proteins in animal hides. This is the process called *tanning*. Tannic and picric acids are sometimes used in the treatment of burns. These acids combine with the protein in the exposed areas to form a leathery coating that excludes air and stops the loss of body fluids. The loss of water and salts is the most significant cause of shock and the fatalities that result from burns. In an emergency, tea can serve as a source of tannic acid for the treatment of severe burns.

f. Denaturation of Hair (Permanent Waving)

In most cases, denaturation is an irreversible process. One important industry, however, owes its existence to a reversible denaturation procedure. The protein of hair (α-keratin) has a high proportion of cysteine, and it is the disulfide linkages between the cysteine moieties that are largely responsible for the shape of the hair. The process of permanent waving (Figure 13.21) involves the cleavage of these *disulfide linkages* by the addition of a reducing agent. The reduced, disordered hair is then placed on curlers and set in the desired pattern. A mild oxidizing agent is added to re-form the disulfide linkages, this time between different cysteine moieties. The effect of this process is the molecular setting of the hair into a new pattern or hairdo. Care must be taken not to oxidize too long, or use too strong an oxidizing agent (such as hypochlorite bleaches). The disulfides can be further oxidized to sulfonic acids and such reactions in hair produce so-called split ends or "frizzies."

---

[9]The danger of lead and/or mercury poisoning has evoked considerable environmental concern. Lead salts are no longer used as pigments in paints, and most of the lead has been removed from gasoline. Mercury compounds still occur in water systems because large quantities of mercury and mercury salts have been dumped by industries into streams and lakes. The mercury is taken in by fish, and then humans eat the fish.

TABLE 13.7 Color Tests for Proteins

Name of Test	Ingredients of Test Reagent	Criterion for Positive Test	Color Produced	Remarks
Ninhydrin	Ninhydrin 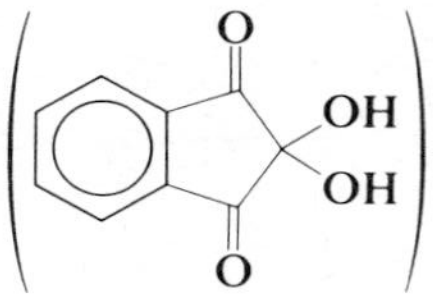	Free —NH$_2$ group	Blue	Positive test given by ammonia and primary amines as well as amino acids, peptides, and proteins. This test is generally used to detect the presence of amino acids. Proline and hydroxyproline give a characteristic yellow color.
Biuret	NaOH, dil CuSO$_4$	Two peptide bonds	Violet	Positive test given by tripeptides, polypeptides, and proteins but not by amino acids and dipeptides.
Xanthoproteic	Conc HNO$_3$	Amino acids containing a benzene ring	Yellow	Positive test given by proteins that contain tryptophan, tyrosine, or phenylalanine. (This test is performed inadvertently by students when they spill nitric acid on their skin.)
Millon	Hg(NO$_3$)$_2$, Hg(NO$_2$)$_2$	Tyrosine	Red	Positive test also given by any phenolic compound.
Hopkins–Cole	Glyoxylic acid 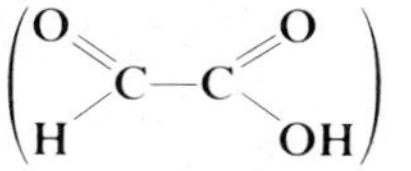 and conc H$_2$SO$_4$	Tryptophan	Violet ring	Positive test also given by any compound containing an indole ring.
Sakaguchi	α-Naphthol 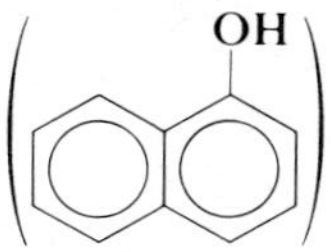 and sodium hypochlorite (NaOCl)	Arginine	Red	Positive test also given by any compound containing the guanidyl group.
Nitroprusside	Sodium nitroprusside [Na$_2$Fe(NO)(CN)$_5$·2H$_2$O]	Cysteine	Red	The presence of sulfur in a protein can also be detected by hydrolyzing the protein with NaOH and then adding lead acetate. A black precipitate, PbS, is produced if sulfur is present.

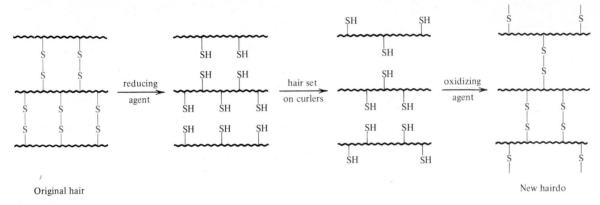

Original hair

New hairdo

FIGURE 13.21   The process of permanent waving.

13.10   *Color Tests for Proteins*

Several color tests for the qualitative detection of proteins are available. Two of them, the *ninhydrin* and *biuret* tests, are general in that they are positive for all proteins. The others are specialized and are positive only for those proteins that contain certain amino acids. These tests are outlined in Table 13.7.

Exercises

13.1 Define, explain, or draw a formula for each of the following.

(a) α-amino acid
(b) amino acid side chain
(c) zwitterion
(d) isoelectric pH
(e) amphoteric
(f) peptide bond
(g) tripeptide
(h) diuretics
(i) diabetes insipidus
(j) sickle cell anemia
(k) oligomeric proteins
(l) prosthetic group
(m) electrophoresis
(n) denaturation

13.2 Distinguish between the terms in each pair.

(a) polypeptide and protein
(b) N-terminal amino acid and C-terminal amino acid
(c) oxytocin and vasopressin
(d) α-helical and β-pleated sheet
(e) primary structure and secondary structure
(f) tertiary structure and quaternary structure
(g) interchain hydrogen bonds and intrachain hydrogen bonds
(h) hemoglobin and myoglobin
(i) α-keratins and β-keratins
(j) simple protein and conjugated protein
(k) globular protein and fibrous protein
(l) albumins and globulins
(m) glycoproteins and lipoproteins
(n) ninhydrin test and biuret test

13.3 List six functions of proteins in the body.

13.4 How does the chemical composition of proteins differ from that of carbohydrates and lipids?

13.5 There are some amino acids found in proteins that are derived from one or another of the 20 fundamental amino acids *after* the latter have been incorporated into the protein chain. 4-Hydroxyproline has been mentioned in Table 13.1. Some others are 5-hydroxylysine, ε-N-methyllysine, 3-methylhistidine, and γ-carboxyglutamic acid. Draw structures for these five amino acids.

13.6 Give the names and write structural formulas for two representatives from each of the four classes of amino acids.

13.7 (a) Give IUPAC names for the amino acids in Exercise 13.6.

(b) Write the formula of a β-amino acid and give its IUPAC name.

13.8 Asparagine (Asn) and glutamine (Gln) are neutral amino acids, whereas aspartic acid and glutamic acid are acidic amino acids. Explain.

13.9 Identify two amino acids that contain more than one chiral carbon atom. Draw all the stereoisomers of (a) alanine, (b) serine, (c) threonine, and (d) isoleucine.

13.10 Identify the amino acid that fits each of the following descriptions.

(a) contains a secondary amino group
(b) contains a heterocyclic ring

(c) contains a benzene ring

(d) not found in proteins

(e) most abundant amino acid in gelatin

(f) abundant in collagen but not found in most other proteins

(g) found in asparagus

(h) most basic of the amino acids

(i) contains a sulfhydryl group

(j) contains a phenolic group

(k) contains a branched chain

(l) disrupts the α-helical conformation

13.11 Choose one amino acid from each class in Table 13.1 and write equations for its reactions with **(a)** an HCl solution and **(b)** a NaOH solution.

13.12 A direct current was passed through a solution containing alanine, histidine, and aspartic acid at a pH of 6.0. One amino acid migrated to the cathode, one migrated to the anode, and one remained stationary. Match the behavior with the correct amino acid.

13.13 (a) The pK_a of the carboxyl group of leucine is 2.36, and the pK_a of its protonated amino group is 9.60. What is the isoelectric pH of leucine?

(b) Using Table 13.3, calculate the isoelectric pH of histidine.

13.14 Draw and name the pairs of dipeptides that can be formed from each of the following.

(a) leucine and phenylalanine

(b) tryptophan and methionine

(c) proline and glycine

(d) cysteine and asparagine

(e) tyrosine and arginine

(f) aspartic acid and histidine

13.15 Let the letters A, B, C, and D represent four different amino acids. Construct all the possible tetrapeptides using each letter only once.

13.16 How many tetrapeptides containing two molecules of alanine and two molecules of valine are possible?

13.17 Which amino acid could be substituted for glutamic acid at position 6 of the β-chain of hemoglobin and not be detrimental to health?

13.18 What would be the products of the acid hydrolysis of the following tetrapeptide?

13.19 In Section 11.2, we mentioned that aspartame (aspartyl-phenylalanine methyl ester) was an artificial sweetener that has taken the place of saccharin in many commercial products. Draw the structure of aspartame.

13.20 Glutathione (γ-glutamylcysteinylglycine) is a tripeptide that is found in all the cells of higher animals. It contains a glutamic acid joined in an unusual peptide linkage involving its γ-carboxyl group. Draw the structure of glutathione.

13.21 Recall (page 200) that the enkephalins are the body's own opiates. Write formulas for the pentapeptide structures of the methionine and leucine enkephalins.

13.22 An octapeptide fragment from a hypothetical protein gave, upon complete hydrolysis, one molecule each of alanine (Ala), glutamic acid (Glu), glycine (Gly), leucine (Leu), methionine (Met), and, valine (Val) and two molecules of cysteine (Cys). The following fragments were isolated after partial hydrolysis: Gly-Cys, Leu-Val-Gly, Cys-Ala, Cys-Glu-Met and Met-Cys. Deduce the amino acid sequence in the octapeptide.

13.23 Glucagon is a polypeptide hormone, produced in the pancreas, that regulates glucose metabolism (see Section 16.3). Its primary structure is His-Ser-Glu-Gly-Thr-Phe-Thr-Ser-Asp-Tyr-Ser-Lys-Tyr-Leu-Asp-Ser-Arg-Arg-Ala-Gln-Asp-Phe-Val-Gln-Trp-Leu-Met-Asn-Thr. Write out the names of the amino acids (1–29) found in glucagon.

13.24 Angiotensin (Asp-Arg-Val-Tyr-Ile-His-Pro-Phe) is an octapeptide that is produced in the kidneys. It functions to increase blood pressure by causing the constriction of blood vessels (vasoconstriction). Draw its primary structure.

13.25 Bradykinin is a nonapeptide that occurs in the blood and functions in the regulation of blood pressure. Its amino acid sequence is Arg-Pro-Pro-Gly-Phe-Ser-Pro-Phe-Arg. Draw its primary structure. Is the secondary structure of bradykinin in the form of an α-helix? Why or why not?

13.26 What is the difference in the type of hydrogen bonding that occurs in secondary and in tertiary structures of proteins?

13.27 What functional groups are found in the side chains of proteins?

13.28 List all of the types of bonds that occur in proteins. Which of these are covalent bonds?

13.29 Consider the following pairs of amino acids and indicate what type of bonding might be expected between them.

(a) Ala and Pro   (b) Asp and Arg

(c) Asp and Ser   (d) Asn and Glu

(e) Glu and Glu   (f) Cys and Cys

(g) Leu and Ile   (h) Lys and Glu

(i) Phe and Phe   (j) Tyr and Asp

13.30 What forces maintain the quaternary structure of **(a)** collagen, **(b)** α-keratin, and **(c)** hemoglobin?

13.31 Give the names of four conjugated proteins and indicate the nature of their prosthetic groups. To what general class of conjugated protein does each belong?

13.32 Give the name of a protein that
  (a) is present in blood   **(b)** is present in milk
  (c) is present in eggs   **(d)** is an antibody
  (e) is present in hair   **(f)** is present in saliva
  (g) is present in bones   **(h)** is an enzyme
  (i) is a hormone
  (j) is present in the digestive tract
  (k) stores oxygen in muscle cells
  (l) transports oxygen in the blood

13.33 Proteins help to maintain the pH of the organism. How can they perform this function?

13.34 Under what conditions does a protein have **(a)** a net positive charge, **(b)** a net negative charge, and **(c)** a net zero charge?

13.35 How do the properties of a protein at its isoelectric pH differ from its properties when in solutions of other pH values?

13.36 What occurs when milk becomes sour?

13.37 Examine the structure of glucagon (Exercise 13.23). Would you expect this polypeptide to have a high, low, or about neutral isoelectric pH? Explain.

13.38 One of the neurotransmitters involved in the sensing of pain information is a polypeptide called substance P. Substance P is an undecapeptide that is released by nerve terminals in response to pain. Its primary structure is   Arg-Pro-Lys-Pro-Gln-Gln-Phe-Phe-Gly-Leu-Met. What would be the net electrical charge ($+$ or $-$) of this polypeptide at pH **(a)** 1.0, **(b)** 6.0, and **(c)** 13.0?

13.39 Briefly describe on the molecular level what occurs during each of the following processes.
  (a) boiling an egg
  (b) sterilization of surgical instruments
  (c) permanent waving
  (d) the use of a dilute solution of silver nitrate as a disinfectant in the eyes of newborn infants

13.40 Why is a 70% alcohol solution more effective as a disinfectant than a 95% alcohol solution?

13.41 In the precipitation of proteins, what is the difference **(a)** between tannic acid and mercuric chloride and **(b)** between acids and alcohol?

13.42 Discuss the use of egg white as an antidote for heavy metal poisoning.

13.43 Name five color tests for proteins and indicate the specific group(s) in the protein molecule responsible for a positive test.

14

Enzymes

The various reactions that occur in biological systems are in many respects identical to those we have discussed previously (for example, hydration of unsaturated bonds, interconversions of alcohols, aldehydes, and acids, hydrolysis of esters and amides). In fact, scientists have been able to duplicate in the test tube (in vitro conditions) many of the reactions commonly carried out in living organisms (in vivo conditions). There is, however, one significant difference and that involves the rate at which the two types of reactions occur. The in vivo reactions take place about 100 to 1 million times faster than the corresponding in vitro reactions. Some enzymatic reactions occur in milliseconds (msec, 10^{-3} sec); others are more rapid, and rates are measured in microseconds (μsec, 10^{-6} sec). For example, it has been estimated that the reactions involved in the transmittal of a nerve impulse take between 1 and 3 millionths of a second. If the in vitro reactions are to take place at an appreciable rate, drastic reaction conditions must be employed. Such conditions, which include high temperatures and the use of potent oxidizing or reducing agents and/or strong acids or bases, are all lethal to living systems. The in vivo reactions are carried out at body temperature ($\sim 37^{\circ}$C) and in the physiological pH range (pH ~ 7). The agents employed by the cell to effect the reactions under these conditions are the highly efficient, highly specific catalysts called enzymes. Life as we know it would be

347

impossible without enzymes because nearly all functions of the cell depend directly or indirectly upon them.

Recall that a **catalyst** is *any substance that increases the rate of a chemical reaction without being consumed in the reaction.* The catalyst does *not* change the position of equilibrium in a reversible reaction; it only increases the rate of attainment of equilibrium. An **enzyme** may be defined as a *complex organic catalyst produced by living cells.* Most enzymes operate within the cell that produces them and are thus termed *intracellular.* A typical human cell contains about 2000 enzymes, and these enzymes are catalyzing over 100 reactions each minute. If the enzyme's usual site of catalytic activity is outside the cell that produces it (as in the case of the digestive enzymes), the enzyme is designated as *extracellular.*

The hydrolysis of sucrose affords a good example by which to distinguish enzyme action from the classical concept of a catalyst. If one were to exclude bacteria and molds, a solution of sucrose in water could be kept indefinitely without undergoing hydrolysis to any appreciable extent. If hydrochloric acid is added and the reaction mixture is heated, hydrolysis takes place, producing glucose and fructose. If the enzyme *invertase* (*sucrase*) were added instead of the acid, the reaction would take place at a greater rate and the solution would not have to be heated at all. Furthermore, hydrochloric acid catalyzes the hydrolysis of lactose and maltose as well as that of sucrose; *invertase* is specific for sucrose alone and will not catalyze the hydrolysis of any other disaccharide.

14.1   *Discovery of Enzymes*

Since ancient times, humans have been aware that yeast cells are capable of fermenting sugars. Louis Pasteur (1860) was one of the first scientists to study systematically such processes as alcoholic fermentation and the souring of milk. He observed that when glucose is stored in sealed, sterile containers, it remains indefinitely unchanged. Furthermore, he observed that the addition of yeast caused the glucose to be fermented to alcohol and carbon dioxide. Pasteur incorrectly believed that the presence of certain intact microorganisms was responsible for the observed catalytic activity and he use the term *ferment* to describe the entire group of catalysts. In 1878, Willy Kuhne proposed that the name enzyme (Greek, *en,* in; *zyme,* yeast) be used to describe these substances.

In 1897, Edward Buchner (Nobel Prize, 1907) prepared a cell-free extract by grinding yeast cells with sand and filtering the resultant material. The cell-free juice was found to be capable of promoting the fermentation of sugars in the same manner as the original intact yeast cells. This experiment is considered to mark the beginning of modern enzyme chemistry. Subsequent studies by many investigators elucidated all of the steps involved in this complicated but efficient process and showed that each step is controlled by a specific enzyme (Chapter 16). However, for 30 years, biochemists vainly attempted to isolate an enzyme in a pure form in order to study its properties. Finally, in 1926, James Sumner (Nobel Prize, 1946) succeeded in isolating a pure crystalline enzyme (*urease*) that he characterized as proteinaceous in nature. Shortly thereafter, other workers succeeded in isolating several different enzymes, all of which were shown to be proteins. Today, over 2000 different enzymes are known; all are proteins of varying complexity, and at least 10% of these have been crystallized.

14.2   *Nomenclature and Classification of Enzymes*

The substance upon which an enzyme acts is known as its **substrate.** Enzymes are most commonly named by adding the suffix *-ase* to the root of the name of the substrate. Thus, for example, *urease* is the enzyme that acts on urea and *sucrase* is the enzyme that acts on sucrose. Sometimes an enzyme is named after the products that are formed as a result of its catalytic activity, as in the case of *invertase,* another name for *sucrase.* Still other enzymes have been given names indicating the type of reaction in which they are involved, such as *alcohol dehydrogenase* and *ascorbic acid oxidase.* Finally, some enzymes still carry the common names given to them at the discretion of their discoverers. *Trypsin,* for example, comes from a Greek word meaning to rub and was selected because the enzyme was first obtained by grinding pancreatic tissue with glycerol. Many digestive enzymes have retained their older names, such as *pepsin, rennin,* and *ptyalin.*

To avoid the continued haphazard naming of enzymes, the International Union of Biochemistry, in 1961, recommended that enzymes be systematically classified according to the general type of reaction that they catalyze. Rules were established for the naming of an enzyme on the basis of both the precise chemical name of the substrate and the nature of the chemical reaction that is catalyzed. For example, the systematic name for *sucrase* is *α-glucopyrano-β-fructofuranohydrolase.* Because the resulting names are so cumbersome, this rational scheme has not been universally accepted, and the common names for enzymes persist in the literature. For this reason, we shall continue to use the common rather than the systematic names for the enzymes. Table 14.1 summarizes the six main divisions recognized in the classification of enzymes. A typical example is given of each type to illustrate the kinds of reactions that we shall deal with in later chapters on metabolism.

14.3   *Characteristics of Enzymes*

All enzymes are globular proteins. Some enzymes, such as *pepsin, trypsin,* and *ribonuclease,* are simple proteins since they consist entirely of amino acid units, but others are known to contain nonprotein portions and are therefore conjugated proteins. Several terms are commonly used in referring to the different components of a conjugated protein enzyme (also called a **holoenzyme**).

The *polypeptide segment of the enzyme* is known as the **apoenzyme.** The *nonprotein organic moiety,* which can frequently be separated from the apoenzyme, is called the **coenzyme.**[1] The apoenzyme is catalytically inactive by itself, but its activity can be restored by the addition of the coenzyme. There are many metalloenzymes in which the metal ion (Mg^{2+}, Mn^{2+}, or Zn^{2+}, for example) is bonded either to the apoenzyme

---

[1] Some authors reserve the term *coenzyme* for those substances that are readily dissociable from the enzyme. If the substance is firmly attached to the protein portion of the enzyme, it is referred to as a *prosthetic group.* We have already defined a prosthetic group as the nonprotein portion of any conjugated protein. In this sense, then, a coenzyme is a specific example of a prosthetic group.

TABLE 14.1 The International Union of Biochemistry Classification of Enzymes

Group Name of Enzyme	Reaction Type Catalyzed	Typical Reaction	Common Names of Enzymes
Oxido-reductase	Oxidation–reduction	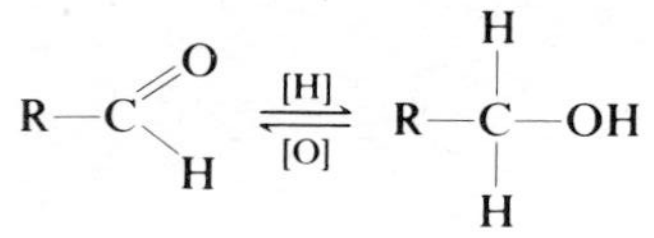	Dehydrogenase, oxidase, peroxidase, reductase
Transferase	Transfer of atoms or groups of atoms from one molecule to another	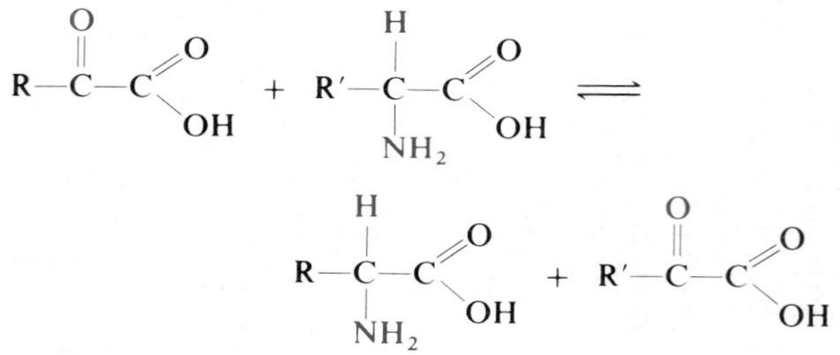	Transaminase, kinase, transacetylase
Hydrolase	Hydrolysis of a variety of substrates by water	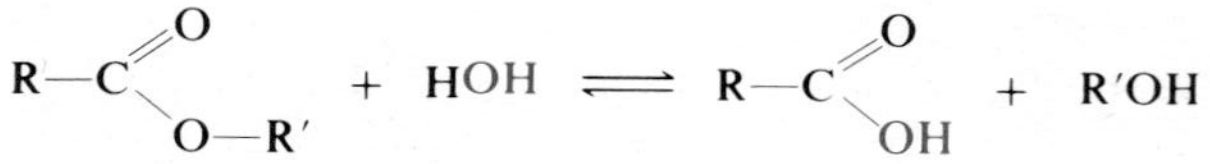	Lipase, phosphatase, amylase, amidase, peptidase
Lyase	Nonhydrolytic addition or removal of substituents from substrates	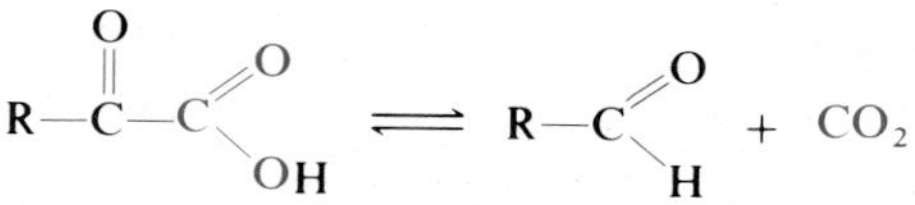	Decarboxylase, fumarase, aldolase, deaminase
Isomerase	Internal rearrangement of certain substituents	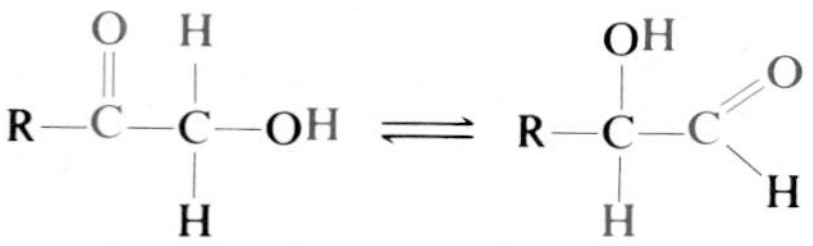	Isomerase, racemase, epimerase, mutase
Ligase	Linking together of two molecules concomitant with the breaking of a high energy bond	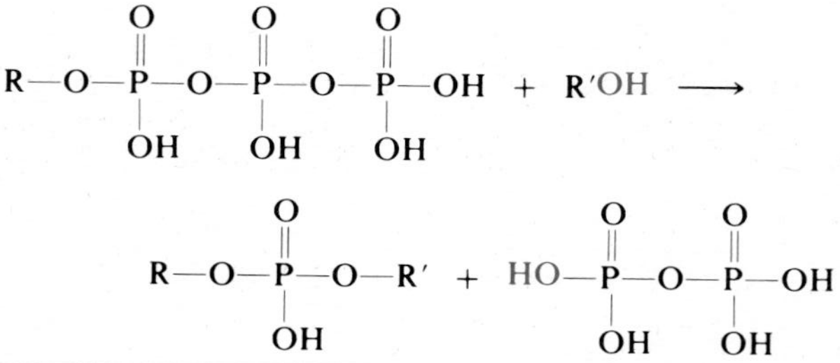	Synthetase, thiokinase, carboxylase

or to the coenzyme. The metal ion is usually designated as the enzyme **activator.** It has been postulated that the metal ion acts to form a coordination complex between the enzyme and the substrate and to activate the substrate by promoting electronic shifts. In our later discussions of metabolic reactions, we cite several examples of enzymes whose catalytic activity is dependent upon the presence of an activator. (*Metal ions and/or coenzymes* are often referred to as **cofactors.**)

Some enzymes, such as the protein-digesting enzymes (*peptidases*) are secreted in larger, inactive forms, known as **zymogens** or **proenzymes.** This is a protective feature that prevents the active form of these enzymes from digesting the proteins in the walls of the digestive tract. *Pepsinogen* and *trypsinogen* are two such compounds that have been carefully studied. These zymogens are converted into the active enzymes *pepsin* and *trypsin* by the action of other enzymes, which remove an inhibitory peptide from the zymogen molecule. For example, *trypsinogen* is synthesized in the pancrease and is secreted into the small intestine, where it is activated in response to the presence of food in the intestine. A hexapeptide is cleaved from the molecule enabling the new protein to attain the conformation essential to its catalytic activity (see Figure 18.1) We shall see that zymogens play an important role in the mechanism of blood clotting (Section 19.5).

14.4 *Vitamins and Coenzymes*

Recall that in Section 12.16, we defined the term *vitamin*, and discussed the fat-soluble vitamins and vitamin C. Since the B vitamins often occur together in foods, they are usually referred to as a group called *B complex*. In this section, we shall see that the major biochemical role of the B complex vitamins is to serve as structural components of coenzymes. Table 14.2 lists the names and structures of some of the coenzymes and their vitamin precursors. In our detailed study of metabolic reactions, we often refer to this table.

a. Thiamine (Vitamin B₁)

Thiamine is necessary for the normal metabolism of carbohydrates. It is converted in the body to the pyrophosphate, which is a coenzyme in the decarboxylation of pyruvic acid to acetyl-CoA (page 443) and α-ketoglutaric acid to succinyl-CoA (page 446). The coenzyme is also involved in the synthesis of ribose, which in turn is used by the body to produce nucleotides and nucleic acids. A deficiency of thiamine in the diet leads to the beriberi syndrome characterized by deterioration of the cardiovascular and nervous systems. This disease is a serious health problem in the Far East because rice, the major food, has a relatively low content of thiamine. Alcoholism is the most common cause of thiamine deficiency in the United States because alcohol is the major caloric contributor to an alcoholic's diet, and therefore there is a low vitamin intake. A severe form of beriberi can also occur in infants of nursing mothers who consume diets deficient in thiamine. Dietary sources of thiamine are pork, lean meat, eggs, whole grain cereals, nuts, and legumes. Synthetic vitamin B₁ is added to enrich the vitamin content of bread and flour. Thiamine is not stored in the body to any significant degree; excesses are excreted in the urine. The vitamin is destroyed in foods that are cooked for prolonged periods at temperatures over 100°C.

b. Riboflavin (Vitamin B₂)

Riboflavin is essential for mammalian cells; a lack in the human diet causes well-defined symptoms such as dermatitis (skin inflammation), glossitis (tongue inflammation), and anemia. Riboflavin is converted in the body to the coenzymes FAD and FMN, and these function in oxidation–reduction reactions occurring in carbohydrate and lipid metabolism. Dietary sources of riboflavin are milk, liver, green vegetables, red meat, fish, and eggs. Riboflavin is destroyed by light and thus does not have a long stability in food products. It is stable at ordinary cooking temperatures.

c. Pyridoxine (Vitamin B₆)

Pyridoxine occurs in the tissues and body fluids of virtually all living organisms. The coenzyme pyridoxal phosphate is required for a wide variety of metabolic transformations of amino acids (Section 18.3-a). Clinical symptoms of vitamin B₆ deficiency include lesions of the skin and mucosa, anemia, irritability, apathy, and neuronal dysfunction including convulsions. Dietary sources of the vitamin are liver, whole grain cereals,

TABLE 14.2 Vitamins and Coenzymes

Vitamin	Coenzyme	Function
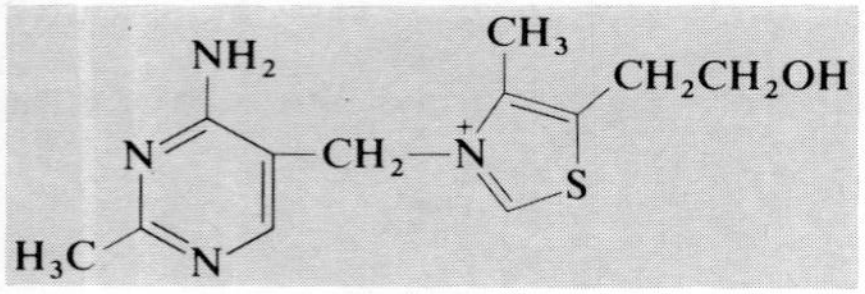 Thiamine (B_1) RDA[a] = 1.5 mg	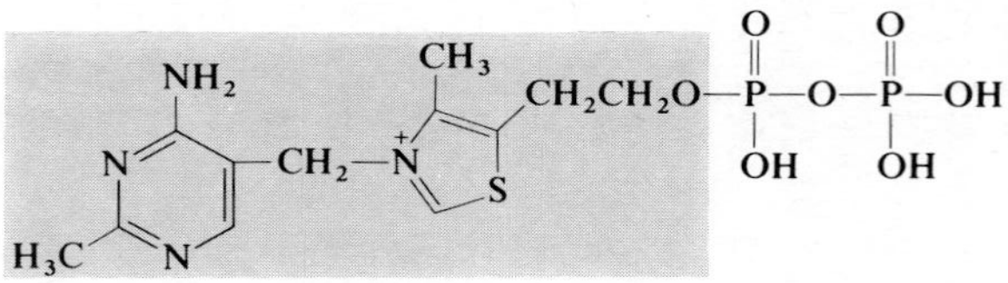 Thiamine pyrophosphate (TPP)	In reactions that remove and/or transfer aldehyde groups
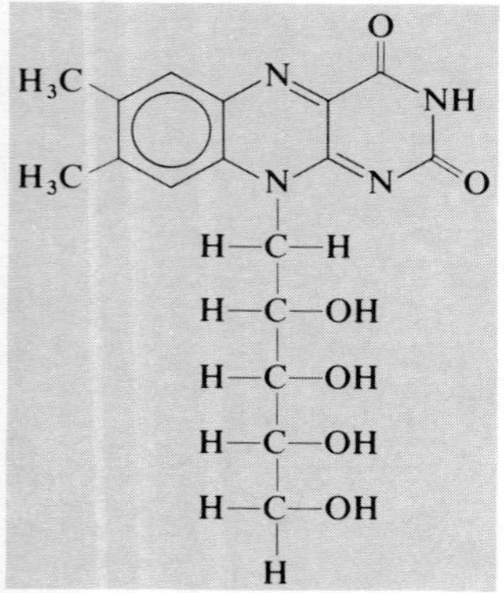 Flavin / Ribitol Riboflavin (B_2) RDA = 1.7 mg	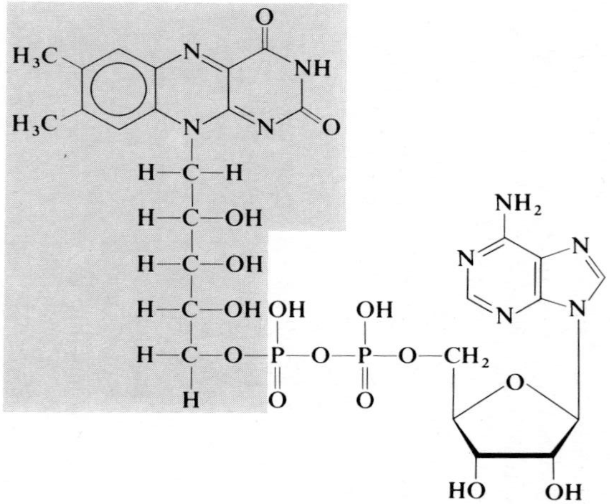 Flavin adenine dinucleotide (FAD)[b]	In oxidation–reduction reactions
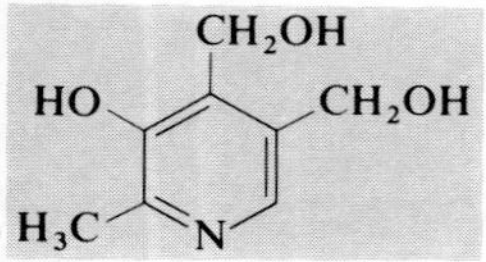 Pyridoxine (B_6) RDA = 2.0 mg	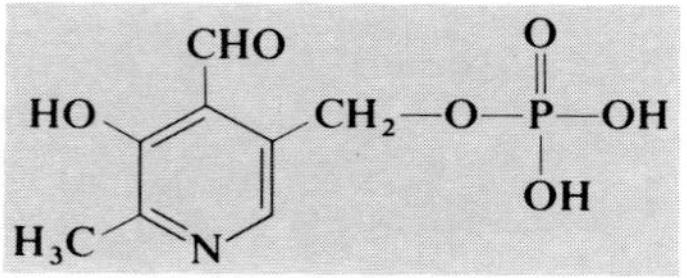 Pyridoxal phosphate	In several reactions of amino acid metabolism: Transamination Decarboxylation Racemization

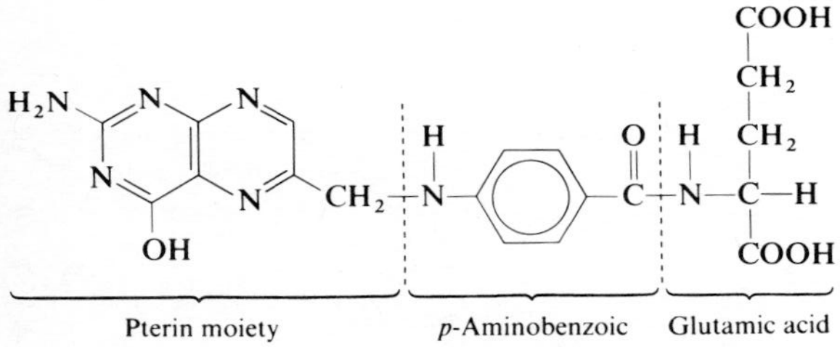

In reactions that transfer single carbon units

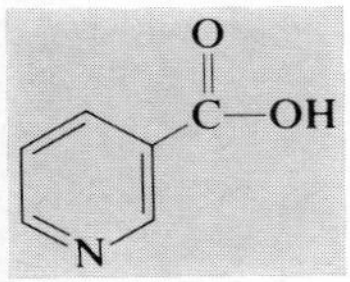

COOH
|
CH$_2$
|
CH$_2$
|
O   H   C—H
‖ | |
C—N   COOH

H   H
| |
H—N
|
CH$_2$
N

H$_2$N   N   OH

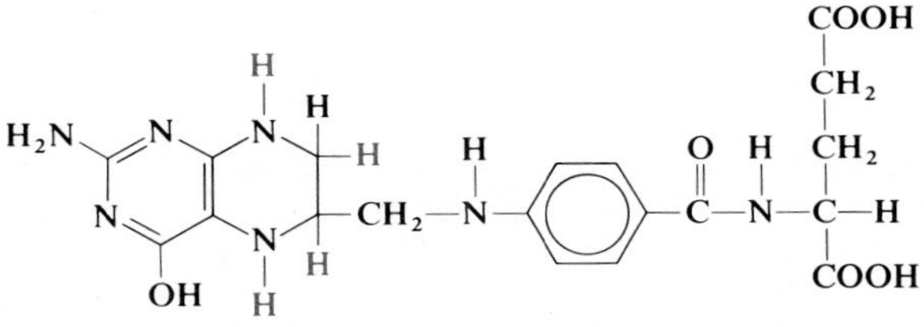

Tetrahydrofolic acid (FH$_4$)

In oxidation–reduction reactions

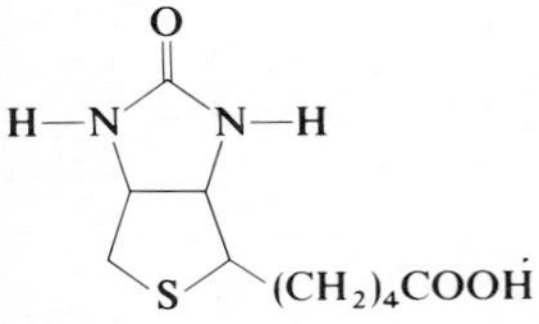

O
‖
C—NH$_2$

N$^+$

Ribose

CH$_2$

OH

HO

O

HO—P=O

O

HO—P=O

O

CH$_2$

O

2'

OH

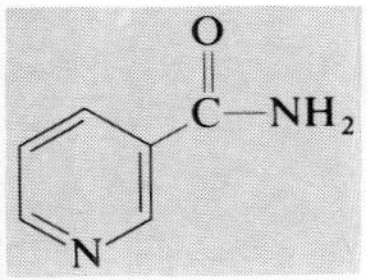

HO

Ribose

NH$_2$

N   N

N   N

Adenine

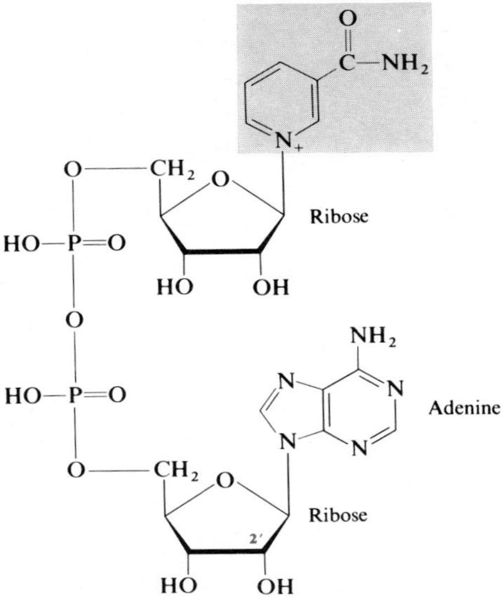

Nicotinamide adenine dinucleotide (NAD$^+$)^c

COOH
|
CH$_2$
|
CH$_2$
|
O   H   C—H
‖ | |
C—N   COOH

Glutamic acid

p-Aminobenzoic acid moiety
Folic acid (F)
RDA = 0.4 mg

H
|
H—N
|
CH$_2$
N

H$_2$N   N   OH

Pterin moiety

O
‖
C—OH

N

Nicotinic acid (Niacin)

O
‖
C—NH$_2$

N

Nicotinamide
RDA = 20 mg

In carboxylation reactions, for the biosynthesis of purines, fatty acids, and urea

(*Table continues*)

Biotin is both a vitamin and a coenzyme

O
‖
C

N—H

H—N

S

(CH$_2$)$_4$COOH

Biotin
RDA = 0.3 mg

TABLE 14.2 (continued)

Vitamin	Coenzyme	Function

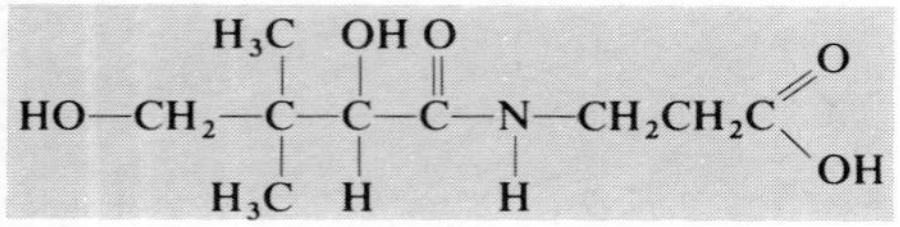

Pantothenic acid

RDA = 10 mg

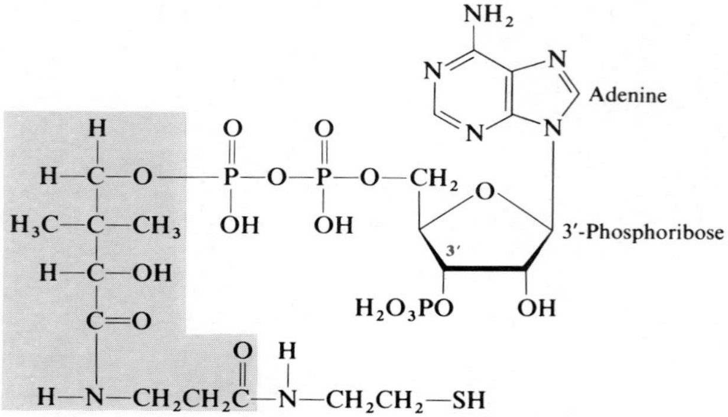

Adenine

3'-Phosphoribose

H_2O_3PO

Pantothenic acid

Coenzyme A (CoA-SH)

β-Mercaptoethylamine

Acyl-group carrier or transfer for the biosynthesis of fatty acids and steroids and for fatty acid oxidation

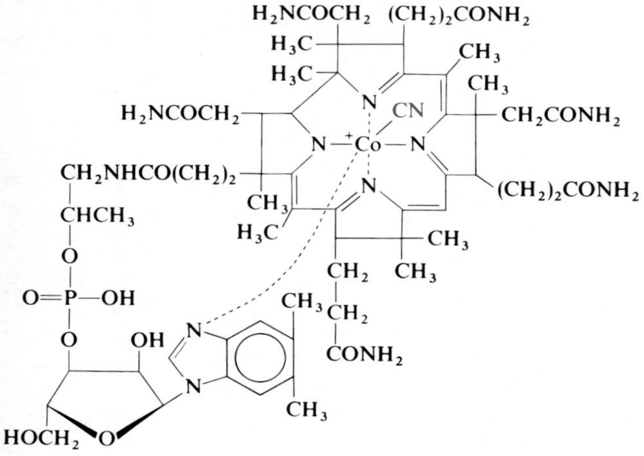

Cyanocobalamin (B_{12})

RDA = 6 µg

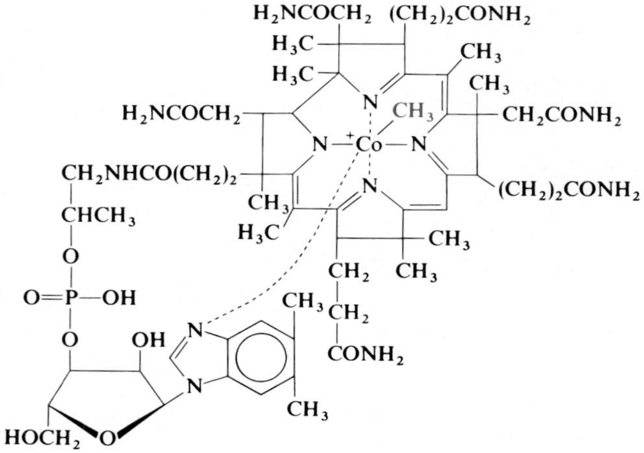

Methylcobalamin

In transmethylation reactions and carbon skeleton rearrangements

bananas, meat, eggs, and milk. Vitamin B_6 enhances the decarboxylation of L-dopa, so the vitamin should be avoided by patients receiving L-dopa to treat Parkinson's disease (see Section 18.3-c). Vitamin B_6 is stable at normal cooking temperatures, but is sensitive to light.

d. Folic Acid

The vitamin is reduced to the coenzyme, tetrahydrofolic acid, which acts as a carrier of one-carbon units (e.g., as formyl or methyl groups) in the formation of compounds such as choline, heme, and nucleic acids. Deficiency of folic acid affects purine biosynthesis, and clinical symptoms include anemia and gastrointestinal disturbances. Folic acid is synthesized by intestinal microorganisms, and it can be absorbed into the general circulation. Dietary sources are liver, yeast, and green vegetables. Folic acid is readily destroyed by cooking. As we shall see in Section 14.9-b, the sulfa drugs act at antimetabolites by interfering with the bacterial biosynthesis of folic acid. The anticancer drug, methotrexate, inhibits the conversion of folic acid to tetrahydrofolic acid. Without the coenzyme, cells cannot grow because they cannot replicate their DNA.

e. Nicotinic Acid (Niacin) and Nicotinamide

Nicotinic acid and its amide are equally effective in supplying human needs. The vitamin is best known for its ability to prevent pellagra in humans. In the early 1900s pellagra was particularly prevalent in the southern United States and was directly associated with low-grade starchy (corn) diets. Pellagra is characterized by loss of appetite and weakness, followed by diarrhea, dermatitis, mental disorders, and death in severe cases. Nicotinamide serves as a component of coenzymes (NAD^+ and $NADP^+$) for a wide variety of enzymes that catalyze oxidation–reduction reactions. Dietary sources of the vitamin are liver, red meat, yeast, fish, and peanuts. Some nicotinic acid (about 10%) is synthesized from tryptophan. The vitamin is not destroyed by cooking, although some is lost to dissolution in the cooking water.

f. Biotin

Biotin is widely distributed as a cell constituent of animal and human tissue. Biochemically, biotin functions as a coenzyme in carboxylation reactions. It is a carbon dioxide carrier in both carbohydrate and lipid metabolism. Dietary sources of biotin are liver, kidney, egg yolks, milk, molasses, and yeast. Because biotin is synthesized by intestinal microorganisms in large quantities, a deficiency of the vitamin seldom occurs in humans. Biotin deficiency can be produced by antibiotics that inhibit the growth of intestinal bacteria. Also, raw egg white contains a protein, avidin, that binds biotin and prevents its absorption from the intestinal tract. An artifically produced deficiency of biotin in humans causes dermatitis, anoreria, nausea, muscle pains, and depression. Biotin is stable at normal cooking temperatures.

g. Pantothenic Acid

Pantothenic acid is a precursor for the biosynthesis of coenzyme A, which is important as a carrier of acyl groups. (The name, coenzyme A, resulted from its involvement in enzymatic acetylation reactions.) Pantothenic (from the Greek word meaning *from everywhere*) acid has a widespread distribution in foods, and deficiency in humans is practically unknown. Symptoms produced by experimental feeding of an antagonist include nausea, fatigue, and burning cramps in the limbs. Dietary sources of pantothenic acid are eggs, yeast, liver, milk, peas, and lean meat. This vitamin is stable at moderate cooking temperatures, but is destroyed at high temperatures.

h. Cyanocobalamin (Vitamin B_{12})

Vitamin B_{12} is a complex cobalt-containing structure that has similarities to the heme group of hemoglobin. The vitamin is converted into two coenzymes, one of which is methylcobalamin (the $-CN$ group is replaced by $-CH_3$). Methylcobalamin acts as a methyl group donor, and it is essential for cell growth and replication and for the maintenance of neural functions (maintaining the myelin sheath). Vitamin B_{12} is formed only by certain bacteria that live in a symbiotic relationship with their hosts. Very small amounts of the vitamin are present in liver, kidney, other meats, milk, fish, eggs, oysters, and clams. Vitamin B_{12} is stored in various tissues, particularly the liver.

Vitamin B_{12} is associated with the disease pernicious anemia. This dietary disease is characterized by the presence of abnormally large, immature, and fragile red blood cells. It is accompanied by gastrointestinal disturbances and lesions of the spinal cord with loss of muscular coordination (ataxia).

Pernicious anemia is usually caused by poor absorption of the vitamin from the intestinal tract rather than by a lack of vitamin B_{12}. Cells of the stomach lining synthesize a glycoprotein, called the *intrinsic factor,* that specifically binds vitamin B_{12} and transports the vitamin into intestinal cells for its subsequent transfer to the blood. Pernicious anemia patients lack or have a deficiency of the intrinsic factor and cannot absorb the ingested vitamin B_{12}. Elderly people often have a decreased synthesis of intrinsic factor and must receive vitamin B_{12} by injection directly into the bloodstream in order to avoid anemia. Since most plants contain little or no vitamin B_{12}, pernicious anemia symptoms are sometimes observed among strict vegetarians. Vitamin B_{12} is stable during most cooking procedures.

14.5 *Mode of Enzyme Action*

In 1888 the Swedish chemist Svante Arrhenius proposed a scheme to account for catalytic activity. He suggested that a catalyst functions to combine with a reactant to form an intermediate compound. This intermediate is more reactive than the initial uncombined species. The formation of an intermediate compound affords a lower energy pathway compared to the uncatalyzed reaction (Figure 14.1). This, in effect, lowers the activation energy of the reaction, accounting for the increased rate of reaction. Enzymes reduce activation energies more effectively than other catalysts, thus enabling biochemical reactions to proceed at relatively low temperatures. Note that the amount of energy absorbed or released in the reaction is not altered by the enzyme.

This scheme applies to all catalytic reactions, whether inorganic, organic, or biochemical. It is generally believed that enzymatic reactions occur in at least two steps. In the first step a molecule of the enzyme **(E)** and a molecule of the substrate **(S)** collide and react to form an intermediate compound, which is called the **enzyme–substrate** complex **(E—S).** (This step is reversible since the complex can break apart yielding the original substrate and the free enzyme.) The enzyme–substrate complex may or may not react with additional substances (water, oxidizing or reducing agents, ATP, etc.) to form products **(P),** which are then released from the surface of the enzyme.

$$S + E \rightleftharpoons E{-}S \qquad \text{Sucrose} + \textit{Sucrase} \rightleftharpoons \text{Sucrose–}\textit{sucrase}\text{ complex}$$

$$E{-}S \longrightarrow P + E \qquad \text{Sucrose–}\textit{sucrase} + H_2O \longrightarrow \text{Glucose} + \text{Fructose} + \textit{Sucrase}$$

The existence of an enzyme–substrate complex has been verified by spectroscopic and kinetic experiments. The bonds that hold the enzyme and the substrate together

FIGURE 14.1 Energy diagram for the progress of a chemical reaction and the effect of an enzyme.

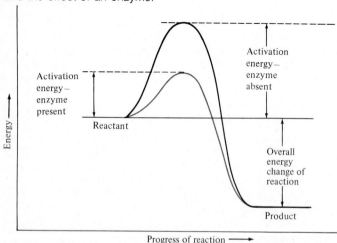

are the same forces involved in maintaining protein structure, that is, electrostatic, covalent, hydrogen bonds, etc. In addition, it has been demonstrated that the structural features or functional groups essential to the formation of the enzyme–substrate complex occur at a specific location on the surface of the enzyme. *This section of the enzyme, which combines with the substrate, and at which transformation from substrate to products occurs,* is called the **active site** of the enzyme. The active site is often a cleft or a crevice in the exterior of the molecule. It possesses a unique conformation (and correctly positioned bonding groups) that is complementary to the structure of the substrate, thus enabling the two molecules to fit together in much the same manner as a key fits into a lock. The **lock and key theory** of enzyme action is illustrated in Figure 14.2. This theory portrays an enzyme as conformationally rigid, and only able to bond to substrates that are structurally suitable. This would explain the high degree of specificity of enzymes for their substrates.

The use of x-ray crystallography in determining the precise three-dimensional structure of enzymes was a major advance in understanding their catalytic activity. It was observed that the binding of some substrates leads to a large conformational change in the enzyme. In 1963 D. E. Koshland, Jr., augmented the older lock and key theory by suggesting that the binding site of an enzyme is not a rigid structure. Instead, some enzymes undergo a change of conformation when they react with a substrate molecule to form an activated complex. The active site has a shape complementary to that of the substrate only *after* the substrate is bound. After catalysis, the enzyme resumes its original structure.

To explain the enzyme's ability to discriminate between similar compounds, a "fit" between the substrate and a portion of the enzyme surface seems essential. However, it appears posssible that this fit is not a static one in which a rigid "positive" substrate fits

FIGURE 14.2 A schematic representation of the interaction of enzyme and substrate.
(a) The active site on the enzyme and the substrate have complementary structures (and complementary bonding groups) and hence fit together as a key fits a lock.
(b) While they are bonded together in the enzyme–substrate complex, the catalytic reaction occurs. (c) The products of the reaction leave the surface of the enzyme, freeing the enzyme to combine with another molecule of substrate.

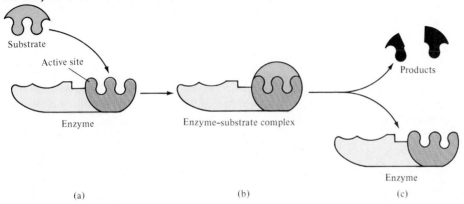

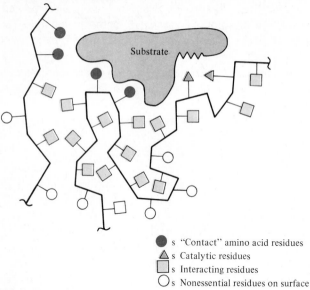

● s "Contact" amino acid residues
▲ s Catalytic residues
▢ s Interacting residues
○ s Nonessential residues on surface

FIGURE 14.3 A schematic representation of an active site. Specificity is determined by the fit of "contact" amino acid residues with the substrate. Catalytic residues act on the substrate bond, which is indicated by the zigzag line. The residues that interact with each other maintain the three-dimensional structure of the enzyme. [Adapted from D. E. Koshland, Jr., *Science*, **142**, 1534 (1963).]

on a rigid "negative" template, but rather, is a dynamic interaction in which the substrate induces a structural change in the enzyme molecule, as a hand changes the shape of a glove.[2]

This **induced-fit theory** is an attractive proposal since it explains several experimental findings that are incompatible with the older theory. Koshland cites examples of compounds that bind to enzymes without undergoing further reaction as well as other compounds that are sterically not suited for the active site but that nevertheless react catalytically. According to Koshland, the active site of an enzyme consists of two components: one is responsible for substrate specificity (*contact groups*), and the other is responsible for catalysis (*catalytic groups*). The active site is a flexible region that can be induced to fit several structurally similar compounds. However, only the proper substrate is capable of correct alignment with the catalytic groups as well (Figures 14.3 and 14.4).

The manner in which an enzyme transforms a substrate into product(s) has been extensively studied. We know that the reactants are brought into proximity as they bind to the enzyme, and this increases the frequency of collisions. Since the enzyme properly aligns each reactant, the effectiveness of each collision will also be increased. As yet, however, the detailed mechanism by which enzymes increase the rate of reactions more efficiently than other catalysts is incompletely understood.

[2]D. E. Koshland, Jr., *Science*, Volume 142, p. 1533 (1963).

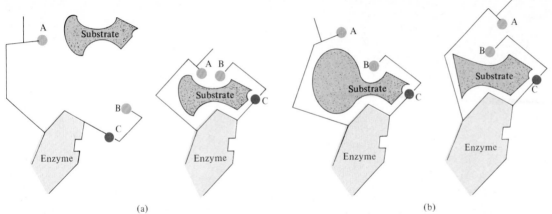

FIGURE 14.4 Schematic representation of a flexible active site. (a) Substrate binding induces proper alignment of catalytic groups A and B so that reaction ensues. (b) Compounds that are either too large or too small are bound, but fail to produce proper alignment of catalytic groups, hence, no reaction. [Adapted from D. E. Koshland, Jr., *Science*, **142,** 1539 (1963).]

We still do not know, for example, whether the full catalytic activity of an enzyme resides in its protein structure as a whole or in a small region associated with the active site. Small peptides have been cleaved from some enzymes (such as *ribonuclease*) without appreciable loss of catalytic activity.

Specific amino acid side chains (such as hydroxyl group of serine, sulfhydryl group of cysteine, imidazole group of histidine) are part of the active sites of various enzymes. The active site probably consists of amino acids from several different positions along the protein chain; these amino acids can be brought into proximity as a result of the folding and bending of the polypeptide chain. The fact that enzymes are inactivated by denaturation, points to the importance of the secondary and tertiary structures in maintaining the active site in a precise three-dimensional arrangement. It is known that two amino acids, histidine-57 and serine-195, are involved in the active site of chymotrypsin (Figure 14.5). The remaining amino acids presumably function in maintaining the active site in the correct geometrical configuration to provide for maximum catalytic activity, as well as in imparting specificity to the molecule.

14.6   *Specificity of Enzymes*

One characteristic that distinguishes an enzyme from all other types of catalysts is its *substrate specificity*. Recall, for example, that hydrogen ions catalyze the hydrolysis of disaccharides, polysaccharides, lipids, and proteins with complete impartiality, whereas different enzymes are required in all four cases. Enzyme specificity is a result of the uniqueness of the active site of each enzyme, and this uniqueness is a function of the chemical nature, electrical charge, and spatial arrangements of the groups located there. A wide range of enzyme specificities exist, and they are arbitrarily grouped as follows.

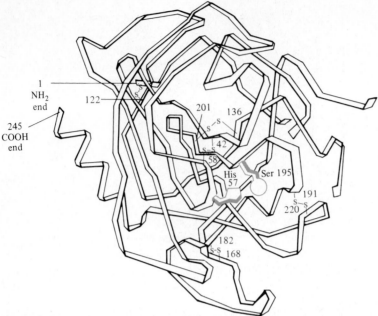

FIGURE 14.5 Model of α-chymotrypsin. The enzyme consists of 245 amino acids in three polypeptide chains, which are interconnected by five disulfide bonds. The active site is in the region of histidine-57 and serine-195. Chymotrypsin is a digestive enzyme that catalyzes the hydrolysis of proteins within the small intestine.

a. Absolute Specificity

Enzymes having absolute specificity catalyze a particular reaction for one particular substrate only and have no catalytic effect on substrates that are closely related. *Urease,* for example, catalyzes the hydrolysis of urea but not of methylurea, thiourea, or biuret.

$$\underset{\text{Urea}}{H_2N-\overset{\displaystyle O}{\overset{\|}{C}}-NH_2} + H_2O \underset{}{\overset{\text{urease}}{\rightleftharpoons}} CO_2 + 2\,NH_3$$

$$\underset{\text{Methylurea}}{H_2N-\overset{\displaystyle O}{\overset{\|}{C}}-NH-CH_3} \qquad \underset{\text{Thiourea}}{H_2N-\overset{\displaystyle S}{\overset{\|}{C}}-NH_2} \qquad \underset{\text{Biuret}}{H_2N-\overset{\displaystyle O}{\overset{\|}{C}}-NH-\overset{\displaystyle O}{\overset{\|}{C}}-NH_2}$$

b. Stereochemical Specificity

Since enzymes are chiral molecules, they show a markedly high degree of specificity toward one stereoisomeric form of the substrate. This is analogous to the binding of monosodium glutamate to the taste buds (see Figure 10.12). L-*Lactic acid dehydrogenase* catalyzes the oxidation of the L-lactic acid found in muscle cells. The D-lactic acid, found in certain microorganisms, does not bind to the enzyme. *Fumarase* adds water to fumaric acid but not to its *cis* isomer, maleic acid.

c. Group Specificity

Enzymes having group specificity are less selective in that they act upon structurally similar molecules having the same functional groups. Many of the *peptidases* fall into this category. *Pepsin* hydrolyzes all peptides having adjacent aromatic amino acids. *Carboxypeptidase* attacks peptides from the carboxyl end of the chain, cleaving the amino acids one at a time (see Section 18.1).

d. Linkage Specificity

Enzymes having linkage specificity are the least specific of all because they attack a particular kind of chemical bond, irrespective of the structural features in the vicinity of the linkage. The *lipases,* which catalyze the hydrolysis of ester linkages in lipids, are an example of this type of enzyme.

14.7 *Factors Influencing Enzyme Activity*

The single most important property of an enzyme is its catalytic activity. Since enzymes are protein catalysts, they are affected by those factors that act upon proteins and upon catalysts in general. The activity of an enzyme may be measured by monitoring the reaction that it catalyzes at fixed time intervals. The rate of the reaction is determined by observing either the rate of disappearance of the substrate or the rate of formation of the product(s). In such experiments, the rate is the only variable; all other experimental conditions are held constant.

a. Concentration of Substrate

The rate of an enzymatic reaction increases as the substrate concentration increases until a limiting rate is reached. At this point, further increase in the substrate concentration produces no significant change in the reaction rate. At excess

FIGURE 14.6 A schematic representation of relative concentrations of enzyme and substrate. (a) Low substrate concentration, (b) adequate substrate concentration, (c) excess substrate concentration.

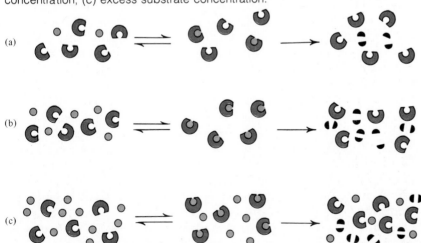

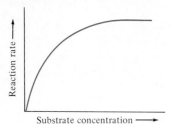

FIGURE 14.7 Effect of substrate concentration on the rate of a reaction that is catalyzed by a fixed amount of enzyme.

substrate concentrations, practically all the enzyme molecules are saturated with the substrate at any given instant. Extra substrate molecules must wait until the enzyme–substrate complexes have dissociated to yield products and the free enzymes before they may undergo reaction. This relationship is illustrated in Figure 14.6 and can be summarized in terms of the two equations given in Section 14.5. At low substrate concentrations the formation of the E—S complex is the rate-determining step, whereas at high substrate concentrations, the slowest step is the dissociation of the E—S complex. Figure 14.7 is a characteristic plot of an enzyme-catalyzed reaction, and it is taken as further evidence of the existence of the enzyme–substrate intermediate.

b. Concentration of Enzyme

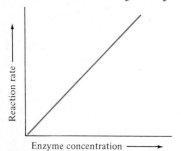

FIGURE 14.8 Effect of enzyme concentration on the rate of a reaction.

Since in essentially all practical cases the concentration of the enzyme is much lower than the concentration of the substrate (the substrate can seldom become saturated with the enzyme), the rate of an enzyme-catalyzed reaction is always directly dependent upon the enzyme concentration (Figure 14.8). This is not a new concept; the reaction rate of any catalytic reaction increases as the concentration of the catalyst is increased. At any given time the concentration of enzyme in a cell is determined by its rate of synthesis and its rate of degradation (i.e., *enzyme turnover*). As we shall see in later chapters, the concentration of enzymes present in a cell can be increased (enzyme induction) or decreased (enzyme suppression) according to the needs of the organism.

c. Temperature

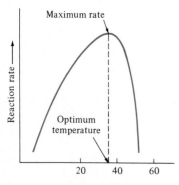

FIGURE 14.9 Effect of temperature on the rate of an enzymatic reaction.

A rule of thumb for most chemical reactions is that a 10°C rise in temperature approximately doubles or triples the reaction rate. (This is due to an increase in the number of molecules that possess sufficient kinetic energy to exceed the activation energy.) To a certain extent, this is true of all enzymatic reactions. After a certain point, however, an increase in temperature causes a decrease in the rate of reaction, as indicated in Figure 14.9. *The temperature that affords maximum activity* is known as the **optimum temperature** for the enzyme in question. Most enzymes of warm-blooded animals have optimum temperatures of about 37°C (98°F). The decrease in rate is a direct consequence of the fact that enzymes are proteins and thus are denatured by heat. Heating disrupts the secondary and tertiary structures of the enzymes, effecting a disorientation of the active site. This disorientation renders the active site inaccessible to the substrate (Figure 14.10).

At temperatures of 0°C and 100°C the rate of enzyme-catalyzed reactions is nearly zero. This fact has several practical applications. We sterilize objects by placing them in boiling water so as to denature the enzymes of any bacteria that may be in contact with them. We refrigerate and freeze our food to slow down the enzyme activity and preserve the food. Animals go into hibernation because of a decrease in

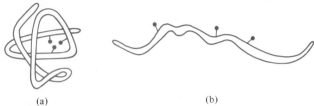

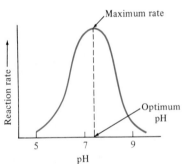

(a)   (b)

FIGURE 14.10 (a) Representation of an active site in an enzyme. (b) Heating denatures the enzyme and the groups of the active site are no longer in close proximity.

their body temperature (in the winter) and as a result, the rate of their metabolic processes decreases. The food required to maintain this lowered metabolic rate is provided by reserves stored in their tissues.

d. Hydrogen Ion Concentration

Again, as a consequence of their protein nature, enzymes are sensitive to changes in the pH of their environments. Extreme values of pH (whether high or low) can cause denaturation of the protein. However, any change in the hydrogen ion concentration alters the degree of ionization of acidic and basic groups both on the enzyme and on the substrate. If any ionizable groups are located at the active site, and if a certain charge is necessary in order for the enzyme to bind its substrate, then an enzyme molecule that has even one of these charges neutralized will lose its catalytic activity. An enzyme will exhibit maximum activity over a narrow pH range in which the molecule exists in its proper charged form. *The median value of this pH range is known as the* **optimum pH** *of the enzyme (Figure 14.11). With the notable exception of gastric juice, most body fluids have pH values between 6 and 8. This is essential because most enzymes exhibit optimal activity in the physiological pH range of 7.0–7.5. However, each enzyme has a characteristic optimum pH, and in a few cases this value is outside the usual physiological range. For example, the optimal pH for pepsin is 2.0 in the stomach, and that for trypsin is 8.0 in the small intestine.

FIGURE 14.11 Effect of pH on the rate of an enzymatic reaction.

14.8 *Enzyme Inhibition*

In the preceding section, we noted that enzymes are inactivated by increased temperatures and by changes in pH. In a sense, then, temperature and hydrogen ion concentration may be considered to be factors that inhibit enzyme activity. In fact, any physical change or chemical reagent that denatures protein will adversely affect the rate of an enzymatic reaction. This type of enzyme inhibition is referred to as nonspecific inhibition since it affects all enzymes in the same manner. In contrast, a specific inhibitor exerts its effect upon a single enzyme or a group of related enzymes. The inhibition of enzyme activity is one of the most important control mechanisms in living organisms. Many poisons act to inhibit specific enzymes (Table 14.3).

TABLE 14.3 Poisons as Enzyme Inhibitors

Poison	Formula	Enzyme Inhibited	Site of Action
Cyanide	CN^-	Cytochrome oxidase, catalase	Binds Fe^{3+} activator
Fluoride	F^-	Enolase	Binds Mg^{2+} activator
Sulfide	S^{2-}	Phenolase	Binds Cu^{2+} activator
Arsenate	$AsO_4{}^{3-}$	Glyceraldehyde 3-phosphate dehydrogenase	Substitutes for phosphate
Iodoacetate	ICH_2COO^-	Triose phosphate dehydrogenase	Binds to cysteine sulfhydryl group
Nerve gas	$F—\overset{\displaystyle O}{\overset{\|}{P}}—OCH(CH_3)_2$ $OCH(CH_3)_2$	Acetylcholinesterase	Binds to serine hydroxyl group

a. Competitive Inhibition

A **competitive inhibitor** is any *compound that bears a close structural resemblance to a particular substrate and competes with that substrate for binding at the same active site on the enzyme.* The inhibitor is not acted upon by the enzyme and so remains bound to the enzyme, preventing the substrate from approaching the active site. The degree of competitive inhibition depends upon the relative concentrations of substrate and inhibitor. If the inhibitor is present in relatively large quantities, it blocks the active sites on all the enzyme molecules and complete inhibition results. However, formation of the inhibitor–enzyme complex is reversible, and increased substrate concentration permits displacement of the inhibitor from the active site. Competitive inhibition can be completely reversed by addition of large excesses of substrate. The reversible nature of competitive inhibition has provided much information about the enzyme–substrate complex and about the specific groups involved at the active sites of various enzymes. Pharmaceutical companies have synthesized drugs that can competitively inhibit metabolic processes in bacteria (Section 14.9) and in cancer cells.

A classic example of competitive inhibition is the effect of malonic acid on the enzyme activity of *succinic acid dehydrogenase.* The former is a homolog of the enzyme's normal substrate, succinic acid. Malonic acid will bind to the active site, since the spacing of its carboxyl groups is not greatly different from that of succinic acid. No catalytic reaction occurs, and malonic acid remains bonded to the enzyme (Figure 14.12). We discuss this reaction again in connection with carbohydrate metabolism (Section 16.8).

Another example of a competitive inhibitor is the compound chlorfenthol, which inhibits the enzyme *DDT dehydrochlorinase* that converts DDT to DDE. You will recall (page 70) that insects become resistant to DDT by a mutation resulting in their ability to biosynthesize the *dehydrochlorinase* enzyme. If chlorfenthol is applied to a field along with DDT, the resistant insects will also be killed off because of the inhibition of their protecting enzyme.

DDT

Chlorfenthol
1,1-Di-(*p*-chlorophenyl)ethanol

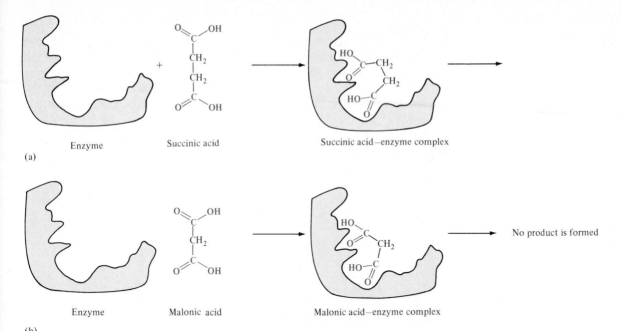

FIGURE 14.12 (a) Succinic acid binds to the enzyme *succinic acid dehydrogenase*. A dehydrogenation reaction occurs and the product (fumaric acid) is released from the enzyme. (b) Malonic acid can also bind to the active site of *succinic acid dehydrogenase*. However, in this case no subsequent reaction occurs, and malonic acid remains bound to the enzyme.

b. Noncompetitive Inhibition

A **noncompetitive inhibitor** is a *substance that can combine with either the free enzyme or the enzyme–substrate complex*. The noncompetitive inhibitor binds to the enzyme at a position relatively remote from the active site and in so doing alters the three-dimensional conformation of the enzyme. This effects a change in the configuration of the active site, so that either the E—S complex does not form at its normal rate, or once formed the E—S complex does not decompose at the normal rate to yield products. Since the inhibitor does not structurally resemble the substrate, the addition of excess substrate does *not* reverse the inhibitory effects.

Many enzymes contain reactive groups, such as —COO$^-$, —NH$_3^+$, —SH, or —OH, that are essential for maintaining the proper three-dimensional conformation of the enzyme. Any chemical reagent that is capable of combining with one or more of these groups will inhibit the enzyme. The heavy metal ions Ag(I), Hg(II), Pb(II) have strong affinities for carboxylate and sulfhydryl groups; we have already discussed their toxic effects with regard to protein denaturation. Iodoacetic acid is another inhibitor that combines specifically with —SH groups.

$$\boxed{\text{Enzyme}}-\text{SH} + \text{I}-\text{CH}_2-\overset{\displaystyle O}{\underset{\displaystyle \text{OH}}{\text{C}}} \longrightarrow \boxed{\text{Enzyme}}-\text{S}-\text{CH}_2-\overset{\displaystyle O}{\underset{\displaystyle \text{OH}}{\text{C}}} + \text{HI}$$

c. Irreversible Inhibition

An **irreversible inhibitor** *inactivates enzymes by forming strong bonds to a particular group at the active site.* The inhibitor does not resemble the substrate, and the inhibitor–enzyme bond is so strong that the inhibition can not be reversed by adding excess substrate. Irreversible inhibition was formerly considered to be another type of noncompetitive inhibition, but it is now recognized as a distinct type of inhibition.

The nerve gases, especially diisopropylfluorophosphate (DFP), irreversibly inhibit biological systems by forming an enzyme–inhibitor complex with a specific hydroxyl group (of serine) situated at the active site of certain enzymes. The proteolytic enzymes *trypsin* and *chymotrypsin* contain such serine groups and are inhibited by DFP. Another enzyme inhibited by organophosphates of this type is *acetylcholinesterase* (recall page 171).

$$\boxed{\text{Enzyme}}\!-\!OH \;+\; F\!-\!\overset{\displaystyle O}{\overset{\|}{P}}\!-\!(OR)_2 \;\longrightarrow\; \boxed{\text{Enzyme}}\!-\!O\!-\!\overset{\displaystyle O}{\overset{\|}{P}}\!-\!(OR)_2 \;+\; HF$$

Metalloenzymes (those that require the presence of a metal ion for activation) are irreversibly inhibited by substances that form strong complexes with the metal. Traces of hydrogen cyanide inactivate iron-containing enzymes such as catalase and cytochrome oxidase. Oxalic and citric acids inhibit blood clotting by forming complexes with calcium ions necessary for the activation of the enzyme *thromboplastin* (see Section 19.5).

d. End-Product Inhibition (Negative Feedback Control)

Some biosynthetic pathways are known in which the enzyme catalyzing the first reaction in the pathway is inhibited by the final product that the pathway produces. Consider the hypothetical series of reactions leading from starting compound A, through the series of intermediates B, C, and D, to the final product E. Each step is catalyzed by a different enzyme as indicated.

$$A \xrightarrow{\text{enzyme a}} B \xrightarrow{\text{enzyme b}} C \xrightarrow{\text{enzyme c}} D \xrightarrow{\text{enzyme d}} E$$
$$\underset{\text{inhibits}}{\longleftarrow\!\!\!-\!\!\!-\!\!\!-\!\!\!-\!\!\!-\!\!\!-\!\!\!-\!\!\!-\!\!\!-}$$

The ultimate product of the sequence, E, serves to prevent formation of its own precursors by noncompetitively inhibiting the action of enzyme a. Thus, an organism is provided with a mechanism for controlling the rate of synthesis of a metabolic intermediate according to its need.

A specific example is the biosynthesis of isoleucine from threonine in *Escherichia coli*. The enzyme *threonine dehydratase*, which catalyzes the first step of this five-step reaction sequence, is inhibited by the end product, isoleucine, when a critical concentration of isoleucine is built up in the cell. When the concentration of isoleucine is sufficiently lowered by its metabolic utilization, its inhibitory effect decreases and biosynthesis of the compound proceeds once again.

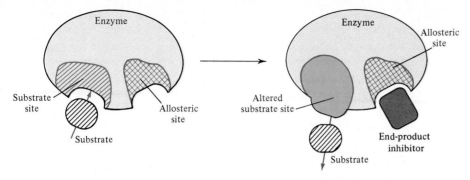

(a) Active enzyme   (b) Inactive enzyme

FIGURE 14.13 A schematic representation of how the binding of an end-product inhibitor inhibits an allosteric enzyme by causing a change in the structural conformation at the substrate-binding site.

It has been shown that in all cases of end-product inhibition, there are at least two distinct binding sites on the enzyme. One is the active site that binds to the substrate, and the other is the *binding site for the inhibitory end product*. This latter site is called the **allosteric site** (other site), and enzymes that contain allosteric sites are referred to as allosteric enzymes.[3] When isoleucine binds to the allosteric site of *threonine dehydratase*, it causes a change in the three-dimensional conformation of the enzyme, distorting the active site sufficiently to render it nonfunctional (Figure 14.13). In most cases of negative feedback control, the first step of a sequence of reactions is inhibited, thereby preventing the cell from wasting energy on the synthesis of useless intermediates.

14.9   *Chemotherapy*

Chemotherapy is the *use of chemicals* (drugs) to *destroy infectious microorganisms (and cancer cells) without damaging the cells of the host*. From bacteria to humans, the metabolic pathways of all living organisms are quite similar, and so the

[3] Allosteric enzymes can be subject to positive as well as negative effects. Certain substances (termed *positive modulators*) can bind to the allosteric site and *increase* the activity of the enzyme by altering its conformation so as to make the substrate-binding *more* accessible to the substrate. All allosteric enzymes consist of two or more polypeptide chains. Usually the active site is on one chain and the allosteric site is on another chain. The change in conformation of one subunit (containing the allosteric site) is amplified and transmitted to effect a conformational change in the other subunit (containing the active site). This change can either increase or decrease the substrate-binding affinity of the active site. Allosteric regulation is probably the most widespread control mechanism for coordinating the biochemical reactions within cells.

discovery of safe and effective chemotherapeutic agents is a formidable task. It is now well established that drugs function through their inhibitory effect on a critical enzyme in the cells of the invading organism.

Chemotherapy is widely used in the treatment of cancer patients. **Antineoplastic** drugs (*substances that inhibit the growth of cancer cells*) such as 5-fluorouracil and 6-mercaptopurine interfere with the production of DNA and RNA in tumor cells by substituting for the pyrimidine and purine bases (see Section 15.1). Use of another drug, however, has engendered a great deal of controversy. Laetrile (amygdalin or "vitamin B_{17}"), a compound extracted from apricot pits (and produced synthetically), was first broadly promoted for cancer treatment in the early 1950s and rose to national prominence in the 1970s. It had been claimed to alleviate pain and prolong life for cancer victims. Laetrile was supposed to function by reacting with the enzyme *β-glucosidase* to release cyanide, which killed cancer cells. Normal cells were said to be protected by an enzyme (*rhodanase*) that detoxifies cyanide by converting it into thiocyanate. The Food and Drug Administration, however, considered the drug unsafe and worthless. In a 1977 report the commissioner of the Food and Drug Administration, Donald Kennedy, stated that "all available evidence indicates that Laetrile is a major health fraud in the United States and there is no documentation of its safety or effectiveness." In 1981 the National Cancer Institute completed extensive clinical testing of laetrile on cancer patients. It was concluded that laetrile offered "no substantive benefit in curing cancer, slowing cancer's advance through the body, or ameliorating its symptoms." Laetrile has been called "the most thoroughly studied failure in the history of medicine," and it is no longer promoted as a cancer cure.

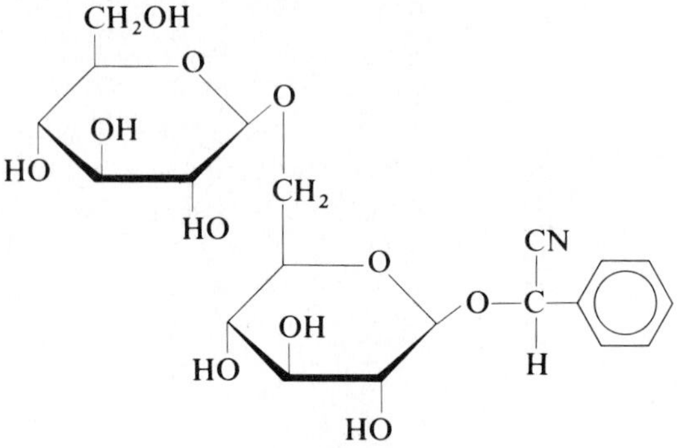

Laetrile
(Amygdalin)
(Mandelonitrile-β-gentobioside)

a. Antimetabolites

An **antimetabolite** is a *substance that possesses a structure closely related to the normal substrate* (the metabolite) *of an enzyme and **competitively** inhibits a significant metabolic reaction.* One of the earliest (1935) and best understood antimetabolites is the synthetic antibacterial agent sulfanilamide. Its effectiveness rests on its structural similarity to *p*-aminobenzoic acid, a compound vital to the growth of many pathogenic bacteria. To reproduce, some bacteria require *p*-aminobenzoic acid for

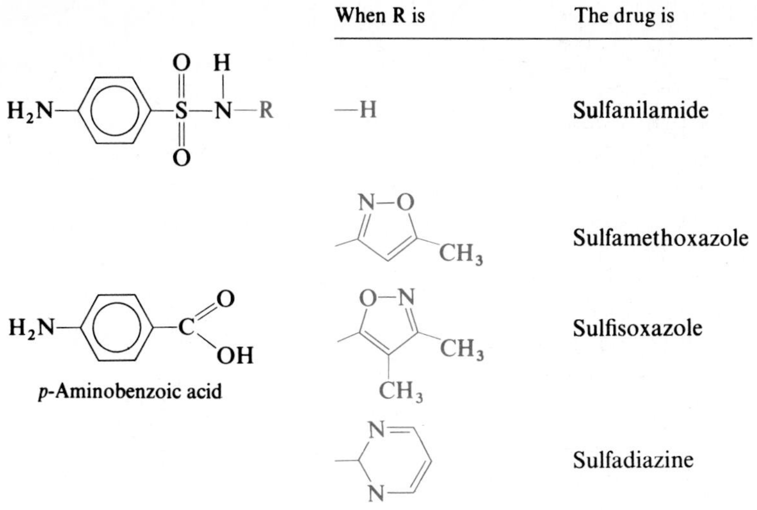

	When R is	The drug is
H₂N—⟨benzene⟩—S(=O)(=O)—N(H)—R	—H	Sulfanilamide
	(isoxazole N—O with CH₃)	Sulfamethoxazole
H₂N—⟨benzene⟩—C(=O)—OH *p*-Aminobenzoic acid	(isoxazole O—N with CH₃, CH₃)	Sulfisoxazole
	(pyrimidine ring with N)	Sulfadiazine

FIGURE 14.14 The sulfa drugs interfere with the normal metabolism of *p*-aminobenzoic acid.

the synthesis of folic acid (a coenzyme precursor—Table 14.2). Sulfanilamide interferes with the enzyme-controlled step involving the incorporation of the *p*-aminobenzoic acid moiety into folic acid. Thus, sulfanilamide (and its derivatives) is *bacteriostatic* and not *bacteriocidal*. The sulfa drugs don't kill bacteria outright but rather prevent their growth and weaken them so they are more readily attacked and destroyed by phagocytes (see page 494).

Sulfanilamide is not harmful to humans (or to other mammals) because we cannot synthesize folic acid but must obtain it, preformed, from our diets. After the drug was recognized as an antibacterial agent, many other sulfanilamide derivatives were synthesized and found to be even more effective in this capacity. Many lives were saved during World War II as a result of these popularly named wonder drugs. Soldiers carried packages of powdered sulfa drugs to sprinkle on open wounds to prevent infection. Unfortunately, prolonged use of sulfa drugs causes a number of side effects, particularly kidney damage, so that they have been largely replaced by the penicillins and other antibiotics. However, they are still prescribed for some specific infections against which they are highly effective, such as bladder and urinary tract infections. Some newer sulfa drugs are used in the treatment of tuberculosis and leprosy, and they are widely used in veterinary medicine. Structures of several of the important sulfa drugs used as urinary antiseptics are given in Figure 14.14, along with the structure of the metabolite *p*-aminobenzoic acid.[4]

[4] It is interesting to note that *p*-aminobenzoic acid (PABA) has become widely known as an ingredient in suntan lotions. PABA acts as a sun filter by absorbing the short-wavelength ultraviolet rays that are responsible for causing sunburn. The most effective sunscreen products are those that contain 5% PABA as the active ingredient. The ethyl ester of PABA is Benzocaine, a local anesthetic found in a wide variety of over-the-counter products including first aid and sunburn sprays, foot powders, cough medicines, and appetite control products.

b. Antibiotics

Although some antibiotics are believed to function as antimetabolites, the terms are not identical. An **antibiotic** is a *compound produced by one microorganism* (bacteria, mold, yeast) *that is toxic to another microorganism.*[5] Antibiotics, many of which can now be synthesized in the laboratory, constitute no well-defined class of chemically related substances but, instead, possess the common property of effectively inhibiting a variety of enzymes essential to bacterial growth. Some antibiotics are bacteriostatic; others are bacteriocidal.

Penicillin, one of the most widely used antibiotics in the world, was fortuitously discovered by Alexander Fleming in 1928. In 1938, Ernst Chain and Howard Florey isolated it in pure form and proved its effectiveness as an antibiotic. (The three

FIGURE 14.15 The penicillins differ only in the nature of their **R** groups. Notice that the amino acids valine and cysteine are incorporated into the penicillin structure.

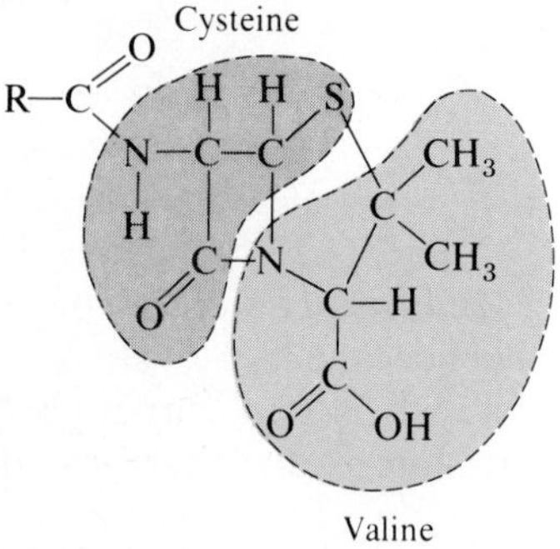

Penicillin	R
F	$-CH_2CH=CHCH_2CH_3$
G	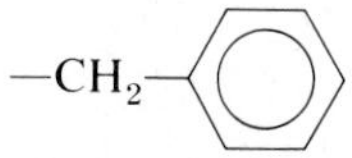 $-CH_2-$⟨phenyl⟩
K	$-(CH_2)_6CH_3$
X	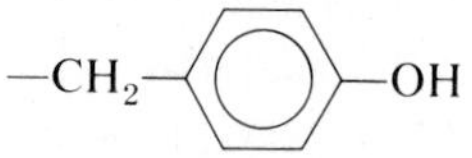 $-CH_2-$⟨phenyl⟩$-OH$
N	$-(CH_2)_3-CH-COOH$ with NH_2
V	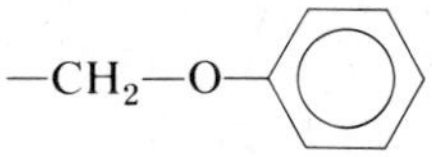 $-CH_2-O-$⟨phenyl⟩
Ampicillin	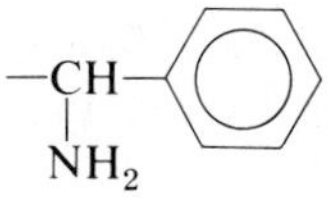 $-CH-$⟨phenyl⟩ with NH_2
Methicillin	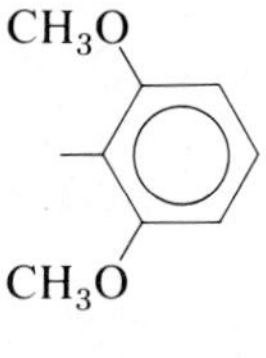 ⟨phenyl with CH_3O substituents⟩

[5] The term *antibiotic* is considered to be synonymous with antibacterial, although many antiviral and anticancer drugs are also antibiotics.

scientists received the Nobel Prize for Physiology and Medicine in 1945.) Penicillin was first introduced into medical practice in 1941. It functions by interfering with the synthesis of cell walls of reproducing bacteria. Penicillin inhibits an enzyme (*transpeptidase*) that catalyzes the last step in bacterial cell wall biosynthesis. This step involves the joining of long polysaccharide chains by short peptide chains. The new cell walls are defective, and subsequently the bacterial cells burst. Since human cells have cell membranes and not cell walls, they are not affected. Several naturally occurring penicillins have been isolated. All have the empirical formula $C_9H_{11}O_4SN_2R$ and contain a four-membered ring fused to a five-membered ring (Figure 14.15). The various R groups are obtained by the addition of the appropriate organic compounds to the culture media.

The penicillins are effective against gram-positive bacteria (bacteria that are stained by Gram's dye) and a few gram-negative bacteria (including *E. coli*). They have proved effective in the treatment of diphtheria, gonorrhea, pneumonia, syphilis, many pus infections, and certain types of boils. Penicillin G was the earliest penicillin to be used on a wide scale. However, it cannot be administered orally for it is quite unstable, and the acid pH of the stomach causes a rearrangement to an inactive derivative. Penicillin V and ampicillin, on the other hand, are acid stable,

Figure 14.16 Structures of some common antibiotics. Notice that aureomycin and tetracycline are similar in all respects except for the absence of a chlorine atom in the latter. Streptomycin is a glycoside containing an amino derivative of glucose and chloramphenicol bears a resemblance to epinephrine. Gramicidin S in a cyclodecapeptide formed from two identical pentapeptide chains. It contains the amino acids D-phenylalanine and ornithine, which are not found in proteins.

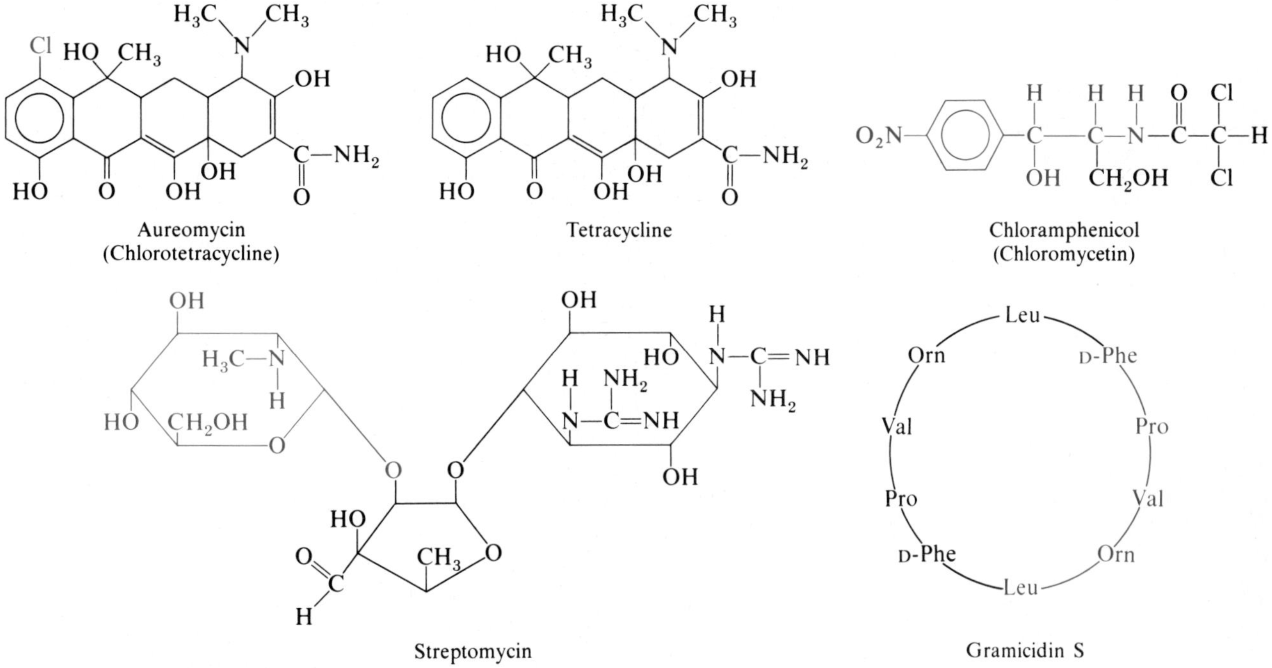

Aureomycin
(Chlorotetracycline)

Tetracycline

Chloramphenicol
(Chloromycetin)

Streptomycin

Gramicidin S

and they are the major oral penicillins. (Ampicillin is a broad-range antibiotic effective against both gram-positive and gram-negative bacteria.) Some strains of bacteria become resistant to penicillin by a mutation that allows them to synthesize an enzyme, *penicillinase,* that breaks down the antibiotic (cleavage of the four-membered ring). To combat these strains, scientists have been able to synthesize penicillin analogs (such as methicillin) that are not inactivated by *penicillinase.*

Some people are allergic to penicillin and therefore must be treated with other antibiotics. (This allergic reaction is so severe that a fatal coma may occur if penicillin is inadvertently administered to a sensitive individual.) Fortunately, a number of antibiotics have been discovered. Most are the complete products of microbial synthesis (e.g., aureomycin, streptomycin). Others are made by chemical modifications of antibiotics (e.g., semisynthetic penicillins, tetracyclines), and some are manufactured entirely by chemical synthesis (e.g., chloramphenicol). They have proved to be as effective as penicillin in destroying infectious microorganisms (Figure 14.16). Many of these antibiotics exert their effects by blocking protein synthesis in microorganisms. (See Table 15.7 for a listing of some of the major bacterial diseases.)

14.10   *Diagnostic Applications of Enzymes*

The measurement of enzyme activity in such body fluids as plasma or serum has become a valuable tool in medical diagnosis. Certain enzymes that function in the plasma, such as the enzymes involved in blood clotting, are continually secreted into the blood by the liver. Most other enzymes, however, are normally present in plasma in very low concentrations. They are derived from the routine destruction of erythrocytes, leukocytes, and other cells. When cells die, their soluble enzymes leak out of the cells and enter the bloodstream. Since not all cells contain the same complement of enzymes, those that are specific to a particular organ can be important in aiding diagnosis. Therefore, an abnormally high level of a particular enzyme in the blood often indicates specific tissue damage, as in hepatitis and myocardial infarction (*myo,* muscle; *cardi,* heart; an *infarct* is an area of dead tissue—see footnote 14 on page 441). Table 14.4 lists the commonly assayed enzymes that are used in clinical diagnosis.

TABLE 14.4   Some Important Enzymes for Clinical Diagnoses

Enzyme Assayed	Organ or Tissue Affected
α-Amylase	Pancreas
Alkaline phosphatase	Bone, liver
Acid phosphatase	Prostate
Creatine kinase (CK)	Muscle, heart
Glutamic oxaloacetic transaminase (GOT)	Heart, liver
Glutamic pyruvic transaminase (GPT)	Liver
Lactic dehydrogenase (LDH)	Heart, liver
Alanine aminotransferase	Liver
Aspartate aminotransferase	Heart, liver

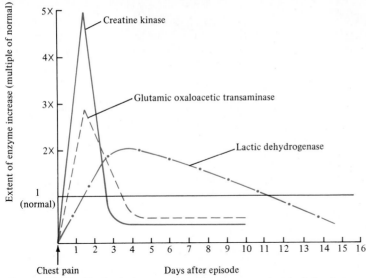

FIGURE 14.17 Typical changes in serum enzyme levels following a heart attack. [Adapted with permission of Macmillan Publishing Company from *Introductory Chemistry for Health Professionals* by Ken Liska and Lucy T. Pryde. Copyright © 1984 by Macmillan Publishing Company.]

Modern medical practices have automated and computerized the assay procedures for most of these serum enzymes. It is important to note that the precise patterns of enzyme changes in certain tissue diseases are characteristic. For example, in a myocardial infarction, the ratio of GOT/GPT is usually high; the reverse is true in liver disease. A summary of changes in serum enzyme levels following a heart attack is illustrated in Figure 14.17.

Exercises

14.1 Define, explain, or draw a formula for each of the following.
 (a) enzyme
 (b) substrate
 (c) holoenzyme
 (d) activator
 (e) metalloenzyme
 (f) zymogen
 (g) vitamin
 (h) enzyme–substrate complex
 (i) active site
 (j) enzyme specificity
 (k) enzyme turnover
 (l) inhibitor
 (m) negative feedback control
 (n) allosteric site
 (o) chemotherapy
 (p) antineoplastic agent
 (q) Laetrile
 (r) sulfa drug
 (s) PABA
 (t) myocardial infarction

14.2 Distinguish between the terms in each of the following pairs.
 (a) in vitro and in vivo
 (b) intracellular enzyme and extracellular enzyme
 (c) sucrose and sucrase
 (d) apoenzyme and coenzyme

(e) peptidase and lipase

(f) trypsin and trypsinogen

(g) contact group and catalytic group

(h) absolute specificity and stereochemical specificity

(i) group specificity and linkage specificity

(j) optimum temperature and optimum pH

(k) competitive inhibitor and noncompetitive inhibitor

(l) irreversible inhibitor and end-product inhibitor

(m) antimetabolite and antibiotic

(n) bacteriostatic and bacteriocidal

14.3 Briefly recount the contributions of Pasteur, Buchner, and Sumner to our understanding of the nature of enzymes.

14.4 In what ways are enzymes similar to ordinary chemical catalysts? How do they differ? What is the effect of an enzyme on an equilibrium system?

14.5 Why are enzymes more specific than inorganic catalysts?

14.6 Animals can digest starch but not cellulose. Explain.

14.7 Identify three intracellular enzymes and three extracellular enzymes.

14.8 Why is it essential that the *peptidases* be secreted as zymogens?

14.9 What type of reaction is catalyzed by each of the following enzymes?

(a) carbohydrase   (b) lipase

(c) esterase   (d) dipeptidase

(e) isomerase   (f) dehydrogenase

(g) decarboxylase   (h) transferase

(i) hydrolase   (j) penicillinase

14.10 Refer to Table 14.1 and give the group name of the enzyme that catalyzes the following reactions. Also give the specific name for those enzymes that you know.

(a) sucrose $\longrightarrow$ glucose + fructose

(b) lactose $\longrightarrow$ galactose + glucose

(c) ascorbic acid $\longrightarrow$ dehydroascorbic acid

(d) ethyl alcohol $\longrightarrow$ acetaldehyde

(e) polypeptides $\longrightarrow$ amino acids

(f) urea $\longrightarrow$ carbon dioxide + ammonia

(g) glucose $\longrightarrow$ fructose

(h) lysine $\longrightarrow$ cadaverine + carbon dioxide

(i) triacylglycerols $\longrightarrow$ fatty acids + glycerol

(j) maltose $\longrightarrow$ 2 glucose

14.11 List the B vitamins and give a function for each. What foods, in general, are a good source of the B vitamins?

14.12 The phrases below refer to the structural changes that occur to vitamins to convert them to coenzymes. Consult Table 14.2 and identify the vitamin(s).

(a) converted into coenzymes that catalyze oxidation–reduction reactions

(b) undergoes no further change (i.e., it is also a coenzyme)

(c) pyrophosphate group is added

(d) oxidized and phosphorylated

(e) reduced

(f) methyl replaces cyanide

14.13 Briefly describe the mode of enzyme action.

14.14 Why is it that only a relatively few enzyme molecules can catalyze the conversion of many molecules of substrate?

14.15 What functional groups are responsible for the formation and maintenance of the enzyme–substrate complex?

14.16 Compare the lock and key theory with the induced-fit theory of Koshland.

14.17 How is it possible that two amino acids that are relatively far apart in the primary structure of the protein chain can be in proximity at the active site?

14.18 Only a small fraction of the enzyme molecule binds to the substrate. What function is served by the remainder of the enzyme molecule?

14.19 Which type of groups (i.e., hydrogen bonding, hydrophobic, etc.) at the active site of an enzyme would bind to the following groups on the substrate?

(a) —COOH   (b) —COO⁻

(c) —NH₂   (d) —NH₃⁺

(e) —OH   (f) —SH

(g) —CH(CH₃)₂   (h)

14.20 Name one enzyme from each of the four categories of enzyme specificity. What is its substrate and what products are obtained from the enzymatic reaction?

14.21 Both wool (protein) and nylon (synthetic polymer) are large molecules in which smaller molecules have been joined by amide linkages. Moths can digest wool, but they can't digest nylon. Explain.

14.22 Would you expect an apoenzyme to bind to more than one type of coenzyme? Explain.

14.23 List the factors that influence the rate of an enzyme reaction. Discuss their effects.

14.24 Explain why enzymes become inactive above and below the (a) optimum temperature and (b) the optimum pH.

14.25 *Amylase* in saliva begins the digestion of starch in the mouth. Would you expect this enzyme to continue to function in the stomach? Explain.

14.26 In Section 13.9, we discussed the denaturation of proteins. What is the effect of denaturing an enzyme? Be specific.

14.27 Briefly discuss the various types of enzyme inhibitors.

14.28 Experimentally, how could you distinguish a competitive inhibitor from a noncompetitive inhibitor?

14.29 Fluoroacetic acid (CH₂FCOOH) is a deadly poison. Suggest a possible mechanism for its toxic effect.

14.30 The nerve gases exert their effects by inhibiting *acetylcho-*

linesterase, the enzyme that hydrolyzes acetylcholine. Can the effects of the nerve gas be overcome by the addition of excess acetylcholine? Explain.

14.31 Is oxaloacetic acid ($HOOCCH_2COCOOH$) an inhibitor of the enzyme *succinic acid dehydrogenase*? What type of inhibitor is it?

14.32 Why are salts of mercury, silver, and lead poisonous to the body?

14.33 Explain how allosteric enzymes function.

14.34 Explain why antimetabolites and antibiotics may both be classified as antiseptics.

14.35 Derivatives of PABA are ubiquitous. Ethyl *p*-amino-benzoate (Benzocaine) is an antiseptic and a local anesthetic. Ointments containing 5–10% Benzocaine are used to treat sunburn and minor scrapes. Draw the structure of Benzocaine.

14.36 List five antibiotics.

15

The Nucleic Acids

In this chapter we consider the nucleic acids, the macromolecules whose structure determines the growth and development of all life forms. The nucleic acids are informational molecules; into their primary structure is encoded a set of directions that ultimately governs the metabolic activities of the living cell. We shall see how differences in primary nucleic acid structure account, on the molecular level, for the factors that differentiate organisms from one another. In this chapter, and in subsequent chapters, we relate the function of biomolecules with the structure and the biological role of the various cell components. When a particular organelle is discussed, you may want to refer back to Figure 12.5-c, which depicts a typical animal cell.

15.1 Structure of Nucleic Acids

Nucleic acids are macromolecular polymers that occur in all living things. There are two types, deoxyribonucleic acid (DNA) and ribonucleic acid (RNA). Their chemical composition is best understood in terms of their hydrolysis products. Complete hydrolysis yields a mixture of heterocyclic amines (purines and pyrimidines), a five-carbon sugar (ribose or 2-deoxyribose), and phosphoric acid. Partial hydrolysis

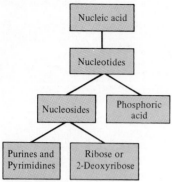

FIGURE 15.1 Products of successive hydrolyses of nucleic acids.

degrades the nucleic acid into somewhat larger subunits, nucleotides and nucleosides. Figure 15.1 shows the successive hydrolytic products of nucleic acids.

a. Pyrimidine Bases

The pyrimidine bases that occur in nucleic acids are substituted derivatives of the parent compound, pyrimidine. Pyrimidine is a heterocyclic six-membered ring compound containing two ring-nitrogen atoms. It does not occur free in nature, but its derivatives _uracil_, _thymine_, and _cytosine_ occur in nucleic acids.

Pyrimidine

Uracil
(2,4-Dioxypyrimidine)

Thymine
(5-Methyl-2,4-dioxypyrimidine)

Cytosine
(4-Amino-2-oxypyrimidine)

Several other pyrimidine derivatives (called modified or minor bases) also are found in various nucleic acids. Among these are 5-methylcytosine and 5-hydroxymethylcytosine (see Exercise 15.5).

b. Purine Bases

The naturally occurring purine bases are derivatives of the parent compound purine, a heterocyclic amine consisting of a pyrimidine ring fused to an imidazole ring. Adenine and guanine are the major purine constituents of nucleic acids.

Purine

Adenine
(6-Aminopurine)

Guanine
(2-Amino-6-oxypurine)

Methylation is the most common form of purine modification, and methylated purines occur in varying amounts in nucleic acids. 6-Methyladenine and 2-methylguanine are two of the minor purine bases that occur in certain nucleic acids (see Exercise 15.5).

c. Nucleosides

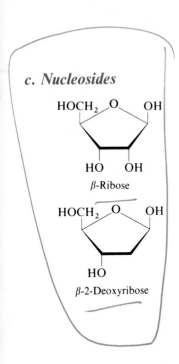

β-Ribose

β-2-Deoxyribose

A purine or a pyrimidine base may be envisioned to combine with a pentose molecule, in a condensation-type reaction, to form a compound known as a nucleoside. The pentose may be D-ribose or 2-deoxy-D-ribose; both sugars exist in the furanose form.

The bond joining the pentose to the nitrogen base is termed an *N-glycosyl linkage,* and it is always beta in naturally occurring nucleosides. The N-glycosyl linkage is formed between carbon-1' of the sugar and nitrogen-1 of the pyrimidine base or nitrogen-9 of the purine base.[1] A molecule of water is eliminated in the process. The following equations are given only to help the student visualize the joining of the sugar to the base; nucleosides are *not* synthesized in this fashion in the cell.

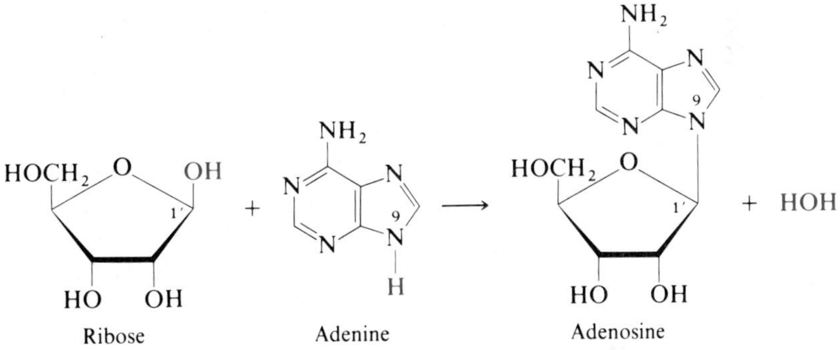

Ribose   +   Adenine ⟶ Adenosine + HOH

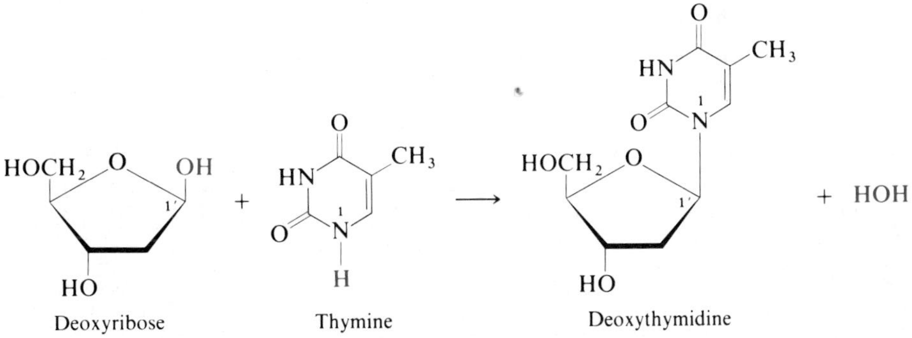

Deoxyribose + Thymine ⟶ Deoxythymidine + HOH

The common names of the ribonucleosides are derived from the names of the nitrogeneous bases. The suffix *-osine* denotes purine nucleosides, and the suffix *-idine*

---

[1] The convention used in numbering is that atoms of the pentose ring are designated by primed numbers, whereas atoms of the purine or pyrimidine ring are designated by unprimed numbers.

TABLE 15.1 The Major Pyrimidine and Purine Nucleosides

Structure	Name	Structure	Name
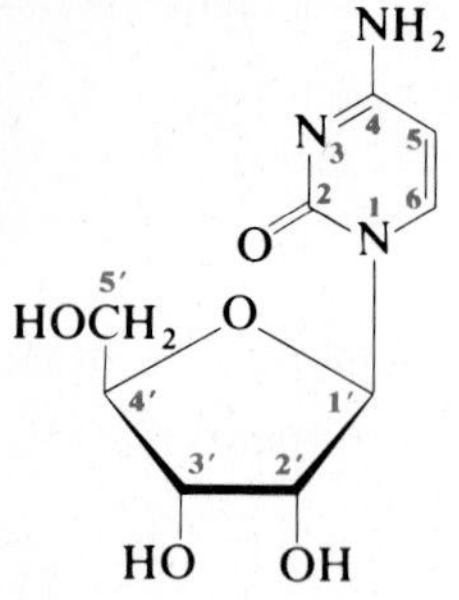	Cytidine	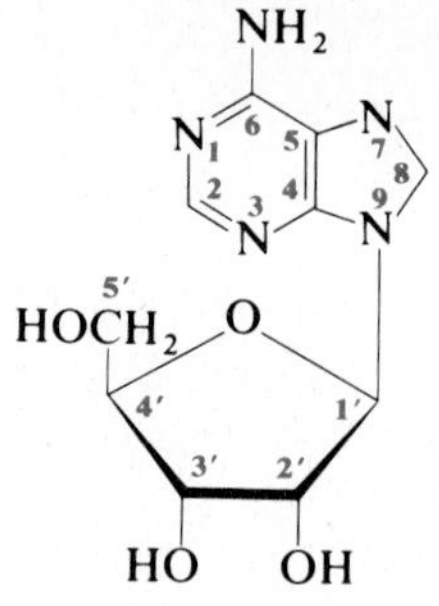	Adenosine
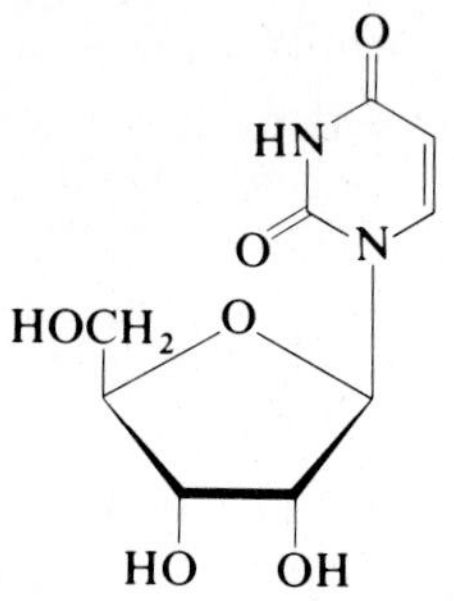	Uridine	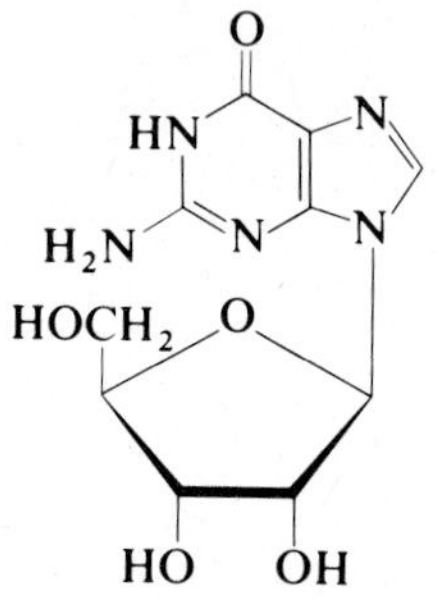	Guanosine
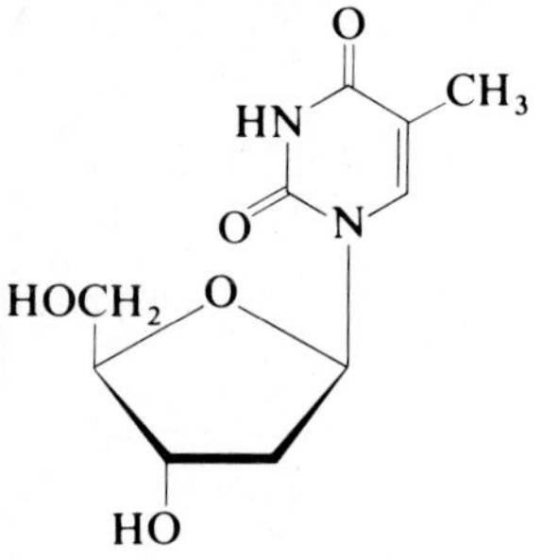	Deoxythymidine		

is used for pyrimidine nucleosides. The prefix *deoxy-* is utilized if the base is combined with deoxyribose (deoxynucleosides): deoxyadenosine, deoxyguanosine, deoxycytidine, and deoxythymidine. Structures and names of the major ribonucleosides and one of the deoxyribonucleosides are given in Table 15.1.

d. Nucleotides

The **nucleotides** are phosphate esters of the nucleosides and may be envisioned to result from the esterification of phosphoric acid with one of the free pentose hydroxyl groups. Note that again we are illustrating the combination of phosphate and a nucleoside. Nucleotides are not formed in this manner.

TABLE 15.2 The Pyrimidine and Purine Nucleotides

Structure	Names	Structure	Names
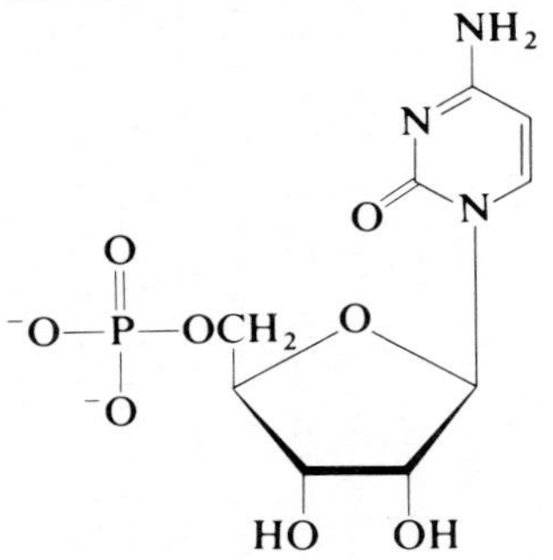	Cytidylic acid Cytidine 5′-monophosphate	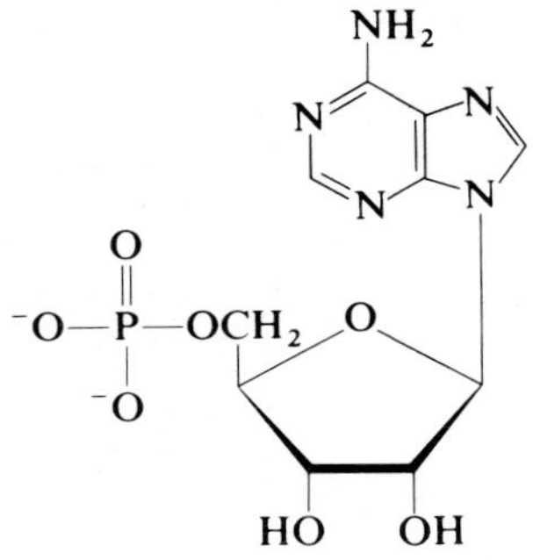	Adenylic acid Adenosine 5′-monophosphate
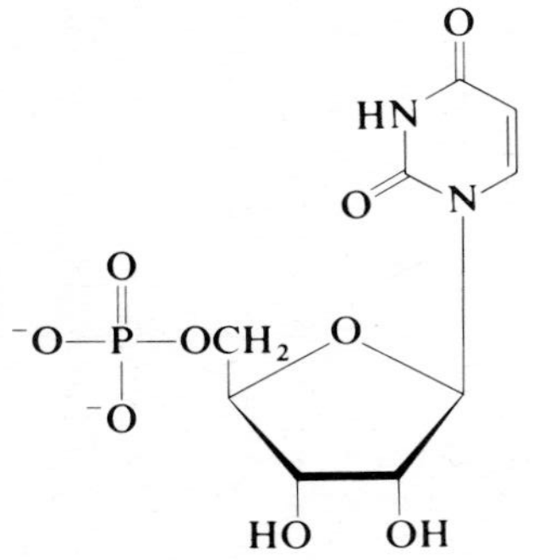	Uridylic acid Uridine 5′-monophosphate	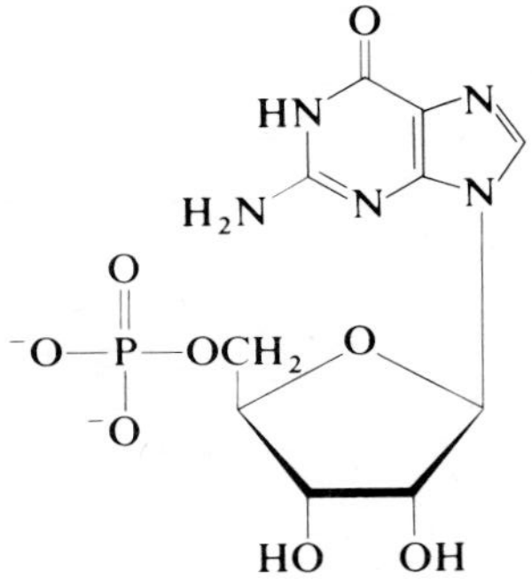	Guanylic acid Guanosine 5′-monophosphate
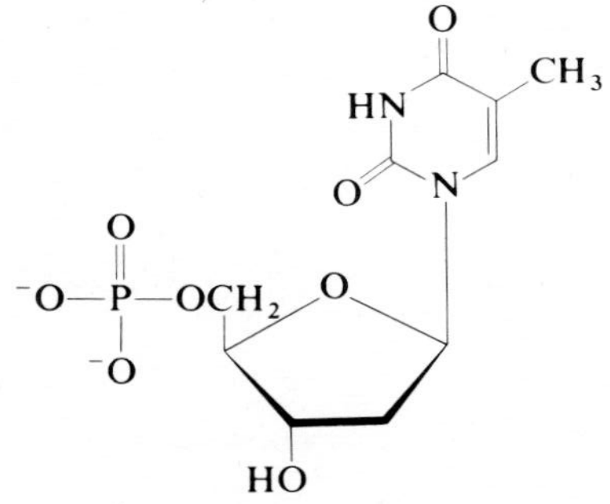	Deoxythymidylic acid Deoxythymidine 5′-monophosphate		

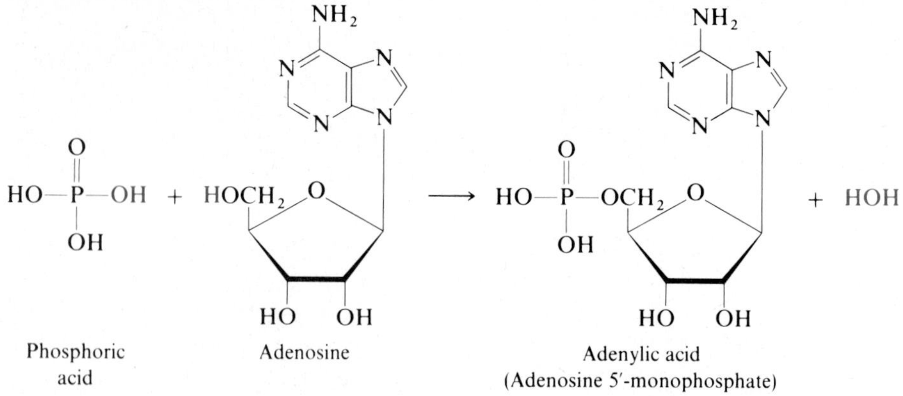

$$\underset{\substack{\text{Phosphoric}\\\text{acid}}}{\text{HO}-\overset{\displaystyle O}{\underset{\displaystyle OH}{\text{P}}}-\text{OH}} + \underset{\text{Adenosine}}{\text{HOCH}_2} \longrightarrow \underset{\substack{\text{Adenylic acid}\\\text{(Adenosine 5'-monophosphate)}}}{\text{HO}-\overset{\displaystyle O}{\underset{\displaystyle OH}{\text{P}}}-\text{OCH}_2} + \text{HOH}$$

The nucleotides are named as nucleoside monophosphates, such as cytidine 5'-monophosphate (CMP) and deoxycytidine 5'-monophosphate (dCMP), or as acids (because of the acidic phosphate group), such as cytidylic acid and deoxycytidylic acid. The names and structures of some nucleotides are given in Table 15.2.

Nucleotides also occur in the free form in all cells. We have already mentioned adenosine monophosphate (AMP) and its derivatives ADP and ATP (see Figures 8.3 and 8.4). In addition, a cyclic 3',5'-phosphate of adenosine occurs in which the phosphate group is bonded to two of the ribose carbons. The compound is adenosine 3',5'-monophosphate (cyclic AMP) and, as we shall see in Section 16.3, it plays a crucial role in metabolism. Certain other nucleotides are structural components of a number of important coenzymes. Refer to Table 14.2 and notice that FAD, NAD^+, and coenzyme A are all adenine nucleotides.

e. Nucleotide Polymers

Nucleic acids are polymers of the nucleotides in much the same way that proteins are polymers of amino acids. Both nucleic acids and proteins are irregular polymers since they are composed of several different types of basic repeating units (four distinct nucleotides in the case of the nucleic acids, and 20 distinct amino acids in the case of the proteins). Biosynthetically, nucleic acids arise from the polymerization of nucleoside triphosphates with the concomitant release of pyrophosphate. Polymerizing enzymes are essential to this process.

When the nucleoside triphosphates condense to form nucleic acids, bonds are formed between the 5'-phosphate of one nucleotide and the 3'-hydroxyl of another nucleotide. The phosphate group can be considered to be the connecting bridge between adjacent nucleotides. This gives rise to a polynucleotide with an alternating sugar–phosphate backbone having a 3',5'-phosphodiester linkage. Nucleic acids, like proteins, have two ends, one called the 3' end (free 3'-hydroxyl) and the other the 5' end (free 5'-hydroxyl). **Nucleic acids,** then, are *polymeric chains with a backbone of repeating sugar units connected by phosphate bridges* (Figure 15.2). Partial structures for DNA and RNA are shown in Figure 15.3.

GTP

ATP

Bond to next
nucleotide

$^-O-\overset{\overset{\displaystyle O}{\|}}{\underset{\underset{\displaystyle O^-}{|}}{P}}-O-\overset{\overset{\displaystyle O}{\|}}{\underset{\underset{\displaystyle O^-}{|}}{P}}-OH$

FIGURE 15.2 General chemical structure of nucleic acids.

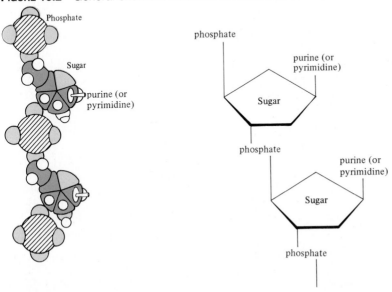

Phosphate

Sugar

purine (or
pyrimidine)

phosphate

purine (or
pyrimidine)

Sugar

phosphate

purine (or
pyrimidine)

Sugar

phosphate

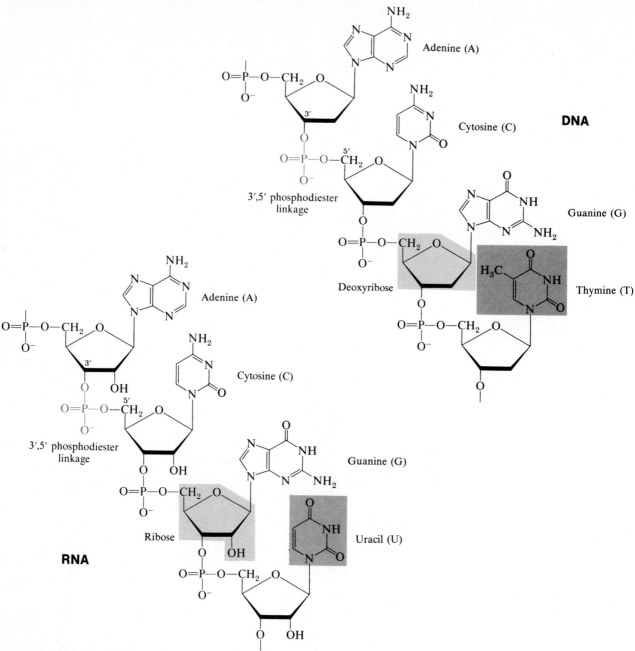

FIGURE 15.3 Partial chemical structures of the strands of DNA and RNA. The sequence of nucleotides differs for each naturally occurring type of DNA or RNA.

15.2 Deoxyribonucleic Acid (DNA)

Deoxyribonucleic acid (DNA) is a high molecular weight nucleic acid that occurs almost exclusively in the nucleus of the cell.[2] It is found in every living organism, although some viruses contain RNA instead (Section 15.7). During mitosis, DNA is apparent in the chromosomes,[3] those structures directly responsible for the transmission of the cellular genetic information. Molecular weight determinations on various types of DNA yield values between 3×10^6 and 3×10^9 daltons. (Recall that the molecular weights of most proteins lie between 10^3 and 10^6 daltons; therefore DNA molecules are several orders of magnitude larger than proteins.) It has been estimated that if all the DNA of a single human cell were stretched out end to end, it would extend more than 2 meters.

DNA is a double-stranded nucleic acid; within its structure two nucleic acid chains are intimately associated with one another by means of hydrogen bonding. The two strands of nucleic acid are oriented toward each other in such a way that the bases projecting from each are in proximity. The structures of the four bases permit hydrogen bonding only between specific base pairs (Figure 15.4). *The purine adenine*

FIGURE 15.4 Pairings of thymine and adenine (a) and of cytosine and guanine (b) by means of hydrogen bonding as in DNA.

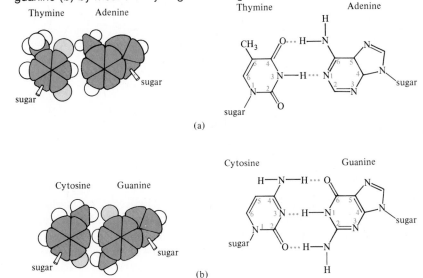

(a)

(b)

[2] Bacteria have no nuclei, and their DNA is packed tightly in a region of the cytosol. DNA is also found in the chloroplasts of green plants and in the mitochondria. For a discussion of the functions of extranuclear DNA, consult an advanced biochemistry text.

[3] Chromosomes are fibers consisting of complex structures of DNA and proteins. Human chromosomes are composed of about 25% DNA and 75% protein. As we shall see, the DNA contains the genetic information, whereas the attached proteins (mainly histones) are a major factor in the regulation of gene expression.

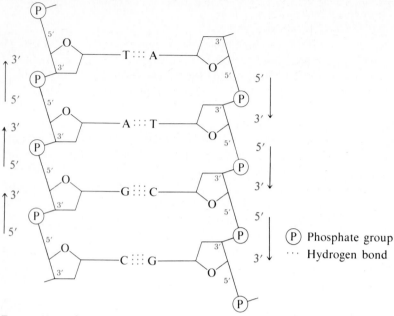

FIGURE 15.5 Secondary structure of a segment of a hypothetical DNA molecule. The two strands are complementary—**not** identical—and are held together through hydrogen bonding of complementary bases. The 3′,5′-phosphodiester linkages are oppositely oriented (antiparallel) in the two chains.

Legend:
(P) Phosphate group
··· Hydrogen bond

always pairs with the pyrimidine thymine, and guanine always pairs with cytosine. These specific base pairs are referred to as **complementary** bases.

This intramolecular hydrogen bonding accounts for the secondary structure of DNA and is the major force holding the two strands of the molecule together. Figure 15.5 shows a schematic representation of a portion of a DNA molecule depicting hydrogen bonding between the bases of the two strands. Owing to the great length of the nucleic acid molecule, there are vast numbers of hydrogen bonds formed between the two chains. It is the additive contribution of all these bonds that imparts great stability to the DNA molecule. Notice that in order for the complementary bases to be adjacent, the two strands of the double helix must be aligned *antiparallel;* that is, their 3′,5′-phosphodiester bridges are in opposite directions.

The tertiary structure of DNA is better appreciated when the student understands some of the experimental data that led to its elucidation. A brief part of that experimental history follows.

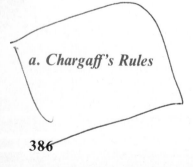

a. Chargaff's Rules

The work of E. Chargaff and others has revealed a number of striking regularities in the "normal" DNA content of a variety of different species. It was Chargaff's insightful discoveries that led J. D. Watson and F. H. C. Crick to postulate their

TABLE 15.3 The Composition of DNA from Various Sources

	Base Composition (%)				Base Ratios			Dissymmetry Ratio
	A	T	G	C	A/T	G/C	Pu/Py	(A + T)/(G + C)
Human	30.9	29.4	19.9	19.8	1.05	1.00	1.04	1.52
Calf	29.0	28.5	21.2	21.2	1.01	1.00	1.01	1.36
Sheep	29.3	28.3	21.4	21.0	1.03	1.02	1.03	1.36
Rat	28.6	28.4	21.4	21.5	1.01	1.00	1.00	1.33
Hen	28.8	29.2	20.5	21.5	1.02	0.95	0.97	1.38
Turtle	29.7	27.9	22.0	21.3	1.05	1.03	1.00	1.33
Salmon	29.7	29.1	20.8	20.4	1.02	1.02	1.02	1.43
Marine crab	47.3	47.3	2.7	2.7	1.00	1.00	1.00	17.52
Sea urchin	32.8	32.1	17.7	17.3	1.02	1.02	1.02	1.85
Yeast	31.3	32.9	18.7	17.1	0.95	1.09	1.00	1.79
Aspergillus niger (mold)	25.0	24.9	25.1	25.0	1.00	1.00	1.00	1.00
Escherichia coli	24.7	23.6	26.0	25.7	1.04	1.01	1.03	0.93
Staphylococcus aureus	30.8	29.2	21.0	19.0	1.05	1.11	1.07	1.50
Sarcina lutea	13.4	12.4	37.1	37.1	1.08	1.00	1.04	0.35

famous molecular model of DNA. Chargaff's findings, now known as **Chargaff's rules,** are summarized as follows.

1. The base composition of the DNA of an organism is constant throughout all the somatic cells of that organism and is characteristic for a given species. (Somatic cells include all body cells other than germ cells.)
2. The base composition of DNA varies greatly from one organism to another. This is clearly expressed by what is known as the *dissymmetry ratio,* (A + T)/(G + C). In other words, differing base compositions among organisms are reflected by variance in their dissymmetry ratios (Table 15.3).
3. The base composition in a given species does not change with age, nutritional state, or changes in the environment.
4. The amount of adenine in the DNA of a given organism is always equal to the amount of thymine (A = T).
5. The amount of guanine in the DNA of a given organism is always equal to the amount of cytosine (G = C).
6. Therefore, the total amount of purine bases in the DNA of a given organism is always equal to the total amount of pyrimidine bases (A + G = T + C).

b. Watson–Crick Model for DNA

Chargaff's findings clearly suggested that the purine and pyrimidine bases of DNA were in some way linked to one another, adenine always pairing with thymine, and guanine always pairing with cytosine. At about the same time that Chargaff and his colleagues published their findings (1950–53), other experimenters, M. Wilkins and R. Franklin, came forward with x-ray diffraction data on the lithium salt of DNA. Wilkins and Franklin found that within the crystal there is a repeat distance of 34 Ångström units (Å) and there are 10 subunits per turn. The repeat distance of the

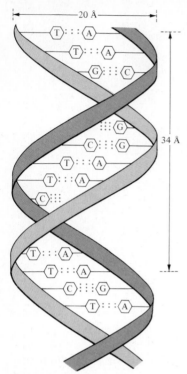

FIGURE 15.6 The Watson–Crick double-stranded helical model for DNA.

subunit is therefore 3.4 Å, which is the same order of magnitude that a single nucleotide might occupy in the chain. These data, coupled with Chargaff's findings, led to the Watson–Crick model for DNA (1953). Watson, Crick, and Wilkins received the Noble Prize in 1962.

According to the model, DNA consists of two right-handed polynucleotide chains that are complementary and coiled about the same axis so that they form a **double helix.** The bases project inward in such a way that a purine of one strand always pairs with a pyrimidine of the other. Only certain base pairs can be spatially accommodated (again, according to Wilkins' x-ray data), and these are the pairs A:::T and G:::C. The bases are associated with each other by means of hydrogen bonding. Such pairing rules demand that the base sequence of one chain be completely determined by the base sequence of the other. The two strands of the double helix are oriented with opposite (antiparallel) polarity. The negatively charged phosphoester backbone, therefore, projects outward from the helix, and the ionized phosphate groups are free to interact with the environment. (In particular, the DNA molecules form complexes with the histones—basic proteins whose protonated amino groups interact electrostatically with the negatively charged phosphate groups.) The helix has a diameter of about 20 Å and contains ten nucleotide pairs in each turn of the helix, which is 34 Å in length. It has been calculated that the total amount of DNA in a typical mammalian cell contains about 5.5×10^9 nucleotides. Figure 15.6 depicts the Watson–Crick model for DNA.

It is not possible to separate the two strands of the DNA molecule without destroying its helical configuration. However, since the strands are complementary and therefore not identical, it should be possible to separate them. This has, in fact, been done in DNA *melting* experiments (Figure 15.7). A solution of DNA is heated to a temperature sufficient to cleave the hydrogen bonds that hold the two strands

FIGURE 15.7 A schematic diagram showing the melting of DNA (a) and the effects of rapid (b) or slow (c) cooling.

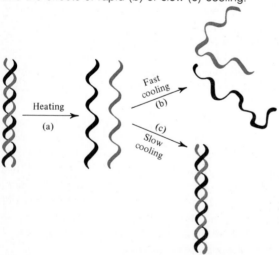

together. (Cleavage and strand separation are indicated by a sharp change in the viscosity and optical density of the solution.) The solution is then cooled rapidly so that the complementary strands do not have the opportunity to realign themselves in their former configuration. The strands may then be separated by any number of physical separation techniques (chromatography, electrophoresis, density gradient centrifugation).

15.3 *DNA as the Hereditary Material*

At the time that Watson and Crick postulated the structure of DNA, biologists throughout the world were beginning to consider the possibility that the nucleic acids were the substances that carried hereditary information. Formerly, scientific speculation favored the proteins because it seemed that more information could be coded in the more diverse protein molecule. However, proteins were known to have a high turnover rate; that is, their metabolic synthesis and breakdown occurred at a high rate. The genetic material, it seemed, should have a relatively high degree of metabolic stability because of its responsibility for transmitting hereditary information from one generation to the next. DNA has the required metabolic stability. In addition, its structure suggests the possibility of **template replication,** a phenomenon that allows for the biosynthesis of two identical daughter molecules from the information contained in one parent molecule. Watson and Crick postulated the current theory of DNA replication according to their own model for the structure of the molecule.

a. DNA Replication—A Semiconservative Mechanism

If the sequence of bases along the DNA strand somehow contains the hereditary information, the biosynthesis of new molecules of DNA must proceed in some fashion that preserves this sequence of bases and hence conserves the information for distribution to progeny cells. Watson and Crick proposed a semiconservative mode of replication for the molecule. They speculated that during DNA replication (which probably occurs during the interphase of mitosis), the two strands uncoil, and each strand acts as a **template** for the formation of a new complementary strand (Figure 15.8). In other words, the deoxynucleoside triphosphates that exist in the free state in the nucleus are attracted (according to the rules of base pairing) to the single-stranded, uncoiling, DNA. In this way, two new strands are formed that are *complementary* to the two parental strands. As the new daughter molecules begin to be formed, they assume the helical configuration of the parent molecule (Figure 15.9). It is because each molecule consists of one old parental strand and one newly synthesized strand that the mechanism is termed *semiconservative.* Thus, sometime prior to nuclear division, the hereditary material is duplicated and its information content is preserved for distribution to progeny cells.

The Watson–Crick hypothesis conformed to the specification that the base composition of the DNA of an organism is constant throughout the cells of the organism. It provided a simple, logical, and likely means by which the hereditary substance could be precisely replicated. It was a satisfactory hypothesis that needed only experimental verification for widespread acceptance.

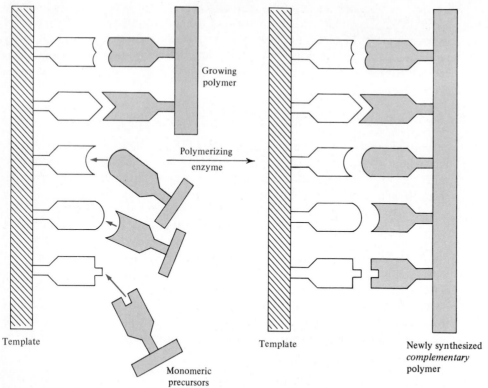

Figure 15.8 A schematic view of the formation of a complementary polymer upon a template surface.

Figure 15.9 A schematic diagram of DNA replication. Replication is assumed to occur by sequential "unzipping" of the double helix. The new nucleotides are positioned (by an enzyme) and phosphate bridges are formed, thus restoring the original double-helical configuration. Each newly formed double helix consists of one old strand and one new strand.

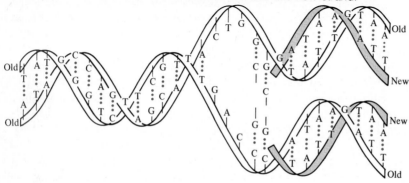

b. DNA Synthesis in Vitro

The work of Meselson and Stahl (1957–58) offered some rather conclusive proof that DNA does indeed replicate by the Watson–Crick semiconservative mechanism. (See an advanced biochemistry text). Further evidence for template replication of DNA was provided by the work of A. Kornberg (Nobel Prize, 1959). He succeeded in isolating an enzyme from the bacterium *Escherichia coli,* which he called *DNA polymerase.* This enzyme was capable of producing a DNA-like molecule having chemical and physical properties almost identical to those of native DNA. Kornberg's DNA replication system included (1) all four deoxynucleoside triphosphates, (2) native DNA (in small amounts), which served both as a template and as a primer, and (3) the *DNA polymerase* enzyme. The DNA that was formed by the system had a base composition nearly identical to the base composition of the native DNA primer, a fact lending support to the theory of template synthesis. In addition, Kornberg found that when the DNA primer was denatured so as to separate the two strands, synthesis proceeded much more efficiently. The equation that describes the system is illustrated.

$$
\begin{array}{c}
n_1\ \text{dATP} \\
+ \\
n_1\ \text{dTTP} \\
+ \\
n_2\ \text{dGTP} \\
+ \\
n_2\ \text{dCTP}
\end{array}
\quad\xrightarrow[\substack{\text{DNA polymerase,}\\ \text{Mg}^{2+}}]{\text{DNA template;}}\quad
\text{DNA}
\begin{bmatrix}
\text{dAMP} \\
\text{dTMP} \\
\text{dGMP} \\
\text{dCMP}
\end{bmatrix}_{2n_1+2n_2}
+\ 2(n_1 + n_2)\ \text{PP}_i
$$

As molecular studies by Kornberg and others continued, it became apparent that the previous views were too simplistic (see footnote 9, page 403). Kornberg's enzyme is now called *DNA polymerase I* because two other *DNA polymerases (II and III)* have since been isolated. It now appears that *DNA polymerase III* is the major enzyme concerned in the replication process. *DNA polymerase I* is also believed to participate in DNA replication, but its main role appears to be in the repair of DNA. (The biological function of *DNA polymerase II* is as yet unknown.)

The *polymerase* enzyme joins nucleotides together *in one direction only,* linking the 5'-nucleoside triphosphates to the 3'-hydroxyl end of a growing polynucleotide chain. Since the two strands of DNA have opposite polarity, and since it is known that the two daughter molecules are synthesized side by side, we are left with the dilemma of postulating a mechanism whereby the strand having the terminal 5'-phosphate is replicated. That is, the antiparallel orientation of complementary DNA strands demands that one daughter strand elongate in the 5' → 3' direction and the other in the 3' → 5' direction. Synthesis in the latter direction would require a polymerizing enzyme with a quite different specificity from the one that operates in the 5' → 3' direction.

One theory about this replication came from the laboratory of R. Okazaki. He found that a large proportion of newly synthesized DNA exists as discontinuous small fragments. These fragments (called *Okazaki fragments*) are about 1000 nucleotides long in prokaryotes and about 100–200 nucleotides long in eukaryotes.[4] Okazaki postulated that only one of the two new strands is synthesized continuously. This strand is called the *leading strand*. The other strand is synthesized discontinuously from fragments that are copied in the $5' \rightarrow 3'$ direction starting at the replication fork (the region of DNA that is unwinding). The developing fragments are covalently joined by the enzyme *DNA ligase*[5] to form a continuous second strand, called the *lagging strand*. Thus, both the leading strand and the lagging strand are synthesized in the $5' \rightarrow 3'$ direction (Figure 15.10).

15.4 *Ribonucleic Acid (RNA)*

Ribonucleic acid (RNA) is a single-stranded nucleic acid having three important structural features that distinguish it from DNA, two of which have already been alluded to.

1. Ribonucleic acid contains the pentose ribose, in contrast to the 2-deoxyribose of DNA.
2. RNA contains the base uracil rather than thymine.[6]
3. Hydrolytic analysis of RNA indicates that its composition does not obey Chargaff's rules. In other words, the purine/pyrimidine ratio in RNA is not 1:1 as it is in DNA.

The absence of systematic hydrogen bonding in RNA results in an irregular and relatively unpredictable structure for the molecule. X-ray diffraction data, however, indicate that portions of the molecule have a helical double-stranded structure. This is probably due to the presence of loops in the molecule formed when part of a strand folds back upon itself. Such regions would be characterized by A–U and G–C pairs, but they must have a limited occurrence since the purine/pyrimidine ratio usually does not approximate 1:1.

All of the cellular RNAs appear to be synthesized from a part of the DNA molecule by a template mechanism that is analogous to DNA replication in many respects. To initiate RNA biosynthesis, the two strands of the DNA molecule begin to uncoil. This occurs at specific sites called *promoters* on the DNA template. The nucleoside triphosphates are attracted to the uncoiling region of the DNA molecule according to the rules of base pairing. They are then enzymatically polymerized; a

---

[4]DNA replication in prokaryotes (recall Figure 12.5) is very rapid; in *E. coli* it occurs at a rate of about 1700 base pairs per second. In eukaryotes, such as animal cells, DNA replication is much slower (about 50 base pairs per second).

[5]*DNA ligases* serve not only to join short lengths of DNA but also to repair broken strands of DNA (see Section 15.6). The intact strand serves as the template for the repair of the damaged strand. *DNA ligases* also are involved in the exchange of sections of DNA fragments that occurs during crossing-over, and scientists make use of these enzymes in recombinant DNA experiments (see Section 15.8).

[6]This latter distinction is not as absolute as once thought, since it is now known that certain transfer RNAs also contain thymine. When thymine is bound to ribose, the nucleoside is called *ribothymidine,* and the nucleotide is *ribothymidylic acid* (see Figure 15.12). Also, there is a bacterial virus (a bacteriophage) that contains deoxyuridine instead of deoxythymidine in its DNA.

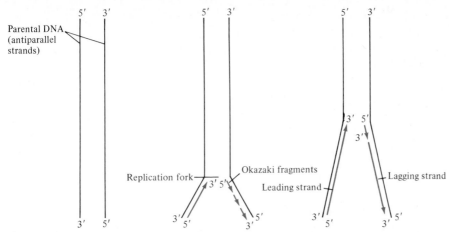

FIGURE 15.10 Both strands of DNA are synthesized in the 5′ → 3′ direction. The leading strand is synthesized continuously, whereas the lagging strand is synthesized in short fragments that are joined by *DNA ligase*.

molecule of RNA is formed, with the concurrent elimination of pyrophosphate. The enzyme that catalyzes the reaction is known as *RNA polymerase*. It can initiate RNA synthesis in the middle of a DNA strand; chain growth proceeds in the 5′ → 3′ direction. The equation for the reaction is illustrated.

$$
\begin{matrix}
\text{ATP} \\
+ \\
\text{UTP} \\
+ \\
\text{GTP} \\
+ \\
\text{CTP}
\end{matrix}
\xrightarrow[\substack{\text{RNA polymerase,} \\ \text{Mg}^{2+}}]{\text{DNA template;}}
\text{RNA}
\begin{bmatrix}
\text{AMP} \\
\text{UMP} \\
\text{GMP} \\
\text{CMP}
\end{bmatrix}
+ \; \text{PP}_i
$$

Because the RNA is a complementary copy of the information contained in the DNA, *the process of RNA biosynthesis* is referred to as **transcription.** We see later (Section 15.5-a) why this process is vital to all growth and development. Figure 15.11 contains a schematic representation of the process.

DNA transcription differs from DNA replication in several ways. In a given DNA region only one of the DNA strands serves as a template and is copied, but in another region of the DNA, the other strand may serve as a template for the biosynthesis of a different RNA. Thus, RNA molecules are much shorter than DNA molecules. Also, RNA molecules do not remain hydrogen bonded to DNA for any length of time. As soon as transcription is completed, the RNA is released and the DNA helix reforms.

Three basic types of cellular RNA are known to exist; the distinctions among them are made primarily on the basis of biochemical function. They do, however, differ also in molecular weight and in secondary structure (Table 15.4).

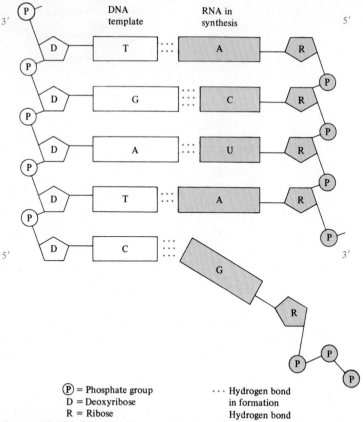

P = Phosphate group
D = Deoxyribose
R = Ribose
··· Hydrogen bond in formation
—— Hydrogen bond

FIGURE 15.11 Transcription–synthesis of RNA upon a DNA template. One strand of the DNA double helix is used as a template for alignment of RNA bases. Once a nucleotide is in position, a phosphate bridge is formed, resulting in a single strand of RNA.

TABLE 15.4 RNA Molecules in *E. coli*

RNA Type	Relative Amount (%)	Average MW (daltons)	Approx. Number of Nucleotides
Messenger RNA (mRNA)	5	4×10^5	1200
Transfer RNA (tRNA)	15	3×10^4	70–90
Ribosomal RNA (rRNA)	80	6×10^5	1800

a. Messenger RNA (mRNA)

Messenger RNA (mRNA) comprises only a few percent of the total amount of RNA within the cell. It has been shown to be complementary to a given segment of the DNA of the organism from which it is isolated. A molecule of mRNA exists for a relatively short time; like proteins, it is continuously being degraded and resynthesized. The rate of mRNA degradation is different from species to species and also from one type of cell to another. In bacteria, one half of the total mRNA is degraded every 2 minutes, whereas in rat liver, the half-life is several days.

The molecular dimensions of the mRNA molecule vary according to the amount of genetic information that the molecule is meant to encode. It is known, however, that there is very little intramolecular hydrogen bonding in this type of RNA and that the molecule exists in a fairly random coil. After transcription, the mRNA passes into the cytosol,[7] carrying the genetic message from DNA to the ribosomes (Section 15.4-c), the sites of protein synthesis. In Section 15.5-a we shall see how mRNA directly governs that synthesis.

b. Transfer RNA (tRNA)

Transfer RNA (tRNA) is a relatively low molecular weight nucleic acid, soluble in solvents commonly used to isolate the higher molecular weight nucleic acids. It functions by attaching itself (with the aid of a specific enzyme—see Section 15.5-a) to a particular amino acid and carrying that amino acid to the site of protein synthesis at the precise moment specified by the genetic code. Each of the 20 amino acids found in proteins has at least one corresponding tRNA, and most amino acids have multiple tRNA molecules (Section 15.5-b). For example, there are two different tRNAs specific for the transfer of lysine, three for isoleucine, four for glycine, and six for serine. The existence of several tRNAs for the same amino acid is termed multiplicity.

A characteristic feature of tRNA molecules is that they contain a rather high percentage ($\sim 10\%$) of modified bases in addition to the common bases A, G, U, and C. Over 40 different modified bases have been discovered in tRNAs, most of which are methylated forms of the common bases. (The bases are modified after the tRNA chain has been synthesized.) Because of this feature, some regions of the cloverleaf diagram have been named for the modified bases that occur in them (Figure 15.12). For example, the T loop is so named because it includes thymine (ribothymidylic acid, T), and the D loop usually includes dihydrouracil (UH_2). Other regions of the cloverleaf are the variable loop, which in different tRNAs has different numbers of nucleotides (ranging from 4 to 21), and the acceptor stem, which accepts the amino acid specific to that particular tRNA. All tRNA molecules have the same terminal sequence —C—C—A at the 3'-end of the acceptor stem. (These bases are added to the end of the tRNA molecule after it has been transcribed.) In addition, the anticodon loop contains a unique sequence of three nucleotides that is different in the tRNAs for different amino acids. This triplet is called the **anticodon** (see Section 15.5-a). Finally notice that the stem regions contain intrachain hydrogen bonding between base pairs. The double-helical configuration occurs by looping of the single-stranded tRNA chain.

[7] Some people use the terms *cytoplasm* and *cytosol* interchangeably. **Cytoplasm** is a term used to refer to *all of the contents of the cell with the exception of the nucleus* (which occupies about 6% of the cell volume). The **cytosol** *consists of the fluid outside the membrane-bound organelles* (see Figure 12.5-c) and generally represents about 55% of the total volume in eukaryotic cells.

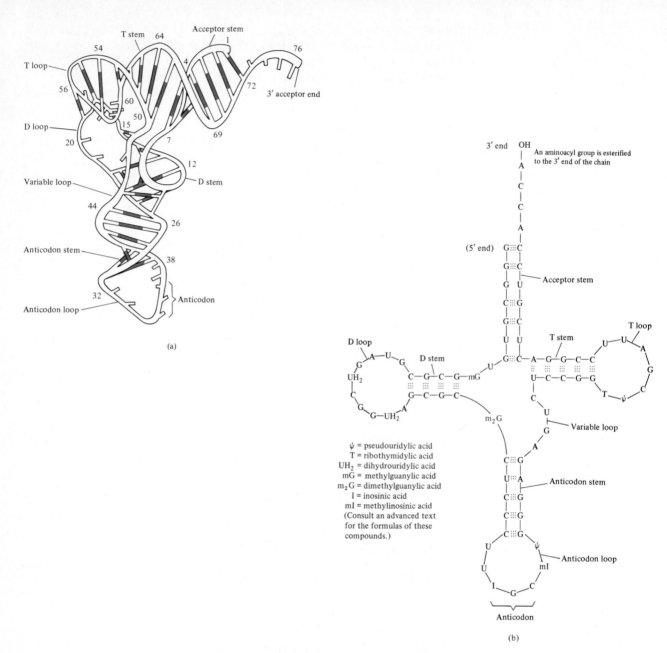

FIGURE 15.12 (a) Folding pattern of the polynucleotide chain in yeast phenylalanine tRNA. The sugar–phosphate backbone of the molecule is represented as a coiled tube with the cross-rungs standing for the nucleotide base pairs in the stem regions. The short rungs indicate bases that are not involved in base-base hydrogen bonding. (b) Cloverleaf diagram of the base sequence in a yeast alanine tRNA molecule. The molecule is a single chain, but in folding back on itself hydrogen bonds are formed and large regions of the molecule are characterized by base pairing. [(a) from S. Kim et al., *Science*, **185**, 436 (1974).]

In 1965 Robert Holley (Nobel Prize, 1968) and his co-workers at Cornell University deduced the entire base sequence of an alanine tRNA isolated from bakers' yeast. The base sequences of over 100 other tRNA molecules have since been elucidated. Although these tRNAs have quite different base sequences, they all can be written in the same cloverleaf pattern. Moreover, they all have a constant number of base pairs in the stem regions: seven in the acceptor stem, five in the T stem, five in the anticodon stem, and three or four in the D stem. These features are maintained in tRNA molecules from plants, animals, bacteria, and viruses. In the discussion of

polypeptide synthesis (Section 15.5-a), we shall use the figure to represent the tRNA molecule.

c. Ribosomal RNA (rRNA)

Ribosomal RNA (rRNA) makes up 80% of the total cellular complement of ribonucleic acid. The **ribosome** is *a cellular substructure that serves as the site for protein synthesis.* Its composition is about 65% rRNA and 35% protein. The ribonucleic acids and the proteins are bonded together by a large number of noncovalent forces such as hydrogen bonds and hydrophobic interactions. Structurally, a ribosome is composed of two spherical particles of unequal size. The smaller of them has a distinct affinity for mRNA; the larger has an attraction for tRNA. In terms of cellular structure, ribosomes are extremely small particles visible only with the aid of an electron microscope. More often than not, they are seen as clusters known as *polyribosomes,* or *polysomes,* bound to the endoplasmic reticulum of animal and plant cells or to the cell membrane of microorganisms. When ribosomes occur in such aggregates, they are held together by strands of mRNA. On the average, five to eight ribosomes are simultaneously synthesizing the same polypeptide from the information in one mRNA strand (large proteins require long strands of mRNA, and as many as 100 individual ribosomes may be attached). The time required for the synthesis of an average size polypeptide (~ 300 amino acids) is about 15 seconds in a bacterial cell and 2 or 3 minutes in a mammalian cell.

In Section 14.9-b we said that many antibiotics exert their effects by interfering with protein synthesis in microorganisms. For example, actinomycin D binds to DNA and prevents transcription by blocking the movement of *RNA polymerase.* Rifamycin interferes with transcription by inhibiting *RNA polymerase.* See also comments on Figure 15.14.

Ribosomal RNA serves as a structural component of the ribosomes, and its 3'-end binds to the 5'-end of mRNA; other biological roles of rRNA are as yet unclear. Students of biochemistry often tend to take the chemical reactions that they study *out* of the cell, thereby taking them out of their proper frame of reference. The cellular substructures and organelles at which the various biochemical processes occur are often actively involved in these same processes. Such is definitely the case with the ribosomes and protein synthesis. We shall see that it is advantageous to regard the ribosome as a type of biological surface catalyst—an enzyme—rather than as some mysterious nonchemical cellular component (or passive workbench). A great deal remains to be discovered about the functions of ribosomal proteins and of rRNA.

15.5 The Genetic Code

Throughout the preceding sections, the role of the nucleic acids in *coding* information has been stressed. We have mentioned that the sequence of bases in the DNA molecule serves to direct protein synthesis. Now let us examine the nature of the code embodied in that sequence.

It is known that the amino acid sequence of any particular cellular protein is vital to the proper biochemical functioning of that protein in the overall scheme of cellular metabolism. For example, the conjugated protein hemoglobin contains some 146 amino acids in each of its two identical β-polypeptide chains. Recall (Section 13.3) that the substitution of just one incorrect amino acid in 146 can result in the formation of nonfunctional hemoglobin molecules, as evidenced by the disease sickle cell anemia. Furthermore, a variety of other diseases (galactosemia, phenylketonuria—see Section 15.6) appear to be related to the synthesis of nonfunctional, defective protein molecules, primarily enzymes. In many cases, there is just one incorrect amino acid in the molecule.

From such data, we may conclude that the primary structure, or amino acid sequence, of the protein molecule is of vital importance to life and that the cell must contain an intricate set of instructions wherein the proper sequence of amino acids for all its proteins is genetically contained. We now know that the code involves the sequence of bases in the DNA. *The segment of a DNA molecule that codes for the biosynthesis of one complete polypeptide chain* is called a **gene.** (If a given protein contains two or more polypeptide chains, each chain is coded by a different gene.) Each gene in a section of DNA contains about 1000–2000 nucleotides.[8] A human cell contains about 100,000–500,000 genes (although there is sufficient DNA to code for 10 times this number of genes).

How can a molecule with just four different monomeric units specify the sequence of the 20 different amino acids that occur in proteins? If each different nucleotide coded for one different amino acid, then obviously the nucleic acids could code for only 4 of the 20 amino acids. Suppose we consider the nucleotides in groups of two. There are 4^2 or 16 different combinations of pairs of the four distinct nucleotides. Such a code is more extensive, but still inadequate. If, however, the nucleotides are considered in groups of three, there are 4^3 or *64 different combinations.* Here we have a code that is extensive enough to govern the primary structure of the protein molecule because it contains more than enough coding units to designate all 20 amino acids. Now we shall see how this code directs protein synthesis.

a. Building a Polypeptide

If the sequence of bases along the DNA strand determines the sequence of amino acids along the polypeptide chain, then the information contained in the DNA must be conveyed from the nucleus to the site of protein synthesis. This is accomplished by the orderly interactions of the nucleic acids with over 100 different enzymes. Recall that mRNA is made from a DNA template and so contains a base sequence

[8] It has been estimated that less than 5% of the DNA of higher organisms conveys the information for the synthesis of proteins. The remainder is involved in the regulation of protein synthesis, in the synthesis of tRNAs and rRNAs, and in maintaining the correct conformation of the DNA. See also footnote 9.

that is complementary to that of the DNA upon whose surface it was synthesized. Once it is formed, the mRNA is transported across the nuclear membrane into the cytosol (and hence to the ribosomes), carrying with it the genetic instructions. *Each group of three bases along the mRNA strand now specifies a particular amino acid, and the sequence of these triplet groups dictates the sequence of the amino acids in the protein.* Because the code involves three bases per coding unit, it is referred to as a **triplet code.** *The coding unit* is called a **codon.**

Now the cell faces the problem of lining up the amino acids according to the sequence called for by the mRNA and of joining them together by means of peptide linkages. Because this process involves *the transfer of the information encoded in the mRNA to the ultimate structure of the protein molecule,* it is often referred to as **translation.**

Before the amino acids may be incorporated into a polypeptide chain they must first be activated. Activation occurs prior to the reaction of the amino acid with its particular tRNA carrier molecule. The amino acid combines with a molecule of ATP, yielding an intermediate compound known as *aminoacyl adenylate* (Figure 15.13). The carboxyl group of the amino acid is bonded to the phosphoryl group of AMP. Each of the 20 amino acids has its own specific activation enzyme, called *aminoacyl–tRNA synthetase,* that is capable of recognizing each amino acid and its corresponding tRNA.

The aminoacyl adenylate remains on the surface of the enzyme and then undergoes reaction with the proper tRNA molecule to form the corresponding aminoacyl–tRNA complex. *Both the enzymes (aminoacyl–tRNA synthetases) and the tRNAs are each highly specific for a particular amino acid.* The high degree of

FIGURE 15.13 The joining of an amino acid to its particular tRNA requires a two-step process that occurs on the surface of the enzyme. (1) The amino acid reacts with ATP to form an activated amino acid. (2) This activated amino acid then can bond to a specific tRNA molecule.

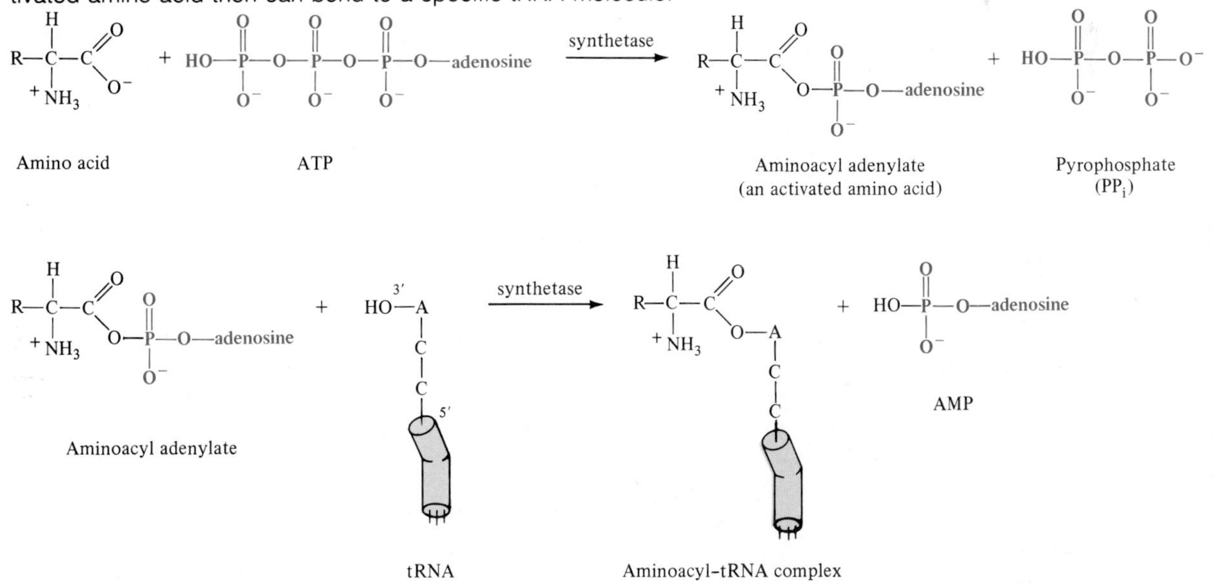

FIGURE 15.14 The elongation steps in protein synthesis.

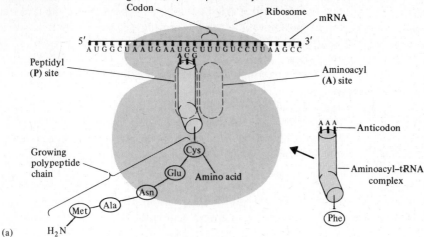

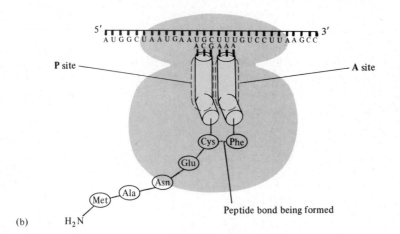

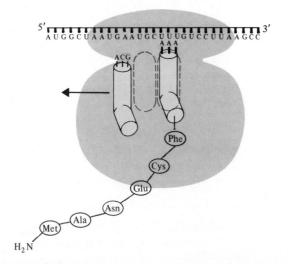

(a) Protein synthesis is already in progress at the ribosome. The growing polypeptide chain is bound to the peptidyl (**P**) site. At this point the aminoacyl (**A**) site is vacant. The codon UUU is lined up above the **A** site. An activated tRNA molecule whose anticodon is AAA approaches the ribosome. (The antibiotic tetracycline blocks the binding of the aminoacyl–tRNA to the **A** site of the ribosome.)

(b) The activated tRNA molecule has become bound to the ribosome at the **A** site. It is also bound to the mRNA molecule by means of base pairing between codon and anticodon. Amino acid Phe is about to be incorporated into the polypeptide chain. The peptide linkage will be formed between the carboxyl group of amino acid Cys and the amino group of amino acid Phe. This is catalyzed by the enzyme *peptidyl transferase*, which is a component of the ribosome. (Chloramphenicol blocks the peptidyl transferase reaction.)

(c) The peptide linkage has been formed, and the growing polypeptide chain is now attached to the **A** site. The tRNA molecule has dissociated from the **P** site and is about to move into the cytosol to pick up another amino acid.

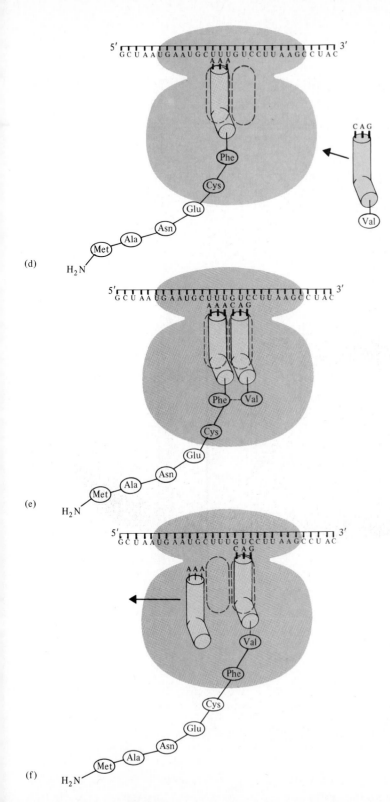

(d) The ribosome moves (**translocates**) from left to right along the mRNA strand. (The energy for translocation is supplied by GTP.) The polypeptide chain, along with the tRNA molecule to which it is bound, is simultaneously shifted from the **A** site to the **P** site. This brings the next codon, GUC, into line over the **A** site. Notice that an activated tRNA molecule (containing the next amino acid to be incorporated into the chain) is moving into position on the surface of the ribosome. Its anticodon is CAG. (Erythromycin blocks the translocation reaction.)

(e) The activated tRNA molecule carrying amino acid Val is now in place on the ribosome. The peptide linkage between the carboxyl group of Phe and the amino group of Val is about to be made.

(f) The peptide linkage has been formed, and the growing chain is attached, through a tRNA molecule, to the **A** site. The polypeptide chain is now seven amino acid units in length. The ribosome will translocate again, and the tRNA–polypeptide complex will be in position at the **P** site. This process will continue until the polypeptide chain is completed (i.e., when one of the three termination codons appears at the **A** site). When the ribosome reaches the end of the message, both it and the polypeptide are released from the mRNA molecule.

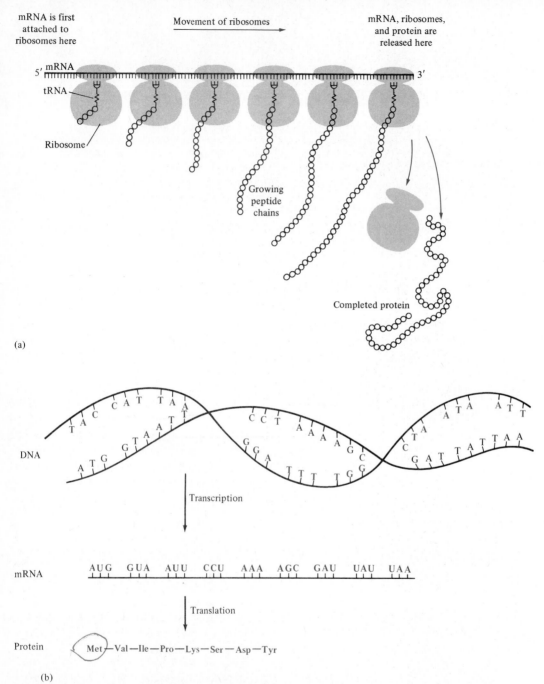

(a)

(b)

FIGURE 15.15 (a) Diagrammatic representation of a polysome. Several ribosomes are attached to mRNA. The ribosomes have progressed varying distances along the mRNA during translation, and each one is associated with a progressively longer protein chain. At the end of the message the mRNA and ribosome separate and the complete protein is released. (b) The relationship between transcription and translation.

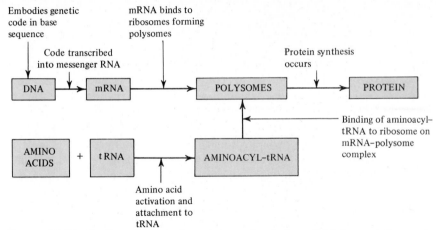

FIGURE 15.16 Outline of events in protein synthesis from DNA transcription and amino acid activation to completed protein.

specificity of the *synthetase* enzymes is absolutely vital to the correct incorporation of amino acids into proteins. Binding of the amino acid to its tRNA occurs at the 3'-position of the terminal adenosine nucleoside. Now that the amino acid molecules have been activated and have undergone reaction with the tRNA carriers, protein synthesis may take place.

Figure 15.14 depicts a schematic stepwise representation of this all-important process. When a certain portion of the mRNA strand has been *read* by a given ribosome, another ribosome may attach itself to the same strand and begin to read it simultaneously. Thus in cells active in protein synthesis we find clusters of ribosomes connected by a single strand of mRNA (Figure 15.15). The amount of any particular protein in a cell depends on the balance between the rate at which it is synthesized (which is largely controlled by the rate at which its mRNA is synthesized in the nucleus) and the rate at which it is degraded. These events are summarized in Figure 15.16.[9]

[9]It is absolutely essential at this point to note that the preceding discussions of DNA replication, RNA synthesis, and protein synthesis are correct, but they are an oversimplification based on the level of this text. Several molecular components (including RNA primers) are required to initiate DNA replication. Additional proteins are required for the elongation and the termination processes. Furthermore, ultraviolet radiation and certain chemicals are known to damage DNA by disrupting the sugar–phosphate backbone and/or by altering the purine and pyrimidine bases. DNA repair enzymes exist that can mend the backbone and correct the base sequence.

Also, the majority of eukaryotic genes occur in pieces, spread out along the DNA. Parts of the gene are expressed (*exons*); the intervening regions (*introns*) are not expressed. Therefore, the RNAs transcribed from these genes have been found to be spliced. That is, the entire length of DNA is copied in a longer transcript that is cut once or several times and hooked back together to give a shorter, functional messenger. In addition, protein synthesis is critically controlled at the level of transcription. We now know that there are structural genes, operator genes, promoter genes, regulatory genes, repressor molecules (which are proteins), plus an involvement of GTP and cyclic AMP. Postribosomal modification of proteins occurs after synthesis on the mRNA–ribosome complex, and involves many types of modifications (e.g., methylation or hydroxylation of amino acid side chains, breakage of peptide bonds to activate a zymogen or a hormone, etc.). The student is directed to an advanced biochemistry text for detailed explanation of the fascinating studies in molecular genetics.

b. Cracking the Code

As we have stated, the sequence of bases on the mRNA directs the precise sequence of the amino acids for each protein. We have indicated that the condon, the unit that codes for a particular amino acid, consists of groups of three adjacent nucleotides on the mRNA. There are 64 possible triplet codons. Early experimenters were faced with the task of determining which codon (or perhaps codons) stood for each of the 20 amino acids. The cracking of the genetic code was the joint accomplishment of several well-known geneticists, notably H. Khorana, M. Nirenberg, P. Leder, and S. Ochoa (1961–64). A genetic dictionary has been compiled and is given in Table 15.5. Of the 64 possible codons, 61 code for amino acids and 3 serve as signals for the termination of polypeptide synthesis (that is, as periods at the end of a sentence). Notice, in the table, that only methionine (AUG) and tryptophan (UGG) have a single codon. All other amino acids have multiple codons.

Further experimentation by Nirenberg has thrown much light on the nature of the genetic code. It now appears that

1. The code is essentially universal—animal, plant, and bacterial cells use the same codons to specify each amino acid. (A few exceptions have recently been discovered.)
2. The code is degenerate—in all but two cases (methionine and tryptophan) more than one triplet codes for a given amino acid.
3. The first two bases of each codon are most significant; the third base often varies. This suggests that a change in the third base by a mutation may still

TABLE 15.5 The Genetic Code

First Base		Second Base				Third Base
	U	C	A	G	
U	UUU UUC } Phe UUA UUG } Leu	UCU UCC UCA UCG } Ser	UAU UAC } Tyr UAA Termination UAG Termination	UGU UGC } Cys UGA Termination UGG Trp	U C A G
C	CUU CUC CUA CUG } Leu	CCU CCC CCA CCG } Pro	CAU CAC } His CAA CAG } Gln	CGU CGC CGA CGG } Arg	U C A G
A	AUU AUC AUA } Ile AUG Met	ACU ACC ACA ACG } Thr	AAU AAC } Asn AAA AAG } Lys	AGU AGC } Ser AGA AGG } Arg	U C A G
G	GUU GUC GUA GUG } Val	GCU GCC GCA GCG } Ala	GAU GAC } Asp GAA GAG } Glu	GGU GGC GGA GGG } Gly	U C A G

permit the correct incorporation of a given amino acid into a protein (see Section 15.6).

4. In general, those codons with C or U as the second base tend to specify the nonpolar amino acids, whereas codons with A or G as the second base specify the polar amino acids (see Table 13.1).

5. The code is continuous and nonoverlapping—there are no special signals and adjacent codons do not overlap (except in the case of a few viruses that do have overlapping genes).

6. There are three codons that do not code for any amino acid. These are the termination codons; they are read by special proteins (called release factors) and signal the end of the translation process.

7. The codon AUG codes for methionine as well as being the initiation codon. Thus methionine is the first amino acid in each newly synthesized polypeptide. Since the vast majority of polypeptides do not begin with methionine, the amino acid is enzymatically removed before the polypeptide chain is completed.

15.6 *Mutations and Genetic Diseases*

We have seen that DNA directs the synthesis of proteins through the intermediary mRNA and that the sequence of bases in the DNA is critical and specific for the proper sequence of amino acids in proteins. On rare occasions, however, the base sequence in DNA may be modified either spontaneously (about 1 in 10 billion) or by exposure to heat, radiation, or certain chemicals. *Any chemical or physical change that alters the sequence of bases in the DNA molecule* is termed a **mutation.** The most common types of mutations are *substitution* (a different base is substituted), *insertion* (addition of a new base), and *deletion* (loss of a base). These changes within the DNA are called **point mutations** because the change occurs at a single nucleotide position (Figure 15.17).

The chemical and/or physical agents that cause mutations are termed **mutagens.** Examples of physical mutagens are ultraviolet and gamma radiation. They exert their mutagenic effects either directly or via free radicals induced by the radiation (Section 2.5). Radiation and free radicals are known to cause covalent modification (often cross-linkage) of bases already incorporated into DNA. For example, upon exposure to UV light, two adjacent thymines on a DNA strand can become covalently linked together producing a thymine dimer (Figure 15.18). If not repaired, the dimer prevents formation of the double helix at the point at which it occurs. The genetic disease *xeroderma pigmentosum* is transmitted as an autosomal recessive trait, and is an example of a defective mechanism for the repair of pyrimidine dimers in DNA. (The enzyme that cuts out the damaged thymine dimers is not synthesized by the cells.) Individuals affected by this condition are abnormally sensitive to light and are more prone to skin cancer than normal individuals. During replication, an abnormal DNA is produced, which apparently has no stop signal and results in the proliferation of cancer cells.

Among the chemical mutagens are two base analogs, 5-bromouracil and 2-aminopurine. They can be incorporated into the new DNA strand, but they exhibit

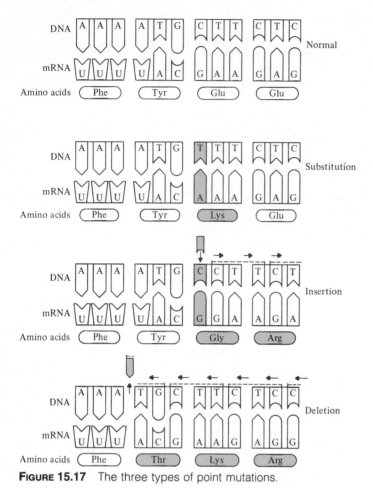

Figure 15.17 The three types of point mutations.

faulty base pairing. 5-Bromouracil is incorporated into DNA in place of thymine, but it can base-pair with guanine (instead of adenine). 2-Aminopurine substitutes for adenine, yet it sometimes base-pairs with cytosine (instead of thymine). Hydroxylamine and nitrous acid are other chemical mutagens. Hydroxylamine (NH_2OH) deaminates cytosine, yielding a product that pairs with adenine instead of guanine. Nitrous acid (HNO_2) can convert cytosine to uracil, which will also bond to adenine instead of guanine. That these compounds are both carcinogenic and mutagenic strongly indicates that disruption of DNA is fundamental to both processes.

Many mutations are perpetuated by replication and thus inherited. Although it is possible for a gene mutation to be beneficial (thus permitting evolutionary advances through natural selection), most mutations are detrimental. If a point mutation occurs at a crucial position, the defective protein will lack biological activity and may result in the death of the cell. In such cases the altered DNA sequence is lost and will not be copied into daughter cells. Nonlethal mutations often lead to metabolic abnormalities or to hereditary diseases. Over 1200 diseases in humans result from

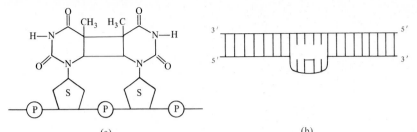

FIGURE 15.18 (a) Structure of a DNA intrastrand thymine dimer. (b) Defect in the double strand produced by the thymine dimer. The dimerization is caused by ultraviolet light. This temporarily stops DNA replication, but the dimer can be enzymatically excised and the strand can be repaired.

mutations in genes; such diseases are called **inborn errors of metabolism** or **genetic diseases.** A partial listing of genetic diseases is presented in Table 15.6, and a few specific conditions are discussed here. In most cases the defective gene results in the failure to synthesize a particular enzyme.

Phenylketonuria (PKU) results when the enzyme *phenylalanine hydroxylase* is absent. A person having PKU cannot convert phenylalanine to tyrosine, which is the precursor of the catecholamines (page 204) as well as the skin pigment melanin.

$$\text{Phenylalanine} + \tfrac{1}{2}O_2 \xrightarrow[\text{hydroxylase}]{\text{phenylalanine}} \text{Tyrosine}$$

Phenylalanine   Tyrosine

In the absence of this step, phenylalanine accumulates and the transamination (see Section 18.3-a) of phenylalanine to phenylpyruvate, normally a very minor process, becomes important.

$$\text{Phenylalanine} \xrightarrow{\text{transaminase}} \text{Phenylpyruvate}$$

Phenylalanine   Phenylpyruvate

Excessive amounts of phenylpyruvate impair normal brain development, causing severe mental retardation. (A mean IQ of 20—about 1% of patients in mental institutions are phenylketonuric.) The lifespan of untreated PKU individuals is significantly shorter than normal (75% are dead by age 30). The disease acquired its name from the high levels of this phenylketone in the urine. PKU may be diagnosed by assaying a sample of blood or urine for phenylalanine or one of its metabolites, and federal law requires that all newborns be tested (usually within the first two weeks). If the condition is detected, mental retardation can be prevented by giving the afflicted infant a diet containing little or no phenylalanine. Because phenylalanine is so prevalent in natural foods, the low phenylalanine diet is composed of a

TABLE 15.6 A Partial Listing of Genetic Disorders in Humans and the Malfunctional or Deficient Protein or Enzyme

Acatalasia	Catalase (red blood cells)
Albinism	Tyrosinase
Alkaptonuria	Homogentistic acid oxidase
Cystathioninuria	Cystathionase
Fabry's disease	α-Galactosidase
Galactosemia	Galactose 1-phosphate uridyl transferase
Gaucher's disease	Glucocerebrosidase
Glycogen storage disease	Different types:
	α-Amylase
	Debranching enzyme
	Glucose 1-phosphatase
	Liver phosphorylase
	Muscle phosphofructokinase
	Muscle phosphorylase
Goiter	Iodotyrosine dehalogenase
Gout and Lesch–Nyhan syndrome	Hypoxanthine–guanine phosphoribosyl transferase
Hemolytic anemias	Different types:
	Glucose 6-phosphate dehydrogenase
	Glutathione reductase
	Phosphoglucoisomerase
	Pyruvate kinase
	Triose phosphate isomerase
Hemophilia	Antihemophilic factor (factor VIII)
Histidinemia	Histidase
Homocystinuria	Cystathionine synthetase
Hyperammonemia	Ornithine transcarbamylase
Hypophosphatasia	Alkaline phosphatase
Isovalericacidemia	Isovaleryl-SCoA dehydrogenase
Maple syrup urine disease	α-Keto acid decarboxylase
McArdle's syndrome	Muscle phosphorylase
Metachromatic leukodystrophy	Sphingolipid sulfatase
Methemoglobinemia	NADPH–methemoglobin reductase and NADH–methemoglobin reductase
Niemann–Pick disease	Sphingomyelinase
Phenylketonuria	Phenylalanine hydroxylase
Pulmonary emphysema	α-Globulin of blood
Sickle cell anemia	Hemoglobin
Tay–Sachs disease	Hexosaminidase A
Tyrosinemia	Hydroxyphenylpyruvate oxidase
Von Gierke's disease	Glucose 6-phosphatase
Wilson's disease	Ceruloplasmin (blood protein)

From R. C. Bohinski, *Modern Concepts in Biochemistry*, 4th ed., Allyn and Bacon, Inc., Boston, 1983. Reprinted by permission.

synthetic protein substitute plus very small measured amounts of natural foods. The diet is maintained until at least three years old, by which time brain development is completed. The incidence of PKU in newborns is about 1 in 15,000.

Galactosemia results from the lack of the enzyme that catalyzes the formation of glucose from galactose. The blood galactose level is markedly elevated, and galactose is found in the urine. This disease may result in impaired liver function, cataracts, mental retardation, and even death. If recognized in early infancy, the effects of galactosemia can be eliminated by removing milk and all other sources of galactose from the diet. As the children grow older, they normally develop an alternate pathway for metabolizing galactose, and thus the need to restrict milk is not permanent. The incidence of galactosemia in the United States is 1 in every 65,000 newborn babies.

There are several genetic diseases that are collectively categorized as lipid storage diseases. As we shall learn in Chapter 17, lipids are constantly being synthesized and broken down. If the enzymes that catalyze lipid decomposition are missing, the lipids tend to accumulate and cause a variety of medical problems. In **Niemann–Pick disease,** a disease of infancy or early childhood, sphingomyelins accumulate in the brain, liver, and spleen because the enzyme *sphingomyelinase* is lacking. The accumulation of the sphingomyelins causes mental retardation and early death.

In **Gaucher's disease** cerebrosides accumulate in the brain and cause severe mental retardation and death. Juvenile and adult forms of this disease are characterized by enlarged spleen and kidneys, hemorrhaging, mild anemia, and fragile bones. This disease is caused by the lack of a specific enzyme called *glucocerebrosidase,* which cleaves glucocerebrosides into glucose and sphingosine.

In the absence of another particular enzyme, *hexosaminidase A,* gangliosides accumulate in brain tissue. The ganglion cells of the brain become greatly enlarged and nonfunctional. This effect, called **Tay–Sachs disease,** results in retardation of development, dementia, paralysis, and blindness. Death usually occurs before the age of three. Tay–Sachs disease can be diagnosed by assaying the amniotic fluid (aminocentesis) for the enzyme. The absence of *hexosaminidase A* in the amniotic fluid allows for a recommendation for a therapeutic abortion since the disease is incurable. Genetic screening can identify Tay–Sachs carriers because they produce only half the normal amount of *hexosaminidase A* (although they do not exhibit symptoms of the disease). Tay–Sachs is prevalent in persons of Eastern European Jewish ancestry.

15.7 *The Nature of Viruses*

A discussion of viruses seems particularly appropriate here since viruses are composed almost entirely of proteins and nucleic acids. The field of virology is a rapidly expanding one, and recent research efforts to establish connections between viruses and cancer have yielded much important and exciting information.

Viruses are a unique group of infectious agents composed of a tightly packed central core of nucleic acids that is enclosed in one or more protein coats (see Figure 13.17). They are divided into two main classes on the basis of the nucleic acid content of the central core. Viruses contain either DNA or RNA *but never both.* (Recall that the cells of higher organisms, from bacteria to humans, contain both kinds of nucleic acids.) Viruses differ from one another in their size[10] and shape. The influenza virus, for

[10]Viruses are exceedingly small, much smaller than bacteria. The tobacco mosaic virus is approximately 300 nm long by 18 nm wide (i.e., 300×10^{-9} m by 18×10^{-9} m) and has a molecular weight of about 40 million daltons. In general, animal viruses are larger than bacterial viruses. Most viruses are visible only under the electron microscope.

example, is about ten times bigger than the polio virus. Viruses may be spherical, rod-shaped, or shaped like threads.

Characteristic of viruses is their infectivity. They are able to inject their nucleic acids into the cells of specific hosts and there replicate or remain apparently latent. In order for viruses to replicate, their nucleic acids must enter the cells of another organism for they do not have the necessary biological machinery to replicate on their own. Outside the host cell a virus is an inert chemical complex; within the host cell it can use the cell's mechanisms to replicate nucleic acids and to use the information contained within these viral nucleic acids to synthesize the viral protein coat. Many copies of the virus can thus be formed within each infected cell. The virus is therefore said to be an *obligatory intracellular parasite*. Viruses have a limited host range, and they may be subclassed accordingly into animal, plant, and bacterial viruses. Within the subclasses, moreover, viruses are able to infect only certain species of organisms; for instance, a virus that can cause disease in a certain species of rodent may not be able to grow at all in a different but related species.

In order to understand the recent theories about *oncogenic* (cancer-causing) viruses, it is necessary to know something of the nature of events that ensue when a virus infects a living cell.

a. Infective Processes

1. Productive Infection Productive infection is said to occur when the virus utilizes the host's cellular apparatus for purposes of making more copies of itself. Somehow the virus is able to interfere with normal host metabolism, and to direct the synthesis of multiple copies of its own nucleic acids. At about the same time, the host's protein synthesis may halt as the synthetic machinery is taken over for the synthesis of multiple copies of virus protein coat. When the two major viral components are synthesized, the coats are assembled around the nucleic acid cores, and the newly formed virus particles (progeny viruses) are

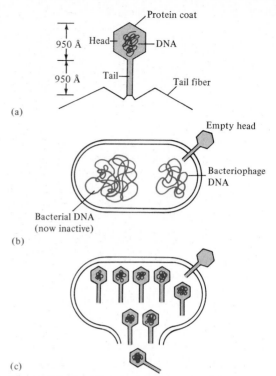

FIGURE 15.19 Life cycle of a bacteriophage. (a) Drawing of T2 bacteriophage. The bacteriophage DNA is located in the head of the particle, surrounded and protected by the protein coat. The end of the tail contains lysozyme, which can hydrolyze the polysaccharide of the bacterial cell wall. (b) When the bacteriophage infects a bacterium, the phage DNA is injected into the bacterial cell. Synthesis of bacterial nucleic acids and protein stops, and the synthetic machinery of the bacterial cells begins to produce phage DNA and protein. (c) As soon as all the phage constituents are synthesized, they begin to form new phage particles. When the bacterial cells becomes filled with phage particles, lysis occurs and the progeny are released. A total of 150–300 progeny viruses are produced in about 30–60 minutes from the infection of one bacterial cell by a single virus.

TABLE 15.7 Human Infectious Diseases

Bacterial Origin	Viral Origin	
Cholera	AIDS	Measles
Diphtheria	Chicken pox	Meningitis
Dysentery	Cold sores	Mumps
Gonorrhea	Common cold	Pneumonia
Plague	Encephalitis	Polio
Syphilis	Gastroenteritis	Rabies
Tetanus	Genital herpes	Shingles
Tuberculosis	German measles	Smallpox
Typhoid fever	Hepatitis	Warts
Whooping cough	Influenza	Yellow fever

released from the cell either by budding or when the cell bursts. In either case, the host cell eventually dies. The progeny viruses so released are free to infect other cells, and so the process is repeated. Figure 15.19 depicts the life cycle of a bacteriophage (a virus that infects bacteria). Cell death and the production of new viruses account for the symptoms of viral infections. As Table 15.7 indicates, a greater number of human diseases are of viral origin than bacterial origin. Infectious diseases of viral origin (especially the common cold and influenza) are among the most significant health problems in our society.

2. Cell Transformation   Certain viruses (among them, the oncogenic viruses) may not cause the death of the host cell upon infection, but rather will alter the properties of the infected cell in such a way that it no longer behaves as it used to in controlling cell division. This type of infection is observed both in vivo and in vitro, and from studies in vitro has come a body of information about the altered properties of these virally *transformed cells*. In order to appreciate the significance of the changes, it is necessary first to know something of the properties of normal (nontransformed) cells in tissue culture (in vitro).

Normal cells grown in a culture dish will continue to divide until their borders come in contact with each other. They usually align themselves in an orderly fashion with respect to one another. This phenomenon is known as *contact inhibition*. Cells infected with transforming viruses lose this characteristic and under the influence of the viral nucleic acid they multiply without restraint. They are seen to climb over one another in tissue culture and to divide so chaotically that three-dimensional clumps of cells are formed rather than single sheets. The normal mechanism that regulates cell division is lost due to some unknown influence exerted by the presence of the viral nucleic acids. This is significant in view of the fact that unregulated and chaotic cell division is typical of cancer in living animals.

Other evidence of altered cellular function is present in colonies of transformed cells. These evidences, often called *viral footprints*, include the appearance of new molecules (virus specific antigens) on the surface of and within the transformed cells, as well as certain enzymes, nonenzymatic proteins, and mRNA, which is strongly suspected to be of viral origin. Viral footprints are not detected in uninfected cells of the same tissue culture line. The finding of footprints is often crucial in establishing that viral infection has taken place because, in many instances, once the infecting virus has entered the cell it seems to disappear and further efforts to detect it (for example, by electron microscopy) will prove futile.

Transformation of cells in tissue culture and induction of cancer in animals are closely related phenomena, as we shall see, since often the inoculation of animals with transformed cells produces tumors.

b. Oncogenic RNA Viruses

The oncogenic RNA viruses are divided into subgroups on the basis of the kind of cancer they produce and the type of animal they infect. Included in this group are the *Rous sarcoma* (cancer in chickens), *Avian myeloblastosis* (leukemia in birds), *Rauscher leukemia* (leukemia in mice), and a virus that produces tumors of the mammary gland in mice. The latter virus is also referred to

as the milk factor or Bittner virus since its main mode of transmission is from mother to offspring through the milk. The virus may be detected in breast tissue long before the appearance of cancer, and it is interesting to note that hormonal factors may play a role in activating the virus so that it may produce the malignant change. In addition, newborn and suckling mice are susceptible to the virus when it is given orally or when injected under the skin or into the abdominal cavity. A viral etiology for human breast cancer is strongly suspected, and research efforts geared toward isolating the agent are underway.

Dr. Howard Temin (University of Wisconsin) and Dr. David Baltimore (MIT), working independently, discovered that the usual order for the transmittal of genetic information can be reversed by certain oncogenic RNA viruses.[11] Working with cultures of *Rous sarcoma* virus, Dr. Temin found evidence that these transformed cells contained new DNA whose sequence of bases were complementary to those of the viral RNA. This could only happen if the viral RNA served as the template for the synthesis of DNA. (Recall from Section 15.4 that this hypothesis is the reverse of the established dogma.) The discovery by Temin and Baltimore that these viruses contained an enzyme (*RNA-dependent DNA polymerase*—commonly called *reverse transcriptase*) capable of transcribing the RNA virus message into DNA was a major breakthrough in the area of virus research. Shortly thereafter, other workers detected this enzyme in a variety of RNA-transformed cancer viruses, including human leukemia.[12] The significance of this research is that if *reverse transcriptase* is required for the cancer-producing effect of the RNA virus, a substance that inhibits this enzyme might prevent leukemia. Various drugs are presently being tested for treatment of leukemia and other forms of cancer in rats and mice.

c. Oncogenic DNA Viruses

Two of the best known of the oncogenic DNA viruses are the polyoma virus (cancer in mice) and simian virus 40 (cancer in monkeys). Both have been widely investigated as agents of cancer induction. The polyoma virus occurs naturally in mice both in the laboratory and in the wild. When introduced experimentally into various animals, it produces a wide range of different types of tumors.

[11] Dr. Baltimore, Dr. Temin, and Dr. Renato Dulbecco of the Imperial Cancer Research Laboratory in London were awarded the 1975 Nobel Prize in Medicine for their discoveries "concerning the interaction between tumor viruses and the genetic material of the cell."

[12] These RNA viruses that contain the enzyme *reverse transcriptase* in their protein coats are referred to as **retroviruses.** It is known that a form of leukemia in humans is caused by the HTLV-I retrovirus. A very similar retrovirus, HTLV-III, is believed to be the virus that causes acquired immune deficiency syndrome (AIDS).

The virus is able to cause productive infection in certain tissue cultures, whereas in others it produces cell transformation. This latter event is detected by the finding of viral footprints as previously described. Cells of certain tissue culture lines, when so transformed, are capable of producing tumors when inoculated into various host animals. In addition, cells taken from the tumors so induced can be transplanted into animals of the same strain and tumor growth is again observed. Finally, preparations of the virus itself are tumorigenic in many species when inoculated under the skin.

Simian virus 40 (SV40) is structurally similar to polyoma virus. It was originally discovered in kidney cultures of a certain species of monkey that were being used to propagate polio virus in the preparation of vaccine. Apparently, SV40 multiplies silently in the kidneys of these species without producing overt disease. However, the virus is tumorigenic in many other animals as well as in particular species of monkeys. Since its discovery as a contaminant in the vaccine preparations, much investigation has taken place, partially due to the finding that many children were observed to excrete the virus in urine for as long as five weeks after ingestion of live polio vaccine. SV40 causes tumors at the site of inoculation into newborn hamsters.

Hamster cells transformed in tissue culture also produce tumors when inoculated into the newborn animals. Cells of the tumors thus induced can be serially passed in animals of the same strain and the tumors to which they give rise are identical to the tumors produced by the virus itself.

From the foregoing discussion, it can be seen that advances in animal models of virally induced tumorigenesis have been great. No comparable advances have been made in the field of human disease thus far. One of the biggest reasons is also the most obvious—experimental tumor induction in humans is not possible. Another stumbling block has been the elimination of contaminating viruses from tissue cultures of human cell lines. In addition, as we have seen, viruses often "disappear" after inducing malignant change, so it is not surprising that efforts to identify viruses from human tumor tissue have not yet been very rewarding. However, research is actively in progress and it appears that before long, definitive results may be forthcoming in the area of certain human cancers for which a viral etiology is strongly suspected. Some of the areas presently under aggressive investigation include certain of the leukemias, lymphomas (tumors of the body's lymphoid tissues), mammary gland cancer, and various bone cancers.

15.8   *Recombinant DNA*

One of the most controversial issues in recent years to arouse the scientific community as well as government officials and the general public is that of gene splicing or recombinant DNA. The term **recombinant DNA** refers to *DNA molecules that have been created by splicing segments of DNA from one organism into pieces of DNA from another organism.* Almost any organism (plants, animals viruses, bacteria) can serve as the DNA donor, but to date the bacterium *Escherichia coli* and certain yeasts have served as the principal DNA acceptors. *E. coli* is particularly suited for this work because its genetic identity has been so thoroughly studied. More is known about the biochemistry and genetics of *E. coli* than any other organism. Most of *E. coli's* genes (about 2000–3000) are contained within a single large, ringed chromosome. In addition, there are *much smaller closed loops of DNA,* called **plasmids,** that carry only a few genes (some of which code for enzymes that confer resistance to various antibiotics). It is these plasmids that serve as the vehicles for the splicing technique. Bacterial plasmids are termed *vectors* because their genetic information allows for the harboring and replication of the recombinant molecule in an *E. coli* cell. The circular DNA of a bacteriophage (phage lambda) has also been used as a vector for the infection of *E. coli,* but these experiments are less common and far less safe. Figure 15.20 illustrates the production of recombinant DNA; the steps are as follow.

1. *E. coli* bacteria are placed in a detergent solution to break open the cells.
2. The plasmids are separated from the chromosomal DNA by differential centrifugation.

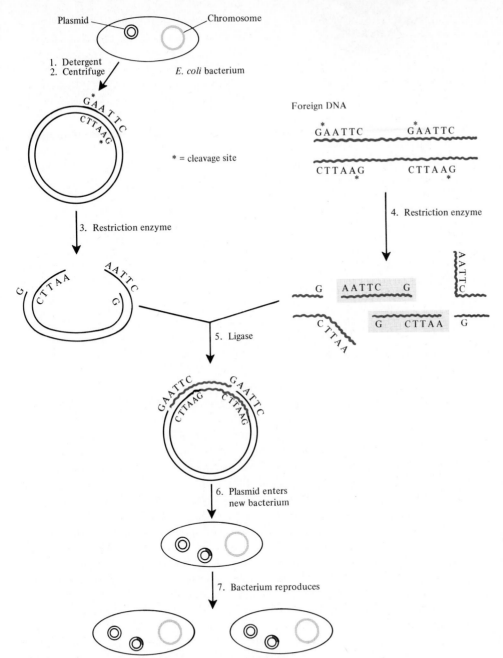

Figure 15.20 Recombinant DNA in *Escherichia coli*.

3. Restriction enzymes (*endonucleases*) are utilized to cleave the plasmid at a specific short sequence in a way that creates overlapping, cohesive (so-called "sticky") ends. Each restriction enzyme can recognize a specific sequence of four to six nucleotides in DNA. For example, the *endonuclease* designated as *Eco*RI cuts at the sequence GAATTC and *Sma*I cuts at CCCGGG. More than 100 restriction enzymes, purified from different species of bacteria, are now commercially available.

4. In vitro combination of the same restriction enzyme with DNA from another organism (foreign DNA) or with synthetic DNA produces segments of DNA having cohesive ends that are complementary to those of the plasmid. (Since different restriction enzymes have different cleavage sites, a given strand of DNA can be separated into many different segments of varying lengths. It is therefore possible to insert almost any foreign gene(s) into *E. coli*.)

5. The enzyme *DNA ligase* seals the foreign DNA segment into place in the plasmid.

6. The resealed plasmid is placed in a solution of calcium chloride containing *E. coli*. When the solution is heated, the bacterium cell membrane becomes permeable, allowing the plasmid to enter.

7. *E. coli* reproduce by dividing (and thus doubling the population) at a rate of about once every 20–30 minutes. The new *E. coli* that are now created have characteristics dictated by their own genes as well as those that have been transplanted from a different species.

Proponents of recombinant DNA research are excited about its great potential benefits. Recombinant techniques are an enormous aid to scientists in mapping and sequencing the position of genes and in determining the functions of different segments of an organism's DNA. The complete DNA sequences of more than 100 mammalian genes have been determined using recombinant DNA technology. An understanding of gene function and gene regulation is a primary goal of scientists working to cure cancer. It is conceivable that recombinant DNA could lead to cures for genetic diseases. If appropriate genes could be successfully inserted in *E. coli,* then the bacteria would become miniature pharmaceutical factories, producing great quantities of insulin, clotting factor for hemophiliacs, missing enzymes, hormones, vitamins, antibodies, vaccines, and so on. DNA-recombinant molecules containing synthetic genes for the production of interferon (a glycoprotein that combats certain viral infections and stimulates the immune system), human insulin, growth hormone, and *urokinase* (a human enzyme used clinically for dissolving blood clots) have already been accomplished in *E. coli*. Vaccines against hepatitis B (human) and hoof and mouth disease (cattle) have been produced. In addition, it may be possible to breed human intestinal bacteria that can digest cellulose, to construct bacterial strains that can clean up oil spills by degrading the compounds in the oil, and to create new plants that can obtain their nitrogen directly from the air rather than from costly petroleum-based fertilizers.

Besides *E. coli,* scientists have used other bacteria as well as yeast and fungi in gene splicing experiments. The enzyme *rennin* has been produced in fungi. *Rennin* is used commercially to coagulate milk into curds in the production of cheese. A bacterial plasmid has been utilized as a vector by plant molecular biologists. The bacterium is

Agrobacterium tumefaciens, which can cause tumors in many plants. Scientists have succeeded in introducing genes for several foreign proteins (including animal protein) into plants by means of Agrobacter, at the same time eliminating its tumor-causing ability. One practical application would be to transfer the gene necessary for the synthesis of a deficient amino acid into a particular plant (e.g., the gene for methionine into soybeans). In 1980 the U.S. Supreme Court, in a landmark decision, decreed that man-made life forms (i.e., the mutant bacteria, yeast, fungi, etc.) are patentable. That decision was expected to increase the vigor with which companies pursued their quest for marketable proteins via genetic manipulation.

Opponents of this research acknowledge the potential benefits, but they are concerned about the considerable risks. They fear that a man-made "Andromeda strain" could be unleashed, which would proliferate uncontrollably, causing mass disease and death.[13] Furthermore, they argue that this research can be misused for political and social purposes. These techniques might be exploited for genetically engineered control of human behavior. Finally, there are some who believe that we should not meddle with evolution by creating new forms of life different from any that exist on earth. Their slogan is "it's not nice to fool Mother Nature."

Concern about the potential biological hazards of recombinant DNA research first arose in 1973 among molecular biologists working in this area, and they voluntarily agreed to halt certain kinds of experiments. In 1976, the NIH issued research guidelines that apply to all federally supported work involving recombinant DNA. These guidelines establish a set of carefully controlled physical and biological containment conditions. The designations P1, P2, P3, and P4 represent increasing levels of physical containment, expressed in more elaborate laboratory facilities and procedures. P1 denotes standard laboratory precautions for experiments in the lowest-risk category (for example, injecting harmless bacterial genes into *E. coli*). P2 and P3 facilities require increasing levels of isolation from the environment. P4 denotes the ultrasecure laboratories for research of greatest risk (such as working with animal tumor viruses or primate cells). No escape of contaminated air, wastes, or untreated materials is permitted. As of 1977, there were only two P4 laboratories in existence in the United States. By 1979, the NIH Guidelines had been eased somewhat because of the safety demonstrated by biological tests. SV40 cancer virus DNAs that were placed in *E. coli* gave cells that did not cause cancer in monkeys. Recall (Section 15.7-c) that the SV40 virus is tumorigenic. Indeed, it is safer to handle cancer virus DNAs as recombinant DNAs in *E. coli*.

[13] A great breakthrough in the area of recombinant DNA research was made by geneticist Roy Curtiss. He was able to develop a strain of *Escherichia coli* that could live only under laboratory conditions (i.e., it needed the nutrients supplied by the laboratory medium to survive). In 1976 the NIH certified this strain for use in recombinant DNA research.

Exercises

15.1 Define, explain, or draw a formula for each of the following.
   (a) *N*-glycosyl linkage
   (b) nucleoside triphosphate
   (c) 3′,5′-phosphodiester linkage
   (d) complementary bases
   (e) dissymmetry ratio
   (f) double helix
   (g) histones
   (h) DNA melting

(i) template replication
(j) semiconservative mechanism
(k) Okazaki fragments
(l) promoters
(m) multiplicity
(n) ribosome
(o) genetic code
(p) translocation
(q) code degeneracy
(r) mutagens
(s) genetic disease
(t) PKU
(u) amniocentesis
(v) oncogenic virus
(w) retrovirus
(x) recombinant DNA
(y) plasmid
(z) restriction endonuclease

15.2 Distinguish between the terms in each of the following pairs.
(a) purine base and pyrimidine base
(b) major base and minor base
(c) ribose and deoxyribose
(d) nucleoside and nucleotide
(e) nucleotide and nucleic acid
(f) somatic cell and germ cell
(g) *DNA polymerase* and *DNA ligase*
(h) T loop and D loop
(i) codon and anticodon
(j) chromosomes and genes
(k) transcription and translation
(l) cytoplasm and cytosol
(m) A site and P site
(n) mutation and point mutation

15.3 Give the names of all the compounds that can be obtained from the complete hydrolysis of (a) DNA and (b) RNA.

15.4 Name and draw the structural formulas of the pyrimidine and purine bases found most commonly in nucleic acids.

15.5 Draw structural formulas for the following minor pyrimidine and purine bases.
(a) 5-methylcytosine
(b) 5-hydroxymethylcytosine
(c) 6-methyladenine
(d) 2-methylguanine
(e) 6-oxypurine
(f) 4-thiouracil
(g) 5,6-dihydrouracil
(h) 1-methylguanine

15.6 Draw structural formulas for (a) deoxythymidine and (b) deoxyadenosine.

15.7 Pseudouridine is a nucleoside found in some tRNAs. The

uracil group is linked to ribose at carbon-5, and at physiological pH the oxy group at carbon-4 is in the keto form, whereas the oxy group at carbon-2 is in the enol form. Draw the structure of pseudouridine.

15.8 The possibilities of hydrogen bond formation are similar for uracil and thymine. Explain.

15.9 Draw structural formulas for the following compounds.
(a) guanosine   (b) AMP   (c) TTP
(d) CDP   (e) $^5{}'G—C^{3'}$ (f) $^3{}'U—A^{5'}$

15.10 Draw structural formulas for dinucleotides composed of (a) cytosine and adenine and (b) thymine and guanine. Show how these dinucleotides might link together by means of hydrogen bonds.

15.11 We shall learn in Section 16.3, that cyclic AMP is a key regulator of cellular metabolism. Its single phosphate group is intramolecularly esterified at both the 3'- and 5'-hydroxyl groups of adenosine. Draw the structure of cyclic AMP.

15.12 (a) What constitutes the backbone of a DNA strand?
(b) How are the two ends of the strand identified?
(c) What forces hold the two strands together?

15.13 DNA can be described as having primary and secondary structure. What does this mean?

15.14 DNA and RNA are termed nucleic acids. What makes them acidic?

15.15 What is meant by the antiparallel chain structure of DNA?

15.16 Hereditary characteristics are encoded in the base sequence of DNA. Explain.

15.17 Show the replication of the following DNA segment.

15.18 How does RNA differ from DNA (a) structurally and (b) in base composition?

15.19 Write the RNA base sequence that would be obtained upon transcription of the lower chain of the following DNA segment.

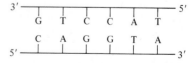

15.20 Briefly outline the functional role of (a) DNA, (b) mRNA, (c) tRNA, and (d) rRNA in protein synthesis.

15.21 (a) How many nucleotide units are present in a codon?
(b) Why is it that a triplet of bases codes for each amino acid?

15.22 What are the two most important sites on tRNA molecules?

15.23 It is essential that each tRNA only bind to a specific amino acid (e.g., leucine and not isoleucine). How is this specificity assured?

15.24 A hypothetical protein has a molecular weight of 60,000 daltons. Assume that the average molecular weight of an amino acid is 120 daltons.
(a) How many amino acids are present in the protein?
(b) How many codons occur in the mRNA that codes for this protein?
(c) How many nucleotide bases are found in the mRNA?

15.25 If the sequence of bases in a section of mRNA is

5′AUGUACCACGGUACGCGGGUAUUGCUA-
GCCGAUGGGUAA³′

what would be the amino acid sequence in the peptide produced from this mRNA? (See Table 15.5.)

15.26 If the base sequence of a gene is

5′TACGAATCTAGAATACTTCCAAAAGTA
TTTTGATACATC³′

what would be the amino acid sequence of the polypeptide that is synthesized?

15.27 The hormone somatostatin, produced in the pancreas, is composed of 14 amino acids. Somatostatin inhibits the release of a variety of hormones including glucagon, insulin, and human growth hormone. It is believed to keep the pancreas's output of insulin and glucagon in a proper balance (see Section 16.3).
(a) Given the structure of somatostatin, postulate a possible base sequence of the mRNA that would direct the synthesis of somatostatin. Include an initiation codon and a termination codon.

```
Ala-Gly-Cys-Lys-Asn-Phe-Phe
              |              |
              S            Trp
              |              |
              S            Lys
              |              |
        Cys-Ser-Thr-Phe-Thr
```

(b) What is the base sequence of the DNA that codes for this mRNA?

15.28 Many antibiotics destroy bacteria by disrupting their protein synthesis. What is the specific effect upon protein synthesis of the following antibiotics?
(a) actinomycin D
(b) tetracycline
(c) chloramphenicol
(d) erythromycin

15.29 Give two examples of physical mutagens and four examples of chemical mutagens.

15.30 Name three genetic diseases and indicate which enzyme is lacking.

15.31 Recall (page 324) that sickle cell hemoglobin differs from normal hemoglobin as a result of the substitution of valine for glutamic acid as the sixth amino acid from the N-terminal end of the polypeptide chain.
(a) What alteration in the base sequence of the DNA could have caused this substitution?
(b) What would be the resulting base sequence in the mRNA?

15.32 The following table contains just a few of the 200 or so known hemoglobin mutants. For each mutation, give the codon for the normal amino acid and then give the codon for the mutant amino acid.

Type	Chain Position		Normal	Mutant
J	α	5	Ala	Asp
I	α	16	Lys	Glu
M	α	58	His	Tyr
D	β	121	Glu	Gln
K	β	136	Gly	Asp

15.33 Consider the following segment of DNA

ACGTTAGCCCCAGCT....

(a) Write the sequence of bases in the corresponding mRNA.
(b) What would be the amino acid sequence formed by translation?
(c) What amino acid sequence would result from (i) replacement of the colored guanine by adenine, (ii) insertion of thymine immediately after the colored guanine, and (iii) deletion of the colored guanine?

15.34 Assume that a segment of a gene that coded for a particular enzyme has the following base sequence.

TACGACGTAACAAGC....

(a) What effect would result from a point mutation in which an adenine substituted for the colored guanine?
(b) What effect would result from a point mutation in which a thymine substituted for the colored adenine?

15.35 Viruses have been described as structures that are at the threshold of life. Comment on this description.

16

Carbohydrate Metabolism

The term **metabolism** is used to describe the *various chemical processes by which food is utilized by a living organism to provide energy, growth substance, and cell repair.* Metabolic reactions include both the *degradation of absorbed food substances into smaller molecules* (**catabolism**) and the *biosynthesis of complex molecules from simpler components* (**anabolism**). In general, catabolic reactions are exothermic, and anabolic reactions are endothermic. *Any chemical compound that is involved in a metabolic reaction* is referred to as a **metabolite**.

In our earlier discussion of carbohydrates, it was mentioned that human existence on this planet is directly dependent upon the plant kingdom. The animal world derives its foodstuffs, and hence its energy, from plant life. Plants obtain water, inorganic salts, and nitrogenous compounds from the soil, and carbon dioxide and oxygen from the atmosphere. With these raw materials, they are able to synthesize carbohydrates, lipids, and proteins. The energy required for these synthetic reactions is obtained from the sun; thus radiant energy (as sunlight) is the ultimate source of biological activity. *The overall process by which glucose is formed from carbon dioxide and water at the expense of solar energy* is termed **photosynthesis.** The general photosynthetic equation can be written as follows.

$$6\,CO_2 \;+\; 6\,H_2O \;+\; 686{,}000\;cal \xrightarrow{\text{sunlight}} \underset{\text{Glucose}}{C_6H_{12}O_6} \;+\; 6\,O_2$$

This synthesis is a distinguishing characteristic of green plants, and from the standpoint of human survival, it represents the most important series of reactions (>100 enzyme-catalyzed steps) that occur on the surface of the earth. Notice that the formation of 1 mole of glucose requires the conservation of 686,000 cal. When compared to the energy requirement of most other endothermic chemical reactions, this is a very large sum. (It is roughly equivalent to the heat required to raise the temperature of 7 quarts of water from 0°C to 100°C.) On the solar scale, however, it is a drop in the bucket.[1]

Carbohydrates are the primary metabolites of the animal kingdom; recall that over half of the food that we ourselves consume is composed of carbohydrates. Carbohydrates serve as the chief fuel of biological systems, supplying living cells with usable energy. Like all fuels, carbohydrates must be burned or oxidized if energy is to be released. The combustion (oxidation) process ultimately results in the conversion of the carbohydrate into carbon dioxide and water; the energy stored in the carbohydrate molecule during photosynthesis is released in the reaction.[2]

$$C_6H_{12}O_6 \;+\; 6\,O_2 \longrightarrow 6\,CO_2 \;+\; 6\,H_2O \;+\; 686{,}000\;cal$$

This equation summarizes the biological combustion of foodstuff molecules by the cell (respiration). The term respiration is often used in a broader sense to include *all metabolic processes by which gaseous oxygen is used to oxidize organic matter to carbon dioxide, water, and energy.*

Both respiration and the combustion of the common fuels (wood, coal, gasoline) use oxygen from the air to break down complex organic substances to carbon dioxide and water. The energy released in the burning of wood, however, is manifested entirely in the form of heat, but excess heat energy is useless and even injurious to the living cell. Organisms conserve almost half of the 686,000 cal by a series of stepwise reactions that liberate small amounts of utilizable energy which is stored in the phosphate bond energy of ATP. The remainder of the energy is used to heat the body and thus to maintain proper body temperature.

---

[1] It has been estimated that on an average sunny day each square centimeter of earth receives about 1 cal of solar radiation every minute. Yet the energy intercepted by the earth is but a tiny part ($2 \times 10^{-7}\%$) of the total energy given off by the sun, and only a small fraction of this intercepted energy ($\sim 1\%$) is utilized in photosynthesis. By far, the major photosynthesizing organisms are the phytoplankton living in the world's oceans. Phytoplankton are the chief source of energy for aquatic organisms.

[2] Plant and animal cells exist in a symbiotic cycle; each requires the products of the other for life. Although H_2O and O_2 are abundant in the atmosphere, CO_2 is present only to the extent of 0.02–0.05%. It has been estimated that if animals were removed from the earth, all the atmospheric CO_2 would be consumed in 1–2 years.

16.1 *Digestion and Absorption of Carbohydrates*

Digestion may be defined as a *hydrolytic process whereby food molecules are broken down into simpler chemical units that can be absorbed by the body*. In humans, digestion takes place in the digestive tract (Figure 16.1-a) and absorption occurs primarily in the small intestine.

Starch is the principal carbohydrate ingested by humans.[3] Its digestion begins in the mouth where the enzyme *ptyalin* (an *amylase* contained in saliva) catalyzes its hydrolysis to a mixture of dextrins plus some maltose and glucose. *Ptyalin* continues to function as food passes through the esophagus, but it is quickly inactivated when it comes in contact with the acidic environment of the stomach. Very little carbohydrate digestion occurs in the stomach; acid-catalyzed hydrolysis proceeds too slowly at body temperature to be effective. The primary site of carbohydrate digestion is the small intestine, where another *amylase*,[4] *amylopsin* (secreted from the

FIGURE 16.1 (a) The human digestive tract. (b) Intestinal villus.

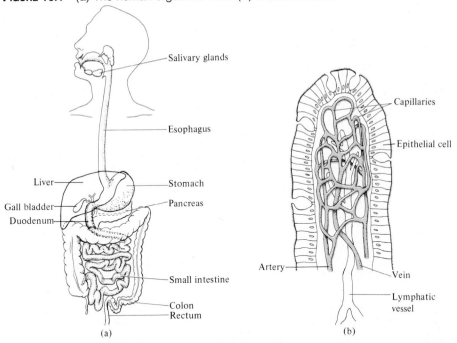

[3] The average American consumes approximately 325 g of carbohydrates each day. This comprises about 160 g of starch, 120 g of sucrose, 30 g of lactose, 10 g of glucose, and 5 g of fructose, and traces of maltose. A typical diet supplies about 50% of total body energy in the form of carbohydrate.

[4] An inhibitor of *amylase* obtained from the red kidney bean has been marketed as a "starch-blocker" for the treatment of obesity. There is, however, considerable controversy as to the efficacy of this enzyme inhibitor as an anti-obesity agent.

pancreas), converts the remaining starch molecules, along with the dextrins, to maltose. Maltose is then cleaved into two glucose molecules by the enzyme *maltase*. Disaccharides such as sucrose and lactose are not digested until they reach the small intestine where they are acted upon by the enzymes *sucrase* and *lactase*. The enzymes that catalyze the hydrolysis of disaccharides are termed *disaccharidases* and are located on the membranes of cells that line the inner surface of the small intestine. Ultimately, the complete hydrolysis of disaccharides and polysaccharides produces three monosaccharide units—glucose, fructose, and galactose. These monosaccharides are then absorbed through the wall of the small intestine into the bloodstream.[5] Absorption of most digested food takes place in the small intestine through the finger-like projections, called *villi,* that line the inner surface (Figure 16.1-b). Each villus is richly supplied with a fine network of blood vessels and a central lymph vessel. Absorption of monosaccharides occurs across the semipermeable membranous wall of each villus into the blood capillaries. However, the absorption does not occur by means of a simple process of osmosis or diffusion through an inert membrane. All cell membranes are selective in their action, a fact that implies that they play an active role in the absorption process. The passage of monosaccharides across the intestinal wall is an energy-requiring process, and so the term *active transport* is used to describe this type of absorption.

Following absorption, the monosaccharides are carried by the portal vein to the liver. Here galactose and fructose are enzymatically converted to glucose or into phosphorylated intermediates to be metabolized by the liver. (Glucose is the only sugar that circulates in the blood, hence the name *blood sugar*.) Some of the glucose passes into the general circulatory system to be transported to other tissues. The glucose in the tissues may be oxidized to CO_2 and H_2O, or it may be converted to muscle glycogen thus serving as a source of readily available energy for the muscle. The majority of the absorbed glucose remains in the liver and is either stored as glycogen (as a reservoir for the maintenance of the normal level of blood glucose) or converted to fat and exported to adipose tissue as lipoprotein complexes (the VLDLs—see page 304). The average person has a sufficient amount of stored glycogen (in liver and muscle cells) to supply energy for about 18 hours.

16.2 *Blood Glucose*

The concentration of glucose in the blood is referred to as the **blood-sugar level.** Under normal circumstances the blood-sugar level remains remarkably constant at about 80 mg/100 mL of blood. However, since individuals differ in their chemical makeup, a normal concentration of glucose may range from 70 to 100 mg/100 mL. (This amounts to a total of about 5–6 g of glucose, or 1 teaspoon, in the entire body.) Soon after eating a meal, the blood-sugar level may rise to 120 mg/100 mL of blood, but this level returns to the normal level within 2 hours.

---

[5] Any polysaccharides or disaccharides that escape hydrolysis by intestinal enzymes cannot be absorbed. Intestinal bacteria metabolize these carbohydrates into lactose, short-chain carboxylic acids, and gases (CO_2, CH_4, and H_2), causing fluid secretion, increased intestinal mobility, and cramps.

As we shall see in Section 16.3, regulation of the blood-sugar level is vital to the well-being of an individual. When the control mechanisms function improperly, dire consequences result. The condition resulting from *a lower than normal concentration of glucose* is called hypoglycemia (usually defined as less than 40 mg/100 mL), and that resulting from *a higher than normal concentration of glucose* is hyperglycemia. In extreme cases of hyperglycemia, the **renal threshold** (160–170 mg/100 mL) is reached, and excess glucose is excreted in the urine (glucosuria). Extreme hypoglycemia, which is usually due to the presence of excessive amounts of insulin hyperinsulinism), can cause unconsciousness, lowered blood pressure, and may result in death. Loss of consciousness is most likely due to the lack of glucose in the brain tissue, which has no capacity for glycogen storage and thus is dependent upon a continuous supply of glucose for its energy requirements. The brain uses about 125 g/day of glucose. At rest the total glucose requirements of all other tissues of the body (heart, liver, kidney, muscle, etc.) is about 200 g/day. The total daily glucose requirements are normally met from the dietary intake of carbohydrate.

The major cause of hyperglycemia is *diabetes mellitus* (see Section 16.3). Over 10 million Americans have diabetes (either mildly or severely), and this disease alone is the third leading cause of death (either outright or through side effects) in the United States. Diabetes is characterized by an abnormal metabolism of carbohydrates as well as of proteins and lipids (see Section 17.4). Because a diabetic is unable to utilize glucose properly, excessive quantities accumulate in the blood and urine. Characteristic symptoms of diabetes are constant hunger, weight loss, extreme thirst, and frequent urination because the kidneys excrete large amounts of water in an attempt to remove excess sugar from the blood. High levels of glucose in the blood can cause

FIGURE 16.2 Glucose tolerance test for a normal patient and for a diabetic.

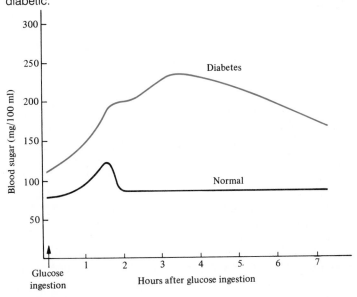

various types of body damage. Among the many complications of diabetes are blindness, cardiovascular disease, gangrene, and kidney disease.

The **glucose tolerance test** is the most important diagnostic test for diabetes mellitus. A patient's blood-sugar level is determined after an overnight fast. Then a known amount (~ 75 g) of glucose is dissolved in about 400 mL of water and administered orally over a period of 5–10 minutes. Blood is drawn from the patient's fingertip at 30-minute intervals after ingestion, and the blood-sugar concentration is determined. The initial rise in blood sugar falls rapidly in a normal individual. In a diabetic, on the other hand, the increase of the blood-sugar level is greater than normal and will remain at the elevated levels for several hours (Figure 16.2).

16.3 *Hormonal Regulation of Blood-Sugar Level*

The most important contributing process in the maintenance of a constant blood-glucose concentration is the synthesis and breakdown of glycogen in the liver. The liver is responsible for removing glucose from the blood when the concentration is too high (after a meal) and releasing it when the blood-sugar level is too low (e.g., between meals). The activity of the liver, in this regard, is controlled by several different hormones; among these are insulin, epinephrine (adrenaline), and glucagon.

Insulin, which is produced by the beta cells of the islets of Langerhans in the pancreas, *is the most important regulator of metabolism in the body.* It is released from the beta cells in response to a high blood-sugar level. In general, insulin promotes anabolic reactions and inhibits catabolic reactions in the liver, muscle, and adipose tissue. Specifically, insulin performs the following functions.

1. It enhances **glycogenesis** (*the formation of glycogen from glucose*) in both the liver and muscle.
2. It promotes the entry of glucose into muscle, liver, and adipose tissue. (Note, however, that red blood cells and cells in the brain, kidney, and intestinal tract do not require insulin for glucose uptake.)
3. It accelerates the conversion of glucose into fatty acids and hence the synthesis and storage of triacylglycerols in adipose tissue (see Section 17.1).
4. It inhibits the breakdown of glycogen and stored fat.
5. It promotes the transport of amino acids into cells and stimulates protein synthesis. It inhibits the intracellular degradation of proteins.
6. It suppresses **gluconeogenesis,** *the synthesis of glucose from noncarbohydrate precursors* (e.g., amino acids, glycerol, lactic acid). Gluconeogenesis occurs primarily in the liver and is particularly important in maintaining a constant blood-sugar level during periods of starvation or strenuous exercise to supply glucose for brain tissue.

Hence, the principal role of insulin is rapidly to remove glucose from the blood thus lowering the blood-sugar level. The mechanism by which insulin accomplishes its prodigous tasks is largely unknown. Insulin does bind to specific receptor proteins in the membranes of target cells. Just how the hormone–receptor interaction is coupled to the physiologic responses is being investigated in many laboratories throughout the world.

If the pancreas does not secrete enough insulin and/or if there are insufficient (or defective) insulin receptors on the cell membranes, **diabetes mellitus** develops. There are two types of this disease. Insulin-dependent diabetics do not produce sufficient amounts of insulin to regulate their blood-sugar level. This type develops early in life and is termed *Type I* or *juvenile-onset diabetes*. It is rapidly reversed by the administration of insulin, and Type I diabetics can lead normal lives provided that they receive insulin as needed. Since insulin is a protein, it cannot be taken orally because it would be digested. Therefore, the Type I diabetic must be treated with daily injections of insulin (either subcutaneously or intramuscularly). Limited success has been achieved by implanting insulin-producing cells from cadavers into adults with Type I diabetes. In some patients, insulin production lasted from 3 to 12 weeks and significantly decreased the patient's needs for injected insulin.

A current theory is that in Type I diabetics the islet cells of the pancreas are destroyed by the body's own immune system. These cells may have been altered by a viral infection (and thus appear as "foreign" to the immune system). Researchers have developed a simple blood test capable of predicting who will develop Type I diabetes several years before the disease becomes apparent. The blood test searches for antibodies that destroy the body's insulin-producing cells. The antibodies require several years before they destroy enough islet cells to cause diabetes.

Type II (adult-onset) diabetes is by far the more common (about 90% of diabetic cases), and it occurs late in life. Type II diabetics produce sufficient amounts of insulin, but the beta cells are not secreting enough of it, or it is not utilized properly (because there is a lack of insulin-receptor proteins on the target cells, or the insulin-receptor proteins are defective). For these people the disease can be controlled with a combination of diet and exercise alone, without insulin injections. Alternatively, there are oral antidiabetic drugs that stimulate the islet cells to secrete insulin (Figure 16.3).

FIGURE 16.3   The antidiabetic drugs are structurally similar to the sulfa drugs (see Figure 14.14). The replacement of the *p*-amino group on the benzene ring with other substituents accounts for the loss of antibacterial properties.

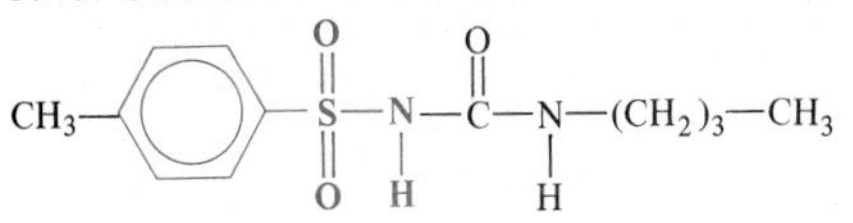

Tolbutamide
(Orinase)

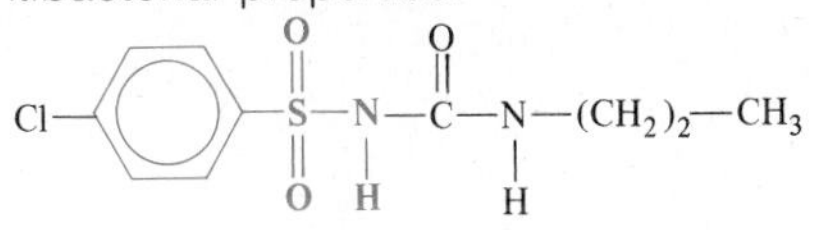

Chlorpropamide
(Diabinese)

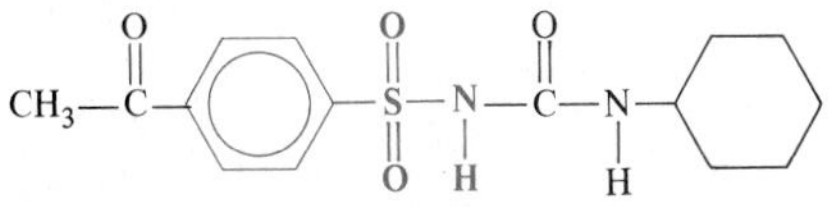

Acetohexamide
(Dymelor)

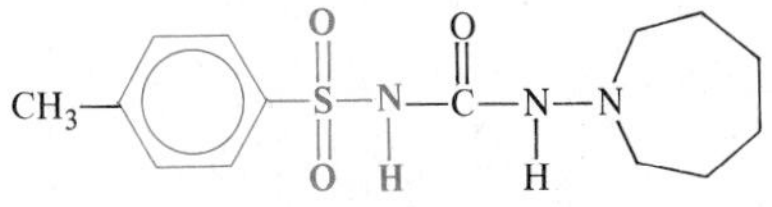

Tolazamide
(Tolinase)

NH₂ structure label:

Adenosine 3′,5′-monophosphate
(Cyclic AMP)

All other hormones that affect glucose metabolism act to raise the blood-sugar level. Both epinephrine and glucagon exert their effects by binding to a receptor protein on the outside of the cell membrane (epinephrine and glucagon each bind to a different specific receptor). These receptor proteins are linked to an enzyme called *adenyl cyclase,* which is bound to the inner membrane. When the outer receptor protein is unbound, the enzyme is not active. In a manner similar to allosteric activation (page 367), the binding of epinephrine or glucagon to their receptors causes conformational changes in other membrane proteins such that *adenyl cyclase* becomes active. The function of the enzyme is to catalyze the conversion of ATP to adenosine 3′,5′-monophosphate (cyclic AMP or cAMP). *Adenyl cyclase* is extremely efficient, and many cyclic AMP molecules can be synthesized by a single activated enzyme.

$$\text{ATP} \xrightarrow[\text{Adenyl cyclase}]{\text{Mg}^{2+}} \text{Cyclic AMP} + \text{PP}_i$$

Cyclic AMP is often referred to as a *second messenger* because it transmits messages (delivered via the blood by the extracellular hormones—the primary messengers) from the cell membrane to enzymes within the cell (Figure 16.4). Cyclic AMP performs its function by binding to and activating certain inactive enzyme precursors, eventually resulting in a cascade of cellular events that results in the stimulation of a wide range of catabolic processes and the inhibition of several anabolic reactions. (See an advanced biochemistry text for a discussion of the cyclic AMP cascade.) It should be noted that cyclic AMP serves as the second messenger for some neurotransmitters and several other hormones in addition to epinephrine

FIGURE 16.4 Binding of an extracellular hormone to a receptor protein activates adenyl cyclase, which catalyzes the synthesis of cAMP.

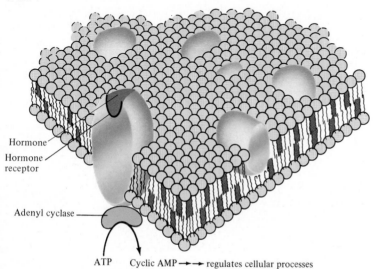

Hormone
Hormone receptor

Adenyl cyclase

ATP   Cyclic AMP → → regulates cellular processes

and glucagon (e.g., norepinephrine—page 204; vasopressin—page 323; FSH and LH—page 297).

Epinephrine (page 204) is secreted by the adrenal medulla in response to a low blood-sugar level. It is also released during exercise or during periods of emotional stress, such as anger or fright, to provide the organism with additional energy. Epinephrine binds to its receptors (called *adrenergic receptors* after its other name, adrenaline), on the membranes primarily of muscle and, to a lesser extent, of the liver. It markedly stimulates **glycogenolysis,** *the breakdown of glycogen to form glucose*. Epinephrine also promotes gluconeogenesis in the liver. Both of these processes increase the amount of glucose in the blood, making that additional glucose available to the body tissues. Epinephrine increases the heart rate and raises the blood pressure.

Glucagon, like insulin, is a polypeptide (29 amino acids—see Exercise 13.23) hormone that is produced by the pancreas (the alpha cells of the islets of Langerhans), but glucagon is an antagonist of insulin. Like epinephrine, it is secreted in response to a low blood-sugar level and its task is to increase the glucose concentration. Glucagon acts primarily on the liver (and adipose tissue) and not on skeletal muscle because there are no receptors for glucagon on muscle-cell membranes. Glucagon stimulates gluconeogenesis and glycogenolysis in the liver, to restore the blood glucose to its normal level. The concentration of glucagon in the blood of a diabetic is above normal, and this may be a significant contributor to the problems associated with diabetes.

Glycogen metabolism is strongly influenced by the ratio of insulin to glucagon in the blood. Higher amounts of insulin lead to glycogen storage after a meal, whereas higher amounts of glucagon favor the breakdown of liver glycogen to add more glucose to the blood. It is the insulin/glucagon ratio that determines the outcome of carbohydrate metabolism. A high ratio leads to carbohydrate anabolism and storage, whereas a low ratio results in carbohydrate catabolism and utilization. The secretion of these hormones is directly governed by the blood-sugar level.

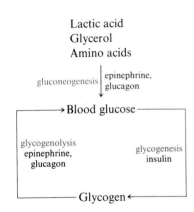

16.4 *Embden–Meyerhof Pathway—Anaerobic Glycolysis*

To appreciate fully the role of carbohydrates in living systems, it is necessary to understand the biochemical nature of their *metabolic pathways*. A metabolic pathway is represented by a type of flow diagram that enables us to explain how an organism converts a certain reactant into a desired end-product. In the 1930s, the German biochemists G. Embden and O. Meyerhof elucidated the sequence of reactions by which glycogen and glucose are degraded in the *absence of oxygen* (**anaerobic** conditions) to pyruvic acid. It was discovered that the two apparently dissimilar processes of alcoholic fermentation in yeast and muscle contraction in animals proceed by the same pathway as far as pyruvic acid. In the absence of oxygen, muscle cells (and certain other animal tissues) convert pyruvic acid into lactic acid, but under similar conditions, the enzymes in yeast convert the pyruvic acid to ethyl alcohol and carbon dioxide. The former process is known as **glycolysis** (glucose splitting), and the latter is **fermentation.**

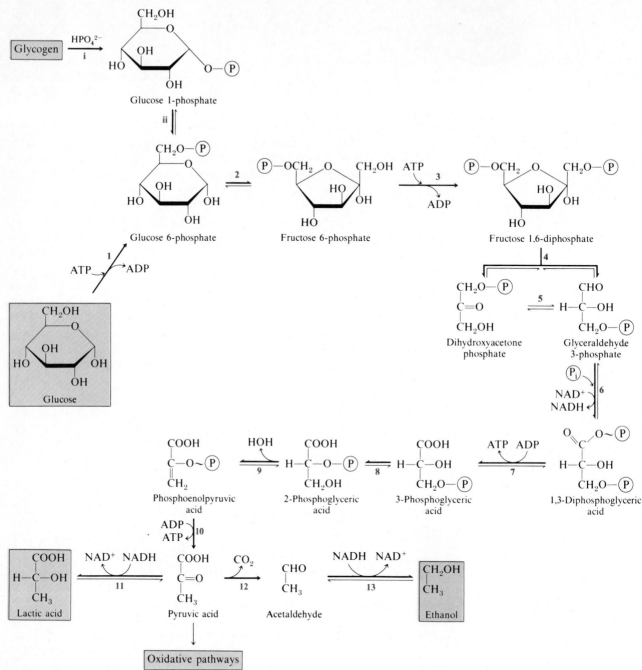

FIGURE 16.5 Embden–Meyerhof pathway.

$$C_6H_{12}O_6 \longrightarrow \longrightarrow \longrightarrow 2\ CH_3-\overset{\displaystyle O}{\overset{\|}{C}}-C\underset{OH}{\overset{\displaystyle O}{\diagup}}$$

Glucose   Pyruvic acid

without oxygen

in yeast — fermentation → $2\ C_2H_5OH\ +\ 2\ CO_2\ +$ Energy

Ethyl alcohol   Carbon dioxide

in muscle — glycolysis → $2\ CH_3CHOHCOOH\ +$ Energy

Lactic acid

It is important to notice that the conversion of glucose to pyruvic acid represents an oxidation reaction (that is, $C_6H_{12}O_6 \rightarrow 2\ CH_3COCOOH + 4\ H^+ + 4\ e^-$), and yet oxygen is not required for the process to occur. We shall see that the outstanding characteristic of the Embden–Meyerhof pathway is the utilization of the coenzyme NAD^+ as the electron acceptor (oxidizing agent). Thus glycolysis and fermentation differ only in the eventual fate of the pyruvic acid, and hence in the means employed for the regeneration of the NAD^+.

A summary of the reactions and the metabolites involved in the Embden–Meyerhof pathway is given in Figure 16.5. This sequence of reactions is probably the best understood of all the metabolic pathways. Over a dozen different, specific enzyme molecules act in such a manner that the product of the first enzyme-catalyzed reaction becomes the substrate of the next. All the reactions of the pathway occur in the cytosol, where the enzymes involved are in solution. The transfer of intermediates from one enzyme to the next occurs by diffusion. Since these enzymes are found in a soluble form in the cell, they were relatively easy to isolate and characterize. Each of the reactions given in Figure 16.5 has been assigned a number that corresponds to the discussion that follows. There is often the tendency to become lost in the complexity of this process and to lose an overall perspective. We are interested in seeing how the principles of organic chemistry apply to these biochemical reactions, but it must always be kept in mind that the central theme of metabolism is the extraction of chemical energy from foodstuff and not merely the degradation of these molecules into simpler substances. In the discussion that follows, the full names of the pertinent enzymes are given in the body of the text. Only the general names are written above or below the arrows of the biochemical equations.

Step i   The liver contains the only store of glycogen that can be converted to glucose and released into the blood for transport to all other tissues. Muscle glycogen, on the other hand, is used solely within the muscle. During stress and exercise, skeletal muscle cells use glycogen as the primary source of energy production.

Liver and muscle cells contain enzymes called *phosphorylases*[6] that catalyze the

---

[6] Recall (Section 16.3) we said that the hormone epinephrine increases the rate of glycolysis by enhancing (via cyclic AMP) the activity of *phosphorylase* enzymes in the liver and in muscle cells. *McArdle's disease* is a genetic disease in which there is a deficiency in, or an absence of, the *phosphorylase* enzyme in muscle cells. Thus, stored glycogen cannot be used to provide energy; the patients have a limited capacity for exercise, and they suffer from painful muscle cramps.

phosphorylitic cleavage of the α-1,4-glucosidic bonds at the nonreducing end of the glycogen chain to produce glucose 1-phosphate. The reaction is analogous to a hydrolytic cleavage, except that here inorganic phosphate takes the place of a water molecule. (Two other enzymes, a *transferase* and a *debranching enzyme,* are required to hydrolytically split off glucose molecules at the branch points of glycogen) Although this reaction is reversible, different enzymes and different reaction sequences are employed for the biosynthesis of glycogen.

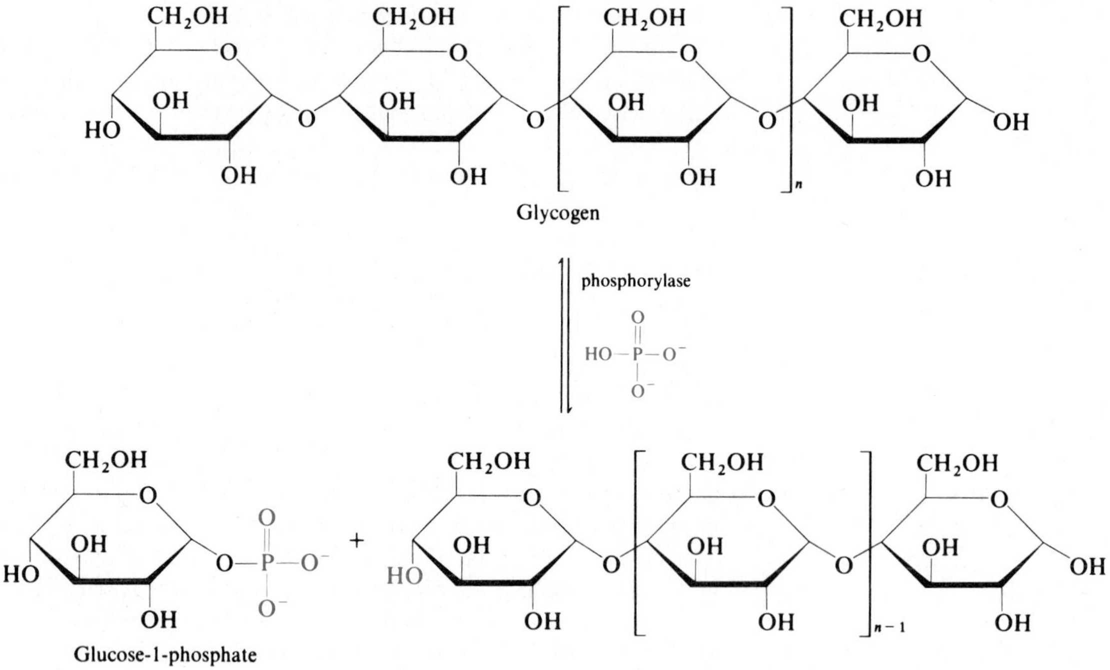

Glycogen

Glucose-1-phosphate

Step ii Glucose 1-phosphate and glucose 6-phosphate are readily interconvertible in the presence of Mg^{2+} ions and the enzyme *phosphoglucomutase.* (In general, a *mutase* is an enzyme that catalyzes the intramolecular transfer of a chemical group.) For simplicity, the isomerization reaction may be thought of as an intramolecular transfer of the phosphate group from carbon-1 to carbon-6 and, although this process is reversible, glucose 6-phosphate formation is favored. (The symbol Ⓟ is a shorthand notation for the phosphite group, PO_3^{2-}).

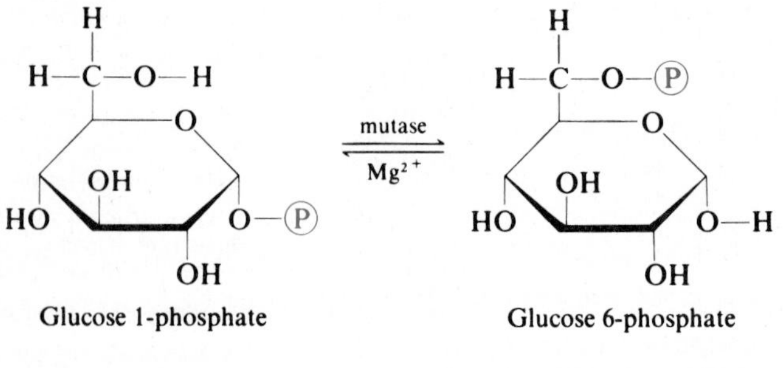

Glucose 1-phosphate   Glucose 6-phosphate

430

Step 1 The initial step, upon entry of glucose into yeast cells and most animal cells, is phosphorylation to glucose 6-phosphate. (With the exception of liver cells, there is very little free glucose inside most cells.) The phosphate donor in this reaction is ATP, and the enzyme, which requires Mg^{2+} ions for its activity, is *hexokinase*.

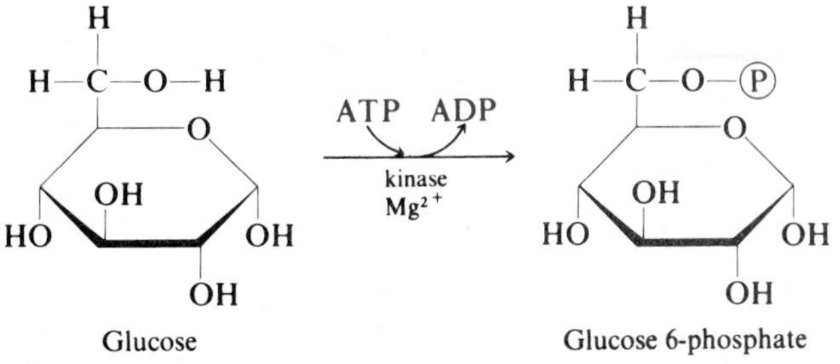

Glucose   Glucose 6-phosphate

The reaction is accompanied by an expenditure of energy, since a molecule of ATP is being utilized rather than synthesized. However, this step is necessary for the activation of the glucose molecule. (We shall see that a second molecule of ATP is expended in step 3. These two initiating molecules of ATP are recovered at a later stage of the pathway.) Reaction 1 is essentially irreversible in the cell; ATP is not formed to an appreciable extent by the reaction of ADP with a simple phosphate ester.

Step 2 Glucose 6-phosphate is isomerized to fructose 6-phosphate by the action of the enzyme *phosphoglucoisomerase*. (In general, an *isomerase* enzyme catalyzes the interconversion of one isomeric form of a sugar to another isomeric form.) *Phosphoglucoisomerase* is highly specific for glucose 6-phosphate.[7]

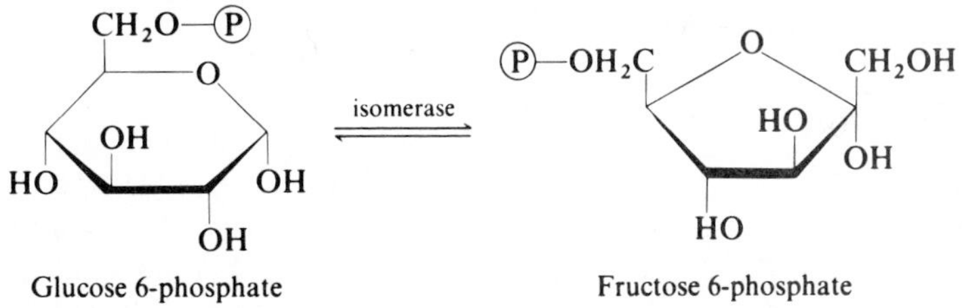

Glucose 6-phosphate   Fructose 6-phosphate

[7] The fate of glucose 6-phosphate differs in muscle cells and in the liver. Since the major role of glycogen in muscle is to provide energy for muscle contraction, the glucose 6-phosphate is immediately converted to fructose 6-phosphate. However, in the liver the chief function of glycogen is to serve as a reservoir of glucose molecules. The liver stores and releases glucose to meet the needs of other tissues. (The liver relies mainly on the oxidation of fatty acids for its energy needs—see Chapter 17.)

When the blood-sugar level decreases, glucose 6-phosphate in the liver is converted into glucose via a hydrolytic reaction catalyzed by *glucose 6-phosphatase*. (Muscle cells lack this enzyme and therefore cannot export glucose.)

$$\text{Glucose 6-phosphate} \;+\; \text{HOH} \; \xrightarrow[\text{(only in the liver)}]{\text{phosphatase}} \; \text{Glucose} \;+\; P_i$$

(Footnote continues)

A mechanism for this aldose–ketose transformation involves the formation of an enediol intermediate and is best understood if the open-chain structures of the sugars are considered (Sections 7.5-b and 11.10-b).

```
      H   O                    H    OH                      H
       \ //                     \  /                        |
        C                        C                      H—C—OH
        |                        ||                         |
   H—C—OH                     C—OH                       C=O
        |                        |                         |
  HO—C—H         ⇌        HO—C—H        ⇌          HO—C—H
        |                        |                         |
   H—C—OH                     H—C—OH                    H—C—OH
        |                        |                         |
   H—C—OH                     H—C—OH                    H—C—OH
        |                        |                         |
     CH₂O—Ⓟ                    CH₂O—Ⓟ                   CH₂O—Ⓟ

  Open-chain form           Enediol              Open-chain form
of glucose 6-phosphate    intermediate        of fructose 6-phosphate
```

$$
\begin{array}{ccc}
\text{Open-chain form of glucose 6-phosphate} & \text{Enediol intermediate} & \text{Open-chain form of fructose 6-phosphate}
\end{array}
$$

Step 3 Next follows another phosphorylation reaction, again involving the utilization of ATP as the phosphate-group donor. The enzyme *phosphofructokinase*[8] is specific for fructose 6-phosphate and, like *hexokinase,* requires Mg^{2+} ions for activity. The reaction is irreversible and necessitates the expenditure of energy from a second molecule of ATP.

```
  Ⓟ—OH₂C    O   CH₂OH              ATP  ADP         Ⓟ—OH₂C    O   CH₂O—Ⓟ
         \  /  \  /                    ↘  ↗                  \  /  \  /
          HO                      ————————————→               HO
           OH                        kinase                     OH
          HO                          Mg²⁺                     HO

    Fructose 6-phosphate                              Fructose 1,6-diphosphate
```

Step 4 Fructose 1,6-diphosphate is then enzymatically cleaved into two molecules of triose phosphate. This cleavage may be regarded as the reverse of an aldol condensation reaction (Section 7.5-a), and hence the enzyme has been named *aldolase.* Again, a better understanding of the reaction is achieved if the fructose 1,6-diphosphate is written in its open-chain form.

The glucose molecules then exit the liver and are transported by the blood for use by other tissues (primarily brain and skeletal muscle). Another genetic disease, *von Gierke's disease,* is caused by a lack of the enzyme *glucose 6-phosphatase* in liver cells. Hypoglycemia occurs because glucose cannot be formed from glucose 6-phosphate.

[8] *Phosphofructokinase* is an allosteric enzyme that is primarily responsible for controlling the rate of glycolysis. This enzyme is strictly regulated by the intracellular concentration of ATP. When there is a high concentration of ATP in the cell, ATP binds to the allosteric site, inhibiting the enzyme, and the pace of glycolysis decreases. When ATP is being utilized to provide energy, the allosteric site is vacant; the enzyme is active, and the glycolytic rate increases.

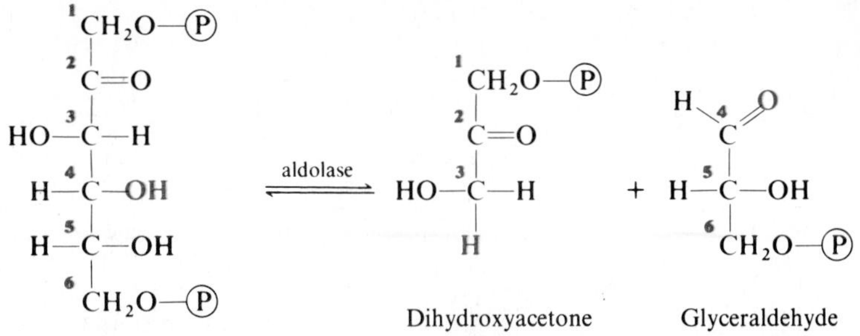

Fructose 1,6-diphosphate   aldolase ⇌   Dihydroxyacetone phosphate   +   Glyceraldehyde 3-phosphate

Step 5 The next step is concerned with the interconversion of the triose phosphates. This is essential since only glyceraldehyde 3-phosphate can be further metabolized by the body. If cells were unable to convert dihydroxyacetone phosphate to glyceraldehyde 3-phosphate, then half of the energy stored in the original glucose molecule would be lost and would accumulate in the cell as the nonmetabolizable ketotriose phosphate. The enzyme for the isomerization reaction is *triose phosphate isomerase,* and the mechanism involves another enediol intermediate.

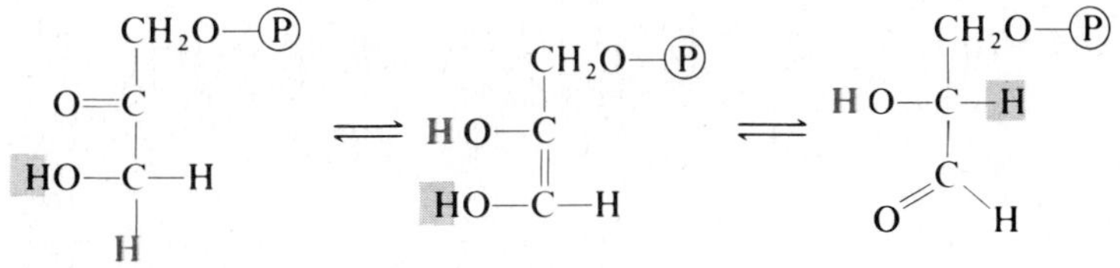

Dihydroxyacetone phosphate   ⇌   Enediol intermediate   ⇌   Glyceraldehyde 3-phosphate

A summation of steps 4 and 5 indicates that the enzymes *aldolase* and *triose phosphate isomerase* have effectively accomplished the conversion of one molecule of fructose 1,6-diphosphate into *two* molecules of glyceraldehyde 3-phosphate. All the remaining steps of glycolysis involve three-carbon compounds. Thus far, the glycolytic pathway has required an energy input in the form of two molecules of ATP but has not yet released any of the energy stored in the glucose.

Step 6 We now come to the first oxidation–reduction reaction in the glycolytic sequence. The conjugated enzyme *glyceraldehyde 3-phosphate dehydrogenase* contains the coenzyme NAD^+. In the process of oxidizing the aldehyde to a carboxylic acid, the coenzyme is reduced to NADH. Notice that the same enzyme also catalyzes a phosphorylation reaction; inorganic phosphate is the phosphate donor. The energy released by the oxidation reaction is utilized for the subsequent phosphorylation reaction. The poisonous effect of iodoacetic acid (recall Table 14.3) in blocking glycolysis occurs at this step. The *dehydrogenase* enzyme contains free sulfhydryl groups at its active site; iodoacetate noncompetitively inhibits the enzyme by covalently bonding to these catalytic groups.

$$
\underset{\substack{\text{Glyceraldehyde}\\\text{3-phosphate}}}{
\begin{array}{c}
\overset{\displaystyle O}{\underset{\displaystyle}{\diagup\!\!\!\diagdown}}{}^{H}\\
\text{C}\\
\text{H}-\text{C}-\text{OH}\\
\text{CH}_2\text{O}-\textcircled{P}
\end{array}}
\quad
\xrightleftharpoons[\text{dehydrogenase}]{\;\;HPO_4^{2-}\quad NAD^+\quad NADH + H^+\;\;}
\quad
\underset{\text{1,3-Diphosphoglyceric acid}}{
\begin{array}{c}
\overset{\displaystyle O}{}{}^{O-\textcircled{P}}\\
\text{C}\\
\text{H}-\text{C}-\text{OH}\\
\text{CH}_2\text{O}-\textcircled{P}
\end{array}}
$$

The reaction is more easily understood if it is considered to occur in two steps: (a) oxidation and (b) phosphorylation. Note, however, that this presentation is an oversimplification that in no way implies the actual enzyme-catalyzed sequence of events.

(a) Oxidation

$$
\begin{array}{c}
\overset{\displaystyle O}{}{}^{H}\\
\text{C}\\
\text{H}-\text{C}-\text{OH}\\
\text{CH}_2\text{O}-\textcircled{P}
\end{array}
\;+\; NAD^+ \;+\; HOH \;\longrightarrow\;
\begin{array}{c}
\overset{\displaystyle O}{}{}^{OH}\\
\text{C}\\
\text{H}-\text{C}-\text{OH}\\
\text{CH}_2\text{O}-\textcircled{P}
\end{array}
\;+\; NADH \;+\; H^+
$$

(b) Phosphorylation

$$
\begin{array}{c}
\overset{\displaystyle O}{}{}^{OH}\\
\text{C}\\
\text{H}-\text{C}-\text{OH}\\
\text{CH}_2\text{O}-\textcircled{P}
\end{array}
\;+\;
\begin{array}{c}
O\\
\parallel\\
HO-P-O^-\\
|\\
O^-
\end{array}
\;\longrightarrow\;
\begin{array}{c}
O\\
\parallel\\
\overset{\displaystyle O}{}{}^{O-P-O^-}\\
\text{C}\qquad |\\
\text{H}-\text{C}-\text{OH}\;\;O^-\\
\text{CH}_2\text{O}-\textcircled{P}
\end{array}
\;+\; HOH
$$

Step 7 1,3-Diphosphoglyceric acid, the product of the reaction in step 6, contains a high-energy acylphosphate bond (recall Table 8.8). It can transfer the phosphate group on carbon-1 directly to a molecule of ADP, thus forming a molecule of ATP. The enzyme that catalyzes the reaction is *phosphoglycerokinase,* and, like all *kinases,* it requires Mg^{2+} ions for activity. It is in this reaction that ATP is first produced in the pathway. Since the ATP is formed by a *direct transfer of a phosphate group from a metabolite to ADP,* the process is referred to as **substrate level phosphorylation** to distinguish it from another ATP-synthesizing process we consider shortly.

$$
\underset{\text{1,3-Diphosphoglyceric acid}}{
\begin{array}{c}
\overset{\displaystyle O}{}{}^{O\sim\textcircled{P}}\\
\text{C}\\
\text{H}-\text{C}-\text{OH}\\
\text{CH}_2\text{O}-\textcircled{P}
\end{array}}
\quad
\xrightleftharpoons[\substack{\text{kinase}\\Mg^{2+}}]{\;\;ADP\quad ATP\;\;}
\quad
\underset{\text{3-Phosphoglyceric acid}}{
\begin{array}{c}
\overset{\displaystyle O}{}{}^{OH}\\
\text{C}\\
\text{H}-\text{C}-\text{OH}\\
\text{CH}_2\text{O}-\textcircled{P}
\end{array}}
$$

Step 8 This step in the pathway is very similar to the intramolecular transfer of phosphate as seen in step ii. The enzyme *phosphoglyceromutase* catalyzes the exchange of a phosphate group from the hydroxyl group of carbon-3 to the hydroxyl group of carbon-2.

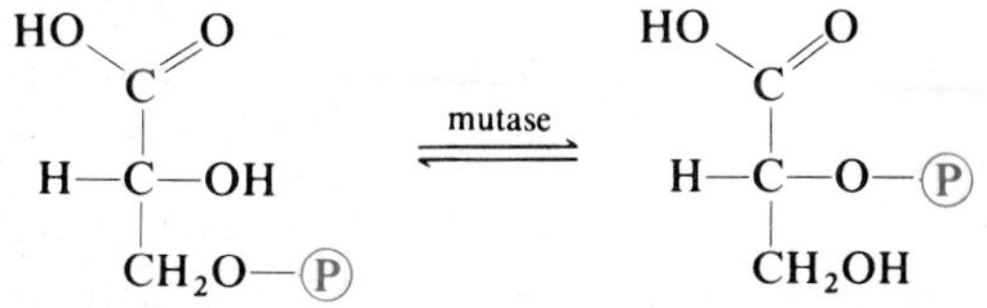

3-Phosphoglyceric acid   2-Phosphoglyceric acid

Step 9 The enzyme *enolase,* which also requires Mg^{2+} ions for activity, catalyzes an alcohol dehydration reaction to produce phosphoenolpyruvic acid, a compound with a high-energy enol phosphate group (recall Table 8.8). Fluoride is an effective poison of the glycolytic pathway because of its inhibition of the enzyme *enolase* (Table 14.3). Fluoride ions bind to the activator Mg^{2+} ions to form a magnesium fluorophosphate complex.

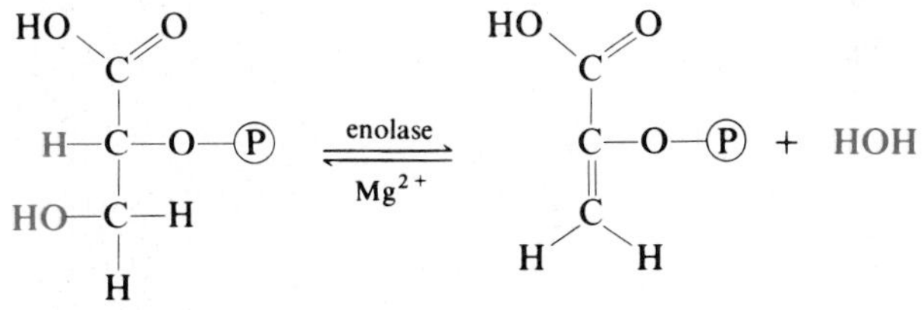

2-Phosphoglyceric acid   Phosphoenolpyruvic acid (PEP)

Step 10 This irreversible step provides a second example of substrate level phosphorylation. The phosphate group of phosphoenolpyruvic acid (PEP) is transferred to ADP; one molecule of ATP is produced per molecule of PEP. It is likely that the reaction proceeds via the enol form of pyruvic acid, which is unstable with respect to tautomeric rearrangement to pyruvic acid. The enzyme *pyruvate kinase* requires both Mg^{2+} and K^+ ions for activity.

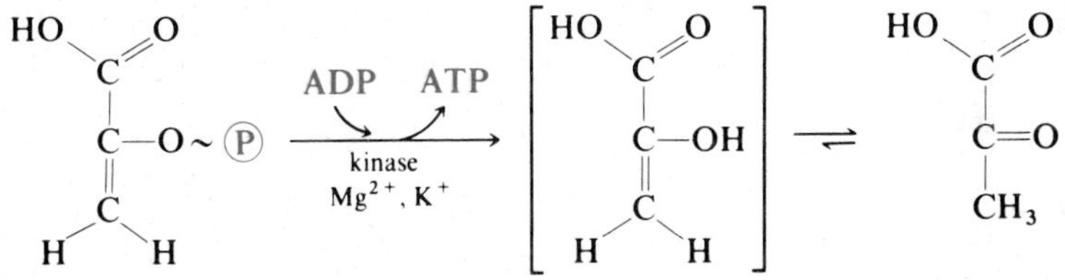

PEP   Enol form of pyruvic acid   Pyruvic acid

Step 11 Steps 1 through 10 are identical for glycolysis and fermentation. Pyruvic acid, as we shall see, is the *crossroads* compound; its metabolic fate is dependent upon the availability of oxygen (anaerobic or aerobic), the organism under

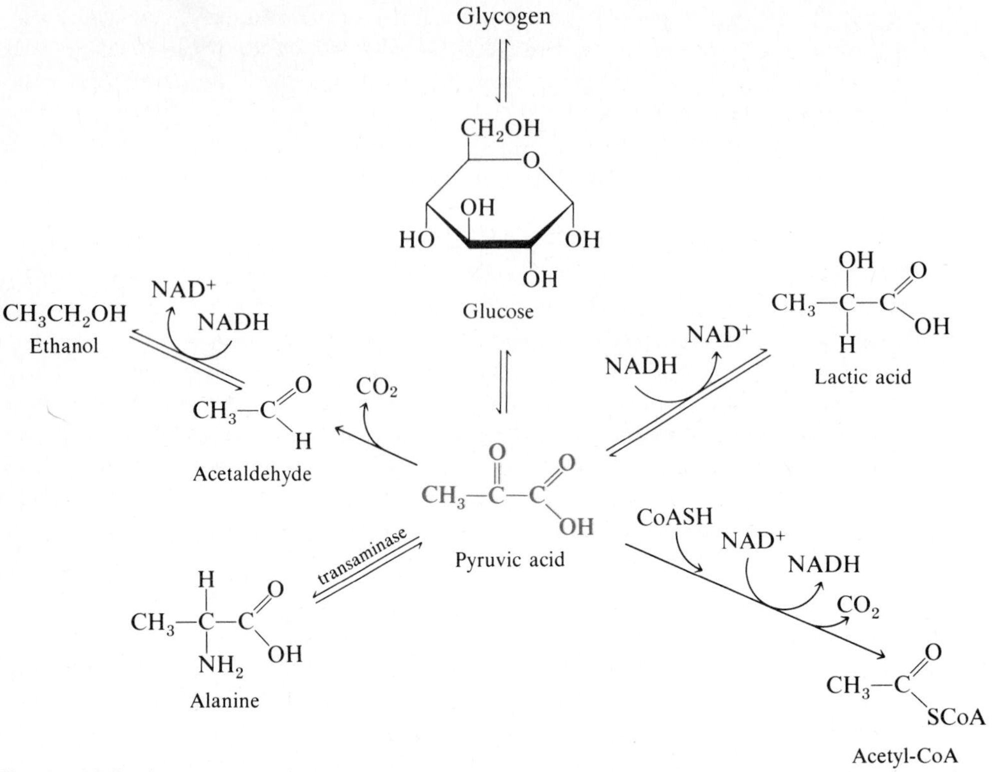

FIGURE 16.6 Importance of pyruvic acid in metabolism.

consideration, and the tissue involved (Figure 16.6). Under usual conditions and during moderate exercise, muscle cells respire aerobically. However, during strenuous exercise, the energy demand on the muscles is enormous, and the respiratory and circulatory systems are unable to deliver oxygen to these cells in sufficient amounts to meet this demand. (This condition is often referred to as *oxygen debt*—not enough oxygen is available for cellular activities.) As a result the muscle cells must obtain their required energy via the anaerobic pathway (glycolysis).

In the presence of NADH the enzyme *lactic acid dehydrogenase*[9] catalyzes the reduction of puruvic acid to lactic acid (ketone to secondary alcohol). This reaction

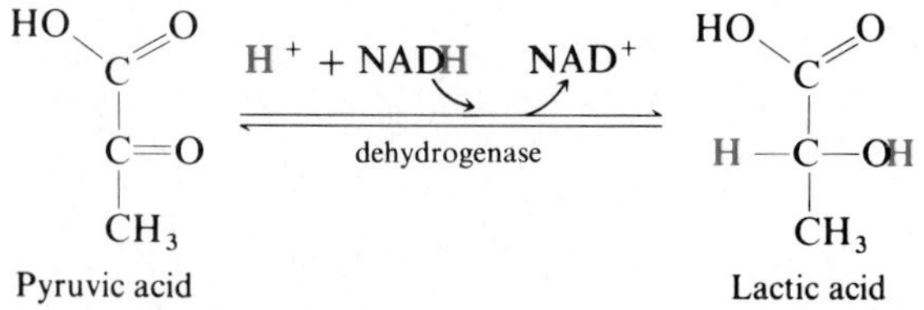

[9]This enzyme derives its name from its catalytic role in reversing this reaction. Recall that enzymes catalyze both the forward and the reverse reactions of an equilibrium mixture and that the enzymes may be named accordingly. An equally descriptive name for this enzyme is *pyruvic acid reductase* or *hydrogenase*.

is essential because it regenerates the NAD^+ that is needed in step 6. If NAD^+ were not replenished, the cell's supply of this coenzyme would be swiftly depleted, anaerobic glycolysis would cease, and glyceraldehyde 3-phosphate would accumulate in the cell.

Lactic acid, then, is the end product of glycolysis,[10] and if there were not some mechanism for its removal, it would accumulate in the muscle cells, raising the level of acidity in these cells. (The intramuscular pH can decrease from a resting value of 7.0 to about 6.5.) Increased acidity impedes muscle performance by causing fatigue (inability of a muscle to contract). How the increase in hydrogen ion concentration leads to muscle fatigue is as yet unclear. Two processes act to maintain a proper level of lactic acid, both of which require oxygen. The heavy breathing that occurs after strenuous exercise helps to supply this oxygen (i.e., repay the oxygen debt).

1. 70–80% of the lactic acid diffuses out of the muscle and is transported to the liver. There it may be oxidized to pyruvic acid and hence to carbon dioxide and water (via the Krebs cycle—see Section 16.8), or converted back into glucose (gluconeogenesis). *The anaerobic catabolism of glucose to lactic acid in muscle cells, the transport of lactic acid via the blood to the liver, and the reconversion of lactic acid to glucose* comprise the **Cori cycle.**

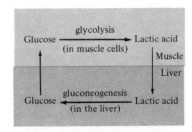

2. The 20–30% that remains within the muscle cells can be reoxidized to pyruvic acid, which then enters the Krebs cycle and is further oxidized to carbon dioxide and water during those periods when the muscle cells are afforded an ample supply of oxygen.

Step 12 Yeast[11] and other microorganisms that can live in a limited supply of oxygen must also have some means of regenerating NAD^+. They do so by first

[10]The average individual, expending normal energy, produces about 120 g of lactic acid each day. (Vigorous exercise increases this value well above 120 g.) It is estimated that one third of this daily lactic acid is produced by tissues that are strictly anaerobic (i.e., erythrocytes, retina, renal medulla, epidermis— these tissues do not have mitochondria).

[11]Yeast and many other organisms (including humans) are referred to as *facultative anaerobes.* Facultative anaerobes may function anaerobically or aerobically depending on the circumstances (i.e., they can utilize either an organic molecule or molecular oxygen as the terminal electron acceptor). In the absence of oxygen, yeasts metabolize glucose to produce carbon dioxide and ethanol, whereas in the presence of oxygen they metabolize glucose to carbon dioxide and water. (This is the reason that air must be excluded from the fermentation vats in the production of alcoholic beverages.) *Strict anaerobes,* such as the bacteria that cause gangrene and botulism, can only use organic molecules as terminal electron acceptors in metabolism and are poisoned in the presence of oxygen.

decarboxylating pyruvic acid to acetaldehyde. This reaction, which is catalyzed by *pyruvic acid decarboxylase,* is responsible for CO_2 production in yeast (the reason yeast is used in certain baking processes). The enzyme requires the coenzyme thiamine pyrophosphate (TPP—see Table 14.2) and Mg^{2+} ions. The decarboxylation reaction is irreversible.

Pyruvic acid   Acetaldehyde

Step 13 The final step in the fermentation process is catalyzed by the enzyme *alcohol dehydrogenase.* Acetaldehyde, the terminal electron acceptor in fermentation, is reduced by NADH to ethyl alcohol,[12] thus regenerating NAD^+. (When yeast

[12]About 3 g/day of ethyl alcohol is produced continuously by microbial fermentation in the large intestine. For many people this amount is insignificant compared with that ingested from alcoholic beverages. In humans the principal route of metabolism of ingested alcohol is believed to be oxidation in the liver, first to acetaldehyde and subsequently to acetic acid. Most of the acetic acid is released by the liver and is transported to other tissues where it is converted to acetyl-CoA. (As we shall see, the acetyl-CoA is metabolized to carbon dioxide and water, or it is utilized in the biosynthesis of fat.)

Alcohol readily crosses the placental membrane and builds up in the fetus because the liver of the fetus does not contain the detoxifying enzymes. Therefore, children of alcoholic mothers have a high risk of *fetal alcohol syndrome* (FAS). FAS consists of facial deformities, growth deficiency, and mental retardation. FAS is believed to be the third most common cause of mental deficiency. (IQs average 35–40 points below normal.)

After heart disease and cancer, alcohol addiction is the third largest health problem in the United States. More than 200,000 people die of it each year (mainly as a result of cirrhosis of the liver, but also cardiovascular disease and cancers of the mouth and larynx), and it is a factor in one of every 10 deaths. People who drive while intoxicated are responsible for about 50% of United States traffic fatalities. Alcohol-impaired driving is the leading cause of death and injury among those under 25 years of age.

One proven method in the treatment of chronic alcoholism is to administer a drug that inhibits the second step of alcohol metabolism. The drug disulfiram (Antabuse) successfully competes with acetaldehyde for the active site on the enzyme *acetaldehyde dehydrogenase.* As a result, when alcohol is ingested by an individual previously treated with disulfiram, the blood acetaldehyde concentrations rise five to ten times higher than in an untreated individual. This effect is accompanied by severe discomfort. Characteristic physiological responses are an increase of the heartbeat and a reduction of the blood pressure. Respiratory difficulties, nausea, sweating, vomiting, chest pains and blurred vision are observed. The effect once elicited lasts between 30 minutes and several hours. To avoid such discomfort, the patients are forced to abstain from alcohol (and the treatments are continued until they can abstain voluntarily). Disulfiram should be administered only by a physician, and therapy is usually commenced in a hospital.

Disulfiram

is used in baking, the alcohol readily evaporates out of the oven.)

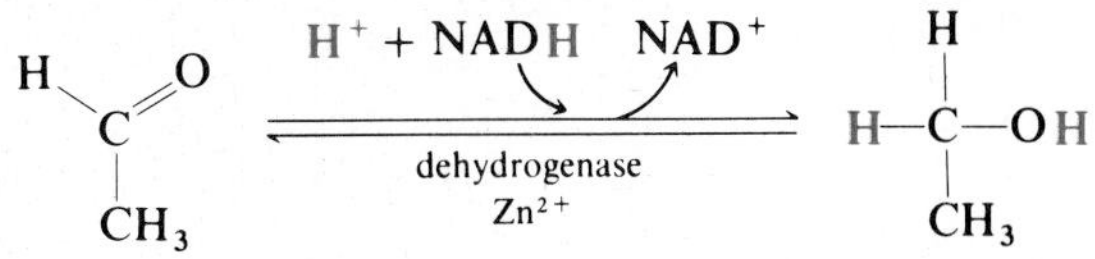

$$
\begin{array}{c}
H \\
\diagdown \\
C = O \\
\diagup \\
CH_3
\end{array}
\quad
\underset{\substack{\text{dehydrogenase} \\ Zn^{2+}}}{\overset{H^+ + NADH \quad NAD^+}{\rightleftharpoons}}
\quad
\begin{array}{c}
H \\
| \\
H - C - OH \\
| \\
CH_3
\end{array}
$$

Acetaldehyde   Ethanol

16.5   *Reversal of Glycolysis and Fermentation*

It has been mentioned that several of the degradative steps of the Embden–Meyerhof pathway are not reversible in vivo. (Single arrows are used in Figure 16.5 to indicate that reactions i, 1, 3, 10, and 12 are irreversible.) This seems to be inconsistent with the fact that lactic acid and any of the other intermediates of the glycolytic sequence can be converted back to glucose and/or glycogen provided an energy source is available. Therefore, other reactions catalyzed by different enzymes must necessarily effect a reversal of steps i, 1, 3, and 10. The irreversibility of parts of a metabolic sequence is a common phenomenon; it will be encountered again in aerobic metabolism as well as in the metabolism of lipids and proteins. Catabolic and anabolic processes may have many of the same intermediates and many of the same enzymes, yet their pathways are not identical. Similarly, more than one pathway may exist for the same process. The separation of degradative and synthetic pathways allows an organism to control one series of reactions without affecting the other. Since we are concerned primarily with reactions that yield energy and produce ATP, we shall not take up the anabolism of carbohydrates. These reactions are found in advanced biochemistry texts.

16.6   *Bioenergetics of Glycolysis and Fermentation*

The net energy yield in the form of high-energy phosphate bonds (moles of ATP) obtained from the anaerobic metabolism of each mole of glucose can be easily calculated from Figure 16.5.

One mole of ATP is expended in the initial phosphorylation of glucose (step 1).

$$
\text{Glucose} \quad \xrightarrow{\quad \overset{ATP \quad ADP}{\searrow \quad \nearrow} \quad} \quad \text{Glucose 6-phosphate}
$$

And a second mole is consumed in the phosphorylation of fructose 6-phosphate (step 3).

$$
\text{Fructose 6-phosphate} \quad \xrightarrow{\quad \overset{ATP \quad ADP}{\searrow \quad \nearrow} \quad} \quad \text{Fructose 1,6-diphosphate}
$$

Steps 4 and 5 are very significant for it is in these reactions that each mole of six-carbon sugar is transformed into 2 moles of the triose phosphate, glyceraldehyde 3-phosphate. In step 7, each mole of 1,3-diphosphoglyceric acid is converted into

1 mole of 3-phosphoglyceric acid; 1 mole of ATP is produced per mole of triose phosphate, or *2 moles of ATP per mole of glucose.*

$$\text{1,3-Diphosphoglyceric acid} \xrightarrow[\quad\quad\quad]{\quad ADP\quad ATP \quad} \text{3-Phosphoglyceric acid}$$

In reaction 10, 1 mole of ATP is generated for each mole of pyruvic acid that is formed from phosphoenolpyruvic acid; again 2 moles of ATP are produced for each mole of glucose that entered the pathway.

$$\text{Phosphoenolpyruvic acid} \xrightarrow[\quad\quad\quad]{\quad ADP\quad ATP \quad} \text{Pyruvic acid}$$

A summation of all these steps reveals that for every mole of glucose degraded, 2 moles of ATP are initially consumed and 4 moles of ATP are ultimately produced. The net production of ATP is thus 2 moles per mole of glucose converted to lactic acid or to ethanol. If, however, glycogen is used as the source of glucose, steps i and ii are operative instead of step 1. It would then be necessary to expend only 1 mole of ATP (in step 3) in order to produce fructose 1,6-diphosphate, and the net yield of ATP would be 3 moles per mole of glucose 1-phosphate. Thus, the utilization of glycogen for anaerobic energy production represents a 50% increase in efficiency compared to glucose.

In Yeast

$$C_6H_{12}O_6 \ + \ 2\,ADP \ + \ 2\,P_i \ \longrightarrow \ 2\,C_2H_5OH \ + \ 2\,CO_2 \ + \ 2\,ATP$$

In Muscle—Starting with Glycogen

$$(C_6H_{12}O_6)_n \ + \ 3\,ADP \ + \ 3\,P_i \ \longrightarrow \ (C_6H_{12}O_6)_{n-1} \ + \ 2\,CH_3CHOHCOOH \ + \ 3\,ATP$$

Recall that about 7300 cal of free energy is conserved per mole of ATP produced (Section 8.14). Recall also that the total amount of energy that can theoretically be obtained from the complete oxidation of 1 mole of glucose is 686,000 cal. The energy conserved by anaerobic metabolism, then, is only a minute amount of the total energy available, that is, either 2.1% (14,600/686,000) or 3.2% (21,900/686,000). Thus, anaerobic cells extract only a small fraction of the total energy of the glucose molecule, yet this amount is sufficient for their survival. Since they are not nearly as efficient as aerobic cells, it is necessary that they utilize more glucose per unit of time to accomplish the same amount of cellular work.

16.7 *Storage of Chemical Energy*

Before continuing on to the aerobic phase of carbohydrate metabolism, we shall consider how muscle cells store and utilize the energy that they extract from carbohydrates (and other compounds). Strenuous muscle activity requires the

$$
\begin{array}{c}
\text{O} \\
\parallel \\
\text{H}-\text{N}\sim\text{P}-\text{O}^- \\
\mid \\
\text{O}^- \\
\mid \\
\text{C}=\text{NH} \\
\mid \\
\text{N}-\text{CH}_3 \\
\mid \\
\text{CH}_2\text{COO}^-
\end{array}
$$

Creatine phosphate
(Phosphocreatine)

$$
\begin{array}{c}
\text{O} \\
\parallel \\
\text{H}-\text{N}\sim\text{P}-\text{O}^- \\
\mid \\
\text{O}^- \\
\mid \\
\text{C}=\text{NH} \\
\mid \\
\text{N}-\text{H} \\
\mid \\
(\text{CH}_2)_3 \\
\mid \\
\text{H}-\text{C}-\text{NH}_3^+ \\
\mid \\
\text{COO}^-
\end{array}
$$

Arginine phosphate
(Phosphoarginine)

continuous production of ATP by glycolysis. *The splitting of ATP to ADP and inorganic phosphate furnishes the energy for muscle contraction.*

In 1927, a compound called *creatine phosphate* was isolated from mammalian muscle and was subsequently demonstrated to be the storage form of energy in the muscles of vertebrates and in nerve tissue. *Arginine phosphate* serves the same purpose in the muscles of invertebrates. These high-energy phosphate compounds, which serve as reservoirs of phosphate-bond energy, are often called *phosphagens.* The high-energy bond of these particular phosphagens occurs between phosphorus and nitrogen rather than between phosphorus and oxygen.

The turnover rate of ATP is very high, and this precludes its use as a storage form of energy. A 70-kg man will hydrolyze and resynthesize about 70 kg of ATP per day. Therefore, the concentration of ATP in muscle is relatively low and cannot meet the demands of muscular exertion for more than a fraction of a second.[13] At rest, mammalian muscle contains four to six times as much creatine phosphate as ATP. As ATP is utilized, creatine phosphate, in the presence of *creatine kinase,*[14] reacts with ADP to produce more ATP and creatine.

$$
\text{Creatine phosphate} \;+\; \text{ADP} \;\underset{\text{Mg}^{2+}}{\overset{\text{kinase}}{\rightleftharpoons}}\; \text{Creatine} \;+\; \text{ATP}
$$

The reaction is readily reversible. When muscular activity is required, the reaction proceeds to the right. When there is an abundant amount of ATP (formed from the catabolism of foodstuffs), the reaction proceeds to the left and creatine phosphate is stored in the muscle cells. Since the concentration of creatine phosphate in the muscle is limited, it is only useful in generating a quick source of utilizable energy. It has been estimated that the creatine phosphate can provide energy for about 20 seconds of strenuous exercise (e.g., long enough to sprint 200 m). Recall we said earlier that energy for prolonged activity is obtained from the anaerobic breakdown of glycogen that is synthesized during long periods of muscular inactivity. The

[13] Trained athletes can exert themselves more strenuously for several reasons: (1) They have larger muscle cells and a greater number of mitochondria in each muscle cell; hence, they produce ATP at a greater rate. (2) Their muscles contain a greater amount of the contractile proteins (actin and myosin). (3) Their cardiac muscles contain an abundant supply of myoglobin, the protein that binds oxygen; this greater amount of stored oxygen allows for prolonged aerobic activity. (4) The physical training stimulates additional capillaries to grow within the muscles, bringing additional oxygen-carrying blood to the muscles.

Conversely, a muscle that is not used to any great extent tends to decrease in size and strength. There is a reduction in the number of mitochondria within each muscle cell and a reduction of the number of capillaries within the muscle fibers. This condition is called *atrophy* and is particularly noticeable in an individual that has removed a cast after several weeks of limb immobilization.

[14] Recall (Section 14.10) that the measurement of *creatine kinase* (CK) activity in the blood is of value in the diagnosis of disorders affecting skeletal and cardiac muscle. A heart attack (myocardial infarction) damages heart muscle cells, and the damaged cells "leak" their enzymes, including CK, into the extracellular fluid and plasma, resulting in elevated serum CK levels. The increase in serum CK concentration is apparent 3–6 hours after a heart attack and reaches its highest extent 18–30 hours after the attack. Thus the CK test is a standard clinical test for the detection of a suspected myocardial infarction (see Figure 14.17; also see page 480 for the GOT test).

glycogen stored in muscle cells is more readily available for energy production than is glucose from the blood. In the days just prior to a marathon, competitors will "load up" on carbohydrates in order to maximize the amount of glycogen stored in muscle cells.

The human body can store roughly 450 g of glucose as glycogen (about 350 g of glycogen in all of the muscle cells and about 100 g of glycogen in the liver). The liver glycogen reservoir allows us to eat intermittently, since it provides the glucose to maintain the blood-sugar level. Stored glycogen, however, does not last long. After a 24-hour fast, very little glycogen remains in the liver. We shall see in Chapter 17 that when carbohydrate intake exceeds the amount required for energy and for storage as glycogen, the excess is converted to fat to be stored in adipose tissue. When the glycogen reserves are depleted, fatty acids provide the bulk of the energy for muscle cells.

16.8 *Aerobic Metabolism*

We have mentioned that the anaerobic phase of glucose catabolism ends with the production of lactic acid or ethanol and that it accounts for only a small fraction of the total available energy of the glucose molecule. Aerobic metabolism is of extreme importance for providing energy to the brain, kidneys, liver, and skeletal muscles. The cells that function aerobically (those that require oxygen to live) utilize the identical series of glycolytic reactions up to the production of pyruvic acid. In aerobic metabolism, however, pyruvic acid is not reduced to lactic acid but is oxidized in a number of discrete enzymatic reactions to carbon dioxide and water.

$$C_6H_{12}O_6 \xrightarrow{\text{glycolysis}} 2\,CH_3CHOHCOOH \;+\; \sim 47{,}000 \text{ cal}$$

$$2\,CH_3COCOOH \;+\; 5\,O_2 \longrightarrow 6\,CO_2 \;+\; 4\,H_2O \;+\; \sim 639{,}000 \text{ cal}$$

From the standpoint of energy production, the oxidation of pyruvic acid is of considerable significance because it liberates most of the energy (93%) stored in the glucose molecule. Much of this energy is stored in a chemical form in the high-energy phosphate bonds of ATP.

A scheme for the complex series of reactions that effects the oxidation of pyruvic acid to carbon dioxide and water was first proposed by Sir Hans Krebs in 1937 (Nobel Prize for Medicine, 1953). Pyruvic acid is transported from the cytosol into the inner compartment of the mitochondrion (see Figure 16.8). Within the mitochondrion, pyruvic acid is first decarboxylated to a two-carbon compound, which then enters a cyclic sequence of reactions known collectively as the **Krebs cycle,** the **tricarboxylic acid cycle,** or the **citric acid cycle.** The Krebs cycle functions to produce ATP and to provide metabolic intermediates for the biosynthesis of needed compounds. As we see in later chapters, the Krebs cycle is not restricted to the metabolism of carbohydrates. It also plays a vital role in the metabolism of lipids and proteins, and thus occurs in almost all cells of higher animals. Every enzyme that participates in the cycle has been identified, and the operation of this metabolic pathway in vivo has been completely verified by the use of isotopic tracers. All of the enzymes involved in the Krebs cycle are located within the mitochondria. Aerobic metabolism is thus separated from the reactions of glycolysis, which occur in the cytosol.

Figure 16.7 is a schematic outline of the Krebs cycle. Each reaction is numbered, and the individual steps of the sequence are now considered in detail.

Step 1 The pyruvic acid formed in the Embden–Meyerhof pathway is not an intermediate in the Krebs cycle. It must first be enzymatically decarboxylated and oxidized (*oxidative decarboxylation*) to yield the extremely important intermediate **acetyl coenzyme A.** The formation of acetyl-CoA from pyruvic acid requires the sequential action of three different enzymes and the participation of five coenzymes: coenzyme A, thiamine pyrophosphate (TPP), lipoic acid, NAD^+, and FAD. (See Table 14.2 for structures of the coenzymes.)

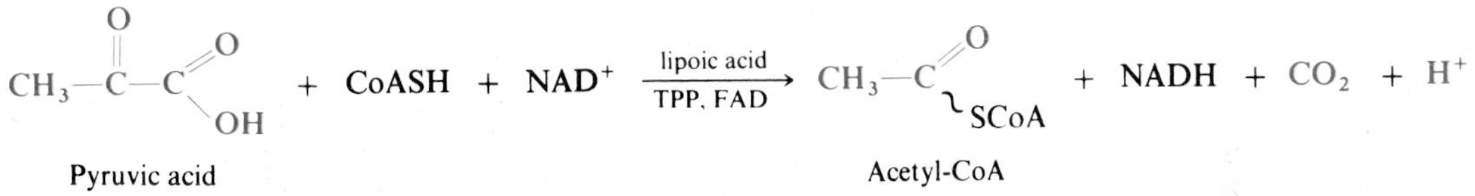

Pyruvic acid   Acetyl-CoA

The name given to this multienzyme system is *pyruvic acid dehydrogenase complex*. The mechanism of the reaction is believed to involve four distinct steps, decarboxylation, reductive acetylation, acetyl transfer, and electron transport. The initial decarboxylation step is irreversible; hence the overall conversion of pyruvic acid to acetyl-CoA is irreversible.

Step 2 Acetyl-CoA is likewise not a true intermediate in the Krebs cycle. It enters the cycle by condensing with the four-carbon dicarboxylic acid oxaloacetic acid, yielding citric acid, the six-carbon tricarboxylic acid for which the cycle was named. Notice that the reaction is very similar to an aldol condensation reaction (page 126) and for this reason the enzyme was originally referred to as the *condensing enzyme*. This enzyme is now called *citric acid synthetase*. The reaction is highly exothermic because of the energy released by the subsequent hydrolysis of the thioester bond of acetyl-CoA (7500 cal/mole). The two carbon atoms that originate from acetyl-CoA are shown in color here and in subsequent reactions. Note that this step regenerates coenzyme A.

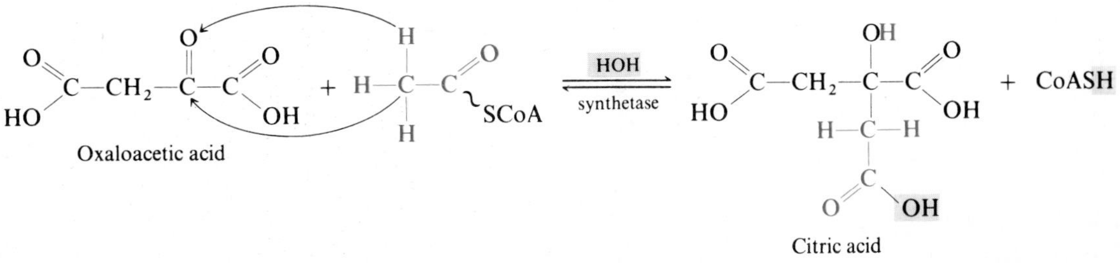

Oxaloacetic acid   Citric acid

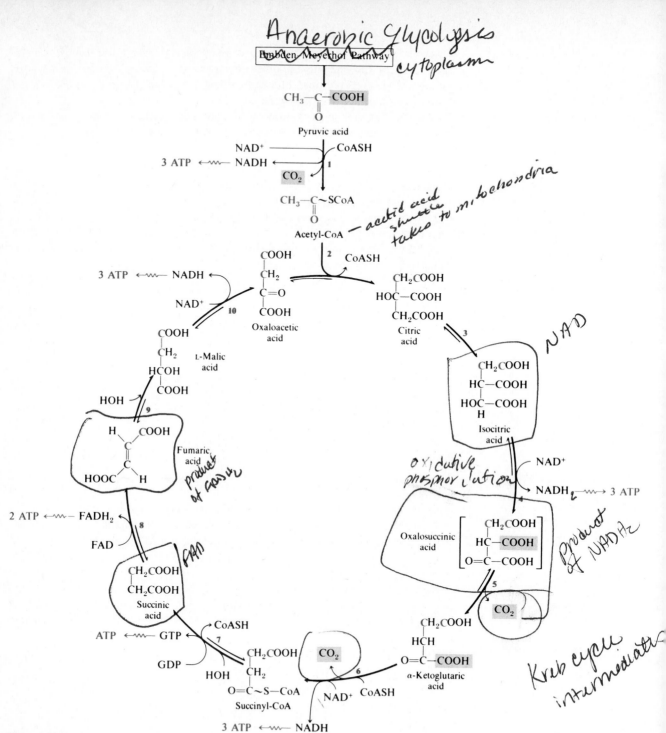

FIGURE 16.7 The Krebs cycle.

The handwritten annotations on the figure read:

Anaerobic Glycolysis
cytoplasm

— acetic acid shuttle takes to mitochondria

oxidative phosphorylation

product of NADH₂

product of FADH₂

Krebs cycle intermediates

NAD

FAD

The printed labels within the cycle diagram:

Embden-Meyerhof Pathway

$CH_3-\overset{O}{\underset{\|}{C}}-COOH$
Pyruvic acid

NAD^+ / NADH → 3 ATP
CoASH
1
CO_2

$CH_3-\overset{O}{\underset{\|}{C}}\sim SCoA$
Acetyl-CoA

2   CoASH

$\begin{array}{l}COOH\\CH_2\\C=O\\COOH\end{array}$
Oxaloacetic acid

3 ATP ← NADH ← 10
NAD^+

$\begin{array}{l}CH_2COOH\\HOC-COOH\\CH_2COOH\end{array}$
Citric acid

$\begin{array}{l}COOH\\CH_2\\HCOH\\COOH\end{array}$
L-Malic acid

3

$\begin{array}{l}CH_2COOH\\HC-COOH\\HOC-COOH\\H\end{array}$
Isocitric acid

HOH
9

$\begin{array}{l}H\quad COOH\\ \quad C\\ \quad C\\HOOC\quad H\end{array}$
Fumaric acid

NAD^+
$NADH_2$ → 3 ATP
4

2 ATP ← FADH₂
FAD
8

Oxalosuccinic acid
$\begin{bmatrix}CH_2COOH\\HC-COOH\\O=C-COOH\end{bmatrix}$

$\begin{array}{l}CH_2COOH\\CH_2COOH\end{array}$
Succinic acid

5
CO_2

ATP ← GTP ←
GDP
CoASH
7

HOH
$\begin{array}{l}CH_2COOH\\CH_2\\O=C\sim S-CoA\end{array}$
Succinyl-CoA

CO_2
6

$\begin{array}{l}CH_2COOH\\HCH\\O=C-COOH\end{array}$
α-Ketoglutaric acid

NAD^+ CoASH

3 ATP ← NADH

Step 3 The next step is sometimes considered to be two separate reactions. The single enzyme *aconitase* catalyzes successive dehydration and hydration reactions. The net result is the isomerization of citric acid to its less symmetrical isomer, isocitric acid. The intermediate is *cis*-aconitic acid, and it normally remains bound to the enzyme. (Notice that the addition of water to the double bond of *cis*-aconitic acid is antiMarkovnikov—Section 3.4-b.)

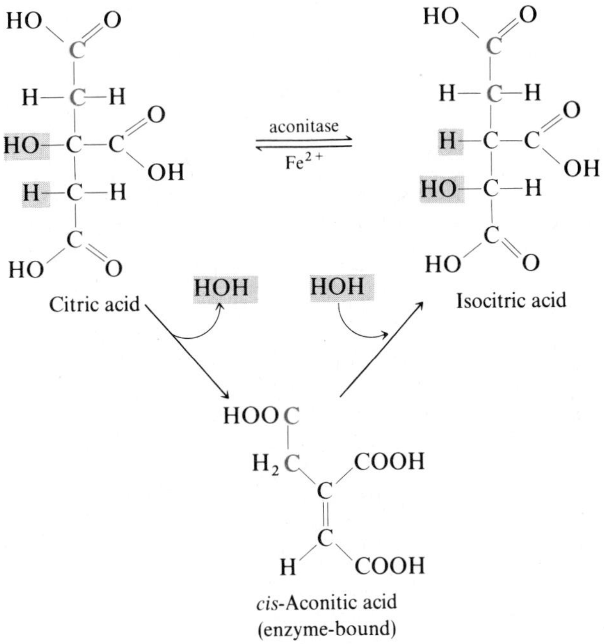

Citric acid

aconitase
Fe^{2+}

HOH   HOH

HOOC

H_2C COOH

C

C

H   COOH

cis-Aconitic acid
(enzyme-bound)

Isocitric acid

Steps 4 and 5 The next two steps are usually discussed together because experimental evidence indicates that the intermediate, oxalosuccinic acid, does not exist free but is firmly bound to the surface of the enzyme. The enzyme *isocitric acid dehydrogenase* catalyzes the oxidative decarboxylation of isocitric acid to α-ketoglutaric acid.

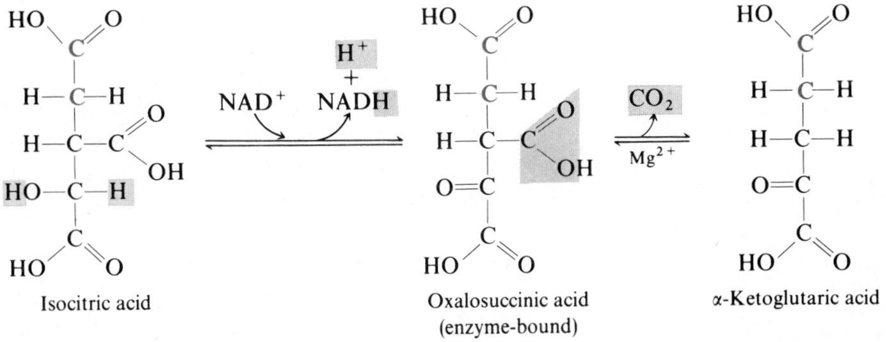

Isocitric acid

NAD^+ H^+ + NADH

Oxalosuccinic acid
(enzyme-bound)

CO_2
Mg^{2+}

α-Ketoglutaric acid

Step 6 This step is analogous to step 1; it is catalyzed by another multienzyme system known as *α-ketoglutaric acid dehydrogenase complex,* which requires the same five coenzymes as *pyruvic acid dehydrogenase.* This is the only nonreversible reaction (because of the decarboxylation step) in the Krebs cycle and, as such, prevents the cycle from operating in the reverse direction.

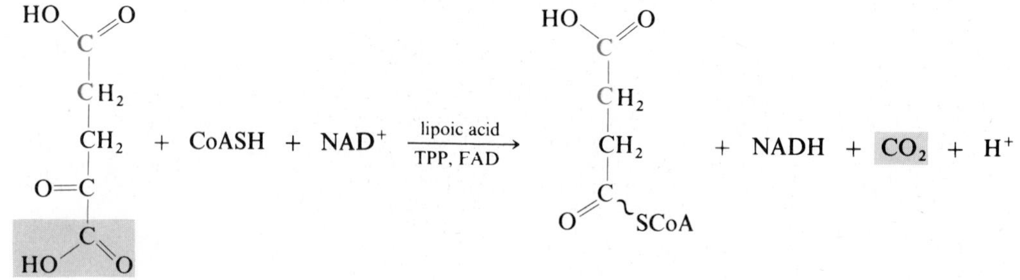

α-Ketoglutaric acid   Succinyl-CoA

Step 7 In this reaction the energy released by the hydrolysis of the high-energy thioester bond of succinyl-CoA ($\sim$8000 cal/mole) is utilized to form guanosine triphosphate (GTP) from guanosine diphosphate (GDP) and inorganic phosphate. This reaction is significant because GTP has a higher free energy of hydrolysis than ATP and can readily transfer its terminal phosphate group to ADP to generate ATP in the presence of the enzyme *nucleoside diphosphokinase.* Here we have another example of substrate level phosphorylation. It is the only reaction in the Krebs cycle that directly involves a high-energy phosphate bond.

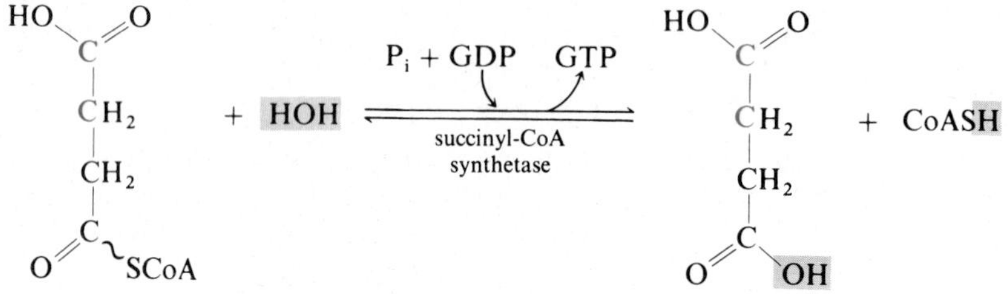

Succinyl-CoA   Succinic acid

$$GTP \ + \ ADP \ \overset{kinase}{\rightleftharpoons} \ GDP \ + \ ATP$$

Step 8 The enzyme *succinic acid dehydrogenase* catalyzes the removal of two hydrogen atoms from succinic acid, thus forming fumaric acid. You will recall that this enzyme is competitively inhibited by malonic acid (Section 14.8-a). This dehydrogenation reaction is the only one in the cycle that utilizes the coenzyme FAD rather than NAD^+. *Succinic acid dehydrogenase* is the only enzyme of the Krebs cycle that is located within the inner mitochondrial membrane. This permits electrons to be transferred directly into the electron transport chain when the $FADH_2$ is reoxidized (see footnote 15, page 451).

446

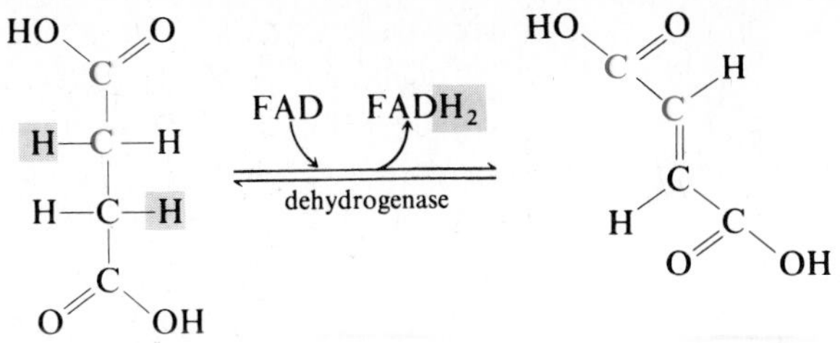

Succinic acid   Fumaric acid

Step 9 The addition of a molecule of water across the double bond of fumaric acid to form L-malic acid is catalyzed by the enzyme *fumarase*. This enzyme is highly stereospecific; *trans* addition occurs so that only the L-isomer of malic acid is produced.

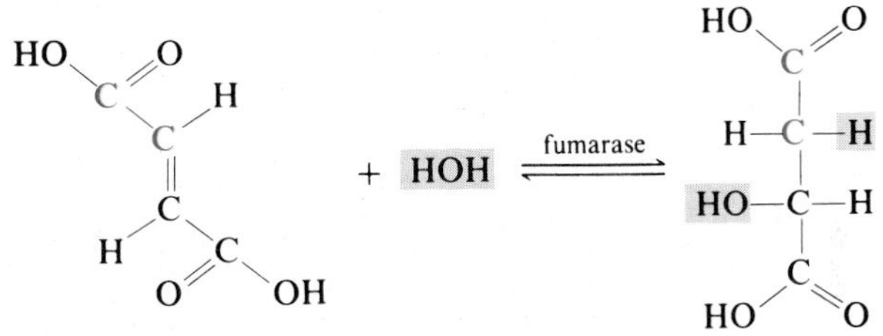

Fumaric acid   L-Malic acid

Step 10 One revolution of the cycle is completed when the oxidation of L-malic acid to oxaloacetic acid is brought about by the enzyme *malic acid dehydrogenase*. This is the fourth oxidation–reduction reaction that utilizes NAD^+ as the oxidizing agent.

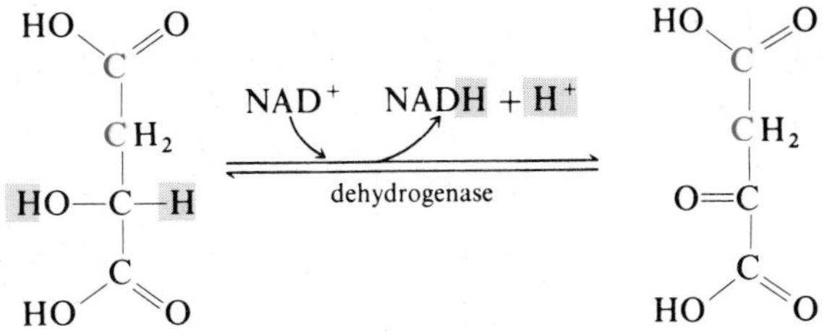

L-Malic acid   Oxaloacetic acid

16.9   *Respiratory Chain (Electron-Transport Chain)*

We have stated that aerobic metabolism occurs only in the presence of molecular oxygen. It has also been mentioned that the major portion of solar energy stored in carbohydrates is conserved in the process. Yet nowhere in the Krebs cycle has the utilization of oxygen or the conservation of energy been indicated. None of the intermediates of the cycle has been shown to be linked to phosphate groups, and no direct synthesis of ATP took place from ADP and inorganic phosphate. The Krebs cycle deals primarily with the fate of the carbon skeleton of pyruvic acid, describing the metabolites involved in its conversion to carbon dioxide. Two carbon atoms enter the cycle as actyl-CoA (step 2), and two different carbon atoms exit the cycle as carbon dioxide (steps 5 and 6). The coenzymes NAD^+ and FAD are reduced to NADH and $FADH_2$; no mechanism has yet been indicated for their regeneration. The reduced coenzymes must be reoxidized if the aerobic phase of carbohydrate metabolism is to continue.

All the enzymes and cofactors that are necessary for the Krebs cycle and for the conservation of energy are localized in the mitochondria, small organelles that are often referred to as the "power plants" of the cell (Figure 16.8). **Mitochondria** are *oval-shaped, dual-membrane structures.* They may be randomly distributed throughout the cytosol, or they may be organized in regular rows or clusters. A cell may contain 100–1000 mitochondria depending on its function. A mitochondrion has two lipid/protein membranes, an *outer membrane* and an *inner membrane* that is extensively folded into a series of internal ridges called *cristae*. Thus there are two compartments in mitochondria: the *intermembrane space* and the *matrix*, which is surrounded by the inner membrane. The outer membrane is permeable, whereas the inner membrane is impermeable to most molecules and ions. (Water, oxygen, and carbon dioxide can freely penetrate both membranes.) The mitochondrial matrix contains all the enzymes of the Krebs cycle, with the exception of *succinic acid dehydrogenase,* which is embedded in the inner membrane. The enzymes that operate to provide energy for the cell are also contained within the inner mitochondrial membrane and are positioned in geometrically specific arrays such that they are capable of functioning as extremely efficient assembly lines. *This sequence of highly organized oxidation–reduction enzymes* is known as the **respiratory chain** or the **electron-transport chain.** As we shall see, the operation of the chain is analogous to a bucket brigade.

Figure 16.9 illustrates the respiratory chain in the mitochondria. The sequence in which the electron carriers of the chain operate is determined by their respective reduction potentials. The reduction potential is a measure of the tendency of a

FIGURE 16.8 Schematic three-dimensional (a) and two-dimensional (b) representations of a liver mitochondrion.

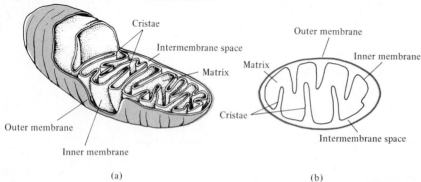

(a)   (b)

Chapter 16 · Carbohydrate Metabolism

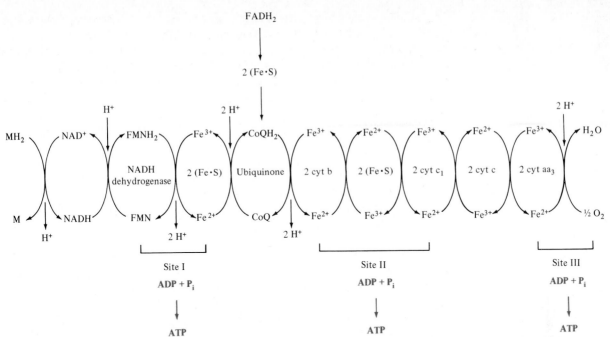

FIGURE 16.9 A schematic diagram of the respiratory chain depicting the oxidized and reduced forms of the carriers, the input and output of protons from the mitochondrial matrix, and the sites along the chain where oxidative phosphorylation occurs.

substance to gain electrons when compared to the standard hydrogen electrode (25°C, 1 atm H_2, and 1 M H^+), which is arbitrarily assigned a value of 0.0 volt. Since in biological systems we are more interested in neutral solutions, a correction is made for changes in pH. At pH 7 the hydrogen electrode has a potential difference of -0.42 volt when measured against the standard hydrogen electrode. The reduction potentials of some of the intermediates of the respiratory chain are listed in Table 16.1.

TABLE 16.1 Standard Reduction Potentials for Some Respiratory Chain Components

System (Oxidant/Reductant)	E (at pH 7) (volts)
$NAD^+/NADH$	-0.32
$FMN/FMNH_2$	-0.30
$CoQ/CoQH_2$	$+0.04$
Cyt b-Fe(III)/cyt b-Fe(II)	$+0.07$
Cyt c_1-Fe(III)/cyt c_1-Fe(II)	$+0.23$
Cyt c-Fe(III)/cyt c-Fe(II)	$+0.25$
Cyt aa_3-Fe(III)/cyt aa_3-Fe(II)	$+0.29$
$\frac{1}{2} O_2/H_2O$	$+0.82$

For each molecule of pyruvic acid that is converted to carbon dioxide and water via the Krebs cycle, five dehydrogenation reactions occur. NAD^+ serves as the electron acceptor in four of these (steps 1, 4, 6, and 10), and FAD is the oxidizing agent in the fifth (step 8). In Figure 16.9, MH_2 symbolizes the reduced metabolites (pyruvic acid, isocitric acid, α-ketoglutaric acid, and malic acid) and M signifies the oxidized metabolites (acetyl-CoA, oxalosuccinic acid, succinyl-CoA, and oxaloacetic acid).

We now examine one of these conversions by writing a balanced half-reaction.

$$HOOC-CH_2-\overset{OH}{\underset{H}{\overset{|}{\underset{|}{C}}}}-COOH \longrightarrow HOOC-CH_2-\overset{O}{\overset{||}{C}}-COOH + 2\,H^+ + 2e^-$$

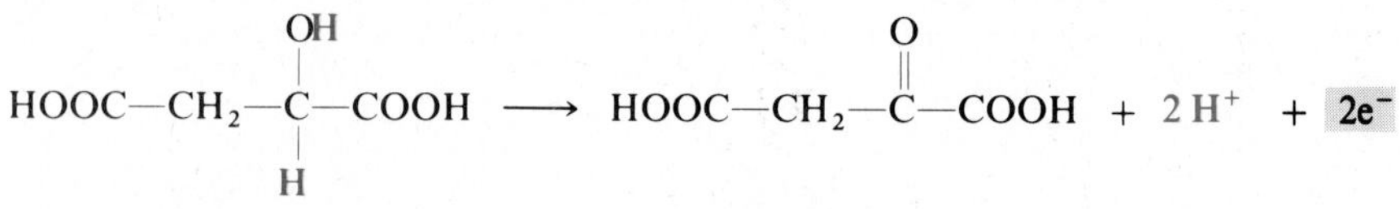

L-Malic acid   Oxaloacetic acid

It has been stated that the function of the coenzyme NAD^+ is to accept this pair of electrons. We can also write a balanced half-reaction for the reduction of the coenzyme; **R**— represents the remaining portion of the NAD^+ molecule (see Table 14.2).

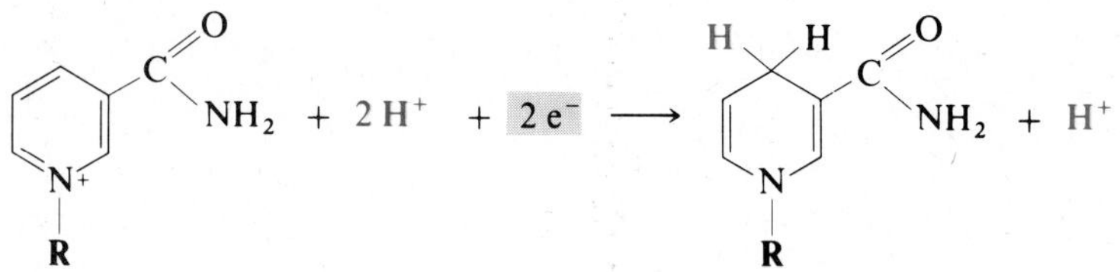

NAD$^+$ NADH

Two hydrogen ions and two electrons are removed from the substrate. The NAD^+ molecule accepts both electrons and one hydrogen ion. The other hydrogen ion is transported from the matrix, across the inner mitochondrial membrane, to the intermembrane space.

In the next step of the respiratory chain, both electrons are passed on to an enzyme called *NADH dehydrogenase* whose coenzyme is FMN (see Table 14.2). By passing the electrons along, NADH is reoxidized back to NAD^+ and FMN is reduced to $FMNH_2$. It is this step that accounts for the regeneration of NAD^+; again we depict only the relevant portion of the FMN molecule.

$$NADH + H^+ \longrightarrow NAD^+ + 2\,H^+ + 2e^-$$

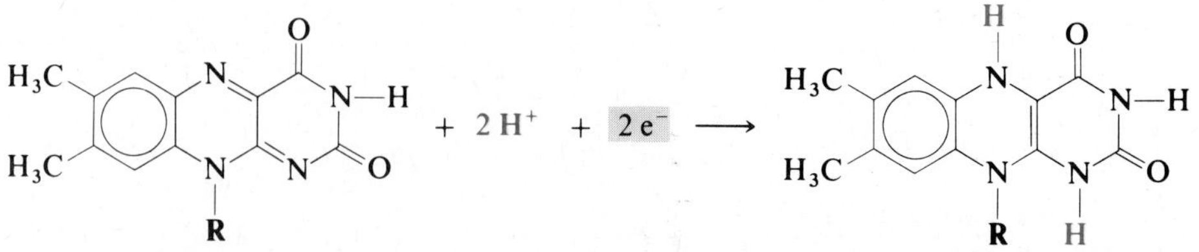

FMN   FMNH$_2$

The electrons are then transferred from $FMNH_2$ to a series of iron–sulfur complexes (abbreviated as Fe·S). A number of iron–sulfur proteins have been identified in the respiratory chain, complexed with other electron carriers. The iron atom in these proteins is in the Fe(III) (ferric) form; by accepting an electron, it is reduced to the Fe(II) (ferrous) form. (Since each iron–sulfur complex can transfer only one electron, two molecules are needed to transport the two electrons and regenerate FMN.)

$$FMNH_2 \longrightarrow FMN + 2H^+ + 2e^-$$
$$2\,Fe(III)\cdot S + 2e^- \longrightarrow 2\,Fe(II)\cdot S$$

The two electrons are next transferred from Fe(II)·S to coenzyme Q (CoQ). This coenzyme is a quinone derivative with a long isoprenoid side chain. Coenzyme Q is also called ubiquinone because it is ubiquitous in all living systems. Several ubiquinones are known, differing only in the number of isoprene units in their side chain. The most common ubiquinone found in the mitochondria of animal tissues contains ten isoprene units and has been designated as CoQ_{10}. The quinone ring is reversibly reducible to a hydroquinone, and so CoQ serves as the electron carrier between the flavin coenzymes and the cytochromes.[15] It is in this step that the oxidized form of the iron–sulfur complex is regenerated.

$$2\,Fe(II)\cdot S \longrightarrow 2\,Fe(III)\cdot S + 2e^-$$

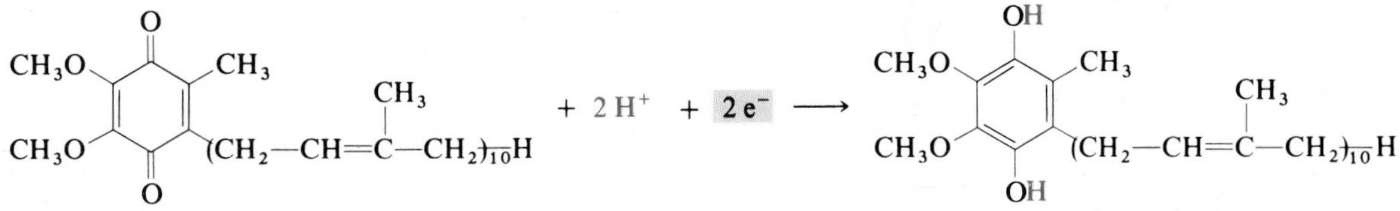

Coenzyme Q_{10} $+ 2H^+ + 2e^- \longrightarrow$ Coenzyme $Q_{10}H_2$

A series of compounds called **cytochromes** were probably the first entities to be associated with electron-transferring reactions. A number of these substances exist, and we have included four of them in Figure 16.9. (Notice that another Fe·S protein is located between two of the cytochromes.) The cytochromes are conjugated

[15] Recall that we said the enzyme *succinic acid dehydrogenase* (which catalyzes the oxidation of succinic acid to fumaric acid—step 8) contains the coenzyme FAD, which is reduced to $FADH_2$. The oxidized form of the coenzyme must be regenerated, and this is accomplished via the respiratory chain. The electrons from $FADH_2$ require a two-step transfer to enter the chain. First they are passed to an iron–sulfur protein,

$$FADH_2 \longrightarrow FAD + 2H^+ + 2e^-$$
$$2\,Fe(III)\cdot S + 2e^- \longrightarrow 2\,Fe(II)\cdot S$$

which in turn passes them on to CoQ (see Figure 16.9).

$$2\,Fe(II)\cdot S \longrightarrow 2\,Fe(III)\cdot S + 2e^-$$
$$CoQ_{10} + 2H^+ + 2e^- \longrightarrow CoQ_{10}H_2$$

protein enzymes. Their prosthetic groups are iron porphyrins, which resemble heme, the pigment in hemoglobin and in myoglobin (see Figure 13.15). The various cytochromes differ with respect to (1) their protein constituents, (2) the manner in which the prophyrin is bound to the protein, and (3) the substituents on the periphery of the porphyrin ring. Such slight differences in structure bestow differences in reduction potential upon the different cytochromes. Like the iron in the Fe·S complexes, the characteristic feature of the cytochromes is the ability of their iron atoms to exist in either the ferrous or the ferric form. Thus each cytochrome in its oxidized form, Fe(III), can accept one electron and be reduced to the Fe(II)-containing form. This change in oxidation state is reversible, and the reduced form can donate its electron to the next cytochrome, and so on. Only the last cytochrome, cytochrome aa_3 (called *cytochrome oxidase*), has the ability to transfer electrons to molecular oxygen. (The combination of cyanide with the ferric ions of cytochrome aa_3 completely inhibits its reduction to the ferrous state, thus blocking the transfer of electrons to molecular oxygen. This accounts for the extreme toxicity of cyanide to living organisms—just 50 mg of HCN constitutes a lethal dose for a human.) Since the Fe(III)/Fe(II) system is only a one-electron exchange, two cytochrome b molecules are necessary to complete the oxidation of coenzyme $Q_{10}H_2$.

$$CoQ_{10}H_2 \longrightarrow CoQ + 2H^+ + 2e^-$$
$$2\,Cyt\,b\text{-}Fe(III) + 2e^- \longrightarrow 2\,Cyt\,b\text{-}Fe(II)$$

Then

$$Cyt\,b \longrightarrow Fe\cdot S \longrightarrow Cyt\,c_1 \longrightarrow Cyt\,c \longrightarrow Cyt\,aa_3$$

In the final step, two molecules of the terminal electron carrier, cytochrome aa_3, pass their electrons to molecular oxygen, the ultimate electron acceptor. It has been estimated that about 95% of the oxygen utilized by cells reacts in this single process.

$$2\,Cyt\,aa_3\text{-}Fe(II) \longrightarrow 2\,Cyt\,aa_3\text{-}Fe(III) + 2e^-$$
$$\tfrac{1}{2}O_2 + 2H^+ + 2e^- \longrightarrow H_2O$$

Each intermediate compound in the respiratory chain is reduced by the addition of electrons in one reaction and is subsequently restored to its original form when it delivers the electrons to the next compound. Thus each pair of electrons that is removed from the substrates of the Krebs cycle utlimately reduces one atom of oxygen.

16.10   Oxidative Phosphorylation

The *process whereby ATP synthesis is linked to the consumption of oxygen in the respiratory chain* is referred to as **oxidative phosphorylation**. A model (the *chemiosmotic hypothesis*) that links the formation of ATP to the operation of the respiratory chain has been proposed by Peter Mitchell (Nobel Prize, 1978—see an advanced biochemistry text for details). The energy required for the production of ATP results from the passage of a pair of electrons from NADH or from $FADH_2$ to oxygen through a series of electron carriers. The electron transport chain can be thought to function as a biochemical battery; that is, energy is obtained from oxidation–reduction reactions.

Electron transport is tightly coupled to oxidative phosphorylation. The reduced forms of the coenzymes, NADH and $FADH_2$, are oxidized by the respiratory chain *only* if ADP is simultaneously phosphorylated to ATP. Within energy-utilizing cells, the turnover of ATP is very high. These cells, then, contain high levels of ADP, and they must consume large quantities of oxygen to continuously phosphorylate the ADP back to ATP. For example, at rest about 20% of an adult's oxygen consumption occurs within brain tissue to supply energy to brain cells. (Much of this energy is used to maintain the concentration gradients across the membranes of the trillions of brain cells—recall Section 8.15). Resting skeletal muscles utilize about 30%, whereas during strenuous exercise these muscles account for almost 90% of the total oxygen consumption of the organism. The enzymes of oxidative phosphorylation are embedded in the inner mitochondrial membrane in association with the enzymes of the respiratory chain. The sites on the respiratory chain at which the oxidative phosphorylations are believed to occur are shown in Figure 16.9. From the data contained in Table 16.1, we can calculate the maximum amount of energy (E) that is made available when a single pair of electrons travels from NADH to oxygen along the chain.

$$
\begin{array}{lr}
 & E \\
NADH \longrightarrow NAD^+ + H^+ + 2\,e^- & +0.32 \\
\frac{1}{2}O_2 + 2\,H^+ + 2\,e^- \longrightarrow H_2O & +0.82 \\
\hline
NADH + \frac{1}{2}O_2 + H^+ \longrightarrow NAD^+ + H_2O & +1.14 \text{ volts}
\end{array}
$$

The energy change for the reaction can be obtained by use of the following equation.[16]

$$\text{Energy change} = -nF\,\Delta E$$

where n = number of electrons transferred
F = Faraday's constant (23,062 cal/volt equivalent)

$$\text{Energy change} = -(2)(23,062)(1.14)$$
$$= -52,600 \text{ cal}$$

This value of 52,600 cal represents a considerable amount of energy. If it were released all at once, much of it would be dissipated as heat and it might prove damaging to the cell. Therefore, the respiratory chain serves as a device for

[16]Since the actual concentrations of the various compounds are unknown, the use of reduction potentials can only yield a rough estimate of the actual energy change for the reaction.

delivering this energy in small increments to be used to phosphorylate ADP. It has been experimentally observed that three molecules of ATP are formed for every molecule of NADH that is oxidized in the chain (but only two ATPs are formed if the primary acceptor is FAD, as is the case when succinic acid serves as a substrate). The net equation for the respiratory chain is

$$NADH + \tfrac{1}{2}O_2 + H^+ + 3\,ADP + 3\,P_i \longrightarrow NAD^+ + H_2O + 3\,ATP$$

Recall that 7300 cal is required for the conversion of 1 mole of ADP to ATP. It can be determined that almost half of the energy released in the electron-transport chain is conserved in the formation of high-energy phosphate bonds. What is not conserved as ATP is liberated as heat and contributes to the maintenance of body temperature in warm-blooded animals.

$$\text{Energy conserved by respiratory chain} = \left(\frac{\text{energy conserved}}{\text{energy available}}\right)(100\%)$$

$$= \frac{(3)(7300)}{(52,600)}(100\%) = 42\%$$

16.11 Energy Yield of Carbohydrate Metabolism

It is now possible to summarize the energy conserved (in ATP production) from the complete oxidation of one molecule of glucose. It must be recalled that, under aerobic conditions, the glycolytic sequence terminates with the formation of two molecules of pyruvic acid. Two molecules of ATP are obtained from substrate level phosphorylation. In addition, since the pyruvic acid is not reduced to lactic acid, there are two molecules of NADH (from step 6) that remain in the cytosol. We know that NAD^+ must be regenerated from NADH for glycolysis to continue. The problem, however, is that NADH cannot pass across the mitochondrial

FIGURE 16.10 The glycerol phosphate shuttle.

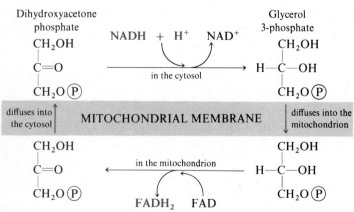

TABLE 16.2 Production of ATP During the Oxidation of One Molecule of Glucose to Carbon Dioxide and Water

Reaction	Type of Phosphorylation	Number of ATP Molecules Formed
Glucose $\longrightarrow$ 2 Pyruvic acid	Substrate level	2
Glyceraldehyde 3-phosphate $\xrightarrow{\text{NAD}^+ \quad \text{NADH} \atop \text{FAD} \quad \text{FADH}_2}$ 1,3-Diphosphoglyceric acid	Oxidative (via glycerol phosphate shuttle)	2×2
Pyruvic acid $\xrightarrow{\text{NAD}^+ \quad \text{NADH}}$ Acetyl-CoA	Oxidative	2×3
Isocitric acid $\xrightarrow{\text{NAD}^+ \quad \text{NADH}}$ Oxalosuccinic acid	Oxidative	2×3
α-Ketoglutaric acid $\xrightarrow{\text{NAD}^+ \quad \text{NADH}}$ Succinyl-CoA	Oxidative	2×3
Succinyl-CoA $\xrightarrow{\text{GDP} \quad \text{GTP}}$ Succinic acid	Substrate level	2×1
Succinic acid $\xrightarrow{\text{FAD} \quad \text{FADH}_2}$ Fumaric acid	Oxidative	2×2
Malic acid $\xrightarrow{\text{NAD}^+ \quad \text{NADH}}$ Oxaloacetic acid	Oxidative	2×3
	Sum	36

membrane to be oxidized by the respiratory chain. A solution is achieved whereby electrons from NADH are transported across the membrane via a shuttle process involving glycerol 3-phosphate and dihydroxyacetone phosphate. These compounds can penetrate the outer mitochondrial membrane.

The first step in this shuttle (Figure 16.10) is the reduction of dihydroxyacetone phosphate by NADH to form glycerol 3-phosphate and to regenerate the NAD^+. This reaction occurs in the cytosol. Glycerol 3-phosphate then enters the mitochondrion where it is reoxidized to dihydroxyacetone phosphate, this time by a *dehydrogenase* enzyme that contains the coenzyme FAD instead of NAD^+. The dihydroxyacetone phosphate diffuses out of the mitochondrion and returns to the cytosol to complete the shuttle. The reduced form of the flavin coenzyme, $FADH_2$, is reoxidized by passing its electrons to the respiratory chain at the level of coenzyme Q. As a result, two molecules of ATP (instead of three) are formed every time a cytosolic NADH molecule is reoxidized via the glycerol phosphate shuttle and the respiratory chain.[17]

[17]The glycerol phosphate shuttle operates in most cells (e.g., skeletal muscle and nerve cells). In certain other tissues, particularly the liver and heart, another type of shuttle exists. This is the malic acid–aspartic acid shuttle. By a quite complex mechanism, it produces three molecules of ATP from the reoxidation of a cytosolic NADH molecule.

The aerobic continuation of glycolysis yields a total of 15 molecules of ATP from each molecule of pyruvic acid, or 30 molecules of ATP per molecule of glucose. Table 16.2 lists the various reactions that result in ATP synthesis.

$$C_6H_{12}O_6 + 6\,O_2 + 36\,ADP + 36\,P_i \longrightarrow 6\,CO_2 + 42\,H_2O + 36\,ATP$$

The above equation summarizes the oxidation of 1 mole of glucose in certain aerobic cells (e.g., skeletal muscle, nerve). The energy released (686,000 cal) is coupled to the synthesis of 36 moles of ATP from ADP and inorganic phosphate. Assuming that 7300 cal is required for the synthesis of each mole of ATP, then $36 \times 7300 = 263,000$ cal is conserved by the cell. The efficiency of conservation therefore is

$$\frac{(263,000)}{(686,000)}(100\%) = 38\%$$

This recovery of energy compares favorably with the efficiency of any manmade machine, and it represents a remarkable achievement on the part of the living organism. In comparison, automobiles are only about 5% efficient in utilizing the energy released in the combustion of gasoline. (Exercise 16.36 asks you to determine the number of ATPs and the energy efficiency in heart and liver cells.)

Exercises

16.1 Define or give an explanation for each of the following terms.
- **(a)** metabolite
- **(b)** villi
- **(c)** blood-sugar level
- **(d)** renal threshold
- **(e)** glucosuria
- **(f)** glucose tolerance test
- **(g)** gluconeogenesis
- **(h)** adenyl cyclase
- **(i)** cyclic AMP
- **(j)** metabolic pathway
- **(k)** Ⓟ
- **(l)** kinase
- **(m)** oxygen debt
- **(n)** Cori cycle
- **(o)** fetal alcohol syndrome
- **(p)** Antabuse
- **(q)** carbohydrate loading
- **(r)** Krebs cycle
- **(s)** oxidative decarboxylation
- **(t)** mitochondria
- **(u)** respiratory chain
- **(v)** cytochromes

16.2 Distinguish between the terms in each of the following pairs.
- **(a)** anabolism and catabolism
- **(b)** photosynthesis and respiration
- **(c)** metabolism and digestion
- **(d)** active transport and passive transport
- **(e)** hypoglycemia and hyperglycemia
- **(f)** glycogenesis and glycogenolysis
- **(g)** anaerobic and aerobic
- **(h)** glycolysis and fermentation
- **(i)** a mutase and an isomerase
- **(j)** a phosphatase and a phosphorylase
- **(k)** facultative anaerobes and strict anaerobes
- **(l)** substrate level phosphorylation and oxidative phosphorylation

16.3 Briefly discuss the digestion and absorption of carbohydrates, enumerating the enzymes involved, and the sites along the gastrointestinal tract where catalysis occurs.

16.4 What is the major factor in maintaining a constant blood-sugar level?

16.5 Name the three hormones that help to control the blood-sugar level. Tell which glands secrete them, and tell how they function in regulating the blood-sugar level.

16.6 Briefly describe the two types of diabetes mellitus.

16.7 How do muscle cells and liver cells differ in their metabolism of glucose 6-phosphate and glycogen?

16.8 When *aldolase* catalyzes the cleavage of ketose mono- and diphosphates, dihydroxyacetone phosphate is always one of the products. Complete the following catalytic reactions.

(a)

$$CH_2O\;\textcircled{P}$$
$$|$$
$$C{=}O$$
$$|$$
$$HO{-}C{-}H$$
$$|$$
$$H{-}C{-}OH$$
$$|$$
$$H{-}C{-}OH$$
$$|$$
$$H{-}C{-}OH$$
$$|$$
$$CH_2O\;\textcircled{P}$$

$\xrightarrow{\text{aldolase}}$

Sedoheptulose
1,7-diphosphate

(b) fructose 1-phosphate $\xrightarrow{\text{aldolase}}$

16.9 Iodoacetic acid and fluoride ions are poisonous because they inhibit enzymes of the Embden–Meyerhof pathway. What enzymes are inhibited?

16.10 What is the fate of the lactic acid formed by muscular activity?

16.11 Draw a molecule of glucose and label carbon-3 and carbon-4.
(a) Which of these two carbon atoms will appear in lactic acid?
(b) Which of these two carbon atoms will appear in ethanol or in carbon dioxide?

16.12 (a) Which is the oxidative step in the Embden–Meyerhof pathway?
(b) How is the reduced coenzyme reoxidized in yeast cells?
(c) How is the reduced coenzyme reoxidized in muscle cells?

16.13 The enzyme *lactic acid dehydrogenase* is not specific for pyruvic acid, but will catalyze the reduction of other keto acids. Write an equation for the enzymatic reduction of phenylpyruvic acid. Does this surprise you in light of what was said about the stereochemical specificity of this enzyme in Section 14.6-b? Comment.

16.14 Write a balanced half-reaction for each of the following conversions.
(a) glucose $\longrightarrow$ pyruvic acid
(b) pyruvic acid $\longrightarrow$ lactic acid
(c) pyruvic acid $\longrightarrow$ ethanol + carbon dioxide

16.15 List four phosphorylated and four nonphosphorylated metabolites of glycolysis and fermentation.

16.16 Assign names to the question marks in the following word equations.

(a) ? $\xrightarrow[\text{dehydrogenase}]{\text{alcohol}}$ Ethanol

(b) Fructose 1,6-diphosphate $\xrightarrow{\text{aldolase}}$? + ?

(c) 2-Phosphoglyceric acid $\xrightarrow{?}$
Phosphoenolpyruvic acid

(d) ? $\xrightarrow{\text{glucose 6-phophatase}}$ Glucose

(e) Glycogen $\xrightarrow{\text{phosphorylase}}$?

(f) Dihydroxyacetone phosphate $\xrightarrow{?}$
Glyceraldehyde 3-phosphate

(g) Glucose $\xrightarrow{\text{hexokinase}}$?

(h) ? $\xrightarrow{\text{lactic acid dehydrogenase}}$ Lactic acid

(i) Fructose 6-phosphate $\xrightarrow{?}$
Fructose 1,6-diphosphate

(j) ? $\xrightarrow{\text{phosphoglucoisomerase}}$ Fructose 6-phosphate

(k) Glyceraldehyde 3-phosphate
$\xrightarrow[\text{dehydrogenase}]{\text{glyceraldehyde 3-phosphate}}$?

(l) 1,3-Diphosphoglyceric acid $\xrightarrow{?}$
3-Phosphoglyceric acid

(m) 3-Phosphoglyceric acid $\xrightarrow{\text{phosphoglyceromutase}}$?

(n) Pyruvic acid $\xrightarrow{?}$ Acetaldehyde

(o) ? $\xrightarrow{\text{pyruvate kinase}}$ Pyruvic acid

(p) Glucose 1-phosphate $\xrightarrow{\text{phosphoglucomutase}}$?

16.17 Refer to Exercise 16.16 and select the equations (by letter) in which the following processes occur.
(a) the expenditure of phosphate-bond energy
(b) the formation of high-energy phosphate bonds
(c) isomerization reactions
(d) a reverse aldol condensation
(e) an oxidation reaction
(f) reduction reactions
(g) a dehydration reaction
(h) a decarboxylation reaction

16.18 The average adult consumes about 65 g of fructose daily (either as the free sugar or as part of sucrose). In the liver, fructose is first phosphorylated to fructose 1-phosphate, which is then split into dihydroxyacetone phosphate and glyceraldehyde. The latter compound is phosphorylated to glyceraldehyde 3-phosphate. Write out equations (using formulas) for these three steps and give the specific names of the enzymes. Indicate which steps utilize ATP.

16.19 The alcohol that we drink is detoxified by enzymes in the liver. Write out the sequence of reactions for the metabolism of alcohol.

   (a) Which enzyme is inhibited by disulfiram?

   (b) Does alcohol supply energy (calories) for the body? Explain.

16.20 Why is it more efficient for anerobic cells to utilize glycogen as a source of energy rather than glucose?

16.21 Write the structural formula of creatine phosphate. What is its role in muscle contraction?

16.22 Why are trained athletes better able to perform strenuous exercise?

16.23 If the carbon atom of the methyl group of pyruvic acid is labeled, where would the label appear after the oxidation of pyruvic acid by one turn of the Krebs cycle?

16.24 Assign names to the question marks in the following word equations.

   (a) ? $\xrightarrow{\text{aconitase}}$ Isocitric acid

   (b) ? $\xrightarrow[\text{synthetase}]{\text{citric acid}}$ Citric acid

   (c) Fumaric acid $\xrightarrow{\text{fumarase}}$?

   (d) Isocitric acid $\xrightarrow{?}$ α-Ketoglutaric acid

   (e) α-Ketoglutaric acid $\xrightarrow{?}$ Succinyl-CoA

   (f) Malic acid $\xrightarrow[\text{dehydrogenase}]{\text{malic acid}}$?

   (g) ? + ? $\xrightarrow[\text{diphosphokinase}]{\text{nucleoside}}$ GTP + ADP

   (h) Pyruvic acid $\xrightarrow{?}$ Acetyl-CoA

   (i) Succinyl-CoA $\xrightarrow[\text{synthetase}]{\text{succinyl-CoA}}$?

   (j) Succinic acid $\xrightarrow[\text{dehydrogenase}]{\text{succinic acid}}$?

16.25 Refer to Exercise 16.24 and select the equations (by letter) in which the following processes occur.

   (a) isomerization
   (b) hydration
   (c) dehydration
   (d) oxidation
   (e) decarboxylation
   (f) phosphorylation

16.26 Fluoroacetic acid (CH_2FCOOH) is the toxic component in a South African plant known as gifblaar (*Dichapetalum cymosum*). Fluoroacetic acid is highly toxic to humans because it can easily enter cells and pass across the mitochondrial membrane. Within the mitochondrion, fluoroacetic acid is converted by *acetyl-CoA synthetase* to fluoroacetyl-CoA. That compound then condenses with oxaloacetic acid to form fluorocitric acid, which is an inhibitor of the enzyme *aconitase*. (Fluorocitric acid is a lethal poison and is used as a rodenticide.) Write equations showing the conversion of fluoroacetic acid to fluorocitric acid. What type of inhibitor (see Section 14.8) is fluorocitric acid?

16.27 Sketch and label the pertinent features of a mitochondrion.

16.28 Outline, in detail, the sequence of reactions that occur in the respiratory chain.

16.29 Write balanced half-reactions for all of the oxidation–reduction reactions that occur in the Krebs cycle and in the respiratory chain.

16.30 How is the energy of carbohydrate metabolism made available to the body cells?

16.31 The complete oxidation of 1 mole of acetic acid in a calorimeter yields about 200 kcal.

   (a) How much energy is stored as ATP when 1 mole of acetic acid is converted to acetyl-CoA and metabolized via the Krebs cycle?

   (b) What is the percentage energy efficiency?

16.32 How many moles of ATP can be formed from each mole of lactic acid that is oxidized to CO_2 and H_2O?

16.33 What are the names of the two shuttle systems that transport NADH from the cytosol into a mitochondrion?

16.34 Only two molecules of ATP are produced by the aerobic conversion of **(a)** succinic acid to fumaric acid and **(b)** glyceraldehyde 3-phosphate to 1,3-diphosphoglyceric acid. Explain.

16.35 Write word equations for all the substrate level phosphorylation reactions that occur during the metabolism of carbohydrates.

16.36 **(a)** How much ATP (in moles) is produced from 1 mole of glucose in a typical liver or heart cell? (Recall footnote 17.)

   (b) What is the efficiency associated with ATP production in a liver or a heart cell?

16.37 Write the net equation for the complete oxidation of 1 mole of glucose in a skeletal muscle cell that is respiring under aerobic conditions.

16.38 **(a)** What purpose do carbohydrates fulfill in metabolism?

   (b) In what respect does carbohydrate metabolism resemble the burning of table sugar in a pan?

   (c) In what way does it differ?

17

Lipid Metabolism

Nearly all the energy required by the animal organism is generated by the oxidation of carbohydrates and lipids. Of all major nutrients, triacylglycerols are the richest energy source. The oxidation of 1 g of a typical lipid liberates about 9500 cal; the oxidation of an equal mass of carbohydrate liberates only about 4200 cal. A lipid molecule has a high proportion of carbon–hydrogen bonds ($-CH_2-$) compared to a carbohydrate molecule ($-CHOH-$). Therefore, lipids have a greater capacity to combine with oxygen and, consequently, have a higher heat content.[1]

Carbohydrates provide a readily available source of energy; lipids function as the principal energy reserve.[2] Men and women differ in their capacity to store lipids. The

[1] A large percentage of the lipid molecule is of the nature of a saturated hydrocarbon (i.e., highly reduced). Thus fats may be thought to be analogous to combustible petroleum products, whereas carbohydrates may be considered to be analogous to alcohols (more oxidized), which are not nearly as effective as fuels. Recall (page 97) our discussion of gasohol. As a food substance, ethyl alcohol liberates about 7100 cal/g during its metabolism (see page 438).

[2] The fat reserves in the average individual provide sufficient energy to survive starvation for 30 days (given sufficient water). In comparison, glycogen in the liver (about 100 g) is depleted within 1 day. From the standpoint of efficiency of fuel storage, fats are far superior. Glycogen, because of its many hydroxyl groups, is extremely hydrated. About 2 g of water is bound to every gram of stored glycogen. Thus in

(*Footnote continues*)

average male stores about 12% of his body weight as fat, whereas about 20% of the weight of an average female is due to body fat. (In contrast, marathon runners average about 5–6% body fat.) Americans consume about 100–125 g of lipids each day. This represents 40% of the daily calorie requirements. The National Cancer Institute has recommended that Americans should reduce their lipid intake, and that only 30% of total calories be provided by lipids. Normally, triacylglycerols constitute about 98% of the total dietary lipids. However, as we shall see, significant quantities can be synthesized from carbohydrates when carbohydrate intake is high and lipid intake is low.

The nutritional aspects of lipids are still not completely understood. Lipids are not dietary necessities; an organism can survive on a lipid-free diet if carbohydrates and proteins are supplied as a source of metabolic energy. Certain lipids, however, are required for normal growth and development. These lipids supply certain fatty acids that the organism cannot synthesize. Prominent among these **essential fatty acids** are *the unsaturated acids containing more than one double bond,* linoleic and linolenic acids. Linoleic acid is utilized by the body to synthesize many of the other unsaturated fatty acids such as arachidonic acid. (Recall from Section 12.15 that arachidonic acid is required as a precursor for the biosynthesis of prostaglandins.) In addition, the essential fatty acids are incorporated into the structures of the membrane lipids (Section 12.11), and they are necessary for the efficient transport and metabolism of cholesterol. The average daily diet should contain about 2–3 g of linoleic acid. The "vitamin F" sold in health food stores is not a vitamin but rather a formulation of the essential fatty acids. Infants lacking essential fatty acids in their diet will lose weight and develop eczema. **Eczema** is *an inflammatory skin disease characterized by scales and crusts on the skin.*

17.1 *Digestion and Absorption of Lipids*

Lipids are not digested by the body until they reach the upper portion of the small intestine (duodenum). In this region, a hormone is secreted that stimulates the gallbladder to discharge bile into the duodenum. In the context of lipid digestion, the principal constituents of the bile are the bile salts (Section 12.12-b). The bile salts act as emulsifiers and serve to disrupt some of the hydrophobic bonds that tend to hold the lipid molecules together. This emulsification process is essential to lipid digestion. It converts the water-insoluble lipids into smaller globules (micelles) that are more susceptible to attack by the water-soluble, fat-splitting enzymes. Another hormone then promotes the secretion of the pancreatic juice. This juice contains the *lipases* that catalyze the digestion of triacylglycerols into 2-monoacylglycerols and fatty acids. Phospholipids and cholesterol esters are also hydrolyzed into their component molecules.

---

comparing stored mass of each fuel in the body, 1 g of fat contains more than six times the energy content of 1 g of hydrated glycogen. The ability to store greater amounts of the more energy-efficient lipids is especially important for migrating birds and terrestrial animals. The camel's hump, for example, is almost all adipose tissue, and migratory birds rely on stored fats to supply the energy for long, sustained flight.

$$
\begin{array}{ccccc}
\underset{\displaystyle \overset{\textstyle O}{\|}}{R-C-O-CH_2} & & HO-CH_2 & & HO-CH_2 \\[2ex]
\underset{\displaystyle \overset{\textstyle O}{\|}}{R'-C-O-CH} & \longrightarrow & \underset{\displaystyle \overset{\textstyle O}{\|}}{R'-C-O-CH} & \longrightarrow & \underset{\displaystyle \overset{\textstyle O}{\|}}{R'-C-O-CH} \\[2ex]
\underset{\displaystyle \overset{\textstyle O}{\|}}{R''-C-O-CH_2} & & \underset{\displaystyle \overset{\textstyle O}{\|}}{R''-C-O-CH_2} & & HO-CH_2 \\
& & + & & + \\
& & R-C\!\!\stackrel{O}{\diagdown}_{OH} & & R''-C\!\!\stackrel{O}{\diagdown}_{OH}
\end{array}
$$

Biochemists do not agree upon the extent to which a lipid must be hydrolyzed in order to be absorbed. It is probable that after emulsification some lipids are absorbed directly through the intestinal membrane before being hydrolyzed. (Any lipid that is not absorbed is eliminated in the feces.) Once the monoacylglycerols, fatty acids, and free cholesterol pass into the cells of the intestinal epithelium, they are immediately resynthesized into triacylglycerols, phospholipids, or cholesterol esters. It has been shown experimentally that the glycerol needed to form these compounds is supplied by the metabolic pool within the intestinal cells and not from the glycerol obtained by lipid hydrolysis. The triacylglycerols then become associated with proteins and form the chylomicrons (recall Section 12.17).

The chylomicrons are transported by the lymphatic system into the blood-stream.[3] Some of the triacylglycerols are carried to the liver, where they are modified and/or utilized to provide energy for liver functions. The remaining lipids (and the triacylglycerols that are synthesized in the liver) are transported to specialized fat-storage cells (via VLDLs), or circulated through the blood as lipoprotein complexes (LDLs and HDLs). The circulating lipids are distributed to the various tissues to be incorporated into membrane lipids or to be oxidized for the generation of energy. After a meal, the lipid content of the blood rises and remains at a high level for several hours; then it gradually decreases to the fasting level. The normal fasting plasma-lipid concentration is about 500 mg/100 mL. Of this, triacylglycerols comprise about 120 mg, phospholipids (mainly lecithins) about 180 mg, and cholesterol (chiefly cholesterol esters) about 200 mg.

An early theory of lipid metabolism held that a portion of ingested food was utilized and that the remainder was stored in adipose tissue as *fat deposits*. The stored fat was believed to be metabolically inactive, excess lipid that remained undisturbed until needed by the body. Through the use of tracer experiments, however, we now know that lipids are in a continuous state of dynamic equilibrium; they are constantly being transported, degraded, utilized, and resynthesized.

[3]Lymph is tissue fluid (water and dissolved substances) that has entered a lymphatic capillary. Lymph vessels transport excess fluid from interstitial spaces and return it to the bloodstream. The blood system and the lymphatic system are separate, but there is a crossover point via the thoracic duct. After a high-fat meal, the lymph changes from a transparent yellow to a milky white because of the emulsified lipid.

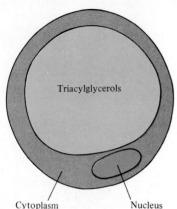

FIGURE 17.1 A typical fat cell found in adipose tissue. Fat cells function to produce and store lipids; they are among the largest cells in the body. Notice that the nucleus and cytoplasm are displaced to the cell periphery by the large lipid droplet.

The human body has a large capacity for the storage of fat, a fact all too evident to many people. **Obesity** is *the condition in which an excessive amount of fat is deposited in the adipose tissue* and the individual becomes overweight. Although in some cases, obesity is due to a derangement of one of the endocrine glands, most individuals are overweight from eating more food than is required for the normal functions of the body and/or from a lack of exercise. (A rough rule of thumb is that 1 g of fat is deposited in adipose tissue for every 9 kcal (9 Cal)[4] of food eaten in excess of the body's requirements. One pound of fat is deposited for every 4100 Cal in excess of intake. To lose that 1 lb of fat requires the energy expended in a brisk 20-mile walk.

Fat deposits, such as those directly beneath the skin (subcutaneous) and in the abdominal region, are specialized tissue in which a relatively large percentage of cytoplasm is replaced by large droplets of triacylglycerols (approximately 90% of the mass of the fat cell—Figure 17.1). Adults have approximately 30–40 billion fat cells. They swell or shrink like a sponge depending on the amount of fat inside them. Fat surrounding the vital organs protects them from mechanical injury, and the subcutaneous layer of fat helps insulate the body against heat loss. Adipose tissue is the only tissue in which free triacylglycerols occur in appreciable amounts.[5] Elsewhere, in cells or in the blood plasma, lipids are bound to proteins as lipoprotein complexes. The fate of lipids in the animal body is outlined in Figure 17.2.

17.2 *Fatty Acid Oxidation*

In order for triacylglycerols to be used for the production of energy, they must first be cleaved to fatty acids and glycerol by the *lipases* contained within the adipose tissue. Fat metabolism, like carbohydrate metabolism, is regulated by hormones. Again, it is the same two energy-invoking hormones, epinephrine and glucagon, along with the neurotransmitter, norepinephrine, that bind to receptor proteins in the cell membranes of adipose tissue. They activate the enzyme *adenyl cyclase* (Section 16.3) to form cyclic AMP from ATP. The cyclic AMP then stimulates the hydrolysis (*lipolysis*) of triacylglycerols (by the lipases) and the release (*mobilization*) of fatty acids from adipose tissue. Insulin, on the other hand, is concerned with the storage of metabolic fuels. It enhances the synthesis of triacylglycerols and the storage of fats in adipose tissue. Insulin will be present in the blood when fatty acids are not needed for energy, that is, when there is a plentiful supply of blood glucose.

---

[4]The nutritional calorie (1 Cal), written with a capital C, is equal to 1 kcal or 1000 cal. The U.S. Recommended Daily Allowance for caloric intake is 2000 Cal for a 128 lb female and 2700 Cal for a 154 lb male.

[5]In addition to the normal white adipose tissue, some organisms contain another type of fatty tissue called **brown fat.** Its brown color is due to the presence of large numbers of mitochondria (stocked with red-brown cytochromes). In these specialized mitochondria, the oxidation of fat is used to generate heat instead of being converted into ATP. Brown fat is particularly abundant in hibernating animals and in the neck and upper back of newborn infants (as well as other mammals that are born hairless). These cells thereby serve as heat factories that revive hibernating animals and maintain the body temperature of young animals. Adults have few brown fat cells because metabolic reactions usually generate more than enough heat (page 454) so that body temperature can be regulated by the dissipation of this heat.

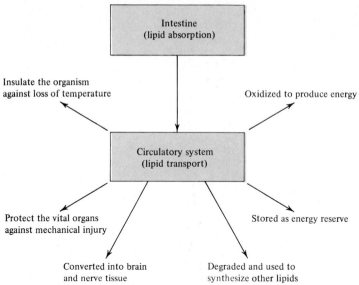

Insulate the organism
against loss of temperature

Oxidized to produce energy

Intestine
(lipid absorption)

Circulatory system
(lipid transport)

Protect the vital organs
against mechanical injury

Stored as energy reserve

Converted into brain
and nerve tissue

Degraded and used to
synthesize other lipids

FIGURE 17.2 Metabolic fate of lipids.

The glycerol obtained from fat hydrolysis is transported to the liver where it is converted to a carbohydrate derivative (see Section 17.5) and is metabolized via the Embden–Meyerhof pathway. The fatty acids are transported by the blood, bound to serum albumins, to other tissues for oxidation. It should be emphasized that in a normal individual, certain organs utilize both glucose and fatty acids (in varying amounts) for their energy needs. The utilization rate of a specific fuel depends on the physiological state of the individual. For example, glucose (or glycogen) is the chief fuel for the immediate energy needs of active skeletal muscle (as in a sprint), whereas fatty acids are the major fuel for resting skeletal muscle. Fatty acids are the major fuel for cardiac muscle, although ketone bodies (see Section 17.4), glucose, and lactic acid are also utilized by the heart. During fasting and prolonged exercise, fatty acids are the preferred fuel for all muscle types. The kidneys and adipose tissue use both glucose and fatty acids, whereas fatty acids are the preferred fuel of the liver. The brain does not use fatty acids because they are bound to albumins and cannot diffuse across the blood–brain barrier. Red blood cells can only use glucose because they lack mitochondria and hence cannot obtain energy via oxidative phosphorylation.

The fatty acids, which contain the bulk of the energy of the lipids, are broken down in a series of sequential reactions accompanied by the gradual release of utilizable energy. Some of these reactions are oxidative and require the same coenzymes (NAD^+, FAD) as those which take part in the oxidation of carbohydrates. The enzymes involved in fatty acid catabolism are localized in the mitochondria along with the enzymes of the Krebs cycle, respiratory chain, and oxidative phosphorylation. This factor (the localization within the mitochondria) is of the utmost importance since it provides for the efficient utilization of the energy stored in the fatty acid molecules.

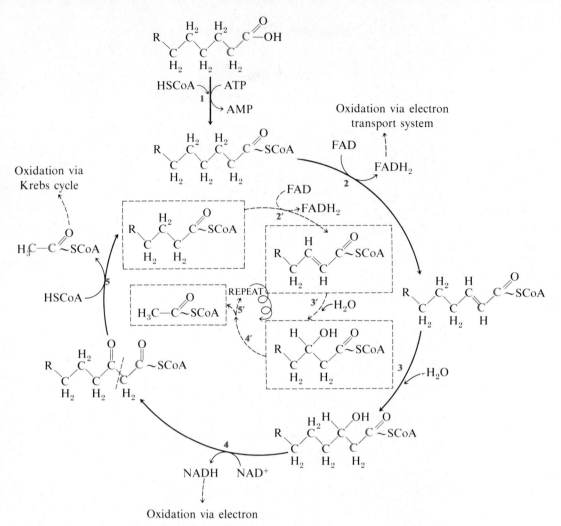

FIGURE 17.3 Fatty acid oxidation (β-oxidation).

In the beginning of the twentieth century, the German biochemist Franz Knoop showed that the breakdown of fatty acids occurs in a stepwise fashion with the removal of two carbon atoms at a time. Subsequent investigations have resulted in the separation and purification of the enzymes involved. The details of the degradation sequence are now fully understood. The reaction scheme is shown in Figure 17.3.

Step 1 The first phase of fatty acid metabolism occurs on the outer mitochondrial membrane. Fatty acids, like carbohydrates and amino acids, are relatively inert and must first be activated by conversion to an energy-rich fatty acid derivative of coenzyme A (called *fatty acyl-CoA*). This activation process is a two-step reaction that is catalyzed by the enzyme *acyl-CoA synthetase*. For each molecule of fatty acid that is activated, one molecule of coenzyme A and one molecule of ATP are used. First the fatty acid reacts with ATP to form a fatty acyl adenylate. (Note the similarity to the activation of amino acids prior to protein synthesis—Section 15.5-a.) Then the sulfhydryl group of coenzyme A attacks the acyl adenylate to yield fatty acyl-CoA and AMP. Finally, the pyrophosphate formed in the initial reaction is hydrolyzed to phosphate ions by the enzyme *pyrophosphatase*. The net effect is the utilization of two high-energy bonds of ATP to activate each molecule of a fatty acid.

$$R-CH_2CH_2CH_2CH_2CH_2C\overset{O}{\underset{OH}{\diagup}} + ATP \underset{}{\overset{synthetase}{\rightleftharpoons}} R-CH_2CH_2CH_2CH_2CH_2C\overset{O}{\underset{AMP}{\diagup}} + PP_i$$

Fatty acid $\qquad\qquad\qquad\qquad\qquad\qquad\qquad\qquad$ Fatty acyl adenylate

$$R-CH_2CH_2CH_2CH_2CH_2C\overset{O}{\underset{AMP}{\diagup}} + HSCoA \underset{}{\overset{synthetase}{\rightleftharpoons}} R-CH_2CH_2CH_2CH_2CH_2C\overset{O}{\underset{SCoA}{\diagup}} + AMP$$

Fatty acyl adenylate $\qquad\qquad$ Coenzyme A $\qquad\qquad\qquad\qquad$ Fatty acyl-CoA

$$PP_i + HOH \overset{pyrophosphatase}{\longrightarrow} 2\,P_i$$

The fatty acyl-CoA combines with a carrier molecule (carnitine—see Exercise 17.20) and is transported into the mitochondrial matrix. Further metabolism of the fatty acyl-CoA occurs entirely within the mitochondrial matrix via a sequence of reactions known as **β-oxidation** (since *the carbon beta to the carboxyl group undergoes successive oxidations*), or the **fatty acid spiral** (because *it involves the progressive removal of two-carbon units from the carboxyl end of the fatty acyl-CoA*).

The reactions involved in steps 2 through 4, dehydrogenation, hydration, and dehydrogenation, are analogous to those involved in the conversion of succinic acid to oxaloacetic acid in the Krebs cycle (succinic → fumaric → malic → oxaloacetic). The functional group at the β-carbon can be visualized as undergoing the following changes: alkane → alkene → secondary alcohol → ketone. The net effect is the introduction of a keto group beta to a carboxyl group.

Step 2 This first oxidation reaction is catalyzed by the enzyme *acyl-CoA dehydrogenase*. The coenzyme FAD accepts two hydrogen atoms from adjacent carbons, one from the α-carbon and one from the β-carbon. The enzyme is stereospecific in that only the *trans* alkene is obtained.

$$R-CH_2CH_2CH_2-\overset{H}{\underset{H}{\overset{|}{\underset{|}{C}}}}-\overset{H}{\underset{H}{\overset{|}{\underset{|}{C}}}}-C\overset{O}{\underset{SCoA}{\diagup}} + FAD \overset{dehydrogenase}{\longrightarrow} R-CH_2CH_2CH_2-\overset{H}{\underset{}{\overset{|}{C}}}=\overset{}{\underset{H}{\overset{|}{C}}}-C\overset{O}{\underset{SCoA}{\diagup}} + FADH_2$$

Fatty acyl-CoA $\qquad\qquad\qquad\qquad\qquad\qquad\qquad\qquad\qquad\qquad$ Enoyl-CoA

Each molecule of $FADH_2$ that is reoxidized via the electron-transport chain supplies energy to form two molecules of ATP.

Step 3 Again, as in the Krebs cycle, only the L-isomer is formed when the stereospecific enzyme *enoyl-CoA hydratase* adds water across the *trans* double bond.

$$R\!-\!CH_2CH_2CH_2\!-\!\underset{\underset{H}{|}}{C}\!=\!\underset{\underset{H}{|}}{C}\!-\!C\!\!\begin{array}{c}\nearrow O\\[-2pt]\searrow SCoA\end{array} + HOH \underset{}{\overset{hydratase}{\rightleftharpoons}} R\!-\!CH_2CH_2CH_2\!-\!\underset{\underset{H}{|}}{\overset{\overset{HO}{|}}{C}}\!-\!\underset{\underset{H}{|}}{\overset{\overset{H}{|}}{C}}\!-\!C\!\!\begin{array}{c}\nearrow O\\[-2pt]\searrow SCoA\end{array}$$

Enoyl-CoA   L-β-Hydroxyacyl-CoA

Step 4 The second oxidation reaction is catalyzed by *β-hydroxyacyl-CoA dehydrogenase,* which exhibits an absolute stereospecificity for the L-isomer. Here the coenzyme NAD^+ is the hydrogen acceptor.

$$R\!-\!CH_2CH_2CH_2\!-\!\underset{\underset{H}{|}}{\overset{\overset{HO}{|}}{C}}\!-\!CH_2C\!\!\begin{array}{c}\nearrow O\\[-2pt]\searrow SCoA\end{array} + NAD^+ \underset{}{\overset{dehydrogenase}{\rightleftharpoons}} R\!-\!CH_2CH_2CH_2\overset{\overset{O}{\|}}{C}CH_2C\!\!\begin{array}{c}\nearrow O\\[-2pt]\searrow SCoA\end{array} + NADH + H^+$$

L-β-Hydroxyacyl-CoA   β-Ketoacyl-CoA

The reoxidation of NADH and the transport of hydrogen ions and electrons through the respiratory chain furnishes three molecules of ATP.

Step 5 The final reaction is a cleavage of the *β-ketoacyl-CoA* by a molecule of coenzyme A. The products of the reaction are acetyl-CoA and a fatty acyl-CoA whose chain length is shortened by two carbon atoms. The enzyme is *acetyl-CoA acetyl transferase (thiolase).*

$$R\!-\!CH_2CH_2CH_2\overset{\overset{O}{\|}}{C}CH_2C\!\!\begin{array}{c}\nearrow O\\[-2pt]\searrow SCoA\end{array} + HSCoA \overset{thiolase}{\longrightarrow} R\!-\!CH_2CH_2CH_2C\!\!\begin{array}{c}\nearrow O\\[-2pt]\searrow SCoA\end{array} + H\!-\!\underset{\underset{H}{|}}{\overset{\overset{H}{|}}{C}}\!-\!C\!\!\begin{array}{c}\nearrow O\\[-2pt]\searrow SCoA\end{array}$$

β-Ketoacyl-CoA   Shortened fatty acyl-CoA   Acetyl-CoA

The newly formed fatty acyl-CoA can then be degraded further by repetition of steps 2 through 5; a molecule of acetyl-CoA is liberated at each turn of the spiral. Normally, the spiral is repeated as many times as is necessary to break down a fatty acid containing an even number of carbon atoms, *n*, into *n*/2 molecules of acetyl-CoA. It should be noted that no further addition of ATP is necessary since the shortened fatty acids already contain the thiol ester. One molecule of ATP is sufficient to activate any fatty acid regardless of the number of carbon atoms in its hydrocarbon chain. The unsaturated fatty acids (which comprise about 50% of the fatty acids found in humans) are also incorporated into the β-oxidation sequence. Two ancillary enzymes are required to convert them into normal substrates of the fatty acid spiral. The overall equation for the β-oxidation of palmitoyl-CoA (16 carbons) is as follows.

$$CH_3(CH_2)_{14}C \overset{O}{\underset{SCoA}{\diagup}} + 7\,FAD + 7\,NAD^+ + 7\,HSCoA + 7\,H_2O \longrightarrow$$

$$8\,CH_3C \overset{O}{\underset{SCoA}{\diagup}} + 7\,FADH_2 + 7\,NADH + 7\,H^+$$

The acetyl-CoA formed in the fatty acid spiral is involved in a myriad of biochemical pathways. It may enter the Krebs cycle to be oxidized to produce energy, or it may be utilized as the starting material for biosynthesis of lipids (triacylglycerols, phospholipids, cholesterol, and other steroids). *When the body ingests more carbohydrate than it needs for energy and for glycogen synthesis, the excess is converted into fatty acids via the common intermediate, acetyl-CoA.* (In normal adults, roughly one third of the total ingested carbohydrates may be converted into fatty acids and stored as fat.) *Fatty acids are built up from two-carbon units that come from acetyl-CoA.* Although we shall not discuss the biosynthetic pathway in this text, it should be pointed out that fatty acid synthesis is not simply the reverse of fatty acid breakdown. In humans, the liver is the major site of fatty acid synthesis. Synthesis occurs in the cytosol, whereas fatty acid oxidation occurs in the mitochondria of various tissues. A completely different set of enzymes is involved. The newly synthesized fatty acids are used to provide energy, or they are esterified with glycerol and stored as triacylglycerols in adipose tissue.

Finally, it is important to recall that the conversion of pyruvic acid to acetyl-CoA is irreversible (page 443) in mammals. Therefore the acetyl-CoA molecules derived from fatty acids cannot be used to synthesize glucose. *Mammals can convert carbohydrates to lipids (via acetyl-CoA). They cannot convert lipids to carbohydrates.*

17.3 *Bioenergetics of Fatty Acid Oxidation*

The combustion of 1 mole of palmitic acid releases a considerable amount of energy, and the precise amount can be determined by conducting the experiment in a calorimeter.

$$C_{16}H_{32}O_2 + 23\,O_2 \longrightarrow 16\,CO_2 + 16\,H_2O + 2,340,000\ cal[6]$$

The amount of energy that is made available to the cell and conserved in the form of ATP is readily calculable.

The breakdown by an organism of 1 mole of palmitic acid, necessitates the utilization of 1 mole of ATP, and 8 moles of acetyl-CoA are formed. You will recall (Figure 16.7) that each mole of acetyl-CoA metabolized by the Krebs cycle yields 12 moles of ATP. Each turn of the fatty acid spiral produces 1 mole of NADH and 1 mole of $FADH_2$. Reoxidation of these compounds by the respiratory chain

[6] Earlier we stated that the high caloric content of lipids is used effectively by migratory birds and many terrestrial animals (e.g., the camel). Note that the oxidation of fatty acids also produces large quantities of water, which sustains the animals for long periods of time.

(within skeletal muscle cells) yields 3 and 2 moles of ATP, respectively. The complete degradation of 1 mole of palmitic acid requires seven turns of the spiral, and a total of 7 moles of $FADH_2$ and NADH is therefore formed. The energy calculations may be summarized as follows.

	Yield of ATP
1 mole of ATP is split to AMP and 2 P_i	−2
8 moles of acetyl CoA formed (8 × 12)	96
7 moles of $FADH_2$ formed (7 × 2)	14
7 moles of NADH formed (7 × 3)	21
Total moles of ATP	129

The percentage of available energy that can theoretically be conserved by the cell in the form of ATP is calculated as follows.

$$\frac{\text{(Energy conserved)}}{\text{(Total energy available)}}(100\%) = \frac{(129)(7300)}{(2,340,000)}(100\%) = 40\%$$

The efficiency with which fatty acids are metabolized is seen to be comparable to that of the carbohydrate metabolism system. Recall we said that carbohydrates are utilized for the normal energy requirements of skeletal muscle, whereas lipids provide energy for prolonged activity. It has been estimated that after 4 hours of sustained exercise, fatty acid oxidation supplies more than 60% of the muscle's energy demands.

17.4 Ketosis

We have previously mentioned abnormal physiological conditions in which fatty acids are utilized to provide the bulk of the energy for most body tissues. These conditions generally occur in conjunction with an impairment of carbohydrate metabolism (e.g., starvation, diabetes mellitus, or zero-carbohydrate "crash" diets). Lacking carbohydrates, the rate of fatty acid oxidation increases and the production of acetyl-CoA increases accordingly. The problem is that most of the acetyl-CoA molecules cannot enter the Krebs cycle. Why? Because there is insufficient oxaloacetic acid (recall step 2, Figure 16.7). If carbohydrate is lacking, oxaloacetic acid levels are reduced because oxaloacetic acid is used to synthesize glucose (via gluconeogenesis) and hence is not available to combine with acetyl-CoA. Consequently, the excess fatty acids released from the adipose tissue are transported to the liver to be metabolized.

When the concentration of acetyl-CoA in the liver mitochondria reaches a certain level, a reversal of the last step of the fatty acid spiral occurs to produce acetoacetyl-CoA. The acetoacetyl-CoA reacts with another molecule of acetyl-CoA and with water to form β-hydroxy-3-methylglutaryl-CoA, which is then cleaved to acetoacetic acid and acetyl-CoA. Some of the acetoacetic acid is reduced to form

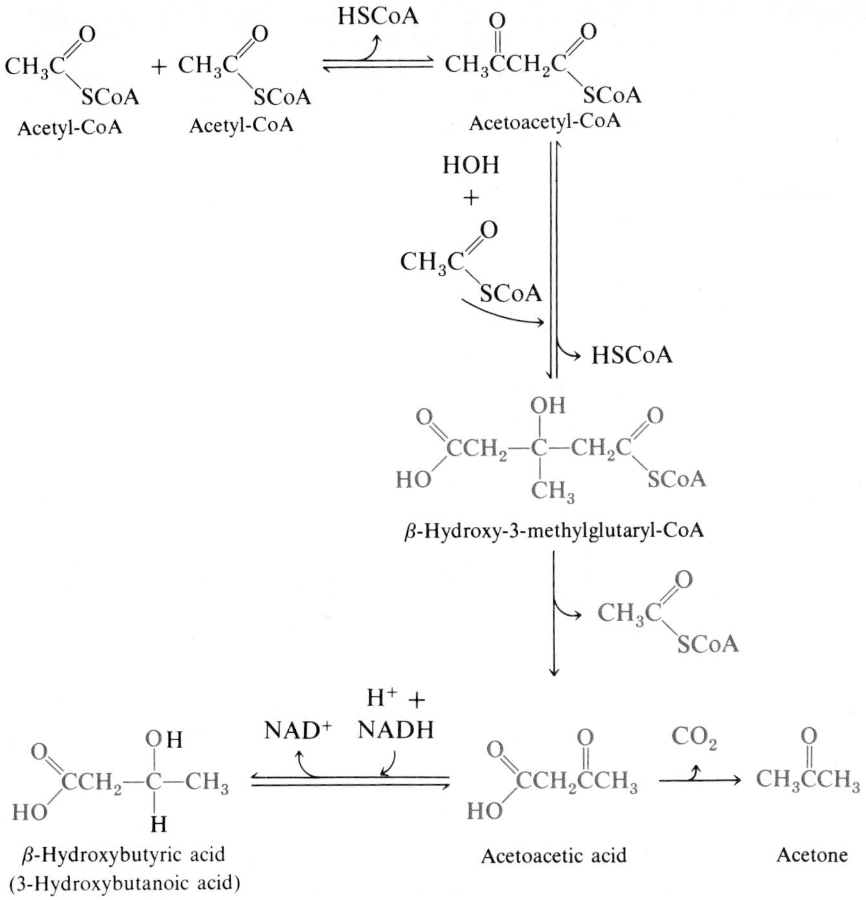

FIGURE 17.4 The formation of ketone bodies.

β-hydroxybutyric acid, and a small fraction of acetoacetic acid is decarboxylated to carbon dioxide and acetone[7] (Figure 17.4).

Acetoacetic acid, β-hydroxybutyric acid and acetone are collectively referred to as **ketone bodies.** Normally the liver synthesizes small amounts of acetoacetic acid and β-hydroxybutyric acid, and releases them into the blood for use as a metabolic fuel (to be converted back into acetyl-CoA molecules) by other aerobic tissues including the heart and the brain. (During prolonged starvation, ketone bodies provide about 70% of the energy requirements of the brain.) However, when the rate of formation of the ketone bodies in the liver greatly exceeds their use by the tissues, excess ketone bodies accumulate in the blood (*ketonemia*—concentrations greater than 3 mg/100 mL) and in the urine (*ketonuria*). These conditions together are referred to as **ketosis.** Since two of the three ketone bodies are acids, their presence in the blood in excessive amounts overwhelms the blood buffers and causes a marked decrease in

---

[7]The acetone is not further metabolized and is transported to the kidneys and the lungs. Since acetone is a fairly volatile compound, it is often expelled in the breath. The sweet smell of acetone is a characteristic of ketosis and is frequently noticed on the breath of severely diabetic patients.

blood pH (to 6.9 from a normal value of 7.4). *This decrease in the pH of the blood leads to a more serious condition known as* **acidosis**.[8] The body tends to lose fluids (become dehydrated) as the kidneys discharge large quantities of water in an attempt to eliminate these excess acids. Acidosis results in interference with the transport of oxygen by the hemoglobin molecule (see Section 19.7), and a fatal coma may result. Any treatment that promotes the utilization of carbohydrates (for example, an injection of insulin) can alleviate both ketosis and acidosis.

17.5 Glycerol Metabolism

One molecule of glycerol is obtained from each molecule of triacylglycerol or phospholipid that is hydrolyzed. Glycerol is transported to the liver where it is readily incorporated into the scheme of carbohydrate metabolism by conversion to dihydroxyacetone phosphate. This two-step process includes the phosphorylation of a primary hydroxyl group followed by the oxidation of a secondary hydroxyl group to a ketone. Dihydroxyacetone phosphate can be utilized to provide energy by conversion to pyruvic acid (via the glycolytic pathway), or it can be transformed into glucose. Since glycerol is the starting compound for glucose synthesis, this is another important example of gluconeogenesis.

$$
\begin{array}{ccccc}
\text{CH}_2\text{OH} & & \text{CH}_2\text{OH} & & \text{CH}_2\text{OH} \\
| & \text{ATP} \quad \text{ADP} & | & \text{NAD}^+ \quad \text{NADH} + \text{H}^+ & | \\
\text{CHOH} & \xrightarrow{\quad\text{kinase}\quad} & \text{CHOH} & \xrightleftharpoons{\quad\text{dehydrogenase}\quad} & \text{C}=\text{O} \\
| & & | & & | \\
\text{CH}_2\text{OH} & & \text{CH}_2\text{O}\,\text{Ⓟ} & & \text{CH}_2\text{O}\,\text{Ⓟ} \\
\text{Glycerol} & & \text{Glycerol} & & \text{Dihydroxyacetone} \\
& & \text{phosphate} & & \text{phosphate}
\end{array}
$$

[8]This is sometimes called metabolic acidosis to distinguish it from respiratory acidosis (see Section 19.8).

Exercises

17.1 Define or give an explanation for each of the following terms.
  (a) essential fatty acid (b) eczema
  (c) bile salts (d) lipases
  (e) chylomicrons (f) lymph
  (g) fat deposits (h) obesity
  (i) lipolysis (j) mobilization
  (k) fatty acyl-CoA (l) β-oxidation
  (m) ketonemia (n) ketonuria

17.2 Compare the energy released when 1 g of carbohydrate, 1 g of lipid, and 1 g of alcohol are oxidized completely in the body.

17.3 It was said (footnote 2) that 1 g of fat contains more than six times the energy content of 1 g of glycogen. Explain.

17.4 What is a characteristic structural feature of the essential fatty acids?

17.5 Briefly discuss the digestion of lipids.

17.6 Show, with equations, the chemical changes that triacylglycerols undergo during digestion.

17.7 Write the formulas for four fatty acids that would be formed during the digestion of the lipids found in peanuts (see Table 12.3).

17.8 What happens to the products of lipid digestion after they pass across the intestinal wall?

17.9 (a) How are lipids transported by the blood?
  (b) Where are they transported?

17.10 Arrange the following in order of decreasing plasma-lipid concentration: triacylglycerols, phospholipids, cholesterol esters.

17.11 What conditions lead to obesity?

17.12 What are the functions of adipose tissue?

17.13 Epinephrine and insulin are usually associated with carbohydrate metabolism. What are the effects of these hormones on lipid metabolism?

17.14 What is the difference between the proteins that transport lipids (triacylglycerols, phospholipids, cholesterol) and the proteins that transport fatty acids?

17.15 What is the chief fuel for each of the following?
(a) brain
(b) liver
(c) cardiac muscle
(d) kidneys
(e) resting skeletal muscle
(f) active skeletal muscle
(g) skeletal muscle (prolonged exercise)
(h) brain (prolonged starvation)

17.16 (a) Why is the system for the oxidation of fatty acids called a spiral instead of a cycle?
(b) Why is it also called β-oxidation?

17.17 Outline in detail the sequence of reactions that occur in the complete oxidation of lauric acid.

17.18 If hexanoic acid, labeled in carbons 2, 4, and 6, were metabolized to acetyl-CoA and butyryl-CoA via β-oxidation, where would the labeled carbons appear?

17.19 How many turns of the fatty acid spiral are necessary to metabolize the following fatty acids?
(a) myristic acid ($C_{13}H_{27}COOH$)
(b) palmitoleic acid ($C_{15}H_{29}COOH$)
(c) cerotic acid ($C_{25}H_{51}COOH$)

17.20 Long-chain fatty acyl-CoA molecules cannot traverse the inner mitochondrial membrane. A special transport mechanism, which uses carnitine as the carrier molecule, is required. The fatty acyl grouping is transferred from its thioester bond with coenzyme A to the hydroxyl group of carnitine to form acyl carnitine plus coenzyme A.

Complete the following equation.

$$RCH_2CH_2CH_2CH_2CH_2C \overset{O}{\underset{SCoA}{\diagup}} \quad +$$

Fatty acyl-CoA

$$CH_3-\overset{+}{\underset{CH_3}{\overset{CH_3}{N}}}-CH_2-\overset{H}{\underset{OH}{C}}-CH_2-C\overset{O}{\underset{OH}{\diagup}} \quad \xrightarrow{\text{acyltransferase}}$$

Carnitine

17.21 Why do the majority of naturally occurring fatty acids contain an even number of carbon atoms?

17.22 The ingestion of excess carbohydrates results in the deposition of fats in adipose tissue. Explain.

17.23 What is the significance of acetyl-CoA in metabolism?

17.24 In which segment of fatty acid catabolism is the most energy made available?

17.25 Calculate the total number of moles of ATP that can be produced from the oxidation of 1 mole of glyceryl stearate to CO_2 and H_2O in a muscle cell that is respiring aerobically.

17.26 What are the ketone bodies? Write equations for their formation from acetyl-CoA.

17.27 Why do diabetics accumulate relatively large amounts of ketone bodies?

17.28 What is the relationship between ketosis and acidosis?

17.29 Show, with equations, how the glycerol obtained from the hydrolysis of lipids is incorporated into the glycolytic pathway.

18

Protein Metabolism

In Chapter 13 we indicated the essential nature of protein molecules and their vital functions in all living organisms. In Chapter 15 the biosynthesis of proteins was discussed in connection with the study of nucleic acids. In this chapter we deal primarily with the catabolic aspects of protein metabolism, with particular reference to the metabolic role of the amino acids.

18.1   Digestion and Absorption of Proteins

Like polysaccharides and lipids, intact proteins cannot normally be absorbed across intestinal membranes. They must first be hydrolyzed into their constituent amino acids. Protein digestion begins in the stomach where the action of the gastric juice effects the hydrolysis of about 10% of the peptide bonds. This bond rupture results in the formation of smaller polypeptides having molecular weights of the order of 600–3000 daltons. Recall that gastric juice has a pH between 1.5 and 2.5, owing to the presence of hydrochloric acid (about 0.5%).[1] This concentration of acid serves to

---

[1] If the concentration of acid in the stomach exceeds this value (hyperacidity), it may be indicative of gastric ulcers, hypertension, or gastritis (inflammation of the stomach lining). Hypoacidity, a lower than normal concentration of acid, is commonly associated with stomach cancer and pernicious anemia.

473

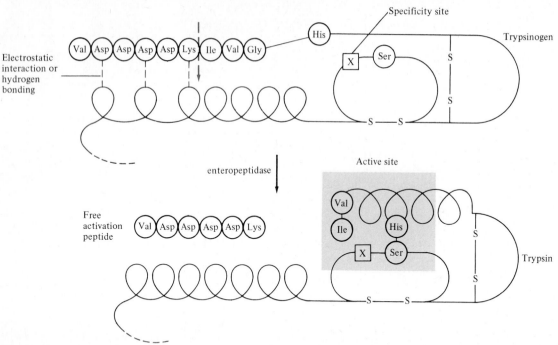

FIGURE 18.1 Schematic representation of the structural changes involved in the activation of trypsinogen. Rupture of the lysyl–isoleucine bond in the N-terminal region (dashed arrow) leads to the liberation of the activation peptide and causes the newly formed N-terminal region of the polypeptide chain to assume a more nearly helical configuration. This in turn permits a histidine and serine side chain to come into juxtaposition so that these catalytic groups are properly aligned. The specificity site of the protein (X) is believed to be pre-existent in the zymogen molecule.

denature the dietary proteins, making them more susceptible to attack by the proteolytic enzymes. The principal digestive component of gastric juice is *pepsinogen,* a zymogen that is produced in secretory cells located in the stomach wall. Pepsinogen is catalytically converted by hydrogen ions into its active form, *pensin.*[2] Once some *pepsin* is formed, it also assists in the activation of the remaining pepsinogen. This phenomenon is known as **autocatalysis,** *the catalysis of a reaction by one of the products of that reaction. Pepsin* is an *endopeptidase* that catalyzes the hydrolysis of peptide linkages within the protein molecule. It has a fairly broad specificity, but acts preferentially on linkages involving the aromatic amino acids tryptophan, tyrosine, and phenylalanine as well as methionine and leucine. The gastric juice of infants contains *rennin,* an enzyme having a specificity very similar to that of *pepsin.*

Protein digestion is completed in the small intestine by the action of the digestive juices of the pancreas and of the intestinal mucosal cells. These juices are sufficiently

---

[2] All of the enzymes that digest proteins (*peptidases*) are secreted in the form of their inactive precursors, probably to protect the tissues from digestion by their own enzymes. A mucous lining protects the digestive tract from hydrolysis by the proteolytic enzymes that function within the tract.

alkaline (pH ~7.5–8.5) to neutralize the acidic material passed on from the stomach. The pancreatic juice contains the zymogens *trypsinogen* and *chymotrypsinogen*. The intestinal mucosal cells secrete the proteolytic enzyme *enteropeptidase*, which converts *trypsinogen* to *trypsin* (Figure 18.1); *trypsin* then activates chymotrypsinogen to *chymotrypsin*. Both of these active enzymes are *endopeptidases*. *Chymotrypsin* preferentially attacks peptide bonds involving the carboxyl groups of the aromatic amino acids (phenylalanine, tryptophan, and tyrosine); *trypsin* attacks peptide bonds involving the carboxyl groups of the basic amino acids (lysine and arginine). Pancreatic juice also contains the zymogen *procarboxypeptidase*, which is cleaved by *trypsin* to *carboxypeptidase*. The latter is an *exopeptidase*; it catalyzes the hydrolysis of the peptide linkages at the free carboxyl end of the peptide chain, resulting in the stepwise liberation of free amino acids from the carboxyl end of the polypeptide.

Two types of *peptidases* are secreted in the intestinal juice: (1) an *exopeptidase, aminopeptidase,* which acts upon the peptide linkages of terminal amino acids possessing a free amino group, and (2) *dipeptidase* and *tripeptidase,* which cleave dipeptides and tripeptides. Figure 18.2 illustrates the specificity of protein-digesting enzymes.

The combined action of the proteolytic enzymes of the gastrointestinal tract results in the hydrolysis not only of the dietary (exogenous) proteins but also of endogenous proteins (the digestive enzymes, other secreted proteins, and dead epithelial cells). (The quantity of endogenous proteins hydrolyzed to amino acids is about 50–70 g each day.) The amino acids are actively transported (an energy-requiring process) across the intestinal wall into the portal circulation[3] and are carried to the liver. The liver is the principal organ responsible for the degradation and synthesis of amino acids. Amino acids have a high turnover rate (half-lives of a few days). The liver is also the site of synthesis of most blood proteins, such as albumins, globulins, fibrinogen, and prothrombin (see Chapter 19). The amino acids synthesized in the liver and the amino acids obtained from protein digestion, along with the amino acids derived from the turnover of tissue proteins, are transported by the blood to all the cells of the body.

FIGURE 18.2 Specificity of peptidase hydrolysis.

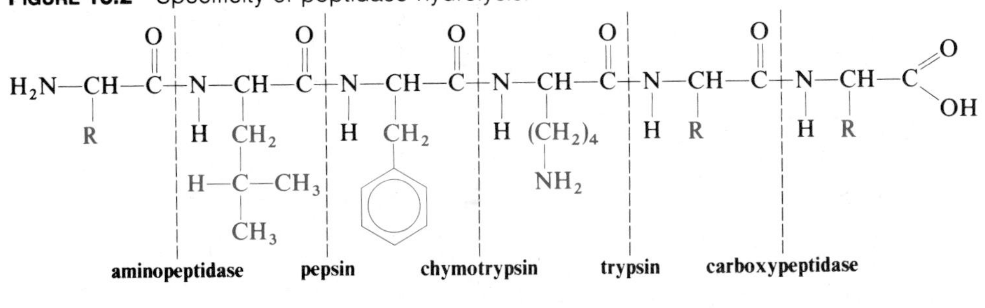

[3]Occasionally, small peptides and polypeptides may be absorbed into the bloodstream. These foreign polypeptides act as *antigens;* that is, they stimulate the formation of specific *antibodies* (see Section 19.2) and are probably responsible for the allergic reactions that some individuals develop for certain foods.

The cellular proteins are in a continual state of flux; they are constantly being degraded and resynthesized from their constituent amino acids. This theory was first validated by the classic study of R. Schoenheimer and D. Rittenberg in 1939. They fed isotopically labeled amino acids to healthy adult rats who had no pressing need for dietary amino acids for growth of tissue protein. They found that almost 60% of the ingested amino acids (labeled with the heavy isotope ^{15}N) were incorporated into tissue proteins. The apparent stability of the adult nongrowing organism is not due, therefore, to metabolic inertness, but rather to a delicate balance between the rates of synthesis and degradation of its constituent compounds. The cell may be visualized as containing a grand mixture, or metabolic pool, of amino acids derived either from the diet or from the degradation of tissue protein. The protein turnover in a normal adult is about 300 g/day. The liver is metabolizing its proteins at such a rate that the equivalent of half the liver tissue is being replaced with fresh protein every 10 days (that is, the half-life of liver protein is 10 days). Proteins of muscle and connective tissue have half-lives of 180 and 1000 days, respectively. The half-life of enzymes varies widely, depending on their metabolic importance and the cell in which they function. Half-lives of enzymes may range from 10 minutes to 6 hours. Hair has no half-life because only synthesis (and not degradation) of hair protein occurs within the ectodermal (skin) cells.

A dietary intake of nitrogen is required to provide for the biosynthesis of the various nitrogenous compounds. Nitrogen is lost in the continuous degradation of tissue protein and in the excretion of certain nitrogen-containing waste materials. Under normal conditions, an *individual's intake of dietary nitrogen is equal to the amount of nitrogen lost in the feces, urine, and sweat.* Such a condition is referred to as **nitrogen balance** or **nitrogen equilibrium.** Organisms are said to be in *positive* nitrogen balance (intake exceeds excretion) whenever tissue is being synthesized, as, for example, during periods of growth, pregnancy, and convalescence from disease. *Negative* nitrogen balance results from (1) an inadequate intake of protein (for example, fasting), (2) fevers, infections, surgery, wasting diseases, or (3) a diet that lacks, or is deficient in, any one of the essential amino acids. These factors cause an accelerated breakdown of tissue protein (in an attempt to supply the missing amino acids), and nitrogen excretion exceeds intake.

Higher plants and certain microorganisms are capable of synthesizing all of their proteins from carbon dioxide, water, and inorganic salts. They obtain their required nitrogen either from soil nitrates or from atmospheric nitrogen (via nitrogen-fixing bacteria). Thus these organisms can grow on a medium that does not contain any performed amino acids. Animals, however, can synthesize only about half of their naturally occurring amino acids; the remainder must be supplied in the diet. All of the amino acids required for the synthesis of a particular protein must be available to the cell at the time of protein synthesis. If just one amino acid is absent, or is present in insufficient quantity, the protein will not be synthesized. An **essential** or *indispensable* amino acid is one that *cannot be synthesized by an organism, from the substrates ordinarily present in its diet, at a rate rapid enough to supply the normal requirements of protein biosynthesis.* A list of essential amino acids is given in

TABLE 18.1 Essential and Nonessential Amino Acids for Humans

Essential	Nonessential
Lysine	Glycine
Leucine	Alanine
Isoleucine	Serine
Methionine	Tyrosine[a]
Threonine	Cysteine
Tryptophan	Aspartic acid
Valine	Asparagine
Phenylalanine	Glutamic acid
Histidine	Glutamine
Arginine[b]	Proline

[a] In the presence of adequate amounts of phenylalanine.

[b] Essential for growing children, not for adults.

Table 18.1. Notice that, as a rule, these compounds contain carbon chains or aromatic rings that are not present as intermediates of carbohydrate or lipid metabolism. The inability to synthesize these amino acids does not arise from a lack of the necessary nitrogen but results rather from the animal's inability to manufacture the correct carbon skeleton. If, for example, an animal is supplied with phenylpyruvic acid, it can readily synthesize the amino analog phenylalanine. Lysine appears to be an exception since the entire preformed amino acid must be supplied. The general effect of a deficiency of one or more essential amino acids is to restrict growth and protein synthesis and produce a negative nitrogen balance.

Essential amino acids are best provided by animal protein. (Casein, the protein from milk, is especially beneficial since it is well balanced in its amino acid distribution.) Proteins that lack an adequate amount of the essential amino acids are termed **incomplete proteins.** Gelatin is an example of an incomplete animal protein because it is deficient in tryptophan.[4] Most plant proteins are deficient in lysine and/or one other essential amino acid. Zein, the protein in corn, is deficient in lysine and tryptophan; rice is low in lysine, methionine, and threonine; wheat is low in lysine; and the legumes are low in methionine and tryptophan. People on vegetarian diets should consume a larger total quantity of protein than would be required from animal protein diets and should include diverse vegetable protein sources to provide the proper amino acid requirements.[5] The average individual requires about 50–60 g (about 2 oz) of protein in the daily diet. This is equivalent to the protein supplied by one quarter-pound hamburger or one regular cheeseburger. (In the industrialized countries, the average daily intake of proteins is about 100 g.)

Prolonged starvation leads to exhaustion of all carbohydrate and fat reserves; the organism is ultimately left with no source of energy except its own tissue proteins and the proteins within the blood. When most of the fat is depleted, a starving human may lose as much as 6% of muscle mass per day. The loss of plasma proteins, especially albumins, occurs to an even greater extent. The Nigerian civil war (1967–68) made the world brutally aware of the serious condition of protein deprivation called *kwashiorkor,* a disease resulting in extreme emaciation, bloated abdomen, mental apathy, diarrhea, lack of pigmentation of the skin and hair, and eventual death. This disease is prevalent in Latin America, Asia, and Africa and is said to be the most severe and widespread nutritional disorder among young children. Kwashiorkor can best be treated by the administration of adequate amounts of well-balanced protein. The problem, of course, is that animal protein is a scarce commodity in many of these areas.

[4] From a health standpoint, "liquid protein" reducing diets are harmful to the individual because the protein is denatured and partially hydrolyzed gelatin. If this is the chief source of protein, the diet lacks the essential amino acid tryptophan.

[5] Amino acids have been touted in articles and advertisements in newspapers and magazines as a cure-all for a variety of ailments including cold sores, depression, fatigue, insomnia, obesity, and pain. For example, there are claims that tryptophan and tyrosine cure depression and insomnia; leucine and phenylalanine are effective as pain relievers. However, the FDA states that it is dangerous for consumers to ingest large amounts of any individual amino acid. Animal studies have shown that amino acid imbalances can be created in the body by abnormal intakes of specific individual amino acids.

18.3 *Metabolism of Amino Acids*

The human body contains about 100 g of *free* (surplus) amino acids. Two amino acids, glutamic acid and glutamine, comprise half of this total, whereas the essential amino acids constitute about 10 g. Most of the amino acids in the metabolic pool are utilized for the synthesis of any of the myriad of proteins necessary to the living organism. This is the major function of amino acids in the body. About 75% of the metabolized amino acids in the normal adult are used for the protein synthesis made necessary by the constant destruction of body proteins by wear and tear. However, amino acids also play an essential role in the metabolism of all nitrogenous compounds. Although the liver is the principal organ for amino acid metabolism, other tissues such as the kidney, intestine, muscle, and adipose tissue are also involved. Generally, the first step in the breakdown of amino acids is the separation

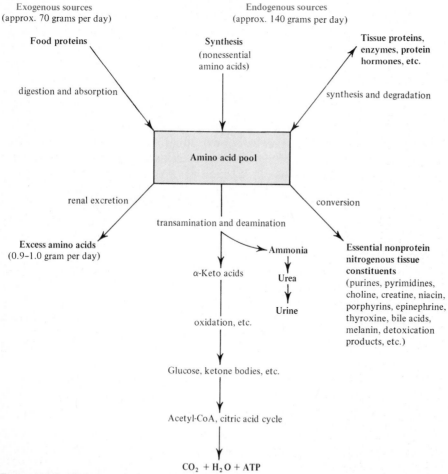

FIGURE 18.3 Scheme of general paths of amino acids in metabolism (average human). [From James M. Orten and Otto W. Neuhaus, *Human Biochemistry*, 9th ed., The C. V. Mosby Co., St. Louis, 1975.]

of the amino group from the carbon skeleton. The amino groups are then incorporated into almost all the other nonprotein nitrogen-containing compounds, for example, neurotransmitters, hormones, nucleic acids, porphyrins, and creatine (Figure 18.3). A discussion of these specific metabolic pathways is beyond the scope of this text.

The carbon skeletons resulting from the deaminated amino acids may be utilized to form either glucose (via gluconeogenesis) or fats (via acetyl-CoA), or they are converted to a metabolic intermediate that can be oxidized by the Krebs cycle. Amino acid catabolism is particularly prevalent during hypoglycemia and fasting (also starvation).[6] Since fatty acids cannot be converted to glucose, proteins must be hydrolyzed to amino acids that can provide the proper carbon skeleton to form the glucose needed by the brain. By the end of the overnight fast, gluconeogenesis is providing blood glucose from amino acids.

a. Transamination

In this reaction, the amino group is transferred from an amino acid to an α-keto acid, resulting in the formation of a new keto acid and a new amino acid.

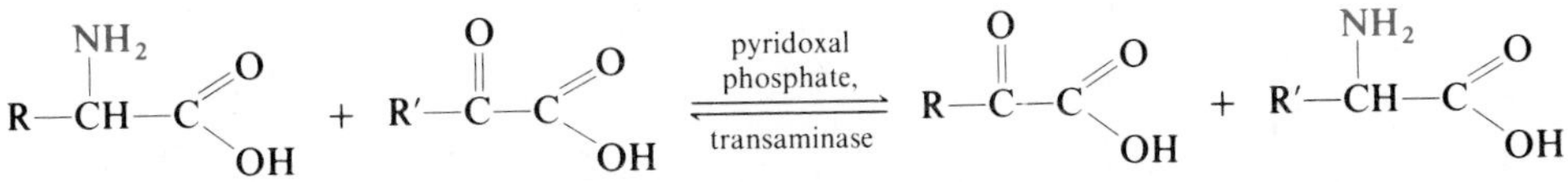

Amino acid α-Keto acid   New α-keto acid   New amino acid

Practically all of the naturally occurring amino acids have been found to undergo this reaction. It is probable that there are a large number of *transaminases* (also called *aminotransferases*), each one catalyzing a reaction between a specific α-amino acid and some α-keto acid. The *transaminases* occur in most tissues and are known to function in conjunction with the coenzyme pyridoxal phosphate (Table 14.2). The first step involves the attachment of the amino acid to the enzyme, which then transfers the amino group to the appropriate α-keto acid, usually α-ketoglutaric acid (or, less frequently, oxaloacetic acid).

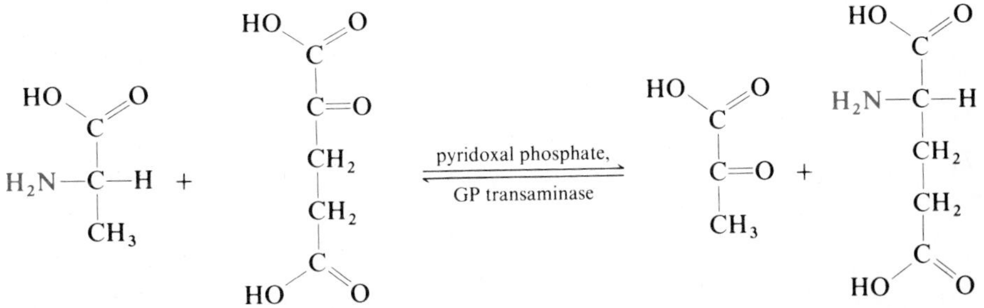

Alanine α-Ketoglutaric acid   Pyruvic acid   Glutamic acid

[6] During fasting or starvation, the liver glycogen is depleted very rapidly and blood glucose must be manufactured from other sources to supply energy for the brain cells. The major source of this blood glucose is the amino acids obtained from the degradation of tissue protein (e.g., proteins from skeletal muscles and from membranes that line the digestive tract). The minimum requirement for carbohydrates is only about 75 g/day because glucose can be synthesized from amino acids.

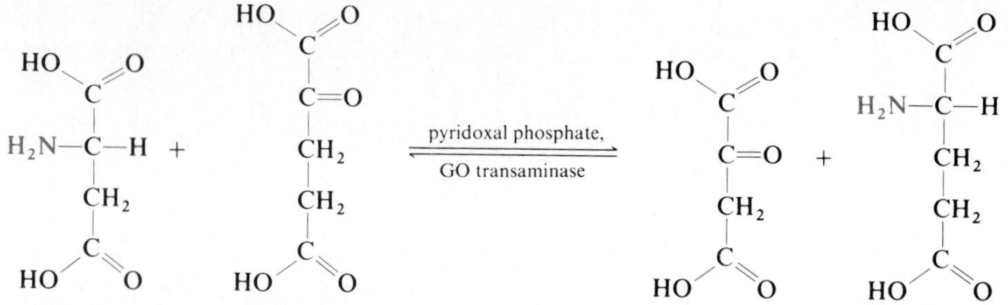

Aspartic acid   α-Ketoglutaric acid   Oxaloacetic acid   Glutamic acid

These two particular transamination reactions are of special clinical interest. Normally the blood contains a low concentration of *transaminase* enzymes. However, when extensive tissue destruction occurs, it is accompanied by rapid and striking increases in the blood *transaminase* levels. *Glutamic–pyruvic transaminase* (GPT) has a particularly high activity in the cytosol of the liver, and an elevated serum level of this enzyme is indicative of liver damage. Heart muscle contains an abundance of *glutamic–oxaloacetic transaminase* (GOT). In the diagnosis of myocardial infarction it is common practice to test blood samples for both GOT and CK (*creatine kinase*—see footnote 14, page 441, and Section 14.10).

It is especially significant that the transamination reactions are reversible, since this reversibility links protein metabolism with the metabolism of carbohydrates and lipids. Those *amino acids that can form any of the metabolites of carbohydrate metabolism* can be converted to glucose (via gluconeogenesis) or oxidized to carbon dioxide, water, and energy; they are referred to as **glucogenic**. *Amino acids that give rise to acetoacetyl-CoA or acetyl-CoA* are precursors for fatty acid synthesis; they are said to be **ketogenic** (because they yield ketone bodies). Certain amino acids fall into both categories; leucine is the only amino acid that is exclusively ketogenic. Table 18.2 classifies the amino acids with regard to the metabolic fate of their carbon skeletons. The products and metabolic routes of the 20 amino acids are summarized in Figure 18.4.

b. Oxidative Deamination

We saw that in transamination reactions, the α-amino group of many amino acids was transferred to α-ketoglutaric acid to form glutamic acid. The latter compound

TABLE 18.2 Glucogenic and Ketogenic Amino Acids

Glucogenic		Ketogenic	Glucogenic and Ketogenic
Alanine	Glycine	Leucine	Isoleucine
Arginine	Histidine		Lysine
Asparagine	Methionine		Phenylalanine
Aspartic acid	Proline		Tyrosine
Cysteine	Serine		Tryptophan
Glutamic acid	Threonine		
Glutamine	Valine		

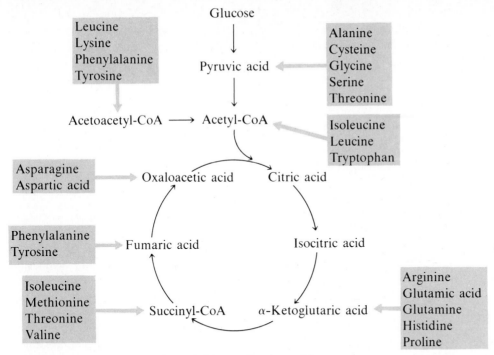

Figure 18.4 Fates of the carbon skeletons of amino acids.

then loses the amino group as ammonia and is oxidized back to α-ketoglutaric acid. This process is termed **oxidative deamination,** and it occurs in the liver mitochondria. The enzyme *glutamic acid dehydrogenase* is unusual in that it uses either NAD^+ or $NADP^+$ as a coenzyme. (The reduced forms of the coenzymes, NADH or NADPH, are ultimately oxidized by the respiratory chain.) The ammonia formed from glutamic acid by oxidative deamination is converted into urea, which is excreted in the urine (see Section 18.5).

The preceding equilibrium reaction is of central importance in linking protein metabolism and carbohydrate metabolism. The reverse reaction is significant in nitrogen metabolism because it is one of the few reactions in animals that can convert inorganic nitrogen (NH_3) into organic nitrogen (amino acids). The amino group in the glutamic acid can be passed on through transamination reactions, producing all the other cellular amino acids, providing the appropriate α-keto acids are available.

c. Decarboxylation

Several *amino acid decarboxylases* eliminate carbon dioxide from amino acids to form primary amines. Pyridoxyl phosphate is the necessary coenzyme.

$$R-\underset{\underset{NH_2}{|}}{\overset{\overset{H}{|}}{C}}-C\underset{OH}{\overset{O}{<}} \xrightarrow[\text{decarboxylase}]{\text{pyridoxyl phosphate,}} R-\underset{\underset{NH_2}{|}}{\overset{\overset{H}{|}}{C}}-H + CO_2$$

These enzymes are found primarily in microorganisms, but they are also found in some animal tissues. Intestinal bacteria are responsible for amino acid decarboxylation, and the foul smell of feces is due in part to the resulting amines. (Recall that cadaverine and putrescine are formed upon decarboxylation of lysine and ornithine—Section 9.8).

Some of the amines produced as a result of decarboxylation have important physiological effects. Many people are allergic to pollen, dust, insect stings, and so on. By a mechanism not completely understood, the body responds to these foreign substances by decarboxylating histidine to *histamine.* (Allergic individuals synthesize abnormally high amounts of histidine.) Histamine dilates blood vessels and thus initiates the inflammatory response (see footnote 10, page 293). The discomfort associated with hay fever and other allergies is due to the inflammation of the eyes, nose, and throat. Furthermore, the expansion of the blood capillaries causes a decrease in the blood pressure that, if severe enough, may induce shock. (Recall that prostaglandins are also released, and they increase the sensitivity to pain associated with inflammation—Section 12.15.) *Antihistamines* are compounds that are structurally similar to histamine. They can occupy the receptor sites[7] normally occupied

[7]There are two types of receptors for histamine, designated as H_1 and H_2. H_1 receptors are found in the walls of capillaries and in the smooth muscle of the respiratory tract. They affect the vascular (dilation and increased permeability of capillaries) and muscular (bronchoconstriction) changes associated with hay fever and asthma. H_1 receptors are blocked by the classical antihistamines (such as Benadryl), which relieve these symptoms. H_2 receptors occur mainly in the wall of the stomach, and their activation causes an increased secretion of hydrochloric acid. Cimetidine blocks H_2 receptors and thus, by reducing acid secretion, is an effective drug for people with ulcers.

Cimetidine
(Tagamet)

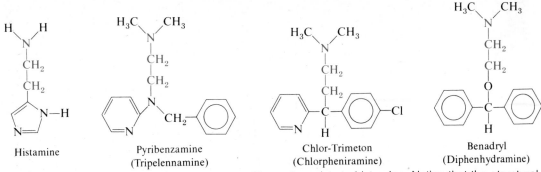

FIGURE 18.5 The antihistamines act as competitive antagonists to histamine. Notice that the structural similarity to histamine is the substituted ethylamine moiety (in color).

by histamine (e.g., on smooth muscles that surround capillaries), thereby preventing the physiological changes produced by histamine (i.e., they are antagonists). Figure 18.5 illustrates some of the common antihistamines found in nasal decongestants, combination pain relievers, and hay fever preparations. The most common side effect of antihistamines is sedation. This can impair one's ability to operate machinery or drive a motor vehicle.

Besides histamine, several other neurotransmitters are synthesized by the decarboxylation of specific amino acids. Serotonin (5-hydroxytryptamine—Section 9.12-f) is formed by the action of a specific *decarboxylase* on 5-hydroxytryptophan. The cell bodies of serotonin-containing neurons are located almost exclusively in the upper brainstem from which axons project to other areas of the central nervous system. Serotonin constricts blood vessels, stimulates smooth muscle, and has a potent inhibitory effect on its postsynaptic neurons. Another

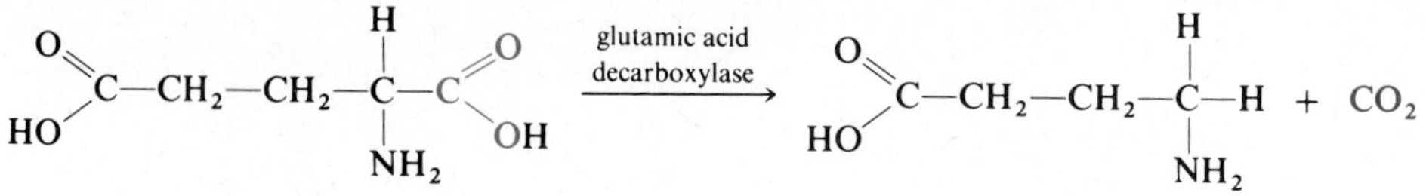

$$O=C(HO)-CH_2-CH_2-\underset{NH_2}{\overset{H}{C}}-C(\overset{O}{\diagdown})OH \xrightarrow[\text{decarboxylase}]{\text{glutamic acid}} O=C(HO)-CH_2-CH_2-\underset{NH_2}{\overset{H}{C}}-H + CO_2$$

Glutamic acid → γ-Aminobutyric acid (GABA)

important inhibitory neurotransmitter, GABA (γ-aminobutyric acid), is formed in the brain and spinal cord from the decarboxylation of glutamic acid. GABA is thought to inhibit dopamine neurons in particular, as well as other neurons throughout the central nervous system. Another *decarboxylase* catalyzes the decarboxylation of 3,4-dihydroxyphenylalanine (L-dopa) to form dopamine. A *deficiency* of dopamine in the brain cells is a primary cause of Parkinson's disease, a disorder of the central nervous system that involves a progressive paralytic rigidity, tremors of the extremities, and unresponsiveness to external stimuli. Dopamine itself cannot be administered because it does not pass across the blood–brain barrier. A major breakthrough in the treatment of Parkinson's disease has been the use of L-dopa. (The L-enantiomer is more effective and less toxic than the D-form of the drug.) Large doses of L-dopa are administered orally; the drug is able to pass from the digestive system into the blood and then cross the blood–brain barrier. L-Dopa is decarboxylated in the brain into the necessary neurotransmitter, dopamine. (Incidentally, numerous studies have linked schizophrenia to an *overabundance* of dopamine in brain cells.)

3,4-Dihydroxyphenylalanine (L-Dopa) $\xrightarrow[\text{decarboxylase}]{\text{dopa}}$ 3,4-Dihydroxyphenylethylamine (Dopamine) + CO_2

18.4 Storage of Nitrogen

In contrast to carbohydrates and lipids, proteins are not stored to an appreciable extent in living organisms. Small amounts of amino acids are excreted by humans, but this is essentially a wasteful process since amino acids contain utilizable energy. (One gram of protein liberates about 4300 cal.) Accordingly, most excess amino acids are deaminated; the resulting carbon skeleton is oxidized to produce energy or is stored as glycogen or fat. Most of the ammonia that is liberated is converted to urea. A relatively small proportion combines with glutamic acid in the presence of *glutamine synthetase* and ATP to form glutamine. This is another example of the conversion of inorganic nitrogen to organic nitrogen.

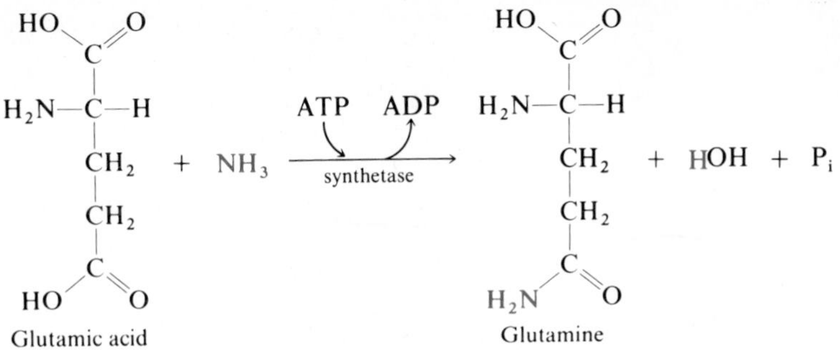

Glutamic acid   Glutamine

Glutamine is present in many tissues and in the blood and serves as a temporary storage and transport form of nitrogen. The formation of glutamine is the major method of disposing of ammonia from the brain.

In the liver and in the kidneys, the amide group of glutamine can be donated in specific reactions to appropriate acceptor molecules to effect the biosynthesis of many nitrogen-containing compounds.

$$\text{Glutamine} + H_2O \xrightarrow{\text{glutaminase}} \text{Glutamic acid} + NH_3$$

18.5   Excretion of Nitrogen

Whenever amino acids are utilized for energy production or for the synthesis of glucose or fat, an amino group is liberated in the form of ammonia. Living organisms must have some mechanism for removing this ammonia from the cell environment because even low concentrations of ammonia are poisonous. Levels of only 5 mg of ammonia per 100 mL of blood (hyperammonemia) are toxic to humans. The normal concentration of this compound is about $1-3$ $\mu g/100$ mL. Organisms differ biochemically in the manner in which they excrete excess nitrogen. Most vertebrates and most adult amphibia excrete nitrogen as urea in the urine. Birds, reptiles, and insects convert nitrogen to uric acid. All marine organisms, from unicellular organisms to fish, excrete free ammonia. The ammonia is very soluble in water and is rapidly diluted in the aqueous environment.

In humans, 95% of the nitrogenous waste products results from amino acid catabolism; the remaining 5% comes from the metabolism of other nitrogen-containing compounds (e.g., creatine, neurotransmitters, pyrimidines). The breakdown of purine bases in the human body results in the production of uric acid, and very small concentrations of this acid are found in the urine and in the body fluids.

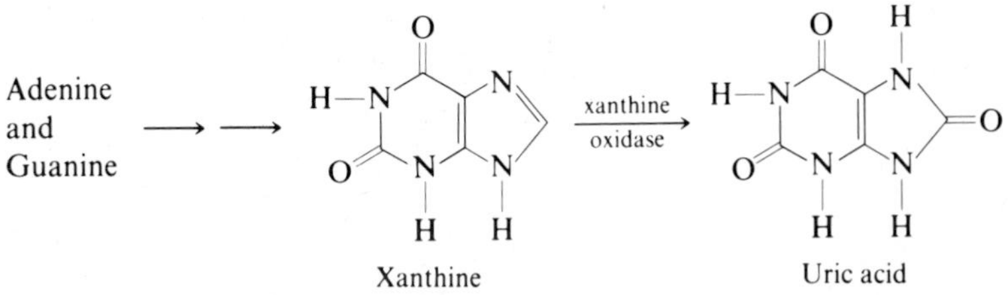

Xanthine   Uric acid

Under certain pathological conditions (impairment of purine metabolism), large quantities of uric acid are produced and the plasma concentration of uric acid becomes abnormally high (>7 mg/100 mL). At physiological pH, the excess uric acid precipitates as the sparingly soluble monosodium salt. When such deposits occur in the joints of the body's digital regions, the surrounding tissue can become inflamed, causing the painful arthritic characteristics of gout. Crystals of monosodium urate may also accumulate in the kidneys (kidney stones) and cause extensive impairment of renal function. Kidney failure is another serious medical consequence of gout. (For some unknown reason, men suffer from gout to a much greater extent than women.)

FIGURE 18.6 Interrelationships of metabolic pathways.

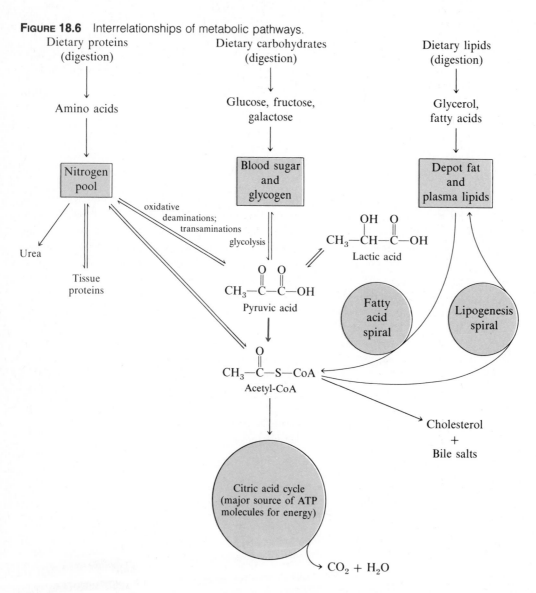

Another serious impairment of purine metabolism results from a genetic defect. In the *Lesch–Nyhan* syndrome, an enzyme necessary for the normal utilization of purines is absent. Affected children are mentally defective and exhibit a compulsive, aggressive behavior toward others (extreme hostility), as well as toward themselves (self-mutilation by chewing their tongue, lips, and fingers). The relationship between the absence of the enzyme and the aberrant behavior is still unknown.

In mammals, the liver is the principal organ concerned with the formation of urea (by a cyclic series of reactions known as the **urea cycle**—see an advanced biochemistry text). Extensive liver damage (e.g., cirrhosis) or a genetic defect of any of the urea cycle enzymes results in hyperammonemia and causes tremor, slurred speech, and blurred vision. A continual rise in ammonia concentrations can lead to coma and death. Once produced, urea is transported by the bloodstream to the kidneys and is eliminated in the urine. The kidneys filter about 100 L of blood each day. The normal individual excretes about 1.5 L of water, containing approximately 30 g of urea daily, although this value varies greatly from day to day and can rise dramatically (to about 100 g of urea) upon ingestion of a high protein diet. A long-term decrease in urea excretion is indicative of liver and/or kidney disease. Loss of nitrogen also occurs through the skin, mainly as sweat (about 2 g of urea) and via the feces (smaller amounts).

18.6 *Relationships Among the Metabolic Pathways*

A great variety of organic compounds can be derived from carbohydrates, lipids, and proteins. All organisms utilize the three major foodstuffs to form acetyl-CoA or the metabolites of the Krebs cycle. These in turn supply energy upon subsequent oxidation by the cycle. A brief summary of the interrelationships of the metabolic pathways is given in Figure 18.6. We should mention once again that in mammals, the conversion of pyruvic acid to acetyl-CoA is irreversible. Thus, an acetyl-CoA molecule derived from fatty acid degradation cannot be directly transformed into pyruvic acid. Therefore, mammals are unable to synthesize carbohydrates from lipids.

Exercises

18.1 Define or give an explanation for each of the following terms.
(a) peptidase
(b) autocatalysis
(c) protein turnover
(d) nitrogen balance
(e) kwashiorkor
(f) pyridoxal phosphate
(g) GABA
(h) Parkinson's disease
(i) L-dopa
(j) hyperammonemia
(k) gout
(l) urea

18.2 Distinguish between the terms in each of the following pairs.
(a) pepsin and pepsinogen
(b) chymotrypsin and trypsin
(c) endopeptidase and exopeptidase
(d) positive and negative nitrogen balance
(e) essential and nonessential amino acid
(f) complete and incomplete protein
(g) transamination and deamination
(h) GPT and GOT
(i) glucogenic and ketogenic amino acids
(j) histamine and antihistamine

18.3 Briefly describe the digestion of proteins. Mention the location and the enzymes involved.

18.4 What is the defense mechanism employed by the body to protect its tissues from the digestive action of the proteolytic enzymes?

NaHCO₃   peptidase

18.5 In the treatment of diabetes, insulin is administered by injection rather than orally. Explain.

18.6 Indicate the expected products from the enzyme action of *pepsin, chymotrypsin,* and *trypsin* on each of the following tripeptides.
(a) Ala-Phe-Tyr
(b) Ile-Tyr-Ser
(c) Phe-Arg-Leu
(d) Thr-Glu-Lys

18.7 Indicate the location in a polypeptide chain that is cleaved by (a) an *aminopeptidase* and (b) a *carboxypeptidase.*

18.8 What was the significance of Schoenheimer's and Rittenberg's experiments with labeled amino acids?

18.9 What is meant by the term "metabolic pool of amino acids"?

18.10 Compare the turnover rate of enzymes and muscle proteins.

18.11 Vegetable protein is converted in the animal body into animal protein. Explain.

18.12 (a) Why are vegetarians instructed to eat a wide variety of vegetables?
(b) What is the best source of essential amino acids?

18.13 (a) What is a unique structural feature of most of the essential amino acids?
(b) What is the effect of a deficiency of one or more of the essential amino acids?

18.14 What are three functions of amino acids in the body?

18.15 What is the significance of the transamination reaction in the metabolism of amino acids?

18.16 The following compound can supply the body with one of the essential amino acids. Show, with an equation, how this is possible.

$$CH_3CH_2CH(CH_3)\overset{O}{\underset{\|}{C}}-C\overset{O}{\underset{OH}{}}$$

18.17 If 1 mole of alanine is converted into pyruvic acid, and the pyruvic acid is metabolized via the Krebs cycle, how many moles of ATP could be produced?

18.18 (a) What toxic compound is formed in the oxidative deamination reaction?
(b) How does the body get rid of this compound?

18.19 (a) How does Benadryl exert its effect?
(b) What is the difference between an H_1 receptor and an H_2 receptor?

18.20 Name the compound that is described.
(a) the amino acid that is exclusively ketogenic
(b) the principal amino group acceptor in transamination reactions
(c) the α-keto acid that is usually formed in oxidative deamination reactions
(d) the compound that is decarboxylated to serotonin
(e) the compound that is decarboxylated to GABA
(f) the compound that serves as the temporary storage form of nitrogen in the body.

18.21 Write equations for each of the following.
(a) oxidative deamination of an amino acid
(b) transamination of an amino acid
(c) decarboxylation of an amino acid

18.22 Write equations for the formation of each of the following compounds from an amino acid.
(a) pyruvic acid   (b) histamine
(c) α-ketoglutaric acid   (d) glutamine
(e) cadaverine   (f) oxaloacetic acid
(g) ethanolamine   (h) dopamine

18.23 (a) What organ is responsible for the synthesis of urea?
(b) What organ is responsible for the excretion of urea?

18.24 Indicate the relationship between the metabolism of carbohydrates, lipids, and proteins.

19

The Blood

The three circulating fluids (extracellular fluids) in the animal organism are *blood, interstitial fluid* (fluid within the tissue spaces—about 17% of body weight), and *lymph* (fluid within the lymph vessels). Blood is by far the most important of these. It is the connecting link between all the tissue cells and the external environment. The circulating blood transports oxygen and nutritive materials to all the cells and metabolic waste products to the excretory organs. It transports hormones from the site of their secretion to the organs that require them. In addition to its transport functions, the blood serves to (1) help maintain osmotic relationships between the tissues (water and electrolyte balance and fluid distribution), (2) protect the organism against infection, (3) regulate the body temperature, and (4) control the pH of the body. The blood contains a number of different types of cells and biochemical compounds that enable it to carry out these varied functions.

19.1 General Characteristics of Mammalian Blood

Circulatory blood consists of a straw-colored fluid portion (the plasma) and the formed elements (blood cells). It accounts for about one twelfth of the body weight of the average individual. The volume of blood varies with body size. A 150 lb (68 kg) man has about 5–6 L of blood; roughly 55–60% of it is plasma. **Plasma** is an extremely complex solution, containing all of the biochemically significant compounds (carbohydrates, amino acids, proteins, lipoproteins, enzymes, hormones, vitamins, and inorganic ions). Although these compounds are continuously entering and leaving the circulatory system, the overall composition of the plasma remains remarkably constant (a state of dynamic equilibrium exists). Approximately 90–92% of plasma is water; the remaining 8–10% consists of dissolved solids, the greatest component of which is the plasma proteins. Dispersed throughout the plasma are the **formed elements,** which account for about 40–45% of the total blood volume. The formed elements of the blood are the **erythrocytes** (red blood cells), **leukocytes** (white blood cells), and **thrombocytes** (platelets). The chemical analysis of blood samples is of great clinical significance. Increased or decreased amounts of certain substances may indicate a particular disease or condition (Table 19.1).

a. Plasma Proteins

The protein content of the blood is about the same as that of the muscle and other tissues. More than 100 proteins have been identified in the blood; most of them are synthesized in the liver. The normal concentration range of blood proteins is 7.0–8.0 g/100 mL of plasma. These proteins remain within the circulatory system and ordinarily are not used as sources of energy. At times of protein deprivation, the plasma protein concentration is maintained at the expense of tissue protein. Plasma proteins are grouped into three main classes on the basis of their solubility properties and methods of isolation.

1. Albumins Albumins are the most abundant proteins ($\sim 55\%$ by mass) in the plasma. Their major function is the maintenance of osmotic pressure (see Section 19.3) by controlling the water balance. They are also important for their buffering capacity and in the transport of fatty acids, certain metal ions, and many drugs (particularly the nonpolar ones). Normal concentrations of albumins range from 3.5 to 4.5 g per 100 mL of plasma.

2. Globulins Globulins account for about 40% by mass of the total plasma proteins. They have higher molecular weights than the albumins (150,000 daltons as compared to 70,000 daltons). Three subclasses of globulins are recognized, and they differ from one another with respect to their rates of movement in an electric field; α-globulins (0.7–1.5 g/100 mL of plasma), β-globulins (0.6–1.1 g/100 mL), and γ-globulins (0.7–1.5 g/100 mL). α-Globulins and β-globulins are synthesized in the liver. They form complexes with the water-insoluble lipids and thus serve to transport these compounds through the aqueous media (i.e., VLDLs, LDLs, and HDLs). Unlike the other plasma proteins the γ-globulins are synthesized by white blood cells in the lymph nodes. They combat certain infectious diseases (such as diphtheria, influenza, measles, mumps, typhoid), and they are referred to as *antibodies* or *immunoglobulins*. Based on molecular weights and chemical properties, five classes of immunoglobulins are recognized (IgG, IgM, IgA, IgD, and IgE). IgG is the most abundant ($\sim 80\%$) immunoglobulin in plasma. IgA accounts for about 13% of the antibodies, IgM about 6%, and the remaining 1% is IgD and IgE. IgA is

TABLE 19.1 Normal Composition of Blood

Determination[a]	Normal Range (per 100 mL)	Clinical Significance	
		Increased in	Decreased in
Calcium (s)	9–10.6 mg (4.5–5.3 mEq/L)[b]	Hyperparathyroidism, Addison's disease, malignant bone tumor, hypervitaminosis D	Hypoparathyroidism, rickets, malnutrition, diarrhea, chronic kidney disease, celiac disease
Cholesterol, total (s)	150–280 mg	Diabetes mellitus, obstructive jaundice, hypothyroidism, pregnancy	Pernicious anemia, hemolytic jaundice, hyperthyroidism, tuberculosis
Uric acid (s)	3–7.5 mg	Gout, leukemia, pneumonia, liver and kidney disease	
Urea nitrogen (b, s)	8–20 mg	Mercury poisoning, acute glomerulonephritis, kidney disease	Pregnancy, low-protein diet, severe hepatic failure
Nonprotein nitrogen (b, s)	15–35 mg	Kidney disease, pregnancy, intestinal obstruction, congestive heart failure	Low-protein diet
Creatine (s)	3–7 mg	Nephritis, renal destruction, biliary obstruction, pregnancy	
Creatinine (b, s)	0.7–1.5 mg	Nephritis, chronic renal disease	
Glucose (b)	70–100 mg	Diabetes mellitus, hyperthyroidism, infections, pregnancy, emotional stress, after meals	Starvation, hyperinsulinism, Addison's disease, hypothyroidism, extensive hepatic damage
Chlorides (s)	100–106 mEq/L	Nephritis, anemia, urinary obstruction	Diabetes, diarrhea, pneumonia, vomiting, burns
Phosphate, inorganic (p)	3–4.5 mg	Hypoparathyroidism, Addison's disease, chronic nephritis	Hyperparathyroidism, diabetes mellitus
Sodium (s)	136–145 mEq/L	Kidney disease, heart disease, pyloric obstruction	Vomiting, diarrhea, Addison's disease, myxedema, pneumonia, diabetes mellitus
Potassium (s)	3.5–5 mEq/L	Addison's disease, oliguria, anuria, tissue breakdown	Vomiting, diabetes acidosis, diarrhea
Carbon dioxide (s, p)	Adults, 24–29 mEq/L Infants, 20–26 mEq/L	Tetany, vomiting, intestinal obstruction, respiratory disease	Acidosis, diarrhea, anesthesia, nephritis
Hemoglobin (b)	Male, 14–18 g Female, 12–16 g	Polycythemia	Anemia

[a]Key to abbreviations ( ):   b = blood   s = serum   p = plasma
[b]mEq/L = milliequivalents per liter.
Reprinted with permission of Macmillan Publishing Company from *Chemistry for the Health Sciences,* 5th ed., by George I. Sackheim and Dennis D. Lehman. Copyright ©1985 by Macmillan Publishing Company.

the principal antibody in external secretions (e.g., mucus, saliva, tears), and thus it represents the initial defense against bacteria and viruses (see Section 19.2).

3. Fibrinogen Fibrinogen is a large protein (MW 340,000 daltons) consisting of six polypeptide chains. It is synthesized in the liver and constitutes about 5% (0.3–0.4 g/100 mL of plasma) of the total blood protein. It functions in blood coagulation (see Section 19.5). The fibrinogen content of plasma increases when inflammatory or infectious processes exist and during menstruation and pregnancy.

b. Formed Elements

Apart from the respiratory function of the hemoglobin molecules in the erythrocytes, the blood cells have specific roles that are not directly concerned with the general metabolic processes. This is in contrast to the plasma, which serves as the metabolic transport medium of the organism.

1. Erythrocytes The red blood cells are formed in the red bone marrow, and they are the most numerous of the formed elements. The blood of the average adult female contains about 5 million of these cells, and that of the average male about 5.5 million, in every cubic millimeter (mm^3) of blood. (One drop of blood is the equivalent of about 100 mm^3, so there are about 500 million erythrocytes in each drop of blood). Any condition that tends to lower the oxygen content of the blood causes an increase in the number of erythrocytes. Persons living at high altitudes generally have a higher erythrocyte count than do those living at sea level. Conversely, increased barometric pressure results in a decrease in the erythrocyte count. The term *hematocrit value* is applied to the volume (in percent) of packed red blood cells in a sample (usually 10 mL) that has been centrifuged under standard conditions. The cells are spun to the bottom of a centrifuge tube and the supernatant liquid (the plasma) is drawn off the top. A blood sample usually contains about 45% red cells; thus 45 is the normal hematocrit value. Any variation from the normal value may be indicative of the existence of certain pathological conditions. When anemia occurs, for instance, *the percentage of erythrocytes (and/or the percentage of hemoglobin) is abnormally low.* Anemia may result from (1) a decreased rate of production of erythrocytes (aplastic anemia), (2) an increased destruction of erythrocytes (hemolytic anemia), or (3) an increased loss of erythrocytes (as in hemorrhaging). *Polycythemia* is the condition arising from an abnormally high percentage of red blood cells.

Red blood cells are disk-shaped with slight depressions at the center, like a solid doughnut (Figure 19.1). Unlike most other cells, erythrocytes contain neither mitochondria nor a nucleus. (The nucleus is lost during the development and maturation of the erythrocyte.) They cannot reproduce, have no aerobic metabolism, and are unable to synthesize carbohydrates, lipids, or proteins. They obtain all of their energy from the pentose phosphate shunt, and from substrate level phosphorylation in the Embden–Meyerhof pathway. The most significant components of the red blood cell are the hemoglobin molecules. An individual's blood type is determined by specific short-chain polysaccharides that are bound to certain proteins (glycoproteins) on the membranes of the red blood cells.

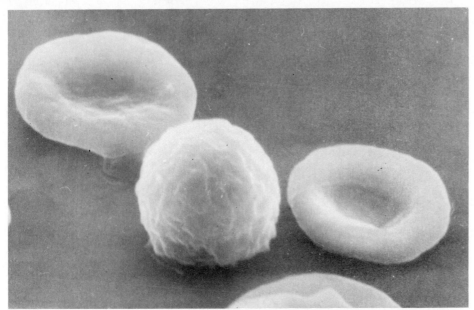

FIGURE 19.1   Red blood cells are doughnut-shaped discs, whereas white blood cells are spherical. The photo was obtained through a scanning electron microscope at 10,000× magnification. [Courtesy of Benjamin Zweifach, Dept. of Bioengineering, Univ. of Calif., San Diego.]

The major function of the erythrocyte is to transport oxygen from the lungs to the cells. Also, it assists in the transport of carbon dioxide from the tissues to the lungs (see Section 19.7). Human red blood cells have a life span of about four months. (In this time interval, each red blood cell makes about 120,000 trips around the body.) During this period, there is no degradation or resynthesis of hemoglobin molecules in the erythrocyte. To maintain a constant level of erythrocytes, new cells are formed in the bone marrow at the same rate that old cells are eliminated by special tissues in the liver and the spleen. It has been estimated that there are approximately 30 trillion red blood cells in an average adult male and about 3 million of these are destroyed every second. Assuming that there are 300 million hemoglobin molecules in each erythrocyte, then 900 trillion molecules of hemoglobin must be synthesized every second (by cells in the bone marrow) in order to maintain a constant supply.

2.   Leukocytes   The composition of white blood cells resembles that of other tissue cells. They are nucleated and they contain glucose, lipids, proteins and other soluble organic substances and inorganic salts. White blood cells constitute the body's primary defenders against foreign organisms (e.g., viruses, bacteria), and the blood is the vehicle that transports them to sites of infections.

The different varieties of leukocytes have specialized functions. *Lymphocytes* are involved in the synthesis and storage of antibodies (see Section 19.2). *Phagocytes* (*macrophages*) can leave the blood by squeezing between the endothelial cells that line the capillary wall. They are attracted toward sites of inflammation by chemicals released from injured tissue. Phagocytes contain lysosomal enzymes, and their function is to engulf and digest the invading organisms.

Billions of leukocytes are produced each day in the bone marrow to replace the ones that die. On the average, there are about 7000 leukocytes per cubic millimeter of blood, but this value is subject to considerable variation. A higher than normal leukocyte count occurs during acute infections such as *appendicitis* (16,000–20,000 per mm^3). High numbers of leukocytes may also appear during emotional disturbances and following vigorous exercise and/or excessive loss of body fluids. Viral diseases such as chickenpox, influenza, measles, mumps, and polio are accompanied by an abnormally low count (< 5000 per mm^3), because in fighting the viruses the leukocytes are killed faster than they can be produced. **Leukemia** is *a cancer that is characterized by the uncontrolled production of leukocytes that fail to mature*. Thus, even though there is a high number of white blood cells, these cells are unable to destroy invading pathogens and the person has a lowered resistance to infections. Invariably some of the cancer cells *metastasize* (spread out) from the bone marrow or lymph nodes to other parts of the body. In these other tissues, the leukemic cells eventually crowd out the normal functioning cells.

3. Thrombocytes There are about 250,000 platelets in every cubic millimeter of blood. These small, non-nucleated cells contain proteins and relatively large amounts of phospholipids, mostly cephalin. The blood platelets liberate species that are instrumental in the mechanism of blood clotting (see Section 19.5). An abnormally low platelet count (< 100,000 per mm^3) is related to a tendency to bleed.

19.2   *The Immune Response*

Immunity *is resistance to specific pathogens* (infectious agents such as viruses, bacteria, fungi, protozoans) *or the toxins secreted by the pathogens*. It involves the **immune response** *in which the lymphocytes and phagocytes recognize foreign substances* (cells or large molecules like proteins) *and act to destroy them*.

Lymphocytes are formed in bone marrow cells, in lymph nodes, and in other lymphatic tissue. Developing, undifferentiated lymphocytes are released from the bone marrow and transported by the blood. Many of them are delivered to the thymus gland (located below the trachea). The majority of the lymphocytes remain inactive within the thymus. However, some of these lymphocytes become differentiated and thereafter are called **T-lymphocytes** (*thymus-derived lymphocytes* or *T-cells*). Some of the T-cells are stored in the thymus; others are transported by the blood to other organs of the lymphatic system (in particular the lymph nodes and spleen). The lymphocytes that do not enter the thymus gland become differentiated in the blood and are called **B-lymphocytes** (*bone marrow-derived lymphocytes*

or *B-cells*). These cells are distributed by the blood to the lymph nodes and spleen along with the T-cells. B-cells and T-cells are similar in size and appearance, but they take part in different forms of the immune response.

We said that the function of leukocytes is to destroy foreign substances. This is accomplished by first being able to distinguish self from non-self. Substances that are foreign to our bodies contain certain proteins, termed **antigens** on their surfaces. These proteins are unique for each foreign substance, yet are markedly different from any of our body proteins. B-cells and T-cells become programmed, or sensitized, to recognize these antigens.

The mechanism whereby the lymphocytes become sensitized to antigens is not fully understood. It is known that T-cells and B-cells respond differently. There are three types of T-cells— cytotoxic T-cells, T-helper cells, and T-suppressor cells. The cytotoxic T-cells can bind directly to the antigen in much the same lock-and-key model as described for enzyme–substrate combinations (Section 14.5). Once attached to the antigen-bearing cell, the T-cell releases a specific toxin that is lethal to the invading pathogen. This type of response is called **cellular immunity.**

B-cells, on the other hand, act indirectly. It is thought that *T-helper* cells must somehow trigger B-cells to produce and secrete the **antibodies** (immunoglobulins), which are then carried by the blood to the site of the pathogenic invasion. Specific antibodies are formed by the B-cells in response to specific antigens and stimulation by T-helper cells. (T-suppressor cells have the opposite effect; they diminish the activity of the B-cells.) The antibodies have complementary groups to those of the antigens, and thus they can bind the antigens to form insoluble precipitates or to cause foreign cells to clump together (*agglutination*). Eventually enzymes and/or phagocytes will arrive to attack and degrade the antigen–antibody complexes. This type of response is called **humoral immunity.**

Immunoglobulins are glycoproteins that interact with antigens on foreign particles (bacteria, viruses, cancer cells) and immobilize these substances so that they can be subsequently destroyed by the phagocytes. Each immunoglobulin molecule is composed of four polypeptide chains: two identical light (L) chains and two identical heavy (H) chains. Light and heavy refer to relative molecular weights. In IgG, the L chains have a molecular weight of about 25,000 daltons ($\sim$ 220 amino acids) and the H chains about 50,000 daltons ($\sim$ 440 amino acids). The quaternary structure of the four polypeptide chains resembles the letter Y. Three disulfide bonds link the four chains together (Figure 19.2), and each of the subunits has both constant and variable amino acid sequences. Moreover, all immunoglobulins contain the same amino acid sequences in their constant region. It is the variable region that distinguishes one immunoglobulin from another, and it is these variable regions that bind to the different antigens. Since each immunoglobulin has two "arms," it can bind two antigens; this results in a large aggregation of particles that either precipitate from the serum or are destroyed by the phagocytes. Binding of antibody to antigen occurs in much the same manner as other protein binding discussed earlier, that is, by hydrogen bonding, hydrophobic bonds, and ionic bonds. Again the structure of the antigen must be able to be recognized and to "fit" the complementary structure of a portion of the immunoglobulin. The large number of antigens that enter our body necessitates the synthesis of numerous different antibodies ($> 10,000$); most of these are variations of IgG.

FIGURE 19.2 Schematic diagram of a Y-shaped immunoglobulin molecule composed of two heavy chains and two light chains cross-linked by three disulfide bonds. The tinted areas represent the variable regions, and they contain the antigen-binding sites—one at the tip of each arm of the Y.

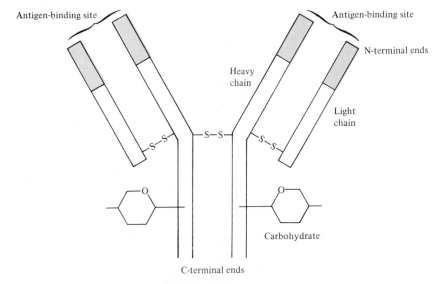

19.3   *Osmotic Pressure*

When a living cell is placed in distilled water, water gradually flows through the semipermeable membrane into the cell. The environment outside the cell is 100% water, while the percentage of water in the internal environment is considerably lower due to the presence of dissolved substances. Since the semipermeable membrane prevents these substances from leaving the cell, equalization of concentration can only be attained by the passage of the small water molecules into the cell. The diffusion of water from a dilute solution (high concentration of water), through a semipermeable membrane, into a more concentrated solution (low concentration of water) is known as the process of *osmosis.*[1] **Osmotic pressure** is defined as *the pressure required to prevent the occurrence of osmosis when two solutions of unequal concentrations are separated by a semipermeable membrane.* The osmotic pressure depends solely upon the concentration of solute particles (either ions or molecules) in the solutions involved.

The concentration of protein in the plasma far exceeds the concentration of protein in the interstitial fluid outside the blood vessels. This concentration gradient results in an osmotic pressure of about 28 mmHg. Therefore, if no external force were applied, fluids would be expected to diffuse from the interstitial fluid into the bloodstream. However, the pumping action of the heart creates the so-called **blood pressure** (see Section 19.4), and this pressure is greater at the arterial end of a capillary (~ 36 mmHg) than at the venous end (~ 21 mmHg). Since the blood pressure at the arterial end is higher than the osmotic pressure, the natural tendency is reversed and there is a net flow *from* the capillary *into* the interstitial fluid. The

FIGURE 19.3   Oxygen and nutrients leave at the arteriole end of the capillaries, whereas carbon dioxide and other cellular waste products enter the venule end of the capillaries.

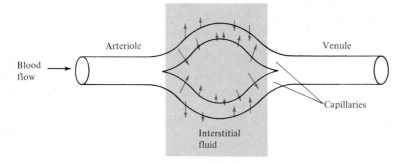

[1] Various terms are used to describe the relative concentrations of cells and their surrounding environments. A **hypotonic** solution is less concentrated than the cell contents. To equalize concentrations, water molecules enter the cell, which swells and may burst. Conversely, a **hypertonic** solution is more concentrated than the contents of the cell. Osmosis occurs in the opposite direction; water diffuses out of the cell, resulting in crenation (a shrinkage in size). An **isotonic** solution contains the same concentration of dissolved particles as the cell. For red blood cells, an isotonic solution is a 0.9% (0.15 *M*) solution of sodium chloride or a 5% solution of glucose.

fluid that leaves the capillary contains the dissolved nutrients, oxygen, hormones, and vitamins needed by the tissue cells. As the blood moves along the capillary branches, the blood pressure decreases until at the venous end the osmotic pressure is greater than the blood pressure and there is a net flow of fluid *from* the interstitial fluid *into* the capillary. The incoming fluid contains the metabolic waste products such as carbon dioxide and excess water (Figure 19.3).

Osmotic pressure, and hence the delicate balance of fluid exchange, is directly related to the concentration of albumins in the plasma. The normal half-life of albumins in the plasma is 20 days. If the albumin level is low, as might be the case from (1) malnutrition (low protein intake), (2) abnormal protein synthesis (liver disease), or (3) the loss of protein in the urine as a result of kidney disease (nephrosis), the osmotic pressure decreases. This results in a net efflux of fluids from the capillaries into the interstitial and cellular regions. This abnormal accumulation of fluids within the interstitial spaces produces noticeable swelling, particularly in the lower extremities. The condition so characterized is known as edema.[2] Recall that children who suffer from the protein-deficiency disease kwashiorkor (page 477) characteristically have bloated abdomens. This swelling is caused by the accumulation of water, which leaves the blood because there are insufficient albumins to maintain the osmotic pressure of the blood.

19.4   *Blood Pressure*

The **blood pressure** is *the force exerted by the blood against the inner walls of the arteries.* The **systolic pressure** is *the maximum pressure achieved during contraction of the heart ventricles.* When the ventricles relax, the blood pressure drops and *the lowest pressure that remains in the arteries before the next ventricular contraction* is called the **diastolic pressure.** The alternate expansion and contraction of the arterial walls can be felt as a *pulse* (when the artery is near the surface of the skin). Blood pressure measurements are reported as a *ratio of systolic pressure* (in mmHg) to *diastolic pressure* (in mmHg), for example, 120/80.

The blood pressure is dependent upon several factors including total volume of blood, heart action, and the smooth muscles that surround the arteries. The blood pressure is directly proportional to the volume of blood. Thus, if there is a loss of blood due to hemorrhaging, the blood pressure drops, but it returns to normal when the lost blood is replaced by a blood transfusion. Contraction of the smooth muscles in the walls of the arteries constricts the vessels (vasoconstriction), causing an increase in blood pressure. Conversely, dilation of the smooth muscles causes vasodilation and a decrease in blood pressure.

One of the major health problems in the United States is high blood pressure or **hypertension.** Over 20% of the world's population is afflicted, and in the United States it is estimated that hypertension occurs in about 25 million people. In order to treat this disorder, it is necessary to attempt to counteract those factors that cause the increased pressure. Blood pressure can be reduced by any of the following

---

[2]Edema may also result from the increased capillary permeability that accompanies an inflammatory reaction.

procedures:

1. Administer diuretics (see page 323) and/or reduce the sodium ion intake. Both of these steps will increase the excretion of urine and so decrease the volume of blood.
2. Negate the stimulating effects of epinephrine on the cardiac muscle with drugs that specifically bind to epinephrine binding sites (so-called *β-blockers*) and/or reduce stress and therefore reduce the levels of epinephrine.
3. Administer vasodilators, drugs that cause relaxation of the smooth muscles in the arterial walls and/or use drugs that block the action of smooth muscle vasoconstrictors.

If hypertension is not brought under control, the left ventricle must contract with a greater force against the increased arterial pressure. As a result of this increased workload, the heart muscle cells require additional oxygen, which cannot be delivered fast enough and in sufficient quantities. Heart cells die, and a myocardial infarction results.

19.5 *Clotting of Blood*

The phenomenon of blood clotting or coagulation is of the utmost importance to the organism. If such a mechanism did not exist, loss of blood would occur whenever a blood vessel was injured. The details of the coagulation process, which is quite complex, have been clarified considerably by research in recent years. The theory accounting for the clotting of blood has evolved from a two-step mechanism to the present model that depicts a multistep cascade of activation of protein factors. We shall restrict our discussion to a general description of the fundamental reactions involved. (See an advanced biochemistry text for a more detailed explanation.)

When blood clots, the soluble plasma protein fibrinogen is converted to the insoluble protein *fibrin*. Fibrin monomers undergo a polymerization reaction that results in the formation of needle-like threads. These threads enmesh the blood cells and effectively seal off the area where the blood vessel has been damaged (Figure 19.4). Once the fibrinogen and the formed elements have been removed from the plasma, the fluid that remains is called the **serum.** The only distinction, therefore between blood plasma and blood serum is the presence of fibrinogen in the plasma. *Because blood serum lacks fibrinogen, it is unable to clot.*

The clotting mechanism becomes operative only when a tissue is cut or injured. Blood platelets and damaged tissue cells are somehow activated and release a group of compounds collectively referred to as *thromboplastin*. In the presence of calcium ions and other cofactors, thromboplastin autocatalytically converts *autoprothrombia III* to *autoprothrombin C. Autoprothrombin C* is the enzyme that acts with calcium ions, phospholipids, and other cofactors to form *thrombin* from *prothrombin*. Autoprothrombin III and prothrombin are globulin plasma proteins that are continually produced by the liver. They are zymogens, and their activation is

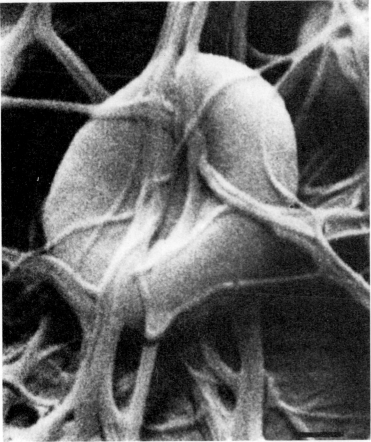

FIGURE 19.4 Scanning electron micrograph of an erythrocyte enmeshed in fibrin fibrils—a part of a typical blood clot. [E. Bernstein and E. Kairinen, *Science,* **173,** Aug. 27, 1971 (cover). Copyright 1971 by the AAAS.]

analogous to the activation of the various digestive enzymes. *Hemophilia* is a sex-linked hereditary disease in which one of the protein factors involved in the formation of *thrombin* is lacking or inactive.[3] *Thrombin* is the actual clotting enzyme. It is believed to effect the conversion of fibrinogen to fibrin by hydrolysis of a peptide fragment from the former. (A peptide bond between an arginine and a glycine is cleaved.) Normally, blood takes 5–8 minutes to form a clot. After the tissue is repaired by the body, the fibrin clot is digested by the enzyme *plasmin* (which circulates in the blood as the zymogen called *plasminogen*). The process involved in blood coagulation may be summarized as follows.

[3] Traditionally, hemophilia has been treated by plasma transfusion. This transfused blood contains the missing factor that can initiate clotting. A more recent development is the use of commercially available plasma concentrate that contains a high percentage of the antihemolytic factor (called factor VIII).

$$\text{Autoprothrombin III} \xrightarrow[\text{Ca}^{2+}]{\text{thromboplastin}} \text{Autoprothrombin C}$$

$$\text{Prothrombin} \xrightarrow[\substack{\text{Ca}^{2+}, \text{ phospholipids,}\\ \text{other factors}}]{\text{autoprothrombin C}} \text{Thrombin}$$

$$\text{Fibrinogen} \xrightarrow{\text{thrombin}} \text{Fibrin}$$

A number of substances, the *anticoagulants,* inhibit the clotting of blood by interfering with one or another of the reactions. Heparin is one of the principal anticoagulating agents. It is a polysaccharide, rich in sulfate ester groups, and is believed to block the catalytic activity of both *thromboplastin* and *thrombin.* Low concentrations of heparin are normally secreted into the circulatory system to prevent **thrombosis,** *the formation of a clot within a blood vessel.* If the clot, or a fragment of the clot, breaks loose and is carried away by the blood to be lodged in small blood vessels elsewhere (e.g., the lungs), then it is called an *embolism.* Certain sodium salts are employed as anticoagulants when blood is collected for clinical purposes. The anions of these salts, citrate, oxalate, and fluoride, form strong complexes with calcium ions, thus preventing them from existing in the free ionic form. If calcium ions are absent in the plasma, blood will not clot.

Two other compounds, vitamin K and dicumarol, are known to affect the clotting process. Vitamin K is a nutritional factor necessary for normal blood clotting (Section 12.16-d). Animal blood deficient in vitamin K has a prolonged coagulation time because of a lack of prothrombin in the plasma. Vitamin K is a coenzyme in the oxygen-dependent carboxylation of the glutamic acid side chains of prothrombin to yield γ-carboxyglutamate residues. These carboxyglutamate groups must be present in prothrombin to enable it to bind to calcium ions during the process of conversion of prothrombin to *thrombin.* Dicumarol is believed to act as a metabolic antagonist of vitamin K. It prevents blood clotting either by repressing prothrombin formation or by inhibiting the enzyme for which vitamin K is a coenzyme. Dicumarol is frequently administered to patients who have suffered heart attacks caused by thrombosis as a preventive measure against further clotting in the blood vessels. Chemists have succeeded in synthesizing new anticoagulants that have a greater potency than dicumarol. One of these is warfarin sodium (Coumadin),[4] which is unique in that it can be administered orally, intravenously, intramuscularly, or rectally. These drugs are usually administered before an operation or after heart attacks to minimize thrombosis.[5]

Dicumarol

Warfarin sodium

[4]Warfarin is also employed as a rat poison. It is safe for use as a rodenticide because regular ingestion of massive doses is fatal to rodents (producing internal hemorrhaging), whereas a single, accidental ingestion by children or pets is harmless.

[5]Some doctors recommend that elderly persons and patients with a history of heart attacks and strokes should take aspirin regularly (two to four tablets daily). Recall (Section 12.15) that among its other physiological effects, aspirin inhibits the synthesis of prostaglandins and thromboxanes, thereby interfering with the aggregation of blood platelets and diminishing the rate at which the blood platelets release the blood-clotting factors (thromboplastin). Hence, aspirin prevents heart attacks by inhibiting thrombosis. The greatest danger from thrombosis is that the clot may become detached and travel through the blood to some vital organ such as the heart or the brain. If the clot becomes lodged in and obstructs the blood vessels to these organs, their tissue cells are starved for oxygen and the cells die. If tissue death occurs in the brain, the condition is termed a *stroke;* if heart muscle tissue is destroyed, the condition is called a *coronary thrombosis* or a *myocardial infarction.*

19.6 *Hemoglobin*

The ability of mammalian blood to transport large quantities of oxygen depends upon the presence of the respiratory pigment hemoglobin (15 g/100 mL of blood). Hemoglobin comprises about 90% of the total protein found in red blood cells. You will recall that hemoglobin is a conjugated protein having a molecular weight of about 68,000 daltons. Upon hydrolysis, it yields the simple protein, globin, and four *heme* groups (iron porphyrins—see Figure 13.15). The heme groups account for about 4% of the total molecular weight. The characteristic red color of blood is due entirely to the presence of hemoglobin or, more precisely, to the presence of the heme groups, which absorb strongly in the blue region of the spectrum ($\sim$ 4000 Å). It is also the heme moiety that combines with molecular oxygen. However, the entire hemoglobin molecule is necessary for oxygen transport. If the globin is removed or replaced by another protein respiration is inhibited.

We have previously mentioned the cytochromes (Section 16.9), heme proteins in which the central iron atom undergoes reversible oxidation–reduction; that is, $Fe^{2+} \rightleftharpoons Fe^{3+}$. The iron of the hemoglobin molecule, however, is in the ferrous state and does not change to the ferric state at any time during normal oxygen transport. When oxygen is bound to the ferrous ion of hemoglobin, the compound is termed *oxyhemoglobin*. The deoxygenated form is sometimes referred to as reduced hemoglobin, but this term is misleading because it implies that the iron in the molecule is reduced. Therefore, we refer to this unoxygenated form simply as hemoglobin. Oxidizing agents such as potassium ferricyanide can oxidize the iron of hemoglobin to the ferric state. The same result is achieved in vivo by the action of nitrites and certain organic compounds (e.g., acetanilide, nitrobenzene, the sulfa drugs). The resulting compound, which contains iron in the +3 oxidation state, is called *methemoglobin,* and is incapable of oxygen transport. Small amounts of methemoglobin are normally present (about 0.3 g/100 mL blood) in the erythrocytes, but appreciable amounts of this substance result in the pathological condition *methemoglobinemia.*[6] The brown color of stale meat and dried blood results from the air oxidation of hemoglobin to methemoglobin.

Carbon monoxide hemoglobin (CO–hemoglobin) is formed by the combination of carbon monoxide with hemoglobin. The ferrous ions in hemoglobin have a much greater affinity for carbon monoxide than they do for oxygen (by a factor of 200) and thus will preferentially combine with any carbon monoxide that is in the blood. CO–hemoglobin will not transport oxygen, because all of the iron-binding sites are tied up by the carbon monoxide molecules. If sufficiently large numbers of hemoglobin molecules become saturated with carbon monoxide (about 60%), death occurs as a result of a failure of the blood to supply the brain with oxygen. Carbon monoxide poisoning is treated by greatly increasing the concentration of oxygen in the blood, either by artificial respiration in fresh air or by breathing in pure oxygen from an oxygen tank. The blood of cigarette smokers contains a relatively high percentage of CO–hemoglobin. The strain of pumping a greater volume of blood to compensate

---

[6]Salami and other preserved meat products contain nitrite salts as preservatives. (Recall that nitrites have been implicated in nitrosoamine formation—page 193.) People who consume relatively large quantities of these foods have a tendency to develop methemoglobinemia.

for insufficient oxygen is likely one of the major contributors to heart disease in smokers.

When red blood cells are destroyed, their hemoglobin molecules are completely catabolized. The porphyrin ring is first cleaved; the globin and the iron are subsequently removed. The globin is then digested, and its amino acids join the others in the metabolic pool. The iron is set free and incorporated into the iron-storage protein *ferritin*. This iron will be reutilized for the synthesis of new hemoglobin in the bone marrow.[7] The porphyrin skeleton is of no further use to the body. It undergoes a series of degradation reactions that lead to the production of the bile pigments, chiefly *biliverdin* and *bilirubin*. These are colored substances that give the bile its yellow color. The degraded pigments are stored in the gall bladder and released into the small intestine. As they travel down the intestinal tract, they undergo additional transformations that result in a darkening of their color, thus accounting for the characteristic colors of feces and urine.

An excess of bilirubin in the blood is responsible for the yellow color of the skin in **jaundice** (French, *jaune,* yellow). Jaundice caused by an excess of bilirubin can arise from (1) infectious hepatitis, where the liver is malfunctioning and cannot remove sufficient bilirubin, (2) the obstruction of bile ducts by gallstones, and (3) an acceleration of erythrocyte destruction in the spleen (hemolytic jaundice). Jaundice occurs in a large percentage of newborn infants because of insufficient synthesis of the liver enzymes that decompose bilirubin. A common treatment of neonatal jaundice is to shine a special fluorescent light onto the baby's skin. The energy of the light is able to decompose some of the bilirubin just beneath the surface of the skin.

19.7   *Respiratory Functions of the Blood*

The human body requires an enormous amount of oxygen to satisfy the demands of the energy-yielding oxidative phosphorylation reactions. The hemoglobin molecule is well suited to meet these demands because of its affinity for oxygen and because the attachment of oxygen to heme is readily reversible. In the alveoli of the lungs, hemoglobin comes into direct contact with a rich supply of oxygen (90–100 mmHg) and is converted to oxyhemoglobin. The oxyhemoglobin is carried by the arterial circulation to the cells in which there is a low oxygen concentration (25–40 mmHg) and a relatively high concentration of carbon dioxide ($\sim$60 mmHg). The oxygen is given up to the cells; the resulting hemoglobin carries some of the carbon dioxide back to the lungs to be expelled, and more oxyhemoglobin is formed.

Oxyhemoglobin does not transfer all of its oxygen to the tissue cells. Normally, every 100 mL of arterial blood combines with about 20 mL of oxygen. In the resting individual, the venous blood carries about 13 mL of oxygen per 100 mL of blood. Therefore, about 65% of the hemoglobin in venous blood is still combined with oxygen. If a person is engaged in strenuous exercise, his oxygen demand is high and the percentage of oxyhemoglobin in the venous blood may fall as low as 25%.

---

[7] An average diet supplies about 12–15 mg of iron per day, but only about 10% of this is absorbed. In order to maintain sufficient iron for hemoglobin synthesis, it is essential that the body retain the 20–25 mg of iron that is released each day.

At the lungs: O_2 pressure relatively high, CO_2 pressure relatively low.

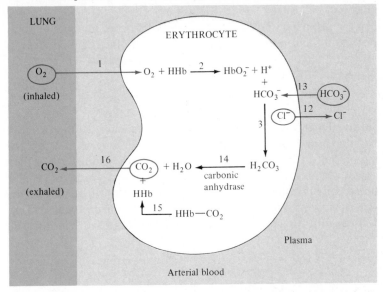

At the cells: CO_2 pressure relatively high, O_2 pressure relatively low.

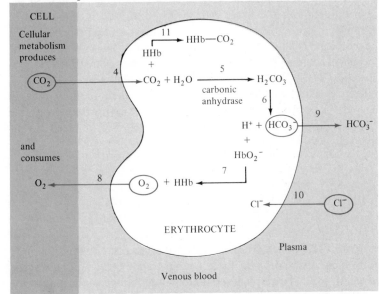

FIGURE 19.5 Oxygen and carbon dioxide transport by the blood.

Arterial blood is crimson in color; venous blood is a darker red, but it is not purple or blue.

The actual mechanism of the respiratory process is more complex than we have implied so far. Figure 19.5 and the following discussion represent an attempt to indicate the essential events that take place as blood traverses the capillaries of the alveoli and the tissues.

Atmospheric oxygen is taken into the lungs where a difference in pressure exists between the alveoli and the capillaries. Oxygen diffuses into the red blood cell (1) where it combines with hemoglobin, HHb, to form oxyhemoglobin, HbO_2^- (2).

$$HHb + O_2 \rightleftharpoons HbO_2^- + H^+$$

Two factors tend to shift this equilibrium to the right. They are the high oxygen concentration in the lungs and the neutralization of hydrogen ions by the bicarbonate ions present in the red blood cells (3). The oxyhemoglobin is then carried to the tissues where carbon dioxide is being produced as a result of cellular metabolism. The carbon dioxide diffuses into the erythrocyte (4) where the enzyme *carbonic anhydrase*[8] catalyzes its combination with water to form carbonic acid (5). Carbonic acid subsequently dissociates into bicarbonate ions and hydrogen ions (6).

$$H_2O + CO_2 \rightleftharpoons H_2CO_3 \rightleftharpoons H^+ + HCO_3^-$$

The major factor that shifts this equilibrium to the right is the neutralization of the hydrogen ions by oxyhemoglobin (7), which is concomitantly split into hemoglobin and oxygen. Because the oxygen pressure is greater in the capillaries than in the tissues, the oxygen diffuses into the tissue cells (8) to be utilized in the oxidative reactions of metabolism. As the bicarbonate ion concentration increases, the ions diffuse out of the red blood cell into the plasma (9). The loss of negative ions by the cell is then balanced by the migration of chloride ions into the cell from the plasma. The process, referred to as the *chloride shift* (10), brings about a reestablishment of electrolyte equilibrium. Most of the carbon dioxide is taken to the lungs as bicarbonate ion ($\sim 90\%$), but some combines with hemoglobin (probably with free amino groups in the globin portion) (11). This species is referred to as *carbaminohemoglobin*.

$$HHbNH_2 + CO_2 \longrightarrow HHbN-C + H^+$$

When the red blood cells return to the lungs, a reversal of events—that is, (12), (13), (3)(14), (15), (16)—sends carbon dioxide into the lungs to be expelled and the cyclic process repeats itself. It should be emphasized that hemoglobin plays the major role in the respiratory process. In addition to its role in oxygen transport, it is also involved in the ability of the blood to carry carbon dioxide without appreciable changes in pH.

---

[8] *Carbonic anhydrase* is one of the most efficient enzymes in the body. It has a turnover number of 600,000 per second. This means that a single enzyme transforms 600,000 molecules of substrate into product each second. (In comparison, the turnover number of *lactic acid dehydrogenase* is 1000 per second.)

19.8   *Blood Buffers*

Recall that the maintenance of pH within narrow limits is vital to the well-being of an organism. Any slight change in hydrogen ion concentration (± 0.2–0.4 pH unit) inhibits oxygen transport and alters the rates of the metabolic processes by decreasing the catalytic efficiency of enzymes. The blood plasma and the erythrocytes contain four buffering systems that maintain the pH of the blood between 7.35 and 7.45. These are (1) the bicarbonate pair, (2) the phosphate pair, (3) the plasma proteins, and (4) the hemoglobin of red blood cells. In Section 8.5 we mentioned the buffering action of the bicarbonate pair H_2CO_3/HCO_3^- and of the phosphate pair $H_2PO_4^-/HPO_4^{2-}$. In Section 13.2 we indicated that the amino acids, in their zwitterionic forms, can neutralize small concentrations of either acids or bases. Since the plasma proteins and the globin of hemoglobin contain both acidic and basic amino acids, they tend to minimize changes in pH by combining with, or liberating, hydrogen ions and thus serve as excellent buffering agents over a wide range of pH values.

The principal *intracellular* buffer of the body cells is the phosphate pair, whereas hemoglobin molecules are the most important buffers within the red blood cells. The major *extracellular* buffer in the blood and interstitial fluid is the bicarbonate pair, owing to its intimate connection with the respiration processes. Under normal conditions the primary metabolic factor that tends to lower the pH of an organism is the continuous production of carbon dioxide and the acidic metabolites (acetoacetic acid, pyruvic acid, lactic acid, α-ketoglutaric acid, etc.). All of these compounds vary in concentration according to metabolic circumstances. When acids enter the blood, they are neutralized by bicarbonate ions, and the slightly dissociable acid, carbonic acid, is formed.

$$H^+ + HCO_3^- \rightleftharpoons H_2CO_3$$

It would seem that this reaction would alter the buffer ratio by decreasing the bicarbonate ion concentration and increasing the concentration of carbonic acid. The excess carbonic acid, however, is readily decomposed to water and carbon dioxide by the enzyme *carbonic anhydrase*. The respiration rate is increased, and the carbon dioxide is eliminated at the lungs, thus preserving the proper buffer ratio.

$$H_2CO_3 \underset{}{\overset{\text{carbonic anhydrase}}{\rightleftharpoons}} H_2O + CO_2$$

The respiration rate is another factor that influences the pH of the blood. *Respiratory acidosis* results from **hypoventilation,** a condition that arises when the rate of breathing is too slow. Hypoventilation is brought on by an obstruction to respiration (e.g., asthma, pneumonia, or pulmonary emphysema), by coronary attack, or by drugs that depress the brain's respiratory center (e.g., morphine, barbiturates). When the respiration rate is very low, carbon dioxide is not expelled from the lungs fast enough, and the preceding equilibrium is shifted to the left. This increases the concentration of carbonic acid and results in a higher than normal H_2CO_3/HCO_3^- ratio and a subsequent decrease in the blood pH.

Hyperventilation, when the rate of breathing is too rapid, causes *respiratory alkalosis.* Hyperventilation arises during strenuous exercise, anxiety, crying, and hysteria. At high altitudes hyperventilation may occur in response to low oxygen pressure. An increased rate of respiration accelerates the removal of carbon dioxide from the lungs, and the equilibrium shifts to the right. The concentration of carbonic acid in the blood decreases and the H_2CO_3/HCO_3^- ratio becomes lower than normal with a subsequent increase in the blood pH.

Exercises

19.1 Define or give an explanation for each of the following terms.
(a) lymph
(b) formed elements
(c) hematocrit value
(d) leukemia
(e) immunity
(f) osmosis
(g) edema
(h) hypertension
(i) hemophilia
(j) anticoagulant
(k) thrombosis
(l) embolism
(m) methemoglobinemia
(n) jaundice

19.2 Distinguish between the terms in each of the following pairs.
(a) aplastic anemia and hemolytic anemia
(b) anemia and polycythemia
(c) lymphocytes and phagocytes
(d) T-cells and B-cells
(e) antibodies and antigens
(f) cellular immunity and humoral immunity
(g) osmotic pressure and blood pressure
(h) systolic pressure and diastolic pressure
(i) vasodilation and vasoconstriction
(j) plasma and serum
(k) stroke and coronary thrombosis
(l) hypoventilation and hyperventilation

19.3 What are the three circulating fluids in the body?

19.4 List five functions of the blood.

19.5 The blood accounts for about 8% of the total body weight. Calculate the volume of blood that your body contains (density of blood = 1.06 g/mL).

19.6 What are some of the nonprotein constituents of the blood?

19.7 Give the formulas for five of the inorganic ions present in the blood.

19.8 What are the functions of (a) albumins, (b) α- and β-globulins, and (c) fibrinogen? Where are they synthesized?

19.9 (a) What are the functions of the γ-globulins?
(b) Where are γ-globulins synthesized?

(c) What are the designations of the five classes of γ-globulins?
(d) Which class is most abundant?

19.10 Name the three formed elements of the blood and give a function for each.

19.11 (a) In what way are red blood cells similar to other cells?
(b) How do they differ?
(c) Where are erythrocytes formed?
(d) Where are they broken down?

19.12 Briefly describe the immune response.

19.13 Explain the processes involved when fluid is exchanged between the plasma and the cells (a) at the arterial end and (b) at the venous end of a capillary.

19.14 What determines the osmotic pressure within the blood vessels?

19.15 How are blood pressure measurements reported?

19.16 Cite three ways of treating hypertension.

19.17 How does each of the following function in the clotting of blood?
(a) prothrombin
(b) thrombin
(c) fibrinogen
(d) thromboplastin
(e) calcium ions
(f) fibrin

19.18 (a) What is the role of vitamin K in blood coagulation? What is the effect of (b) heparin and (c) dicumarol in preventing thrombosis?
(d) How do anions such as oxalates, fluorides, and citrates prevent blood coagulation?
(e) Write a structural formula of γ-carboxyglutamic acid.

19.19 Describe the chemical structure of hemoglobin.

19.20 Contrast the structural features of the following pairs of compounds.
(a) heme and hemoglobin
(b) hemoglobin and myoglobin
(c) hemoglobin and oxyhemoglobin
(d) oxyhemoglobin and CO–hemoglobin
(e) oxyhemoglobin and methemoglobin
(f) CO–hemoglobin and carbaminohemoglobin

19.21 What is the oxidation state of iron in (a) hemoglobin,

(b) oxyhemoglobin, (c) methemoglobin, and (d) CO–hemoglobin?

19.22 (a) Why is carbon monoxide such a deadly poison?
(b) How is carbon monoxide poisoning treated?

19.23 Name three heme-containing compounds.

19.24 Name the two bile pigments that are formed from the porphyrin skeleton of hemoglobin.

19.25 What are the causes of jaundice?

19.26 (a) Contrast the contents of arterial blood and venous blood.
(b) What are the colors of arterial blood and venous blood?

19.27 Briefly explain the processes of oxygen transport from lungs to cells and of carbon dioxide transport from cells to lungs. Write equations for the reactions that occur at the cells and at the lungs.

19.28 What is the function of *carbonic anhydrase?*

19.29 What is the normal pH range of the blood?

19.30 What is the relationship between acidosis and oxygen transport?

19.31 Illustrate, with equations, how the blood buffers prevent the change in pH of the blood when small amounts of acid or base are produced during metabolic reactions.

19.32 Name the two blood buffer pairs. Why is the bicarbonate buffer pair effective in spite of the fact that the ratio of acid to anion is 1:10?

19.33 Make use of the following equilibrium to explain what occurs to cause (a) respiratory acidosis and (b) respiratory alkalosis.

$$H^+ + HCO_3^- \rightleftharpoons H_2O + CO_2$$

Index

A

C

Cadaverine, 187
Calorie, defined, 462n
Cancer
    chemotherapy, 368
    estrogen therapy, 297n
    leukemia, 494
    oncogenic viruses, 411–12, 415
    tar, 60
    vitamins and, 300
Carbaminohemoglobin, 504
Carbocation, 41–42, 90
Carbohydrate metabolism, 419
    chemical energy storage, 440–42
    digestion and absorption, 421–22
    energy yield, 454–56
    See also Krebs cycle
Carbohydrates, 241–42
    classification, 242
    defined, 241
    sweeteners, 242–43
    tests, 263–66
    See also Disaccharides,
        Monosaccharides, Polysaccharides,
        Sugars
Carbolic acid, 62
Carbon
    α-, 127, 149
    chiral, 219
    compounds, 5–6
    penultimate, 245
    tetrahedral atom, 217
Carbon–carbon bonds, 4t, 122
Carbonic acid, 504–506
Carbon monoxide, toxicity, 501
Carbonyl compounds, reduction, 132
Carbonyl group, 115, 121–22, 252
Carboxylic acids, 137
    acidity, 143–44, 149–50
    ammonia addition, 179
    buffers, 147–49
    chemical properties, 150–54
    hydrogen ion concentration, 144–46
    nomenclature, 138–39, 151
    oxidation, 140–41
    physical properties, 142–43
    preparation, 140–42
    reaction with sodium bicarbonate, 151
    resonance, 149
Carboxypeptidases, 326, 474

Carcinogens, 60
Cardiovascular disease, 291
    anticoagulants, 500
    cholesterol and, 291, 303–305
    coronary thrombosis, 500n
    estrogen therapy and, 297n
    myocardial infarction, 372–73, 441n, 480,
        498, 500n
    serum enzyme levels, 373
β-Carotene, 46, 48f
Casein, 325t, 339
Catabolism, defined, 419
Catalyst, 40, 348
Catalytic groups, 358
Catechol, 100
Catecholamines, 204
Cell membranes, 25, 284f, 288–89
Cellobiose, 257–58
Cell types, 284f
Cellular immunity, 495
Cellulose, 262–63, 309
Cephalin, 286
Cerebrosides, 287
Chain reaction, 24
Chair conformation, 29–30
Chargaff's rules, 386–87
Chemical bonds, defined, 3
Chemical formulas, 7–9
Chemiosmotic hypothesis, 452
Chemotherapy, 367–72
    antibiotics, 370–72
    antimetabolites, 368–69
    antineoplastic drugs, 68
    defined, 367
Chinese restaurant syndrome, 232n
Chirality, 219
Chloral hydrate, 123n
Chlordane, 76f, 78
Chlordiazepoxide, 207–208
Chloride shift, 504
Chlorination, alkanes, 24
Chlorpromazine, 207–208
Cholecalciferol, 300–302
Cholesterol, 290–91, 303–305
Choline, 285
Cholinergic nerves, 171n
Chromoprotein, 336
Chromosome, 385
Chylomicron, 304, 461
Chymotrypsin, 359–60, 366, 474
Chymotrypsinogen, 474

Cinnamaldehyde, 127
Cis-trans isomerism, 229–31, 234
Citric acid, 157
Citric acid cycle, *see* Krebs cycle
Citric acid synthetase, 443
Coagulation, 498–500
Coal tar, 60
Cocaine, 106n, 205
Codeine, 198
Codon, 399
Coenzyme, 349, 352–54t
Coenzyme A, 110, 354–55
Coenzyme Q (COQ), 101, 451
Cofactors, 351
Collagen, 335
Competitive inhibitor, defined, 364
Complementary bases, nucleic acid, 386
Condensation polymers, 44, 180
Condensation reaction, defined, 126
Condensed structural formula, 9
Configurational isomers, defined, 214
Conformational isomers, 29n, 214
Conjugated diene, defined, 46
Contact groups, 358
Contact inhibition, 411
Cori cycle, 437
Corpus luteum, 297
Cortisol, 293
Covalent bonds, 3–5
"Crack," 205
Cracking, defined, 37
Creatine kinase (CK) 372t, 373f, 441, 480
Creatine phosphate, 441
Cresols, 99
C-terminal end, peptide, 319
Curare, 171n
Cyanides, reduction, 185–86
Cyanocobalamin, 354–55
Cyanohydrins, 125
Cyclamates, 242
Cyclic AMP, 168, 382, 426, 462
Cycloalkanes, 28–30
Cycloalkenes, 36, 230–31
Cyclobutane, 28–29
Cycloheptane, 28
Cyclohexane, 28–30
Cyclopentane, 28–29
Cyclopropane, 28–29
Cysteine, 313t
Cytochromes, 325t, 449, 451–52

Cytoplasm, 395n
Cytosol, 395

D

Dacron, 160–61
Dalmane, 202
Dalton, 310n
Darvon, 198f, 200
DDT, 76–78
DDT dehydrochlorinase, 70, 364
Deamination, oxidative, 480–82
Debranching enzyme, 430
Decarboxylation, 155, 443, 482–84
n-Decyl alcohol, 87
Dehydration, 37, 90–92, 155
7-Dehydrocholesterol, 300–301
Dehydrogenation, 119, 155, 450
Dehydrohalogenation, 38, 69–70
Delaney Clause, 242–43
Demerol, 198f, 200
Denaturation
  acids or bases, 340
  alkaloid reagents, 341
  defined, 339
  hair, 341, 343f
  heat, 339
  heavy metal ion salts, 340–41
  organic compounds treatment, 339
  proteins, 339–41
  ultraviolet radiation, 339
Denatured alcohol, 96
Deoxyribonucleic acid, *see* DNA
Deoxyribose, 252
Designer drugs, 208–209
Detergent, 281–83
Dextrins, 260
Dextrorotatory, 216
Diabetes, 323, 423, 425
Diastereomers, 223–25
Diastolic pressure, defined, 497
Diazepam, 207–208
Dicarboxylic acids, aliphatic, 139
Dichlorodiphenyltrichloroethane (DDT)
  76–78
2,6-Dichlorophenol, 68
Dicumarol, coagulation and, 500
Dieldrin, 76f, 78
Diet, liquid protein, 477n

M

Macrophages, 494
Malathion, 170*f*, 172
Maleic acid, geometric isomers, 230–31
L-Malic acid, formation, 447
Malic acid dehydrogenase, 447
Malonic acid, 139*t*, 364, 365*f*
Maltase, 422
Maltose, 256–58
Malt sugar, 256
Malvalic acid, 271*n*
D-Mannose, 245, 247, 249–50
Marijuana, 202–204
Markovnikov's rule, 41–42
McArdle's disease, 429*n*
MDMA, 209
Melting points, 21–22
Membranes
  cell, 284*f*, 288–89
  mitochondrial, 448
Menopause, 297
Meperidine, analogs, 209
Mercaptan, 108
Mercury compounds, 341*n*
Meso compounds, racemic mixtures and, 225–26
Messenger RNA, 394–95
Meta, 58
Metabolic pathways, 427
  Embden-Meyerhof, 428*f*
  fatty acid spiral, 464*f*
  Krebs cycle, 444*f*
  interrelationships, 486–87
Metabolism
  defined, 419
  importance of pyruvic acid, 436*f*
  See also Carbohydrate metabolism, Lipid metabolism, Protein metabolism
Metabolite, defined, 419
Metalloenzymes, 366
Methadone, 198–99
Methamphetamine, 204–205
Methanal, 95, 117*t*, 121*t*, 123*n*, 133
Methane
  structure, 5–7
  tetrahedral configuration, 217–18
Methanol, *see* Methyl alcohol
Methaqualone, 202
Methemoglobin, 501
Methionine, 312*t*

Methyl alcohol, 26, 82*t*, 94–95, 140–41
  acidity, 89
2-Methyl-1,3-butanediol, stereoisomers, 223
2-Methyl-1-butanol, enantiomers, 221–22
Methylcobalamin, 354*t*, 355
Methylfentanyl, 208–209
Methyl salicylate, 163–64
Micelles, 280–81
Milk sugar, 258
Mineralocorticoid, 293
Minimum daily requirement, nutrient, 300*n*
Mitochondrion(ia), 448, 462*n*, 463
Molecular formula, defined, 7
Molecular models, 7
Molisch test, 264
Monosaccharides, 242
  absorption, 422
  specific rotation, 250–51
  stereochemistry, 244–47
  terminology, 243
Monosodium urate, 486
L-Monosodium glutamate, 232–33
Morphine, 197–98
MPPP, 209
Multistep synthesis, 125
Mutagens, 405
Mutarotation, 250–52, 257
Mutations, 405–407
Myasthenia gravis, 171*n*
Myoglobin, 325*t*, 332–33, 334

N

NADH dehydrogenase, 450
Naloxone, 200
Naltrexone, 200
Naphthalene, 60
Narcotic antagonists, 200
Narcotics, 199
Natural gas, 21, 26
Nembutal, 201
Nerve gases, 170, 366
Nerve transmission, phosphate esters and, 170–72
Neutralization, 150–52
Neuron, 171*f*
Neurotransmitters, 170, 204, 207
Niacin, 196, 355
Nicotinamide, 353*t*, 355

Nicotinamide adenine dinucleotide
(NAD⁺/NADH), 353*t*, 437–38, 447,
448, 450, 454–55, 481
Nicotine, 196
Nicotinic acid, 196, 355
Niemann–Pick disease, 409
Ninhydrin test, 342–43
Nitration, 61
Nitriles, 142*n*, 185–86
Nitrite salts, as preservatives, 501*n*
Nitrobenzene, uses, 62
Nitro compounds, reduction, 185
Nitrogen
balance, 476–77
equilibrium, 476–77
excretion, 485–87
heterocycles, 194–97
storage, 484–85
Nitroglycerin, 97–99
Nitrosation, amines, 192–93
Nitrosoamines, 193, 501*n*
Nitrous acid, 192–93, 406
Nitrous acid esters, 94
Nonbarbiturate sleep-inducing drugs, 202*f*
Nonsaponifiable lipids, 270
Norepinephrine, 204–205, 462
NTA, 283
N-terminal end, peptide, 319
Nucleic acids, 382, 384*f*
hydrolysis, 377–78
nucleosides, 379–80
nucleotides, 380–82, 384*f*
purine and pyrimidine bases, 378–79
structure, 377–84
See also DNA, RNA
Nucleophiles, 71
Nucleoproteins, 336
Nucleoside diphosphokinase, 446
Nucleosides, 379–80
Nucleotides, 380–82, 384*f*
Nylon, 180

O

Obesity, 462
Obligatory intracellular parasite, 410
Octane rating, 27
Odors, primary, 234*t*, 236–37
Oil, barrel, 27*n*
Oils, 46, 270–71, 273
Okazaki fragments, 391–92

Olefins, 33
Olfactory receptor sites, 236*f*
Oncogenic viruses, 411–12
Opium, 197–98
Opium alkaloids, 197–201
Opsin, 234–35
Optical activity, 215, 224*t*
defined, 216
structure and, 216–23
Optical rotation, 216, 245
Optimum temperature, 362
Oral contraceptives, 295–97
Organic acids, *see* Carboxylic acids
Organic compounds, 8*t*
optically active and inactive, 224*t*
properties, 3*t*
Organic halides
aryl, defined, 65
chemical properties, 69–71
preparation, 67–68
substitution, 71
toxic, 74–75*t*
uses, 72–76
See also Alkyl halides
Organophosphates, 167–68, 170–72
Orlon, 45*t*
Ortho, 58
Osmotic pressure, 496–97
Osteomalacia, 300
Oxalic acid, 139*t*, 156
Oxaloacetic acid, 447, 480
Oxidation
alcohols, 92–93, 119, 140
alkenes, 43–44
carboxylic acids, 140–41
ketones, 130–31
lipoic acid, 109
phenols, 101
thiols, 109
Oxonium ion, 89, 91, 105
Oxygen debt, 436
Oxyhemoglobin, 501–502, 504
Oxytocin, 319, 322*f*
Ozone, 25, 73

P

PABA, 369*n*
Palmitic acid, breakdown, 467–68
Pantothenic acid, 354–55
Para, 58

R,S convention, 226–27
  structure and optical activity, 216–23
Steroids, 289–90, 292, 294–95
Stomach, pH, 473
Strict anaerobes, 437n
Stroke, 500n
Structural formula, defined, 7
Structural isomers, 213
Substitution, 23, 61, 71
Substrate, defined, 349
Substrate level phosphorylation, 434
Sucaryl sodium, 242
Succinic acid, 139t
Succinic acid dehydrogenase, 364–65, 446, 451n
Succinylcholine, 171n
Sucrase, 422
Sucrose, 216, 254–56, 348
Sugars, 247–49, 255, 265
D-Sugars, 244–45
L-Sugars, 244
Sulfanilamide, 368–69
Sulfhydryl group, 107
Sulfonation, 62
Sulfone, 109
Sulfoxide, 109
Sunscreen products, 369n
Sweeteners, 242–43
Sympathomimetic amines, 204
Synapses, 170–71
Syndets, 281
Systolic pressure, defined, 497

T

Tannic acid, 341
Tar cancer, 60
Tartaric acid, 157, 225–26
Tautomerism, defined, 128
Tay–Sachs disease, 409
T-cells, 494–95
Teflon, 45t
Template replication, 389
Teratogenesis, 206–207
Terpenes, 46
Tertiary alcohol, 82, 90
Tertiary structure, protein, 328–32
Testosterone, 294–95
Tetraethyllead, 27
Tetrahydrocannabinol, 202–204
Tetrapeptide, 321f

Thiamine, 31, 352t
Thioesters, 110
Thioethers, nomenclature, 108
Thiols, 107–10
Threonine, 312t
Threonine dehydratase, 366–67
Threose, 244
Thrombin, 498–99
Thrombocytes, 490, 494
Thromboplastin, 366, 498
Thrombosis, 500
Thromboxanes, 298
T loop, RNA, 395–96
T-lymphocytes, 494–95
Tobacco mosaic virus, 325t, 334f, 335, 409n
Tocopherol, 25–26, 301–302
Tollens' reagent, 131
Toluene, 59, 62, 141n
Tranquilizers, 207–208
Transaminases, 479–80
Transamination, amino acids, 479–80
Transcription, 393–94, 402f
Transferase, 430
Transfer RNA (tRNA), 395–97, 399–403
Translation, 399, 402f
Translocation, 401f
Triacylglycerols, 270, 272, 460, 461
  chemical properties, 274–77
  drying oils, 278
  halogenation, 274–75
  hydrogenation, 275–77
  physical properties, 274
  rancidity, 277
  saponification, 274
Tricarboxylic acid cycle, see Krebs cycle
Triglycerides, 270n
Trimedlure, 233
2,4,6-Trinitrophenol, 100–101, 341
Trinitrotoluene, 62
Triose phosphate isomerase, 433
Trioses, 252
Tripalmitin, 22f
Tripeptidase, 474
Tripeptide, 321f
Triple bond, 4
Triplet code, 399
Tripolyphosphoric acid, 165
Trypsin, 366, 474–75
Trypsinogen, 474f, 475
Tryptophan, 63, 311t
Tyrosine, 63, 313t

Periodic Table of the Elements

s block	p block	d block	f block

Period

Transition Elements

IA	IIA		IIIB	IVB	VB	VIB	VIIB			
1 **H** 1.008										
3 **Li** 6.941	**4** **Be** 9.012									
11 **Na** 22.990	**12** **Mg** 24.305									
19 **K** 39.10	**20** **Ca** 40.08		**21** **Sc** 44.96	**22** **Ti** 47.90	**23** **V** 50.94	**24** **Cr** 51.996	**25** **Mn** 54.94	**26** **Fe** 55.85	**27** **Co** 58.93	
37 **Rb** 85.47	**38** **Sr** 87.62		**39** **Y** 88.91	**40** **Zr** 91.22	**41** **Nb** 92.91	**42** **Mo** 95.94	**43** **Tc** 98.91	**44** **Ru** 101.07	**45** **Rh** 102.91	
55 **Cs** 132.91	**56** **Ba** 137.34		**57–71** **La-Lu**	**72** **Hf** 178.49	**73** **Ta** 180.95	**74** **W** 183.85	**75** **Re** 186.2	**76** **Os** 190.2	**77** **Ir** 192.22	
87 **Fr** (223)	**88** **Ra** 226.03		**89–103** **Ac-Lr**	**104** **Unq*** (261)	**105** **Unp** (262)	**106** **Unh** (263)	**107** **Uns** (262)	**108** **Uno**	**109** **Une**	

Lanthanoid Series

57 **La** 138.91	**58** **Ce** 140.12	**59** **Pr** 140.91	**60** **Nd** 144.24	**61** **Pm** (145)	**62** **Sm** 150.36

Actinoid Series

89 **Ac** 227.03	**90** **Th** 232.04	**91** **Pa** 231.04	**92** **U** 238.03	**93** **Np** 237.05	**94** **Pu** (244)